Classics in Mathematics

Lars Hörmander

The Analysis of Linear Partial Differential Operators III

Born on January 24, 1931, on the southern coast of Sweden, Lars Hörmander did his secondary schooling as well as his undergraduate and doctoral studies in Lund. His principle teacher and adviser at the University of Lund was Marcel Riesz until he retired, then Lars Gårding. In 1956 he worked in the USA, at the universities of Chicago, Kansas, Minnesota and New York, before returning to a chair at the University of Stockholm. He remained a frequent visitor to the US, particularly to Stanford and was Professor at the IAS, Princeton from 1964 to 1968. In 1968 he accepted a chair at the University of Lund, Sweden, where, today he is Emeritus Professor.

Hörmander's lifetime work has been devoted to the study of partial differential equations and its applications in complex analysis. In 1962 he was awarded the Fields Medal for his contributions to the general theory of linear partial differential operators. His book Linear Partial Differential Operators, published 1963 by Springer in the Grundlehren series, was the first major account of this theory. His four volume text *The Analysis of Linear Partial Differential Operators*, published in the same series 20 years later, illustrates the vast expansion of the subject in that period.

Lars Hörmander

The Analysis of Linear Partial Differential Operators III

Pseudo-Differential Operators

Reprint of the 1994 Edition

Springer

Lars Hörmander
University of Lund
Department of Mathematics
Box 118
SE-22100 Lund
Sweden
email: lvh@maths.lth.se

Originally published as Vol. 274 in the series:
Grundlehren der Mathematischen Wissenschaften

Library of Congress Control Number: 2007900647

Mathematics Subject Classification (2000):
35A, G, H, J, L, M, P, S, 47G, 58G

ISSN 1431-0821
ISBN 978-3-540-49937-4 Springer Berlin Heidelberg New York

Springer is a part of Springer Science+Business Media

springer.com

Production: LE-TEX Jelonek, Schmidt & Vöckler GbR, Leipzig
Cover-design: WMX Design GmbH, Heidelberg

SPIN 11912330 41/3100YL - 5 4 3 2 1 0 Printed on acid-free paper

Grundlehren der mathematischen Wissenschaften 274

A Series of Comprehensive Studies in Mathematics

Lars Hörmander

The Analysis of Linear Partial Differential Operators III

Pseudo-Differential Operators

Springer-Verlag
Berlin Heidelberg NewYork Tokyo

Lars Hörmander
Department of Mathematics
University of Lund
Box 118
S-221 00 Lund, Sweden

Corrected Second Printing 1994
With 7 Figures

AMS Subject Classification (1980): 35A, G, H, J, L, M, P, S; 47G; 58G

ISBN 3-540-13828-5
Springer-Verlag Berlin Heidelberg New York Tokyo
ISBN 0-387-13828-5
Springer-Verlag New York Heidelberg Berlin Tokyo

Library of Congress Cataloging in Publication Data
(Revised for vol. 3)
Hörmander, Lars.
The analysis of linear partial differential operators.
(Grundlehren der mathematischen Wissenschaften; 256-)
Expanded version of the author's 1 vol. work: Linear partial differential operators.
Includes bibliographies and indexes.
Contents: 1. Distribution theory and Fourier analysis - 2. Differential operators with constant coefficients - 3. Pseudo-differential operators.
1. Differential equations, Partial. 2. Partial differential operators. I. Title. II. Series.
QA377. H578 1983 515.7'242 83-616
ISBN 0-387-12104-8 (U.S.: v. 1)

SPIN 10427377 41/3140 – 5 4 3 2 1 – Printed on acid-free paper

Preface

to Volumes III and IV

The first two volumes of this monograph can be regarded as an expansion and updating of my book "Linear partial differential operators" published in the Grundlehren series in 1963. However, volumes III and IV are almost entirely new. In fact they are mainly devoted to the theory of linear differential operators as it has developed after 1963. Thus the main topics are pseudo-differential and Fourier integral operators with the underlying symplectic geometry. The contents will be discussed in greater detail in the introduction.

I wish to express here my gratitude to many friends and colleagues who have contributed to this work in various ways. First I wish to mention Richard Melrose. For a while we planned to write these volumes together, and we spent a week in December 1980 discussing what they should contain. Although the plan to write the books jointly was abandoned and the contents have been modified and somewhat contracted, much remains of our discussions then. Shmuel Agmon visited Lund in the fall of 1981 and generously explained to me all the details of his work on long range scattering outlined in the Goulaouic-Schwartz seminars 1978/79. His ideas are crucial in Chapter XXX. When the amount of work involved in writing this book was getting overwhelming Anders Melin lifted my spirits by offering to go through the entire manuscript. His detailed and constructive criticism has been invaluable to me; I as well as the readers of the book owe him a great debt. Bogdan Ziemian's careful proofreading has eliminated numerous typographical flaws. Many others have also helped me in my work, and I thank them all.

Some material intended for this monograph has already been included in various papers of mine. Usually it has been necessary to rewrite these papers completely for the book, but selected passages have been kept from a few of them. I wish to thank the following publishers holding the copyright for granting permission to do so, namely:
Marcel Dekker, Inc. for parts of [41] included in Section 17.2;
Princeton University Press for parts of [38] included in Chapter XXVII;
D. Reidel Publishing Company for parts of [40] included in Section 26.4;
John Wiley & Sons Inc. for parts of [39] included in Chapter XVIII.
(Here [N] refers to Hörmander [N] in the bibliography.)

Finally I wish to thank the Springer-Verlag for all the support I have received during my work on this monograph.

Djursholm in November, 1984 — Lars Hörmander

Contents

Introduction

to Volumes III and IV

A great variety of techniques have been developed during the long history of the theory of linear differential equations with variable coefficients. In this book we shall concentrate on those which have dominated during the latest phase. As a reminder that other earlier techniques are sometimes available and that they may occasionally be preferable, we have devoted the introductory Chapter XVII mainly to such methods in the theory of second order differential equations. Apart from that Volumes III and IV are intended to develop systematically, with typical applications, the three basic tools in the recent theory. These are the theory of pseudo-differential operators (Chapter XVIII), Fourier integral operators and Lagrangian distributions (Chapter XXV), and the underlying symplectic geometry (Chapter XXI). In the choice of applications we have been motivated mainly by the historical development. In addition we have devoted considerable space and effort to questions where these tools have proved their worth by giving fairly complete answers.

Pseudo-differential operators developed from the theory of singular integral operators. In spite of a long tradition these played a very modest role in the theory of differential equations until the appearance of Calderón's uniqueness theorem at the end of the 1950's and the Atiyah-Singer-Bott index theorems in the early 1960's. Thus we have devoted Chapter XXVIII and Chapters XIX, XX to these topics. The early work of Petrowsky on hyperbolic operators might be considered as a precursor of pseudo-differential operator theory. In Chapter XXIII we discuss the Cauchy problem using the improvements of the even older energy integral method given by the calculus of pseudo-differential operators.

The connections between geometrical and wave optics, classical mechanics and quantum mechanics, have a long tradition consisting in part of heuristic arguments. These ideas were developed more systematically by a number of people in the 1960's and early 1970's. Chapter XXV is devoted to the theory of Fourier integral operators which emerged from this. One of its first applications was to the study of asymptotic properties of eigenvalues (eigenfunctions) of higher order elliptic operators. It is therefore discussed in Chapter XXIX here together with a number of later developments which give beautiful proofs of the power of the tool. The study by Lax of the propagation of singularities of solutions to the Cauchy problem was one of the forerunners of the theory. We prove such results using only pseudo-differential operators in Chapter XXIII. In Chapter XXVI the propagation of singularities is discussed at great length for operators of principal type. It is the only known approach to general existence theorems for such operators. The completeness of the results obtained has been the reason for the inclusion of this chapter and the following one on subelliptic operators. In addition to Fourier integral operators one needs a fair amount of symplectic geometry then. This topic, discussed in Chapter XXI, has deep roots in classical mechanics but is now equally indispensible in the theory of linear differential operators. Additional symplectic geometry is provided in the discussion of the mixed problem in Chapter XXIV, which is otherwise based only on pseudo-differential operator theory. The same is true of Chapter XXX which is devoted to long range scattering theory. There too the geometry is a perfect guide to the analytical constructs required.

The most conspicuous omission in these books is perhaps the study of analytic singularities and existence theory for hyperfunction solutions. This would have required another volume – and another author. Very little is also included concerning operators with double characteristics apart from a discussion of hypoellipticity in Chapter XXII. The reason for this is in part shortage of space, in part the fact that few questions concerning such operators have so far obtained complete answers although the total volume of results is large. Finally, we have mainly discussed single operators acting on scalar functions or occasionally determined systems. The extensive work done on for example first order systems of vector fields has not been covered at all.

Chapter XVII. Second Order Elliptic Operators

Summary

The study of differential operators with variable coefficients has led to the development of quite elaborate techniques which will be exposed in the following chapters. However, much simpler classical methods will often work in the second order case, and some results are in fact only valid then. Moreover, second order operators (or rather related first order systems) play an important role in many geometrical contexts, so it seems natural to exploit the simplifications which are possible for them. However, the well motivated reader aiming for the most high powered machinery can very well skip this chapter altogether.

Elliptic operators are of constant strength so the results proved in Chapter XIII are applicable to them. The perturbation arguments used in Chapter XIII are recalled in Section 17.1 in the context of elliptic operators with low regularity assumptions on the coefficients and with L^p or Hölder conditions on the solutions. However, we shall not aim for such refinements later on since their main interest comes from the theory of non-linear differential equations which is beyond the scope of this book.

Section 17.2 is mainly devoted to the Aronszajn-Cordes uniqueness theorem stating in particular that if

$$\sum_{|\alpha|\leq 2} a_\alpha(x) D^\alpha u = 0$$

is an elliptic equation where a_α are real valued Lipschitz continuous functions for $|\alpha|=2$ and a_α are bounded for $|\alpha|<2$, then u vanishes identically if u vanishes of infinite order at some point. No such result is true for operators of higher order than two although there are weaker uniqueness theorems concerning solutions vanishing in an open set (see also Chapter XXVIII). In this context we also return to the uniqueness theorems of Section 14.7 where we now allow first order perturbations.

In Section 17.3 we study the simplest classical boundary problem, the Dirichlet problem, consisting in finding a solution of $\Delta u = f$ with given boundary values. When the coefficients are constant and the boundary is flat, a reduction to the results of Section 17.1 is obtained by a simple reflection argument. As in Section 17.1 we can then use perturbation methods to handle variable coefficients and a curved boundary. Thus the boundary is flattened, coefficients are frozen at a boundary point, the norm of the error then committed is estimated, and a Neuman series is applied. Obviously no good information on the singularities of solutions can be obtained in that way. In Section 17.4 we therefore present the Hadamard parametrix method which exploits the simple form of a second order operator in geodesic coordinates to describe the singularities of the fundamental solution with arbitrarily high precision. This method is in fact applicable to all second order operators with real non-degenerate principal symbol. It can also be applied to the Dirichlet problem although with considerable limitations due to the possible occurence of tangential or multiply reflected geodesics.

In Section 17.5 we combine the results of Sections 17.3 and 17.4 to a study of the asymptotic properties of eigenfunctions and eigenvalues of the Dirichlet problem. First we prove the precise error estimate of Avakumovič away from the boundary. A fairly precise analogue at the boundary is given, but one component of the proof cannot be completed until Chapter XXIV. Further refinements will be given in Chapter XXIX.

17.1. Interior Regularity and Local Existence Theorems

Despite the title of the chapter we shall here study a differential operator

$$P(x,D)=\sum_{|\alpha|\leqq m} a_\alpha(x)D^\alpha$$

of arbitrary order m in an open set $X\subset\mathbb{R}^n$. We assume that for some $p\in(1,\infty)$

(i) a_α is continuous when $|\alpha|=m$;

(ii) $P_m(0,D)=\sum_{|\alpha|=m} a_\alpha(0)\,D^\alpha$ is elliptic;

(iii) $a_\alpha\in L^{n/(m-|\alpha|)}_{\text{loc}}$ if $m-|\alpha|<n/p$, $a_\alpha\in L^{p+\varepsilon}_{\text{loc}}$ for some $\varepsilon>0$ if

$$m-|\alpha|=n/p,\quad a_\alpha\in L^p_{\text{loc}} \text{ if } m-|\alpha|>n/p.$$

We can then supplement Theorem 13.2.1 as follows:

Theorem 17.1.1. *If* (i)–(iii) *are fulfilled and* X *is a sufficiently small neighborhood of* 0, *then there is a linear operator* E *in* $L^p(X)$ *such that*

(17.1.1) $L^p(X)\ni f\mapsto D^\alpha Ef\in L^q(X)$ *is continuous if* $p\leqq q\leqq\infty$ *and* $1/q\geqq 1/p-(m-|\alpha|)/n$ *with strict inequality if* $q=\infty$;

(17.1.2) $$P(x,D)Ef=f,\quad f\in L^p(X),$$

(17.1.3) $$EP(x,D)u=u \text{ if } u\in C_0^\infty(X).$$

Proof. Let $p(D)=P_m(0,D)$ and choose $F_0\in\mathscr{S}'$ according to Theorem 7.1.22 so that $\hat{F}_0(\xi)=1/p(\xi)$ when $|\xi|\geqq 1$ and $\hat{F}_0\in C^\infty$. Then it follows from Theorems

7.9.5 and 4.5.9 that

$$\|D^\alpha F_0 * g\|_{L^q} \leqq C\|g\|_{L^p} \quad \text{if } g\in L^p\cap\mathscr{E}',\ 1/q=1/p-(m-|\alpha|)/n,\ q<\infty. \tag{17.1.4}$$

Moreover, $D^\alpha F_0 \in L^r_{\text{loc}}$ if $m-|\alpha|>n(1-1/r)$, for $D^\alpha F_0$ is essentially homogeneous of degree $m-|\alpha|-n>-n/r$.

Let E_0 be a fundamental solution of $p(D)$. Then $F_0-E_0\in C^\infty$. If $g\in L^p(X)$ we define $g_0=g$ in X and $g_0=0$ in $\complement X$, and set $E_0 g=E_0 * g_{0|X}$. From (17.1.4) and the subsequent observations it follows if X is contained in the unit ball that

$$\|D^\alpha E_0 g\|_{L^q(X)} \leqq C\|g\|_{L^p(X)},\quad g\in L^p(X). \tag{17.1.5}$$

Here $1/q=1/p-(m-|\alpha|)/n$ when $m-|\alpha|<n/p$, we choose $q=p(p+\varepsilon)/\varepsilon$ with ε as in condition (iii) when $m-|\alpha|=n/p$, and $q=\infty$ when $m-|\alpha|>n/p$ (take $1/r+1/p=1$). Now

$$P(x,D)E_0 g=p(D)E_0 g+(P(x,D)-p(D))E_0 g=g+Rg,$$
$$Rg=\sum_{|\alpha|=m}(a_\alpha(x)-a_\alpha(0))\,D^\alpha E_0 g+\sum_{|\alpha|<m}a_\alpha(x)\,D^\alpha E_0 g.$$

By Hölder's inequality, (17.1.5) and conditions (i) and (iii), we have

$$\|Rg\|_{L^p(X)}\leqq \tfrac12\|g\|_{L^p(X)},\qquad g\in L^p(X),$$

if X is sufficiently small. Thus $I+R$ is then invertible, and $E=E_0(I+R)^{-1}$ has properties (17.1.1) and (17.1.2) by (17.1.5) and the fact that

$$P(x,D)\,Ef=(I+R)(I+R)^{-1}f=f.$$

Finally, if $f=P(x,D)u$, $u\in C_0^\infty(X)$, then the unique solution of the equation $g+Rg=f$ is $g=p(D)u$, for $E_0 g=u$, hence

$$p(D)u+Rp(D)u=p(D)u+\sum_{|\alpha|=m}(a_\alpha(x)-a_\alpha(0))D^\alpha u+\sum_{|\alpha|<m}a_\alpha(x)\,D^\alpha u$$

is equal to $P(x,D)u$ in X. This completes the proof.

If one replaces the L^p conditions by Hölder conditions one obtains the following theorem instead:

Theorem 17.1.1′. *Assume that for some $\gamma\in(0,1)$ the coefficients of $P(x,D)$ are in C^γ in a neighborhood of 0, and that $P_m(0,D)$ is elliptic. If X is a sufficiently small ball with center at 0 then there exists a linear operator E in $\bar C^\gamma(X)$ such that*

$$\bar C^\gamma(X)\ni f\mapsto D^\alpha Ef\in\bar C^\gamma(X)\quad \textit{is continuous if } |\alpha|\leqq m; \tag{17.1.1$'$}$$

$$P(x,D)\,Ef=f,\quad f\in\bar C^\gamma(X); \tag{17.1.2$'$}$$

$$EP(x,D)u=u\quad \textit{if } u\in C_0^\infty(X). \tag{17.1.3$'$}$$

Here $\bar{C}^\gamma(X)$ is the set of all continuous functions in X such that the norm

$$\sup_{x\in X}|g(x)| + \sup_{x,y\in X}|g(x)-g(y)|/|x-y|^\gamma$$

is finite. If X has radius r, then a C^γ extension to the whole space is given by

$$\begin{aligned} g_0(x) &= g(x), && x\in X;\\ g_0(x) &= g(rx/|x|)(2-|x|/r), && r\leqq |x|\leqq 2r;\\ g_0(x) &= 0, && |x|>2r. \end{aligned}$$

The proof of Theorem 17.1.1′ is identical to that of Theorem 17.1.1 except that g_0 is defined in this way and that (17.1.4) is replaced by the continuity in C^γ when $|\alpha|=m$, which follows from Theorem 7.9.6. We leave the details for the reader since the result will never be used here.

By a slight twist of the proof of Theorem 17.1.1 one can prove a logarithmic convexity theorem for the L^p norms of the derivatives which will be useful later on. To shorten the proofs we exclude lower order terms now. First we prove a lemma.

Lemma 17.1.2. *If $P(D)$ is homogeneous and elliptic of order m, then*

$$\sum_{|\alpha|\leqq m} A^{m-|\alpha|}\|D^\alpha v\|_{L^p} \leqq C(\|P(D)v\|_{L^p} + A^m\|v\|_{L^p}) \tag{17.1.6}$$

if $A>0$ and $D^\alpha v\in L^p$, $|\alpha|\leqq m$.

Proof. Introducing Ax as a new variable instead of x makes A disappear in (17.1.6) so we may assume in the proof that $A=1$. We define F_0 as in the proof of Theorem 17.1.1, thus $P(D)F_0=\delta+\omega$ where $\omega\in\mathscr{S}$. Then we have

$$D^\alpha v = D^\alpha F_0 * P(D)v - (D^\alpha\omega)*v,$$

and (17.1.6) follows since $D^\alpha\omega\in L^1$ and $D^\alpha F_0$ satisfies the hypotheses of Theorem 7.9.5.

Remark. It follows from the proof that C can be taken independent of P if P varies in a compact set of elliptic polynomials of degree m.

Theorem 17.1.3. *Assume that $P_m(x,D)$ satisfies the hypotheses* (i) *and* (ii) *above in a compact neighborhood K of* 0. *Let $X\subset K$ be an open set, and denote by $d(x)$ the distance from $x\in X$ to $\complement X$. If $D^\alpha u\in L^p(X)$, $|\alpha|\leqq m$, it follows then that*

$$\|d(x)^{|\alpha|}D^\alpha u\|_{L^p(X)} \leqq C(\|d(x)^m P_m(x,D)u\|_{L^p(X)} + \|u\|_{L^p(X)})^{|\alpha|/m}\|u\|_{L^p(X)}^{1-|\alpha|/m}, \tag{17.1.7}$$

where C is independent of X.

Proof. Let $B=B(y,R)$ be a ball with radius R and center $y\in X$ with $d(y)\geqq 2R$. Set $\chi_B(x)=\chi((x-y)/R)$ with a fixed $\chi\in C_0^\infty(B(0,1))$ which is equal to 1 in $B(0,\frac12)$. Applying (17.1.6) to $P(D)=P_m(y,D)$ and $v=\chi_B u$ gives with another C

$$\begin{aligned}&\sum_{|\alpha|\leqq m} A^{p(m-|\alpha|)}\int_{B(y,\frac12 R)}|D^\alpha u|^p\,dx\\&\leqq C\Big(\int_{B(y,R)}|P_m(x,D)u|^p\,dx+\varepsilon(R)\sum_{|\alpha|=m}\int_{B(y,R)}|D^\alpha u|^p\,dx\\&\qquad+\sum_{|\alpha|<m}R^{-p(m-|\alpha|)}\int_{B(y,R)}|D^\alpha u|^p\,dx+A^{pm}\int_{B(y,R)}|u|^p\,dx\Big).\end{aligned}$$

Here we have expanded $P(D)(\chi_B u)$ by Leibniz' formula and estimated $\chi_B(P_m(y,D)-P_m(x,D))u(x)$ by means of the modulus of continuity ε of the coefficients. Thus $\varepsilon(R)\to 0$ when $R\to 0$. Now we take $A=M/R$ where M is a large constant and multiply by R^{pm}. This gives

$$\begin{aligned}&\sum_{|\alpha|\leqq m} M^{p(m-|\alpha|)}R^{p|\alpha|}\int_{B(y,\frac12 R)}|D^\alpha u|^p\,dx\\&\leqq C\Big(\int_{B(y,R)}|R^m P_m(x,D)u|^p\,dx+\varepsilon(R)\sum_{|\alpha|=m}\int_{B(y,R)}|R^m D^\alpha u|^p\,dx\\&\qquad+\sum_{|\alpha|<m}R^{p|\alpha|}\int_{B(y,R)}|D^\alpha u|^p\,dx+M^{pm}\int_{B(y,R)}|u|^p\,dx\Big).\end{aligned}$$

With some small R_0 to be chosen later we define

$$R(y)=\min(R_0,d(y)/2)$$

and integrate with respect to $R(y)^{-n}dy$ over X. Since $|R(x)-R(y)|\leqq|x-y|/2$ it follows if $|x-y|<R(y)$ that $|R(y)-R(x)|<R(y)/2$, hence

$$R(y)/2<R(x)<3R(y)/2.$$

On the other hand, if $|x-y|<2R(x)/5$ then $|R(y)-R(x)|<R(x)/5$ so

$$4R(x)/5<R(y)<6R(x)/5.$$

Hence

$$\int_{x\in B(y,R(y))} dy/R(y)^n\leqq(3/2)^n\int_{|x-y|<2R(x)} dy/R(x)^n=3^n\int_{|y|<1}dy,$$

$$\int_{x\in B(y,\frac12 R(y))} dy/R(y)^n\geqq(5/6)^n\int_{|x-y|<2R(x)/5} dy/R(x)^n=3^{-n}\int_{|y|<1}dy.$$

With a new constant C independent of R_0 it follows that

$$\begin{aligned}&\sum_{|\alpha|\leqq m} M^{p(m-|\alpha|)}\int|R(x)^{|\alpha|}D^\alpha u|^p\,dx\\&\leqq C\Big(\int|R(x)^m P_m(x,D)u|^p\,dx+\varepsilon(R_0)\sum_{|\alpha|=m}\int|R(x)^m D^\alpha u|^p\,dx\\&\qquad+\sum_{|\alpha|<m}\int|R(x)^{|\alpha|}D^\alpha u|^p\,dx+M^{pm}\int|u|^p\,dx\Big).\end{aligned}$$

Choose R_0 so small that $C\varepsilon(R_0)<\frac{1}{2}$. When $M\geqq M_0$, say, we can then cancel the two sums on the right-hand side against half of the left-hand side and obtain

$$M^{m-|\alpha|}\|R(x)^{|\alpha|}D^\alpha u\|_{L^p}\leqq C(\|R(x)^m P_m(x,D)u\|_{L^p}+M^m\|u\|_{L^p}).$$

We choose $M=M_0$ if $\|R(x)^m P_m(x,D)u\|_{L^p}<M_0^m\|u\|_{L^p}$; otherwise we take M so that

$$M^m\|u\|_{L^p}=\|R(x)^m P_m(x,D)u\|_{L^p},$$

which gives (17.1.7).

Corollary 17.1.4. *Assume that* P_m *satisfies the hypotheses* (i), (ii) *in a neighborhood* K *of* 0. *If* $D^\alpha u\in L^p$ *in* $K\setminus\{0\}$ *for* $|\alpha|\leqq m$ *and*

$$\int_{R<|x|<2R}|u|^p\,dx=O(R^N),\qquad R\to 0, \tag{17.1.8}$$

$$|P_m(x,D)u|\leqq C\sum_{|\alpha|<m}|D^\alpha u|\,|x|^{|\alpha|-m}\quad \textit{in } K\setminus\{0\} \tag{17.1.9}$$

then it follows if $|\alpha|\leqq m$ *that*

$$\int_{R<|x|<2R}|R^{|\alpha|}D^\alpha u|^p\,dx=O(R^N),\qquad R\to 0. \tag{17.1.10}$$

Proof. We can apply Theorem 17.1.3 with $X=B(0,2R)\setminus B(0,R)$ if R is small. Then

$$d(x)^m|P_m(x,D)u|\leqq C\sum_{|\alpha|<m}d(x)^{|\alpha|}|D^\alpha u|$$

because $d(x)\leqq R\leqq|x|$. Hence it follows from (17.1.7) that

$$S=\sum_{|\alpha|<m}\|d^{|\alpha|}D^\alpha u\|_{L^p(X)}\leqq C_1 S^{(m-1)/m}\|u\|_{L^p(X)}^{1/m}.$$

Thus

$$\sum_{|\alpha|<m}\int_{4R<3|x|<5R}|(R/3)^{|\alpha|}D^\alpha u|^p\,dx\leqq\sum_{|\alpha|<m}\|d^{|\alpha|}D^\alpha u\|_{L^p(X)}^p$$
$$\leqq C_1^{mp}\|u\|_{L^p(X)}^p=O(R^N),$$

which proves (17.1.10) for $|\alpha|<m$. Another application of (17.1.7) gives (17.1.10) when $|\alpha|=m$ also.

With applications to global existence theory in mind we shall discuss in Section 17.2 whether a solution u of a differential equation with principal symbol P_m must be zero when (17.1.8) is valid for all N (or, equivalently, if (17.1.10) is valid for all α with $|\alpha|<m$ and all N). We shall then have to assume that the coefficients of P_m are Lipschitz continuous, that is, $|a_\alpha(x)-a_\alpha(y)|\leqq C|x-y|$, $|\alpha|=m$. Then we can define $P_m(x,D)u$ in the distribu-

tion sense if $D^\alpha u \in L^p$, $|\alpha|<m$, and Theorem 17.1.3 as well as Corollary 17.1.4 can be improved by means of Friedrichs' lemma:

Lemma 17.1.5. *Let $v\in L^p(\mathbb{R}^n)$ and let $|a(x)-a(y)|\leqq M|x-y|$ if $x,y\in\mathbb{R}^n$. If $\phi\in C_0^\infty$ and $\phi_\varepsilon(x)=\phi(x/\varepsilon)\,\varepsilon^{-n}$, then*

$$\|(aD_j v)*\phi_\varepsilon - a(D_j v*\phi_\varepsilon)\|_{L^p} \leqq M\|v\|_{L^p}\int(|\phi|+|y||D_j\phi|)\,dy. \tag{17.1.11}$$

For fixed v the left-hand side tends to 0 when $\varepsilon\to 0$.

Proof. Since C_0^∞ is dense in L^p we may assume that $v\in C_0^\infty$, and it suffices to prove (17.1.11) since it is then obvious that the limit is 0. The quantity to estimate is

$$\begin{aligned}&|\int(a(x-y)-a(x))(D_j v)(x-y)\,\phi_\varepsilon(y)\,dy|\\ &=|\int(a(x-y)-a(x))\,v(x-y)\,D_j\phi_\varepsilon(y)\,dy-\int(D_j a)(x-y)\,v(x-y)\,\phi_\varepsilon(y)\,dy|\\ &\leqq M\int|v(x-y)|\,(|y|\,|D_j\phi_\varepsilon(y)|+|\phi_\varepsilon(y)|)\,dy.\end{aligned}$$

(17.1.11) follows now from Minkowski's inequality since

$$\int(|\phi_\varepsilon(y)|+|y|\,|D_j\phi_\varepsilon(y)|)\,dy$$

is independent of ε.

Let us now return to Theorem 17.1.3 assuming only that $D^\alpha u\in L^p(X)$, $|\alpha|<m$, but that a_α are Lipschitz continuous and that $P_m(x,D)u\in L^p(X)$. Let $\chi_0, \chi_1\in C_0^\infty(X)$, $\chi_1=1$ in a neighborhood of supp χ_0, and set $v=\chi_0 u$. Then $v\in\mathscr{E}'(X)$ and $D^\alpha v\in L^p$, $|\alpha|<m$, $P_m(x,D)v\in L^p$. Choose $\phi\in C_0^\infty$ with $\int\phi\,dx=1$ and set $v_\varepsilon=v*\phi_\varepsilon$ where $\phi_\varepsilon(x)=\phi(x/\varepsilon)/\varepsilon^n$. Then $v_\varepsilon\in C_0^\infty$ and if $b_\alpha=\chi_1 a_\alpha$ we have for small ε

$$P_m(x,D)\,v_\varepsilon=\sum_{|\alpha|=m} b_\alpha D^\alpha v_\varepsilon\to P_m(x,D)\,v \text{ in } L^p$$

by Lemma 17.1.5 since $P_m(x,D)\,v=\sum b_\alpha D^\alpha v$. Hence we can apply (17.1.7) to $v_\varepsilon-v_\delta$ and conclude that $D^\alpha v_\varepsilon$ has a limit in L^p when $\varepsilon\to 0$ if $|\alpha|=m$. Hence $D^\alpha u\in L^p_{\text{loc}}(X)$ when $|\alpha|\leqq m$. The estimate (17.1.7) is therefore true if X is replaced by $\{x\in X; d(x)>\rho\}$. Letting $\rho\to 0$ we obtain (17.1.7) as it stands. Thus Theorem 17.1.3 and Corollary 17.1.4 are valid when a_α are Lipschitz continuous and $D^\alpha u\in L^p$, $|\alpha|<m$.

17.2. Unique Continuation Theorems

We shall begin with a unique continuation theorem similar to Theorem 8.6.5 where operators of higher order are allowed. Let

$$P_m(x,D)=\sum_{|\alpha|=m} a_\alpha(x)\,D^\alpha$$

be defined in an open set $X \subset \mathbb{R}^n$ and assume

(i) a_α is Lipschitz continuous in X,

(ii) P_m is elliptic in X.

By Σ we denote the closed conic set

$$\Sigma = \{(x,N) \in T^*(X) \setminus 0;\ P_m(x,\xi+\tau N) \text{ has a zero } \tau \text{ of multiplicity} \geqq 2 \text{ with } \xi+\tau N \neq 0 \text{ for some } \xi \in \mathbb{R}^n\}. \tag{17.2.1}$$

Of course τ cannot be real then.

Theorem 17.2.1. *If $D^\alpha u \in L^2_{\text{loc}}(X)$, $|\alpha|<m$, and $P_m(x,D)u \in L^2_{\text{loc}}(X)$,*

$$|P_m(x,D)u| \leqq C \sum_{|\alpha|<m} |D^\alpha u| \quad \text{in } X \tag{17.2.2}$$

then $\bar{N}(\operatorname{supp} u) \subset \Sigma$, where Σ is defined by (17.2.1).

For the notation $\bar{N}$ and the global uniqueness results which follow from Theorem 17.2.1 we refer to Sections 8.5 and 8.6. The definitions of Σ and of $\bar{N}$ are both local and invariant under local diffeomorphisms so it is sufficient to prove that if $0 \in X$ and $(0,N) \notin \Sigma$, $N=(0,\dots,0,1)$ then $u=0$ in a neighborhood of 0 if $\operatorname{supp} u \cap \{x; x_n \geqq 0\} \subset \{0\}$. This will be done by means of estimates with respect to high powers of a weight function with maximum in the support of u taken at 0 only.

Set $p(\xi)=P_m(0,\xi)$. Then the hypothesis $(0,N) \notin \Sigma$ means that $p(\xi+i\tau N)$ and $p^{(n)}(\xi+i\tau N)=\partial p(\xi+i\tau N)/\partial \xi_n$ have no common zero $(\xi,\tau) \in \mathbb{R}^{n+1} \setminus \{0\}$. Thus

$$\sum_{|\alpha| \leqq m} \tau^{2(m-|\alpha|)} |\xi^\alpha|^2 \leqq C(|p(\xi+i\tau N)|^2 + \tau^2 |p^{(n)}(\xi+i\tau N)|^2); \quad (\xi,\tau) \in \mathbb{R}^{n+1}; \tag{17.2.3}$$

for both sides are homogeneous of degree $2m$ and can only vanish if $\tau=0$ and $p(\xi)=0$, that is $\xi=0$. Next we need an identity of Treves which is closely related to the commutation relations.

Lemma 17.2.2. *Let $Q(x)=\sum a_j x_j + \sum b_j x_j^2/2$ be a real quadratic polynomial in $\mathbb{R}^n$ and let $P(D)$ be a differential operator with constant coefficients. If $u \in C_0^\infty(\mathbb{R}^n)$ and $v=u e^{Q/2}$ then*

$$\int |P(D)u|^2 e^Q dx = \int |P(D+iQ'/2)v|^2 dx = \int \sum_\alpha \left|\bar{P}^{(\alpha)}(D-iQ'/2)v\right|^2 b^\alpha/\alpha!\, dx. \tag{17.2.4}$$

Proof. The first equality is obvious since $D_j u = e^{-Q/2}(D_j + i\partial_j Q/2)v$. The adjoint of $D_j + i\partial_j Q/2$ is $D_j - i\partial_j Q/2$ so we must show that

$$\bar{P}(D-iQ'/2)P(D+iQ'/2) = \sum P^{(\alpha)}(D+iQ'/2)\bar{P}^{(\alpha)}(D-iQ'/2) b^\alpha/\alpha!. \tag{17.2.5}$$

Now the commutators

$$[D_j - i\partial_j Q/2, D_k + i\partial_k Q/2] = \partial_j \partial_k Q = b_j \delta_{jk}$$

are the same as the commutators of ∂_j and $b_k x_k$. Since as operators

$$\bar{P}(\partial)\, P(b\,x) = \sum (\partial^\alpha P(b\,x))\, \bar{P}^{(\alpha)}(\partial)/\alpha!$$

by Leibniz' rule and this is a purely algebraic consequence of the commutation relations, it follows that (17.2.5) holds.

The following is the crucial estimate in the proof of Theorem 17.2.1.

Proposition 17.2.3. *Let $P_m(x,D)$ satisfy conditions* (i) *and* (ii) *above in a neighborhood of 0 and assume that $(0,N)\notin\Sigma$. Then there is a neighborhood $X_0 \subset X$ of 0 such that with $\phi(x) = x_n + x_n^2/2$ we have for small $\varepsilon>0$ and large $\tau>0$*

$$\sum_{|\alpha|\le m} \tau^{2(m-|\alpha|)-1} \int |D^\alpha u|^2 e^{2\tau\phi}\,dx \le C\int |P_m(\varepsilon x, D)u|^2 e^{2\tau\phi}\,dx, \quad u\in C_0^\infty(X_0). \tag{17.2.6}$$

Proof. If we write $v(x) = u(x)\,e^{\tau\phi(x)}$ then

$$Du = e^{-\tau\phi}(D + i\tau\phi')v \quad \text{and} \quad Dv = e^{\tau\phi}(D - i\tau\phi')u.$$

Apart from the size of the constant, (17.2.6) is therefore equivalent to

$$\sum_{|\alpha|\le m} \tau^{2(m-|\alpha|)-1} \int |D^\alpha v|^2\,dx \le C\int |P_m(\varepsilon x, D + i\tau\phi')v|^2\,dx, \quad v\in C_0^\infty(X_0). \tag{17.2.6$'$}$$

Assume first that the coefficients of P_m are *constant*, thus $P_m = p$. If we apply (17.2.4) with $P = p$ and $Q = 2\tau\phi$ it follows that

$$\int |\bar{p}(D - i\tau\phi')v|^2\,dx + 2\tau\int |\bar{p}^{(n)}(D - i\tau\phi')v|^2\,dx \le \int |p(D + i\tau\phi')v|^2\,dx. \tag{17.2.7}$$

By (17.2.3) and Parseval's formula we have for all $v\in C_0^\infty(\mathbb{R}^n)$

$$\sum_{|\alpha|\le m} \tau^{2(m-|\alpha|)} \int |D^\alpha v|^2\,dx \le C\Big(\int |\bar{p}(D - i\tau N)v|^2\,dx + \tau^2\int |\bar{p}^{(n)}(D - i\tau N)v|^2\,dx\Big) \tag{17.2.3$'$}$$

If $v\in C_0^\infty(X_0)$ and $|x|<\delta$ in X_0, it follows from (17.2.3)$'$ that

$$\sum_{|\alpha|\le m} \tau^{2(m-|\alpha|)} \int |D^\alpha v|^2\,dx \le 2C\Big(\int |\bar{p}(D - i\tau\phi')v|^2\,dx + \tau^2\int |\bar{p}^{(n)}(D - i\tau\phi')v|^2\,dx\Big) + C'(1+\delta^2\tau^2) \sum_{|\alpha|\le m} \tau^{2(m-1-|\alpha|)} \int |D^\alpha v|^2\,dx. \tag{17.2.8}$$

When δ is small and τ is large we have $C'(1+\delta^2\tau^2)<\tau^2/2$ which allows us to cancel the last sum against half of the left hand side. (17.2.6) is then a consequence of (17.2.8) and (17.2.7).

To complete the proof we need an elementary lemma which allows us to handle variable coefficients. We denote the L^2 norm simply by $\| \ \|$.

Lemma 17.2.4. *Let $X\subset\mathbb{R}^n$ be an open set, and let A be a Lipschitz continuous function with $|A(x)-A(y)|\leqq L|x-y|$ for $x, y\in X$. Then*

$$|\int A(x)(D^\alpha u(x)\overline{D^\beta v(x)}-D^\beta u(x)\overline{D^\alpha v(x)})\,dx|\leqq|\alpha+\beta|\,LM$$

if $u, v\in C_0^\infty(X)$ and

$$\|D^{\alpha'} u\|\,\|D^{\beta'} v\|\leqq M \qquad when\ |\alpha'+\beta'|<|\alpha+\beta|,\quad \max(|\alpha'|,|\beta'|)\leqq\max(|\alpha|,|\beta|).$$

Also the last inequality can be taken strict when $|\alpha|\neq|\beta|$.

Proof. This is obvious when $\alpha+\beta=0$. If $|\alpha+\beta|=1$ we just have to note that

$$\int A(x)(D_j u(x)\overline{v(x)}-u(x)\overline{D_j v(x)})\,dx=-\int D_j A(x)\,u(x)\overline{v(x)}\,dx.$$

An integration by parts also gives the statement when $|\alpha|=|\beta|=1$,

$$\int A(x)(D_j u(x)\overline{D_k v(x)}-D_k u(x)\overline{D_j v(x)})\,dx=$$
$$-\int u(x)(D_jA(x)\overline{D_k v(x)}-D_kA(x)\overline{D_j v(x)})\,dx.$$

These two identities allow us to exchange indices between α and β and transfer excess derivatives at a cost of LM for each index affected.

End of Proof of Proposition 17.2.3. Writing $P_m(0,D)=p(D)$ and $r(x,D)=P_m(x,D)-p(D)$ now, we know by hypothesis that the coefficients of $r(\varepsilon x,D)$ and their Lipschitz constants are $O(\varepsilon)$ in X. With the notation in the first part of the proof we form

$$\int|P_m(\varepsilon x,D+i\tau\phi')v|^2\,dx-\int|\overline{P}_m(\varepsilon x,D-i\tau\phi')v|^2\,dx.$$

Inserting $P_m=p+r$ we first obtain the terms

$$\int|p(D+i\tau\phi')v|^2\,dx-\int|\bar{p}(D-i\tau\phi')v|^2\,dx\geqq2\tau\int|\bar{p}^{(n)}(D-i\tau\phi')v|^2\,dx.$$

The other terms where no derivative falls on ϕ' are of the form

$$\tau^{2m-|\alpha|-|\beta|}\int A(x)(D^\alpha v(x)\overline{D^\beta v(x)}-D^\beta v(x)\overline{D^\alpha v(x)})\,dx;\qquad|\alpha|\leqq m,\ |\beta|\leqq m;$$

where the Lipschitz constant of A is $O(\varepsilon)$. These terms can be estimated by means of Lemma 17.2.4. In addition there are terms of the form

$$\tau^\nu\int A(x)\,D^\alpha v(x)\overline{D^\beta v(x)}\,dx;\qquad\nu+|\alpha|+|\beta|<2m,\qquad|\alpha|\leqq m,\ |\beta|\leqq m;$$

where $\sup|A| = O(\varepsilon)$. Thus

$$\|\bar{P}_m(\varepsilon x, D - i\tau\phi')v\|^2 + \tau\|\bar{p}^{(n)}(D - i\tau\phi')v\|^2 \leqq \|P_m(\varepsilon x, D + i\tau\phi')v\|^2 + C\varepsilon \sum_{|\alpha|\leqq m} \tau^{2(m-|\alpha|)-1} \|D^\alpha v\|^2.$$

If we observe that (17.2.8) remains valid with $\bar{p}(D - i\tau\phi')$ replaced by $\bar{P}_m(\varepsilon x, D - i\tau\phi')$ provided that $\varepsilon < \delta$, we complete the proof of (17.2.6) just as in the constant coefficient case.

Proof of Theorem 17.2.1. We recall that it suffices to prove that if $0 \in X$ and $(0, N) \notin \Sigma$, $N = (0, \dots, 0, 1)$ then $u = 0$ in a neighborhood of 0 if $\operatorname{supp} u \cap \{x; x_n \geqq 0\} \subset \{0\}$. In doing so we set $u_\varepsilon(x) = u(\varepsilon x)$ where ε is chosen so small that (17.2.6) is valid for a neighborhood $X_0 \subset X/\varepsilon$ of 0. Let $\chi \in C_0^\infty(X_0)$ be equal to 1 in a neighborhood V of 0, and set $U = \chi u_\varepsilon$. If $P_m(x, D)u = f$ then

$$P_m(\varepsilon x, D)\, U = \varepsilon^m \chi(x) f(\varepsilon x) + \sum_{0<|\alpha|\leqq m} D^\alpha \chi\, P_m^{(\alpha)}(\varepsilon x, D)\, u_\varepsilon/\alpha!$$

which implies that $P_m(\varepsilon x, D)\, U \in L^2$ and that, by (17.2.2),

$$|P_m(\varepsilon x, D)\, U| \leqq C \sum_{|\alpha|<m} \varepsilon^{m-|\alpha|} |D^\alpha U| \quad \text{in } V.$$

By the remarks at the end of Section 17.1 we have $D^\alpha U \in L^2$ when $|\alpha| \leqq m$, so it is clear that (17.2.6) may be applied to U. If $\operatorname{supp}\chi$ is small enough we have $\phi \leqq -c$ for some $c > 0$ in $\operatorname{supp} U \setminus V$. Hence we obtain using (17.2.6)

$$\tau^{\frac{1}{2}} \sum_{|\alpha|<m} \|e^{\tau\phi} D^\alpha U\| \leqq C\|e^{\tau\phi} P_m(\varepsilon x, D)\, U\| \leqq C' \sum_{|\alpha|<m} \|e^{\tau\phi} D^\alpha U\| + C'' e^{-c\tau}.$$

For large τ it follows that

$$\tau^{\frac{1}{2}} \sum_{|\alpha|<m} \|e^{\tau\phi} D^\alpha U\| \leqq 2C'' e^{-c\tau}.$$

Hence $U = 0$ when $\phi > -c$, which proves the theorem.

In the second order case the following lemma shows that the set Σ has a very simple description:

Lemma 17.2.5. *Let p be a quadratic form in $\mathbb{R}^n$ with complex coefficients which is elliptic, that is, $p(\xi) \neq 0$ when $0 \neq \xi \in \mathbb{R}^n$. If $N \in \mathbb{R}^n \setminus 0$ and $\xi \in \mathbb{R}^n \setminus \mathbb{R}N$, $n \neq 2$, it follows that the equation $p(\xi + \tau N) = 0$ has one root with $\operatorname{Im}\tau > 0$ and one with $\operatorname{Im}\tau < 0$. When $n = 2$ the roots are distinct unless p is the square of a linear form.*

Proof. $\mathbb{R}^n \setminus \mathbb{R}N$ is connected if $n > 2$. Since $p(\xi + \tau N)$ has no real zero if $\xi \in \mathbb{R}^n \setminus \mathbb{R}N$ it follows that the number of zeros with $\operatorname{Im}\tau > 0$ is independent of ξ. Replacing ξ by $-\xi$ changes the sign of τ also so there must be one zero in each half plane. When $n=2$ there is a factorization $p(\xi)=L_1(\xi)L_2(\xi)$ with linear factors L_1 and L_2. They must be proportional if they have a common zero; and then they can be chosen equal.

If $m=2$ it follows that Σ is empty when $n>2$ and that $\Sigma=\bigcup(T_x^*\setminus 0)$ for all x such that $P_m(x,\xi)$ is the square of a linear form when $n=2$. If X is connected and $P_m(x,\xi)$ is real for some x then Σ is empty, for the two zeros of $P_m(x,\xi+\tau N)$ must remain in different half planes for reasons of continuity.

In what follows we shall only consider the second order case and shall then use the notation $p(x,D)$ instead of $P_m(x,D)$. We shall prove that if u satisfies a weakened form of (17.2.2) and vanishes of infinite order at a point where the coefficients are real, then u is equal to 0.

Theorem 17.2.6. *Let $p(x,D)=\sum a_{jk}(x)D_jD_k$ be an elliptic operator in a connected neighborhood X of 0 such that $a_{jk}(0)$ is real, a_{jk} is continuous in X, Lipschitz continuous in $X\setminus\{0\}$, and $|a'_{jk}|\leq C|x|^{\delta-1}$ for some $\delta>0$. If $D^\alpha u\in L^2_{\mathrm{loc}}$, $|\alpha|\leq 1$, and*

$$|p(x,D)u|\leq C\sum_{|\alpha|\leq 1}|x|^{\delta+|\alpha|-2}|D^\alpha u|, \tag{17.2.2)'}$$

$$\int_{|x|<\varepsilon}|u|^2\,dx=O(\varepsilon^N),\quad \varepsilon\to 0, \tag{17.2.9}$$

for every N, then $u=0$ in X.

Proof. Since (17.2.2)′ implies (17.1.9) it follows from Corollary 17.1.4 in the extended form discussed at the end of Section 17.1 that for $|\alpha|\leq 2$ and all N

$$\int_{\varepsilon<|x|<2\varepsilon}|D^\alpha u|^2\,dx=O(\varepsilon^N),\quad \varepsilon\to 0. \tag{17.2.9)'}$$

Hence u is the sum of a function in $H^{\mathrm{loc}}_{(2)}(X)$ and a distribution with support at 0. However, no distribution with support at 0 is in L^2_{loc} so it follows that $u\in H^{\mathrm{loc}}_{(2)}(X)$. By Theorem 17.2.1 it suffices to show that $u=0$ in a neighborhood of 0. Without restriction we may assume that $p(0,D)=\sum D_j^2$.

As in the proof of Proposition 14.7.1 we introduce polar coordinates in $\mathbb{R}^n\setminus\{0\}$ by writing $x=e^t\omega$ where $t\in\mathbb{R}$ and $\omega\in S^{n-1}$. Then we have

$$\partial/\partial x_j=e^{-t}(\omega_j\partial/\partial t+\Omega_j)$$

where Ω_j is a vector field in S^{n-1}. With the notation $p(x,D)=\sum a_{jk}(x)D_jD_k$ it follows that

$$p(x,D)=-e^{-2t}\sum a_{jk}(e^t\omega)(\omega_j(\partial/\partial t-1)+\Omega_j)(\omega_k\partial/\partial t+\Omega_k).$$

With $U(t,\omega)=u(e^t\omega)$ the inequality (17.2.2)′ can be written

$$\Big|\sum a_{jk}(e^t\omega)(\omega_j(\partial/\partial t-1)+\Omega_j)(\omega_k\partial/\partial t+\Omega_k)U\Big|\leq C\sum_{|\alpha|\leq 1}e^{\delta t}|U_\alpha| \tag{17.2.2)''}$$

where $U_\alpha=(\omega\,\partial/\partial t+\Omega)^\alpha U$. By assumption we have $a_{jk}(e^t\omega)=\delta_{jk}+O(e^{\delta t})$ as $t\to-\infty$, first order derivatives are $O(e^{\delta t})$, and

$$\sum(\omega_j(\partial/\partial t-1)+\Omega_j)(\omega_j\,\partial/\partial t+\Omega_j)=\partial^2/\partial t^2+(n-2)\,\partial/\partial t+\sum\Omega_j^2$$

since $\sum\omega_j\Omega_j=0$ and $\sum\Omega_j\omega_j=\sum r\,\partial\omega_j/\partial x_j=\sum r\,\partial(x_j/r)/\partial x_j=n-1$. The operator $\sum\Omega_j^2$ is the Laplace-Beltrami operator Δ_ω in the unit sphere. The adjoint of Ω_j as an operator in $L^2(S^{n-1})$ is $(n-1)\,\omega_j-\Omega_j$. In fact,

$$\begin{aligned}\int(\Omega_j u)\,v\,dx+\int u\,\Omega_j v\,dx=\int\Omega_j(u\,v)\,dx&=\int|x|\,\partial(u\,v)/\partial x_j\,dx-\iint\omega_j\,\partial(u\,v)/\partial r\,r^n\,d\omega\,dr\\&=-\int\omega_j u\,v\,dx+n\int\omega_j u\,v\,r^{n-1}\,d\omega\,dr\\&=(n-1)\int\omega_j u\,v\,dx.\end{aligned}$$

In spite of this Δ_ω is of course self-adjoint; indeed, we have

$$\sum((n-1)\,\omega_j-\Omega_j)^2=(n-1)^2-(n-1)\sum\Omega_j\omega_j+\sum\Omega_j^2=\sum\Omega_j^2.$$

In the proof of Theorem 17.2.1 the essential estimate (17.2.7) was obtained from (17.2.4) thanks to the positivity of b_j, that is, the convexity of the exponent ϕ. To obtain a similar effect we introduce for some ε with $0<\varepsilon<\delta$ a new variable T instead of t,

$$t=T+e^{\varepsilon T};\qquad dt/dT=1+\varepsilon\,e^{\varepsilon T}>0.$$

Note that $T<t<T+1<T/2$ if $T<-2$. After multiplication by $(1+\varepsilon e^{\varepsilon T})^2$ the operator in the left-hand side of (17.2.2)'' becomes

$$Q=\partial^2/\partial T^2+c(T)\,\partial/\partial T+(1+\varepsilon\,e^{\varepsilon T})^2\sum\Omega_j^2+\sum_{|\alpha|+j\leqq 2}c_{\alpha,j}(T,\omega)(\partial/\partial T)^j\,\Omega^\alpha.$$

Here $c(T)=(n-2)(1+\varepsilon\,e^{\varepsilon T})-\varepsilon^2\,e^{\varepsilon T}/(1+\varepsilon\,e^{\varepsilon T})$ is close to $n-2$ at $-\infty$, and

(17.2.10) $$c_{\alpha,j}=O(e^{\delta t}),\quad dc_{\alpha,j}=O(e^{\delta t})\ \text{as}\ T\to-\infty.$$

(Note that this change of variables is not smooth in the original variables.) We shall prove that for some T_0

(17.2.11) $$\begin{aligned}\sum_{j+|\alpha|\leqq 2}\tau^{3-2(j+|\alpha|)}\iint|(\partial/\partial T)^j\,\Omega^\alpha\,U|^2\,e^{-(2\tau-\varepsilon)T}\,d\omega\,dT\\ \leqq C\iint|QU|^2\,e^{-2\tau T}\,d\omega\,dT,\qquad U\in C_0^\infty((-\infty,T_0)\times S^{n-1}).\end{aligned}$$

(When $|\alpha|=2$ we define Ω^α for example as a product $\Omega_j\Omega_k$ with $j\leqq k$.) This will serve the same purpose as (17.2.6) did in the proof of Theorem 17.2.1.

Proof of (17.2.11). Set $U=e^{\tau T}\,V$ and

$$Q_\tau V=e^{-\tau T}\,Q(e^{\tau T}\,V).$$

Thus Q_τ is obtained from Q when $\partial/\partial T$ is replaced by $\partial/\partial T+\tau$. Then (17.2.11) is equivalent to

(17.2.11)' $$\begin{aligned}\sum_{j+|\alpha|\leqq 2}\tau^{3-2(j+|\alpha|)}\iint|(\partial/\partial T)^j\,\Omega^\alpha\,V|^2\,e^{\varepsilon T}\,d\omega\,dT\\ \leqq C\iint|Q_\tau V|^2\,d\omega\,dT,\qquad V\in C_0^\infty((-\infty,T_0)\times S^{n-1}).\end{aligned}$$

Let Q_τ^- be the operator obtained from Q_τ when $\partial/\partial T$ and Ω_j are replaced by $-\partial/\partial T$ and $-\Omega_j$ while $c_{\alpha,j}$ is replaced by $\overline{c_{\alpha,j}}$. (With our present notation this is essentially equivalent to the complex conjugation in the proof of Proposition 17.2.3.) We shall examine the difference

$$\iint |Q_\tau V|^2\, d\omega\, dT - \iint |Q_\tau^- V|^2\, d\omega\, dT. \tag{17.2.12}$$

In addition to paying attention to the powers of τ and orders of differentiation as in the proof of Proposition 17.2.3 we must now take the exponential decrease at $-\infty$ into account. It will follow from (17.2.10) that the terms involving $c_{\alpha,j}$ are not important, so we first consider the other terms in Q_τ and Q_τ^-,

$$(\pm\partial/\partial T+\tau)^2 + c(T)(\pm\partial/\partial T+\tau) + (1+\varepsilon e^{\varepsilon T})^2 \sum \Omega_j^2.$$

The corresponding contribution to (17.2.12) is

$$\begin{aligned} &4\,\mathrm{Re}((\partial^2/\partial T^2+\tau^2+c(T)\tau+(1+\varepsilon e^{\varepsilon T})^2\textstyle\sum\Omega_j^2)V,\ (2\tau+c)\partial V/\partial T)\\ &\quad = -2\,\mathrm{Re}(c'(T)\partial V/\partial T,\ \partial V/\partial T) - 2((3\tau^2+2c\tau)c'V,V) + 2\textstyle\sum(h\Omega_j V,\Omega_j V).\end{aligned}$$

Here we have used that $\sum\omega_j\Omega_j=0$, and

$$\begin{aligned} h=\frac{d}{dT}(1+\varepsilon e^{\varepsilon T})^2(2\tau+c) &= ((2\tau+c)2\varepsilon^2 e^{\varepsilon T} + (1+\varepsilon e^{\varepsilon T})c')(1+\varepsilon e^{\varepsilon T})\\ &\geqq 2\varepsilon^2\tau e^{\varepsilon T}\end{aligned}$$

when τ is large enough. All other terms in (17.2.12) are of the form

$$\begin{aligned} &\tau^j \iint a(T,\omega)\big((\partial/\partial T)^{\alpha_0}\Omega^\alpha V\,\overline{(\partial/\partial T)^{\beta_0}\Omega^\beta V}\\ &\quad -((-\partial/\partial T)^{\beta_0}(-\Omega)^\beta V\,\overline{(-\partial/\partial T)^{\alpha_0}(-\Omega)^\alpha V}\big)\,d\omega\, dT\end{aligned}$$

with $j+\alpha_0+|\alpha|+\beta_0+|\beta|\leqq 4$, $\alpha_0+|\alpha|\leqq 2$, $\beta_0+|\beta|\leqq 2$ and $a=O(e^{\delta T})$, $a'=O(e^{\delta T})$. We can estimate them by an obvious modification of the proof of Lemma 17.2.4, which can also be used directly after decomposition by an appropriate partition of unity. (Recall that the adjoint of Ω_j differs from $-\Omega_j$ by an operator of order 0.) Hence we obtain

$$\begin{aligned} &\|Q_\tau^- V\|^2 + 4\varepsilon^2\tau\textstyle\sum\iint|\Omega_j V|^2 e^{\varepsilon T}\, d\omega\, dT \leqq \|Q_\tau V\|^2\\ &\quad + C\sum_{j+|\alpha|\leqq 2}\tau^{3-2(j+|\alpha|)}\iint|(\partial/\partial T)^j\Omega^\alpha V|^2(e^{\delta T}+\tau^{-1}e^{\varepsilon T})\,d\omega\, dT.\end{aligned} \tag{17.2.13}$$

This is an adequate substitute for (17.2.7). Instead of (17.2.8) we shall prove that for some T_0

$$\begin{aligned} &\sum_{j+|\alpha|\leqq 2}\tau^{4-2(j+|\alpha|)}\iint|(\partial/\partial T)^j\Omega^\alpha V|^2 e^{\varepsilon T}\, d\omega\, dT\\ &\quad\leqq C\iint(|Q_\tau V|^2+|Q_\tau^- V|^2+\tau^2\textstyle\sum|\Omega_j V|^2)e^{\varepsilon T}\,d\omega\, dT,\\ &V\in C_0^\infty((-\infty,T_0)\times S^{n-1}),\end{aligned} \tag{17.2.14}$$

if τ is large enough. If we introduce $Ve^{\varepsilon T/2}$ as a new dependent variable, freeze the coefficients at $-\infty$, and drop terms of lower order, we find that it is sufficient to prove that

$$\begin{aligned}(17.2.15)\qquad &\sum_{j+|\alpha|\leqq 2} \tau^{4-2(j+|\alpha|)} \iint |(\partial/\partial T)^j \Omega^\alpha V|^2\, d\omega\, dT \\ &\leqq C \iint (|((\partial/\partial T+\tau)^2+\Delta_\omega) V|^2 + |((\partial/\partial T-\tau)^2+\Delta_\omega) V|^2 \\ &\quad +\tau^2 \sum |\Omega_j V|^2)\, d\omega\, dT.\end{aligned}$$

The integral in the right-hand side is equal to

$$\iint (2|(\partial^2/\partial T^2+\tau^2+\Delta_\omega) V|^2 + 8\tau^2 |\partial V/\partial T|^2 + \tau^2 \sum |\Omega_j V|^2)\, d\omega\, dT.$$

Furthermore

$$\begin{aligned}\|(\partial^2/\partial T^2+\tau^2+\Delta_\omega) V\|^2 &= \|\partial^2 V/\partial T^2\|^2 + \tau^4 \|V\|^2 + \|\Delta_\omega V\|^2 \\ &\quad + 2\sum \|\partial/\partial T \Omega_j V\|^2 - 2\tau^2 (\sum \|\Omega_j V\|^2 + \|\partial V/\partial T\|^2).\end{aligned}$$

By the ellipticity of Δ_ω we have

$$\sum_{|\alpha|=2} \|\Omega^\alpha V\|^2 \leqq C(\|\Delta_\omega V\|^2 + \sum_{|\alpha|\leqq 1} \|\Omega^\alpha V\|^2).$$

If we combine these estimates we obtain (17.2.15), hence (17.2.14). Using (17.2.13) to estimate the right hand side of (17.2.14) we obtain when τ is large enough

$$\begin{aligned}&\sum_{j+|\alpha|\leqq 2} \tau^{4-2(j+|\alpha|)} \iint |(\partial/\partial T)^j \Omega^\alpha V|^2 (e^{\varepsilon T} - C'(e^{\delta T} + \tau^{-1} e^{\varepsilon T})/\varepsilon^2)\, d\omega\, dT \\ &\leqq C\tau/2\varepsilon^2 \|Q_\tau V\|^2.\end{aligned}$$

When τ is sufficiently large and e^{T_0} is sufficiently small, the estimate (17.2.11)′ follows.

End of Proof of Theorem 17.2.6. First recall that the function u in the theorem satisfies the differential inequality (17.2.2)″ when considered as a function U of ω and t. When we take $t=T+e^{\varepsilon T}$ the inequality is replaced by

$$(17.2.2)''' \qquad |QU| \leqq C' \sum_{|\alpha|\leqq 1} e^{\delta T} U_\alpha$$

where $U_0=|U|$ and $U_\alpha=|\partial U/\partial T|+|\Omega^\alpha U|$ when $|\alpha|=1$. Choose $\psi\in C^\infty(\mathbb{R})$ equal to 1 in $(-\infty, T_0-1)$ and 0 in (T_0, ∞), and set

$$U^\psi(T,\omega)=\psi(T)\, U(T,\omega).$$

It follows from (17.2.9)′ that the derivatives of U^ψ of order $\leqq 2$ multiplied by e^{-NT} are in L^2 for any N. By cutting u off for large negative T and regularizing we conclude that (17.2.11) is valid for U^ψ. The right-hand side can then be estimated by

$$C''(e^{-2\tau(T_0-1)}+\sum_{|\alpha|\leqq 1}\iint |U_\alpha^\psi|^2 e^{-2T(\tau-\delta)}\,d\omega\,dT)$$

where $U_0^\psi=|U^\psi|$ and $U_\alpha^\psi=|\partial U^\psi/\partial T|+|\Omega^\alpha U^\psi|$ if $|\alpha|=1$. Hence (17.2.11) gives for large τ, since $\varepsilon\leqq 2\delta$,

$$\sum_{j+|\alpha|\leqq 2}\tau^{3-2(j+|\alpha|)}\iint |(\partial/\partial T)^j\Omega^\alpha U^\psi|^2 e^{-(2\tau-\varepsilon)T}\,d\omega\,dT\leqq 2C''e^{-2\tau}(T_0-1).$$

When $\tau\to\infty$ it follows that $U=0$ when $T<T_0-1$. Thus the function u in Theorem 17.2.6 vanishes in a neighborhood of 0. The proof is complete.

As we saw in Chapters X and XIII a major application of uniqueness theorems is the proof of global existence theorems. We shall give another example here using Theorem 17.2.1.

Theorem 17.2.7. *Let $a_{jk}(x)$ be Lipschitz continuous in an open set $X\subset\mathbb{R}^n$ (or a C^2 manifold), $a_{jk}=a_{kj}$, and assume that $(\operatorname{Re}a_{jk}(x))$ is positive definite. Then*

$$\sum D_j(a_{jk}D_k u)=f \tag{17.2.16}$$

implies $u\in H_{(2)}^{\mathrm{loc}}(X)$ if $u\in L_{\mathrm{loc}}^2(X)$, which makes the equation defined, and $f\in L_{\mathrm{loc}}^2(X)$. Moreover, the equation has a solution $u\in H_{(2)}^{\mathrm{loc}}(X)$ for every $f\in L_{\mathrm{loc}}^2(X)$.

Proof. 1) To prove the regularity statement we first show that

$$u\in L_{\mathrm{loc}}^2(X),\quad \sum D_j(a_{jk}D_k u)\in H_{(-1)}^{\mathrm{loc}}(X)\Rightarrow u\in H_{(1)}^{\mathrm{loc}}(X).$$

This statement has the advantage that the hypotheses remain valid if we replace u by χu where $\chi\in C_0^\infty(X)$. In the proof we may therefore assume that $u\in\mathscr{E}'(X)$. Let $K\subset X$ be a compact neighborhood of supp u. If $v\in C_0^\infty(K)$ then

$$\begin{aligned}\sum\int |D_j v|^2\,dx &\leqq C\operatorname{Re}\int\sum a_{jk}(x)\,D_k v\,\overline{D_j v}\,dx\\ &= C\operatorname{Re}\int\sum (D_j a_{jk}(x)\,D_k v)\bar v\,dx.\end{aligned} \tag{17.2.17}$$

As at the end of Section 17.1 we choose $v=u_\varepsilon=u*\phi_\varepsilon$ where $\phi\in C_0^\infty$, $\int\phi\,dx=1$ and $\phi_\varepsilon(x)=\phi(x/\varepsilon)\varepsilon^{-n}$. By Lemma 17.1.5 we know that

$$a_{jk}D_k u_\varepsilon-(a_{jk}D_k u)*\phi_\varepsilon$$

is bounded in L^2 when $\varepsilon\to 0$. Hence

$$f_\varepsilon=\sum D_j(a_{jk}D_k u_\varepsilon)=\sum (D_j a_{jk}D_k u)*\phi_\varepsilon+\sum D_j(\sum a_{jk}D_k u_\varepsilon-(\sum a_{jk}D_k u)*\phi_\varepsilon)$$

is bounded in $H_{(-1)}$. By (17.2.17) it follows that

$$\|u_\varepsilon\|_{(1)}^2\leqq C_1\sum\|D_j u_\varepsilon\|^2\leqq C_2\|f_\varepsilon\|_{(-1)}\|u_\varepsilon\|_{(1)}$$

which implies that $\|u_\varepsilon\|_{(1)}\leqq C_2\|f_\varepsilon\|_{(-1)}$ is bounded. Thus $u\in H_{(1)}$. If $f\in L^2$ it follows as we saw at the end of Section 17.1 that $u\in H_{(2)}$.

2) To construct u it is enough to show that

$$|(f,\phi)| \leqq \|M \sum D_j \bar{a}_{jk} D_k \phi\|_{L^2}, \quad \phi \in C_0^\infty(X) \tag{17.2.18}$$

for some positive continuous function M. In fact, by the Hahn-Banach theorem it follows then that for some $g \in L^2$

$$(f,\phi) = (g, M \sum D_j \bar{a}_{jk} D_k \phi), \quad \phi \in C_0^\infty(X),$$

which means that $u = Mg$ satisfies the equation (17.2.16); by the first part of the proof u is in $H_{(2)}^{\text{loc}}$. If K is a compact subset of X we also know from the first part of the proof that

$$\|\phi\|_{(1)} \leqq C \| \sum D_j \bar{a}_{jk} D_k \phi \|_{(-1)}, \quad \phi \in C_0^\infty(K)$$

which of course implies (17.2.18) if $M > C \|f\|_{L^2(K)}$ and supp $\phi \subset K$. As in the proof of Theorem 10.7.8 the only problem is to increase the compact set K without increasing M much on a somewhat smaller set. Let $K_1, K_2, \dots$ be a sequence of compact sets with union X, each contained in the interior of the following one and chosen so that $X \setminus K_j$ has no component which is precompact in X. Let M be a function such that (17.2.18) holds when $\phi \in C_0^\infty(K_j)$ for some $j > 2$, and let $\varepsilon > 0$. Then we claim that (17.2.18) remains valid when $\phi \in C_0^\infty(K_{j+1})$ for some $\tilde{M}$ such that $\tilde{M} = (1+\varepsilon)M$ in K_{j-2}. Taking a sequence ε_j with $\prod(1+\varepsilon_j) < \infty$ we conclude from this that there is a function M such that (17.2.18) is valid.

If the claim were false then (17.2.18) would be false with $\tilde{M} \geqq (1+\varepsilon)M$ everywhere and $\tilde{M}$ very large on $\complement K_{j-1}$, so we can find a sequence $\phi_N \in C_0^\infty(K_{j+1})$ with

$$(f, \phi_N) = 1, \quad (1+\varepsilon)\|M \sum D_j \bar{a}_{jk} D_k \phi_N\|_{L^2} \leqq 1,$$
$$\|\sum D_j \bar{a}_{jk} D_k \phi_N\|_{L^2(\complement K_{j-1})} \leqq 1/N.$$

By the first part of the proof the sequence ϕ_N remains bounded in $H_{(2)}$ as $N \to \infty$ so it has a subsequence converging in $H_{(1)}$ to a limit Φ with

$$(f, \Phi) = 1, \quad (1+\varepsilon)\|M \sum D_j \bar{a}_{jk} D_k \Phi\|_{L^2} \leqq 1,$$
$$\operatorname{supp} \Phi \subset K_{j+1}, \quad \sum D_j \bar{a}_{jk} D_k \Phi = 0 \quad \text{in } \complement K_{j-1}$$

By hypothesis every component of $X \setminus K_{j-1}$ contains points outside the compact set K_{j+1}. Hence it follows from Theorem 17.2.1 that $\Phi = 0$ there, so supp $\Phi \subset K_{j-1}$. By the first part of the proof $\Phi \in H_{(2)}$. If we regularize Φ we obtain functions in $C_0^\infty(K_j)$ violating the assumption that (17.2.18) is valid in $C_0^\infty(K_j)$, which completes the proof.

We shall now prove an extension of Theorem 14.7.2 which gives a uniqueness theorem for a perturbation of the Laplacean at infinity rather than at 0. After passage to polar coordinates the problem will be very close to that discussed in Theorem 17.2.6. We assume now that a_{jk} are Lipschitz continuous and that

$$|a_{jk}(x)-\delta_{jk}| \leqq C/|x|^{1+\delta}, \quad |a'_{jk}(x)| \leqq C/|x|^{2+\delta} \tag{17.2.19}$$

in a neighborhood of infinity. Thus $p(x,D)=\sum a_{jk}(x)D_jD_k$ approaches minus the Laplacean at infinity; any other homogeneous elliptic operator of second order with *real* coefficients could of course be used as well.

Theorem 17.2.8. *Let X be a connected neighborhood of ∞ in $\mathbb{R}^n$ where (17.2.19) is valid, and let $(1+|x|)^\tau D^\alpha u \in L^2(X)$ for all τ when $|\alpha| \leqq 1$. If $\lambda>0$ and*

$$|p(x,D)u-\lambda u| \leqq C|x|^{-1} \sum_{|\alpha|\leqq 1} |D^\alpha u|, \quad x\in X, \tag{17.2.20}$$

it follows that $u=0$.

Proof. As in the proof of Theorem 17.2.6 we introduce polar coordinates $x=e^t\omega$ where $t\in\mathbb{R}$ and $\omega\in S^{n-1}$, and obtain with $U(t,\omega)=u(e^t\omega)$

$$|\sum a_{jk}(e^t\omega)(\omega_j(\partial/\partial t-1)+\Omega_j)(\omega_k\partial/\partial t+\dot{\Omega}_k)U+\lambda e^{2t}U| \leqq C\sum_{|\alpha|\leqq 1} e^{(1-|\alpha|)t}|U_\alpha| \tag{17.2.20$'$}$$

where $U_\alpha=(\omega\partial/\partial t+\Omega)^\alpha U$. By assumption we have $a_{jk}(e^t\omega)=\delta_{jk}+O(e^{-t-\delta t})$ as $t\to+\infty$, first order derivatives are also $O(e^{-t-\delta t})$, and at infinity the sum is

$$(\partial^2/\partial t^2+(n-2)\partial/\partial t+\Delta_\omega)U$$

as before. Since we work with large positive t now we set $t=T-e^{-\varepsilon T}$ for some $\varepsilon\in(0,\delta)$, which is legitimate since $dt/dT=1+\varepsilon e^{-\varepsilon T}>0$. When $t>0$ we have $T>0$, hence $|t-T|<1$. After multiplication by $(dt/dT)^2$ the operator in the left-hand side of (17.2.20)′ becomes

$$Q=\partial^2/\partial T^2+c(T)\partial/\partial T+(1+\varepsilon e^{-\varepsilon T})^2\Delta_\omega+\sum_{|\alpha|+j\leqq 2} c_{\alpha,j}(T,\omega)(\partial/\partial T)^j\Omega^\alpha+\lambda(T)$$

where $c_{\alpha,j}=O(e^{-T-\delta T})$, $dc_{\alpha,j}=O(e^{-T-\delta T})$, and

$$c(T)=(n-2)(1+\varepsilon e^{-\varepsilon T})+\varepsilon^2 e^{-\varepsilon T}/(1+\varepsilon e^{-\varepsilon T}),$$
$$\lambda(T)=\lambda e^{2T}e^{-2e^{-\varepsilon T}}(1+\varepsilon e^{-\varepsilon T})^2.$$

Note that $\lambda'(T)\geqq\lambda e^{2T}$ for large T. We shall prove that for large T_0 and τ

$$\sum_{|\alpha|\leqq 1}\|e^{T(\tau+1-|\alpha|)}D^\alpha U\|^2 \leqq C(e^{-T_0/2}+\tau^{-1})\|e^{\tau T}QU\|^2, \quad U\in C_0^\infty((T_0,\infty)\times S^{n-1}). \tag{17.2.21}$$

Here D^α denotes any product of $|\alpha|$ factors Ω_j and $\partial/\partial T$. In view of Friedrichs' lemma it follows as in the proof of Theorem 14.7.2 that (17.2.21) may be applied to the function U in (17.2.20)′ multiplied by a cutoff function of T which is 0 for $T<T_0+1$ and 1 for $T>T_0+2$. When T_0 is chosen large enough we obtain as before when $\tau\to\infty$ that $U=0$ when $T>T_0+2$; thus $u=0$ in X by Theorem 17.2.1. What remains is therefore to prove the estimate (17.2.21).

Set $V=Ue^{\tau T}$ and

$$Q_\tau V = e^{\tau T} Q(Ve^{-\tau T})$$

which is the operator obtained when $\partial/\partial T$ is replaced by $\partial/\partial T-\tau$ in Q. Then (17.2.21) is a consequence of the estimate

$$\sum_{|\alpha|\leqq 1}\|(\tau+e^T)^{1-|\alpha|}D^\alpha V\|^2 \leqq C(e^{-T_0/2}+\tau^{-1})\|Q_\tau V\|^2,\quad V\in C_0^\infty((T_0,\infty)\times S^{n-1}), \tag{17.2.21$'$}$$

apart from the size of the constant. Denoting by Q_τ^-, the operator obtained from Q_τ when $\partial/\partial T$ and Ω_j are replaced by $-\partial/\partial T$ and $-\Omega_j$ while $c_{\alpha,j}$ is replaced by $\overline{c_{\alpha,j}}$, we shall again consider the difference $\|Q_\tau V\|^2-\|Q_\tau^- V\|^2$ in order to prove the following analogue of (17.2.13)

$$\|Q_\tau^- V\|^2+4\varepsilon^2\tau\sum\|e^{-\varepsilon T/2}\Omega_j V\|^2+\tau\lambda\|e^T V\|^2\leqq\|Q_\tau V\|^2+CR(V), \tag{17.2.22}$$

where $V\in C_0^\infty((T_0,\infty)\times S^{n-1})$, T_0 is large, and

$$R(V)=\sum_{|\alpha|=2}\|e^{-T}D^\alpha V\|^2/\tau+\sum_{|\alpha|\leqq 1}(\|e^{-\delta T}D^\alpha V\|^2\tau+\|D^\alpha V\|^2)\tau^{2-2|\alpha|}. \tag{17.2.23}$$

First assume that all $c_{\alpha,j}$ vanish. Then $Q_\tau=L_1+L_2$, $Q_\tau^-=L_1-L_2$ where

$$L_1=\partial^2/\partial T^2+\tau^2-c(T)\tau+(1+\varepsilon e^{-\varepsilon T})^2\Delta_\omega+\lambda(T);\quad L_2=-(2\tau-c(T))\partial/\partial T.$$

L_1 is symmetric and the adjoint of L_2 is $-L_2-c'(T)$. Hence

$$\|Q_\tau V\|^2-\|Q_\tau^- V\|^2=2(L_1V,L_2V)+2(L_2V,L_1V)=2([L_1,L_2]V,V)\\ -2(c'L_1V,V).$$

Computing the commutator $[L_1,L_2]$ and using the definition of L_1 we find that the right-hand side can be estimated by $CR(V)$ apart from the terms

$$(-4\tau\varepsilon^2e^{-\varepsilon T}(1+\varepsilon e^{-\varepsilon T})\Delta_\omega V,V)+(((2\tau-c(T))\lambda'(T)-2c'\lambda(T))V,V).$$

Since $\lambda'(T)\geqq\lambda e^{2T}$ for large T we obtain (17.2.22) in this case even with λ replaced by $3\lambda/2$ say in the last term on the left. If the coefficients $c_{\alpha,j}$ are not all 0 then we have three other types of terms to consider:

a) The crossproducts involving $\lambda(T)$ and $c_{\alpha,j}$ are

$$2\,\mathrm{Re}\sum((c_{\alpha,j}(\partial/\partial T-\tau)^j\Omega^\alpha V,\lambda(T)V)-(\lambda(T)V,\overline{c_{\alpha,j}}(-\partial/\partial T-\tau)^j(-\Omega)^\alpha V)).$$

After an integration by parts as in Lemma 17.2.4 we find that these can be estimated by

$$C\iint e^{T-\delta T}(\tau|V|^2+\sum_{|\alpha|=1}|D^\alpha V||V|)\,d\omega\,dT\leqq\|e^TV\|^2+C'\sum_{|\alpha|\leqq 1}\tau^{2-2|\alpha|}\|D^\alpha V\|^2.$$

The last sum can be estimated by means of R and the preceding term can be cancelled against the extra $\lambda\tau\|e^T V\|^2/2$ which we had in the left-hand side in the "unperturbed case".

b) After a similar integration by parts, crossproducts between $c_{\alpha,j}$ and some other term than $\lambda(T)$ in the unperturbed operator can be estimated by

$$C\sum\int e^{-T(1+\delta)}\tau^j|D^\alpha V||D^\beta V|\,d\omega\,dT$$

where $j+|\alpha|+|\beta|\leqq 3$ and $|\alpha|\leqq 2$, $|\beta|\leqq 2$ in the sum. If $|\alpha|=2$ we can write the integrand as $\tau^{-1/2}|e^{-T}D^\alpha V|\tau^{j+1/2}|e^{-\delta T}D^\beta V|$, where $j+|\beta|\leqq 1$, so the integral can be estimated by $CR(V)$. The same is obviously true when $|\beta|=2$. If $|\alpha|\leqq 1$ and $|\beta|\leqq 1$ we can estimate the integrand by $\tau\tau^{1-|\alpha|}e^{-\delta T}|D^\alpha V|\tau^{1-|\beta|}e^{-\delta T}|D^\beta V|$, for δ may be assumed <1, and the integral can therefore again be estimated by $CR(V)$.

c) Terms containing two coefficients $c_{\alpha,j}$ are smaller for large T than those examined in b) so the same estimates hold for them. This completes the proof of (17.2.22).

Next we shall estimate the T derivative of V which is missing in the left-hand side of (17.2.22). To do so we observe that

$$(Q_\tau - Q_\tau^-)V/2 = L_2 V = -(2\tau - c(T))\,\partial V/\partial T$$

when all $c_{\alpha,j}$ are equal to 0; otherwise terms which can be estimated by $\sum e^{-T(1+\delta)}\tau^{2-|\alpha|}|D^\alpha V|$ may occur. Hence

$$\tau^2\|\partial V/\partial T\|^2 \leqq \|Q_\tau V\|^2 + \|Q_\tau^- V\|^2 + C\sum_{|\alpha|\leqq 2}\|e^{-T(1+\delta)}D^\alpha V\|^2\tau^{4-2|\alpha|}.$$

If we divide by τ and add to (17.2.22) we obtain, with a new constant of course,

$$\varepsilon^2\tau\sum\|e^{-\varepsilon T/2}\Omega_j V\|^2 + \tau\|\partial V/\partial T\|^2 + \lambda\tau\|e^T V\|^2 \leqq C(\|Q_\tau V\|^2 + R(V)). \tag{17.2.24}$$

Finally we shall prove an analogue of (17.2.3)'. It is complicated by the fact that Q_τ contains the exponentially large term $\lambda(T)$ which has to be cut down to a size which can be controlled by the term $\lambda\tau\|e^T V\|^2$ in (17.2.24). In the following lemma the right-hand side contains also the other terms from the left-hand side of (17.2.24), with A playing the role of e^T. The left-hand side contains what is needed to introduce variable coefficients, estimate the essential contribution to the error term R in (17.2.24), and finish the proof of (17.2.21)'. We denote by Δ' the Laplace operator in $x' = (x_1, \dots, x_{n-1}) \in \mathbb{R}^{n-1}$.

Lemma 17.2.9. *For $u \in C_0^\infty(\mathbb{R}^n)$ and large positive A and τ we have*

$$K^{-1}\sum_{|\alpha|\leqq 2}\tau^{4-2|\alpha|}\|D^\alpha u\|^2 + A^{-\varepsilon}\tau^3\|u\|^2 + M\sum_{|\alpha|\leqq 1}\|D^\alpha u\|^2\tau^{2-2|\alpha|}$$
$$\leqq C\left(K^{-1}\|(\Delta' + (\partial/\partial x_n - \tau)^2)u\|^2 + \tau A^{-\varepsilon}\sum_1^{n-1}\|D_j u\|^2 + \tau A^2\|u\|^2\right) \tag{17.2.25}$$

provided that

$$A^{-\varepsilon}\tau^2 \leqq \tau^3/K + A^2, \quad \tau^3/K \leqq A^{-\varepsilon}\tau^2 + A^2, \quad M^2 K \leqq \tau A^2, \quad M \leqq A^{1-\varepsilon/2}. \tag{17.2.26}$$

Proof. By Parseval's formula the estimate is equivalent to one of the form

$$(17.2.25)' \quad (|\xi|^2+\tau^2)^2/K+A^{-\varepsilon}\tau^3+M(|\xi|^2+\tau^2) \\ \leqq C'(||\xi|^2-\tau^2+2i\tau\xi_n|^2/K+\tau A^{-\varepsilon}|\xi'|^2+\tau A^2).$$

We distinguish two cases:

a) If $||\xi'/\tau|^2-1|\geqq 1/2$ or $|\xi_n/\tau|\geqq 1/2$ then

$$(|\xi|^2+\tau^2)/||\xi|^2-\tau^2+2i\tau\xi_n|=(|\xi/\tau|^2+1)/||\xi/\tau|^2-1+2i\xi_n/\tau|$$

is bounded so the estimate of the first term in (17.2.25)′ is obvious. The estimate of the second term follows since $A^{-\varepsilon}\tau^3\leqq\tau^4/K+\tau A^2$, and that of the third term follows from the inequality between geometric and arithmetic means since $M^2\leqq\tau A^2/K$.

b) If $||\xi'/\tau|^2-1|\leqq 1/2$ and $|\xi_n/\tau|\leqq 1/2$ then $|\xi'|^2\geqq\tau^2/2$ and the estimate is equivalent to

$$\tau^4/K+A^{-\varepsilon}\tau^3+M\tau^2\leqq C''(\tau^3A^{-\varepsilon}+\tau A^2).$$

The estimate of the first term follows from the second inequality (17.2.26), that of the second term is trivial, and the estimate of the third term follows from the last condition (17.2.26). The proof is complete.

Proof of Theorem 17.2.8 continued. If we take $\omega_1,\dots,\omega_{n-1}$ as coordinates on the unit sphere in a neighborhood of $(0,\dots,0,1)$ then Q_τ differs from $\Delta'_{\omega'}+(\partial/\partial T-\tau)^2$ by $\lambda(T)$, by an operator with coefficients $O(|\omega'|+e^{-\varepsilon T})$, and by an operator of first order in τ and D. If $1/e<|e^T/A|<e$, $T\geqq T_0$, and $|\omega'|$ is small enough in supp V, it follows from (17.2.25) that

$$(17.2.27) \quad \sum_{|\alpha|=2}\|K^{-1/2}D^\alpha V\|^2+\tau^3\|e^{-\varepsilon T/2}V\|^2+M\sum_{|\alpha|\leqq 1}\|D^\alpha V\|^2\tau^{2-2|\alpha|} \\ \leqq C'(\|K^{-1/2}Q_\tau V\|^2+\tau\sum\|e^{-\varepsilon T/2}\Omega_j V\|^2+\tau\|e^T V\|^2)$$

if in addition to (17.2.26) we have

$$(17.2.28) \qquad A^2/K\leqq\tau$$

which implies that the term $\lambda(T)V$ in $Q_\tau V$ can be absorbed in the last term. For the proof we just have to observe that the other perturbing terms in Q_τ can be cancelled against half of the first sum in the left-hand side of (17.2.25) with $|\alpha|\leqq 2$ still. To satisfy (17.2.28) and respect the lower bounds in (17.2.26) with K as small as possible we choose

$$(17.2.29) \qquad \begin{aligned} K&=A^\varepsilon\tau && \text{if } A^{2+\varepsilon}\leqq\tau^2,\\ K&=\tau^3/A^2 && \text{if } A^2\leqq\tau^2\leqq A^{2+\varepsilon},\\ K&=A^2/\tau && \text{if } \tau^2\leqq A^2. \end{aligned}$$

The first two inequalities (17.2.26) are then fulfilled, and $\tau A^2/K=A^{2-\varepsilon}$ in the first case, $\tau A^2/K=A^4/\tau^2\geqq A^{2-\varepsilon}$ in the second case, while $\tau A^2/K=\tau^2$ in the third case. The remaining conditions are therefore satisfied if

$M \leqq \min(A^{1-\varepsilon/2}, \tau)$. We take $M = (e^{-T_0/2} + 1/\tau)^{-1}$ in what follows. Since $K \geqq A$ in all cases we can replace $K^{-1/2}$ by $e^{-T/2}$ in the right-hand side of (17.2.27), and since $K^{-1/2} \geqq MA^{-1}\tau^{-1/2}$ by (17.2.26) we replace $K^{-1/2}$ by $Me^{-T}\tau^{-1/2}$ in the left-hand side. This gives with a new constant

(17.2.27)′
$$M\Big(\sum_{|\alpha|=2} \|e^{-T}D^\alpha V\|^2/\tau + \sum_{|\alpha|\leqq 1} \|D^\alpha V\|^2 \tau^{2-2|\alpha|}\Big) + \tau^3 \|e^{-\varepsilon T/2} V\|^2$$
$$\leqq C(\|e^{-T/2} Q_\tau V\|^2 + \tau \sum \|e^{-\varepsilon T/2} \Omega_j V\|^2 + \tau \|e^T V\|^2).$$

The point $(0, \dots, 0, 1)$ on the unit sphere has no special properties. We can therefore choose a partition of unity $1 = \sum \phi_j$ on the unit sphere and a partition of unity $1 = \sum \psi(T-k)$ on $\mathbb{R}$ such that $\operatorname{supp} \psi \subset (-1, 1)$ and (17.2.27)′ is applicable to $V_{jk} = \phi_j(\omega)\psi(T-k)V(T, \omega)$ for all j and k when $V \in C_0^\infty((T_0, \infty) \times S^{n-1})$. If we sum over j and k we find that (17.2.27)′ is valid for V itself with a larger C. In the left-hand side we just use that at most a fixed number of supports can overlap, and in the right-hand side we use the later terms in (17.2.27)′ to take care of terms where a derivative falls on ϕ_j or on ψ.

Fix ε now so that $0 < \varepsilon < \delta$. Since

$$R(V) \leqq \sum_{|\alpha|=2} \|e^{-T} D^\alpha V\|^2/\tau + \sum_{|\alpha|\leqq 1} \|D^\alpha V\|^2 \tau^{2-2|\alpha|}$$
$$+ C_1 e^{(\varepsilon - 2\delta)T_0} \sum_{|\alpha|\leqq 1} \|e^{-\varepsilon T/2} D^\alpha V\|^2 \tau^{3-2|\alpha|},$$

it follows from (17.2.27)′ if T_0 and τ are large enough that the term $CR(V)$ in (17.2.24) is less than $\|Q_\tau V\|^2$ plus one half of the left hand side. Thus

$$\varepsilon^2 \tau \sum \|e^{-\varepsilon T/2} \Omega_j V\|^2 + \lambda\tau \|e^T V\|^2 \leqq 2(C+1)\|Q_\tau V\|^2.$$

If this estimate is used in the right-hand side of (17.2.27)′ we obtain (17.2.21)′ which completes the proof.

17.3. The Dirichlet Problem

In the study of subharmonic functions in Section 16.1 we discussed the Dirichlet problem for the Laplacean in a half space. We shall continue the study here, adding L^2 estimates and some variable coefficient theory. This will allow us to sketch with a minimum of technicalities various methods which have been used in the study of general boundary problems for elliptic differential equations but which will not be covered by this book.

First we shall just study the Dirichlet problem for the Laplacean in

$$\mathbb{R}^n_+ = \{x \in \mathbb{R}^n; x_n > 0\}.$$

It consists in finding for given f and ϕ a solution u of

(17.3.1)
$$\Delta u = f \text{ in } \mathbb{R}^n_+; \quad u = \phi \text{ on } \partial\mathbb{R}^n_+.$$

If ϕ is defined in $\overline{\mathbb{R}}^n_+$ and sufficiently smooth, then replacing $u-\phi$ by u and $f-\Delta\phi$ by f reduces (17.3.1) to the Dirichlet problem with homogeneous boundary data

$$\Delta u=f \text{ in } \mathbb{R}^n_+; \quad u=0 \text{ in } \partial\mathbb{R}^n_+, \tag{17.3.1)$'$}$$

and we shall study it in that form.

We recall from Appendix B the notation $\overline{H}_{(1)}(\mathbb{R}^n_+)$ for the set of restrictions to $\mathbb{R}^n_+$ of functions in $H_{(1)}(\mathbb{R}^n)$. By Corollary B.2.5 we have

$$\overline{H}_{(1)}(\mathbb{R}^n_+)=\{u\in L^2(\mathbb{R}^n_+);\ D^\alpha u\in L^2(\mathbb{R}^n_+),\ |\alpha|=1\}. \tag{17.3.2}$$

It follows from Theorem B.2.7 that every $u\in\overline{H}_{(1)}(\mathbb{R}^n_+)$ can be considered as a bounded continuous function of $x_n\in\overline{\mathbb{R}}_+$ with values in $H_{(1/2)}(\mathbb{R}^{n-1})$. If $u_0=u$ when $x_n\geqq 0$ and $u_0=0$ when $x_n<0$ then $u_0\in L^2(\mathbb{R}^n)$ and

$$\partial u_0/\partial x_n=u(.,0)\otimes\delta(x_n)+(\partial u/\partial x_n)_0, \quad \partial u_0/\partial x_j=(\partial u/\partial x_j)_0 \quad \text{if } j\neq n.$$

Thus $u_0\in H_{(1)}(\mathbb{R}^n)$ if and only if $u(.,0)=0$.

The following is an analogue of Theorem 13.2.1, and the proof is similar.

Theorem 17.3.1. *Let $P(x,D)=\sum_{|\alpha|\leqq 2} a_\alpha(x)D^\alpha$ have continuous coefficients in a neighborhood of* 0, *and assume that $p(\xi)=\sum_{|\alpha|=2} a_\alpha(0)\xi^\alpha$ is elliptic with real coefficients. If X is a sufficiently small neighborhood of* 0 *in $\mathbb{R}^n$ and $X_+=X\cap\mathbb{R}^n_+$, $X_0=X\cap\partial\mathbb{R}^n_+$, there is a linear operator E in $L^2(X_+)$ such that*

(17.3.3) $\quad P(x,D)Ef=f,\quad Ef=0$ *in* X_0 *if* $f\in L^2(X_+)$;

(17.3.4) $\quad EP(x,D)u_+=u_+$ *if* u_+ *is the restriction to* X_+ *of some* $u\in C_0^\infty(X)$ *with* $u=0$ *in* X_0;

(17.3.5) $\quad D^\alpha E$ *is a bounded linear operator in* $L^2(X_+)$ *if* $|\alpha|\leqq 2$.

By (17.3.5) *we have* $D^\alpha Ef\in L^2(X_+)$ *for* $|\alpha|\leqq 2$ *if* $f\in L^2(X_+)$. *Thus* $\chi Ef\in\overline{H}_{(1)}(\mathbb{R}^n_+)$ *for every* $\chi\in C_0^\infty(X)$, *so boundary values of* Ef *are defined in* X_0.

Proof. We may assume that $p(\xi)=|\xi|^2$ for this can be achieved by a linear change of variables respecting the boundary plane. Let e be a fundamental solution of $p(D)$ which is even in x_n, for example the Newton kernel (Theorem 3.3.2). If $f\in L^2(X_+)$ we define $Tf=f$ in X_+, $Tf(x',-x_n)=-f(x',x_n)$ if $(x',x_n)\in X_+$, and $Tf=0$ elsewhere. Then $Tf\in L^2_{\text{comp}}(\mathbb{R}^n)$ if X is bounded, as we assume, and Tf is odd as a function of x_n. Hence $e*Tf\in H^{\text{loc}}_{(2)}(\mathbb{R}^n)$ is also odd in x_n so the restriction to the plane $x_n=0$ must vanish. Let E_0f be the restriction to X_+. Then

(17.3.3)$'$ $\quad p(D)E_0f=f$ *in* X_+, $\quad E_0f=0$ in X_0 if $f\in L^2(X_+)$;

(17.3.4)′ $E_0 p(D)u_+ = u_+$ if u_+ is the restriction to X_+ of some $u \in C_0^\infty(X)$ with $u=0$ in X_0;

(17.3.5)′ $D^\alpha E_0$ is bounded in $L^2(X_+)$ if $|\alpha| \leqq 2$.

We have already verified (17.3.3)′. To prove (17.3.4)′ we first observe that Tu_+ is continuous. Hence $D_n Tu_+$ is the even extension of $D_n u_+$, which is continuous, and $D_n^2 Tu_+$ is the odd extension $TD_n^2 u_+$ of $D_n^2 u_+$. This means that $Tp(D)u_+ = p(D)\,Tu_+$ so $e * Tp(D)u_+ = e * p(D)\,Tu_+ = Tu_+$ which proves (17.3.4)′. (Here it is important that $p(D)$ has no term which is odd in D_n.) Condition (17.3.5)′ follows from Theorem 10.3.1 for example.

Now we just copy the proof of Theorem 13.2.1. Writing

$$P(x,D) = p(D) + \sum_{|\alpha| \leqq 2} b_\alpha(x) D^\alpha,$$

where $b_\alpha(0)=0$ when $|\alpha|=2$, we look for a solution of the equation

$$P(x,D)u = p(D)u + \sum b_\alpha(x)\, D^\alpha u = f \in L^2(X_+)$$

which is of the form $u = E_0 g$, $g \in L^2(X_+)$. By (17.3.3)′ this guarantees the Dirichlet condition for u, so we only have to solve the equation

$$g + Ag = f; \quad Ag = \sum b_\alpha D^\alpha E_0 g.$$

The operator A in $L^2(X_+)$ has norm $<\frac{1}{2}$ if $\sum \sup|b_\alpha|$ is small enough. Then we define

$$E = E_0(I+A)^{-1}$$

and deduce (17.3.3)–(17.3.5) from (17.3.3)′–(17.3.5)′ exactly as in the proof of Theorem 13.2.1. If the coefficients b_α are not small we can apply this conclusion with b_α replaced by $b_\alpha(\varepsilon x)\,\varepsilon^{2-|\alpha|}$ if ε is small enough. Taking εx as a new variable we then obtain Theorem 17.3.1 with X replaced by εX. The proof is complete.

Remarks. 1. If $a_\alpha \in C^\infty$ we can adapt the proof of Theorem 13.3.3 to show that there is a linear map $E\colon L^2(\mathbb{R}^n_+) \to \bar{H}_{(2)}(\mathbb{R}^n_+)$ such that

$$P(x,D)Ef = f \text{ in } X_+, \quad Ef = 0 \text{ in } X_0 \text{ if } f \in L^2(\mathbb{R}^n_+);$$

$$EP(x,D)u_+ = u_+ \text{ in } X_+ \text{ if } u_+ \text{ is the restriction to } X_+ \text{ of some } u \in C_0^\infty(X) \text{ with } u=0 \text{ in } X_0;$$

$$D^\alpha Ef \in L^2(\mathbb{R}^n_+), \quad |\alpha| \leqq s+2, \text{ if } D^\alpha f \in L^2(\mathbb{R}^n_+), \quad |\alpha| \leqq s.$$

Here s is any integer $\geqq 0$.

2. In Theorem 17.3.1 we assumed $p(D)$ real just to have a simple explicit solution of the constant coefficient Dirichlet problem (17.3.3)′ given by the reflection argument. However, this is by no means essential. If $n>2$ or $n=2$ and the zeros of $p(1,\tau)$ are in opposite half planes $\operatorname{Im}\tau \gtrless 0$ then we can easily construct E_0 by taking Fourier transforms with respect to x' and solving an ordinary differential equation. The same approach is in fact

applicable to elliptic operators or systems of arbitrary order with quite general boundary conditions. To obtain a sufficiently general framework for perturbation arguments one should then start by solving the constant coefficient boundary problem with inhomogeneous boundary data too. This approach was used systematically in Chapter X of "Linear partial differential operators", but in the study of boundary problems in Chapter XX below we shall use another more constructive method.

(17.3.3) and (17.3.5) give a local existence theorem for solutions of the Dirichlet problem (17.3.1)′ such that $D^\alpha u\in L^2$, $|\alpha|\leqq 2$, if $f\in L^2$. One can pass to a global existence theorem with the methods used in the proof of Theorem 17.2.7 for example. We shall not do so here but show instead how to use (17.3.4)′ and (17.3.5)′ to prove a regularity theorem analogous to one for interior regularity at the end of Section 17.1.

Theorem 17.3.2. *Assume in addition to the hypotheses in Theorem 17.3.1 that a_α is Lipschitz continuous when $|\alpha|=2$. If $D^\alpha u\in L^2(X_+)$, $|\alpha|\leqq 1$, and*

$$P(x,D)u=f\in L^2(X_+),\quad u=0 \text{ in } X_0,$$

it follows that $D^\alpha u\in L^2(Y_+)$, $|\alpha|\leqq 2$, for every $Y\Subset X$, and that

$$\sum_{|\alpha|=2}\|D^\alpha u\|_{L^2(Y_+)}\leqq C_Y(\|P(x,D)u\|_{L^2(X_+)}+\sum_{|\alpha|\leqq 1}\|D^\alpha u\|_{L^2(X_+)}). \tag{17.3.6}$$

Proof. Choose $\chi\in C_0^\infty(X)$ equal to 1 in Y, and set $v=\chi u$. Then $P(x,D)v=g\in L^2(X_+)$, $D^\alpha v\in L^2(X_+)$ when $|\alpha|\leqq 1$, $v=0$ in X_0, and $v=0$ outside a compact subset of X. Choose $\phi\in C_0^\infty(\mathbb{R}^{n-1})$ with $\phi\geqq 0$, $\int\phi(x')\,dx'=1$, and set

$$v_\varepsilon(x)=\int v(x'-\varepsilon y',x_n)\phi(y')\,dy'.$$

By Minkowski's inequality we have with $\|\ \|$ denoting the norm in $L^2(X_+)$

$$\|D^\alpha v_\varepsilon\|\leqq\|D^\alpha v\|,\quad |\alpha|\leqq 1,$$

and $D^\alpha v_\varepsilon\in L^2(X_+)$ if D^α has at most one factor D_n, for we can let the others act on ϕ. Since $P(x,D)$ can be divided by the coefficient c of D_n^2, we may assume that $c=1$. Then

$$P(x,D)v_\varepsilon-g_\varepsilon=\sum_{\alpha_n<2}(a_\alpha D^\alpha v_\varepsilon-(a_\alpha D^\alpha v)_\varepsilon).$$

When $|\alpha|=2$ we can write $D^\alpha=D_jD_k$ for some $j\neq n$ and apply Friedrichs' lemma (Lemma 17.1.5) for every fixed x_n. This gives

$$\|P(x,D)v_\varepsilon-g_\varepsilon\|\to 0,\quad \varepsilon\to 0.$$

In particular it follows that $D_n^2 v_\varepsilon\in L^2(X_+)$.

By (17.3.4) and (17.3.5) we have

$$\sum_{|\alpha|\leqq 2}\|D^\alpha w_+\|\leqq C\|P(x,D)w_+\| \tag{17.3.6$'$}$$

if w_+ is the restriction to X_+ of a function $w \in C_0^\infty(X)$ vanishing in X_0. If Tv_ε is the odd continuation of v_ε from $\mathbb{R}^n_+$ to $\mathbb{R}^n$ then $D^\alpha Tv_\varepsilon \in L^2$ if $\alpha_n \leqq 2$ (see the proof of Theorem 17.3.1). If we apply (17.3.6)′ to the regularizations w of Tv_ε by convolution with even functions, it follows at once that (17.3.6)′ can be applied to $w_+ = v_\varepsilon$. When $\varepsilon \to 0$ we conclude that $D^\alpha v \in L^2$, $|\alpha| \leqq 2$, and that (17.3.6) is valid. The proof is complete.

The Dirichlet problem can be solved with great ease and in great generality by means of *Dirichlet's principle*. Consider a differential operator of divergence form

$$P(x,D)u = \sum D_j(a_{jk} D_k u)$$

where $(a_{jk}(x))$ is a real positive definite symmetric matrix which is a continuous function of x in the closure of an open bounded subset X of $\mathbb{R}^n$. If $P(x,D)u = f$ in X the equation is equivalent to the weak form

$$(f,v) = \int f\bar{v}\,dx = \int \sum a_{jk} D_k u \overline{D_j v}\,dx, \quad v \in C_0^\infty(X). \tag{17.3.7}$$

This condition makes sense if $D^\alpha u \in L^2(X)$, $|\alpha| \leqq 1$, and remains valid then for all v in the closure $\mathring{H}$ of $C_0^\infty(X)$ in $H_{(1)}(\mathbb{R}^n)$. If $\chi \in C_0^\infty(\mathbb{R}^n)$ it is clear that $\chi u \in \mathring{H}$ for every $u \in \mathring{H}$. If $\partial X \in C^1$ at $x_0 \in \partial X$ we can choose a C^1 map ψ of a neighborhood of 0 in $\mathbb{R}^n$ on a neighborhood X_0 of x_0 mapping $\mathbb{R}^n_+$ to X and conclude if $\chi \in C_0^\infty(X_0)$ that $\psi^*(\chi u) \in \bar{H}_{(1)}(\mathbb{R}^n_+)$ and that $\psi^*(\chi u) = 0$ on $\partial \mathbb{R}^n_+$. Conversely, if $\partial X \in C^1$ and this condition is fulfilled at every boundary point then $u \in \mathring{H}$. (We could also identify $\mathring{H}$ with the space $\dot{H}_{(1)}(\bar{X})$ of distributions in $H_{(1)}(\mathbb{R}^n)$ with support in $\bar{X}$, for no such distributions have support in ∂X.) The condition $u \in \mathring{H}$ is therefore a generalization to an arbitrary domain of our previous statement of the homogeneous Dirichlet condition. The Dirichlet problem can thus be restated as follows: Find $u \in \mathring{H}$ satisfying (17.3.7) for all $v \in \mathring{H}$. This is Dirichlet's principle.

The solution is extremely simple. Write

$$Q(u,v) = \sum a_{jk} D_j u \overline{D_k v}\,dx; \quad u, v \in C_0^\infty(X);$$

which is a non-negative hermitian symmetric form. With $\| \ \|$ denoting the L^2 norm we have

$$\sum \|D_j u\|^2 \leqq C Q(u,u), \quad u \in C_0^\infty(X).$$

The left-hand side is equivalent to $\|u\|^2_{(1)}$ since for example

$$\int |u|^2\,dx = -2\,\mathrm{Re} \int x_1 u\, \partial\bar{u}/\partial x_1\,dx \leqq \tfrac{1}{2} \int |u|^2\,dx + 2 \int |x_1 D_1 u|^2\,dx$$

and this implies

$$\|u\|^2 \leqq 4\,\|x_1 D_1 u\|^2.$$

Hence $(Q(u,u))^{\frac{1}{2}}$ is a norm equivalent to $\|u\|_{(1)}$ on $C_0^\infty(X)$ and therefore on $\mathring{H}$. For every $f \in L^2(X)$ we have

$$|(f,v)| \leqq \|f\|\,\|v\| \leqq C'\,Q(v)^{\frac{1}{2}}, \quad v \in \mathring{H},$$

which proves that there is a unique $u \in \mathring{H}$ satisfying (17.3.7) for every $v \in \mathring{H}$. Taking $v \in C_0^\infty$ we obtain $P(x,D)u=f$, and $u \in \mathring{H}$ means that the homogeneous Dirichlet boundary conditions are fulfilled. We can strengthen the conclusion if $\partial X \in C^2$ and a_{jk} is Lipschitz continuous. In fact, at any boundary point we can then take a C^2 diffeomorphism ψ as above which flattens out the boundary in a neighborhood. Using Theorem 17.3.2 we obtain $D^\alpha u \in L^2(X)$ when $|\alpha| \leqq 2$, and

$$\sum_{|\alpha| \leqq 2} \|D^\alpha u\|_{L^2(X)} \leqq C \|f\|_{L^2(X)}.$$

If a_{jk}, f and ∂X are smooth it is not hard to deduce that u is also smooth.

We shall finally outline another classical method for solving the Dirichlet problem and indicate how it will be modified in Chapter XX. For a change we shall now emphasize the role of the boundary data. Thus we assume that $X \subset \mathbb{R}^n$ is bounded and that $\partial X \in C^\infty$, and we consider the Dirichlet problem for the homogeneous Laplace equation

$$\Delta u = 0 \text{ in } X, \quad u = \phi \text{ on } \partial X. \tag{17.3.8}$$

If X were a half space, then the solution given in Section 16.1 would be

$$u(x) = 2 \int_{\partial X} \phi(y)\, \partial E(x-y)/\partial n_y \, dS(y)$$

where E is the fundamental solution given by the Newton kernel (Theorem 3.3.2), n is the exterior unit normal and dS the surface area on ∂X. This is just another way of stating the reflection method used to prove Theorem 17.3.1. In the general case one therefore tries to find ψ so that the double layer potential

$$u(x) = 2 \int \psi(y)\, \partial E(x-y)/\partial n_y \, dS(y)$$

will satisfy the boundary condition in (17.3.8); u is automatically harmonic in X. This gives an equation of the form

$$\psi + K\psi = \phi$$

where K is a compact (Fredholm) integral operator. The reason for this is that K would be 0 if the boundary were flat so K just expresses the deviation from that case. Fredholm theory was developed precisely to solve the preceding equation.

If instead of the Dirichlet problem one is interested in the Neumann problem, that is, the boundary condition $du/dn = \phi$, one obtains in the half space case the solution

$$u(x) = -2 \int \phi(y)\, E(x-y)\, dS(y).$$

In the general case one therefore tries a simple layer potential

$$u(x) = -2 \int \psi(y)\, E(x-y)\, dS(y),$$

and the problem again becomes a Fredholm equation $\psi + K\psi = \phi$.

To carry these arguments over to general operators and boundary conditions is rather cumbersome. First one must solve constant coefficient problems in all tangential half spaces to find appropriate kernels for the problem at hand. Then one makes an "ansatz" as above and has to show that it leads to a Fredholm equation. However, there is a useful modification of this approach. By Green's formula

$$u(x)=\int_{\partial X} u_0(y)\,dE(x-y)/dn_y\,dS-\int_{\partial X} u_1(y)\,E(x-y)\,dS \tag{17.3.9}$$

if u_0 and u_1 are the boundary values and the normal derivative of u respectively. Thus we know u if u_0 and u_1 are known. The formula (17.3.9) defines a harmonic function for arbitrary u_0 and u_1 but it need not have boundary values and normal derivative equal to u_0 and u_1. Indeed, we know that a harmonic function is determined by its boundary values so $u_1 = A u_0$ for some operator A if u_0 and u_1 are boundary values and normal derivatives of a harmonic function. The operator calculus which we shall develop in Chapter XVIII will give a quite explicit representation of operators such as A. By using the differential equation $\Delta u=0$ we can write any differential boundary condition in the form $B_0 u_0 + B_1 u_1 = \phi$ where B_0 and B_1 are differential operators in ∂X. Thus the solution of the boundary problem is reduced to solving the equation $(B_0+B_1 A)u_0=\phi$ in the manifold without boundary ∂X. The operator $B_0+B_1 A$ is not a differential operator but it belongs to a closely related class of operators to which the theory of elliptic differential operators, for example, is easily extended.

These remarks are admittedly and purposely vague. We just hope to convince the sceptical reader that there are good reasons for the introduction of a fairly large machinery in Chapter XVIII, and that the effort it requires will be rewarded when we return to boundary problems in Chapter XX.

17.4. The Hadamard Parametrix Construction

We have seen in Chapter XIII and again in Sections 17.1 and 17.3 how to extend results on constant coefficient elliptic operators to variable coefficient ones. However, this gives only the existence of fundamental solutions and information on their continuity as operators. One is often interested in the precise form of their singularities. We shall therefore present a remarkably simple and precise construction due to J. Hadamard, which gives the singularities of the fundamental solution with any desired precision. It is also applicable in non-elliptic cases where the methods of Chapter XIII and Sections 17.1, 17.3 fail completely. In the second half of the section we shall indicate this in a special case needed in Section 17.5. The extension to general second order hyperbolic equations should cause the reader no difficulty. (See also the notes to Chapter XXIII.)

Let P be a second order differential operator of the form

$$P=-\sum \partial/\partial x_j(g^{jk}\,\partial/\partial x_k)+\sum b^j\,\partial/\partial x_j+c$$

where g^{jk}, b^j, c are C^∞ functions in an open set $X\subset\mathbb{R}^n$ and (g^{jk}) is a real positive definite matrix. We wish to construct a right parametrix for P, or rather for $P-z$ when $z\in\mathbb{C}\setminus\overline{\mathbb{R}}_+$ since this will eliminate some irrelevant difficulties in the constant coefficient case. First we assume that P is equal to minus the Laplacean Δ,

$$\Delta=\sum \partial^2/\partial x_j^2.$$

The inverse Fourier transform of $(|\xi|^2-z)^{-1}$ is a fundamental solution of $-\Delta-z$. We shall also introduce the powers, for these occur inevitably when one makes a perturbation (cf. Section 12.5). Thus we set (in the distribution sense) for $\nu=0,1,\dots$

$$F_\nu(x)=\nu!\,(2\pi)^{-n}\int e^{i\langle x,\xi\rangle}(|\xi|^2-z)^{-\nu-1}\,d\xi. \tag{17.4.1}$$

It follows from Theorem 7.1.22 that $F_\nu\in C^\infty(\mathbb{R}^n\setminus 0)$, and $D^\alpha F_\nu$ is a locally integrable function also at 0 when $|\alpha|\leqq 2$ unless $\nu=0$ and $|\alpha|=2$; then we also have a term which is homogeneous of degree $-n$. It is obvious that F_ν is a function of $|x|=(x_1^2+\dots+x_n^2)^{\frac12}$. It would be easy to give an expression for F_ν in terms of Bessel functions, but that would just suggest lengthy proofs for the essential properties which are easily seen from (17.4.1). They are first of all

$$(-\Delta-z)F_\nu=\nu F_{\nu-1},\quad \nu>0;\quad (-\Delta-z)F_0=\delta_0; \tag{17.4.2}$$

$$-2\,\partial F_\nu/\partial x=x F_{\nu-1},\quad \nu>0. \tag{17.4.3}$$

(17.4.2) is obvious, and (17.4.3) follows from the fact that the Fourier transform of $-2\,\partial F_\nu/\partial x_k$ is

$$-2\nu!\,i\,\xi_k(|\xi|^2-z)^{-\nu-1}=(\nu-1)!\,(-D_k)(|\xi|^2-z)^{-\nu}.$$

We shall keep z fixed in the following discussion and have therefore not indicated in the notation that F_ν depends on z also.

As a first step to an extension of (17.4.2), (17.4.3) we pull back the identities by a linear bijection T in $\mathbb{R}^n$. If $y=Tx$ then ${}^tT\partial/\partial y=\partial/\partial x$ so

$$\Delta_y=\sum \partial/\partial x_j\, g^{jk}\,\partial/\partial x_k \quad \text{if } (g^{jk})=T^{-1}\,{}^tT^{-1}.$$

Since F_ν is rotationally symmetric we may by some abuse of notation write $F_\nu(x)=F_\nu(|x|)$, and then we have $F_\nu(y)=F_\nu(|x|_g)$ if

$$|x|_g^2=\sum g_{jk}\,x_j x_k$$

where (g_{jk}) is the inverse of (g^{jk}). Recalling Example 6.1.3 we obtain since $(\det T)^{-2}=\det(g^{jk})$

$$(17.4.2)' \qquad (-\sum \partial_j g^{jk} \partial_k - z) F_\nu(|x|_g) = \nu F_{\nu-1}(|x|_g), \quad \nu > 0;$$
$$= (\det g^{jk})^{\frac{1}{2}} \delta_0 \quad \text{if } \nu = 0;$$

$$(17.4.3)' \qquad -2\sum g^{jk} \partial_k F_\nu(|x|_g) = x_j F_{\nu-1}(|x|_g), \quad \nu > 0.$$

The preceding formulas are valid for any symmetric positive definite matrix (g^{jk}) but the whole point is that they are applicable also if P has variable coefficients provided that the coordinates are well chosen. Recall that the principal symbol $\sum g^{jk}(x)\xi_j\xi_k$ of P is invariantly defined in the cotangent bundle. The dual quadratic form $\sum g_{jk}(x)\,dx_j\,dx_k$ in the tangent bundle defines a Riemannian metric. As proved for example in Appendix C.5 we can for every point in X introduce geodesic normal coordinates which vanish there and satisfy the condition

$$(17.4.4) \qquad \sum_k g_{jk}(x)\,x_k = \sum_k g_{jk}(0)\,x_k, \quad j=1,\dots,n.$$

This means that the rays through 0 are geodesics with arc length equal to the distance in the Euclidean metric

$$|x|_g = |x|_{g,0} = (\sum g_{jk}(0)\,x_j x_k)^{\frac{1}{2}}.$$

We have $g_{jk}(x) - g_{jk}(0) = O(|x|^2)$ and similarly for g^{jk}. Usually one requires that $g_{jk}(0) = \delta_{jk}$ which can of course be achieved by an additional linear coordinate change. However, it will be more convenient for us later on to require only (17.4.4).

From (17.4.4) we obtain if $f \in C^1$

$$\sum g^{jk}(x)\,\partial_k f(|x|_g^2) = \sum g^{jk}(0)\,\partial_k f(|x|_g^2), \quad j=1,\dots,n.$$

This shows that (17.4.2)′ remains valid when $x \neq 0$ with $g^{jk}(0)$ replaced by $g^{jk}(x)$. The same is true at 0 in the distribution sense, for replacing $g^{jk}(0)$ by $g^{jk}(x)$ can only add locally integrable terms since the difference is $O(|x|^2)$. By the product rule and (17.4.2)′, (17.4.3)′ we obtain if $u_\nu \in C^\infty$ and $\nu > 0$

$$(P-z)(u_\nu F_\nu) = \nu u_\nu F_{\nu-1} + (P u_\nu) F_\nu - (h u_\nu - 2\langle x, \partial u_\nu/\partial x\rangle) F_{\nu-1}/2,$$
$$(17.4.5) \qquad h(x) = \sum g_{jk}(0)\,b^j(x)\,x_k = \sum g_{jk}(x)\,b^j(x)\,x_k.$$

Similarly we obtain if $F_0(x) = f(|x|^2)$

$$(P-z)(u_0 F_0) = u_0(0)(\det g^{jk})^{\frac{1}{2}}\delta_0 + (P u_0) F_0 - 2(h u_0 - 2\langle x, \partial u_0/\partial x\rangle) f'.$$

When we add for $\nu = 0, \dots, N$, it follows that

$$(17.4.6) \qquad (P-z)\sum_0^N u_\nu F_\nu = u_0(0)(\det g^{jk})^{\frac{1}{2}}\delta_0 + (P u_N) F_N$$

if u_ν are chosen so that

$$(17.4.7) \qquad 2\nu u_\nu - h u_\nu + 2\langle x, \partial u_\nu/\partial x\rangle + 2 P u_{\nu-1} = 0, \quad \nu = 0, \dots, N,$$

for this makes the coefficients of f' and of $F_0, \dots, F_{N-1}$ all equal to 0. (When $\nu = 0$ one should interpret $u_{\nu-1}$ as 0 in (17.4.7).)

We shall now prove that the equations (17.4.7) have a unique smooth solution with $u_0(0)=1$; no boundary condition is required for u_ν when $\nu\neq 0$.

Lemma 17.4.1. *Let X be an open subset of $\mathbb{R}^n$ which is starshaped with respect to 0, that is, $x\in X \Rightarrow tx\in X$ if $0\leq t\leq 1$. If $h\in C^\infty(X)$ and $h(0)=0$ then the equation*

$$h u_0 = 2\langle x, \partial u_0/\partial x\rangle \tag{17.4.8}$$

has a unique solution $u_0\in C^\infty(X)$ with $u_0(0)=1$. If $f\in C^\infty(X)$ and $\nu>0$ then the equation

$$(2\nu-h)u_\nu + 2\langle x, \partial u_\nu/\partial x\rangle = f \tag{17.4.9}$$

has a unique solution $u_\nu\in C^\infty(X)$.

Proof. If we introduce polar coordinates $x=r\omega$ with $\omega\in S^{n-1}$ and $r>0$, then (17.4.8) means for $x\neq 0$ that $\partial u_0/\partial r = h u_0/2r$. If $u_0(0)=1$ we obtain

$$u_0(x)=\exp\Big(\int_0^r h(s\omega)\,ds/2s\Big)=\exp\Big(\int_0^1 h(tx)\,dt/2t\Big).$$

Since $h(0)=0$ the quotient $h(tx)/t$ is a C^∞ function of $(x,t)\in X\times[0,1]$, so $u_0\in C^\infty(X)$ and $u_0\neq 0$ in X. To solve (17.4.9) we set $u_\nu=u_0 v$ and obtain the equation

$$\nu v + r\,\partial v/\partial r = g,\qquad g=f/2u_0.$$

Thus $\partial(r^\nu v)/\partial r = r^{\nu-1} g$, which gives

$$r^\nu v(r\omega)=\int_0^r s^{\nu-1} g(s\omega)\,ds = r^\nu \int_0^1 t^{\nu-1} g(tr\omega)\,dt.$$

It follows that

$$u_\nu(x)=u_0(x)\int_0^1 t^{\nu-1} f(tx)/2u_0(tx)\,dt$$

is the only solution of (17.4.9); that $u_\nu\in C^\infty(X)$ is obvious.

Remark. If b^j, c and therefore h are square matrices, the preceding constructions work with no essential modification. Only the formula for u_0 becomes less explicit. This makes the Hadamard construction applicable to systems with principal symbol $\sum g^{jk}(x)\xi_j\xi_k I$ where I is the identity matrix. This observation is quite useful, for the Laplacean on forms defined in a Riemannian manifold is of this type.

Let us now just assume that we are given a second order operator P in an open set $X\subset\mathbb{R}^n$ with C^∞ coefficients and positive definite principal symbol $p(x,\xi)=\sum g^{jk}(x)\xi_j\xi_k$. Then it follows from Corollary C.5.2 that there is a neighborhood V of $\{0\}\times X$ in $\mathbb{R}^n\times X$, a neighborhood W of the diagonal in $X \times X$, and a uniquely defined diffeomorphism

$$V \ni (\tilde{x}, y) \mapsto (\gamma(\tilde{x}, y), y) \in W$$

with $\gamma(0, y) = y$, $\gamma'_{\tilde{x}}(0, y)$ equal to the identity, and the principal symbol in the $\tilde{x}$ coordinates

$$\sum \tilde{g}^{jk}(\tilde{x}, y)\, \xi_j \xi_k = p(\gamma(\tilde{x}, y), {}^t\gamma'_{\tilde{x}}(\tilde{x}, y)^{-1} \xi)$$

satisfying (17.4.4). We choose V so that $\{\tilde{x}; (\tilde{x}, y) \in V\}$ is convex for every $y \in X$. If $(x, y) \in W$ we have a well defined Riemannian distance $s(x, y)$,

$$s(\gamma(\tilde{x}, y), y) = \left(\sum \tilde{g}_{jk}(0, y)\, \tilde{x}_j \tilde{x}_k\right)^{\frac{1}{2}}.$$

The square is in $C^\infty(W)$. Pulling the functions $u_\nu(\tilde{x}, y)$ defined by (17.4.7) back to W from V, we obtain uniquely defined $U_\nu \in C^\infty(W)$ such that

$$(17.4.6)' \qquad (P(x, D) - z) \sum_0^N U_\nu(x, y)\, F_\nu(s(x, y))$$
$$= (\det g^{jk}(y))^{\frac{1}{2}}\, \delta_y(x) + (P(x, D)\, U_N(x, y))\, F_N(s(x, y)).$$

This works of course equally well on a manifold. Note that $\delta_y(x)$ is a distribution density on X which becomes a distribution when divided by the natural Riemannian density $(\det g^{jk})^{-\frac{1}{2}}\, dx$.

Since $F_N(s(x, y)) \in C^{2N+1-n}$ the error term on the right-hand side of (17.4.6)′ is as smooth as we please when N is large. All terms are C^∞ off the diagonal. If we choose $\chi \in C^\infty(X \times X)$ with support in W so that $\chi = 1$ in some neighborhood of the diagonal and set

$$F(x, y) = \chi(x, y) \sum_0^N U_\nu(x, y)\, F_\nu(s(x, y)),$$

it follows that

$$(P(x, D) - z)\, F(x, y) = (\det g^{jk}(y))^{\frac{1}{2}}\, \delta_y(x) + R(x, y)$$

where $R \in C^{2N+1-n}$. The operator

$$\mathscr{F} f(x) = \int F(x, y) f(y)\, (\det g^{jk}(y))^{-\frac{1}{2}}\, dy$$

maps $\mathscr{E}'(X)$ to $\mathscr{D}'(X)$ and preserves wave front sets. With $\mathscr{R}$ defined similarly with F replaced by R, we have

$$(P - z)\, \mathscr{F} = I + \mathscr{R}$$

so $\mathscr{F}$ is a right parametrix in the sense that $\mathscr{R}$ is as smooth as we wish if N is large. Taking the adjoint of a right parametrix for the adjoint of $P - z$ we obtain a left parametrix for $P - z$ also. (A simple argument which we postpone until Section 18.1 shows that they have essentially the same singularities.) From these facts it is easy to derive for P the results on existence and regularity of solutions proved in Chapter XIII for general elliptic differential equations. (In Section 18.1 we shall also give a simple general method for the construction of a parametrix for an arbitrary elliptic operator.)

The preceding construction can also be applied to the wave equation

$$\partial^2/\partial t^2 + P$$

in $\mathbb{R}^{n+1}$ associated with P, and this will be essential in Section 17.5. If we replace z by τ^2 in (17.4.1) and take the inverse Fourier-Laplace transform defined by

$$(17.4.1)' \quad E_\nu(t,x) = \nu!(2\pi)^{-n-1} \int_{\operatorname{Im}\tau = c < 0} e^{i(\langle x,\xi\rangle + t\tau)}(|\xi|^2 - \tau^2)^{-\nu-1}\, d\xi\, d\tau$$

in the sense of distribution theory, we obtain a distribution with support in the forward light cone $\{(t,x);\, t \geqq |x|\}$ (cf. (7.4.7)). We have

$$(17.4.2)'' \quad (\partial^2/\partial t^2 - \Delta) E_\nu = \nu E_{\nu-1}, \quad \nu > 0; \quad (\partial^2/\partial t^2 - \Delta) E_0 = \delta_{0,0};$$

$$(17.4.3)'' \quad -2\partial E_\nu/\partial x = x E_{\nu-1}, \quad \nu > 0.$$

In the following lemma we collect some further properties which will be required in this section and in Section 17.5.

Lemma 17.4.2. *E_ν is a homogeneous distribution of degree $2\nu+1-n$ with support in the forward light cone,*

$$(17.4.10) \quad E_\nu = 2^{-2\nu-1}\pi^{(1-n)/2}\chi_+^{\nu+(1-n)/2}(t^2 - |x|^2), \quad t > 0.$$

E_ν is a C^∞ function of t with values in $\mathscr{D}'(\mathbb{R}^n)$ when $t \geqq 0$, and

$$\partial_t^k E_\nu(+0,.) = 0 \text{ when } k \leqq 2\nu, \quad \partial_t^{2\nu+1} E_\nu(+0,.) = \nu!\,\delta_0.$$

Furthermore,

$$(17.4.11) \quad WF(E_\nu - \check{E}_\nu) = \{(t,x;\tau,\xi);\, t^2 = |x|^2,\ \tau^2 = |\xi|^2,\ \tau x + t\xi = 0\},$$

and $E_\nu - \check{E}_\nu$, $\partial_t(E_\nu - \check{E}_\nu)$ are continuous functions of x with values in $\mathscr{D}'^{2k}(\mathbb{R})$ if k is an integer with $k \geqq (n-1)/2 - \nu$. If $k = (n-1)/2 - \nu$ then for $x = 0$

$$\partial_t(E_\nu - \check{E}_\nu) = 2^{2k-n+1}\pi^{(1-n)/2}(-1)^k k!\,\delta^{(2k)}(t)/(2k)!.$$

$\partial_t(E_0 - \check{E}_0)$ is the Fourier transform of $\frac{1}{2}\operatorname{sgn}\tau\, de_0(x,\tau^2)$, where

$$(17.4.12) \quad e_0(x,\tau^2) = (2\pi)^{-n} \int_{|\xi| < |\tau|} e^{i\langle x,\xi\rangle}\, d\xi.$$

Proof. (17.4.10) was proved in Section 6.2 when $\nu = 0$. Since

$$(\partial^2/\partial t^2 - \Delta)\chi_+^a(t^2 - |x|^2) = (4a + 2n - 2)\chi_+^{a-1}(t^2 - |x|^2)$$

by the computations preceding Theorem 6.2.1, the recursion formula (17.4.2)'' follows from (17.4.10). By Theorem 6.2.3 the equation (17.4.2)'' has only one solution E_ν with support in the forward light cone when $E_{\nu-1}$ is known. Hence (17.4.10) follows by induction. The statement on the wave front set follows at once from Theorem 8.2.4 when $t \neq 0$. When $x=t=0$ we just have to note that the wave front set is closed and that

$$(\partial^2/\partial t^2 - \Delta)^{\nu+1}(E_\nu - \check{E}_\nu) = \nu!(\delta_{0,0} - \delta_{0,0}) = 0;$$

by Theorem 8.3.1 it follows that $\tau^2 = |\xi|^2$ in the wave front set.

In particular (17.4.11) implies that $E_\nu - \check{E}_\nu$ is a C^∞ function of x with values in $\mathscr{D}'$. Set $a=\nu+(1-n)/2$ and let $\delta=|x|^2>0$. Then $\partial_t(E_\nu-\check{E}_\nu)$ is a constant times the distribution

$$e_{\delta,a} = \chi_+^{a-1}(t^2-\delta)\, 2|t| = t/|t| \frac{d}{dt} \chi_+^a(t^2-\delta).$$

If ϕ is an odd test function then $\langle e_{\delta,a}, \phi\rangle = 0$, and if ϕ is an even test function then

$$\langle e_{\delta,a}, \phi\rangle = -\langle e_{\delta,a+1}, \psi\rangle, \quad \psi(t) = \phi'(t)/2t.$$

ψ is also an even test function, and by Taylor's formula

$$\sup|\psi^{(j)}| \leq \tfrac{1}{2}\sup|\phi^{(j+2)}|.$$

Since $e_{\delta,0}$ and $e_{\delta,\frac{1}{2}}$ are obviously continuous with values in $\mathscr{D}'^0$, it follows that $e_{\delta,a}$ is continuous with values in $\mathscr{D}'^{2k}$ if k and $2a$ are integers and $a+k \geq 0$. We have $e_{0,-k} = 2(-1)^k k!\, \delta^{(2k)}/(2k)!$

E_0 is the limit when $\varepsilon \to 0$ of $E_{0,\varepsilon}$,

$$E_{0,\varepsilon}(t,x) = (2\pi)^{-n-1} \int_{\operatorname{Im}\tau=-1} e^{i(\langle x,\xi\rangle + t\tau) - \varepsilon|\xi|^2} (|\xi|^2 - \tau^2)^{-1}\, d\xi\, d\tau.$$

Since

$$\int e^{it\tau}\, i\tau/(|\xi|^2-\tau^2)\, d\tau = 2\pi i^2/(-2)(e^{it|\xi|} + e^{-it|\xi|}),$$

if the integral is taken from $-i-\infty$ to $-i+\infty$ and from $i+\infty$ to $i-\infty$, it follows that $\partial(E_{0,\varepsilon} - \check{E}_{0,\varepsilon})/\partial t$ is the Fourier transform of $\frac{1}{2}\operatorname{sgn}\tau\, e^{-\varepsilon\tau^2}\, de_0(x,\tau^2)$. When $\varepsilon \to 0$ we obtain the last statement in Lemma 17.4.2. The proof is complete.

With some abuse of the notation we shall write $E_\nu(t,|x|)$ instead of $E_\nu(t,x)$ in what follows; when $t=0$ this should be interpreted as the limit when $t \to +0$. If the coordinates are geodesic in the convex neighborhood X_0 of 0, it follows from (17.4.6) with $u_0(0)=1$ that in X_0

$$(17.4.6)'' \qquad (\partial^2/\partial t^2 + P(x,D)) \sum_0^N u_\nu(x) E_\nu(t,|x|_g) = (\det g^{jk})^{\frac{1}{2}} \delta_{0,0} + (P(x,D)\, u_N(x))\, E_N(t,|x|_g).$$

The error term is in C^k if $k < N-(n-1)/2$. If $|x|_g < c$ implies $x \in X_0$ then $E_\nu(t,x)=0$ for $t<c$ in a neighborhood of $\{x; |x|_g \geq c\}$, so (17.4.6)″ remains valid in $(-\infty,c)\times X$ for any $X \supset X_0$ if u_ν is extended arbitrarily to X.

As in the elliptic case we can extend the construction to general coordinates by taking the pullback by the inverse of the map $(t,\tilde{x},y) \mapsto (t,\gamma(\tilde{x},y),y)$ to geodesic coordinates. Then we obtain:

Proposition 17.4.3. *For any open* $Y \Subset X \subset \mathbb{R}^n$ *one can choose* $c>0$ *and* $U_j \in C^\infty(X \times Y)$ *such that, with* $s(x,y)$ *denoting the geodesic distance from x to y, we have in* $(-\infty, c) \times X \times Y$

$$(17.4.6)''' \quad (\partial^2/\partial t^2 + P(x,D)) \sum_0^N U_\nu(x,y) E_\nu(t, s(x,y))$$
$$= (\det g^{jk})^{\frac{1}{2}} \delta_{0,y} + (P(x,D) U_N(x,y)) E_N(t, s(x,y)).$$

When $s(x,y) \leqq c$ *the coefficients* U_j *are defined by integrating the equations* (17.4.7) *in geodesic coordinates, and when* $s(x,y)>c$ *their definition is irrelevant.*

If the coefficients of P are in $C^\infty(\overline{X})$ and P remains elliptic in $\overline{X}$, we can extend the coefficients to a neighborhood of $\overline{X}$ and then take $Y=X$. However, the situation is much more difficult when we want to construct a parametrix for the mixed problem for the wave operator $\partial^2/\partial t^2 + P$ in $\mathbb{R} \times X$ with Dirichlet data on ∂X. Then Proposition 17.4.3 can only be used if $t<d(y)$, where $d(y)$ is the geodesic distance from y to ∂X. If $P=-\Delta$ and X is a half space, then the "Green's function" with pole at $(0,y)$ introduced in Theorem 12.9.12 is just $E_0(t, x-y) - E_0(t, x-y^*)$ where y^* is the reflection of y in ∂X. We shall prove that this construction can be modified in a manner completely analogous to (17.4.6)''' when t is not larger than a fixed multiple of the distance $d(y)$ of y from the boundary. It is possible to extend the construction to all t which are small compared to $d^{\frac{1}{2}}$, which has to be assumed in order to guarantee that no light rays emanating from y arrive tangentially at the boundary. However, this is more laborious and will not be essential in the application in Section 17.5.

Before proceeding it is useful to rephrase the existence of geodesic normal coordinates as the existence of an *exponential map*. Given

$$P = -\sum \partial/\partial x_j (g^{jk} \partial/\partial x_k) + \sum b^j \partial/\partial x_j + c$$

in $X \subset \mathbb{R}^n$ we define again the Riemannian metric $\sum g_{jk}(x)\, dx_j\, dx_k$ where (g_{jk}) is the inverse of (g^{jk}). When $y \in X$ and s is a tangent vector at y with small norm,

$$|s|_{g,y} = |s|_y = (\sum g_{jk}(y) s_j s_k)^{\frac{1}{2}}$$

then one defines $\exp_y s$ as the point at distance $|s|_y$ from y on the geodesic with tangent vector s at y. If we introduce geodesic normal coordinates at y, then this becomes the identity map. Hence it follows that

$$(y,s) \mapsto (y, \exp_y s)$$

is a diffeomorphism of a neighborhood of the zero section in the tangent bundle of X on a neighborhood of the diagonal in $X \times X$. (The right setting for this is of course a Riemannian manifold.)

Let us now consider a Riemannian manifold with boundary, or for simplicity an open bounded subset X of $\mathbb{R}^n$ with C^∞ boundary ∂X. We assume that $g^{jk} \in C^\infty(\overline{X})$ and choose some extension of g^{jk} to a neighborhood of $\overline{X}$. When the geodesic distance $d(y)$ from y to ∂X is sufficiently small and $|s|_y < 4d(y)$, say, we shall then define a reflected exponential map as follows. First we start from y on the geodesic with direction s. If the boundary ∂X is encountered at a distance $<|s|_y$ we shall prove that the intersection is transversal and that continuing on the geodesic with direction obtained by the usual law of reflection to a total arc length $|s|_y$ we obtain a point $\exp_y^r s$ with the standard properties of the exponential map. (In the flat case discussed above it will be the point $(y+s)^*$.) In particular, (17.4.4) is valid for the metric in X when these coordinates are used. When $|s|_y = 2d(y)$ and s has the direction of the geodesic from y minimizing the boundary distance, we obtain $\exp_y^r s = y$.

To simplify the discussion of the reflected exponential map we assume that X is defined by $x_n \geqq 0$ in a neighborhood of $0 \in \mathbb{R}^n$ and that the coordinates are geodesic with respect to the boundary plane ∂X defined by $x_n = 0$ (Corollary C.5.3). Thus

$$g_{jn} = 0 \quad \text{for } j \neq n, \; g_{nn} = 1.$$

Let K_0 be a compact subset of ∂X. Translating and changing scales we obtain the metric G^y defined by

$$\sum g_{jk}((y',0) + y_n x) S_j S_k = |S|_x^{y\,2}.$$

If $(y',0) \in K_0$ this is defined in any desired bounded subset of $\mathbb{R}^n$ when y_n is small enough, and when $y_n = 0$ we obtain a flat metric with no cross products between dx_n and dx_j for $j \neq n$. The point y corresponds to the fixed point $\Pi = (0, \dots, 0, 1)$. The exponential map

$$B^y = \{S \in \mathbb{R}^n;\ |S|_\Pi^y < 4\} \ni S \mapsto \exp_\Pi^y S$$

corresponding to G^y is a C^∞ function of (S, y). For $y_n = 0$ it reduces to $S \mapsto S + \Pi$, so for fixed small y_n it is a diffeomorphism in a neighborhood of $\overline{B}^{(y',0)}$. The nth coordinate $F(S,y)$ of $\exp_\Pi^y S$ is equal to $1 + S_n$ when $y_n = 0$. Hence the zero set of this function is a C^∞ hypersurface close to the plane $S_n = -1$ which divides B^y into the starshaped inverse image B_+^y of X and the complement $B_-^y = B^y \setminus B_+^y$ mapped to $\complement X$. When $S \in B_-^y$ the equation $F(\lambda S, y) = 0$ has a unique solution $\lambda(S,y) \in [0,1]$; it is a C^∞ function equal to $-1/S_n$ when $y_n = 0$. The point of reflection

$$x(S,y) = \exp_\Pi^y(\lambda(S,y) S)$$

is therefore also a C^∞ function of S and y, and so is the tangent $T(S,y)$ of the geodesic there. When $y_n = 0$ it is equal to S. Let $\check{T}(S,y)$ be the reflection of $T(S,y)$ in the tangent plane of ∂X at $x(S,y)$, which just means a change of sign for the nth coordinate. Then our reflected exponential map is defined by

$$\exp_\Pi^{r,y} S = \exp_{x(S,y)}^y((1 - \lambda(S,y))\, \check{T}(S,y)), \qquad S \in B_-^y.$$

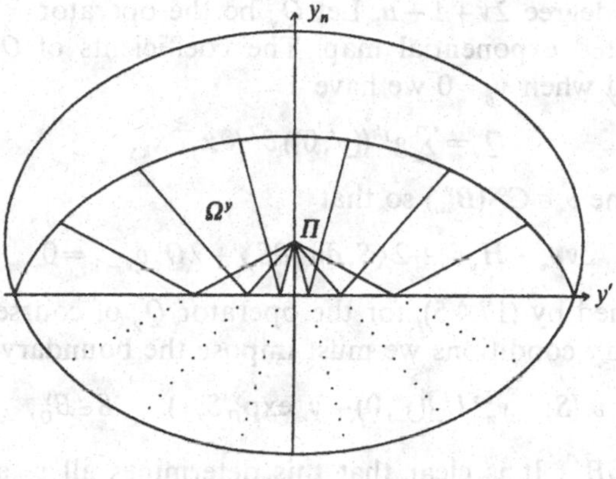

Fig. 1

For small y_n this is a C^∞ function of (S, y) and it reduces to

$$S \mapsto (S_1, \dots, S_{n-1}, -1-S_n); \quad S_n \leqq -1, \ |S|_\Pi^y < 4;$$

when $y_n = 0$. Note that the reflected exponential map gives another coordinate system in a neighborhood of Π. (See Fig. 1.)

The differential with respect to S of the total length $|S|_\Pi^y$ of the broken geodesic from Π to $x(S, y)$ and on to $Y(S, y) = \exp_\Pi^{r,y} S$ is equal to the differential of the length of the geodesic from $x(S, y)$ to $Y(S, y)$ when $x(S, y)$ is left fixed, for the reflection law means that it is stationary with respect to variations of $x(S, y)$. Thus the surface $\{\exp_\Pi^{r,y} S; |S|_\Pi^y = R\}$ is orthogonal to the geodesic from $x(S, y)$ to $Y(S, y)$. For the metric $G^{r,y}$ in B_-^y obtained by pulling the metric G^y back from X by the reflected exponential map $\exp_\Pi^{r,y}$, the rays through 0 are therefore geodesics with arc length $|S|_\Pi^y$ and they are orthogonal to the spheres $|S|_\Pi^y = \text{constant}$. But this means precisely that $G^{r,y}$ satisfies (17.4.4). We can therefore use the Hadamard construction with these coordinates in $\Omega^y = \exp_\Pi^{r,y} B_-^y$ to construct the reflection of the parametrix of P at y.

To do so we first rewrite (17.4.6)''' in terms of the stretched coordinates. Set

$$\begin{aligned} P_y(x, D) = & -\sum \partial/\partial x_j (g^{jk}((y', 0) + y_n x) \partial/\partial x_k) \\ & + y_n \sum b^j((y', 0) + y_n x) \, \partial/\partial x_j + y_n^2 c((y', 0) + y_n x)). \end{aligned}$$

Then (17.4.6)''' means that

$$\begin{aligned} & (\partial^2/\partial t^2 + P_y(x, D)) \sum_0^N U_v((y', 0) + y_n x, y) \, E_v(t, s^y(x, \Pi)) \, y_n^{2v} \\ & = (\det g^{jk}(y))^{\frac{1}{2}} \, \delta_{0,\Pi} + (P_y(x, D) \, U_N((y', 0) + y_n x, y)) \, E_N(t, s^y(x, \Pi)) \, y_n^{2N} \end{aligned}$$

Here $s^y(x, \Pi)$ is the G^y distance from Π to x, and we have used that the Dirac measure in $\mathbb{R}^{n+1}$ is homogeneous of degree $-n-1$ and that E_v is homogeneous of degree $2\nu+1-n$. Let Q_y be the operator P_y pulled back to B^y_- by the reflected exponential map. The coefficients of Q_y are C^∞ functions of (x, y), and when $y_n=0$ we have

$$Q_y=\sum g^{jk}((y',0))\,\partial^2/\partial x_j\,\partial x_k.$$

Now we determine $v_\nu\in C^\infty(B^y_-)$ so that

$$2\nu v_\nu - H_y v_\nu + 2\langle S, \partial v_\nu/\partial S\rangle + 2Q_y v_{\nu-1}=0 \tag{17.4.7$'$}$$

where H_y is defined by (17.4.5), for the operator Q_y of course. To satisfy the Dirichlet boundary conditions we must impose the boundary conditions

$$v_\nu(S)=y_n^{2\nu}\,U_\nu((y',0)+y_n\exp^y_\Pi S, y), \qquad S\in B^y_0, \tag{17.4.13$'$}$$

where $B^y_0=B^y\cap\partial B^y_-$. It is clear that this determines all v_ν as C^∞ functions of (S, y). When $y_n=0$ we obtain $v_0=1$ and $v_\nu=0$ for $\nu\neq 0$. Define $V_\nu(.,y)$ in Ω^y so that the pullback by $\exp^{r,y}_\Pi$ is v_ν, and set

$$\begin{aligned}\mathscr{E}^N(t,x,y)=\sum_0^N(&U_\nu((y',0)+y_n x, y)\,y_n^{2\nu}\,E_\nu(t, s^y(x,\Pi))\\ &-V_\nu(x,y)\,E_\nu(t, s^{r,y}(x,\Pi))), \qquad x\in\Omega^y,\end{aligned} \tag{17.4.14}$$

where $s^{r,y}(x,\Pi)$ is the reflected G^y distance from Π to x. Let Ω be the set of all (x, y) with $x\in\Omega^y$ and y in a small neighborhood ω of $K\subset\partial X$. Then we have $\mathscr{E}^N(t,x,y)=0$ in the sense of distribution theory when $x_n=0$, and

$$\begin{aligned}&(\partial^2/\partial t^2+P_y(x,D))\,\mathscr{E}^N(t,x,y)\\ &=(\det g^{jk}(y))^{\frac12}\delta_{0,y}+(P_y(x,D)\,U_N((y',0)+y_n x, y))\,y_n^{2N}E_N(t, s^y(x,\Pi))\\ &\quad-(P_y(x,D)\,V_N(x,y))\,E_N(t, s^{r,y}(x,\Pi)); \qquad (x,y)\in\Omega.\end{aligned} \tag{17.4.15}$$

Note that the error term becomes arbitrarily smooth when N is large. Summing up, we have proved

Proposition 17.4.4. *Let K_0 be a compact subset of ∂X. Then K_0 has a neighborhood ω in X such that* (17.4.14) *for every N defines a parametrix $\mathscr{E}^N$ for the Dirichlet-Cauchy mixed problem in $\Omega=\{(x,y);\ y\in\omega,\ s^{r,y}(x,\Pi)<4\}$ when $t<4$; this means that $\mathscr{E}^N(t,x,y)=0$ when $x_n=0$, in the distribution sense, and that*

$$(\partial^2/\partial t^2+P_y(x,D))\,\mathscr{E}^N(t,x,y)=(\det g^{jk}(y))^{\frac12}\delta_{0,\Pi}+R_N(t,x,y) \tag{17.4.15$'$}$$

where $R_N\in C^\mu$ when $\mu<N-(n-1)/2$. The coefficients $V_\nu(x,y)$ are C^∞ functions of $(x,y)\in\Omega$, and $V_0(x,y)=1$, $V_\nu(x,y)=0$ for $\nu\neq 0$ when $y_n=0$.

Since the coefficients of the incident wave in (17.4.14) vanish as $y_n^{2\nu}$ when $y_n=0$, and differentiation with respect to x increases the order of the zero, one might ask if the last statement can be improved. However, this is not possible. To see this we shall carry out the proof of Proposition 17.4.4 to first order in $y_n=\delta$ when $y'=0$. We may assume that the metric is Euclidean at 0 and geodesic inside the boundary, hence

$$G^y \equiv \sum dx_j^2 - 2\delta x_n H(dx')$$

where $\equiv$ means equality mod $O(\delta^2)$. Here the quadratic form H is the second fundamental form of ∂X at 0 with normal oriented towards X. Since only the second fundamental form at 0 is important, we can as well do the calculation in $\mathbb{R}^n$ with the metric $\sum dx_j^2$ and X defined by $x_n > \delta H(x')/2$, for this gives rise to a metric G^y of the preceding form when one takes geodesic coordinates with respect to ∂X. We may assume that $H(x') = \sum H_j x_j^2$ where H_j are the principal curvatures, and we shall write $H(x') = \langle x', Hx' \rangle$ with H diagonal.

The first step is to find λ so that $(\lambda S', 1+\lambda S_n) \in \partial X$, that is,

$$1+\lambda S_n = \delta H(\lambda S')/2.$$

Thus $\lambda S_n = -1 + O(\delta)$ and $\lambda S_n \equiv -1 + \delta H(S'/S_n)/2$. The reflection takes place at $x = (\lambda S', \delta H(\lambda S')/2)$. The normal is there $(-\delta H x', 1) \equiv (\delta H S'/S_n, 1)$, with length $\equiv 1$. The reflection $\check{S}$ of S in the tangent plane is

$$\check{S} \equiv S - 2(\delta H(S')/S_n + S_n)(\delta H S'/S_n, 1) \equiv (S' - 2\delta H S', -S_n - 2\delta H(S')/S_n).$$

Thus

$$\exp_{\Pi}^{y} S \equiv (S' - 2(1+1/S_n)\delta H S', -1 - S_n - \delta H(S')(1/S_n^2 + 2/S_n)).$$

The pullback Q of Δ by this map is equal to the pullback by the map $S \mapsto S + \delta\psi(S)$ where

$$\psi(S) = (-2(1+1/S_n) H S', (1/S_n^2 + 2/S_n) H(S')),$$

hence

$$Q \equiv \Delta_S - \delta \sum \partial/\partial S_j \psi_{jk} \partial/\partial S_k + \delta \sum \partial^2 \psi_j / \partial S_j \partial S_k \, \partial/\partial S_k$$

where $\psi_{jk}(S) = \partial\psi_k/\partial S_j + \partial\psi_j/\partial S_k$. The function h in (17.4.5) becomes

$$\delta \sum S_k \partial^2 \psi_j / \partial S_j \partial S_k = 2\delta(\operatorname{Tr} H + H(S'/S_n))/S_n.$$

The first transport equation is in polar coordinates

$$2r\,\partial v_0/\partial r = h v_0$$

with boundary condition $v_0 = 1$ at λS. Thus $v_0 \equiv 1 + \delta w_0$ where $2r\,\partial w_0/\partial r = 2(\operatorname{Tr} H + H(S'/S_n))/S_n$, $w_0 = 0$ when $S_n = -1$. This gives

$$w_0 = -(\operatorname{Tr} H + H(S'/S_n))(1+1/S_n),$$

$$Q v_0 \equiv \delta \Delta w_0 = \delta(-2\operatorname{Tr} H(S_n+2)/S_n^3 - 6H(S')(S_n+2)/S_n^5).$$

The next transport equation $(2r\partial/\partial r + 2)v_1 - h v_1 = -2Q v_0$ with boundary condition $v_1 = 0$ when $S_n = -1$ has a solution $O(\delta)$ so we can drop the term $h v_1$ and integrate explicitly to get

$$v_1 \equiv -2\delta(\operatorname{Tr} H + 3H(S'/S_n))(1+S_n)/S_n^3.$$

Inductively we obtain for $v>0$

(17.4.16) $\quad v_\nu \equiv -\delta\, 2^\nu(2\nu-1)!!(\operatorname{Tr} H+(2\nu+1)\,H(S'/S_n))(1+S_n)/S_n^{2\nu+1}.$

In particular, taking $S'=0$ and $S_n=-2$ we obtain

Proposition 17.4.5. *For the reflection coefficients V_ν in* (17.4.14) *we have when* $y_n=0$

$$\partial V_\nu(\Pi, y)/\partial y_n = -2^{-\nu-1}(2\nu-1)!!\,F(y')$$

where $F(y')$ is the mean curvature of ∂X at $(y',0)$. Here $(-1)!!$ should be interpreted as 1.

17.5. Asymptotic Properties of Eigenvalues and Eigenfunctions

As in Section 17.4 we denote by $P(x,D)$ a second order differential operator of the form

$$P(x,D) = -\sum \partial/\partial x_j(g^{jk}\partial/\partial x_k) + \sum b^j \partial/\partial x_j + c$$

with coefficients in $\bar{C}^\infty(X)$ where X is a bounded open subset of $\mathbb{R}^n$ with C^∞ boundary. We assume that (g^{jk}) is real and positive definite in $\bar{X}$ and that P is symmetric with respect to a density Υdx with $0<\Upsilon\in\bar{C}^\infty(X)$,

(17.5.1) $\qquad (P(x,D)u, v)_\Upsilon = (u, P(x,D)v)_\Upsilon; \quad u,\ v\in C_0^\infty(X),$

where

$$(u,v)_\Upsilon = \int u\bar{v}\,\Upsilon dx.$$

(The density simplifies coordinate changes but should be disregarded as otherwise irrelevant.) Let $\mathscr{P}$ be the operator defined by $P(x,D)$ in $L^2_\Upsilon(X)$ with Dirichlet boundary conditions. As explained in Section 17.3 we have $u\in\mathscr{D}_{\mathscr{P}}$ if and only if $P(x,D)u=f\in L^2$ and u is in the closure $\mathring{H}$ of $C_0^\infty(X)$ in $H_{(1)}(X)$. Then we have

$$\sum (g^{jk}D_k u, D_j v) + \sum (b^j\partial_j u, v) + (cu, v) = (f, v); \quad v\in\mathring{H}.$$

If $v\in\mathscr{D}_{\mathscr{P}}$ we conclude using (17.5.1) that $(\mathscr{P}u,v)_\Upsilon = (u,\mathscr{P}v)_\Upsilon$. With $v=\Upsilon u$ we obtain

$$\sum \|D_j u\|^2 + \|u\|^2 \leqq C_1(\mathscr{P}u,u)_\Upsilon + C_2\|u\|^2; \quad u\in\mathscr{D}_{\mathscr{P}};$$

where $\|\ \|$ denotes the L^2 norm in X. Hence $\mathscr{P}$ is closed. As in Section 17.3 it follows that $P+C_2/C_1$ is bijective from $\mathscr{D}_{\mathscr{P}}$ to L^2, which proves that $\mathscr{P}$ is self-adjoint. If $P(x,D)$ is replaced by $P(x,D)+C_2/C_1$ we do not affect any of the statements on the eigenfunctions proved below, and then we have

(17.5.2) $\qquad \|u\|_{(1)}^2 \leqq C_1(\mathscr{P}u,u); \quad u\in\mathscr{D}_{\mathscr{P}};$

where we have used the notation

$$\|u\|_{(k)}^2 = \sum_{|\alpha| \leq k} \|D^\alpha u\|^2.$$

From now on we assume that (17.5.2) is valid. We shall also write $\overline{H}_{(k)}(X)$ for the set of all u with $D^\alpha u \in L^2(X)$, $|\alpha| \leq k$, and with the norm just defined. By Cauchy-Schwarz' inequality and (17.5.2) we have

$$\|u\|_{(1)} \leq C_1 \|\mathscr{P}u\|, \quad u \in \mathscr{D}_{\mathscr{P}}. \tag{17.5.2$'$}$$

In Section 17.3 we proved the stronger estimate (17.3.6), and we shall elaborate it further as follows.

Lemma 17.5.1. *If $u \in \mathscr{D}_{\mathscr{P}}$ and $\mathscr{P}u \in \overline{H}_{(k)}(X)$ where k is a non-negative integer, then $u \in \overline{H}_{(k+2)}(X)$ and*

$$\|u\|_{(k+2)} \leq C_k \|\mathscr{P}u\|_{(k)}, \quad u \in \mathscr{D}_{\mathscr{P}}. \tag{17.5.3}$$

Proof. For $k=0$ this follows from Theorem 17.3.2. When $k>0$ we may assume that the lemma has already been proved with k replaced by $k-1$. If $\phi \in C_0^\infty(\mathbb{R}^n)$ we then obtain

$$\|\mathscr{P}(\phi u)\|_{(k)} \leq \sup|\phi| \, \|\mathscr{P}u\|_{(k)} + C\|u\|_{(k+1)} \leq C' \|\mathscr{P}u\|_{(k)}.$$

Hence it suffices to prove (17.5.3) when u has support in a coordinate patch. We can then assume the coordinates chosen so that X is there equal to the half space $x_n > 0$. If $j < n$ then $D_j u \in \overset{\circ}{H}$ and $PD_j u = D_j Pu + iP_{(j)}u$ where $P_{(j)}$ is also a second order operator. Hence (17.5.3) with k replaced by $k-1$ gives

$$\|D_j u\|_{(k+1)} \leq C_{k-1}(\|D_j \mathscr{P}u\|_{(k-1)} + \|u\|_{(k+1)}) \leq C' \|\mathscr{P}u\|_{(k)}.$$

This gives the desired estimate for all $D^\alpha u$, $|\alpha| = k+2$, except $D_n^{k+2}u$. Now the differential equation gives

$$D_n^2 u = (g^{nn})^{-1}\Big(-\sum_{j,k \neq n} D_j(g^{jk} D_k u) + \ldots + \mathscr{P}u\Big)$$

where dots denote lower order terms, so the estimate of $D_n^{k+2}u$ follows at once. The proof is complete.

A useful consequence of (17.5.3) is that, with new constants,

$$\|u\|_{(2k)} \leq C_k \|\mathscr{P}^k u\|, \quad u \in \mathscr{D}_{\mathscr{P}^k}. \tag{17.5.4}$$

This is clear if $k=1$. If $k>1$ and (17.5.4) is proved for smaller values of k, then

$$\|\mathscr{P}u\|_{(2k-2)} \leq C_k \|\mathscr{P}^k u\|, \quad u \in \mathscr{D}_{\mathscr{P}^k},$$

and (17.5.4) follows from (17.5.3).

Already the estimate (17.5.2)$'$ implies that $\mathscr{P}$ has a discrete spectrum, and $\mathscr{P}$ is positive by (17.5.2). In fact, if E_λ is the spectral family and $u = E_\lambda u$, then

$\|\mathscr{P}u\| \leqq \lambda \|u\|$ so $\|u\|_{(1)} \leqq C_1 \lambda \|u\|$ by (17.5.2)′. The set of all such u with $\|u\| \leqq 1$ is compact in L^2 by Theorem 10.1.10. Thus $E_\lambda L^2(X)$ is a finite dimensional space (see e.g. Lemma 19.1.4). Let $\phi_1, \phi_2, \dots$ be an orthonormal basis in L^2_Υ for the eigenfunctions such that $0 < \lambda_1 \leqq \lambda_2 \leqq \lambda_3 \leqq \dots$ for the corresponding eigenvalues. Then the kernel of E_λ is $\Upsilon(y)$ times

$$e(x, y, \lambda) = \sum_{\lambda_j \leqq \lambda} \phi_j(x) \overline{\phi_j(y)}, \tag{17.5.5}$$

and $\phi_j \in \bar{C}^\infty(X)$ since $\phi_j \in \bar{H}_{(k)}(X)$ for every k. Here we have used a supplement to the Sobolev lemma (Lemma 7.6.3) which will also give an estimate for the *spectral function* e:

Lemma 17.5.2. *If* $\nu + n/2 < k$ *then* $u \in \bar{H}_{(k)}(X)$ *implies* $u \in \bar{C}^\nu(X)$ *and*

$$\lambda^{k-\nu-\frac{1}{2}n} \sum_{|\alpha| \leqq \nu} \sup |D^\alpha u|^2 \leqq C(\|u\|^2_{(k)} + \lambda^k \|u\|^2), \qquad \lambda \geqq 1. \tag{17.5.6}$$

Proof. Decomposition of u by a partition of unity shows that it is sufficient to prove this when u has support in a coordinate patch, so we may assume that $X = \mathbb{R}^n_+$ and that u has compact support. By Theorem B.2.1 it suffices to prove (17.5.6) when $u \in \bar{C}^\infty_0(\mathbb{R}^n_+)$. Then we shall even prove that

$$\lambda^{k-|\alpha|-\frac{1}{2}n} \sup |D^\alpha u|^2 \leqq C\Big(\sum_{|\beta|=k} \|D^\beta u\|^2 + \lambda^k \|u\|^2\Big), \qquad |\alpha| \leqq \nu. \tag{17.5.6$'$}$$

Taking $\lambda^{\frac{1}{2}} x$ as a new variable we see that it suffices to prove the estimate when $\lambda = 1$. Then we first prove that

$$\sum_0^k \|D_n^j u\|^2 \leqq C(\|D_n^k u\|^2 + \|u\|^2). \tag{17.5.7}$$

This is clear when $k=1$. To prove (17.5.7) for $k=2$ we observe that

$$\int_0^\infty (|v''|^2 - |v'|^2 + |v|^2)\,dt = \int_0^\infty |v'' + v' + v|^2\,dt + |v(0) + v'(0)|^2, \qquad v \in \bar{C}^\infty_0(\mathbb{R}_+).$$

It suffices to prove this when v is real valued, and then we have

$$v''^2 - v'^2 + v^2 - (v'' + v' + v)^2 = (v'' + v')(v'' - v' - v'' - v' - 2v) = -2(v'' + v')(v + v').$$

(a deeper reason for this identity is that $\xi^4 - \xi^2 + 1 = |\xi^2 - i\xi - 1|^2$, $\xi \in \mathbb{R}$, where $\xi^2 - i\xi - 1$ has its zeros in the upper half plane.) Hence we obtain (17.5.7) with $k=2$ and $C=2$. Changing scales we have with L^2 norms on $(0, \infty)$

$$\|v'\|^2 \leqq \varepsilon \|v''\|^2 + \varepsilon^{-1} \|v\|^2, \qquad \varepsilon > 0,$$

which proves that $\|v'\|^2 \leqq 2\|v''\|\,\|v\|$. If now (17.5.7) is proved for a certain value of k, we obtain when applying it to $D_n u$

$$\sum_1^{k+1} \|D_n^j u\|^2 \leqq C(\|D_n^{k+1} u\|^2 + \|D_n u\|^2) \leqq C(\|D_n^{k+1} u\|^2 + 2\|u\|\,\|D_n^2 u\|).$$

Since $2C\|u\|\,\|D_n^2u\|\leq\|D_n^2u\|^2/2+2C^2\|u\|^2$, we obtain (17.5.7) with k replaced by $k+1$ and another C after cancellation of $\|D_n^2u\|^2/2$.

Using Parseval's formula for fixed x_n we can strengthen (17.5.7) to

$$(17.5.7)' \qquad \sum_{|\alpha|\leq k}\|D^\alpha u\|^2\leq C\big(\sum_{|\alpha|=k}\|D^\alpha u\|^2+\|u\|^2\big).$$

This is obvious if $\alpha_n=0$ in the two sums and follows without such a restriction if we then use (17.5.7). The proof of (17.5.6)′ is now reduced to proving the estimate

$$\sup_{x_n>0}|u|^2\leq C\|u\|_{(k)}^2,\qquad u\in\bar{H}_{(k)}(\mathbb{R}^n_+),$$

if $k>n/2$. By Corollary B.2.6 it follows from the same result in $H_{(k)}(\mathbb{R}^n)$ which is precisely Lemma 7.6.3. The proof is complete.

We can now prove a crude but useful estimate for the spectral function.

Theorem 17.5.3. *There are constants C_α such that*

$$(17.5.8) \qquad |D^\alpha_{x,y}e(x,y,\lambda)|\leq C_\alpha\lambda^{(n+|\alpha|)/2};\qquad x,\ y\in X.$$

Proof. Let $f\in L^2$ and apply (17.5.6) to $u=E_\lambda f$ with $2k>|\alpha|+n/2$. Since

$$\|u\|_{(2k)}\leq C_k\|\mathscr{P}^k u\|\leq C_k\lambda^k\|f\|$$

by (17.5.4), it follows from (17.5.6) that

$$|D^\alpha E_\lambda f(x)|^2\leq C_\alpha\lambda^{|\alpha|+n/2}\|f\|^2.$$

Here

$$D^\alpha E_\lambda f(x)=(f,g)_\Upsilon=(E_\lambda f,g)_\Upsilon,\qquad \bar{g}(y)=D^\alpha_x e(x,y,\lambda).$$

Thus $E_\lambda g=g$ and $\|g\|^2\leq C_\alpha\lambda^{|\alpha|+n/2}$. Now $\|\mathscr{P}^k g\|\leq\lambda^k\|g\|$ so (17.5.6) gives

$$|D^\beta_y g(y)|^2\leq C_{\alpha\beta}\lambda^{|\alpha|+|\beta|+n}$$

which completes the proof.

An immediate consequence of (17.5.8) is that if $N(\lambda)$ is the number of eigenvalues $\leq\lambda$, counted with multiplicities, then

$$N(\lambda)=\operatorname{Tr}E_\lambda=\int e(x,x,\lambda)\Upsilon(x)\,dx=O(\lambda^{n/2}).$$

where Tr denotes the trace. (We just have to integrate the definition (17.5.5).) The goal of this section is to prove rather precise asymptotic properties of $N(\lambda)$ by estimating $e(x,x,\lambda)$ first in compact subsets of X and then at the boundary.

Remark. If the coefficients of $P(x,D)$ are just in $C^\infty(X)$ and $\mathscr{P}$ is any self adjoint extension of $P(x,D)$ with domain C_0^∞ which is bounded from below, then a slight modification of the proof of Theorem 17.5.3 shows that E_λ has a C^∞ kernel satisfying (17.5.8) on compact subsets of $X\times X$. If ∂X has a smooth part ω, the coefficients of $P(x,D)$ remain smooth in $X\cup\omega$, and the domain of $\mathscr{P}$ contains the smooth functions with support in $X\cup\omega$ vanishing on ω, then (17.5.8) remains valid on compact subsets of $(X\cup\omega)\times(X\cup\omega)$. In these circumstances $\mathscr{P}$ may not have a discrete spectrum so (17.5.5) is no longer meaningful. However, the results proved below on the spectral function remain valid.

Although the aim is to study the kernel of E_λ we shall first examine the kernel of another function of $\mathscr{P}$ which is easier to approach. A particularly convenient choice is the cosine transform of the spectral measure

$$\cos(t\sqrt{\mathscr{P}})=\int_0^\infty \cos(t\sqrt{\lambda})\,dE_\lambda.$$

To identify the kernel of this operator we take $\psi\in\mathscr{S}(\mathbb{R})$ and $f\in C_0^\infty(X)$ which gives

$$\int_{-\infty}^{\infty}(\cos(t\sqrt{\mathscr{P}})f,f)_\Upsilon\,\psi(t)\,dt=\int_{-\infty}^{\infty}\psi(t)\,dt\int\cos(t\sqrt{\lambda})\,de(f,f,\lambda)$$

where

$$e(f,f,\lambda)=(E_\lambda f,f)_\Upsilon=\iint e(x,y,\lambda)\,\overline{f(x)}\,f(y)\,\Upsilon(x)\,\Upsilon(y)\,dx\,dy$$

is an increasing function of λ bounded by $\|f\|^2$. Interchanging the orders of integration we obtain

$$\int_{-\infty}^{\infty}(\cos(t\sqrt{\mathscr{P}})f,f)_\Upsilon\,\psi(t)\,dt$$

$$=\tfrac12\int(\hat\psi(\sqrt{\lambda})+\hat\psi(-\sqrt{\lambda}))\,de(f,f,\lambda)=\tfrac12\int_0^\infty(\hat\psi(\tau)+\hat\psi(-\tau))\,de(f,f,\tau^2)$$

$$=\iint\overline{f(x)}\,f(y)\,\Upsilon(x)\,\Upsilon(y)\,dx\,dy\int_0^\infty(\hat\psi(\tau)+\hat\psi(-\tau))\,d_\tau e(x,y,\tau^2)/2.$$

The last interchange of orders of integration is obviously justified if $\hat\psi\in C_0^\infty$. Hence the distribution kernel $K(t,x,y)\in\mathscr{D}'(\mathbb{R}\times X\times X)$ of $f\mapsto\cos(t\sqrt{\mathscr{P}})f$ is the Fourier transform with respect to τ of the temperate measure $dm(x,y,\tau)$,

$$m(x,y,\tau)=\Upsilon(y)(\operatorname{sgn}\tau)\,e(x,y,\tau^2)/2. \tag{17.5.9}$$

That this is a temperate measure follows by polarization from the fact that

$$|a|^2e(x,x,\lambda)+a\bar b\,e(x,y,\lambda)+\bar a b\,e(y,x,\lambda)+|b|^2e(y,y,\lambda)$$

is increasing and $O(\lambda^{n/2})$ (by Lemma 17.5.3) for arbitrary complex a and b. If we can get good information on $\cos(t\sqrt{\mathscr{P}})$ then inversion of the Fourier transformation should give the desired results concerning $e(x,y,\lambda)$.

If $f \in C_0^\infty$ then $u(t,x) = \cos(t\sqrt{\mathscr{P}}) f(x)$ is equal to f when $t=0$, we have $\partial u/\partial t = 0$ when $t=0$, and

$$\|(\partial/\partial t)^j \mathscr{P}^k u\| \leqq \|\mathscr{P}^{\frac{1}{2}j+k} f\|$$

for arbitrary positive integers j and k. In view of (17.5.4) and Lemma 17.5.2 it follows that $u \in \bar{C}^\infty(\mathbb{R} \times X)$ and that $u=0$ on $\mathbb{R} \times \partial X$. Furthermore, we have

$$(P + \partial^2/\partial t^2) u = 0.$$

Using the parametrices constructed in Section 17.4 and the initial conditions $u = f$ and $\partial u/\partial t = 0$ when $t=0$ we can therefore reconstruct u approximately. Before writing down the approximation we prove a lemma which will allow us to estimate the error.

Lemma 17.5.4. *Let $v \in C^\infty([0,T] \times \bar{X})$ be a solution of the mixed problem*

$$\text{(17.5.10)} \qquad \begin{aligned} (P + \partial^2/\partial t^2) v &= h \quad \text{in } [0,T] \times \bar{X}; \\ v &= 0 \quad \text{in } [0,T] \times \partial X; \\ v = \partial v/\partial t &= 0 \quad \text{if } t = 0. \end{aligned}$$

Assume that $\partial^j h/\partial t^j = 0$ when $t=0$ if $j<k$. Then it follows that

$$\text{(17.5.11)} \qquad \sum_0^{k+1} \|D_t^{k+1-j} v(t,.)\|_{(j)} \leqq C_k \left(\int_0^t \|D_s^k h(s,.)\|\, ds + \sum_0^{k-1} \|D_t^{k-1-j} h(t,.)\|_{(j)} \right).$$

Proof. If $k=0$ the assertion is a standard energy estimate proved as follows. Since $(P\partial v/\partial t, v)_Y = (\partial v/\partial t, Pv)_Y$, we have

$$2\,\mathrm{Re}(h, \partial v/\partial t)_Y = \partial/\partial t(\|\partial v/\partial t\|_Y^2 + (Pv, v)_Y).$$

Integration from 0 to t gives in view of (17.5.2)

$$\|\partial v/\partial t\|_Y^2 + \|v\|_{(1)}^2/C_1 \leqq 2M \int_0^t \|h(s,.)\|_Y\, ds,$$

$$M^2 = \sup_{0 \leqq s \leqq t} \|\partial v(s,.)/\partial s\|_Y^2 + \|v(s,.)\|_{(1)}^2/C_1.$$

The same estimate holds for smaller values of t, so

$$M^2 \leqq 2M \int_0^t \|h(s,.)\|_Y\, ds.$$

which proves (17.5.11) when $k=0$. If $k>0$ we have $\partial^2 v/\partial t^2 = h - Pv = 0$ when $t=0$. If (17.5.11) is proved for smaller values of k we can therefore apply (17.5.11) to $\partial v/\partial t$ and obtain (17.5.11) with the term for $j=k+1$ missing in the left-hand side. Now it follows from (17.5.3) that

$$\|v(t,.)\|_{(k+1)} \leqq C\|Pv(t,.)\|_{(k-1)} \leqq C(\|h(t,.)\|_{(k-1)} + \|D_t^2 v(t,.)\|_{(k-1)})$$

which completes the proof.

Choose $c>0$ so that (17.4.6)''' holds for all $(x,y)\in X\times X$ at geodesic distance $s(x,y)<c$. Let $d(y)$ denote the geodesic distance from y to ∂X and set $X_\rho=\{y\in X; d(y)>\rho\}$ for some $\rho<c$. If $y\in X_\rho$ and $t<\rho$ the parametrix in (17.4.6)'''

$$\mathscr{E}(t,x,y)=\sum_0^N U_\nu(x,y)\,E_\nu(t,s(x,y))$$

is well defined when $x\in X$ and vanishes near ∂X. If $f\in C_0^\infty(X_\rho)$ then

$$u(t,x)=\int \mathscr{E}(t,x,y)(\det g^{jk}(y))^{-\frac{1}{2}} f(y)\,dy$$

is in $\bar{C}^\infty([0,\rho]\times X)$; $u=0$ in $[0,\rho]\times\partial X$ and $u=0$, $\partial u/\partial t=f$ when $t=0$. Moreover, u depends continuously on f in the C^∞ topology. This follows at once from Lemma 17.4.2 if we introduce instead of y a new variable z with $\exp_y z=x$, for then we obtain a sum of terms with $E_\nu(t,z)$ acting on a C^∞ function of z and x, as a function of z. Thus

$$v=\cos(t\sqrt{\mathscr{P}})f-\partial u/\partial t$$

has zero Cauchy and Dirichlet data, and by (17.4.6)'''

$$\begin{aligned} h(t,x)&=(P(x,D)+\partial^2/\partial t^2)\,v(t,x)\\ &=-\int \partial R_N(t,x,y)/\partial t(\det g^{jk}(y))^{-\frac{1}{2}} f(y)\,dy,\\ R_N(t,x,y)&=(P(x,D)U_N(x,y))E_N(t,s(x,y)).\end{aligned}$$

By Lemma 17.4.2 we have $R_N\in C^{k+1}$ if $N>k+(n+1)/2$, and for $|\alpha|\leqq k$ we then obtain

$$|D^\alpha_{t,x,y}\,\partial R_N(t,x,y)/\partial t|\leqq C\,t^{2N-n-|\alpha|}.$$

Since $s(x,y)\leqq t$ in the support it follows that

$$\int_0^t \|D_s^k h(s,.)\|\,ds+\sum_0^{k-1}\|D_t^{k-1-j}h(t,.)\|_{(j)}\leqq C\,t^{2N+1-\frac{1}{2}n-k}\,\|f\|_{L^1}.$$

Hence Lemma 17.5.4 gives

$$\|v(t,.)\|_{(k+1)}+\|D_t v(t,.)\|_{(k)}\leqq C\,t^{2N+1-\frac{1}{2}n-k}\,\|f\|_{L^1},\qquad t\geqq 0.$$

We may replace k by 0 here. If $|\alpha|+\frac{1}{2}n<k+1$, hence if $|\alpha|+n+1<N$, we obtain using Lemma 17.5.2 with $\lambda=t^{-2}$ that $D^\alpha_{x,t}v$ is continuous and that

$$|D^\alpha_{x,t}v|\leqq C t^{2N-|\alpha|-n}\|f\|_{L^1}, \tag{17.5.12}$$

provided that at most one derivative is taken with respect to t. This restriction is removed at once by means of the equation $\partial^2 v/\partial t^2=h-Pv$.

Set

$$K_N(t,x,y)=\widehat{dm}(x,y,t)-\mathscr{E}'_t(t,x,y)(\det g^{jk}(y))^{-\frac{1}{2}},\qquad t>0,$$

which is a continuous function of t with values in $\mathscr{D}'(X\times X_0)$. If we replace f by $D^\beta f$ in the proof of (17.5.12) we obtain

$$|D^\alpha_{x,t}\langle K_N(t,x,y), D^\beta f(y)\rangle|\leqq C\,t^{2N-|\alpha+\beta|-n}\,\|f\|_{L^1}$$

if $|\alpha+\beta|+n+1<N$. Choose $\chi\in C_0^\infty(\mathbb{R}^n)$ with $\int\chi\,dx=1$ and set $\chi_\varepsilon(x)=\chi(x/\varepsilon)/\varepsilon^n$. Taking $f(y)=\chi_\varepsilon(z-y)$ we then obtain

$$|D^\alpha_{x,z,t}\langle K_N(t,x,.),\chi_\varepsilon(z-.)\rangle|\leqq Ct^{2N-|\alpha|-n},\quad |\alpha|+n+1<N.$$

Letting $\varepsilon\to 0$ we conclude that $K_N\in C^{N-n-3}$ and that

$$|D^\alpha_{x,y,t}K_N(t,x,y)|\leqq Ct^{2N-|\alpha|-n},\quad |\alpha|\leqq N-n-3.$$

Since $\widehat{dm}$ is even we obtain the same result when $t\in\mathbb{R}$ with $K_N(t,x,y)$ replaced by

$$\widehat{dm}(x,y,t)-\partial_t(\mathscr{E}(-t,x,y)-\check{\mathscr{E}}(-t,x,y))(\det g^{jk}(y))^{-\frac12}.$$

Here all terms are continuous functions of (x,y) with values in $\mathscr{D}'(\mathbb{R})$ so we can put $x=y$ and obtain

Theorem 17.5.5. *Let $\Omega=\{(t,x)\in\mathbb{R}\times X; |t|<\min(d(x),c)\}$. Then*

$$\widehat{dm}(x,x,t)-\sum_{2\nu<n}\partial_t(E_\nu(t,0)-\check{E}_\nu(t,0))\,U_\nu(x,x)(\det g^{jk}(x))^{-\frac12}$$

is in $C^\infty(\Omega)$ if n is even, and in $C^\infty(\Omega)$ after division by $|t|$ if n is odd. All derivatives are bounded in Ω. For $t\geqq 0$ the Taylor expansion with respect to t is in both cases

$$\sum_{2\nu\geqq n}\partial_t(E_\nu(t,0)-\check{E}_\nu(t,0))\,U_\nu(x,x)(\det g^{jk}(x))^{-\frac12}.$$

Here $U_\nu(x,x)$ are the coefficients in Proposition 17.4.3 restricted to the diagonal, so they are polynomials in the coefficients of P, their derivatives, and $(\det g^{jk}(x))^{-1}$.

It follows from Lemma 17.4.2 with the notation (3.2.10)′ that

$$\partial_t(E_\nu(t,0)-\check{E}_\nu(t,0))=2^{-2\nu}\pi^{(1-n)/2}\,\underline{t}^{2\nu-n}/\Gamma(\nu+(1-n)/2)$$

if n is even; when n is odd we just have to replace $\underline{t}$ by $|t|$ unless $2\nu=n-1-2k$ where k is an integer $\geqq 0$; then we have

$$\partial_t(E_\nu(t,0)-\check{E}_\nu(t,0))=2^{-2\nu}\pi^{(1-n)/2}(-1)^k k!\,\delta^{(2k)}/(2k)!.$$

This is clear when $t\neq 0$ and follows at 0 since the left-hand side is homogeneous of degree $2\nu-n$ and even in t, so a distribution with support at 0 can only occur in the last case which was discussed explicitly in Lemma 17.4.2.

If the Fourier transform of a measure $d\mu$ is known in an interval I, then we can compute the convolution of $d\mu$ with any function ϕ such that $\operatorname{supp}\hat\phi\subset I$. The following lemma will give some estimates for $d\mu$ which can then be obtained. We denote by ϕ a positive function in $\mathscr{S}(\mathbb{R})$ with $\int\phi(\tau)\,d\tau=1$ and $\operatorname{supp}\hat\phi\subset(-1,1)$, such as $\phi=|\psi|^2*|\psi|^2$ where $\hat\psi\in C_0^\infty(-\frac12,\frac12)$ and the L^2 norm of ψ is 1. Set $\phi_a(\tau)=\phi(\tau/a)/a$, $a>0$, so that $\hat\phi_a(t)=\hat\phi(a\,t)$ has support in $(-1/a,1/a)$.

Lemma 17.5.6. *If μ is an increasing temperate function with $\mu(0)=0$, and ν is a function of locally bounded variation such that $\nu(0)=0$ and*

$$|d\nu(\tau)| \leqq M_0(|\tau|+a_0)^{n-1}\,d\tau, \tag{17.5.13}$$
$$|(d\mu-d\nu)*\phi_a(\tau)| \leqq M_1(|\tau|+a_1)^{\kappa}, \quad \tau\in\mathbb{R},$$

for some $\kappa\in[0,n-1]$ and $a_0\geqq a$, $a_1\geqq a$, then

$$|\mu(\tau)-\nu(\tau)| \leqq C(M_0\,a(|\tau|+a_0)^{n-1}+M_1(|\tau|+a)(|\tau|+a_1)^{\kappa}) \tag{17.5.14}$$

where C only depends on κ and n.

Proof. Choose $c_0>0$ so that $\phi>c_0$ in $(-\frac{1}{2},\frac{1}{2})$. Then

$$c_0\,a^{-1}\int_{\tau-\frac{1}{2}a}^{\tau+\frac{1}{2}a} d\mu \leqq d\mu*\phi_a(\tau) \leqq M_1(|\tau|+a_1)^{\kappa}+C_1\,M_0(|\tau|+a_0)^{n-1},$$

since

$$\int\phi_a(s)(|\tau-s|+a_0)^{n-1}\,ds=\int\phi(s)(|\tau-a\,s|+a_0)^{n-1}\,ds$$
$$\leqq(|\tau|+a_0)^{n-1}\int\phi(s)(1+|s|)^{n-1}\,ds.$$

From this estimate it follows that

$$c_0|\mu(\tau)-\mu(\tau-a\,s)| \leqq a(|s|+1)(C_1\,M_0(|\tau|+a_0+a\,|s|)^{n-1}$$
$$+M_1(|\tau|+a_1+a\,|s|)^{\kappa})$$

if we divide $(0,s)$ into $\leqq|s|+1$ intervals of length $\leqq 1$. Multiplication by $\phi(s)$ and integration yields

$$|\mu(\tau)-\mu*\phi_a(\tau)| \leqq C_2\,a(M_0(|\tau|+a_0)^{n-1}+M_1(|\tau|+a_1)^{\kappa}).$$

The first part of (17.5.13) gives such an estimate with μ replaced by ν and M_1 replaced by 0. Since integration of the second part of (17.5.13) from 0 to τ proves that

$$|(\mu-\nu)*\phi_a(\tau)-(\mu-\nu)*\phi_a(0)| \leqq M_1\,|\tau|(|\tau|+a_1)^{\kappa},$$

the estimate (17.5.14) follows.

In the following theorem $e_0(0,\lambda)$ is defined by (17.4.12), hence equal to $(2\pi)^{-n}\,\lambda^{n/2}$ times the volume of the unit ball.

Theorem 17.5.7. *For the spectral function $e(x,y,\lambda)$ of the Dirichlet problem in X we have, if $d(x)$ is the distance from x to ∂X,*

$$|e(x,x,\lambda)\,\Upsilon(x)-e_0(0,\lambda)(\det g^{jk}(x))^{-\frac{1}{2}}| \leqq C\,\lambda^{n/2}/(1+d(x)\,\lambda^{\frac{1}{2}}). \tag{17.5.15}$$

Proof. We shall apply Lemma 17.5.6 with $a=1/\min(d(x),c)$ and

$$\mu(\tau)=m(x,x,\tau)=\Upsilon(x)\operatorname{sgn}\tau\, e(x,x,\tau^2)/2$$
$$\nu(\tau)=\operatorname{sgn}\tau\, e_0(0,\tau^2)(\det g^{jk}(x))^{-\frac{1}{2}}/2.$$

By Lemma 17.4.2 the leading term in the expansion of $\widehat{d\mu}$ given by Theorem 17.5.5 is the Fourier transform of dv; the other singular terms are Fourier transforms of $|t|^{n-1-2\iota}$ times smooth functions of x for $0<\iota\leqq(n-1)/2$. Hence $(d\mu-dv)*\phi_a$ is the sum of the regularizations of these functions and a bounded function, so (17.5.13) is valid with $\kappa=\max(n-3,0)$. Hence (17.5.14) is valid which proves (17.5.15) when $d(x)\lambda^{\frac{1}{2}}>1$. Since (17.5.15) is a consequence of (17.5.8) when $d(x)\lambda^{\frac{1}{2}}\leqq 1$, the proof is complete.

Corollary 17.5.8. *If $N(\lambda)$ is the number of eigenvalues $\leqq\lambda$ of the Dirichlet problem in X, then*

$$|N(\lambda)-(2\pi)^{-n}C_n\operatorname{vol}(X)\lambda^{n/2}|\leqq C\lambda^{(n-1)/2}\log\lambda \tag{17.5.16}$$

where $\operatorname{vol}(X)$ *is the Riemannian volume of X and C_n is the Euclidean volume of the unit ball.*

Proof. Since

$$N(\lambda)=\int_X e(x,x,\lambda)\,\Upsilon(x)\,dx$$

we obtain (17.5.16) by integration of (17.5.15).

(17.5.16) is a vast improvement compared to our earlier bound $N(\lambda)=O(\lambda^{n/2})$. However, we shall show now that the factor $\log\lambda$ can also be eliminated. The interest of this improvement is that an example below shows that the error term $O(\lambda^{(n-1)/2})$ is sometimes optimal. We shall also show that the boundary ∂X has an influence proportional to $\lambda^{(n-1)/2}$ and to its Riemannian volume.

The main point is an extension of Theorem 17.5.5 which gives better information at ∂X. To state it we first introduce the map

$$\partial X\times[0,c]\ni(x',\delta)\mapsto x(x',\delta)\in\overline{X}$$

such that $\delta\mapsto x(x',\delta)$ is the geodesic with arc length δ normal to ∂X at x' when $\delta=0$. For small c this is a diffeomorphism on a neighborhood X_c of ∂X in $\overline{X}$. We shall denote the inverse by $x\mapsto(\gamma(x),d(x))$. Thus $d(x)$ denotes the geodesic distance from x to ∂X as before, and $\gamma(x)$ is the point on ∂X with geodesic distance $d(x)$ to x.

Theorem 17.5.9. *If c is sufficiently small and $x\in X_c$, $0<t<c$, then*

$$t^n\widehat{dm}(x,x,t)=I_1(x,t)-I_2(\gamma(x),d(x)/t,t) \tag{17.5.17}$$

where

(i) *$I_1\in C^\infty(\overline{X}\times[0,c))$ and the Taylor expansion at $t=0$ is given by*

$$\sum 2^{-2\nu}\pi^{(1-n)/2}t^{2\nu}(\Gamma(\nu+(1-n)/2))^{-1}U_\nu(x,x)(\det g^{jk}(x))^{-\frac{1}{2}}$$

with U_ν defined in Proposition 17.4.3;

(ii) *$2\theta \leq 1$ in the support of $I_2(x',\theta,t)$ and $2\theta=1$ in the singular support,*

$$I_2(x',\theta,t)-\sum_0^{\mu-1} V_\nu(\Pi,x(x',\theta t))(\det g^{jk}(x(x',\theta t)))^{-\frac{1}{2}}\theta^{-2\nu}E'_\nu(1,2\theta)$$
$$\in C^k(\partial X\times[\tfrac{1}{3},1]\times[0,c]),\quad k<\mu-(n-1)/2,$$

with V_ν as in Proposition 17.4.4 and $E'_\nu(t,x)=\partial E_\nu(t,x)/\partial t$. Moreover, $I_2(x',\theta,0) = E'_0(1,2\theta)(\det g^{jk}(x',0))^{-\frac{1}{2}}$.

Before discussing the proof, which we cannot complete until Chapter XXIV, we shall apply Lemma 17.5.6 to draw conclusions concerning the spectral function.

Theorem 17.5.10. *For the spectral function $e(x,y,\lambda)$ of the Dirichlet problem in X we have*

$$|e(x,x,\lambda)\Upsilon(x)-(e_0(0,\lambda)-e_0(2d(x),\lambda))(\det g^{jk}(x))^{-\frac{1}{2}}| \leq C\lambda^{\frac{1}{2}}(\lambda^{\frac{1}{2}}+d(x)^{-1})^{n-2}. \tag{17.5.18}$$

Here and below we sometimes write $e_0(\delta,\lambda)$ instead of $e_0(y,\lambda)$ when $|y|=\delta$.

Proof. When $d(x)>c>0$ this is a consequence of Theorem 17.5.7, for

$$|e_0(2d(x),\lambda)|\leq C\lambda^{n/2}(1+d(x)\lambda^{\frac{1}{2}})^{-(n+1)/2} \tag{17.5.19}$$

by Corollary 7.7.15, and the right-hand side is $O(\lambda^{(n-1)/2}/d(x))$. When $d(x)\leq c$ we shall use Lemma 17.5.6 with $a=1/c$ fixed and

$$\mu(\tau)=m(x,x,\tau)=\tfrac{1}{2}\Upsilon(x)\operatorname{sgn}\tau\, e(x,x,\tau^2),$$
$$\nu(\tau)=\tfrac{1}{2}\operatorname{sgn}\tau(e_0(0,\tau^2)-e_0(2d(x),\tau^2)V_0(\Pi,x))(\det g^{jk}(x))^{-\frac{1}{2}}.$$

From (17.4.12) it follows that

$$|de_0(x,\tau^2)/d\tau|=(2\pi)^{-n}\Big|\int_{|\xi|=|\tau|} e^{i\langle x,\xi\rangle}\,dS\Big|$$

where dS is the Euclidean surface measure. Hence

$$|de_0(x,\tau^2)|\leq C|\tau|^{n-1}\,d\tau$$

where $(2\pi)^n C$ is the area of the unit sphere. This proves the first part of (17.5.13). The theorem will follow if we prove that

$$|(d\mu-d\nu)*\phi_a(\tau)|\leq C(|\tau|+d(x)^{-1})^{n-2}. \tag{17.5.20}$$

In fact, $V_0(\Pi,x)-1=O(d(x))$, and $d(x)e_0(2d(x),\tau^2)=O(\tau^{n-1})$ by (17.5.19).

The proof of Theorem 17.5.7 gives the bound $C(1+|\tau|)^{n-2}$ for the inverse Fourier transform of

$$\hat\phi(t/c)(\widehat{dm}(x,x,t)-t^{-n}I_2(\gamma(x),d(x)/t,t)$$
$$-(E'_0(t,0)+E'_0(-t,0))(\det g^{jk})^{-\frac{1}{2}}).$$

What remains is to estimate the inverse Fourier transform of

$$\hat\phi(t/c)t^{-n}(I_2(\gamma(x),d(x)/t,t)-V_0(\Pi,x)E'_0(1,2d(x)/t)(\det g^{jk})^{-\frac{1}{2}}).$$

When $0 \leqq \theta \leqq \frac{1}{3}$ and $x_n = d(x)$ then

$$t^n \widehat{dm}(x,x,t) - (E_0'(1,0) - V_0(\Pi,x)\, E_0'(1,2\theta))(\det g^{jk}(x))^{-\frac{1}{2}}$$

has uniformly bounded derivatives with respect to (x',θ,t) for small t and vanishes when $\theta = 0$ or $t = 0$, so this difference is $O(\theta t) = O(d(x))$. Hence we have, with $d = d(x)$

$$|I_2(x',\theta,t) - V_0(\Pi,x)\, E_0'(1,2\theta)(\det g^{jk}(x))^{-\frac{1}{2}}| \leqq t^2 + d, \quad 0 \leqq \theta \leqq \tfrac{1}{3}.$$

When $\frac{1}{3} < \theta < 1$ it follows from (ii) in Theorem 17.5.9 that such an estimate holds after subtraction of sufficiently many singular terms. For a function $\leqq C|\hat{\phi}(t/c)|((t^2+d)/(t+d)^n + 1)$ the inverse Fourier transform with respect to t can be estimated by

$$C(c + d\int_0^c (t+d)^{-n}\, dt + \int_0^c (t+d)^{2-n}\, dt) \leqq C(c + d^{2-n}(1/(n-1) + c)).$$

Choose $\psi \in C_0^\infty((-4,4))$ even, with $\psi = 1$ in $(1,3)$ and $\psi = 0$ in $(0,\frac{1}{2})$. What remains is to estimate the inverse Fourier transform of the singular terms in I_2 with $\nu \neq 0$ cut off by a factor $\psi(t/d)$. First we prove that the inverse Fourier transform with respect to t of $\psi(t/d)\, E_\nu'(t/d,2)/d^n$ can be estimated by $C d^{1-n}(1+|\tau|\, d)^{(n-1)/2-\nu}$. Introducing t/d as a new variable we reduce the proof to the case where $d = 1$. Then

$$\psi(t)\, E_\nu'(t,2) - C_\nu \psi(t)\, t(t+2)^{\nu-(n+1)/2}\, \chi_+^{\nu-(n+1)/2}(t-2).$$

The inverse Fourier transform of the last factor is apart from a constant factor $e^{2it}(\tau + i0)^{(n-1)/2-\nu}$ (Example 7.1.17), so the statement follows since the inverse Fourier transform of the other factor is in $\mathscr{S}$. Recall that $V_\nu = O(d)$. Now

$$d^{2-n}(1+|\tau|\, d)^{(n-1)/2-\nu} \leqq C(d^{2-n} + |\tau|^{n-2}),$$

for $(n-1)/2 - \nu \leqq n-2$ when $\nu \neq 0$. This proves (17.5.20) and completes the proof of the theorem.

We shall now prove that the logarithm can be dropped in (17.5.16). To prepare for a refinement in Chapter XXIX we shall in fact prove somewhat more:

Corollary 17.5.11. *If $\psi \in C^1(\overline{X})$ then*

(17.5.21)
$$\begin{aligned}&\overline{\lim_{\lambda\to\infty}}\, \lambda^{(1-n)/2} \Big| \int \psi(x)\, e(x,x,\lambda)\, \Upsilon(x)\, dx - \\ &- \lambda^{n/2}(2\pi)^{-n} C_n \int (\det g^{jk}(y))^{-\frac{1}{2}} \psi(y)\, dy + \\ &+ 4^{-1}\lambda^{(n-1)/2}(2\pi)^{1-n} C_{n-1} \int \psi\, dS \Big| \leqq C \int_X |\psi|\, \Upsilon\, dx\end{aligned}$$

where dS is the Riemannian volume element in ∂X and C_ν is the volume of the unit ball in $\mathbb{R}^\nu$; C is independent of ψ.

Proof. Let c be so small that $\partial X \times [0,c] \ni (x',\delta) \mapsto x(x',\delta)$ is a diffeomorphism. If $d(x)>c/2$ in $\operatorname{supp}\psi$ then (17.5.21) follows from Theorem 17.5.7 so we may assume that $d(x)<c$ in $\operatorname{supp}\psi$. By (17.5.8) and (17.5.19) we may replace the right-hand side of (17.5.18) by $C\lambda^{n/2}$, and this is a better estimate if $d(x)^{2-n}>\lambda^{(n-1)/2}$. After modifying (17.5.18) in this way then, we multiply by $\lambda^{(1-n)/2}|\psi(x)|$ and integrate over X. If $n=2$ the integral of the right-hand side is obviously bounded by the L^1 norm of ψ. If $n>2$ this remains true for the integral over the set where $d(x)\lambda^{\frac{1}{2}}>1$. The measure of the set where $d(x)^{2-n}>\lambda^{(n-1)/2}$ is $O(\lambda^{\frac{1}{2}(n-1)/(2-n)})$, and the integral of $\lambda^{\frac{1}{2}}$ over this set is therefore $O(\lambda^{1/(4-2n)})$. The integral of $\lambda^{(2-n)/2}d(x)^{2-n}$ over the set where $d(x)^{2-n}<\lambda^{(n-1)/2}$ but $d(x)\lambda^{\frac{1}{2}}<1$ has a bound of the same type, multiplied by $\log\lambda$ if $n=3$. Hence the integral over the set where $d(x)\lambda^{\frac{1}{2}}<1$ is $O(\lambda^{1/(4-2n)}\log\lambda)\to 0$ as $\lambda\to\infty$.

It remains to show that

$$\text{(17.5.22)} \quad \lambda^{(1-n)/2}\Big(\int \psi(x)\, e_0(2d(x),\lambda)(\det g^{jk}(x))^{-\frac{1}{2}}\,dx - \\ -4^{-1}\lambda^{(n-1)/2}(2\pi)^{1-n}C_{n-1}\int \psi\, dS\Big)\to 0 \text{ as } \lambda\to\infty.$$

To do so we shall use the coordinates (x',δ). Then the Riemannian volume element $(\det g^{jk}(x))^{-\frac{1}{2}}dx$ becomes $a(x',\delta)\,dS\,d\delta$ where $a(x',0)=1$. Thus

$$|\psi(x(x',\delta))\,a(x',\delta)-\psi(x',0)|\leqq C\delta,$$

and

$$|e_0(2\delta,\lambda)|\leqq C\lambda^{n/2}(1+\delta\lambda^{\frac{1}{2}})^{-(n+1)/2}$$

by (17.5.19). We have

$$\lambda^{\frac{1}{2}}\int_0^c \delta(1+\delta\lambda^{\frac{1}{2}})^{-(n+1)/2}\,d\delta=\lambda^{-\frac{1}{2}}\int_0^{c\lambda^{\frac{1}{2}}} t(1+t)^{-(n+1)/2}\,dt\to 0$$

for the integral is $O(\lambda^{\frac{1}{2}})$ since $n\geqq 2$. We also have

$$\lambda^{\frac{1}{2}}\int_c^\infty (1+\delta\lambda^{\frac{1}{2}})^{-(n+1)/2}\,d\delta=\int_{c\lambda^{\frac{1}{2}}}^\infty (1+t)^{-(n+1)/2}\,dt=O(\lambda^{-\frac{1}{2}})$$

so we may replace the first integral in (17.5.22) by

$$\lambda^{(1-n)/2}\int_{\partial X} dS\int_0^\infty \psi(x',0)\,e_0(2\delta,\lambda)\,d\delta.$$

Now

$$\int_0^\infty e_0(2\delta,\lambda)\,d\delta=\tfrac{1}{4}\int_{-\infty}^\infty e_0(\delta,\lambda)\,d\delta.$$

With the notation $\xi'=(\xi_2,\dots,\xi_n)$ we have

$$e_0(\delta,\lambda)=(2\pi)^{-n}\int e^{i\delta\xi_1}\,d\xi_1\int_{|\xi'|^2<\lambda-\xi_1^2} d\xi',$$

so Fourier's inversion formula gives

$$\int_{-\infty}^{\infty} e_0(\delta,\lambda)\,d\delta=(2\pi)^{1-n}\int_{|\xi'|^2<\lambda} d\xi'=(2\pi)^{1-n}C_{n-1}\lambda^{(n-1)/2}.$$

This completes the proof of (17.5.22) and of Corollary 17.5.11.

(17.5.21) with $\psi=1$ gives in particular

$$(17.5.16)' \qquad |N(\lambda)-(2\pi)^{-n}C_n\operatorname{vol}(X)\lambda^{n/2}|\leqq C\lambda^{(n-1)/2}.$$

We can also choose ψ equal to 1 in a neighborhood of ∂X so that the right-hand side of (17.5.21) is as small as we wish. If it is possible to determine $e(x,x,\lambda)$ on compact subsets of X with an error $o(\lambda^{(n-1)/2})$ we shall therefore be able to determine $N(\lambda)$ too with such an error; the boundary contribution will be

$$-4^{-1}\lambda^{(n-1)/2}(2\pi)^{1-n}C_{n-1}\operatorname{vol}(\partial X).$$

This will be discussed in Section 29.3 after the required technical tools have been developed. However, we shall now discuss examples which show that some additional hypothesis must then be made.

Our first example is the Laplace-Beltrami operator on the sphere $S^n\subset\mathbb{R}^{n+1}$. The discussion so far in this section has been restricted to open sets $X\subset\mathbb{R}^n$. This was only done for the sake of readers who might not feel familiar with Riemannian geometry. All arguments were local and can therefore be carried over to an arbitrary compact Riemannian manifold with boundary; we shall consider the unit sphere S^n.

In $\mathbb{R}^{n+1}$ we shall use the polar coordinates $x=r\omega$ where $r>0$ and $\omega\in S^n$. Then the Laplace operator assumes the form

$$\Delta=r^{-2}\Delta_\omega+\partial^2/\partial r^2+n/r\,\partial/\partial r$$

outside the origin. Here Δ_ω is the Laplace-Beltrami operator in S^n. If $u(x)=r^\mu v(\omega)$ is a homogeneous function of degree μ, it follows that outside the origin

$$\Delta u=r^{\mu-2}(\Delta_\omega v+\mu(\mu+n-1)v).$$

Hence $\Delta u=0$ outside the origin if and only if v is an eigenfunction of $-\Delta_\omega$ with eigenvalue $\lambda=\mu(\mu+n-1)$. Since λ assumes all values $\geqq 0$ when $\mu\leqq 1-n$, we obtain all eigenfunctions of $-\Delta_\omega$ by restricting to S^n all distributions u in $\mathbb{R}^{n+1}$ which are harmonic and homogeneous of degree $\mu\leqq 1-n$ outside the origin. (Cf. Section 3.2.) Then Δu has to be a linear combination of the derivatives of the Dirac measure at 0, so we find that $\mu=1-n-k$ where k is an integer $\geqq 0$, and that

$$u(x)=\sum_{|\alpha|=k} a_\alpha D^\alpha E$$

where E is the fundamental solution of Δ and a_α are constants. (When $k=0$ and $n=1$ the logarithmic potential E must be replaced by a constant. Otherwise there is no harmonic function homogeneous of degree μ in $\mathbb{R}^{n+1}$.) The Fourier transform of u is $\sum a_\alpha \xi^\alpha |\xi|^{-2}$ so it follows that u is supported by the origin if and only if $|\xi|^2$ divides the polynomial $\sum a_\alpha \xi^\alpha$. Let N_k be the dimension of the space of homogeneous polynomials of degree k in $n+1$ variables,

$$N_k = \binom{n+k}{n} = k^n(1+O(1/k))/n!.$$

We define $N_k=0$ for $k<0$. Then it follows that the multiplicity of the eigenvalue $\lambda_k = k(k+n-1)$ of $-\Delta_\omega$ is $N_k - N_{k-2}$ for $k=0,1,\dots$. If $N(\lambda)$ is the number of eigenvalues $\leqq \lambda$ of $-\Delta_\omega$, it follows that

$$N(\lambda_k+0) - N(\lambda_k-0) = N_k - N_{k-2} = 2k^{n-1}(1+O(1/k))/(n-1)!.$$

Thus

$$\lambda_k^{(1-n)/2}(N(\lambda_k+0) - N(\lambda_k-0)) \to 2/(n-1)!, \quad k\to\infty,$$

which proves that it is impossible to find an asymptotic formula for $N(\lambda)$ with continuous main term and error $o(\lambda^{(n-1)/2})$.

If X is the hemisphere where $x_{n+1}>0$ then the eigenfunctions of the Dirichlet problem in X are the eigenfunctions in S^n which are odd with respect to x_{n+1}. Let N_k^o and N_k^e be the dimensions of the spaces of homogeneous polynomials of degree k in $n+1$ variables which are odd and even respectively with respect to x_{n+1}. Clearly

$$N_k = N_k^o + N_k^e, \quad N_k^o = N_{k-1}^e$$

since the odd polynomials are products of x_{n+1} and even polynomials. It follows that

$$N_k - N_{k-2} = N_k^o - N_{k-2}^o + N_k^e - N_{k-2}^e = (N_k^o - N_{k-2}^o) + (N_{k+1}^o - N_{k-1}^o).$$

Now $N_k^o - N_{k-2}^o$ is the dimension of the eigenspace of the Dirichlet problem with the eigenvalue λ_k, so we obtain again jumps of the order of magnitude $\lambda^{(n-1)/2}$. The conclusion is the same as for S^n, and now we can of course realize X by stereographic projection as a ball in $\mathbb{R}^n$ with a Riemannian metric conformal to the Euclidean one.

Before embarking on the proof of Theorem 17.5.9 we need two additions to Lemma 17.5.4. The first is the finite propagation speed for solutions of the mixed Dirichlet-Cauchy problem.

Lemma 17.5.12. *Let $v\in C^\infty([0,\infty)\times\overline{X})$ be a solution of the mixed problem*

$$(\partial^2/\partial t^2 + P)v = h \quad \text{in } [0,\infty)\times\overline{X};$$

$$v=0 \text{ in } [0,\infty)\times\partial X; \quad v=v_0 \text{ and } \partial v/\partial t = v_1 \text{ when } t=0.$$

Let $s(x,y)$ be the geodesic distance from x to y and set

$$\psi(x)=\min(\min\{t+s(x,y);\ (t,y)\in\operatorname{supp} h\},$$
$$\min\{s(x,y);\ y\in\operatorname{supp} v_0\cup\operatorname{supp} v_1\}).$$

Then $v(t,x)=0$ if $t<\psi(x)$.

Proof. By the triangle inequality for s we have

$$|\psi(x)-\psi(y)|\leqq s(x,y); \quad x,y\in\overline{X}.$$

For $0<\varepsilon<1$ we can approximate $(1-\varepsilon)\psi$ by $\psi_\varepsilon\in C^\infty(\overline{X})$ so that $\psi_\varepsilon\to\psi$ uniformly in $\overline{X}$ when $\varepsilon\to 0$, $\psi_\varepsilon\leqq\psi$, and $|\psi_\varepsilon(x)-\psi_\varepsilon(y)|\leqq(1-\frac{1}{2}\varepsilon)\,s(x,y)$, that is,

$$\sum g^{jk}(x)\,\partial\psi_\varepsilon(x)/\partial x_j\,\partial\psi_\varepsilon(x)/\partial x_k\leqq 1-\tfrac{1}{2}\varepsilon.$$

Thus the surfaces $S_a=\{(t,x);\ t=\psi_\varepsilon(x)+a,\ x\in\overline{X}\}$ are non-characteristic, and h, v_0 and v_1 vanish below S_0. Let A be the supremum of all $a\leqq 0$ such that $v=0$ below S_a. If $A<0$ then it follows from the uniqueness theorem for the Cauchy problem (Theorem 23.2.7) or the Dirichlet-Cauchy mixed problem (Theorem 24.1.4) that $v=0$ in a neighborhood of S_A. (The proofs are just slight variations of the energy identity used to prove Lemma 17.5.4 but would not be convenient to give here.) Hence A must be 0 so $v=0$ when $t<\psi_\varepsilon(x)$. When $\varepsilon\to 0$ the lemma follows.

In particular we note that if $f\in C_0^\infty(X)$ and $s(z,y)<d(y)$ when $z\in\operatorname{supp} f$, then $s(x,y)<d(y)+t$ in $\operatorname{supp}(\cos(t\sqrt{\mathscr{P}})f)$, for the triangle inequality gives $s(x,y)\leqq\psi(x)+d(y)$ for the function in Lemma 17.5.12. When $t<3d(y)$ we obtain $s(x,y)<4d(y)$. Thus we can study $\cos(t\sqrt{\mathscr{P}})f$ then by blowing up a neighborhood of y with diameter proportional to $d(y)$ to a fixed size, as we did in the proof of Proposition 17.4.4. Before doing so we must prove another supplement to Lemma 17.5.4 where parameter dependence is allowed.

By $P_z(x,D)$ we shall denote an elliptic operator of the same form as before except that it now depends on a parameter $z\in Z$ where Z is a compact convex body in $\mathbb{R}^N$ for some N. We assume that the coefficients of P_z are in $C^\infty(\overline{X}\times Z)$ and that (17.5.2) is valid uniformly for all P_z.

Lemma 17.5.13. *Let $v_z\in C^\infty([0,T]\times\overline{X}\times Z)$ be a solution of the mixed problem*

$$(P_z+\partial^2/\partial t^2)\,v_z=h_z \quad \text{in } [0,T]\times\overline{X}\times Z; \tag{17.5.10}'$$
$$v_z=0 \text{ in } [0,T]\times\partial X\times Z; \quad v_z=\partial v_z/\partial t=0 \text{ if } t=0.$$

Assume that $\partial^j h_z/\partial t^j=0$ when $t=0$ if $j<k$. Then it follows that with supremum taken over $[0,T]\times\overline{X}\times Z$

$$\sum_{|\alpha|+\frac{1}{2}n<k}\sup|D^\alpha_{t,x,z}v_z(t,x)|\leqq C\sum_{|\alpha|\leqq k}\sup|D^\alpha_{t,x,z}h_z(t,x)|. \tag{17.5.11}'$$

Proof. Let the sum in the right-hand side be equal to 1. Then it follows from (17.5.11) that we have a uniform bound for $\|D_t^{k+1-j} v_z(t,.)\|_{(j)}$ if $j \leqq k+1$. Differentiating the differential equation $(\partial^2/\partial t^2 + P_z) v_z = h_z$ with respect to z_ν we obtain

$$(\partial^2/\partial t^2 + P_z)\, \partial v/\partial z_\nu = \partial h_z/\partial z_\nu - \partial P_z/\partial z_\nu\, v_z = H.$$

We have a bound for $\|D_t^{k-1-j} H\|_{(j)}$ if $j \leqq k-1$ and conclude using (17.5.11) again that there is a uniform bound for $\|D_t^{k-j} \partial v/\partial z_\nu\|_{(j)}$ if $j \leqq k$. Continuing in this way we obtain uniform bounds for $\|D_t^{k+1-|\alpha|-j} \partial_z^\alpha v_z(t,.)\|_{(j)}$ when $j \leqq k + 1 - |\alpha|$, $|\alpha| \leqq k$, and the lemma is therefore a consequence of Lemma 17.5.2. The proof is complete.

Remark. In this lemma it is not necessary to assume that $\overline{X}$ is bounded; it suffices to assume that all v_z vanish outside $[0, T] \times K$ for some fixed compact set $K \subset \overline{X}$.

Proof of Theorem 17.5.9. Let K_0 be a compact part of ∂X as in Proposition 17.4.4, defined by $x_n = 0$ in local coordinates. Let $z \in X$ be close to K_0. From (17.4.15)′ we can obtain a parametrix for P_z when y is close to $\Pi = (0, \dots, 0, 1)$. In fact, we can first pull (17.4.15)′ back by the inverse of the map

$$(t, x, y) \mapsto (y_n t, (y', 0) + y_n x, y),$$

that is

$$(t, x, y) \mapsto (t/y_n, (x - (y', 0))/y_n, y),$$

to construct a parametrix for $\partial^2/\partial t^2 + P$ with pole at $(0, y)$. Then the pull-back by the map

$$(t, x, y) \mapsto (z_n t, (z', 0) + z_n x, (z', 0) + z_n y)$$

gives a parametrix for $\partial^2/\partial t^2 + P_z$ with pole at $(0, y)$. The composition of the maps is

$$(t, x, y) \mapsto (t/y_n, (x - (y', 0))/y_n, (z', 0) + z_n y).$$

The conclusion is that with the notation $y_z = (z', 0) + z_n y$ we have

$$\begin{aligned} &y_n^2 (\partial^2/\partial t^2 + P_z(x, D))\, \mathscr{E}^N(t/y_n, (x - (y', 0))/y_n, y_z) \\ &\quad = (\det g^{jk}(y_z))^{\frac{1}{2}}\, \delta_{0,y}\, y_n^{n+1} + R_N(t/y_n, (x - (y', 0))/y_n, y_z) \end{aligned}$$

when y is close to Π, $|t| < 4$ and $s^{r,z}(x, y) < 4$. We can now argue as in the proof of Theorem 17.5.5, but with

$$\begin{aligned} K_N(t, x, y, z) = {} & z_n^n\, \widehat{dm}(x_z, y_z, z_n t) \\ & - \frac{\partial}{\partial t} \mathscr{E}^N(t/y_n, (x - (y', 0))/y_n, y_z) (\det g^{jk}(y_z))^{-\frac{1}{2}}\, y_n^{1-n}. \end{aligned}$$

For $f \in C_0^\infty(\mathbb{R}^n)$ with support close to Π we have

$$v_z(t, x) = \int K_N(t, x, y, z)\, D^\beta f(y)\, dy \in C^\infty,$$

the Cauchy and Dirichlet data are 0, and

$$h_z(t,x) = (\partial^2/\partial t^2 + P_z)v_z(t,x)$$

has $N-n-2-|\beta|$ uniformly bounded derivatives with respect to x,t,z when $\|f\|_{L^1} \leqq 1$ and $|t|<3$. By Lemma 17.5.12 we have $s^z(x,\Pi)<\frac{7}{2}$ in the support. Thus we can apply Lemma 17.5.13, by the remark following its proof, and conclude as in the proof of Theorem 17.5.5 that for z in a neighborhood of K_0

$$K_N(t,x,y,z) \in C^{N-n-4} \text{ for } x \text{ and } y \text{ close to } \Pi.$$

In particular the restriction to $x=y=\Pi$ has this property which means that

$$z_n^n \widehat{dm}(z,z,z_n t) - \sum_0^N (U_\nu(z)\, z_n^{2\nu}\, E'_\nu(t,0) - V_\nu(\Pi,z)\, E'_\nu(t,2))(\det g^{jk}(z))^{-\frac{1}{2}}$$

is in C^{N-n-4} when $0<t<3$ and z is in a neighborhood of K in X; all derivatives have uniform bounds.

Let M be the function corresponding to m when X is replaced by a neighborhood Y of $\overline{X}$ and the coefficients of P are smoothly extended to Y. Then we know from Theorem 17.5.5 that

$$\widehat{dM}(z,z,t) - \sum_0^N U_\nu(z)\, E'_\nu(t,0)(\det g^{jk}(z))^{-\frac{1}{2}} \in C^{N-n-4}$$

for small t, hence

$$z_n^n \widehat{dM}(z,z,z_n t) - \sum_0^N U_\nu(z)\, z_n^{2\nu}\, E'_\nu(t,0)(\det g^{jk}(z))^{-\frac{1}{2}}$$

is in C^{N-n-4}. If we put

$$I_1(z,t) = t^n \widehat{dM}(z,z,t)$$

then (i) of Theorem 17.5.9 holds and, with I_2 defined by (17.5.17),

$$t^{-n} I_2(z', 1/t, t z_n) - \sum_0^N V_\nu(\Pi,z)\, E'_\nu(t,2)(\det g^{jk}(z))^{-\frac{1}{2}}$$

has bounded derivatives of order $N-n-4$ when $t \leqq 3$. It follows from Lemma 17.5.12 that $t \geqq 2$ in the support. Now introduce new variables $\theta = 1/t \in (\frac{1}{3}, 1)$ and $s = t z_n$. Then we have $t = 1/\theta$ and $z_n = s\theta$ which shows that

$$I_2(z',\theta,s) - \sum_0^N V_\nu(\Pi,(z',s\theta))\, \theta^{-2\nu}\, E'_\nu(1,2\theta)(\det g^{jk}(z',s\theta))^{-\frac{1}{2}}$$

has bounded derivatives with respect to z',θ,s of order $N-n-4$. This proves condition (ii) in Theorem 17.5.9 when $\theta \geqq \frac{1}{3}$. The same proof is applicable when $\theta \geqq \delta$ for some fixed $\delta>0$, but the proof for small θ requires another argument.

$I_1(x', t\theta, t)$ is a C^∞ function equal to $E_0'(1,0)(\det g^{jk}(x',0))^{-\frac{1}{2}}$ when $t=0$. Thus the remaining statement in the theorem is that $t^n \widehat{dm}(x,x,t)$ when $x_n = t\theta$ and $0 \leqq \theta \leqq \frac{2}{5}$ is a C^∞ function of (x', θ, t) for small t, and equal to $(E_0'(1,0) - E_0'(1, 2\theta))(\det g^{jk}(x',0))^{-\frac{1}{2}}$ when $t=0$. For the proof we shall study the kernel

$$K(t,x,y,z) = z_n^n \widehat{dm}(x_z, y_z, z_n t);$$

no approximation is attempted now. If $f \in C_0^\infty(\mathbb{R}^n)$ has support where $s^z(x, \Pi) < 2$, say, and L^1 norm $\leqq 1$, then

(17.5.23) $$v_z(t,x) = \int K(t,x,y,z) D^\beta f(y)\, dy$$

is a C^∞ function with $(\partial^2 \partial t^2 + P_z) v_z(t,x) = 0$, the Dirichlet data are 0 and $v_z = D^\beta f$, $\partial v_z / \partial t = 0$ when $t = 0$. When $t < 2$ we have $s^z(x, \Pi) < 4$ in the support by Lemma 17.5.12. We wish to prove uniform bounds for all derivatives with respect to t, x, z when $t = 1$ and $s^{r,z}(x,y) < \frac{4}{5}$ when $y \in \operatorname{supp} f$. Assuming for a moment that this has been done, we conclude as in the proof of Theorem 17.5.5 that $K \in C^\infty$ and, taking $t=1$, $x' = y' = z'$, $x_n = y_n = \theta < \frac{2}{5}$, that

$$z_n^n\, dm(x,x,z_n), \qquad x = (z', \theta z_n)$$

has uniformly bounded derivatives with respect to z', θ, z_n when $0 \leqq \theta < \frac{2}{5}$ and z_n is small. When $z_n = 0$ we obtain

$$K(t,x,y,z) = (E_0'(t, s^z(x,y)) - E_0'(t, s^{r,z}(x,y))(\det g^{jk}(z))^{-\frac{1}{2}}$$

and the parenthesis reduces to $(E_0'(1,0) - E_0'(1, 2\theta))$ for the preceding choice of t, x, y. Changing notation so that z_n is called t we shall therefore have proved the theorem.

It is easy to see that with v_z defined by (17.5.23) and any k the map $(t,z) \mapsto v_z(t,.)$ is k times differentiable for $t \leqq 2$ with values in $\bar{\mathscr{D}}'^{\mu}$, and that derivatives of order $\leqq k$ are uniformly bounded in $\bar{\mathscr{D}}'^{\mu}$, if $\mu > |\beta| + k + n/2$. To do so we observe that P_z is self adjoint with respect to a C^∞ density Υ_z, hence

$$(v_z(t,.), \phi)_{\Upsilon_z} + (D^\beta f, \partial \Phi_z(t,0,.)/\partial s)_{\Upsilon_z} = 0$$

if $\Phi_z(t,s,x)$ is the solution of the mixed problem

$$(\partial^2/\partial s^2 + P_z)\Phi_z(t,s,x) = 0;$$

$$\Phi_z(t,s,x) = 0, \text{ if } x \in \partial X; \quad \partial \Phi_z(t,s,x)/\partial s = -\phi(x)$$

$$\text{and } \Phi_z(t,s,x) = 0 \text{ if } x \in X, \ s = t.$$

Thus

$$\Phi_z(t,s,x) = ((\sin(t-s)\sqrt{\mathscr{P}_z})/\sqrt{\mathscr{P}_z})\,\phi(x).$$

It follows at once from the spectral theorem, (17.5.4) and Lemma 17.5.2, that we have uniform estimates for the (x,t) derivatives of Φ_z of order $< \mu + 1 - n/2$ if ϕ is bounded in C_0^μ. As in the proof of Lemma 17.5.13 we obtain by successive differentiations with respect to z that this is true for all (x,t,z) derivatives of order $< \mu - n/2$ also. If $\mu > |\beta| + k + n/2$ we conclude that v_z has the stated differentiability properties. The proof will therefore be completed by the following lemma.

Lemma 17.5.14. *Let $v_z(t,x)$ be a solution of*

$$(\partial^2/\partial t^2 + P_z)\, v_z(t,x) = 0$$

when $|t|<3$ which is a C^∞ function of (z,t,x) when $z_n>0$ and z is close to a compact set $K_0 \subset \{z \in \mathbb{R}^n; z_n = 0\}$. Assume that

(i) *$v_z(t,x)=0$ when $x_n=0$ (Dirichlet condition)*

(ii) *$s^z(y,(z',0))<\frac{2}{5}$ when $(0,y)\in \operatorname{supp} v_z$ (Cauchy data condition)*

(iii) *for every α one can find μ such that $D_z^\alpha v_z(t,.)$ belongs to a fixed bounded subset of $\bar{\mathscr{D}}'^{\mu}(\mathbb{R}^n_+)$.*

Then there are fixed bounds for the derivatives with respect to z,t,x of $v_z(t,x)$ when z is in a neighborhood K' of K_0, $1\leqq t\leqq 2$ and $s^z(x,(z',0))<\frac{4}{5}$.

Note that when $z_n=0$ this is quite clear since v_z is equal to the solution of a constant coefficient Cauchy problem with data satisfying (ii) together with the reflection in the boundary plane $x_n=0$. There can be no singularities in the considered set then since it cannot be reached by light rays from the Cauchy data. The proof of the lemma will be given in Section 24.7 after a systematic study of the propagation of singularities of solutions to the mixed problem. It will show that the constant coefficient situation remains true for small perturbations.

Notes

As pointed out in the notes to Chapter XIII the methods used in Sections 17.1 and 17.3 have a long history. For the elliptic case we might add references to Agmon-Douglis-Nirenberg [1,2], Agmon [1,5], Browder [1], Gårding [2], Lions and Magenes [1], Schechter [1] although this list is far from complete. Lemma 17.1.5 comes from Friedrichs [2]. Uniqueness theorems such as Theorem 17.2.1 were first proved by Carleman [1] in the two dimensional case. It was he who introduced the idea of using norms containing powers of a weight function. It has dominated all later work in the field. For second order operators in several variables Müller [1], Heinz [1], Aronszajn [2], Cordes [1], Aronszajn-Krzywcki-Szarski [1] and Agmon [4] proved Theorem 17.2.6 in increasing generality for real a_{jk}. Alinhac and Baouendi [1] showed that it is then only necessary to assume the coefficients real at the distinguished point. Counterexamples due to Alinhac [2] show essentially that this condition is necessary and also that such strong uniqueness theorems do not hold for operators of higher order. We have here followed the exposition in Hörmander [41] where weaker conditions on the lower order terms are also discussed. (Jerison and Kenig [1] have shown that they are not optimal.) Calderón [1] was the first to prove general uniqueness theorems such as Theorem 17.2.1 for operators of higher order in more than two variables. Here we have just included his result in the elliptic case assuming only that the coefficients are Lipschitz continuous. Pliš [3] has proved that this condition cannot be replaced by any weaker Hölder condition. For the background of Theorem 17.2.8 we refer to the notes of Chapter XIV.

The parametrix construction of Hadamard discussed in Section 17.4 is essentially taken from Hadamard [1] although the arguments used here in some ways are closer to those of M. Riesz [1]. Seely [3] studied a parametrix $E(t,x,y)$ with three terms for the mixed Dirichlet problem for the wave equation in $\mathbb{R}^4$ defined when the square of the distance between x and y is small compared to the boundary distance of y. He applied it to prove Corollary 17.5.1 for the Laplace operator in $X \subset \mathbb{R}^3$. In Seeley [4] the results were extended to higher dimensions, and Pham The Lai [1] did so too about the same time. In the latter paper it is evident that it is the Hadamard construction which is being used, and it is clear that it is not essential to start from the constant coefficient Laplace operator. (Some estimates seem incorrect but the end results are valid.)

Carleman [3] found a way to determine the asymptotic behavior of the spectral function of a second order elliptic operator with Dirichlet boundary condition. (The corresponding results on the eigenvalues go back to Weyl [4].) His idea was to study the kernel of the resolvent and then apply a Tauberian theorem; this approach was extended to general elliptic operators by Gårding [7]. Minakshisundaram and Pleijel [1] showed that one can study the Laplace transform of the spectral function as well, for this is related to the Green's function of a heat operator. These methods do not give precise error bounds, but Levitan [1,2] and Avakumovič [1] discovered that taking the cosine transform one obtains optimal result result by means of the properties of the corresponding wave equation. Thus Theorem 17.5.7 is due to Avakumovič [1] but our proof follows to some extent Hörmander [22] where the results were extended to higher order operators (See Chapter XXIX.) As mentioned above, precise results on the spectral function close to the boundary were first obtained by Seeley [3, 4] and Pham The Lai [1]. They only considered the Laplace operator. Ivrii [3] showed by a quite different method that the asymptotic formula for the eigenvalues is valid without this restriction, with the error term given by Seeley and Pham The Lai. He also showed that there is a better asymptotic formula with a second term when there are not too many closed multiply reflected geodesics. The proof was greatly simplified by Melrose [7] who derived Ivrii's result from the basic facts on propagation of singularities. We have combined these ideas in Section 17.5 to give improved estimates for the spectral function near the boundary. The Hadamard type of construction from Seeley [3,4] and Pham The Lai [1] is only used in the simplest situation to approximate $E(t,x,y)$ when t is less than a constant times the distance from y to the boundary. We have presented the arguments in Section 29.3 so that they prepare for the full proof of Ivrii's results in Section 29.3.

Chapter XVIII. Pseudo-Differential Operators

Summary

There is no major difficulty in extending the study of second order elliptic operators in Sections 17.1 and 17.3 to higher order operators. From Theorem 7.1.22 we know that an elliptic operator $P(D)$ with constant coefficients has a parametrix E of the form

$$Ef(x)=(2\pi)^{-n}\int e^{i\langle x,\xi\rangle}a(\xi)\hat{f}(\xi)\,d\xi, \quad f\in\mathscr{S},$$

where $a(\xi)=1/P(\xi)$ for large $|\xi|$. As in Chapter XIII an elliptic operator $P(x,D)$ with variable coefficients (cf. Section 8.3) can then be regarded near a point x_0 as a perturbation of the constant coefficient operator $P(x_0,D)$ obtained by freezing the coefficients at x_0. The proof of Theorem 13.3.3 gives a local fundamental solution E of $P(x,D)$ as a norm convergent series. However, the smoothness of the terms in the series does not grow in general, so it is not suitable for a precise description of the singularities of E. The situation can be improved by taking as a first approximation for E the operator A defined by

$$Af(x)=(2\pi)^{-n}\int e^{i\langle x,\xi\rangle}a(x,\xi)\hat{f}(\xi)\,d\xi, \quad f\in\mathscr{S}, \tag{18.1}$$

where $a(x,\xi)=1/P(x,\xi)$ for large $|\xi|$. This means that we define $Af(x)=E_x f(x)$ where E_x is a parametrix for P with coefficients frozen at x. Differentiation under the integral sign in (18.1) gives

$$P(x,D)Af(x)=(2\pi)^{-n}\int e^{i\langle x,\xi\rangle}P(x,\xi+D_x)a(x,\xi)\hat{f}(\xi)\,d\xi, \quad f\in\mathscr{S}.$$

Here

$$\begin{aligned}P(x,\xi+D_x)a(x,\xi)&=P(x,\xi)a(x,\xi)+\sum_{\alpha\neq 0}P^{(\alpha)}(x,\xi)D_x^\alpha a(x,\xi)/\alpha!\\&=1+O(1/|\xi|)\end{aligned}$$

which suggests that $P(x,D)A=I+R$ where I is the identity operator and R is an operator improving differentiability by one unit. A right parametrix should then be given by $A(I+R)^{-1}=A-AR+AR^2-\dots$ where the smoothness of the terms increases indefinitely.

The preceding somewhat formal argument can actually be justified, but it is preferable to move the successive approximation to the construction of a function a such that

$$P(x,\xi+D_x)\,a(x,\xi)=1+b(x,\xi) \tag{18.2}$$

with b rapidly decreasing as $\xi\to\infty$. If

$$P(x,\xi)=P_m(x,\xi)+P_{m-1}(x,\xi)+\dots$$

is a decomposition of $P(x,\xi)$ in homogeneous terms with respect to ξ, then it is natural to expect a to have an asymptotic expansion

$$a(x,\xi)\sim a_{-m}(x,\xi)+a_{-m-1}(x,\xi)+\dots,\qquad \xi\to\infty, \tag{18.3}$$

where a_k is homogeneous of degree k. The condition (18.2) then gives if we equate terms of equal homogeneity

$$\begin{aligned} &P_m(x,\xi)\,a_{-m}(x,\xi)=1,\\ &P_m(x,\xi)\,a_{-m-1}(x,\xi)+P_{m-1}(x,\xi)\,a_{-m}(x,\xi)\\ &\qquad+\sum P_m^{(j)}(x,\xi)\,D_{x_j}a_{-m}(x,\xi)=0, \end{aligned} \tag{18.4}$$

and a sequence of equations expressing $P_m(x,\xi)\,a_{-m-k}(x,\xi)$ for any $k>0$ in terms of $a_{-m},\dots,a_{-m-k+1}$. Since $P_m(x,\xi)\neq 0$, $\xi\neq 0$, if P is elliptic, the sequence a_{-m}, $a_{-m-1},\dots$ is uniquely determined and gives a solution of (18.2) with b decreasing as rapidly as we please. (Compare the proof of Theorem 8.3.1 which is based on the same principle.)

The preceding argument indicates that any elliptic operator has a parametrix of the form (18.1) with a admitting an asymptotic expansion of the form (18.3). Such operators A are called *pseudo-differential operators*. The reason why they are so useful is that they form an algebra, invariant under passage to adjoints and change of variables, so that it can also be defined on a C^∞ manifold. The correspondence between the operator A and the function a in (18.1), called the *symbol* of A, allows one to give formulas for these operations in terms of the symbols which are as simple as those for differential operators. We shall develop these facts in Section 18.1.

The Schwartz kernel $K(x,y)$ of a pseudo-differential operator is singular only at the diagonal. There it is singular essentially as a homogeneous function of $x-y$, so application of a first order differential operator which is tangential to the diagonal does not make K more singular. In Section 18.2 we study the space of distributions $I(X,Y)\subset\mathscr{D}'(X)$ with the same relation to a C^∞ submanifold Y of the C^∞ manifold X. For these distributions the wave front set is contained in the conormal bundle $N(Y)$ of Y; accordingly we call them conormal distributions for Y. They play an important role also in the development of the calculus of totally characteristic pseudo-differential operators in a manifold with boundary presented in Section 18.3. In particular, the invariance under coordinate transformations of these operators follows from a characterization of their kernels as conormal distributions on a modification of $X\times X$. Section 18.3 is intended to give a solid framework for the study of boundary problems, in particular a discussion of spaces of distributions in a manifold with boundary and their wave front sets. However, little use will be made of the results in the later chapters where we use more conventional but non-invariant techniques. The notions introduced in Section 18.3 seem so natural though that they are bound to play an increasingly important role.

The calculus built up in Sections 18.1, 18.2 and 18.3 is based on the study of Gaussian convolutions in Section 7.6. In Section 18.4 we resume their study in preparation for the calculus of operators with general symbols developed in Sections 18.5 and 18.6. In addition to operators of the form (18.1) with general symbols we also discuss the Weyl calculus which has important advantages due to various symmetry properties. For example, a real valued symbol always gives rise to a self-adjoint operator; this is why it was originally introduced in quantum mechanics. The main theorems in the calculus are proved in Section 18.5, and Section 18.6 is devoted to estimates for such operators. They will be needed in parts of Chapters XXVI and XXVIII, but apart from that it will usually suffice to have absorbed the material in Section 18.1 in order to read the following chapters.

18.1. The Basic Calculus

When studying operators of the form (18.1) we must first specify the conditions which a must satisfy. We definitely want to accept all smooth functions $a(x,\xi)$ which are homogeneous in ξ for large $|\xi|$ and also their linear combinations. The following is a slightly wider class which is technically more convenient:

Definition 18.1.1. If m is a real number then $S^m = S^m(\mathbb{R}^n \times \mathbb{R}^n)$ is the set of all $a \in C^\infty(\mathbb{R}^n \times \mathbb{R}^n)$ such that for all α, β the derivative $a^{(\alpha)}_{(\beta)}(x,\xi) = \partial^\alpha_\xi \partial^\beta_x a(x,\xi)$ has the bound

(18.1.1) $$|a^{(\alpha)}_{(\beta)}(x,\xi)| \leq C_{\alpha,\beta}(1+|\xi|)^{m-|\alpha|}; \quad x, \xi \in \mathbb{R}^n.$$

S^m is called the space of symbols of order m. We write $S^{-\infty} = \bigcap S^m$, $S^\infty = \bigcup S^m$.

Definition 18.1.1 is a global version of the special case of Definition 7.8.1 with $\delta = 0$ and $\rho = 1$. Later on we shall localize S^m and also consider general ρ and δ.

It is clear that S^m is a Fréchet space with semi-norms given by the smallest constants which can be used in (18.1.1). One advantage of not insisting on homogeneity is that $a(x,\xi) = \chi(\xi)$ is in S^0 if $\chi \in \mathscr{S}$. The following proposition is mainly used in this case where it is closely related to the regularization which was introduced already in Section 1.3.

Proposition 18.1.2. *Let $a\in S^0(\mathbb{R}^n\times\mathbb{R}^n)$ and set $a_\varepsilon(x,\xi)=a(x,\varepsilon\xi)$. Then a_ε is bounded in S^0, $0\leqq\varepsilon\leqq 1$, and $a_\varepsilon\to a_0$ in S^m for every $m>0$ when $\varepsilon\to 0$.*

Proof. Since $a_0\in S^0$ the statement follows if we show that for $0\leqq m\leqq 1$

$$(1+|\xi|)^{|\alpha|-m}|\partial_\xi^\alpha\partial_x^\beta(a_\varepsilon(x,\xi)-a(x,0))|\leqq C_{\alpha\beta}\varepsilon^m,\qquad 0\leqq\varepsilon\leqq 1.$$

When $\alpha=0$ this follows since by Taylor's formula

$$|a_{(\beta)}(x,\varepsilon\xi)-a_{(\beta)}(x,0)|\leqq C_\beta|\varepsilon\xi|^m.$$

When $\alpha\neq 0$ we just have to use that

$$(1+|\xi|)^{|\alpha|-m}(1+|\varepsilon\xi|)^{-|\alpha|}\varepsilon^{|\alpha|-m}\leqq((1+|\xi|)/(\varepsilon^{-1}+|\xi|))^{|\alpha|-m}\leqq 1.$$

When working with the spaces S^m it is useful to keep in mind that

$$a\in S^m\Rightarrow a_{(\beta)}^{(\alpha)}\in S^{m-|\alpha|};\quad a\in S^m \text{ and } b\in S^{m'}\Rightarrow ab\in S^{m+m'}. \tag{18.1.2}$$

The proofs are obvious and show for example the continuity of the bilinear map $(a,b)\mapsto ab$.

We shall now prove a simple result which gives a precise meaning to asymptotic sums like (18.3).

Proposition 18.1.3. *Let $a_j\in S^{m_j}$, $j=0,1,\dots$ and assume that $m_j\to-\infty$ as $j\to\infty$. Set $m_k'=\max_{j\geqq k}m_j$. Then one can find $a\in S^{m_0'}$ such that $\operatorname{supp}a\subset\bigcup\operatorname{supp}a_j$ and for every k*

$$a-\sum_{j<k}a_j\in S^{m_k'}. \tag{18.1.3}$$

The function a is uniquely determined modulo $S^{-\infty}$ and has the same property relative to any rearrangement of the series $\sum a_j$; we write

$$a\sim\sum a_j.$$

Proof. The uniqueness follows immediately from (18.1.3) and so does the invariance under rearrangements. To prove the existence we choose $\chi\in C_0^\infty$ equal to 1 in a neighborhood of 0. By Proposition 18.1.2 we can find a positive sequence ε_j converging to 0 so rapidly that

$$|\partial_\xi^\alpha\partial_x^\beta((1-\chi(\varepsilon_j\xi))\,a_j(x,\xi))|<2^{-j}(1+|\xi|)^{m_j+1-|\alpha|},\qquad |\alpha|+|\beta|\leqq j,$$

for $1-\chi(\varepsilon.)\to 0$ in S^1 when $\varepsilon\to 0$. Set $A_j(x,\xi)=(1-\chi(\varepsilon_j\xi))\,a_j(x,\xi)$. Then the sum $a=\sum A_j$ is locally finite, hence $a\in C^\infty$. Given α,β,k we can choose N so large that $N\geqq|\alpha|+|\beta|$ and $m_N'+1\leqq m_k'$. Then we obtain

$$|\partial_\xi^\alpha\partial_x^\beta(a(x,\xi)-\sum_{j<N}A_j(x,\xi))|\leqq(1+|\xi|)^{m_k'-|\alpha|}.$$

Since $a_j - A_j \in S^{-\infty}$ and $A_j \in S^{m'_k}$ when $j \geq k$, it follows that

$$(18.1.3)' \qquad |\partial_\xi^\alpha \partial_x^\beta (a(x,\xi) - \sum_{j<k} a_j(x,\xi))| \leq C_{\alpha\beta k}(1+|\xi|)^{m'_k - |\alpha|}.$$

This is the explicit meaning of (18.1.3), so the proof is complete.

When one has a good candidate for a, the verification of (18.1.3)′ is often simplified by the following observation which shows that little attention has to be paid to the derivatives.

Proposition 18.1.4. *Let $a_j \in S^{m_j}$, $j=0,1,\ldots$ and assume that $m_j \to -\infty$ when $j \to \infty$. Let $a \in C^\infty(\mathbb{R}^n \times \mathbb{R}^n)$ and assume that for all α, β we have for some C and μ depending on α and β*

$$(18.1.4) \qquad |a_{(\beta)}^{(\alpha)}(x,\xi)| \leq C(1+|\xi|)^\mu; \qquad x,\ \xi \in \mathbb{R}^n.$$

If there is a sequence $\mu_k \to -\infty$ such that

$$(18.1.3)'' \qquad |a(x,\xi) - \sum_{j<k} a_j(x,\xi)| < C_k(1+|\xi|)^{\mu_k},$$

it follows that $a \in S^m$, $m = \sup m_j$, and that $a \sim \sum a_j$.

Proof. Subtraction from a of some $A \in S^m$ with $A \sim \sum a_j$ reduces the proof to the case where all a_j are equal to 0. The hypothesis is then that a is rapidly decreasing when $\xi \to \infty$ and that (18.1.4) is valid. We have to show that all derivatives of a are rapidly decreasing when $\xi \to \infty$. It suffices to do so for the first order derivatives and iterate the conclusion. If η is a unit vector we obtain using Taylor's formula and (18.1.4) that

$$|a(x,\xi+\varepsilon\eta) - a(x,\xi) - \langle a'_\xi(x,\xi), \varepsilon\eta\rangle| \leq C\varepsilon^2(1+|\xi|)^\mu, \qquad x \in K,\ 0<\varepsilon<1,$$

for some C, μ. Hence

$$|\langle a'_\xi(x,\xi), \eta\rangle| \leq C\varepsilon(1+|\xi|)^\mu + |a(x,\xi) - a(x,\xi+\varepsilon\eta)|/\varepsilon,$$

which gives $|a'_\xi(x,\xi)| < C'(1+|\xi|)^{\mu-N}$ if we take $\varepsilon = (1+|\xi|)^{-N}$. The derivatives with respect to x can be discussed in the same way, and this completes the proof.

Definition 18.1.5. If $m \in \mathbb{C}$ and h is the reciprocal of a positive integer, the set of all $a \in S^{\mathrm{Re}\, m}$ such that

$$a(x,\xi) \sim \sum_0^\infty a_j(x,\xi)$$

where a_j is homogeneous of degree $m - jh$ when $|\xi| > 1$, will be called polyhomogeneous of degree m and step h. We write $a \in S^{m,h}_{\mathrm{phg}}$. When $h=1$ the step size will be omitted.

The homogeneity in the definition means that

$$a_j(x,t\xi) = t^{m-jh} a_j(x,\xi); \qquad |\xi|>1,\ t>1.$$

This implies homogeneity of degree $m-jh-|\alpha|$ for $a_{j(\beta)}^{(\alpha)}$ so a_j is automatically in $S^{\text{Re}\,m-jh}$ if $a_j \in C^\infty$ and vanishes for large x. That $a_j \in C^\infty$ excludes of course in general that a_j is homogeneous for all $\xi \neq 0$, but since the formula $a \sim \sum a_j$ is only a condition for large $|\xi|$ and the proof of Proposition 18.1.3 cuts away singularities of a_j for small $|\xi|$ we shall use this notation even if a_j is just in C^∞ when $\xi \neq 0$ and homogeneous there.

The symbols in S^m_{phg} are those to which we were led in the summary apart from the fact that the discussion there was local in the x variable. We shall later on localize S^m in the x variables and sometimes in the ξ variables also. However, before doing so we shall discuss pseudo-differential operators in $\mathbb{R}^n$ with symbols in S^m, for the results are then stronger than in the local case and the proofs are cleaner.

Theorem 18.1.6. *If* $a \in S^m$ *and* $u \in \mathscr{S}$, *then*

$$a(x,D)\,u(x) = (2\pi)^{-n} \int e^{i\langle x,\xi\rangle}\, a(x,\xi)\, \hat{u}(\xi)\, d\xi \tag{18.1.5}$$

defines a function $a(x,D)u \in \mathscr{S}$, *and the bilinear map* $(a,u) \mapsto a(x,D)u$ *is continuous. The commutators with* D_j *and multiplication by* x_j *are*

$$[a(x,D), D_j] = i a_{(j)}(x,D); \qquad [a(x,D), x_j] = -i a^{(j)}(x,D). \tag{18.1.6}$$

One calls $a(x,D)$ *a pseudo-differential operator of order* m.

The notation (18.1.5) is justified by the fact that if a is a polynomial in ξ then $a(x,D)$ is obtained from $a(x,\xi)$ by replacing ξ with $D = -i\partial/\partial x$ put to the right of the coefficients. This follows from Fourier's inversion formula. Sometimes we write Op a instead of $a(x,D)$.

Proof. Since $\hat{u} \in \mathscr{S}$ it is clear that (18.1.5) defines a continuous function with

$$|a(x,D)u| \leq (2\pi)^{-n} \int (1+|\xi|)^m\, |\hat{u}(\xi)|\, d\xi \,\sup |a(x,\xi)|(1+|\xi|)^{-m}.$$

The first relation (18.1.6) states that

$$D_j a(x,D)\, u(x) = a(x,D)\, D_j u(x) - i a_{(j)}(x,D)\, u(x).$$

It follows by differentiation of (18.1.5) under the integral sign since $\xi_j \hat{u}(\xi)$ is the Fourier transform of $D_j u$. The Fourier transform of $x_j u(x)$ is $-D_j \hat{u}(\xi)$ so an integration by parts gives the second relation (18.1.6),

$$a(x,D)(x_j u) = x_j a(x,D)u - i a^{(j)}(x,D)\, u.$$

Repeated use of (18.1.6) shows that $x^\alpha D^\beta a(x,D)u$ is a linear combination of

$$a^{(\alpha')}_{(\beta')}(x,D)\, x^{\alpha''} D^{\beta''} u; \qquad \alpha' + \alpha'' = \alpha, \qquad \beta' + \beta'' = \beta.$$

Hence $x^\alpha D^\beta a(x,D)u$ is bounded by the product of a semi-norm of u in $\mathscr{S}$ and a semi-norm of a in S^m. The proof is complete.

Remark. If we introduce the definition of $\hat{u}$ in (18.1.5), it follows that the Schwartz kernel K of $a(x,D)$ is given by

$$K(x,y)=(2\pi)^{-n}\int e^{i\langle x-y,\xi\rangle} a(x,\xi)\,d\xi, \tag{18.1.7}$$

which exists as an oscillatory integral (see section 7.8). Alternatively we can interpret (18.1.7) as $\hat{a}(x,y-x)/(2\pi)^n$ where $\hat{a}$ is the Fourier transform of $a(x,\xi)$ with respect to the ξ variable, defined by an obvious modification of Definition 7.1.9. Hence Fourier's inversion formula gives

$$a(x,\xi)=\int K(x,x-y)e^{-i\langle y,\xi\rangle}\,dy \tag{18.1.8}$$

which should again be read as the Fourier transform of $K(x,x-y)$ with respect to y. The formulas (18.1.7) and (18.1.8) establish a bijection $a\leftrightarrow K$ between distributions in $\mathscr{S}'(\mathbb{R}^{2n})$. Now a variant of Schwartz' kernel theorem (Theorem 5.2.1), which we leave for the reader to verify, states that the maps with kernel in $\mathscr{S}'(\mathbb{R}^{n_1+n_2})$ are precisely the continuous linear maps from $\mathscr{S}(\mathbb{R}^{n_2})$ to $\mathscr{S}'(\mathbb{R}^{n_1})$. For any $a\in\mathscr{S}'(\mathbb{R}^{2n})$ we can thus interpret (18.1.5) as a continuous map $a(x,D)$: $\mathscr{S}(\mathbb{R}^n)\to\mathscr{S}'(\mathbb{R}^n)$. The meaning of Theorem 18.1.6 is that it maps $\mathscr{S}(\mathbb{R}^n)$ into itself when $a\in S^m$. Later on this will also be proved under much weaker hypotheses on a.

We shall now determine the adjoint of $a(x,D)$ with respect to the sesquilinear scalar product

$$(u,v)=\int u\bar{v}\,dx;\quad u,\ v\in\mathscr{S}.$$

Assume first that $a\in\mathscr{S}$. The kernel K of $a(x,D)$, given by (18.1.7), is then in $\mathscr{S}$ and so is the kernel K^* of the adjoint,

$$K^*(x,y)=\overline{K(y,x)}=(2\pi)^{-n}\int e^{i\langle x-y,\eta\rangle}\,\overline{a(y,\eta)}\,d\eta.$$

Since $K^*(x,x-y)=(2\pi)^{-n}\int e^{i\langle y,\eta\rangle}\,\bar{a}(x-y,\eta)\,d\eta$ it follows from (18.1.8) that K^* is the kernel of $b(x,D)$ where $b\in\mathscr{S}$ and

$$\begin{aligned} b(x,\xi)&=(2\pi)^{-n}\int e^{i\langle y,\eta-\xi\rangle}\,\bar{a}(x-y,\eta)\,dy\,d\eta\\ &=(2\pi)^{-n}\int e^{-i\langle y,\eta\rangle}\,\bar{a}(x-y,\xi-\eta)\,dy\,d\eta\end{aligned}$$

Since the quadratic form $(y,\eta)\mapsto 2\langle y,\eta\rangle$ in $\mathbb{R}^{2n}$ has signature 0, determinant $(-1)^n$, and is its own dual, the Fourier transform of $(2\pi)^{-n}e^{-i\langle y,\eta\rangle}$ is by Theorem 7.6.1 equal to $e^{i\langle\hat{y},\hat{\eta}\rangle}$ if $\hat{y}$, $\hat{\eta}$ are the dual variables of y and η. Thus

$$b(x,\xi)=e^{i\langle D_x,D_\xi\rangle}\,\bar{a}(x,\xi) \tag{18.1.9}$$

in the sense that the Fourier transform of b is equal to that of $\bar{a}$ multiplied by $e^{i\langle\hat{x},\hat{\xi}\rangle}$ (see Section 7.6). The map $a\mapsto b$ defined by (18.1.9) is continuous in $\mathscr{S}'$, so it follows that

$$(a(x,D)u,v)=(u,b(x,D)v);\qquad u,v\in\mathscr{S}; \tag{18.1.10}$$

for every $a\in\mathscr{S}'$ if $b\in\mathscr{S}'$ is defined by (18.1.9).

Theorem 18.1.7. *If* $a\in S^m$ *then* $b(x,\xi)=e^{i\langle D_x, D_\xi\rangle}\,\bar a(x,\xi)\in S^m$,

$$(18.1.11)\qquad b(x,\xi)\sim\sum_0^\infty (i\langle D_x, D_\xi\rangle)^k\,\bar a(x,\xi)/k! = \sum_\alpha \partial_\xi^\alpha D_x^\alpha \bar a(x,\xi)/\alpha!.$$

(18.1.10) is valid and shows that $a(x,D)$ *can be extended to a continuous map from* $\mathscr{S}'$ *to* $\mathscr{S}'$, *as the adjoint of* $b(x,D)$.

Proof. Choose $\chi\in C_0^\infty(\mathbb{R}^n)$ so that $\chi(\xi)=1$ when $|\xi|<\frac12$ and $\chi(\xi)=0$ when $|\xi|>1$, and set for integers $\nu\geqq 0$

$$a_\nu(x,\xi)=\chi(\xi/2^\nu)\,\bar a(x,\xi),\qquad b_\nu(x,\xi)=e^{i\langle D_x, D_\xi\rangle}\,a_\nu(x,\xi).$$

In view of Proposition 18.1.2 we have for all ν, if $(x,\xi)\in\operatorname{supp} a_\nu$,

$$|a^{(\alpha)}_{\nu(\beta)}(x,\xi)|\leqq C_{\alpha\beta}(1+|\xi|)^{m-|\alpha|}\leqq C_{\alpha\beta}(1+2^\nu)^{|m|}.$$

Hence Theorem 7.6.5 gives

$$(18.1.12)\qquad |b_\nu(x,\xi)-\sum_{j<k}(i\langle D_x, D_\xi\rangle)^j\,a_\nu(x,\xi)/j!|<C_k\,2^{|m|\nu}.$$

If $|\xi|>2^{\nu+1}$ then the distance from (x,ξ) to $\operatorname{supp} a_\nu$ is at least equal to $|\xi|-2^\nu\geqq|\xi|/2\geqq(1+|\xi|)/4$ so the sum drops out and we are allowed to insert a factor $(1+|\xi|)^{-k}$ in the right-hand side. Given ξ denote by μ the smallest integer $\geqq 0$ with $|\xi|\leqq 2^{\mu+2}$. Then either $\mu=0$ and $|\xi|\leqq 4$ or $\mu>0$ and $2^{\mu+1}<|\xi|\leqq 2^{\mu+2}$. In both cases (18.1.12) with the improvement just mentioned gives

$$(18.1.12)'\qquad |b_\mu(x,\xi)-\sum_{j<k}(i\langle D_x, D_\xi\rangle)^j\,a_\mu(x,\xi)/j!|<C_k'(1+|\xi|)^{|m|-k}.$$

To estimate

$$b(x,\xi)-b_\mu(x,\xi)=\sum_\mu^\infty(b_{\nu+1}(x,\xi)-b_\nu(x,\xi))$$

we set $A_\nu(x,\xi)=a_{\nu+1}(x,\xi)-a_\nu(x,\xi)$, $B_\nu(x,\xi)=b_{\nu+1}(x,\xi)-b_\nu(x,\xi)$, and observe that

$$|A^{(\alpha)}_{\nu(\beta)}(x,\xi)|\leqq C'_{\alpha\beta}\,2^{\nu(m-|\alpha|)}$$

since $2^{\nu-1}\leqq|\xi|\leqq 2^{\nu+1}$ in $\operatorname{supp} A_\nu$. Hence

$$|\partial_\xi^\alpha\partial_x^\beta A_\nu(x,2^\nu\xi)|\leqq C'_{\alpha\beta}\,2^{m\nu}.$$

Since

$$B_\nu(x,2^\nu\xi)=e^{i\langle D_x, D_\xi\rangle/2^\nu}\,A_\nu(x,2^\nu\xi)$$

we obtain from Theorem 7.6.5 where the factor $\|A\|$ is now important

$$|B_\nu(x,2^\nu\xi)-\sum_{j<k}(i\langle D_x, D_\xi\rangle/2^\nu)^j\,A_\nu(x,2^\nu\xi)/j!|\leqq C_k''\,2^{(m-k)\nu},$$

that is,

$$|B_\nu(x,\xi)-\sum_{j<k}(i\langle D_x, D_\xi\rangle)^j\,A_\nu(x,\xi)/j!|\leqq C_k''\,2^{(m-k)\nu}.$$

Let $k \geqq m+1$ and sum for $\nu=\mu$ to ∞. The sum on the right-hand side is at most twice the first term, and $(1+|\xi|)/2^\mu$ lies between fixed bounds. Hence we obtain

$$\left|\sum_\mu^\infty (B_\nu(x,\xi) - \sum_{j<k} (i\langle D_x, D_\xi\rangle)^j A_\nu(x,\xi)/j!)\right| \leqq C_k'''(1+|\xi|)^{m-k}$$

which combined with (18.1.12)′ proves that with a new constant C_k

$$|b(x,\xi) - \sum_{j<k} (i\langle D_x, D_\xi\rangle)^j \bar{a}(x,\xi)/j!| \leqq C_k(1+|\xi|)^{|m|-k}. \tag{18.1.11$'$}$$

Since

$$b^{(\alpha)}_{(\beta)}(x,\xi) = e^{i\langle D_x, D_\xi\rangle} \bar{a}^{(\alpha)}_{(\beta)}(x,\xi)$$

a similar expansion is valid for the derivatives of b, which completes the proof of (18.1.11) and of the theorem.

Remark. The proof gives (18.1.11)′ with a constant C_k depending only on a finite number of semi-norms of a in S^m. (Since such statements can also be recovered from the qualitative statement in Theorem 18.1.7 by means of the closed graph theorem we shall usually omit them in what follows.) If a^ν is a bounded sequence in S^m which tends to 0 in C^∞ it follows that $e^{i\langle D_x, D_\xi\rangle} a^\nu(x,\xi)$ is bounded in S^m and tends to 0 in $\mathscr{S}'$, hence in C^∞.

Next we study the composition of operators.

Theorem 18.1.8. *If $a_j \in S^{m_j}$, $j=1,2$, then as operators in $\mathscr{S}$ or $\mathscr{S}'$*

$$a_1(x,D)\, a_2(x,D) = b(x,D) \tag{18.1.13}$$

where $b \in S^{m_1+m_2}$ is given by

$$b(x,\xi) = e^{i\langle D_y, D_\eta\rangle} a_1(x,\eta)\, a_2(y,\xi)|_{\eta=\xi, y=x} \tag{18.1.14}$$

and has the asymptotic expansion

$$\begin{aligned} b(x,\xi) &\sim \sum (i\langle D_y, D_\eta\rangle)^j a_1(x,\eta)\, a_2(y,\xi)|_{\eta=\xi, y=x}/j! \\ &= \sum_\alpha a_1^{(\alpha)}(x,\xi)\, D_x^\alpha a_2(x,\xi)/\alpha!. \end{aligned} \tag{18.1.15}$$

Proof. Assume first that $a_j \in \mathscr{S}$. If $u \in \mathscr{S}$ then $a_2(x,D)u \in \mathscr{S}$ and the Fourier transform is

$$\eta \mapsto (2\pi)^{-n} \int e^{i\langle y, \xi-\eta\rangle} a_2(y,\xi)\, \hat{u}(\xi)\, d\xi\, dy.$$

Hence

$$a_1(x,D)\, a_2(x,D)\, u(x) = (2\pi)^{-2n} \iint e^{i\langle x,\eta\rangle + i\langle y, \xi-\eta\rangle} a_1(x,\eta)\, a_2(y,\xi)\, \hat{u}(\xi)\, d\xi\, dy\, d\eta$$

so (18.1.13) is valid with

$$b(x,\xi) = (2\pi)^{-n} \iint e^{-i\langle x-y, \xi-\eta\rangle} a_1(x,\eta)\, a_2(y,\xi)\, dy\, d\eta$$

which is equivalent to (18.1.14) as shown before the statement of Theorem 18.1.7. Now consider for arbitrary $a_j \in S^{m_j}$

$$B(x,\xi,y,\eta)=e^{i\langle D_y, D_\eta\rangle} a_1(x,\eta)\, a_2(y,\xi).$$

Since

$$|\partial_\eta^\alpha \partial_y^\beta a_1(x,\eta)\, a_2(y,\xi)| \leqq C_{\alpha\beta}(1+|\eta|)^{m_1-|\alpha|}(1+|\xi|)^{m_2}$$

it follows from Theorem 18.1.7 and the remark after its proof that

$$\begin{aligned}|B(x,\xi,y,\eta)-\sum_{j<k}(i\langle D_y, D_\eta\rangle)^j a_1(x,\eta)\, a_2(y,\xi)/j!|\\ \leqq C_k(1+|\eta|)^{m_1-k}(1+|\xi|)^{m_2}.\end{aligned}$$

Here C_k can be estimated by a finite sum of products of semi-norms of a_1 in S^{m_1} and a_2 in S^{m_2}. Since differentiations commute with $e^{i\langle D_y, D_\eta\rangle}$ we have more generally

$$\begin{aligned}|\partial_\xi^\alpha \partial_x^\beta \partial_\eta^{\alpha'} \partial_y^{\beta'}(B(x,\xi,y,\eta)-\sum_{j<k}(i\langle D_y, D_\eta\rangle)^j a_1(x,\eta)\, a_2(y,\xi)/j!)|\\ \leqq C_{k,\alpha,\alpha',\beta,\beta'}(1+|\eta|)^{m_1-k-|\alpha'|}(1+|\xi|)^{m_2-|\alpha|},\end{aligned}$$

where $C_{k,\alpha,\alpha',\beta,\beta'}$ has a similar bound. Hence $b(x,\xi)=B(x,\xi,x,\xi)\in S^{m_1+m_2}$ and the bilinear map $(a_1,a_2)\mapsto b$ is continuous from $S^{m_1}\times S^{m_2}$ to $S^{m_1+m_2}$. It remains to verify (18.1.13) in general.

Choose $\chi\in C_0^\infty$ equal to 1 in a neighborhood of 0 and set

$$a_j^\nu(x,\xi)=\chi(x/\nu)\,\chi(\xi/\nu)\, a_j(x,\xi).$$

Since $\chi(\xi/\nu)\, a_j(x,\xi)\to a_j(x,\xi)$ in S^{m_j+1} when $\nu\to\infty$ (Proposition 18.1.2) we have $a_2^\nu(x,D)u\to a_2(x,D)u$ in $\mathscr{S}$ as $\nu\to\infty$, if $u\in\mathscr{S}$. Hence

$$a_1^\nu(x,D)\, a_2^\nu(x,D)u\to a_1(x,D)\, a_2(x,D)u \quad \text{in } \mathscr{S}.$$

If $b^\nu(x,\xi)$ is defined by (18.1.14) with a_j replaced by a_j^ν then b^ν is bounded in $S^{m_1+m_2}$ and converges to b pointwise, hence in $\mathscr{S}'$, so $b^\nu(x,D)u\to b(x,D)u$ in $\mathscr{S}'$ as $\nu\to\infty$. Since $b^\nu(x,D)u=a_1^\nu(x,D)\, a_2^\nu(x,D)u$ it follows that (18.1.13) holds. The proof is complete.

Theorem 18.1.8 permits us to give a precise form to the discussion of inversion of elliptic operators in the summary.

Theorem 18.1.9. *Let $a\in S^m$ and $b\in S^{-m}$. Then the conditions*

(i) $$a(x,D)\, b(x,D)-I\in \operatorname{Op} S^{-\infty}$$

(ii) $$b(x,D)\, a(x,D)-I\in \operatorname{Op} S^{-\infty}$$

are equivalent, and a determines b mod $S^{-\infty}$. *(I is the identity operator* Op 1.) *They imply*

(iii) $$a(x,\xi)\, b(x,\xi)-1\in S^{-1}$$

which in turn implies that for some positive constants c and C

(iv) $$|a(x,\xi)|>c|\xi|^m \quad \text{if } |\xi|>C.$$

Conversely, if (iv) *is fulfilled one can find* $b\in S^{-m}$ *satisfying* (i), (ii), (iii).

Proof. (i) and (ii) both imply (iii) by Theorem 18.1.8. From (iii) it follows that $|a(x,\xi)\,b(x,\xi)-1|<\frac{1}{2}$ for $|\xi|>C$, hence

$$\tfrac{1}{2}<|a(x,\xi)\,b(x,\xi)|<C'\,|a(x,\xi)|\,|\xi|^{-m}, \qquad |\xi|>C,$$

which proves (iv). From (iii) it also follows that

$$a(x,D)\,b(x,D)=I-r(x,D), \qquad r\in S^{-1}.$$

We want to invert $I-r(x,D)$ by the Neuman series, so we set

$$b(x,D)\,r(x,D)^k=b_k(x,D), \qquad b_k\in S^{-m-k}.$$

With $b'\sim\sum_0^\infty b_j$ we obtain

$$a(x,D)\,b'(x,D)-I=a(x,D)(b'(x,D)-\sum_{j<k} b_j(x,D))-r(x,D)^k\in \operatorname{Op} S^{-k}$$

for every k. This proves that (i) is valid with b replaced by b'. In the same way we can find $b''\in S^{-m}$ such that (ii) is fulfilled with b replaced by b''. When

$$a(x,D)\,b'(x,D)-I\in \operatorname{Op} S^{-\infty} \quad \text{and} \quad b''(x,D)\,a(x,D)-I\in \operatorname{Op} S^{-\infty}$$

then

$$\begin{aligned} b''(x,D)-b'(x,D)&=b''(x,D)(I-a(x,D)\,b'(x,D))\\ &\quad+(b''(x,D)\,a(x,D)-I)\,b'(x,D)\end{aligned}$$

is in $\operatorname{Op} S^{-\infty}$ so b' and b'' satisfy both (i) and (ii). This proves the equivalence of (i) and (ii) and also that a determines $b \bmod S^{-\infty}$. It remains to prove that (iv) $\Rightarrow$ (iii). The proof is reduced to the case $m=0$ if we introduce $a(x,\xi)(1+|\xi|^2)^{-m/2}$ and $b(x,\xi)(1+|\xi|^2)^{m/2}$ instead of a and b. Then it is a consequence of the following:

Lemma 18.1.10. *If* $a_1,\dots,a_k\in S^0$ *and* $F\in C^\infty(\mathbb{C}^k)$ *then* $F(a_1,\dots,a_k)\in S^0$.

Proof. Since $\operatorname{Re} a_\nu$, $\operatorname{Im} a_\nu\in S^0$ we may assume that a_j are real valued and that $F\in C^\infty(\mathbb{R}^k)$. We have

$$\partial F(a)/\partial x_j=\sum \partial F/\partial a_\nu\, a_{\nu(j)}, \qquad \partial F(a)/\partial \xi_j=\sum \partial F/\partial a_\nu\, a_\nu^{(j)},$$

and $a_{\nu(j)}\in S^0$, $a_\nu^{(j)}\in S^{-1}$. Hence it follows by induction with respect to $|\alpha|+|\beta|$ that the derivatives of $F(a)$ satisfy (18.1.1).

End of Proof of Theorem 18.1.9. Assume that (iv) is fulfilled and that $m=0$. Choose $F\in C^\infty(\mathbb{C})$ so that $F(z)=1/z$ when $|z|>c$. Then $b=F(a)\in S^0$ and

$a(x,\xi)\,b(x,\xi)=1$ when $|\xi|>C$, which strengthens (iii) to

$$\text{(iii)}' \qquad a(x,\xi)\,b(x,\xi)=1 \quad \text{when } |\xi|>C.$$

This completes the proof.

Theorems 18.1.7 to 18.1.9 are the core of the calculus of pseudodifferential operators. They lead to improved continuity properties:

Theorem 18.1.11. *If $a\in S^0$ then $a(x,D)$ is bounded in $L^2(\mathbb{R}^n)$.*

For the proof we need a classical lemma of Schur:

Lemma 18.1.12. *If K is a continuous function in $\mathbb{R}^n\times\mathbb{R}^n$ and*

$$\sup_y \int |K(x,y)|\,dx \leqq C, \qquad \sup_x \int |K(x,y)|\,dy \leqq C,$$

then the integral operator with kernel K has norm $\leqq C$ in $L^2(\mathbb{R}^n)$.

Proof. Cauchy-Schwarz' inequality gives

$$|K u(x)|^2 \leqq \int |K(x,y)|\,|u(y)|^2\,dy \int |K(x,y)|\,dy.$$

If the last integral is estimated by C, an integration with respect to x gives

$$\int |K u(x)|^2\,dx \leqq C \iint |K(x,y)|\,|u(y)|^2\,dx\,dy \leqq C^2 \int |u(y)|^2\,dy.$$

Proof of Theorem 18.1.11. Assume first that $a\in S^{-n-1}$. Then the kernel K of $a(x,D)$ is continuous and

$$|K(x,y)| \leqq (2\pi)^{-n} \int |a(x,\xi)|\,d\xi \leqq C.$$

Now $(x-y)^\alpha\,K(x,y)$ is the kernel of the commutator

$$[x_1,[x_1,\ldots,[x_n,a(x,D)]]]=i^{|\alpha|}\,a^{(\alpha)}(x,D),$$

by (18.1.6), so this is also a bounded function. Hence

$$(1+|x-y|)^{n+1}\,|K(x,y)| \leqq C,$$

and the L^2 continuity of $a(x,D)$ follows from Lemma 18.1.12.

Next we prove by induction that $a(x,D)$ is L^2 continuous if $a\in S^k$ and $k\leqq -1$. To do so we form for $u\in\mathscr{S}$

$$\|a(x,D)\,u\|^2=(a(x,D)\,u, a(x,D)\,u)=(b(x,D)\,u, u)$$

where $b(x,D)=a^*(x,D)\,a(x,D)\in \operatorname{Op} S^{2k}$. The continuity of $a(x,D)$ is therefore a consequence of that of $b(x,D)$,

$$\|a(x,D)\,u\|^2 \leqq \|b(x,D)\,u\|\;\|u\| \leqq \|b(x,D)\|\;\|u\|^2.$$

From the first part of the proof the continuity of $a(x,D)$ for all $a\in S^k$ now follows successively for $k\leqq -(n+1)/2$, $k\leqq -(n+1)/4,\ldots,$ hence for $k\leqq -1$

after a finite number of steps. Assume now that $a\in S^0$ and choose $M>2\sup|a(x,\xi)|^2$. Then

$$c(x,\xi)=(M-|a(x,\xi)|^2)^{\frac{1}{2}}\in S^0$$

by Lemma 18.1.10 since $M/2\leqq M-|a(x,\xi)|^2$ and we can choose $F\in C^\infty(\mathbb{R})$ with $F(t)=t^{\frac{1}{2}}$ when $t\geqq M/2$. Now Theorems 18.1.7 and 18.1.8 show that

$$c(x,D)^*\,c(x,D)=M-a(x,D)^*\,a(x,D)+r(x,D),$$

where $r\in S^{-1}$. Hence

$$\|a(x,D)\,u\|^2\leqq M\,\|u\|^2+(r(x,D)\,u,u)$$

which completes the proof since r is already known to be L^2 continuous.

It follows from the proof that the norm of $a(x,D)$ can be estimated by a semi-norm of a in S^0. There is a very simple proof of L^2 continuity which requires no smoothness at all in ξ but instead some decay as $x\to\infty$.

Theorem 18.1.11′. *Let $a(x,\xi)$ be a measurable function which is $n+1$ times continuously differentiable with respect to x for fixed ξ. If*

$$\sum_{|\alpha|\leqq n+1}\int|D_x^\alpha a(x,\xi)|\,dx\leqq M,\qquad \xi\in\mathbb{R}^n,$$

for some $M<\infty$, it follows that $a(x,D)$ is bounded in $L^2(\mathbb{R}^n)$ with norm $\leqq CM$.

Proof. If $\hat{u}\in C_0^\infty$ the Fourier transform of $a(x,D)\,u$ is

$$\eta\mapsto\int A(\eta-\xi,\xi)\,\hat{u}(\xi)\,d\xi$$

where

$$A(\eta,\xi)=(2\pi)^{-n}\int a(x,\xi)\,e^{-i\langle x,\eta\rangle}\,dx.$$

By hypothesis

$$(1+|\eta|)^{n+1}\,|A(\eta,\xi)|\leqq C_1\,M$$

which implies that

$$\int|A(\eta-\xi,\xi)|\,d\eta\leqq CM,\qquad \int|A(\eta-\xi,\xi)|\,d\xi\leqq CM.$$

In view of Lemma 18.1.12 it follows that the L^2 norm of the Fourier transform of $a(x,D)\,u$ is at most $CM\,\|\hat{u}\|$, which completes the proof.

Let us now reconsider the spaces $H_{(s)}$ introduced in Definition 7.9.1: we have $u\in H_{(s)}$ if $u\in\mathscr{S}'$ and $\hat{u}\in L^2_{\text{loc}}$,

$$\|u\|_{(s)}=((2\pi)^{-n}\int|\hat{u}(\xi)|^2(1+|\xi|^2)^s\,d\xi)^{\frac{1}{2}}<\infty.$$

If $E_s(\xi)=(1+|\xi|^2)^{s/2}$, then $E_s\in S^s$ and the Fourier transform of $E_s(D)\,u$ is $E_s(\xi)\,\hat{u}(\xi)$, so the definition means exactly that $E_s(D)\,u\in L^2$. This gives a precise meaning to the idea that $H_{(s)}$ consists of the distributions with derivatives of order s in L^2, and we have

Theorem 18.1.13. *If* $a \in S^m$ *then* $a(x,D)$ *is a continuous operator from* $H_{(s)}$ *to* $H_{(s-m)}$ *for every* s.

Proof. If $u \in H_{(s)}$ then $v = E_s(D)u \in L^2$ and

$$E_{s-m}(D)\,a(x,D)\,u = E_{s-m}(D)\,a(x,D)\,E_{-s}(D)\,v \in L^2$$

by Theorem 18.1.11 since $E_{s-m}(D)\,a(x,D)\,E_{-s}(D) \in \operatorname{Op} S^0$.

Remark. $a(x,D)$ is also continuous ${}^pH_{(s)} \to {}^pH_{(s-m)}$ by Corollary B.1.6.

The proof of continuity in Theorem 18.1.11 was based on estimating $a(x,D)^*\,a(x,D)$ from above. We shall now prove a stronger one sided estimate which is often important. It is usually called the *sharp Gårding inequality*. Various improvements of it will be given later on, and a shorter proof will be possible when we have developed stronger techniques. However, we give a direct and in principle elementary proof here for the benefit of readers who do not wish to go through Sections 18.4 to 18.6.

Theorem 18.1.14. *If* $a \in S^{2m+1}$ *and* $\operatorname{Re} a \geqq 0$ *then*

$$\operatorname{Re}(a(x,D)\,u,u) \geqq -C\,\|u\|^2_{(m)}, \quad u \in \mathscr{S}. \tag{18.1.16}$$

Proof. (18.1.16) follows from Theorem 18.1.13 if $a \in S^{2m}$. Since

$$(a(x,D) + a(x,D)^*)/2 - (\operatorname{Re} a)(x,D) \in \operatorname{Op} S^{2m}$$

it is therefore sufficient to prove (18.1.16) with a replaced by $\operatorname{Re} a$. Thus we assume $a \geqq 0$ in what follows. To prove (18.1.16) we shall then write $a(x,D)$ as a superposition of positive operators with an error of order $2m$. We start by choosing an even function $\phi \in C_0^\infty(\mathbb{R}^{2n})$ with L^2 norm one and define $\psi \in \mathscr{S}$ by $\psi(x,D) = \phi(x,D)^*\,\phi(x,D)$. Then ψ is even by (18.1.14) and (18.1.9), and we claim that

$$\iint \psi(y,\eta)\,dy\,d\eta = 1. \tag{18.1.17}$$

For the proof we observe that if K_ϕ and K_ψ are the kernels of $\phi(x,D)$ and of $\psi(x,D)$ then

$$\begin{aligned}(2\pi)^{-n} \iint \psi(x,\xi)\,dx\,d\xi &= \int K_\psi(x,x)\,dx = \iint |K_\phi(x,y)|^2\,dx\,dy \\ &= (2\pi)^{-n} \iint |\phi(x,\xi)|^2\,dx\,d\xi\end{aligned}$$

where the last equality follows from (18.1.7) and Parseval's formula.

Having proved (18.1.17) we now set $a = a_0 + a_1$ where

$$\begin{aligned} a_1(x,\xi) &= \iint \psi((x-y)\,q(\eta),\,(\xi-\eta)/q(\eta))\,a(y,\eta)\,dy\,d\eta, \\ q(\eta) &= (1+|\eta|^2)^{\frac14}. \end{aligned} \tag{18.1.18}$$

Since $(\psi(x,D)\,u,u) = \|\phi(x,D)\,u\|^2 \geqq 0$ we obtain $(\psi(tx,D/t)\,u,u) \geqq 0$ if we replace u by $u(x/t)$. Hence it follows that

$$(\psi(t(x-y)),(D-\eta)/t)\,u,u) \geqq 0, \quad u \in \mathscr{S},\ (y,\eta) \in \mathbb{R}^{2n},$$

if u is replaced by $u(x+y)e^{-i\langle x,\eta\rangle}$. It is therefore clear that

$$(a_1(x,D)u,u)\geqq 0, \quad u\in\mathscr{S},$$

so (18.1.16) will follow if we prove that $a_0\in S^{2m}$.

When we differentiate with respect to x_j under the integral sign in (18.1.18), the derivative can be replaced by $-\partial/\partial y_j$. After an integration by parts the net result is that a has been differentiated with respect to y_j. Differentiation of a_1 with respect to ξ_j is more complicated since

$$(\partial/\partial\xi_j+\partial/\partial\eta_j)\psi((x-y)q(\eta),(\xi-\eta)/q(\eta))$$
$$=\psi'((x-y)q(\eta),(\xi-\eta)/q(\eta))F_j(\eta),$$

where $F_j(\eta)=q(\eta)^{-1}\partial q(\eta)/\partial\eta_j\in S^{-1}$ and

(18.1.19) $$\psi'(x,\xi)=\frac{\partial}{\partial t}\psi(tx,\xi/t)|_{t=1}=\langle x,\partial\psi/\partial x\rangle-\langle\xi,\partial\psi/\partial\xi\rangle$$

is another even function in $\mathscr{S}$. Note that $\iint\psi'(x,\xi)\,dx\,d\xi=0$. Differentiation of a_1 with respect to ξ_j therefore gives one term of the form (18.1.18) with a replaced by $a^{(j)}$ and one where ψ is replaced by ψ' and a factor $F_j\in S^{-1}$ is introduced. Inductively it follows that $a_{0(\beta)}^{(\alpha)}=a_{(\beta)}^{(\alpha)}-a_{1(\beta)}^{(\alpha)}$ is a finite sum of terms of the form

(18.1.20) $$\begin{aligned}b_0(x,\xi)=&\iint\psi_1((x-y)q(\eta),(\xi-\eta)/q(\eta))b(y,\eta)\,dy\,d\eta\\&-b(x,\xi)\iint\psi_1(y,\eta)\,dy\,d\eta\end{aligned}$$

where $\psi_1\in\mathscr{S}$ is even and $b\in S^{2m+1-|\alpha|}$. The theorem will be proved if we show that $b\in S^{\mu}$ implies $|b_0(x,\xi)|\leqq C(1+|\xi|)^{\mu-1}$.

a) When $|\xi-\eta|\geqq(1+|\xi|)/2$ we have $1+|\eta|\leqq 3|\xi-\eta|$ and

$$(1+|\xi|)/(1+|\eta|)\leqq 1+|\xi-\eta|, \quad (1+|\eta|)/(1+|\xi|)\leqq 1+|\xi-\eta|.$$

The factor ψ_1 in the first integral in (18.1.20) can be estimated by

$$C_N(1+|x-y|q(\eta)+|\xi-\eta|/q(\eta))^{-2n-1}(1+|\xi-\eta|^2/q(\eta)^2)^{-N}$$

for any N. Since $q(\eta)^2\leqq 1+|\eta|\leqq 3|\xi-\eta|$ the last factor may be replaced by $(1+|\xi-\eta|/3)^{-N}$. Hence the integrand in the first term in (18.1.20) has the bound

$$C'_N(1+|x-y|q(\xi)+|\xi-\eta|/q(\xi))^{-2n-1}(1+|\xi|)^{|\mu|+n+1-N}$$

when $|\xi-\eta|\geqq(1+|\xi|)/2$ and $N>|\mu|+n+1$. The integral of the first factor with respect to y and η over $\mathbb{R}^{2n}$ is finite and independent of (x,ξ), so we obtain the desired bound for the corresponding part of the first term in (18.1.20) if N is large enough.

b) When $|\xi-\eta|<(1+|\xi|)/2$ we have

$$(1+|\xi|)<2(1+|\theta|)<3(1+|\xi|)$$

for every θ on the line segment between ξ and η. Hence Taylor's formula gives

$$B(x,y,\xi,\eta)=|b(y,\eta)-\sum_{|\alpha+\beta|\leqq 2} b^{(\alpha)}_{(\beta)}(x,\xi)(y-x)^\beta(\eta-\xi)^\alpha/\alpha!\beta!|$$
$$\leqq C(|x-y|\,q(\xi)+|\xi-\eta|/q(\xi))^3(1+|\xi|)^{\mu-\frac{3}{2}}.$$

(Note that the scales in ψ were chosen to match in this estimate.) Thus

$$\iint_{2|\xi-\eta|<1+|\xi|} B(x,y,\xi,\eta)|\psi_1((x-y)\,q(\eta),(\xi-\eta)/q(\eta))|\,dy\,d\eta\leqq C(1+|\xi|)^{\mu-\frac{3}{2}}.$$

It remains to evaluate the integrals

$$\iint_{2|\xi-\eta|<1+|\xi|} \psi_1((x-y)\,q(\eta),(\xi-\eta)/q(\eta))(y-x)^\beta(\eta-\xi)^\alpha\,dy\,d\eta.$$

To do so we would like to replace $q(\eta)$ by $q(\xi)$. With the notation ψ_1' in (18.1.19) we have for any N when t and $1/t$ have a fixed bound

$$|\psi_1(tz,\theta/t)-\psi_1(z,\theta)-(t-1)\,\psi_1'(z,\theta)|\leqq C_N(1+|z|+|\theta|)^{-N}(t-1)^2.$$

Assuming still that $|\xi-\eta|<(1+|\xi|)/2$ we shall take $t=q(\eta)/q(\xi)$. Then

$$q(\xi)(t-1)=q(\eta)-q(\xi)=\langle q'(\xi),\eta-\xi\rangle+O(|\eta-\xi|/q(\xi))^2/q(\xi)$$

so replacing (z,θ) by $((x-y)\,q(\xi),\ (\xi-\eta)/q(\xi))$ we conclude that

$$|\psi_1((x-y)\,q(\eta),(\xi-\eta)/q(\eta))-\psi_1((x-y)\,q(\xi),(\xi-\eta)/q(\xi))$$
$$-\langle q'(\xi),(\eta-\xi)/q(\xi)\rangle\,\psi_1'((x-y)\,q(\xi),(\xi-\eta)/q(\xi))|$$
$$\leqq C_N'(1+|x-y|\,q(\xi)+|\xi-\eta|/q(\xi))^{2-N}(1+|\xi|)^{-1}.$$

Since $2|\xi-\eta|>1+|\xi|$ implies $|\xi-\eta|/q(\xi)>(1+|\xi|)/(2q(\xi))$ we obtain

$$\iint_{2|\xi-\eta|<1+|\xi|} \psi_1((x-y)\,q(\eta),(\xi-\eta)/q(\eta))(y-x)^\beta(\eta-\xi)^\alpha\,dy\,d\eta$$
$$=q(\xi)^{|\alpha|-|\beta|}\Big[\iint\psi_1(y,\eta)\,y^\beta\,\eta^\alpha\,dy\,d\eta$$
$$+\iint\psi_1'(y,\eta)\,y^\beta\,\eta^\alpha\langle\eta,q'(\xi)\rangle\,dy\,d\eta+O((1+|\xi|)^{-1})\Big].$$

The first integral on the right vanishes when $|\alpha+\beta|=1$ and the second vanishes when $\alpha+\beta=0$. When $|\alpha+\beta|=2$ we get the bound $C\,q(\xi)^{|\alpha|-|\beta|}=C\,q(\xi)^{2|\alpha|-2}$, when $|\alpha+\beta|=1$ we get the bound $C\,q(\xi)^{|\alpha|-|\beta|-1}=C\,q(\xi)^{2|\alpha|-2}$, and when $\alpha+\beta=0$ the integral is $\iint\psi_1(y,\eta)\,dy\,d\eta+O((1+|\xi|)^{-1})$. If we multiply by $b^{(\alpha)}_{(\beta)}(x,\xi)/\alpha!\,\beta!$ and sum, we obtain after subtraction of the second integral in (18.1.20)

$$|b_0(x,\xi)|\leqq C_1\sum_{|\alpha+\beta|\leqq 2}|b^{(\alpha)}_{(\beta)}(x,\xi)|(1+|\xi|)^{|\alpha|-1}+C_2(1+|\xi|)^{\mu-\frac{3}{2}}.$$

This completes the proof.

Remark 1. In Section 22.3 we shall prove some refinements of Theorem 18.1.14. It will then be important that the preceding proof shows that if

$0 \leqq a \in S^2$ then one can find $a_0 \in S^1$ such that $a(x,D)-a_0(x,D)$ is non-negative and

$$|a_0(x,\xi)| \leqq C\Big(\sum_{|\alpha+\beta|\leqq 2} (1+|\xi|)^{|\alpha|-1}\,|a_{(\beta)}^{(\alpha)}(x,\xi)| + (1+|\xi|)^{\frac{1}{2}}\Big). \tag{18.1.21}$$

This is a special case of the preceding estimate of b_0.

Remark 2. The calculus of pseudo-differential operators works equally well if we allow symbols $a \in S^m$ with values in $\mathscr{L}(B_1, B_2)$ where B_1 and B_2 are reflexive Banach spaces. Thus $a(x,D)$ maps $\mathscr{S}(\mathbb{R}^n, B_1)$ to $\mathscr{S}(\mathbb{R}^n, B_2)$ and $\mathscr{S}'(\mathbb{R}^n, B_1)$ to $\mathscr{S}'(\mathbb{R}^n, B_2)$. We just have to replace absolute values by norms in the arguments above. When B_1 and B_2 are Hilbert spaces the results on L^2 continuity remain valid and the proofs only require obvious modifications. Also the sharp Gårding inequality (Theorem 18.1.14) is valid if $a \in S^{2m+1}$ has values in $\mathscr{L}(H,H)$ where H is a Hilbert space, and the function u in (18.1.16) takes its values in H. In fact, if A is a positive bounded operator in H then

$$(\psi((x-y)/t, (D-\eta)/t)\,Au, u) \geqq 0; \quad y,\ \eta \in \mathbb{R}^n,\ t>0;$$

when ψ is defined as in the proof of Theorem 18.1.14, for the spectral theorem reduces the proof to the scalar case. Thus the first part of the proof remains valid, and the second part requires essentially no change at all

The following simple consequence of Theorem 18.1.14 gives Theorem 18.1.11 a more precise form which is often important.

Theorem 18.1.15. *For every bounded subset A of S^0 there is a constant C such that*

$$\operatorname{Re}(a(x,D)u, u) \geqq -C\delta\,\|u\|^2, \quad u \in \mathscr{S}, \tag{18.1.22}$$

if $a \in A$, $a \geqq 0$, $0<\delta<1$, and $a(x,\xi)=0$ when $|\xi\delta|<1$. We also have

$$\|a(x,D)\| \leqq \sup|a| + (C\delta)^{\frac{1}{2}}, \quad u \in \mathscr{S}, \tag{18.1.23}$$

if $a \in A$, $0<\delta<1$, and $a(x,\xi)=0$ when $|\xi\delta|<1$.

Proof. Let $A_1 = \{a/\delta;\, a \in A$ and $a(x,\xi)=0$ when $|\xi\delta|<1,\ 0<\delta<1\}$. Since $\delta^{-1} \leqq 1+|\xi|$ in supp a it follows that A_1 is a bounded subset of S^1. Hence (18.1.16) is valid uniformly with $m=0$ for all $a \in A_1$, which proves (18.1.22). Choose $\chi \in C^\infty(\mathbb{R}^n)$ with $0 \leqq \chi \leqq 1$, $\chi(\xi)=1$ when $|\xi|>1$ and $\chi(\xi)=0$ when $|\xi|<\frac{1}{2}$. Then we can apply (18.1.22) to

$$M^2\chi(\delta\xi)^2 - |a(x,\xi)|^2$$

if $a \in A$, $M = \sup|a|$ and $a(x,\xi)=0$ when $|\delta\xi|<1$ for some $\delta \in (0,1)$. This gives with another C

$$M^2\|u\|^2 - \operatorname{Re}(|a|^2(x,D)u, u) \geqq -C\delta\,\|u\|^2, \quad u \in \mathscr{S}.$$

We have $a^*(x,D)\,a(x,D)=|a|^2(x,D)+b(x,D)$ where $b(x,\xi)$ is bounded in S^{-1} and vanishes when $|\delta\xi|<1$. Hence $b(x,\xi)/\delta$ belongs to a bounded set in S^0 so $\|b(x,D)/\delta\|\leqq C'$ by Theorem 18.1.11 and

$$M^2\,\|u\|^2-\|a(x,D)\,u\|^2\geqq -(C+C')\,\delta\,\|u\|^2.$$

This completes the proof of (18.1.23).

In the following theorem we sum up some smoothness properties of the kernel of a pseudo-differential operator, partly encountered already.

Theorem 18.1.16. *Let $a\in S^m$ and denote by $K\in\mathscr{S}'(\mathbb{R}^n\times\mathbb{R}^n)$ the Schwartz kernel of $a(x,D)$ defined by (18.1.7). Then $K\in C^j(\mathbb{R}^n\times\mathbb{R}^n)$ if $m+j+n<0$, and $K\in C^\infty(\mathbb{R}^n\times\mathbb{R}^n\setminus\Delta)$ for any m if Δ is the diagonal $\{(x,x);\,x\in\mathbb{R}^n\}$. More precisely*

$$WF(K)\subset\{(x,x,\theta,-\theta);\,x,\theta\in\mathbb{R}^n\} \tag{18.1.24}$$

which is the conormal bundle of Δ. We have

$$WF(a(x,D)\,u)\subset WF(u),\quad u\in\mathscr{S}', \tag{18.1.25}$$

$$\operatorname{sing\,supp} a(x,D)\,u\subset\operatorname{sing\,supp} u,\quad u\in\mathscr{S}'. \tag{18.1.26}$$

If $a\in S^{-\infty}$ then $a(x,D)\,\mathscr{S}'\subset C^\infty$.

Proof. (18.1.7) is absolutely convergent and remains so after j differentiations if $m+j+n<0$. This proves the first statement. To prove the second one we observe that if $\chi,\psi\in C_0^\infty$ then $\chi(x)\,K(x,y)\,\psi(y)$ is the kernel of the operator

$$u\mapsto\chi\,a(x,D)\,\psi\,u,\qquad u\in\mathscr{S},$$

which is in $\operatorname{Op}S^{-\infty}$ by Theorem 18.1.8 if $\operatorname{supp}\chi\cap\operatorname{supp}\psi=\emptyset$. Hence the kernel is in C^∞ then. The more precise result (18.1.24) follows at once from Theorem 8.1.9 with $\phi(x,y,\theta)=\langle x-y,\theta\rangle$. If $u\in\mathscr{E}'$ then Theorem 8.2.13 gives (18.1.25) which implies (18.1.26). (It is also easy to prove (18.1.26) directly when $u\in\mathscr{E}'$, for $\operatorname{sing\,supp} a(x,D)\,u\subset\operatorname{supp} u$ since $K\in C^\infty(\mathbb{R}^n\times\mathbb{R}^n\setminus\Delta)$, and we can write $u=u_1+u_2$ with $u_2\in C_0^\infty$, hence $a(x,D)\,u_2\in C^\infty$, and $\operatorname{supp} u_1$ close to $\operatorname{sing\,supp} u$.) If $a\in S^{-\infty}$ and $u\in\mathscr{S}'$ then

$$D_x^\alpha\,a(x,D)\,u(x)=(2\pi)^{-n}\langle\hat{u},D_x^\alpha(e^{i\langle x,\xi\rangle}\,a(x,\xi))\rangle,$$

for this is true when $u\in\mathscr{S}$ and the right-hand side is a continuous map from $\mathscr{S}'$ to C^0. Hence $a(x,D)\,\mathscr{S}'\subset C^\infty$. If $\chi,\psi\in C_0^\infty$ and $\chi=1$ in a neighborhood of $\operatorname{supp}\psi$, it follows that

$$\psi\,a(x,D)\,u-\psi\,a(x,D)(\chi\,u)=\psi\,a(x,D)(1-\chi)\,u\in C^\infty.$$

Thus $WF(\psi\,a(x,D)\,u)\subset WF(\chi\,u)$ which yields (18.1.25) for any $u\in\mathscr{S}'$.

Remark. (18.1.26) is often called the pseudo-local property while (18.1.25) is referred to as the microlocal property of $a(x,D)$.

Before discussing the effect of a change of variables on $\operatorname{Op} S^m$ we observe that if $a \in S^m$ then a direct computation gives

$$a(x, D)(e^{i\langle x, \xi\rangle} u) = e^{i\langle x, \xi\rangle} a(x, D+\xi) u; \quad u \in \mathscr{S}', \ \xi \in \mathbb{R}^n.$$

If we take $u(x) = v(\varepsilon x)$ where $\hat{v} \in C_0^\infty$ and $v(0) = 1$, it follows when $\varepsilon \to 0$ that

$$a(x, D)\, e^{i\langle x, \xi\rangle} = a(x, \xi)\, e^{i\langle x, \xi\rangle}. \tag{18.1.27}$$

This gives a convenient way to recover the symbol from the operator.

Theorem 18.1.17. *Let X and X_κ be open subsets of $\mathbb{R}^n$ and $\kappa\colon X \to X_\kappa$ a diffeomorphism. If $a \in S^m$ and the kernel of $a(x, D)$ has compact support in $X \times X$ then*

$$a_\kappa(\kappa(x), \eta) = e^{-i\langle \kappa(x), \eta\rangle} a(x, D)\, e^{i\langle \kappa(x), \eta\rangle}, \tag{18.1.28}$$

$a_\kappa(y, \xi) = 0$ if $y \notin X_\kappa$, defines a function $a_\kappa \in S^m$ such that the kernel of $a_\kappa(x, D)$ has compact support in $X_\kappa \times X_\kappa$ and

$$(a_\kappa(x, D)\, u) \circ \kappa = a(x, D)(u \circ \kappa), \quad u \in \mathscr{S}'. \tag{18.1.29}$$

For a_κ we have the asymptotic expansion

$$a_\kappa(\kappa(x), \eta) \sim \sum a^{(\alpha)}(x, {}^t\kappa'(x)\eta)\, D_y^\alpha e^{i\langle \rho_x(y), \eta\rangle}/\alpha!\,|_{y=x}, \tag{18.1.30}$$

where $\rho_x(y) = \kappa(y) - \kappa(x) - \kappa'(x)(y-x)$ vanishes of second order at x. The terms in the series are in $S^{m-|\alpha|/2}$.

Proof. If we show that $a_\kappa \in S^m$ then (18.1.28) means precisely that (18.1.29) is valid when $u(x) = e^{i\langle x, \eta\rangle}$. This proves (18.1.29) since both sides are continuous from $\mathscr{S}'$ to $\mathscr{E}'(X)$ and linear combinations of exponential functions are dense in $\mathscr{S}'$. Before the proof that $a_\kappa \in S^m$ and that (18.1.30) is valid we observe that

$$\phi_\alpha(x, \eta) = D_y^\alpha e^{i\langle \rho_x(y), \eta\rangle}|_{y=x} \tag{18.1.31}$$

is a polynomial in η of degree $\leqq |\alpha|/2$ with C^∞ coefficients. In fact, a differentiation producing a factor η_j also brings out a derivative of $\rho_x(y)$ vanishing at x. If k derivatives bring out a coordinate of η each, we must get a zero term unless $|\alpha| - k \geqq k$, that is, $2k \leqq |\alpha|$, so that there are enough derivatives left to remove these zeros. This shows that the terms in (18.1.30) are indeed in $S^{m-|\alpha|/2}$ so that the asymptotic series is well defined. Note the values of the first few polynomials ϕ_α:

$$\phi_0 = 1; \quad \phi_\alpha = 0, \quad |\alpha| = 1; \tag{18.1.31)'$$

$$\phi_\alpha(x, \eta) = D_x^\alpha i\langle \kappa(x), \eta\rangle, \quad |\alpha| = 2 \text{ or } 3.$$

By Proposition 18.1.4 we shall obtain that $a_\kappa(\kappa(x), \eta) \in S^m$, hence that $a_\kappa \in S^m$, and that (18.1.30) is valid, if we just prove estimates of the form (18.1.3)″. In fact, differentiation of (18.1.28) gives a finite sum of monomials in η multiplied by a similar expression with another a, so (18.1.4) follows.

Choose $\phi \in C_0^\infty(X)$ so that $\phi(x)=\phi(y)=1$ for (x,y) in a neighborhood of the support of the kernel of $a(x,D)$. Then

(18.1.28)′ $a_\kappa(\kappa(x),\eta)=\phi(x)\,e^{-i\langle\kappa(x),\eta\rangle}\,a(x,D)(\phi(x)\,e^{i\langle\kappa(x),\eta\rangle})$,

which shows at once that $a_\kappa \in C^\infty$. To study a_κ for large η we introduce the Fourier transform

$$\Phi(\xi,\eta)=\int \phi(y)\,e^{i\langle\kappa(y),\eta\rangle-i\langle y,\xi\rangle}\,dy.$$

The differential of the phase is ${}^t\kappa'(y)\eta-\xi$. If $|\kappa'(y)|\leqq C$ and $|\kappa'(y)^{-1}|\leqq C$, $y\in\operatorname{supp}\phi$, then

$$|{}^t\kappa'(y)\eta-\xi|\geqq|\xi|/2 \quad \text{if } |\xi|/2\geqq C|\eta|,$$
$$|{}^t\kappa'(y)\eta-\xi|\geqq C^{-1}|\eta-{}^t\kappa'(y)^{-1}\xi|\geqq|\eta|/2C \quad \text{if } C|\xi|<|\eta|/2.$$

In both cases we can normalize the exponent by writing

$$\langle\kappa(y),\eta\rangle-\langle y,\xi\rangle=\omega(\langle\kappa(y),\eta/\omega\rangle-\langle y,\xi/\omega\rangle), \quad \omega=|\xi|+|\eta|$$

and obtain using Theorem 7.7.1 for any N

$$|\Phi(\xi,\eta)|\leqq C_N(1+|\xi|+|\eta|)^{-N} \quad \text{unless } |\eta|/2C<|\xi|<2C|\eta|.$$

Now choose $\chi\in C_0^\infty(\mathbb{R}^n)$ equal to 1 when $1/2C<|\xi|<2C$ and equal to 0 when $|\xi|<1/4C$. Then

$$a(x,D)(\phi(x)\,e^{i\langle\kappa(x),\eta\rangle})=I_1(x,\eta)+I_2(x,\eta),$$

where

$$I_1(x,\eta)=(2\pi)^{-n}\int e^{i\langle x,\xi\rangle}\,a(x,\xi)\,\Phi(\xi,\eta)(1-\chi(\xi/|\eta|))\,d\xi$$

decreases faster than any power of $1/(1+|\eta|)$ as $\eta\to\infty$ and

$$I_2(x,\eta)=(\omega/2\pi)^n\int e^{i\omega(\langle x-y,\xi\rangle+\langle\kappa(y),\eta/\omega\rangle)}\,a(x,\omega\xi)\,\chi(\xi)\,\phi(y)\,dy\,d\xi.$$

Here $\omega=|\eta|$. The integral I_2 is of the form studied in Theorem 7.7.7 if $y-x$ is taken as a new variable instead of y, the roles of x and y are interchanged, and another parameter η/ω occurs. Note that $\chi(\xi)=1$ near the critical point $\xi={}^t\kappa'(x)\eta/\omega$. Hence we obtain for I_2 the asymptotic expansion

$$e^{i\langle\kappa(x),\eta\rangle}\sum\langle iD_y,D_\xi/\omega\rangle^j\,e^{i\langle\rho_x(y),\eta\rangle}\,a(x,\omega\xi)/j!\,|_{y=x,\,\xi={}^t\kappa'(x)\eta/\omega},$$

if we observe that all derivatives of $a(x,\omega\xi)$ can be estimated by ω^m in the support of the integrand, and that $\phi(x)=1$ in a neighborhood of the support of $a_\kappa(\kappa(x),\eta)$. The proof is complete.

The simplest consequence of (18.1.30) is by (18.1.31)′ that

$$a_\kappa(\kappa(x),\eta)-a(x,{}^t\kappa'(x)\eta)\in S^{m-1}.$$

In particular, if a is polyhomogeneous with principal symbol a^0, that is, if a^0 is the homogeneous term of highest degree, then a_κ is polyhomogeneous with principal symbol a_κ^0 satisfying

$$a_\kappa^0(\kappa(x),\eta)=a^0(x,{}^t\kappa'(x)\eta).$$

The *principal symbol* therefore transforms as a function invariantly defined on the cotangent bundle, just as in the case of differential operators discussed in Section 6.4. For the next term a^1 resp. a_κ^1 we obtain (assuming that the step is 1)

$$a_\kappa^1(\kappa(x),\eta)=a^1(x,{}^t\kappa'(x)\eta)+\sum_{|\alpha|=2} a^{0(\alpha)}(x,{}^t\kappa'(x)\eta)\,D_x^\alpha\langle i\,\kappa(x),\eta\rangle/\alpha!.$$

A simpler transformation law is valid for the *subprincipal symbol* defined by

$$a^{1s}(x,\xi)=a^1(x,\xi)+i/2\sum a_{(j)}^{0(j)}(x,\xi). \tag{18.1.32}$$

To compute it for a_κ we use that

$$a_\kappa^0(y,\eta)=a^0(x,{}^t\kappa'(x)\eta),\; y=\kappa(x),$$

where $({}^t\kappa'(x)\eta)_l=\partial\langle y,\eta\rangle/\partial x_l$, which gives

$$\begin{aligned}\sum a_{\kappa(j)}^{0(j)}&=\sum \partial x_k/\partial y_j\,\partial/\partial x_k \sum a^{0(l)}\,\partial y_j/\partial x_l\\
&=\sum a_{(k)}^{0(l)}\,\partial x_k/\partial y_j\,\partial y_j/\partial x_l+\sum \partial x_k/\partial y_j(a^{0(l)}\,\partial^2 y_j/\partial x_l\,\partial x_k\\
&\quad+\sum a^{0(l\,p)}\,\partial^2\langle y,\eta\rangle/\partial x_k\,\partial x_p\,\partial y_j/\partial x_l).\end{aligned}$$

If $J=\det(\partial y_j/\partial x_k)$ then $J(\partial x_k/\partial y_j)$ is the cofactor matrix of the matrix $(\partial y_j/\partial x_k)$ so $J^{-1}\partial J/\partial x_l=\sum \partial^2 y_j/\partial x_k\,\partial x_l\,\partial x_k/\partial y_j$. Hence

$$a_\kappa^{1s}(\kappa(x),\eta)=a^{1s}(x,{}^t\kappa'(x)\eta)-\tfrac12\sum a^{0(l)}(x,{}^t\kappa'(x)\eta)(D_l J)/J. \tag{18.1.33}$$

This may not look much simpler at first sight, but we note that a^{1s} is at least invariantly defined at the points in the cotangent bundle where a^0 vanishes of second order. It is also an invariant under measure preserving changes of variables, and we shall see later that a slight modification of our point of view gives a complete invariant. A more conceptual motivation for the notion of subprincipal symbol will be given in Section 18.5.

Our next aim is to define pseudo-differential operators on a manifold. First we must discuss symbols. If $X\subset\mathbb{R}^n$ is open we define $S_{\mathrm{loc}}^m(X\times\mathbb{R}^n)$ as the set of all $a\in C^\infty(X\times\mathbb{R}^n)$ such that $\phi(x)a(x,\xi)\in S^m(\mathbb{R}^n\times\mathbb{R}^n)$ for every $\phi\in C_0^\infty(X)$. This means that for every compact set $K\subset X$ we can find $C_{\alpha\beta K}$ such that

$$|a_{(\beta)}^{(\alpha)}(x,\xi)|\le C_{\alpha\beta K}(1+|\xi|)^{m-|\alpha|};\quad x\in K,\ \xi\in\mathbb{R}^n. \tag{18.1.1'}$$

By means of a partition of unity we can immediately extend Proposition 18.1.3 to this local case.

More generally, let $\Gamma\subset\mathbb{R}^n\times\mathbb{R}^n$ be open and conic with respect to the second variable, that is, $(x,\xi)\in\Gamma\Rightarrow(x,t\xi)\in\Gamma$ if $t>0$. Then we define $S_{\mathrm{loc}}^m(\Gamma)$ as the set of all $a\in C^\infty(\Gamma)$ such that for every compact set $K'\subset\Gamma$ the estimate (18.1.1)′ is valid in $\{(x,t\xi);\ (x,\xi)\in K',\ t\ge 1\}$. Taking $K'=$

$K\times\{\xi;|\xi|\leq 1\}$ we find that this agrees with the earlier definition when $\Gamma=X\times\mathbb{R}^n$. We shall usually simplify notation by writing $S^m(\Gamma)$ instead of $S^m_{\mathrm{loc}}(\Gamma)$ since there can be no ambiguity unless $\Gamma=\mathbb{R}^n\times\mathbb{R}^n$. Symbols behave well under a change of variables:

Lemma 18.1.18. *Let X_1 and X_2 be open sets in $\mathbb{R}^n$ and let $\phi\colon X_1\to X_2$ and $\Phi\colon X_1\to GL(n,\mathbb{R})$ (the group of invertible $n\times n$ matrices) be C^∞ maps. If $\Gamma_1\subset X_1\times\mathbb{R}^n$ and $\Gamma_2\subset X_2\times\mathbb{R}^n$ are open and conic, $(\phi(x),\Phi(x)\xi)\in\Gamma_2$ when $(x,\xi)\in\Gamma_1$, then*

$$a_1(x,\xi)=a_2(\phi(x),\Phi(x)\xi)$$

is in $S^m(\Gamma_1)$ for every $a_2\in S^m(\Gamma_2)$.

Proof. $K_2=\{(\phi(x),\Phi(x)\xi);(x,\xi)\in K_1\}$ is a compact set in Γ_2 if K_1 is a compact set in Γ_1. Since $|\Phi(x)\xi|/|\xi|$ is bounded from above and below when $(x,\xi)\in K_1$ and

$$a_1^{(j)}=\sum a_2^{(k)}\Phi_{kj},\qquad a_{1(j)}=\sum a_{2(k)}\partial\phi_k/\partial x_j+\sum a_2^{(k)}\partial\Phi_{k\nu}/\partial x_j\xi_\nu$$

the required estimates (18.1.1)′ follow inductively.

If $X\subset\mathbb{R}^n$ is open and $a\in S^m(X\times\mathbb{R}^n)$, then an operator

$$a(x,D)\colon \mathscr{S}'(\mathbb{R}^n)\to\mathscr{D}'(X)$$

is still defined by (18.1.5), and it restricts to an operator $\mathscr{E}'(X)\to\mathscr{D}'(X)$ or $C_0^\infty(X)\to C^\infty(X)$. In fact, if $\phi\in C_0^\infty(X)$ then $\phi(x)a(x,\xi)\in S^m$ so the corresponding operator has already been discussed. It follows also that $a(x,D)$ is a continuous map $H_{(s)}\to H^{\mathrm{loc}}_{(s-m)}(X)$. However, there is some lack of symmetry between right and left multiplication which must be removed before one can take adjoints. The situation is clarified by the following

Proposition 18.1.19. *If $A\colon C_0^\infty(X)\to C^\infty(X)$ is a continuous linear map and for all $\phi,\psi\in C_0^\infty(X)$ the operator*

$$\mathscr{S}\ni u\mapsto\phi A\psi u$$

is in $\operatorname{Op}S^m$, then one can find $a\in S^m(X\times\mathbb{R}^n)$ such that

$$A=a(x,D)+A_0$$

where the kernel of A_0 is in $C^\infty(X\times X)$. Here a is uniquely determined modulo $S^{-\infty}(X\times\mathbb{R}^n)$.

Proof. Let $1=\sum\psi_j(y)$ be a locally finite partition of unity in X. Then $\psi_jA\psi_ku=a_{jk}(x,D)u$, $u\in\mathscr{S}$, where $a_{jk}\in S^m$ and $a_{jk}(x,\xi)=0$ when $x\notin\operatorname{supp}\psi_j$. Set

$$a(x,\xi)=\sum{}'a_{jk}(x,\xi)$$

where we sum over all j and k for which $\operatorname{supp}\psi_j\cap\operatorname{supp}\psi_k\neq\emptyset$. The sum is locally finite since any compact subset of X meets only finitely many

$\operatorname{supp}\psi_j$ and they meet only finitely many $\operatorname{supp}\psi_k$. Hence $a\in S^m(X\times\mathbb{R}^n)$. If K is the kernel of A then the kernel of $A-a(x,D)$ is the sum

$$\sum{}'' \psi_j(x)\,K(x,y)\,\psi_k(y)$$

taken over the indices for which $\operatorname{supp}\psi_j\cap\operatorname{supp}\psi_k=\emptyset$. It is in $C^\infty(X\times X)$ since the sum is locally finite and the terms are in C^∞ by Theorem 18.1.16. Now if $b\in S^m$ and the kernel is in $C_0^\infty(\mathbb{R}^n\times\mathbb{R}^n)$, then

$$b(x,\xi)=e^{-i\langle x,\xi\rangle}\,b(x,D)\,e^{i\langle x,\xi\rangle}$$

is rapidly decreasing when $\xi\to\infty$ so $b\in S^{-\infty}$. If $A=0$ we can apply this conclusion to $b(x,D)u=\phi\,a(x,D)\psi u$ taking $\phi=\psi=1$ near any given point in X. This proves that $a\in S^{-\infty}(X\times\mathbb{R}^n)$ which completes the proof.

We are now ready to define pseudo-differential operators on manifolds:

Definition 18.1.20. A pseudo-differential operator of order m on a C^∞ manifold X is a continuous linear map $A\colon C_0^\infty(X)\to C^\infty(X)$ such that for every local coordinate patch $X_\kappa\subset X$ with coordinates $X_\kappa\ni x\mapsto\kappa(x)=(x_1,\dots,x_n)\in\tilde X_\kappa\subset\mathbb{R}^n$ and all $\phi,\psi\in C_0^\infty(\tilde X_\kappa)$ the map

$$\mathscr{S}'(\mathbb{R}^n)\ni u\mapsto\phi(\kappa^{-1})^*A\,\kappa^*(\psi u)$$

is in $\operatorname{Op}S^m$. We shall then write $A\in\Psi^m(X)$ and extend A to a map $\mathscr{E}'(X)\to\mathscr{D}'(X)$.

If $X\subset\mathbb{R}^n$ it follows from Proposition 18.1.19 that A must be the sum of an operator $a(x,D)$ with $a\in S^m(X\times\mathbb{R}^n)$ and an operator with kernel in $C^\infty(X\times X)$. Theorem 18.1.17 shows that conversely every such operator is in $\Psi^m(X)$. Definition 18.1.20 just means that the restriction of A to each coordinate patch is of the preceding form in the local coordinates. It is of course sufficient to know this for so many coordinate patches X_κ in X that the products $X_\kappa\times X_\kappa$ form an atlas for $X\times X$. It suffices to use an atlas for X if one requires in addition that the kernel of A is smooth outside the diagonal. In particular, if $a_\kappa\in S^m$ and the kernel of $a_\kappa(x,D)$ has compact support in $\tilde X_\kappa\times\tilde X_\kappa$ then we can define $A\in\Psi^m(X)$ by

$$Au=\kappa^*\,a_\kappa(x,D)(\kappa^{-1})^*u\quad\text{in } X_\kappa,\qquad Au\in\mathscr{E}'(X_\kappa).$$

If A is polyhomogeneous it follows from Theorem 18.1.17 that a principal symbol a^0 is invariantly defined on $T^*(X)\smallsetminus 0$, where 0 denotes the 0 section. It is obtained by just pulling the principal symbol back from $T^*(\tilde X_\kappa)\smallsetminus 0$ to $T^*(X_\kappa)\smallsetminus 0$. To define the principal symbol for any $A\in\Psi^m$ we first define $S^m(T^*(X))$ as the set of all $a\in C^\infty(T^*(X))$ such that the pullback to $T^*(\tilde X_\kappa)=\tilde X_\kappa\times\mathbb{R}^n$ is in $S^m(\tilde X_\kappa\times\mathbb{R}^n)$ for every coordinate patch. By Lemma 18.1.18 it is enough to require this for an atlas, and the definition agrees with our earlier one if $X\subset\mathbb{R}^n$. If $A\in\Psi^m$ then the restriction of A to X_κ identified with $\tilde X_\kappa$ defines a symbol in $S^m(\tilde X_\kappa\times\mathbb{R}^n)/S^{-\infty}(\tilde X_\kappa\times\mathbb{R}^n)$ by Proposition 18.1.19. If $a_\kappa \in S^m(T^*(X_\kappa))$ is the pullback of a representative then $a_\kappa - a_{\kappa'} \in S^{m-1}(T^*(X_\kappa \cap X_{\kappa'}))$ by Theorem 18.1.17 for every pair of coordinate patches. With a partition of unity $\{\psi_j\}$ subordinate to a covering by coordinate patches X_{κ_j} we set

$$a = \sum \psi_j a_{\kappa_j} \in S^m(T^*(X))$$

and obtain $a - a_\kappa \in S^{m-1}(T^*(X_\kappa))$ for every κ. This determines a modulo S^{m-1} so we obtain a *principal symbol* isomorphism

$$\Psi^m(X)/\Psi^{m-1}(X) \cong S^m(T^*(X))/S^{m-1}(T^*(X)).$$

To prove surjectivity we take ψ_j now with $\sum \psi_j^2 = 1$ and set for $a \in S^m(T^*(X))$

$$A_j u = \psi_j \kappa_j^* a_j(x, D)(\kappa_j^{-1})^*(\psi_j u), \qquad u \in C^\infty(X),$$

where a_j is the pullback of a to $T^*(\tilde{X}_{\kappa_j})$. Then $A = \sum A_j$ has the principal symbol a. We also have an isomorphism between $\Psi^{-\infty}(X)$ and the operators with C^∞ kernel, that is, operators mapping $\mathscr{E}'(X)$ into $C^\infty(X)$.

To be able to compose operators freely one needs to have some information on the support of the distribution kernel, which is a distribution in $X \times X$ with values in $1 \boxtimes \Omega$, that is, a distribution density in the second variable.

Definition 18.1.21. The (pseudo-differential) operator A in X is said to be properly supported if both projections from the support of the kernel in $X \times X$ to X are proper maps, that is, for every compact set $K \subset X$ there is a compact set $K' \subset X$ such that

$$\operatorname{supp} u \subset K \Rightarrow \operatorname{supp} Au \subset K'; \; u = 0 \text{ at } K' \Rightarrow Au = 0 \text{ at } K.$$

Note that A can then be extended to a map $\mathscr{D}'(X) \to \mathscr{D}'(X)$ so that the last property is preserved.

Proposition 18.1.22. *Every $A \in \Psi^m$ can be written as a sum $A = A_1 + A_0$ where $A_1 \in \Psi^m$ is properly supported and the kernel of A_0 is in C^∞.*

Proof. As in the proof of Proposition 18.1.19 we take a partition of unity $1 = \sum \psi_j$ in X and set

$$A_1 u = \sum{}' \psi_j A(\psi_k u)$$

with the sum taken over all j and k such that $\operatorname{supp} \psi_j \cap \operatorname{supp} \psi_k \neq \emptyset$. The same proof shows that A_1 is properly supported and that $A_0 = A - A_1$ has a C^∞ kernel.

Using the preceding decomposition and Proposition 18.1.3 it is easy to show that if $A_j \in \Psi^{m_j}(X)$, $m_j \downarrow -\infty$, then one can find $A \in \Psi^{m_0}$ with

$$A - \sum_{j<k} A_j \in \Psi^{m_k} \quad \text{for every } k.$$

The details of the proof are left for the reader.

Since the asymptotic formulas in the calculus of pseudo-differential operators only allow one to recognize them modulo $\Psi^{-\infty}$, we shall usually be working with $\Psi^m/\Psi^{-\infty}$ instead of Ψ^m. By Proposition 18.1.22 it is then always possible to pick a properly supported representative for the class considered. It is therefore no essential restriction that we require proper supports in the following immediate consequence of Theorem 18.1.8.

Theorem 18.1.23. *If $A_j \in \Psi^{m_j}(X)$ are properly supported, $j=1,2$, then $A = A_1 A_2 \in \Psi^{m_1+m_2}(X)$ is properly supported and the principal symbol is the product of those of A_1 and of A_2.*

Proof. Let $\phi, \psi \in C_0^\infty(Y)$ where Y is a coordinate patch, and choose $\chi \in C_0^\infty(Y)$ equal to 1 in a neighborhood of $\operatorname{supp} \psi$. Then

$$\phi A_1 A_2 \psi = (\phi A_1 \chi)(\chi A_2 \psi) + \phi A_1 (1-\chi^2) A_2 \psi.$$

The first term on the right is in Ψ^m by Theorem 18.1.8 and the other has a C^∞ kernel.

The proof that (iii) $\Rightarrow$ (i), (ii) in Theorem 18.1.9 gives with no change an extension of Theorem 7.1.22:

Theorem 18.1.24. *If $A \in \Psi^m$ is properly supported and elliptic in the sense that the principal symbol $a \in S^m(T^*(X))/S^{m-1}(T^*(X))$ has an inverse in $S^{-m}(T^*(X))/S^{-m-1}(T^*(X))$ then one can find $B \in \Psi^{-m}$ properly supported such that*

$$BA - I \in \Psi^{-\infty}, \quad AB - I \in \Psi^{-\infty}.$$

One calls B a parametrix for A.

In Chapter XIX we shall discuss the existence theory for elliptic operators which follows from the existence of a parametrix. Here we proceed to discuss local versions of Theorem 18.1.24.

Definition 18.1.25. If $a \in S^m(T^*(X))$ is a principal symbol of $A \in \Psi^m$ then A is said to be non-characteristic at $(x_0, \xi_0) \in T^*(X) \smallsetminus 0$ if $ab - 1 \in S^{-1}$ in a conic neighborhood of (x_0, ξ_0) for some $b \in S^{-m}$. The set of characteristic points is denoted by Char A.

The definition is of course independent of the choice of a. The proof of the equivalence of (iii) and (iv) in Theorem 18.1.9 shows that in terms of local coordinates an equivalent condition is that $|a(x,\xi)| \geq c|\xi|^m$ for large $|\xi|$ in a conic neighborhood of (x_0, ξ_0). If A has a homogeneous principal symbol a, the condition is equivalent to $a(x_0, \xi_0) \neq 0$ so our present definition of Char A coincides with (8.3.4) for differential operators.

If $A \in \Psi^m$ and $k < m$ we shall say that A is in Ψ^k, or of order k, at $(x_0, \xi_0) \in T^*(X) \smallsetminus 0$ if for the complete symbol $a(x, \xi)$ of A restricted to a coordinate patch containing x_0 we have $a \in S^k$ in a conic neighborhood of (x_0, ξ_0). By Proposition 18.1.19 and Theorem 18.1.17 this condition is independent of the choice of a and of the local coordinates. The case $k = -\infty$ is of particular importance:

Proposition 18.1.26. *If $A \in \Psi^m(X)$ and Γ is a closed conic subset of $T^*(X) \setminus 0$, the following conditions are equivalent, $\mathscr{A}$ denoting the kernel of A:*

(i) *A is of order $-\infty$ in $T^*(X) \setminus 0 \setminus \Gamma$.*
(ii) *$WF'(\mathscr{A}) \subset \{(\gamma, \gamma); \gamma \in \Gamma\}$.*
(iii) *$WF(Au) \subset \Gamma \cap WF(u), u \in \mathscr{E}'(X)$.*

Proof. The statements are local so we may assume that $X \subset \mathbb{R}^n$ and that $A = a(x, D)$, where $a \in S^m$ and $a(x, \xi) = 0$, $|\xi| < 1$. If a is rapidly decreasing in a conic neighborhood V of (x_0, ξ_0) we can choose $q \in C^\infty(\mathbb{R}^n \times (\mathbb{R}^n \setminus 0))$ with support in V so that $q(x, \xi)$ is homogeneous in ξ of degree 0 and equal to 1 in a neighborhood of (x_0, ξ_0). Then the kernel of $(aq)(x, D)$ is in C^∞, and $(x_0, \xi_0, x_0, -\xi_0)$ is not in the wave front set of the kernel of $(a(1-q))(x, D)$ by Theorem 8.1.9. In view of (18.1.24) it follows that (i) $\Rightarrow$ (ii). That (ii) $\Rightarrow$ (iii) is a consequence of Theorem 8.2.13. Finally assume that (iii) is valid, and let $(x_0, \xi_0) \notin \Gamma$. Choose $q \in S^0$ with support in a closed cone Γ_1 with $\Gamma_1 \cap \Gamma = \emptyset$ and equal to 1 at ∞ in a conic neighborhood of (x_0, ξ_0). Then $WF(q(x, D)u) \subset \Gamma_1$ for every $u \in \mathscr{E}'$ since (i) $\Rightarrow$ (iii), so $WF(a(x, D)q(x, D)u) \subset \Gamma \cap \Gamma_1 = \emptyset$. Thus $a(x, D)q(x, D)$ has a C^∞ kernel K. Choose $\phi \in C_0^\infty$ equal to 1 in a neighborhood of x_0 and set $a(x, D)q(x, D)\phi u = b(x, D)u$. By (18.1.27) we have

$$b(x, \xi) = e^{-i\langle x, \xi\rangle} \int K(x, y)\phi(y) e^{i\langle y, \xi\rangle} dy$$

and the right-hand side is rapidly decreasing when $\xi \to \infty$. The same is true for the derivatives so $b \in S^{-\infty}$ in a neighborhood of x_0. But $b - a \in S^{-\infty}$ in a conic neighborhood of (x_0, ξ_0) so it follows that $a \in S^{-\infty}$ in a conic neighborhood of (x_0, ξ_0). Thus (iii) $\Rightarrow$ (i) which completes the proof.

Since $WF'(\mathscr{A})$ is contained in the diagonal of $T^*(X) \setminus 0 \times T^*(X) \setminus 0$ it is natural to identify it with a conic subset of $T^*(X) \setminus 0$. We shall write

$$WF(A) = \{\gamma \in T^*(X) \setminus 0, (\gamma, \gamma) \in WF'(\mathscr{A})\}. \tag{18.1.34}$$

By Proposition 18.1.26 $WF(A)$ is the smallest conic set such that A is of order $-\infty$ in the complement, we have

$$WF(Au) \subset WF(A) \cap WF(u), \quad u \in \mathscr{E}'(X), \tag{18.1.35}$$

and no smaller set than $WF(A)$ can be used in the right-hand side. It is clear that $WF(AB) \subset WF(A) \cap WF(B)$ if $A \in \Psi^m(X)$, $B \in \Psi^{m'}(X)$.

We can now state a microlocal version of Theorem 18.1.24:

Theorem 18.1.24'. *If $A \in \Psi^m$ is properly supported and $(x_0, \xi_0) \notin \operatorname{Char} A$ then one can find a properly supported $B \in \Psi^{-m}$ such that $(x_0, \xi_0) \notin WF(BA - I)$ and $(x_0, \xi_0) \notin WF(AB - I)$; these conditions are equivalent.*

Proof. By Definition 18.1.25 we can choose $B_1 \in \Psi^{-m}$ properly supported (like all operators in what follows) so that $AB_1 - I$ is of order -1 at (x_0, ξ_0). This means that $AB_1 = I + R_1 + R_2$ where $R_1 \in \Psi^{-1}$ and $(x_0, \xi_0) \notin WF(R_2)$. By Theorem 18.1.24 we can find $B_2 \in \Psi^0$ with $(I + R_1) B_2 - I \in \Psi^{-\infty}$. Since $(x_0, \xi_0) \notin WF(R_2 B_2)$ it follows that $(x_0, \xi_0) \notin WF(AB - I)$ if $B = B_1 B_2$. Similarly we find B' with $(x_0, \xi_0) \notin WF(B'A - I)$. Since

$$B' - B = (B'A - I) B - B'(AB - I)$$

it follows that $(x_0, \xi_0) \notin WF(B' - B)$. Hence $(x_0, \xi_0) \notin WF(BA - I)$, and the proof is complete.

Theorem 18.1.24′ allows us to give an alternative description of the wave front set of a distribution.

Theorem 18.1.27. *If $u \in \mathscr{D}'(X)$ we have for every $m \in \mathbb{R}$*

$$WF(u) = \bigcap \operatorname{Char} A \tag{18.1.36}$$

where the intersection is taken over all properly supported $A \in \Psi^m(X)$ such that $Au \in C^\infty(X)$.

Proof. Assume that $(x_0, \xi_0) \notin WF(u)$. Then we can choose $A \in \Psi^m$ with $WF(A) \cap WF(u) = \emptyset$ and $(x_0, \xi_0) \notin \operatorname{Char} A$, by just working in a coordinate patch containing x_0. This proves that $\bigcap \operatorname{Char} A \subset WF(u)$. On the other hand, assume that $A \in \Psi^m$, $Au \in C^\infty$ and that $(x_0, \xi_0) \notin \operatorname{Char} A$. We must then prove that $(x_0, \xi_0) \notin WF(u)$. Choose $B \in \Psi^{-m}$ using Theorem 18.1.24′ so that $(x_0, \xi_0) \notin WF(BA - I)$. Then

$$u = BAu + (I - BA) u$$

where $BAu \in C^\infty$ and $(x_0, \xi_0) \notin WF((I - BA) u)$ by Proposition 18.1.26. The proof is complete.

From now on we shall always fall back on (18.1.36) as our *definition* of $WF(u)$. With this definition the analogue of Theorem 8.3.1 is obvious:

Theorem 18.1.28. *If $A \in \Psi^m(X)$ is properly supported and $u \in \mathscr{D}'(X)$ then*

$$WF(u) \subset WF(Au) \cup \operatorname{Char} A. \tag{18.1.37}$$

Proof. If $(x, \xi) \notin WF(Au)$ we can find $B \in \Psi^0$ with $BAu \in C^\infty$ and $(x, \xi) \notin \operatorname{Char} B$. If $(x, \xi) \notin \operatorname{Char} A$ then $(x, \xi) \notin \operatorname{Char} BA$ so $(x, \xi) \notin WF(u)$.

Remark. If Γ is a closed cone $\subset T^*(X) \setminus 0$ we introduced in Section 8.2 the space

$$\mathscr{D}'_\Gamma(X) = \{u \in \mathscr{D}'(X),\ WF(u) \subset \Gamma\}.$$

In Definition 8.2.2 we introduced a notion of convergence for sequences in $\mathscr{D}'_\Gamma(X)$. The preceding arguments show with little change that $u_j \to u$ in

$\mathscr{D}'_\Gamma(X)$ is equivalent to $u_j \to u$ in $\mathscr{D}'(X)$ and $Au_j \to Au$ in $C^\infty(X)$ for every properly supported A with $\Gamma \cap WF(A) = \emptyset$. We leave the details for the reader.

In appendix B we have defined the space $H^{loc}_{(s)}(X)$ of distributions u on X such that $(\kappa^{-1})^*(\phi u) \in H_{(s)}(\mathbb{R}^n)$ for every local coordinate system $\kappa: X_\kappa \to \tilde{X}_\kappa \subset \mathbb{R}^n$ and every $\phi \in C_0^\infty(X_\kappa)$. The main point was the invariance of $H^{comp}_{(s)}(\tilde{X}_\kappa)$ under changes of variables proved in Theorem B.1.8. This is also an immediate consequence of Theorem 18.1.13 and Theorem 18.1.17. If $A \in \Psi^m$ we have $A\phi u \in C^\infty(X \setminus \operatorname{supp} \phi)$, and $A\phi u \in H^{loc}_{(s-m)}(X_\kappa)$ by Theorem 18.1.13, if $u \in H^{loc}_{(s)}(X)$, so we obtain, using Theorem 18.1.24 to prove the converse,

Theorem 18.1.29. *If $u \in H^{loc}_{(s)}(X)$ (resp. $H^{comp}_{(s)}(X)$) and $A \in \Psi^m$ is properly supported then $Au \in H^{loc}_{(s-m)}(X)$ (resp. $Au \in H^{comp}_{(s-m)}(X)$). The converse is true if A is elliptic.*

Thus one can define $H^{loc}_{(s)}(X)$ as the set of all $u \in \mathscr{D}'(X)$ mapped to $L^2_{loc}(X)$ by every (some elliptic) properly supported operator in $\Psi^s(X)$. The preceding discussion can be localized with the following terminology:

Definition 18.1.30. If $u \in \mathscr{D}'(X)$ then $u \in H^{loc}_{(s)}$ at $x_0 \in X$ if $u = u_1 + u_0$ with $u_1 \in H^{loc}_{(s)}(X)$ and $u_0 \in C^\infty$ in a neighborhood of x_0. If $(x_0, \xi_0) \in T^*(X) \setminus 0$ we say that $u \in H^{loc}_{(s)}$ at (x_0, ξ_0) if $u = u_1 + u_0$ with $u_1 \in H^{loc}_{(s)}(X)$ and $(x_0, \xi_0) \notin WF(u_0)$.

It is obvious that $u \in H^{loc}_{(s)}$ at x_0 if and only if $\phi u \in H^{loc}_{(s)}(X)$ for some $\phi \in C^\infty(X)$ with $\phi(x_0) \neq 0$. The condition $u \in H^{loc}_{(s)}$ at (x_0, ξ_0) can be expressed in a similar way with pseudo-differential operators replacing cutoff functions:

Theorem 18.1.31. *If $u \in \mathscr{D}'(X)$ and $A \in \Psi^m(X)$ is properly supported then*

$$u \in H^{loc}_{(s)} \text{ at } (x_0, \xi_0) \Rightarrow Au \in H^{loc}_{(s-m)} \text{ at } (x_0, \xi_0). \tag{18.1.38}$$

One can choose A so that $Au \in H^{loc}_{(s-m)}(X)$ and $(x_0, \xi_0) \notin \operatorname{Char} A$. On the other hand,

$$Au \in H^{loc}_{(s-m)} \text{ at } (x_0, \xi_0) \text{ and } (x_0, \xi_0) \notin \operatorname{Char} A \Rightarrow u \in H^{loc}_{(s)} \text{ at } (x_0, \xi_0). \tag{18.1.39}$$

If $u \in H^{loc}_{(s)}$ at (x_0, ξ_0) for every $\xi_0 \in T^*_{x_0} \setminus 0$ then $u \in H^{loc}_{(s)}$ at x_0.

Proof. If $u = u_1 + u_0$ and $u_1 \in H^{loc}_{(s)}(X)$ then $Au_1 \in H^{loc}_{(s-m)}(X)$ by Theorem 18.1.29, and if $(x_0, \xi_0) \notin WF(u_0)$ then $(x_0, \xi_0) \notin WF(Au_0)$. This proves (18.1.38) and also that $Au \in H^{loc}_{(s-m)}(X)$ if $WF(u_0) \cap WF(A) = \emptyset$. If $(x_0, \xi_0) \notin \operatorname{Char} A$ we choose $B \in \Psi^{-m}(X)$ according to Theorem 18.1.24′ and obtain if $Au \in H^{loc}_{(s-m)}$ at (x_0, ξ_0) that $BAu \in H^{loc}_{(s)}$ at (x_0, ξ_0), and $WF(u - BAu)$

$\subset WF(I-BA)$ which does not contain (x_0, ξ_0). Hence (18.1.39) follows. Assume now that $u \in H^{\mathrm{loc}}_{(s)}$ at (x_0, ξ) for every $\xi \in T^*_{x_0} \setminus 0$. Since the unit sphere in $T^*_{x_0}$ is compact and Char A is closed, we can choose finitely many $A_j \in \Psi^0(X)$ such that $A_j u \in H^{\mathrm{loc}}_{(s)}(X)$ and

$$T^*_{x_0} \cap \operatorname{Char} A_1 \cap \ldots \cap \operatorname{Char} A_J = \emptyset.$$

Let $\bar{A}_j \in \Psi^0$ have a principal symbol which is complex conjugate to that of A_j and set $A = \sum \bar{A}_j A_j$. Then $A \in \Psi^0$ and A is elliptic in a neighborhood of x_0. Using Theorem 18.1.24 we can choose $B \in \Psi^0$ such that $BA - I$ is of order $-\infty$ in a neighborhood of x_0. Hence $BAu - u \in C^\infty$ in a neighborhood of x_0, and since $BAu \in H^{\mathrm{loc}}_{(s)}(X)$ it follows that $u \in H^{\mathrm{loc}}_{(s)}$ in a neighborhood of x_0.

Remark. The discussion above is equally valid for ${}^pH_{(s)}$ for any p.

Occasionally it is useful to introduce functions in X or in $T^*(X) \setminus 0$ which measure the smoothness of u by means of $H_{(s)}$ spaces,

(18.1.40) $$s_u(x) = \sup\{s; u \in H^{\mathrm{loc}}_{(s)} \text{ at } x\}$$

(18.1.41) $$s^*_u(x, \xi) = \sup\{s; u \in H^{\mathrm{loc}}_{(s)} \text{ at } (x, \xi)\}.$$

These are obviously lower semi-continuous functions with $s_u(x) \leqq s^*_u(x, \xi)$. If $s < s^*_u(x, \xi)$ for every $\xi \in T^*_x \setminus 0$ then $u \in H^{\mathrm{loc}}_{(s)}$ at (x, ξ) for every ξ, so $u \in H^{\mathrm{loc}}_{(s)}$ at x, hence

(18.1.42) $$s_u(x) = \inf_\xi s^*_u(x, \xi).$$

By Theorem 18.1.31 we have $s^*_{Au}(x, \xi) \geqq s^*_u(x, \xi) - m$ if $A \in \Psi^m$, and there is equality if $(x, \xi) \notin \operatorname{Char} A$.

We have postponed until now the discussion of adjoints of pseudo-differential operators. The reason is that the dual objects of functions are densities, as we saw in Section 6.3. The adjoint of a pseudo-differential operator is therefore a pseudo-differential operator from densities to densities, unless a positive density is distinguished which allows identification of functions and densities. We must therefore make some comments on pseudo-differential operators between sections of vector bundles E and F over X. This will also be important in Chapters XIX and XX.

Definition 18.1.32. Let E and F be complex C^∞ vector bundles over the C^∞ manifold X. Then a pseudo-differential operator of order m from sections of E to sections of F is a continuous linear map

$$A\colon\ C_0^\infty(X, E) \to C^\infty(X, F)$$

such that for every open $Y \subset X$ where E and F are trivialized by

$$\phi_E\colon E|_Y \to Y \times \mathbb{C}^e, \quad \phi_F\colon F|_Y \to Y \times \mathbb{C}^f,$$

there is a $f\times e$ matrix of pseudo-differential operators $A_{ij}\in\Psi^m(Y)$ such that

$$(\phi_F(Au)|_Y)_i=\sum A_{ij}(\phi_E u)_j,\quad u\in C_0^\infty(Y,E).$$

We shall then write $A\in\Psi^m(X;E,F)$.

Naturally it suffices to assume that there is a covering of $X\times X$ by coordinate patches $Y\times Y$ such that A_{ij} can be expressed as an operator in $\operatorname{Op}S^m$ modulo C^∞ in the local coordinates. We leave as an exercise for the reader to verify that the principal symbol of A is well defined as an element in

$$S^m(T^*(X);\operatorname{Hom}(E,F))/S^{m-1}(T^*(X);\operatorname{Hom}(E,F))$$

where $\operatorname{Hom}(E,F)$ is the vector bundle with fiber at (x,ξ) consisting of the linear maps from E_x to F_x. (See also Section 6.4 for the case of differential operators.) We also leave for the reader to convince himself that spaces $H_{(s)}^{\text{loc}}(X,E)$ of sections of E can be defined as in the scalar case and that $A\in\Psi^m(X;E,F)$ implies that $A\colon H_{(s)}^{\text{loc}}(X,E)\to H_{(s-m)}^{\text{loc}}(X,F)$ is continuous if A is properly supported. The obvious extension of Theorem 18.1.23 to operators between sections of bundles is also left for the reader.

In Section 6.4 we defined the density bundle Ω on X: a section of Ω expressed in local coordinates $x_1,\dots,x_n$ is a function u such that the measure $u|dx|$ is independent of how they are chosen, $|dx|$ denoting the Lebesgue measure in the local coordinates. For the representation u' in the local coordinates x' we therefore have

$$u'|dx'|=u|dx|.$$

We can define the powers Ω^a of Ω for any $a\in\mathbb{C}$ by just changing the transformation law to

$$u'|dx'|^a=u|dx|^a$$

or, more formally, we take the transition functions

$$g_{\kappa\kappa'}=|\det(\kappa\circ\kappa'^{-1})'|^a\circ\kappa'\qquad\text{in } X_\kappa\cap X_{\kappa'}$$

if κ and κ' are arbitrary local coordinates with coordinate patches X_κ and $X_{\kappa'}$. We shall now work out the transformation law for the second term in the symbol of a polyhomogeneous operator acting on half densities, that is, sections of $\Omega^{\frac12}$. This means that (18.1.29) must be replaced by

$$((a_\kappa(x,D)u)\circ\kappa)|J|^{\frac12}=a(x,D)((u\circ\kappa)|J|^{\frac12})\tag{18.1.29$'$}$$

where $J=\det\kappa'(x)$. Now

$$|J|^{-\frac12}a(x,D)(v|J|^{\frac12})=b(x,D)v$$

where

$$\begin{aligned}b(x,\xi)&\sim\sum|J|^{-\frac12}a^{(\alpha)}(x,\xi)D^\alpha|J|^{\frac12}/\alpha!\\&=a(x,\xi)+\tfrac12\sum a^{(k)}(x,\xi)D_kJ/J \mod S^{m-2}.\end{aligned}$$

The sum cancels that in (18.1.33). For the subprincipal symbol we now have the simple transformation law

$$a_\kappa^{1s}(\kappa(x),\eta)=a^{1s}(x,{}^t\kappa'(x)\eta) \tag{18.1.33$'$}$$

so it is an invariantly defined function on $T^*(X)\setminus 0$. The same calculation can be applied to any symbol in S^m. If we repeat the argument in the definition of the principal symbol we obtain

Theorem 18.1.33. *If $A\in\Psi^m(X;\Omega^{\frac12},\Omega^{\frac12})$ then A has a refined principal symbol $\sigma(A)\in S^m(T^*(X))/S^{m-2}(T^*(X))$ such that if A is defined by $a(x,D)$ in a local coordinate system then*

$$\sigma(A)-a-i/2\sum a_{(j)}^{(j)}\in S^{m-2}.$$

When A is polyhomogeneous this means that the subprincipal symbol (18.1.32) *is invariantly defined on $T^*(X)\setminus 0$.*

The product of two half densities is a density so the (anti-)dual space of $C_0^\infty(X,\Omega^{\frac12})$ is $\mathscr{D}'(X,\Omega^{\frac12})$. Hence the adjoint of a continuous linear operator $C_0^\infty(X,\Omega^{\frac12})\to C^\infty(X,\Omega^{\frac12})$ is a map $\mathscr{E}'(X,\Omega^{\frac12})\to\mathscr{D}'(X,\Omega^{\frac12})$. From Theorem 18.1.7 we now obtain:

Theorem 18.1.34. *Every $A\in\Psi^m(X;\Omega^{\frac12},\Omega^{\frac12})$ has an adjoint $A^*\in\Psi^m(X;\Omega^{\frac12},\Omega^{\frac12})$, thus*

$$(Au,v)=(u,A^*v);\quad u,\ v\in C_0^\infty(X,\Omega^{\frac12}).$$

If a is a (refined) principal symbol for A, then $\bar a$ is one for A^*.

The (anti-)dual of a complex vector bundle E over X is defined so that the fibers are the (anti-)duals of those of E. Thus the transition matrices g_{ij} of E are replaced by ${}^tg_{ij}^{-1}$ for the dual and g_{ij}^{*-1} for the anti-dual E^*. We also define $E\otimes\Omega^{\frac12}$ as the vector bundle with transition matrices obtained by multiplying those of E and of $\Omega^{\frac12}$, the latter being scalars. If $u\in C^\infty(X,E\otimes\Omega^{\frac12})$ and $v\in C^\infty(X,E^*\otimes\Omega^{\frac12})$, then $(u(x),v(x))\in C^\infty(X,\Omega)$ and can be integrated over X if the support is compact. The following is an obvious extension of Theorem 18.1.34:

Theorem 18.1.34′. *If E, F are complex vector bundles, then every $A\in\Psi^m(X;E\otimes\Omega^{\frac12},F\otimes\Omega^{\frac12})$ has an adjoint $A^*\in\Psi^m(X;F^*\otimes\Omega^{\frac12},E^*\otimes\Omega^{\frac12})$,*

$$(Au,v)=(u,A^*v);\quad u\in C_0^\infty(X,E\otimes\Omega^{\frac12}),\quad v\in C_0^\infty(X,F^*\otimes\Omega^{\frac12}).$$

If a is a principal symbol for A then a^* is one for A^*.

Another advantage of always having a half density bundle factored out is seen in the form that the Schwartz kernel theorem (Theorem 5.2.1) takes for manifolds: Every continuous linear map $C_0^\infty(X,E\otimes\Omega_X^{\frac12})\to\mathscr{D}'(Y,F\otimes\Omega_Y^{\frac12})$ has a kernel $\in\mathscr{D}'(Y\times X,\mathrm{Hom}(E,F)\otimes\Omega_{Y\times X}^{\frac12})$ where $(\mathrm{Hom}(E,F))_{y,x}$ is the space of linear maps from E_x to F_y. The verification is obvious once one satisfies oneself that the converse is true. It was only to avoid being involved in such discussions that we stated Theorem 8.2.12 and the following results in Section 8.2 only for open subsets of $\mathbb{R}^n$.

The results proved in this section are valid for more general symbol spaces. In particular one can use some of the spaces $S^m_{\rho,\delta}$ obtained when (18.1.1) is replaced by

$$|a^{(\alpha)}_{(\beta)}(x,\xi)| \leqq C_{\alpha,\beta}(1+|\xi|)^{m-\rho|\alpha|+\delta|\beta|}; \quad x,\ \xi\in\mathbb{R}^n. \tag{18.1.1)''}$$

Here $0<\rho\leqq 1$ and $0\leqq\delta<1$. These spaces were already introduced in Section 7.8. The basic reason for their interest is that by Theorem 11.1.3 we have estimates of the form (18.1.1)″ for $1/P(\xi)$ if P is hypoelliptic and $|\xi|$ is large. Since $S^m_{\rho,\delta}\supset S^m_{1,0}=S^m$, Proposition 18.1.3 and 18.1.4 are valid with no change, and so is Theorem 18.1.6. The asymptotic series in (18.1.11) is only defined when $\delta<\rho$, but Theorem 18.1.7 is valid when $\delta\leqq\rho$ apart from the breakdown of (18.1.11) when $\delta=\rho$. The same is true for Theorem 18.1.8 whereas in Theorem 18.1.9 we must assume that $\delta<\rho$. Theorems 18.1.11 and 18.1.13 are valid when $\delta\leqq\rho$ but the proofs given are only applicable when $\delta<\rho$. In Theorem 18.1.14 the hypothesis should be replaced by $a\in S^{2m+\rho-\delta}_{\rho,\delta}$. Changing variables requires an additional condition, for Lemma 18.1.18 is only valid when $1-\rho\leqq\delta$, that is, $\rho\geqq 1-\delta$. When $\delta\leqq\rho$ this implies $\rho\geqq\frac{1}{2}$ with equality only when $\delta=\frac{1}{2}$ also. Now Theorem 18.1.17 remains valid when $1-\rho\leqq\delta\leqq\rho$ except for the asymptotic expansion which we only have when $\delta<\rho$. The rest of the section is really just formal and requires no change. The notation $\Psi^m_{\rho,\delta}(X;E,F)$ is used for the pseudo-differential operators based on $S^m_{\rho,\delta}$.

We shall not carry out the proofs of the preceding statements. They may be supplied by a reader wanting to consolidate his grasp of the material in this chapter. Alternatively, the classes $S^m_{\rho,\delta}$ may be regarded as very special cases of the general classes of pseudo-differential operators discussed later on in this chapter. (See the end of Section 18.4.) However, we shall prove a technically useful result concerning products of pseudo-differential operators in $\mathbb{R}^n$ and pseudo-differential operators in $x'=(x_1,\dots,x_{n-1})\in\mathbb{R}^{n-1}$ depending on the parameter x_n.

Theorem 18.1.35. *Let $a\in S^m(\mathbb{R}^n\times\mathbb{R}^n)$, $b\in S^{m'}(\mathbb{R}^n\times\mathbb{R}^{n-1})$, and assume that for some $\varepsilon>0$ we have*

$$a(x,\xi)=0 \text{ if } \varepsilon|\xi_n|>1 \text{ and } |\xi'|\leqq\varepsilon|\xi_n|. \tag{18.1.43}$$

Then $a(x,D)\,b(x,D')$ and $b(x,D')\,a(x,D)$ are in $\operatorname{Op}(S^{m+m'})$, and the asymptotic expansion of the symbols can be obtained from (18.1.15).

Note that in the estimate

$$|D^\beta_x D^\alpha_\xi b(x,\xi')| \leqq C_{\alpha\beta}(1+|\xi'|)^{m-|\alpha|}$$

we can replace $|\xi'|$ by $|\xi|$ in the set where $\varepsilon|\xi_n|\leqq|\xi'|$, for

$$1+|\xi'|\leqq 1+|\xi|\leqq(1+|\xi'|)(1+1/\varepsilon).$$

The asymptotic series (18.1.15) for the compositions are thus well defined.

Proof of Theorem 18.1.35. Choose $\chi\in C^\infty(\mathbb{R}^n)$ homogeneous of degree 0 outside a compact set so that $|\varepsilon\xi_n|>1$ and $|\xi'|<|\varepsilon\xi_n|$ if $\xi\in\operatorname{supp}\chi$, and $\chi(\xi)=1$ when $|\varepsilon\xi_n|\geqq 2$ and $2|\xi'|\leqq|\varepsilon\xi_n|$. Then

$$b_1(x,\xi)=b(x,\xi')(1-\chi(\xi))\in S^{m'}$$

and $b_1=b$ in a neighborhood of $\operatorname{supp} a$. Hence $b_1(x,D)a(x,D)\in\operatorname{Op}S^{m+m'}$ and the asymptotic expansion of the symbol is given by the usual formula for $b(x,D)a(x,D)$. By (18.1.43) we have $\chi(D)a(x,D)=c(x,D)$ where $c\in S^{-\infty}$. Now

$$\begin{aligned} b(x,D')c(x,D)e^{i\langle x,\xi\rangle}&=b(x,D')e^{i\langle x,\xi\rangle}c(x,\xi)\\ &=e^{ix_n\xi_n}b(x,D')c(x,D',\xi_n)e^{i\langle x',\xi'\rangle}\\ &=e^{i\langle x,\xi\rangle}r(x,\xi)\end{aligned}$$

where $r(x,D',\xi_n)=b(x,D')c(x,D',\xi_n)$ is a composition of pseudo-differential operators in $n-1$ variables containing x_n and ξ_n as parameters. Since $D_{x_n}^j b(x',x_n,\xi')$ is uniformly bounded in $S^{m'}(\mathbb{R}^{n-1}\times\mathbb{R}^{n-1})$ when $x_n\in\mathbb{R}$, for any j, and $(1+|\xi_n|)^N D_{x_n}^j D_{\xi_n}^k c(x',x_n,\xi',\xi_n)$ is also uniformly bounded in $S^{-N}(\mathbb{R}^{n-1}\times\mathbb{R}^{n-1})$ when $(x_n,\xi_n)\in\mathbb{R}^2$, for any j,k and N, it follows that $r\in S^{-\infty}(\mathbb{R}^n\times\mathbb{R}^n)$. We have

$$b(x,D')c(x,D)u=r(x,D)u,\quad u\in\mathscr{S}',$$

since both sides are continuous in $\mathscr{S}'$ and the equality holds for exponentials. Hence

$$b(x,D')a(x,D)=b_1(x,D)a(x,D)+r(x,D)\in\operatorname{Op}S^{m+m'}$$

and the symbol has the usual asymptotic expansion. Thus $b(x,D')^*a(x,D)^*$ belongs to $\operatorname{Op}S^{m+m'}$, which implies that $a(x,D)b(x,D')\in\operatorname{Op}S^{m+m'}$; the symbol has the usual asymptotic expansion.

The kernel of $b(x,D')$ is equal to $\delta(x_n-y_n)K(x,y')$ where K is defined by the oscillatory integral

$$K(x,y')=(2\pi)^{1-n}\int e^{i\langle x'-y',\xi'\rangle}b(x,\xi')\,d\xi'.$$

Even if b is of order $-\infty$ so that $K\in C^\infty$ we can only be sure that the wave front set is in the conormal bundle of $\{(x,y);\,x_n=y_n\}$ (Theorem 8.1.5). However, these are always the only singularities which can occur besides those for pseudo-differential operators:

Theorem 18.1.36. *If $b\in S^m(\mathbb{R}^n\times\mathbb{R}^{n-1})$ is of order $-\infty$ outside the closed cone $\Gamma\subset\mathbb{R}^n\times(\mathbb{R}^{n-1}\smallsetminus 0)$, then the wave front set of the kernel of $b(x,D')$ is* *contained in the union of* $\{(x,x,\xi,-\xi);\ (x,\xi')\in\Gamma\}$ *and* $\{(x,y,\xi,-\xi);\ x_n=y_n,\ \xi'=0\}$. *Thus we have for* $u\in\mathscr{E}'$

$$WF(b(x,D')u)\subset\{(x,\xi)\in WF(u);\ (x,\xi')\in\Gamma\} \cup\{(x,0,\xi_n);\ (y',x_n,0,\xi_n)\in WF(u) \text{ for some } y'\}. \tag{18.1.44}$$

On the other hand,

$$WF(u)\subset WF(b(x,D')u) \cup\{(x,\xi);\ \xi'=0 \text{ or } (x,\xi')\in\operatorname{Char} b\}. \tag{18.1.45}$$

Proof. By Theorems 8.1.9 and 8.1.5

$$WF(K)\subset\{(x,y,\xi,-\xi);\ x'=y',\ \xi_n=0,\ (x,\xi')\in\Gamma\},$$
$$WF(\delta(x_n-y_n))=\{(x,y,\xi,-\xi);\ x_n=y_n,\ \xi'=0\}.$$

The statement on the wave front set of the kernel of $b(x,D')$ is therefore a consequence of Theorem 8.2.10, and (18.1.44) follows from Theorem 8.2.13. To prove (18.1.45) we choose $a\in S^0(\mathbb{R}^n\times\mathbb{R}^n)$ with support in a compactly generated cone which does not intersect the right-hand side of (18.1.45). Then $a(x,D)\,b(x,D')u\in C^\infty$ by Proposition 18.1.26, and $a(x,D)\,b(x,D')$ is a pseudo-differential operator which is non-characteristic where a is. Hence $WF(u)\subset\operatorname{Char} a$, which proves (18.1.45).

18.2. Conormal Distributions

By (18.1.7) the Schwartz kernel of an operator in $\operatorname{Op} S^m$ is an oscillatory integral of the form

$$K(x,y)=(2\pi)^{-n}\int e^{i\langle x-y,\xi\rangle}a(x,\xi)\,d\xi;\qquad x,y\in\mathbb{R}^n; \tag{18.2.1}$$

where $a\in S^m(\mathbb{R}^n\times\mathbb{R}^n)$. It is singular only at the diagonal in $\mathbb{R}^n\times\mathbb{R}^n$ where the wave front set is contained in the conormal bundle (Theorem 18.1.16). In this section we shall discuss a corresponding class of distributions associated with an arbitrary C^∞ submanifold Y of a manifold X, starting with the case of a vector space and a linear subspace. To see where we should aim we introduce in (18.2.1) new variables $x'=x-y$, $x''=x$, so that the diagonal is defined by $x'=0$, and obtain the distribution

$$(2\pi)^{-n}\int e^{i\langle x',\xi\rangle}a(x'',\xi)\,d\xi. \tag{18.2.2}$$

Another choice of x'' variable, say $x''=y$ would have made a dependent on both x' and x'' though. We change notation now so that $\mathbb{R}^{2n}$ becomes $\mathbb{R}^n$ with the variables $x=(x_1,\dots,x_n)$ split into two groups $x'=(x_1,\dots,x_k)$ and $x''=(x_{k+1},\dots,x_n)$. First we show that it does not really matter if one allows a in (18.2.2) to depend on x' or not. In doing so we also note that in

Definition 18.1.1 and the following statements of the properties of symbols it is of course irrelevant that there are as many x variables as ξ variables.

Lemma 18.2.1. *If $a\in S^m(\mathbb{R}^n\times\mathbb{R}^k)$ and u is defined by the oscillatory integral*

$$u(x)=\int e^{i\langle x',\xi'\rangle}a(x,\xi')\,d\xi' \tag{18.2.3}$$

then we also have

$$u(x)=\int e^{i\langle x',\xi'\rangle}\tilde a(x'',\xi')\,d\xi' \tag{18.2.3$'$}$$

where $\tilde a\in S^m(\mathbb{R}^{n-k}\times\mathbb{R}^k)$ is defined by

$$\tilde a(x'',\xi')=e^{-i\langle D_{x'},D_{\xi'}\rangle}a(x,\xi')|_{x'=0} \tag{18.2.4}$$

and has the asymptotic expansion

$$\tilde a(x'',\xi')\sim\sum\langle -iD_{x'},D_{\xi'}\rangle^j a(x,\xi')/j!|_{x'=0}. \tag{18.2.5}$$

Proof. Assume first that $a\in\mathscr{S}$. Then $u\in\mathscr{S}$ and (18.2.3)$'$ means that $(2\pi)^k\tilde a$ is the Fourier transform of u with respect to x',

$$\begin{aligned}\tilde a(x'',\xi')&=(2\pi)^{-k}\iint e^{i\langle x',\theta-\xi'\rangle}a(x,\theta)\,d\theta\,dx'\\&=e^{-i\langle D_{x'},D_{\xi'}\rangle}a(x,\xi')|_{x'=0}.\end{aligned}$$

The last equality follows from the discussion preceding Theorem 18.1.7. Now the map

$$a\mapsto e^{-i\langle D_{x'},D_{\xi'}\rangle}a(x,\xi')$$

is continuous in $\mathscr{S}'$ and maps bounded sets in S^m to bounded sets; this follows from Theorem 18.1.7 since the presence of the parameters x'' is obviously immaterial. For a general $a\in S^m$ we can take a sequence $a_\nu\in\mathscr{S}$ which is bounded in S^m and converges to a in $\mathscr{S}'$, that is, uniformly on every compact set, and conclude that (18.2.3)$'$ is always valid with $\tilde a$ defined by (18.2.4). The asymptotic expansion also follows from Theorem 18.1.7.

We shall now determine the precise regularity properties of distributions of the form (18.2.3)$'$ with $a\in S^m$. Noting that a density on the subspace defined by $x'=0$ can be written in the form (18.2.3)$'$, with $\tilde a$ independent of ξ', we first prove an extension of Theorem 7.1.28.

Proposition 18.2.2. *Let $\phi\in C_0(\mathbb{R}^n)$ be equal to 0 in a neighborhood of 0, let $a(x'',\xi')\in S^m(\mathbb{R}^{n-k}\times\mathbb{R}^k)$ vanish for x'' outside a compact set, and let u be defined by the oscillatory integral*

$$u(x)=\int e^{i\langle x',\xi'\rangle}a(x'',\xi')\,d\xi'. \tag{18.2.6}$$

Then $\hat u\in L^2_{\mathrm{loc}}$ and

$$\int|\hat u(\xi)|^2|\phi(\xi/R)|\,d\xi\le CR^{k+2m},\quad R>1. \tag{18.2.7}$$

If $a\in S^m_{\mathrm{phg}}(\mathbb{R}^{n-k}\times\mathbb{R}^k)$ *and* a_0 *is the principal symbol,* $m'=\operatorname{Re} m$, *then*

(18.2.8)
$$\lim_{R\to\infty} R^{-k-2m'}(2\pi)^{-n-k}\int|\hat{u}(\xi)|^2\,\phi(\xi/R)\,d\xi$$
$$=\iint|a_0(x'',\xi')|^2\,\phi(\xi',0)\,dx''\,d\xi'.$$

Proof. Let $\hat{a}$ be the Fourier transform of a with respect to x''. Then $\hat{u}(\xi)=(2\pi)^k\,\hat{a}(\xi'',\xi')$, so

$$\int|\hat{u}(\xi)|^2\,\phi(\xi/R)\,d\xi=(2\pi)^{2k}\int|\hat{a}(\xi'',\xi')|^2\,\phi(\xi/R)\,d\xi$$
$$=(2\pi)^{2k}R^k\int|\hat{a}(\xi'',R\xi')|^2\,\phi(\xi',\xi''/R)\,d\xi.$$

Since $|x''|$ is bounded in the support of a we have for every N

$$|\hat{a}(\xi'',\xi')|\leqq C_N(1+|\xi''|)^{-N}(1+|\xi'|)^m.$$

We can write $\phi=\phi_1+\phi_2$ where $\phi_j\in C_0(\mathbb{R}^n)$ and $|\xi'|\geqq c$ in $\operatorname{supp}\phi_1$, $|\xi''|\geqq c$ in $\operatorname{supp}\phi_2$, for some $c>0$. When ϕ is replaced by $|\phi_2|$ we have $|\xi''|>cR$ in the support of the integrand and the integral is rapidly decreasing as $R\to\infty$. When ϕ is replaced by $|\phi_1|$ then $|\xi'|\geqq c$ in the support of the integrand, and (18.2.7) follows. If a is polyhomogeneous then

$$R^{-2m'}\int|\hat{a}(\xi'',R\xi')|^2\,\phi_1(\xi',\xi''/R)\,d\xi$$
$$=\int|\hat{a}(\xi'',R\xi')/R^m|^2\,\phi_1(\xi',\xi''/R)\,d\xi$$
$$\to\int|\hat{a}_0(\xi'',\xi')|^2\,\phi_1(\xi',0)\,d\xi$$
$$=(2\pi)^{n-k}\iint|a_0(x'',\xi')|^2\,\phi(\xi',0)\,dx''\,d\xi'$$

by dominated convergence and Parseval's formula. This proves (18.2.8).

The singularities of u lie in the plane $x'=0$:

Proposition 18.2.3. *If the hypotheses of Proposition* 18.2.2 *are fulfilled and* $\chi\in C_0^\infty(\mathbb{R}^k)$ *is equal to* 1 *in a neighborhood of* 0, *then* $(1-\chi(x'))u\in\mathscr{S}(\mathbb{R}^n)$.

Proof. If $\alpha=(\alpha_1,\dots,\alpha_k,0,\dots,0)$ then

$$x'^\alpha u(x)=\int e^{i\langle x',\xi'\rangle}(-D_{\xi'})^\alpha a(x'',\xi')\,d\xi'$$

is a bounded continuous function if $m-|\alpha|<-k$ so that the integral is absolutely convergent. If $m-|\alpha|<-k-\nu$ then the derivatives of order $\leqq\nu$ are also bounded and continuous. This proves the statement, for $(1-\chi(x'))\,|x'|^{-2N}$ has bounded derivatives of all orders for any N.

From Proposition 18.2.3 we see that the behavior of $\hat{u}$ at infinity examined in Proposition 18.2.2 depends only on the regularity properties of u when $x'=0$. Using Definition B.1.1 of the Besov spaces ${}^\infty H_{(s)}$ we can state (18.2.7) in the form

(18.2.7)′
$$u\in{}^\infty H_{(-m-k/2)}(\mathbb{R}^n)\quad\text{if } a\in S^m.$$

(Note in particular that if u is the kernel of an operator in $\operatorname{Op} S^m$ in $n/2$ variables, then $u \in {}^\infty H_{(-m-n/4)}$.) The regularity property (18.2.7)′ is preserved under suitable differentiations of u. In fact, if u is defined by (18.2.3) then

$$D_j u(x) = \int e^{i\langle x',\xi'\rangle} D_{x_j} a(x,\xi')\, d\xi'; \quad j>k;$$

$$x_i D_{x_j} u(x) = \int e^{i\langle x',\xi'\rangle} (x_i D_{x_j} - D_{\xi_i} \xi_j)\, a(x,\xi')\, d\xi'; \quad i,j \leq k;$$

so these operations preserve the form (18.2.3) and the order of the symbol a. (The order of the factors x and D is not important since $x_l D_j - D_j x_l = i\delta_{jl}$.) We shall now prove a converse:

Lemma 18.2.4. *Let $u \in \mathscr{E}'(\mathbb{R}^n)$ and assume that*

$$x^\alpha D^\beta u \in {}^\infty H_{(-m-k/2)}$$

for all α and β with $|\alpha'| \geq |\beta'|$. Then u is of the form (18.2.3)′ with an amplitude $\tilde{a} \in S^m(\mathbb{R}^{n-k} \times \mathbb{R}^k)$.

Proof. By hypothesis the Fourier transform $\hat{u}$ is in C^∞ and

$$\int_{R/2<|\xi|<2R} |\xi^\beta D^\alpha \hat{u}(\xi)|^2\, d\xi \leq C_{\alpha\beta} R^{2m+k}, \quad R \geq 1, \ |\alpha'| \geq |\beta'|,$$

for the order of x^α and D^β is irrelevant, as just pointed out. Taking $|\beta'| = 0$ and $|\beta''| > N + m + k/2$ we conclude that for any N

$$\int_{|\xi-\eta|<1} |D^\alpha \hat{u}(\eta)|^2\, d\eta \leq C_{\alpha,N} |\xi|^{-2N} \quad \text{if } |\xi''| \geq |\xi'|.$$

By Lemma 7.6.3 for example it follows that $|D^\alpha \hat{u}(\xi)| \leq C'_{\alpha,N} |\xi|^{-N}$, if $|\xi''| > |\xi'|$. To deal with the opposite case we introduce

$$U_R(\xi) = \hat{u}(R\xi', \xi'')/R^m$$

and observe that

$$\int_{E_R} |\xi^\beta D^\alpha U_R(\xi)|^2\, d\xi \leq C_{\alpha\beta}, \quad |\alpha'| = |\beta'|$$

if E_R is the ellipsoidal annulus defined by $\frac{1}{4} < |\xi'|^2 + |\xi''/R|^2 < 4$. Since the maximum of $|\xi'^{\beta'}|$ when $|\beta'| = |\alpha'|$ is bounded from below in a neighborhood of the unit sphere in $\mathbb{R}^k$, we obtain for all α and N using Lemma 7.6.3

$$|D^\alpha U_R(\xi)| \leq C'_{\alpha,N} (1 + |\xi''|)^{-N}, \quad |\xi'| = 1, \quad |\xi''| < R.$$

Returning to the original scales we have proved the estimate

$$|D^\alpha \hat{u}(\xi)| \leq C''_{\alpha,N} (1 + |\xi'|)^{m-|\alpha|} (1 + |\xi''|)^{-N}$$

for all ξ. Hence

$$\tilde{a}(x'', \xi') = (2\pi)^{-n} \int \hat{u}(\xi)\, e^{i\langle x'',\xi''\rangle}\, d\xi''$$

is in $S^m(\mathbb{R}^{n-k} \times \mathbb{R}^k)$ as claimed.

The vector fields considered in Lemma 18.2.4 are tangential to the plane defined by $x'=0$, and they generate all such vector fields:

Lemma 18.2.5. *Any C^∞ vector field in $\mathbb{R}^n$ which is tangential to the subspace defined by $x'=0$ can be written in the form*

$$\sum_{i,j\leqq k} a_{ij}(x)\,x_i\,\partial/\partial x_j + \sum_{k+1}^{n} a_j(x)\,\partial/\partial x_j,$$

where a_{ij} and a_j are in C^∞.

Proof. Let $\sum a_j(x)\,\partial/\partial x_j$ be a C^∞ vector field which is tangent to the subspace defined by $x'=0$. This means that $a_j(0,x'')=0$ when $j\leqq k$. Hence Theorem 1.1.9 (with parameters) gives that for some $a_{ij}\in C^\infty$

$$a_j(x)=\sum_{i\leqq k} a_{ij}(x)\,x_i,\quad j\leqq k,$$

which proves the lemma.

Let X be an open set in $\mathbb{R}^n$ and let $u\in\mathscr{D}'(X)$ be defined by (18.2.3) where $a\in S^m(X\times\mathbb{R}^k)$. Then it follows from Lemma 18.2.5 and the remarks preceding Lemma 18.2.4 that $L_1\dots L_N u\in{}^\infty H^{\mathrm{loc}}_{(-m-k/2)}(X)$ if $L_1,\dots,L_N$ are any number of first order differential operators tangential to the plane $x'=0$. The converse follows from Lemma 18.2.4. In fact, if $\psi_j\in C_0^\infty(X)$ have locally finite supports and $\sum\psi_j^2=1$, then Lemma 18.2.4 gives

$$\psi_j u(x)=\int e^{i\langle x',\xi'\rangle}\,\tilde a_j(x'',\xi')\,d\xi''$$

where $\tilde a_j\in S^m(\mathbb{R}^{n-k}\times\mathbb{R}^k)$, and (18.2.3) follows with $a=\sum\psi_j\tilde a_j$. Thus we are led to the following definition.

Definition 18.2.6. Let X be a C^∞ manifold, E a C^∞ (complex) vector bundle over X and Y a closed C^∞ submanifold of X. Then the space $I^m(X,Y;E)$ of distribution sections of E, conormal with respect to Y and of degree m, is defined as the set of all $u\in\mathscr{D}'(X,E)$ such that

$$L_1\dots L_N u\in{}^\infty H^{\mathrm{loc}}_{(-m-n/4)}(X,E),\qquad n=\dim X,\tag{18.2.9}$$

for all N and all first order differential operators L_j between distribution sections of E with C^∞ coefficients tangential to Y. The topology is the weakest one which makes the maps $u\mapsto L_1\dots L_N u\in{}^\infty H^{\mathrm{loc}}_{(-m-n/4)}$ continuous.

Recall that the principal symbol of L_j is a linear function on the fibers $T_x^*(X)$ with values in the linear transformations in E_x. The condition in Definition 18.2.6 is that it vanishes on the conormals of Y. The normalization in (18.2.9) has been chosen so that the kernel of a pseudo-differential operator in $\Psi^m(X;\Omega^{\frac12}\otimes E,\Omega^{\frac12}\otimes F)$ is in $I^m(X\times X,\Delta;\Omega^{\frac12}\otimes\mathrm{Hom}(E,F))$, where Δ is the diagonal in $X\times X$.

It is sufficient to take the operators L_j in (18.2.9) among a system of generators, that is, a set $M_1, \dots, M_\nu$ of first order differential operators tangential to Y such that every tangential operator is of the form

$$L = \sum_1^\nu a_j M_j + a_0 \tag{18.2.10}$$

where $a_j \in C^\infty(X, \operatorname{Hom}(E, E))$. In fact, we can replace L_N by such a sum in (18.2.9). This gives a sum with one factor L_j less if we replace L_{N-1} by the product $L_{N-1} a_j$. Continuing in this way we obtain at last a sum of terms of the form $a M_{i_1} \dots M_{i_N} u$ with $a \in C^\infty(X, \operatorname{Hom}(E, E))$. In particular, we can always take all M_j with principal symbols proportional to the identity.

Theorem 18.2.7. *If $u \in I^m(X, Y; E)$ and $A \in \Psi^{m'}(X; E, F)$ is properly supported, then $Au \in I^{m+m'}(X, Y; F)$.*

Proof. Let L_j be first order differential operators between sections of F which are tangential to Y and have principal symbols proportional to the identity. We have to show that

$$L_1 \dots L_N A u \in {}^\infty H^{\mathrm{loc}}_{(-m-m'-n/4)}(X, F).$$

If $N=0$ this follows from the continuity of A from ${}^\infty H^{\mathrm{loc}}_{(-m-n/4)}$ to ${}^\infty H^{\mathrm{loc}}_{(-m-m'-n/4)}$. If $N>0$ we choose a first order operator L'_N on sections of E whose principal symbol is the same multiple of the identity as that of L_N. Then

$$L_N A = A L'_N + A_0$$

where $A_0 \in \Psi^{m'}(X; E, F)$. Since $L'_N u \in I^m(X, Y; E)$ the proof is reduced to a smaller value of N and therefore it follows by induction.

We can now show that $I^m(X, Y; E)$ does in fact consist of the distributions which are locally of the form we first set out to study.

Theorem 18.2.8. *$u \in I^m(X, Y; E)$ if and only if $\psi_j u \in I^m(X, Y; E)$ for every ψ_j in a partition of unity on X. If X is an open set in $\mathbb{R}^n$ and Y is defined by $x' = (x_1, \dots, x_k) = 0$ while $E = X \times \mathbb{C}^N$, then any $u \in I^m(X, Y; E)$ with compact support is of the form*

$$u(x) = \int e^{i\langle x', \xi' \rangle} a(x'', \xi') \, d\xi' \tag{18.2.11}$$

where $a \in S^{m+(n-2k)/4}(\mathbb{R}^{n-k} \times \mathbb{R}^k; \mathbb{C}^N)$. Conversely, every u of this form is in $I^m(X, Y; E)$.

Proof. The first statement is an immediate consequence of Theorem 18.2.7. The second one follows from Lemma 18.2.4, and the final statement is a consequence of Proposition 18.2.2 and Lemma 18.2.5. The proof is complete.

Next we shall introduce a principal symbol for elements in $I^m(X,Y;E)$. In doing so we start with the simplest case where $E=\Omega^{\frac{1}{2}}$, the half density bundle on X, and we take $X\subset\mathbb{R}^n$. If u has compact support and is of the form (18.2.11) with $a\in S^{m+(n-2k)/4}_{\mathrm{phg}}(\mathbb{R}^{n-k}\times\mathbb{R}^k)$ and principal symbol a_0, then (18.2.8) gives if $m'=\operatorname{Re} m$

(18.2.8)′
$$\lim_{R\to\infty} R^{-2m'-n/2}(2\pi)^{-n}\int|\hat{u}(\xi)|^2\,\phi(\xi/R)\,d\xi$$
$$=(2\pi)^k\int|a_0(x'',\xi')|^2\,\phi(\xi',0)\,dx''\,d\xi'.$$

Since u is a half density in $\mathbb{R}^n$ it follows from Parseval's formula that $\hat{u}$ transforms as a half density under linear changes of variables. The left-hand side is therefore invariant under such changes of variables. Now the normal bundle of the plane $x'=0$ is parametrized by

(18.2.12)
$$(x'',\xi')\mapsto(0,x'',\xi',0)$$

so it is natural to expect $a_0(x'',\xi')$ to define invariantly a half density there, making the right-hand side of (18.2.8)′ also invariant. Since the codimension k does not occur in the left-hand side we want to make it disappear in the right-hand side by including a factor $(2\pi)^{k/2}$ in the principal symbol, or rather inserting a factor $(2\pi)^{-k/2}$ in (18.2.11). If u is the kernel of a pseudo-differential operator of order m in $n/2$ variables this does not quite give the customary factor $(2\pi)^{-n}$ in (18.1.7) so we take an additional factor $(2\pi)^{-n/4}$ to get agreement. Thus we change (18.2.11) to

(18.2.11)′
$$u(x)=(2\pi)^{-(n+2k)/4}\int e^{i\langle x',\xi'\rangle}\,a(x'',\xi')\,d\xi',$$
$$a\in S^{m+(n-2k)/4}.$$

We want to show that the half density

$$a(x'',\xi')\,|dx''|^{\frac{1}{2}}\,|d\xi'|^{\frac{1}{2}}$$

which this defines on the normal bundle of the plane $x'=0$ with the parametrization (18.2.12) is invariantly defined modulo symbols one degree lower.

Theorem 18.2.9. *Let X and X_κ be open subsets of $\mathbb{R}^n$ and let $\kappa\colon X\to X_\kappa$ be a diffeomorphism preserving the plane $Y=\{x;x'=0\}$. Let $u_\kappa\in\mathscr{E}'(X_\kappa)$ and let $u=|\det\kappa'|^{\frac{1}{2}}\,\kappa^* u_\kappa\in\mathscr{E}'(X)$ be the pullback to X as a half density. If $u_\kappa\in I^m(X_\kappa,Y)$,*

(18.2.11)″
$$u_\kappa(x)=(2\pi)^{-(n+2k)/4}\int e^{i\langle x',\xi'\rangle}\,a_\kappa(x'',\xi')\,d\xi',$$

then $u\in I^m(X,Y)$ is of the form (18.2.11)′ *with*

(18.2.13)
$$a(x'',\xi')-a_\kappa(\kappa_2(0,x''),{}^t\kappa'_{11}(0,x'')^{-1}\xi')|\det\kappa'_{11}(0,x'')|^{-\frac{1}{2}}$$
$$\cdot|\det\kappa'_{22}(0,x'')|^{\frac{1}{2}}\in S^{m+(n-2k)/4-1}.$$

Here $\kappa=(\kappa_1,\kappa_2)$ *and*

$$\kappa'=\begin{pmatrix}\kappa'_{11} & \kappa'_{12}\\ \kappa'_{21} & \kappa'_{22}\end{pmatrix}$$

are the splittings of κ *and of* κ' *corresponding to the splitting of the variables; we have* $\kappa'_{12}(0,x'')=0$ *since* $\kappa_1(0,x'')=0$.

Proof. Since $\kappa_1(0,x'')=0$ it follows from Theorem 1.1.9 that we can choose a C^∞ function ψ with $k\times k$ matrix values so that

$$\kappa(x',x'')=(\psi(x)x',\kappa_2(x)).$$

Then $\psi(0,x'')=\kappa'_{11}(0,x'')$ is non-singular. Since u_κ and u are in C^∞ when $x'\neq 0$ we may shrink X so that this is true in all of X. Now we have

$$\begin{aligned}u(x)&=|\det\kappa'(x)|^{\frac12}(2\pi)^{-(n+2k)/4}\int e^{i\langle\psi(x)x',\eta'\rangle}a_\kappa(\kappa_2(x),\eta')\,d\eta'\\ &=(2\pi)^{-(n+2k)/4}\int e^{i\langle x',\xi'\rangle}a_\kappa(\kappa_2(x),{}^t\psi(x)^{-1}\xi')|\det\kappa'(x)|^{\frac12}/|\det\psi(x)|\,d\xi'.\end{aligned}$$

Here we have put $\eta'={}^t\psi(x)^{-1}\xi'$ in the oscillatory integral. It follows from Lemmas 18.2.1 and 18.1.18 that u is of the form (18.2.11)′ with

$$a(x'',\xi')=e^{-i\langle D_{x'},D_{\xi'}\rangle}a_\kappa(\kappa_2(x),{}^t\psi(x)^{-1}\xi')|\det\kappa'(x)|^{\frac12}/|\det\psi(x)||_{x'=0}. \tag{18.2.14}$$

Modulo $S^{m+(n-2k)/4-1}$ this is given by the first term in (18.2.5), that is,

$$a_\kappa(\kappa_2(0,x''),{}^t\psi(0,x'')^{-1}\xi')|\det\kappa'(0,x'')|^{\frac12}|\det\psi(0,x'')|^{-1}.$$

Since $|\det\kappa'(0,x'')|=|\det\psi(0,x'')||\det\kappa'_{22}(0,x'')|$ because $\kappa'(0,x'')$ is triangular, we obtain (18.2.13).

The half density $a(x'',\xi')|dx''|^{\frac12}|d\xi'|^{\frac12}$ on the conormal bundle $N(Y)$ of Y should be considered to have order $m+n/4$ when $a\in S^{m+n/4-k/2}$. In fact, a *function* f on $N(Y)$ is homogeneous of degree μ if $M_t^*f=t^\mu f$, $t>0$, where $M_t(x'',\xi')=(x'',t\xi')$ denotes multiplication by t in the fibers. If we define homogeneity of a half density f by the same condition then the half density $|dx''|^{\frac12}|d\xi'|^{\frac12}$ corresponding to $a=1$ is homogeneous of degree $k/2$ where k is the number of ξ' variables.

Definition 18.2.10. Let V be a real C^∞ vector bundle of fiber dimension k over a C^∞ manifold Y. Then the space $S^\mu(V,\Omega^{\frac12})$ of half density valued symbols on V of order μ is the space of half densities which in a local coordinate patch κ: $Y_\kappa\to\tilde Y_\kappa\subset\mathbb{R}^d$ with local coordinates $y\in\mathbb{R}^d$ take the form $a(y,\eta)|dy|^{\frac12}|d\eta|^{\frac12}$ with $a\in S^{\mu-k/2}(\tilde Y_\kappa\times\mathbb{R}^k)$ if V is identified with $\tilde Y_\kappa\times\mathbb{R}^k$.

With this terminology Theorem 18.2.9 means that the correspondence $a\mapsto u$ given by (18.2.11)′ gives rise to an isomorphism

$$S^{m+n/4}(N(Y); \Omega^{\frac{1}{2}}_{N(Y)})/S^{m+n/4-1}(N(Y); \Omega^{\frac{1}{2}}_{N(Y)})$$
$$\to I^m(X, Y; \Omega^{\frac{1}{2}}_X)/I^{m-1}(X, Y; \Omega^{\frac{1}{2}}_X).$$

There is no difficulty in extending this result to distributions with values in other vector bundles than $\Omega^{\frac{1}{2}}_X$ because multiplication of u by a transition matrix will just affect the symbol in the same way. Thus we have:

Theorem 18.2.11. *Let X be a C^∞ manifold, Y a closed C^∞ submanifold and E a C^∞ complex vector bundle over X. Then there is an isomorphism*

$$S^{m+n/4}(N(Y), \Omega^{\frac{1}{2}}_{N(Y)} \otimes \hat{E})/S^{m+n/4-1}(N(Y), \Omega^{\frac{1}{2}}_{N(Y)} \otimes \hat{E})$$
$$\to I^m(X, Y; \Omega^{\frac{1}{2}}_X \otimes E)/I^{m-1}(X, Y; \Omega^{\frac{1}{2}}_X \otimes E)$$

defined locally by (18.2.11)′. *Here $\hat{E}$ is the lifting of the bundle E to $N(Y)$ (the fiber of $\hat{E}$ at (y, η) is equal to E_y). The image under the inverse map is called the principal symbol.*

Note that we have half density bundles on different spaces here. This is why it is convenient to factor out a half density bundle from the beginning. Also note that the codimension of Y has now disappeared. The presence of the term $n/4$ in the degree of the symbol is caused by our insistence on agreement with the degree of pseudo-differential operators. In that case we did not have a half density as symbol. However, the normal bundle of the diagonal in $X \times X$ is isomorphic to the cotangent bundle in X so it has a natural density defined invariantly by the symplectic form and given in local coordinates by $|dx||d\xi|$. Hence there is a natural half density $|dx|^{\frac{1}{2}}|d\xi|^{\frac{1}{2}}$ of order $(\dim X)/2 = \dim(X \times X)/4$. When the function defined as principal symbol in Section 18.1 is multiplied by this half density the order is raised as in Theorem 18.2.11. In general there is no natural half density in $N(Y)$ permitting us to identify half densities with functions.

By Theorem 18.2.7 we know how pseudo-differential operators act on $I^m(X, Y; E)$. We shall now determine this operation more explicitly on the symbols. To do so it is of course sufficient to work locally, with a trivial bundle.

Thus we assume that $u \in \mathscr{E}'(\mathbb{R}^n)$ is of the form (18.2.11)′. Let p be in $S^{m'}(\mathbb{R}^n \times \mathbb{R}^n)$. To compute $p(x, D)u$ we first observe that Fourier's inversion formula gives

$$\hat{u}(\xi) = (2\pi)^{k-(n+2k)/4} \int a(y'', \xi') e^{-i\langle y'', \xi''\rangle} dy''.$$

Hence we obtain, taking p and a in $\mathscr{S}$ at first,

$$p(x, D)u(x) = (2\pi)^{k-n-(n+2k)/4} \iint e^{i\langle x, \xi\rangle} p(x, \xi) a(y'', \xi') e^{-i\langle y'', \xi''\rangle} d\xi\, dy''$$
$$= (2\pi)^{-(n+2k)/4} \int e^{i\langle x', \xi'\rangle} a_1(x, \xi') d\xi'$$

where

$$a_1(x,\xi') = (2\pi)^{k-n} \iint e^{i\langle x''-y'',\xi''\rangle} p(x,\xi)\,a(y'',\xi')\,d\xi''\,dy''$$
$$= e^{i\langle D_{y''},D_{\xi''}\rangle} p(x,\xi)\,a(y'',\xi')|_{y''=x'',\xi''=0}$$

by the calculations preceding Theorem 18.1.7. The proof of Theorem 18.1.7 gives with no essential change that $a_1 \in S^{m+m'+n/4-k/2}$ and that

$$a_1(x,\xi') \sim \sum \langle iD_{y''},D_{\xi''}\rangle^j p(x,\xi)\,a(y'',\xi')/j!|_{y''=x'',\xi''=0}.$$

Hence it follows from Lemma 18.2.1 that

$$p(x,D)\,u(x) = (2\pi)^{-(n+2k)/4} \int e^{i\langle x',\xi'\rangle} b(x'',\xi')\,d\xi', \tag{18.2.15}$$

$$b(x'',\xi') = e^{i(\langle D_{y''},D_{\xi''}\rangle - \langle D_{x'},D_{\xi'}\rangle)} p(x,\xi)\,a(y'',\xi')|_{y''=x'',x'=\xi''=0} \tag{18.2.16}$$
$$\sim \sum (\langle iD_{y''},D_{\xi''}\rangle - \langle iD_{x'},D_{\xi'}\rangle)^j p(x,\xi)\,a(y'',\xi')/j!|_{y''=x'',x'=\xi''=0}.$$

We have now proved

Theorem 18.2.12. *If $u \in I^m(X,Y;E)$ and $P \in \Psi^{m'}(X;E,F)$ is properly supported, then $Pu \in I^{m+m'}(X,Y;F)$ and the principal symbol is that of u multiplied by the restriction to $N(Y)$ of the principal symbol of P. If u has compact support in a local coordinate patch where Y is defined by $x'=0$ and u is given by (18.2.11)', then the complete symbol of Pu is given by (18.2.15), (18.2.16) where p is the complete symbol of P.*

So far we have made no comments on the polyhomogeneous case. However, it is perfectly clear from the formulas (18.2.14) and (18.2.16) that everything said is applicable then since the step is the reciprocal of an integer. We shall need this remark in what follows.

In the theory of boundary problems one encounters the following situation. X is a C^∞ manifold, Y is the closure of an open subset of X with C^∞ boundary ∂Y. Let E, F be vector bundles on X and P a properly supported operator in $\Psi^m_{\text{phg}}(X;E,F)$. We want P to induce a map $C^\infty(Y,E) \to C^\infty(Y,F)$ which is not always the case.

Definition 18.2.13. P is said to satisfy the transmission condition with respect to Y if for every $u \in C^\infty(Y,E)$ the restriction of Pu_0 to the interior Y° of Y is in $\bar{C}^\infty(Y^\circ,E)$, that is, has a C^∞ extension to X. Here $u_0 = u$ in Y and $u_0 = 0$ in $X \setminus Y$.

We want to determine the conditions on the symbol required for P to satisfy the transmission condition. The question is obviously local so we consider the case where $X \subset \mathbb{R}^n$, Y is defined by $x_1 \geq 0$, and E, F are trivial bundles. We assume also that the support of u is compact in Y. Then

$$u_0(x) = (2\pi)^{-1} \int e^{ix_1\xi_1} a(x'',\xi_1)\,d\xi_1 \tag{18.2.17}$$

where $x''=(x_2,\dots,x_n)$ and

$$a(x'',\xi_1)=\int_0^\infty u(x)e^{-ix_1\xi_1}\,dx_1 \sim -i\sum_0^\infty \xi_1^{-1-k} D_1^k u(0,x'') \tag{18.2.18}$$

is in $S^{-1}_{\text{phg}}(\mathbb{R}^{n-1}\times\mathbb{R}^1)$. The asymptotic expansion follows by repeated partial integrations of course. Thus $u_0\in I^{-(n+2)/4}_{\text{phg}}(X,\partial Y)$. In the symbol a the terms of odd (even) order are odd (even), and by appropriate choice of u we can get modulo $S^{-\infty}$ any symbol $a\in S^{-1}_{\text{phg}}(\mathbb{R}^{n-1}\times\mathbb{R}^1)$ with this property and compact support in x''. Now assume that $P\in\Psi^m_{\text{phg}}$ where m is an integer. Then Pu is of the form (18.2.17) with a new amplitude $b\in S^{m-1}_{\text{phg}}$ given by (18.2.16). Without affecting the asymptotic expansion of the symbol we can multiply by a function in $C_0^\infty(X)$ which is 1 in a neighborhood of $\operatorname{supp} u$ to make the support compact. We shall now prove a lemma which is closely related to the Paley-Wiener-Schwartz theorem (cf. Theorem 7.4.3).

Lemma 18.2.14. *If $v\in\mathscr{E}'(X)$, $X\subset\mathbb{R}^n$, and*

$$v(x)=\int e^{ix_1\xi_1} b(x'',\xi_1)\,d\xi_1$$

with $b\in S^\mu_{\text{phg}}(\mathbb{R}^{n-1}\times\mathbb{R}^1)$ for some $\mu\in\mathbb{C}$, that is,

$$b\sim\sum_0^\infty b_j(x'',\xi_1)$$

where b_j is homogeneous of degree $\mu-j$, then $v|_{x_1>0}$ has a C^∞ extension to the closed half space $x_1\geqq 0$ if and only if for every j

$$b_j(x'',-1)=b_j(x'',1)\,e^{\pi i(\mu-j)}. \tag{18.2.19}$$

This means that b_j is the restriction to $\mathbb{R}\setminus 0$ of $b_j(x'',1)\zeta_1^{\mu-j}$ with $\zeta_1^{\mu-j}$ equal to 1 at 1 and analytic in the upper half plane.

Proof. a) Sufficiency. Let Γ be the curve in $\mathbb{C}$ consisting of the real axis with $(-1,1)$ replaced by the half unit circle in the upper half plane. Then

$$v-\sum_{j<N}\int_\Gamma e^{ix_1\zeta_1} b_j(x'',1)\,\zeta_1^{\mu-j}\,d\zeta_1\in C^\nu$$

if $N>\operatorname{Re}\mu+\nu+1$. Writing the terms in the sum as

$$D_1^k\int_\Gamma e^{ix_1\zeta_1} b_j(x'',1)\,\zeta_1^{\mu-j-k}\,d\zeta_1$$

with $\operatorname{Re}\mu-j-k<-1$ and using Cauchy's integral formula we find that they vanish when $x_1>0$. Hence all derivatives of v are bounded when $x_1>0$.

b) Necessity. By Borel's theorem (Theorem 1.2.6) we can find $w\in C_0^\infty(X)$ equal to v when $x_1>0$. Then

$$b_\phi(\zeta_1)=(2\pi)^{-1}\langle v-w, e^{-ix_1\zeta_1}\phi(x'')\rangle,\qquad \phi\in C_0^\infty(\{x\in X;\, x_1=0\}),$$

is an entire function and

$$|b_\phi(\zeta_1)| \leqq C(1+|\zeta|)^M, \quad \operatorname{Im}\zeta_1 \geqq 0,$$

by the Paley-Wiener-Schwartz theorem (Theorem 7.3.1). On the real axis we have $b_\phi(t\xi_1)t^{-\mu} \to b_\phi^0(\xi_1)$, $t \to +\infty$, where

$$b_\phi^0(\xi_1) = \int b_0(x'', \xi_1)\phi(x'')\,dx''.$$

Hence

$$b_\phi(\zeta_1)\zeta_1^{-\mu}(i+\varepsilon\zeta_1)^{-M-|\mu|-1}$$

tends to 0 at ∞ in the upper half plane and has a bound on Γ which is independent of ε. By the maximum principle it follows that there is also a fixed bound above Γ. Thus

$$|b_\phi(\zeta_1)| \leqq C|\zeta_1|^{\operatorname{Re}\mu}, \quad |\zeta_1|>1, \quad \operatorname{Im}\zeta_1 \geqq 0.$$

Now we can choose $t_\nu \to +\infty$ so that

$$B(\zeta_1) = \lim b_\phi(t_\nu\zeta_1)t_\nu^{-\mu}$$

exists and is analytic in the upper half plane. The boundary values on the real axis are $b_\phi^0(\xi_1)$ in view of say (3.1.13) with $N=0$. Hence $B(\zeta_1) = b_\phi^0(1)\zeta_1^\mu$, $0 \leqq \arg\zeta_1 \leqq \pi$, which proves that b_0 satisfies (18.2.19). By part a) of the proof we can now subtract a distribution corresponding to b_0 and vanishing when $x_1 > 0$ and then conclude that b_1 satisfies (18.2.19) and so on. The proof is complete.

Suppose now that $P \in \Psi_{\mathrm{phg}}^m(X)$ and that the symbol has the expansion

$$\sum_0^\infty p_j$$

where p_j is homogeneous of degree $m-j$. Then Pu_0 is of the form (18.2.15) with b given by (18.2.16) and a defined by (18.2.18) If P satisfies the transmission condition with respect to the half space $x_1 \geqq 0$ it follows from Lemma 18.2.14 that the principal part $p_0(0,x'',\xi_1,0)u(0,x'')/i\xi_1$ must satisfy (18.2.19), so

$$p_0(0,x'',-1,0) = e^{\pi i m} p_0(0,x'',1,0).$$

Note that this condition would still have followed if we had weakened the transmission condition by demanding in Definition 18.2.13 that u vanishes of some fixed order on ∂Y. This weakened transmission condition remains valid if P is multiplied to the right or to the left by any differential operator, for $D_{x_1}^k u_0$ will not contain any terms supported by the plane $x_1=0$ if u vanishes at least of order k. In particular, the commutators of P with D_j and x_j any number of times satisfy this weakened transmission condition, so it follows in view of (18.1.6) that for arbitrary α, β we have

$$p_{0(\beta)}^{(\alpha)}(0,x'',-1,0) = e^{\pi i(m-|\alpha|)} p_{0(\beta)}^{(\alpha)}(0,x'',1,0). \tag{18.2.20}$$

Conversely, (18.2.20) guarantees in view of (18.2.16) and Lemma 18.2.14 that the operator with symbol p_0 satisfies the transmission condition. Subtracting it from P we conclude that p_1 has the same property and so on. Hence

$$p_{j(\beta)}^{(\alpha)}(0,x'',-1,0)=e^{\pi i(m-|\alpha|-j)}\, p_{j(\beta)}^{(\alpha)}(0,x'',1,0) \tag{18.2.20$'$}$$

is a necessary and sufficient condition for P to satisfy the transmission condition.

To put (18.2.20)′ in an invariant form we introduce for $p\in S^m_{\mathrm{phg}}(X\times\mathbb{R}^n)$,

$$p\sim\sum_0^\infty p_j,$$

where p_j is homogeneous of degree $m-j$, a new symbol $\check{p}$ such that

$$\check{p}(x,\xi)\sim\sum_0^\infty e^{-\pi i(m-j)}\, p_j(x,-\xi). \tag{18.2.21}$$

Clearly $\check{\check{p}}-e^{-2\pi i m}p\in S^{-\infty}$ so $p\mapsto e^{\pi i m}\check{p}$ defines an involution of $S^m_{\mathrm{phg}}(X\times\mathbb{R}^n)/S^{-\infty}(X\times\mathbb{R}^n)$. It gives rise to an involution of $\Psi^m_{\mathrm{phg}}(X)/\Psi^{-\infty}(X)$ for any C^∞ manifold. Indeed, since $\check{q}=q$ if q is any polynomial, we obtain with the notation of Theorem 18.1.17

$$(p_\kappa)\check{}\,(\kappa(x),\eta)\sim\sum p_j^{(\alpha)}(x,-{}^t\kappa'(x)\eta)\,e^{-\pi i(m-j-|\alpha|)}\,\phi_\alpha(x,\eta)/\alpha!$$

which is equal to the expansion of $(\check{p})_\kappa(\kappa(x),\eta)$. A similar calculation using (18.1.15) shows that we have an involution of the *algebra* $\Psi^0_{\mathrm{phg}}(X)/\Psi^{-\infty}(X)$ and that it commutes with taking adjoints as well. No change is required if bundles are present. We have now proved

Theorem 18.2.15. *A properly supported pseudo-differential operator $P\in\Psi^m_{\mathrm{phg}}(X;E,F)$ satisfies the transmission condition with respect to the closure Y of an open subset with C^∞ boundary if and only if*

(18.2.22) *the symbol of $\check{P}-P$ vanishes of infinite order on the interior conormal bundle of ∂Y.*

The vanishing condition must of course be worked out in local coordinates but it is independent of how they are chosen. Since $\check{\check{P}}=e^{-2\pi i m}P$ we obtain from (18.2.22) the equivalent condition

(18.2.22)′ the symbol of $P-e^{2\pi i m}\check{P}$ vanishes of infinite order on the exterior conormal bundle of ∂Y.

The orientation of ∂Y as a boundary is therefore irrelevant when m is an integer.

The proof of the sufficiency of (18.2.20)′ shows that Pu admits a C^∞ extension from Y° to Y also if u is a smooth simple (or multiple) layer on ∂Y. We shall now study the boundary values of Pu in that case. Again we may work locally so we assume that $X=\mathbb{R}^n$, that Y is defined by $x_1\geqq 0$ and that $P=p(x,D)$ where $p\sim\sum p_j$ for some p_j homogeneous of degree $m-j$ satisfying (18.2.20)′. Let $u=\delta(x_1)\otimes v(x'')$ where $v\in C_0^\infty(\mathbb{R}^{n-1})$. To compute Pu we take $\phi\in C_0^\infty((-1,1))$ with $\int\phi(t)\,dt=1$ and note that $Pu=\lim Pu_\varepsilon$ if $u_\varepsilon(x)=\phi(x_1/\varepsilon)\,v(x'')/\varepsilon$. Now

$$Pu_\varepsilon(x)=(2\pi)^{-n}\int e^{i\langle x,\xi\rangle}p(x,\xi)\,\hat\phi(\varepsilon\xi_1)\,\hat v(\xi'')\,d\xi,$$

and we shall prove that the integral with respect to ξ_1 has a limit when $\varepsilon\to 0$.

Lemma 18.2.16. *Let $q(t)$, $t\in\mathbb{R}$, be a continuous function and assume that there is an analytic function $Q(t)$ in $\Omega_R=\{t\in\mathbb{C};\ \operatorname{Im}t\geqq 0,\ |t|\geqq R\}$ for some R, such that $Q(t)=O(t^N)$ for some N when $t\to\infty$ in Ω_R, and $q(t)-Q(t)=O(t^{-2})$ when $t\to\infty$ on $\mathbb{R}$. Then*

$$\int^+ q(t)\,dt=\int_{|t|>R}(q(t)-Q(t))\,dt+\int_{-R}^{R}q(t)\,dt-\int_\pi^0 Q(Re^{i\theta})\,Rie^{i\theta}\,d\theta \tag{18.2.23}$$

is independent of the choice of Q. If $F(t,s)$ is an analytic function of t when $\operatorname{Im}t\geqq 0$, for $0\leqq s\leqq 1$, and F is a bounded continuous function of (s,t) then $\int^+ q(t)\,F(t,s)\,dt$ is a continuous function of s.

Proof. If $q=0$ it follows from the maximum principle that

$$\sup_{\Omega_R}|t^2Q(t)(1-i\varepsilon t)^{-N-3}|\leqq\sup_{\partial\Omega_R}|t^2Q(t)|,$$

for $|1-i\varepsilon t|\geqq 1+\varepsilon\operatorname{Im}t\geqq 1$ in $\partial\Omega_R$. Letting $\varepsilon\to 0$ we conclude that $t^2Q(t)$ is bounded in Ω_R, so it follows from Cauchy's integral formula that

$$\int_{\partial\Omega_R}Q\,dt=0.$$

This proves that (18.2.23) gives a unique definition of $\int^+ q(t)\,dt$. The last statement follows at once by dominated convergence after $Q(t)F(t,s)$ has been subtracted from $q(t)F(t,s)$.

Remark. If q is a rational function with no real pole, then the hypothesis of the lemma is fulfilled with Q equal to the sum of the terms of degree $\geqq -1$ in the Laurent expansion at infinity. Since $q(t)-Q(t)=O(t^{-2})$ at infinity in $\mathbb{C}$, it follows from Cauchy's integral formula that $\int^+ q(t)\,dt$ is equal to $2\pi i$ times the sum of the residues of q in the upper half plane.

Let us now return to the boundary values of Pu. For fixed ξ'' and large ξ_1 we have when $\xi_1\to\infty$

$$p(x,\xi)-\sum\xi_1^{m-j-|\alpha|}\,p_j^{(\alpha)}(x,1,0)\,\xi''^\alpha/\alpha!=O(\xi_1^{-2})\cdot$$

if we sum over all j and α with $\alpha_1=0$ and $j+|\alpha|<2+\operatorname{Re} m$. This follows if we write $p_j(x,\xi)=|\xi_1|^{m-j}p_j(x,\xi_1/|\xi_1|,\ \xi''/|\xi_1|)$, expand by Taylor's formula and apply (18.2.20)'. We have

$$|e^{ix_1\xi_1}\hat{\phi}(\varepsilon\xi_1)|\leqq\int|\phi(t)|\,dt,\qquad \operatorname{Im}\xi_1\geqq 0,$$

provided that $\varepsilon\leqq x_1$. Hence it follows that

$$\int e^{ix_1\xi_1}p(x,\xi)\,\hat{\phi}(\varepsilon\xi_1)\,d\xi_1\to\int^+ e^{ix_1\xi_1}p(x,\xi)\,d\xi_1$$

when $\varepsilon\to 0$, and the integral can be bounded by a power of $(1+|\xi''|)$ independent of ε and x_1. Letting $x_1\to 0$ also now, we conclude that $Pu(x)\to q(x'',D'')\,v(x'')$ when $x_1\to 0$, where

$$q(x'',\xi'')=(2\pi)^{-1}\int^+ p(0,x'',\xi_1,\xi'')\,d\xi_1.$$

We have $q\sim\sum q_j$ where

$$q_j(x'',\xi'')=(2\pi)^{-1}\int^+ p_j(0,x'',\xi_1,\xi'')\,d\xi_1 \tag{18.2.24}$$

is homogeneous of degree $m+1-j$. Indeed, choose N so that $N>\operatorname{Re} m+2$ and set

$$p=\sum_{j<N}p_j+R_N.$$

Then we have $R_N\in S^{\operatorname{Re} m-N}$ when $|\xi''|>1$, say, and it follows that

$$\begin{aligned}|D^\beta_{x''}D^\alpha_{\xi''}\int R_N(0,x'',\xi_1,\xi'')\,d\xi_1|&\leqq C_{\alpha\beta}\int(1+|\xi_1|+|\xi''|)^{\operatorname{Re} m-N-|\alpha|}\,d\xi_1\\&\leqq C'_{\alpha\beta}(1+|\xi''|)^{\operatorname{Re} m+1-N-|\alpha|}.\end{aligned}$$

This proves that $q\sim\sum q_j$. Summing up, we have

Theorem 18.2.17. *Assume that $p\sim\sum p_j\in S^m_{\text{phg}}(\mathbb{R}^n\times\mathbb{R}^n)$ and that the transmission conditions (18.2.20)' are fulfilled. If $v\in C_0^\infty(\mathbb{R}^{n-1})$ and $u=\delta(x_1)\otimes v(x'')$ then $p(x,D)u$ has a C^∞ extension from the half space $\{x\in\mathbb{R}^n; x_1>0\}$ to its closure, and*

$$\lim_{x_1\downarrow 0}p(x,D)\,u=q(x'',D'')\,v(x'')$$

where $q\sim\sum q_j\in S^{m+1}_{\text{phg}}(\mathbb{R}^{n-1}\times\mathbb{R}^{n-1})$ with q_j defined by (18.2.24).

The boundary values of $D_1^k\,p(x,D)(D_1^l\,u)$ are of course also given by pseudo-differential operators acting on v; we just have to apply the theorem to the pseudo-differential operator $D_1^k\,p(x,D)\,D_1^l$, the symbol of which is given by the calculus.

A crucial point in the proof of Theorem 18.2.15 was the fact that functions in X with support in Y and with C^∞ restriction to Y can be identified with the elements in $I_{\text{phg}}^{-(n+2)/4}(X,\partial Y;E)$ with support in Y. For any complex number μ we can define the closely related space

$$C^\infty_\mu(Y,E)=\{u\in I_{\text{phg}}^{\mu-(n-2)/4}(X,\partial Y;E),\ \operatorname{supp}u\subset Y\}. \tag{18.2.25}$$

(We have omitted X in the notation since it does not matter how the manifold with boundary Y is extended to an open manifold X.) If u has support in a local coordinate patch where Y is defined by $x_1 \geqq 0$ and E is trivial, then we can write

$$u(x)=\int e^{ix_1\xi_1} a(x'',\xi_1)\,d\xi_1 \tag{18.2.26}$$

where

$$a\sim\sum_0^\infty a_j(x'',\xi_1) \tag{18.2.27}$$

and a_j is homogeneous of degree $\mu-j$. If we change the signs of x_1 and of ξ_1 it follows from Lemma 18.2.14 that $a_j(x'',\xi_1)$ can be extended to a homogeneous analytic function of ξ_1 in the half plane $\operatorname{Im}\xi_1<0$. In view of Example 7.1.17 it follows when μ is not an integer that u has an asymptotic expansion

$$u(x)\sim\sum_0^\infty u_j(x'')\otimes x_{1+}^{j-\mu-1};\qquad u_j\in C^\infty(\mathbb{R}^{n-1}).$$

Here we have used the notation of Section 3.2 and the expansion means that the difference between u and a partial sum of high order is as smooth as we please. If μ is an integer $\leqq -1$ we obtain the functions vanishing when $x_1<0$ which are in C^∞ and $O(x_1^{-\mu-1})$ when $x_1\geqq 0$. Finally, when μ is an integer $\geqq 0$ then u is the sum of a function U, which is in C^∞ when $x_1\geqq 0$ and vanishes when $x_1<0$, and a multiple layer

$$\sum_{j\leqq\mu} u_j(x'')\otimes\delta^{(j)}(x_1),$$

where $u_j\in C^\infty(\mathbb{R}^{n-1})$. There are boundary problems for which one expects the solutions to behave as in one of these cases.

With Y still denoting the closure of an open subset of the C^∞ manifold X with $\partial Y\in C^\infty$, we can now extend Theorem 18.2.15 as follows.

Theorem 18.2.18. *Let $P\in\Psi^m_{\text{phg}}(X;E,F)$ be properly supported and assume that for every $u\in C^\infty_\mu(Y,E)$ the restriction of Pu to the interior Y° of Y is in $\bar{C}^\infty(Y^\circ,F)$. Then the symbol of*

$$\check{P}-e^{2\pi i\mu}P \tag{18.2.28}$$

vanishes of infinite order on the interior normal bundle of ∂Y and conversely.

Proof. We can follow the proof of Theorem 18.2.15 closely, working again in local coordinates. If u is of the form (18.2.26), (18.2.27) with a_j homogeneous of degree $\mu-j$, then

$$a_j(x'',1)=e^{\pi i(\mu-j)}a_j(x'',-1)$$

by Lemma 18.2.14 with the sign of x_1 changed, for u vanishes when $x_1<0$. Now Pu is of the form (18.2.26) with a replaced by a polyhomogeneous symbol with the leading term $p_0(0,x'',\xi_1,0)\,a_0(x'',\xi_1)$. Hence it follows from Lemma 18.2.14 that if $Pu\in\bar{C}^\infty(Y^\circ)$ then

$$\begin{aligned} p_0(0,x'',-1,0)\,a_0(x'',-1) &= e^{\pi i(m+\mu)}\,p_0(0,x'',1,0)\,a_0(x'',1)\\ &= e^{\pi i(m+2\mu)}\,p_0(0,x'',1,0)\,a_0(x'',-1). \end{aligned}$$

Since we can choose a_0 so that $a_0(x'',-1)\neq 0$ at any point, it follows that

$$p_0(0,x'',-1,0)=e^{\pi i(m+2\mu)}\,p_0(0,x'',1,0).$$

Following the proof of Theorem 18.2.15 we can now obtain the corresponding condition for arbitrary derivatives of p_0 or of the lower order terms. The repetition of the details is left for the reader.

Remark. It follows from Theorem 18.2.18 that the modified transmission condition examined in the theorem only depends on the residue class of μ in $\mathbb{C}/\mathbb{Z}$. If (18.2.28) is valid then the symbol of

$$e^{-2\pi i(m+\mu)}\,P-\check{P}$$

vanishes of infinite order on the exterior normal bundle. If

$$m+\mu+\mu'\in\mathbb{Z} \tag{18.2.29}$$

it follows that

$$\check{P}-e^{2\pi i\mu'}\,P$$

vanishes of infinite order on the exterior normal bundle. Hence we may replace Y by $Y'=\overline{\complement Y}$ if μ is replaced at the same time by some μ' satisfying (18.2.29).

18.3. Totally Characteristic Operators

This section is devoted to the study of a class of pseudo-differential operators in a C^∞ manifold X with boundary ∂X. (This notion is defined in section B.2 of Appendix B.) In Section 18.2 we introduced the transmission condition with respect to X for a pseudo-differential operator P in an open manifold of the same dimension containing X. This condition guarantees that P defines a map from $C_0^\infty(X)$ to $C^\infty(X)$ but P does not restrict to an operator from $C_0^\infty(\partial X)$ to $C^\infty(\partial X)$ which is sometimes desirable in the study of boundary problems. A first order differential operator L has this property if and only if it is tangential to ∂X (cf. Definition 18.2.6). The algebra of operators which we shall define is built up from such first order differential operators in the same way that standard pseudo-differential operators are built up from general first order differential operators.

As a model for a manifold with boundary we shall use the closure $\overline{\mathbb{R}}^n_+$ of the half space $\mathbb{R}^n_+=\{x\in\mathbb{R}^n; x_n>0\}$. As explained in Section B.2, if F is a space of distributions in $\mathbb{R}^n$ we shall use the notation $\bar{F}(\mathbb{R}^n_+)$ for the space of restrictions to $\mathbb{R}^n_+$ of elements in F and we shall write $\dot{F}(\bar{\mathbb{R}}^n_+)$ for the set of distributions in F supported by $\bar{\mathbb{R}}^n_+$. The space $\bar{C}^\infty(\mathbb{R}^n_+)$ is by Theorem 1.2.6 identical to the space $C^\infty(\bar{\mathbb{R}}^n_+)$ of C^∞ functions in $\bar{\mathbb{R}}^n_+$ and we shall use both notations.

According to Lemma 18.2.5 the first order differential operators in $\bar{\mathbb{R}}^n_+$ which are tangential to $\partial\bar{\mathbb{R}}^n_+$ are generated by the operators $\partial/\partial x_j, j<n$, and $x_n\partial/\partial x_n$. We can extend Lemma 18.2.5 as follows:

Lemma 18.3.1. *The algebra* $\operatorname{Diff}_b(\bar{\mathbb{R}}^n_+)$ *generated by the first order differential operators with coefficients in* $\bar{C}^\infty(\mathbb{R}^n_+)$ *tangential to* $\partial\bar{\mathbb{R}}^n_+$ *consists of the operators of the form*

$$P=\sum a_\alpha(x)\,x_n^{\alpha_n}D^\alpha$$

where $a_\alpha\in\bar{C}^\infty(\mathbb{R}^n_+)$.

Proof. Since $x_nD_n=D_nx_n+i$, the operators $(x_nD_n)^j$, $j\leqq k$, are linear combinations of the operators $x_n^jD_n^j$, $j\leqq k$, and vice versa. The lemma is therefore a consequence of Lemma 18.2.5.

The elements of $\operatorname{Diff}_b(\bar{\mathbb{R}}^n_+)$ will be called *totally characteristic*. Lemma 18.3.1 suggests extending the subspace $\operatorname{Diff}_b^m(\bar{\mathbb{R}}^n_+)$ of operators of order $\leqq m$ to a class of operators defined by

$$\tilde{a}(x,D)u=(2\pi)^{-n}\int e^{i\langle x,\xi\rangle}\tilde{a}(x,\xi)\,\hat{u}(\xi)\,d\xi,\qquad u\in C_0^\infty(\mathbb{R}^n),\tag{18.3.1}$$

$$\tilde{a}(x,\xi)=a(x,\xi',x_n\xi_n),\tag{18.3.2}$$
$$x_n\geqq 0,\quad \xi'=(\xi_1,\dots,\xi_{n-1});\quad \tilde{a}(x,\xi)=0,\quad x_n<0.$$

Since our primary concern is the behavior as $x_n\to 0$ we shall choose a in a symbol class S^m_+ defined so that $\tilde{a}$ will satisfy (18.1.1) for $x_n>1$:

Definition 18.3.2. By S^m_+ we shall denote the set of all $a\in C^\infty(\bar{\mathbb{R}}^n_+\times\mathbb{R}^n)$ such that for all multi-indices α, β and all integers $\nu\geqq 0$

$$|a^{(\alpha)}_{(\beta)}(x,\xi)|\leqq C_{\alpha,\beta,\nu}(1+|\xi|)^{m-|\alpha|}(1+x_n)^{-\nu},\qquad x\in\bar{\mathbb{R}}^n_+,\quad \xi\in\mathbb{R}^n.\tag{18.3.3}$$

The kernel of (18.3.1) is the inverse Fourier transform

$$K(x,y)=(2\pi)^{-n}\int e^{i\langle x-y,\xi\rangle}\tilde{a}(x,\xi)\,d\xi;\qquad (x,y)\in\mathbb{R}^{2n};\tag{18.3.4}$$

defined in the sense of Schwartz. It is a continuous function of x with values in $\mathscr{S}'(\mathbb{R}^n_y)$ when $x_n\geqq 0$ and vanishes when $x_n<0$. We want (18.3.1) to depend only on the restriction of u to $\mathbb{R}^n_+$ so we must require that $y_n\geqq 0$ in supp K. This means that the oscillatory integral

$$\int e^{i(x_n-y_n)\xi_n}a(x,\xi',x_n\xi_n)\,d\xi_n$$

which is well defined when $x_n \geqq 0$ and $y_n < 0$ must vanish then. Taking $x_n \xi_n$ as a new integration variable when $x_n > 0$ we write this condition in the form

$$\int e^{-it\xi_n} a(x, \xi', \xi_n)\, d\xi_n = 0 \text{ if } t \leqq -1 \text{ and } x_n \geqq 0, \tag{18.3.5}$$

that is, the Fourier transform with respect to ξ_n must vanish when $t < -1$.

Definition 18.3.3. We shall say that $a \in S_+^m$ is lacunary or satisfies the lacunary condition if (18.3.5) is fulfilled. The set S_{la}^m of all $a \in S_+^m$ satisfying (18.3.5) is a closed subspace of S_+^m, thus a Fréchet space.

The lacunary condition is not very restrictive, for just like the condition of proper support it only affects the residual part of the symbol:

Lemma 18.3.4. *Let $\rho \in \mathscr{S}(\mathbb{R})$, $\hat\rho = 1$ in a neighborhood of 0 and $\operatorname{supp} \hat\rho \subset (-\frac{1}{2}, 1)$. If $a \in S_+^m$ it follows that*

$$a_\rho(x, \xi) = \int_{-\infty}^{\infty} a(x, \xi', \xi_n - t)\, \rho(t)\, dt$$

is in S_{la}^m, we have $a - a_\rho \in S_+^{-\infty}$, the map $S_+^m \ni a \mapsto a - a_\rho \in S_+^{-\infty}$ is continuous, and $x_n/2 \leqq y_n \leqq 2x_n$ in the support of the kernel of $\tilde a_\rho(x, D)$.

Proof. The Fourier transform of a_ρ with respect to ξ_n is the product of $\hat\rho$ and that of a, so we obtain not only (18.3.5) but a stronger condition

$$\int e^{-it\xi_n} a_\rho(x, \xi', \xi_n)\, d\xi_n = 0 \quad \text{if } t \notin (-\tfrac{1}{2}, 1). \tag{18.3.5$'$}$$

This implies that $x_n/2 \leqq y_n \leqq 2x_n$ in the support of the kernel of $\tilde a_\rho(x, D)$. Now the condition on ρ implies that $\int \rho\, dt = 1$, $\int t^j \rho\, dt = 0$, $j > 0$. Hence

$$a_\rho(x, \xi) - a(x, \xi) = \int_{-\infty}^{\infty} \Big(a(x, \xi', \xi_n - t) - \sum_{j<N} \partial^j a(x, \xi)/\partial \xi_n^j (-t)^j/j!\Big)\, \rho(t)\, dt$$

for any N. By Taylor's formula the integrand can be estimated by

$$C_{\nu,N} t^N (1+|\xi|)^{m-N} (1+x_n)^{-\nu} |\rho(t)| \qquad \text{if } |t| < |\xi|/2.$$

Evaluating each term separately we obtain the bound

$$\begin{aligned} &C_{\nu,N}(1+|t|)^{|m|+N}(1+x_n)^{-\nu}|\rho(t)| \\ &\quad \leqq C_{\nu,N}(1+|\xi|)^{m-N}(1+x_n)^{-\nu}(1+2|t|)^{2(N+|m|)}|\rho(t)|, \qquad |t| > |\xi|/2. \end{aligned}$$

Since $\rho \in \mathscr{S}$ and N is arbitrary this shows that $a_\rho(x, \xi) - a(x, \xi)$ can be estimated by any power of $(1+|\xi|)^{-1}(1+x_n)^{-1}$; the constants obtained are semi-norms in S_+^m. This is also true for all derivatives of $a_\rho - a$ since convolution with ρ commutes with them. The proof is complete.

From the lemma it follows that

$$S_{la}^m / S_{la}^{-\infty} = S_+^m / S_+^{-\infty}.$$

The restriction to lacunary symbols will therefore have no effect on the symbol calculus which we shall develop along the lines of Section 18.1. First we give an analogue of Theorem 18.1.6.

Theorem 18.3.5. *If $a \in S^m_{la}$ and $u \in \bar{\mathscr{S}}(\overline{\mathbb{R}}^n_+)$ then (18.3.1) applied to any extension of u in $\mathscr{S}(\mathbb{R}^n)$ defines $\tilde{a}(x,D)u \in \bar{\mathscr{S}}(\overline{\mathbb{R}}^n_+)$, and the bilinear map $(a,u) \mapsto \tilde{a}(x,D)u$ is continuous in these spaces. We have*

$$[\tilde{a}(x,D), D_j] = i\,\widetilde{a_{(j)}}(x,D) + i\,\delta_{jn}\,\widetilde{a^{(n)}}(x,D)\,D_n, \tag{18.3.6}$$
$$[\tilde{a}(x,D), x_j] = -i\,\widetilde{a^{(j)}}(x,D) - i\,\delta_{jn}\,x_n\,\widetilde{a^{(n)}}(x,D).$$

For any integer $k \geqq 0$

$$D^k_n \tilde{a}(x,D)u|_{x_n=0} = \sum_{j \leqq k} \binom{k}{j} a_{kj}(x',D')(D^j_n u|_{x_n=0}), \tag{18.3.7}$$

$$a_{kj}(x',\xi') = \sum_{i \leqq j} \binom{j}{i} D^{k-j}_{x_n} D^i_{\xi_n} a(x',0,\xi',0) \in S^m(\mathbb{R}^{n-1} \times \mathbb{R}^{n-1}).$$

Proof. The lacunary condition guarantees that $y_n \geqq 0$ in the support of the kernel K. Hence (18.3.1) is independent of the extension chosen. Since $\hat{u} \in \mathscr{S}$ it is clear that $\tilde{a}(x,D)u$ is a C^∞ function with all derivatives bounded. Differentiation under the integral sign or integration by parts gives (18.3.6). As in the proof of Theorem 18.1.6 it follows then that the map $S^m_{la} \times \bar{\mathscr{S}}(\overline{\mathbb{R}}^n_+) \ni (a,u) \mapsto \tilde{a}(x,D)u$ is continuous with values in $\bar{\mathscr{S}}(\overline{\mathbb{R}}^n_+)$. When we differentiate with respect to x_n under the integral sign in (18.3.1) a factor ξ_n appears when the derivative falls on the exponential function or the last argument of $a(x,\xi',x_n\xi_n)$. When there are j such derivatives altogether we obtain (18.3.7) since

$$(2\pi)^{-1} \int \xi^j_n \hat{u}(\xi',\xi_n)\,d\xi_n$$

is the Fourier transform of $D^j_n u(x',0)$ with respect to x'.

(18.3.7) shows that the purpose of the definition of the operators $\tilde{a}(x,D)$ has been achieved: all normal derivatives of $\tilde{a}(x,D)u$ on the boundary can be calculated by letting pseudo-differential operators in the boundary act on the normal derivatives of at most the same order. We shall now show that the main structure of the calculus of pseudo-differential operators is preserved in spite of the fact that $\tilde{a}(x,\xi)$, defined by (18.3.2), has rather bad symbol properties with respect to x_n. One of the main differences is that the kernel of $\tilde{a}(x,D)$ may have some singularities even when $a \in S^{-\infty}$. We shall therefore examine such operators now, using the notation

$$Q = \{(x,y) \in \mathbb{R}^{2n};\ x_n \geqq 0,\ y_n \geqq 0\}, \qquad \partial_2 Q = \{(x,y) \in \mathbb{R}^{2n},\ x_n = y_n = 0\}$$

for the quaterspace containing the support of the kernel K and for its distinguished boundary. In Q it is convenient to use the symmetric singular

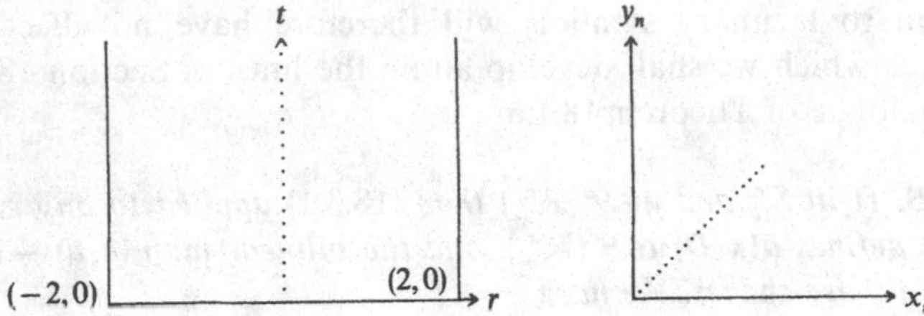

Fig. 2

(polar) coordinates with respect to x_n and y_n given by (see Fig. 2)

(18.3.8) $$t=(x_n+y_n)/2, \quad r=(x_n-y_n)/t=2(x_n-y_n)/(x_n+y_n),$$

that is, $x_n=t(1+r/2)$, $y_n=t(1-r/2)$. Note that $t\geqq 0$ and $|r|\leqq 2$ if $(x,y)\in Q$.

Theorem 18.3.6. *If $a\in S_{la}^{-\infty}$ then the kernel K of $\tilde{a}(x,D)$ is in $L^1_{\text{loc}}(\mathbb{R}^{2n})$ and in $C^\infty(\mathbb{R}^{2n}\setminus\partial_2 Q)$, $\operatorname{supp} K\subset Q$, and*

$$F(x',y',t,r)=t\,K(x',t(1+r/2),y',t(1-r/2))$$

is in C^∞ when $t\geqq 0$, equal to 0 when $|r|\geqq 2$. For all $\alpha,\beta,\tau,\rho,\nu$ we have

(18.3.9) $$|D_{x'}^\alpha D_{y'}^\beta D_t^\tau D_r^\rho F(x',y',t,r)|\leqq C_{\alpha\beta\tau\rho\nu}(1+|x'-y'|+t)^{-\nu}.$$

Conversely, if $K\in L^1_{\text{loc}}(\mathbb{R}^{2n})$, $\operatorname{supp} K\subset Q$, and F has these properties, then K is the kernel of $\tilde{a}(x,D)$ for some $a\in S_{la}^{-\infty}$.

Proof. The inverse Fourier transform

(18.3.10) $$A(x,y)=(2\pi)^{-n}\int e^{i\langle y,\xi\rangle}a(x,\xi)\,d\xi$$

is a C^∞ function when $x_n\geqq 0$. The proof of Lemma 7.1.3 gives

(18.3.11) $$|D_x^\alpha D_y^\beta A(x,y)|\leqq C_{\alpha\beta N}(1+|y|)^{-N}(1+x_n)^{-N}, \quad x_n\geqq 0,$$

for all α,β,N and shows that conversely every A satisfying (18.3.11) with $A(x,y)=0$ for $y_n>1$ is of the form (18.3.10) with $a\in S_{la}^{-\infty}$. The kernel K is a continuous function of x with values in $\mathscr{S}'$ when $x_n\geqq 0$, and (18.3.4) gives

(18.3.12) $$K(x,y)=A(x,x'-y',(x_n-y_n)/x_n)/x_n, \quad x_n>0.$$

Hence $K\in L^1_{\text{loc}}(\mathbb{R}^{2n})$, $\operatorname{supp} K\subset Q$ and $K\in C^\infty$ when $x_n>0$. If $t\geqq 0$ and $r>-2$ then

(18.3.13) $$F(x',y',t,r)=2A(x',t(1+r/2),x'-y',2r/(2+r))/(2+r).$$

This is a C^∞ function. With $x_n=t(1+r/2)$ and $y_n=2r/(2+r)$ we have

$$1+|y_n|\geqq 2/(2+r), \quad 1+|y_n|+x_n\geqq 2t^{\frac{1}{2}}.$$

We can therefore estimate F by any power of $(1+r/2)(1+|x'-y'|+t)^{-1}$. The same is true for all derivatives of F since these are sums of terms of the same form as F but with A replaced by a derivative and powers of t and $1/(2+r)$ as factors. If we set $F(x',y',t,r)=0$ when $r\leqq -2$ it follows that $F\in C^\infty$ when $t\geqq 0$ and that (18.3.9) holds.

Conversely, assume that $K\in L^1_{\text{loc}}(\mathbb{R}^{2n})$, that $\operatorname{supp} K\subset Q$ and that $F\in C^\infty$ when $t\geqq 0$, the estimates (18.3.9) being valid. Since $t(1+r/2)=x_n$ and $2r/(2+r)=y_n$ implies $r=2y_n/(2-y_n)$, $t=x_n(2-y_n)/2$ if $x_n>0$ and $y_n<2$, we set

$$A(x,y)=2F(x',x'-y',x_n(2-y_n)/2,2y_n/(2-y_n))/(2-y_n), \tag{18.3.13$'$}$$

$y_n<2$, and $A(x,y)=0$ when $y_n>1$. The two definitions agree in the common domain since $2y_n/(2-y_n)>2$ when $1<y_n<2$. When $y_n\leqq 1$ we have

$$2y_n/(2-y_n)+2=4/(2-y_n)\leqq 8/(1+|y_n|),$$
$$1+x_n(2-y_n)/2\geqq 1+x_n/2.$$

Note that since $F=0$ when $r<-2$ we can use Taylor's formula to strengthen (18.3.9) to

$$|D^\alpha_{x'}D^\beta_{y'}D^\tau_t D^\rho_r F(x',y',t,r)| \leqq C_{\alpha\beta\tau\rho\nu}(1+|x'-y'|+t)^{-\nu}(r+2)^\nu. \tag{18.3.9$'$}$$

This shows that (18.3.11) follows from (18.3.9) and completes the proof

Remarks. 1. In terms of the notions introduced in Section 18.2 the condition $F\in C^\infty$ means precisely that $K\in I^{-n/2}_{\text{phg}}(\mathbb{R}^{2n},\partial_2 Q)$. We leave the simple verification for the reader.

2. The proof shows that if $a\in S^{-n-2}_{la}$ then

$$|K(x,y)|\leqq C(1+|x'-y'|)^{-n}|x_n y_n|/(|x_n|+|y_n|)^3. \tag{18.3.14}$$

In fact, since $(1+|y'|)^n(1+|y_n|)^2|A(x,y)|\leqq C$ we have

$$(1+|x'-y'|)^n|K(x,y)|\leqq C(1+|x_n-y_n|/x_n)^{-2}/x_n=Cx_n/(x_n+y_n-x_n)^2$$
$$\leqq 8Cx_n y_n/(x_n+y_n)^3 \quad \text{if } 0<x_n\leqq y_n.$$

In addition $(1+|y'|)^n|A(x,y)|\leqq C|1-y_n|$, for $A(x,y)=0$ when $y_n>1$ and $(1+|y'|)^n|\partial A(x,y)/\partial y_n|\leqq C$. Hence

$$(1+|x'-y'|)^n|K(x,y)|\leqq C y_n/x_n^2\leqq 8Cx_n y_n/(x_n+y_n)^3 \quad \text{if } 0<y_n<x_n.$$

This proves (18.3.14) which will be a convenient starting point for the proof of L^2 estimates below.

It follows from Theorem 18.3.6 that if $a\in S^{-\infty}_{la}$ then $\tilde a(x,D)^*=\tilde b(x,D)$ where $b\in S^{-\infty}_{la}$. To prepare for the proof of the analogous result for S^m_{la} we shall now give an analytic expression for b when $a\in S^{-\infty}_{la}$ and the strong lacunary condition

$$\int e^{it\xi_n}a(x,\xi',\xi_n)\,d\xi_n=0 \quad \text{if } t\notin(-1,\tfrac12) \tag{18.3.5$''$}$$

is fulfilled. With the notation (18.3.10) this means that $A(x,y)=0$ when $y_n \notin (-1,\frac{1}{2})$, thus $\frac{1}{2} \leqq |x_n/y_n| \leqq 2$ if $(x,y) \in \operatorname{supp} K$. By (18.1.9) and (18.1.10) we have

(18.3.15) $$(\tilde{a}(x,D)u,v)=(u,\tilde{b}(x,D)v); \quad u,v \in \mathscr{S}(\mathbb{R}^n);$$

where $\tilde{b}=e^{i\langle D_x,D_\xi\rangle}\bar{\tilde{a}}(x,\xi) \in \mathscr{S}'$ and $x_n \geqq 0$ in $\operatorname{supp}\tilde{b}$. If $\phi \in C^\infty(\mathbb{R})$ and $\phi=0$ in $(-\infty,1)$, $\phi=1$ in $(2,\infty)$, then $\phi(x_n/\varepsilon)\bar{\tilde{a}}(x,\xi)=\phi(x_n/\varepsilon)\bar{a}(x,\xi',x_n\xi_n) \in S^{-\infty}$ and converges to $\bar{\tilde{a}}$ in $\mathscr{S}'$ as $\varepsilon \to 0$. Hence $\tilde{b}$ is the $\mathscr{S}'$ limit as $\varepsilon \to 0$ of

$$(x,\xi) \mapsto (2\pi)^{-n}\iint e^{-i\langle x-y,\xi-\eta\rangle}\phi(y_n/\varepsilon)\bar{a}(y,\eta',y_n\eta_n)\,dy\,d\eta$$

defined by Theorem 7.6.5 or interpreted as an integral first with respect to η, then with respect to y. The integral with respect to η_n vanishes by (18.3.5)″ unless $(y_n-x_n)/y_n \in (-1,\frac{1}{2})$, that is, $\frac{1}{2}<y_n/x_n<2$. Choose $\chi \in C_0^\infty(\frac{1}{3},3)$ equal to 1 in $(\frac{1}{2},2)$. Without changing the integral we can insert a factor $\chi(y_n/x_n)$. Letting $\varepsilon \to 0$ we then obtain when $x_n>0$

$$\begin{aligned} b(x,\xi',x_n\xi_n)&=\tilde{b}(x,\xi)\\ &=(2\pi)^{-n}\iint_{y_n<x_n} e^{-i\langle y,\eta\rangle}\bar{a}(x-y,\xi'-\eta',(x_n-y_n)(\xi_n-\eta_n))\,\chi((x_n-y_n)/x_n)\,dy\,d\eta. \end{aligned}$$

The integral exists as a repeated integral. Replacing y_n by $x_n y_n$ and η_n by η_n/x_n we conclude that

(18.3.16) $$\begin{aligned} b(x,\xi)&\\ &=(2\pi)^{-n}\iint e^{-i\langle y,\eta\rangle}\bar{a}(x'-y',x_n(1-y_n),\xi'-\eta',(1-y_n)(\xi_n-\eta_n))\,\chi(1-y_n)\,dy\,d\eta\\ &=e^{i\langle D_y,D_\eta\rangle}(\bar{a}(y',x_ny_n,\eta',y_n\eta_n)\chi(y_n))|_{y=(x',1),\eta=\xi}. \end{aligned}$$

We shall now study this formula when $a \in S_+^m$ for some finite m.

Lemma 18.3.7. *If $a \in S_+^m$ and $\chi \in C_0^\infty(0,\infty)$ is equal to 1 in a neighborhood of 1, then (18.3.16) defines a symbol $b \in S_{la}^m$ such that*

(18.3.17) $$\begin{aligned} b(x,\xi)&\sim\sum\frac{1}{j!}\langle D_y,iD_\eta\rangle^j\bar{a}(y',x_ny_n,\eta',y_n\eta_n)|_{y'=x',y_n=1,\eta=\xi}\\ &=\sum\frac{1}{j!}\langle D_y,iD_\eta\rangle^j\bar{a}(y,\eta',y_n\eta_n)|_{y=x,\eta'=\xi',\eta_n=\xi_n/x_n}. \end{aligned}$$

The map $S_+^m \ni a \mapsto b \in S_{la}^m$ is continuous.

The asymptotic sum is well defined in the sense of Proposition 18.1.3 for if a y derivative falls on the argument $y_n\eta_n$ it causes the degree of a to decrease by one unit which compensates for the appearance of a factor η_n.

Proof. There is a constant M such that $\operatorname{supp}\chi \subset (M^{-1},M)$. If we set

$$c_{x_n}(y,\eta)=\bar{a}(y',x_ny_n,\eta',y_n\eta_n)\chi(y_n)$$

it follows from the bounds $1/M \leqq y_n \leqq M$ in the support that $(1+x_n)^\nu c_{x_n}$ belongs to a bounded set in S^m for every ν and $x_n \geqq 0$. By Theorem 18.1.7

$$C(y,\eta,x_n)=e^{i\langle D_y,D_\eta\rangle}c_{x_n}(y,\eta)$$

is a C^∞ function of (y,η), and we have for all α,β,ν

$$|D_\eta^\alpha D_y^\beta C(y,\eta,x_n)| \leqq C_{\alpha\beta\nu}(1+|\eta|)^{m-|\alpha|}(1+x_n)^{-\nu}.$$

Since the derivatives of c_{x_n} with respect to x_n are of the same form it follows that C is a C^∞ function of all variables and that

$$|D_{x_n}^j D_\eta^\alpha D_y^\beta C(y,\eta,x_n)| \leqq C_{\alpha\beta\nu j}(1+|\eta|)^{m-|\alpha|}(1+x_n)^{-\nu}.$$

Hence $b(x,\xi)=C(x',1,\xi,x_n)$ is in S_+^m. The asymptotic expansion also follows at once from Theorem 18.1.7 since $\chi=1$ in a neighborhood of 1. If $a\in S_+^{-\infty}$ we obtain by taking $\xi_n-\eta_n$ as new integration variable in the first form of (18.3.16)

$$\int b(x,\xi)\,e^{it\xi_n}\,d\xi_n$$
$$=(2\pi)^{1-n}\iint e^{-i\langle y',\eta'\rangle+it\eta_n}\tilde a(x'-y',x_n(1-t),\xi'-\eta',(1-t)\eta_n)\chi(1-t)\,dy'\,d\eta$$

which is equal to 0 unless $M^{-1}\leqq 1-t\leqq M$, that is, $-1-M\leqq t\leqq 1-M^{-1}$. If $a\in S_+^m$ we obtain the same conclusion by taking $a_j\in S_+^{-\infty}$ converging to a in S_+^{m+1}. Thus $b\in S_{la}^m$ which completes the proof.

Remark. That the lacunary condition holds is no surprise since χ has the effect of cutting off the kernel of $\tilde a(x,D)$ by a factor $\chi(x_n/y_n)$.

Theorem 18.3.8. *For every $a\in S_{la}^m$ the adjoint of $\tilde a(x,D)$ is equal to $\tilde b(x,D)$ for some $b\in S_{la}^m$ in the sense that* (18.3.15) *is valid. If a satisfies the strong lacunary condition* (18.3.5)″ *and $\chi\in C_0^\infty(0,\infty)$ is equal to 1 in $(\frac{1}{2},2)$ then b is given by* (18.3.16) *and has the asymptotic expansion* (18.3.17).

Proof. If $a\in S_{la}^{-\infty}$ the first statement follows from Theorem 18.3.6 for $(x,y)\mapsto \overline{K(y,x)}$ has the same properties as K. Thus the first statement follows from the second one and Lemma 18.3.4. If $a\in S_{la}^m$ and (18.3.5)″ is fulfilled we choose $a_j\in S_+^{-\infty}$ so that $a_j\to a$ in S_+^{m+1} as $j\to\infty$ and define $a_{j\rho}$ as in Lemma 18.3.4. Then $a_{j\rho}\in S_{la}^{-\infty}$ satisfies (18.3.5)″ and $a_{j\rho}\to a_\rho$ in S_+^{m+1} as $j\to\infty$. Let $b_j\in S_{la}^{-\infty}$ and $b_0\in S_{la}^m$ be defined by (18.3.16) with a replaced by $a_{j\rho}$ and by a_ρ. Then

$$(\tilde a_{j\rho}(x,D)u,v)=(u,\tilde b_j(x,D)v);\qquad u,v\in\mathscr{S};$$

and $b_j\to b_0$ in S_{la}^{m+1} as $j\to\infty$. Hence it follows from Theorem 18.3.5 that

$$(\tilde a_\rho(x,D)u,v)=(u,\tilde b_0(x,D)v);\qquad u,v\in\mathscr{S}.$$

Since $a-a_\rho\in S_{la}^{-\infty}$ satisfies (18.3.5)″ we also have with $b-b_0\in S_{la}^{-\infty}$

$$((\tilde a(x,D)-\tilde a_\rho(x,D))u,v)=(u,(\tilde b(x,D)-\tilde b_0(x,D))v);\qquad u,v\in\mathscr{S};$$

which gives (18.3.15) and completes the proof.

By Theorem 18.3.5 $\tilde{a}(x,D)$ is a continuous map $\bar{\mathscr{S}}(\mathbb{R}^n_+)\to\bar{\mathscr{S}}(\mathbb{R}^n_+)$, and the same is true for the adjoint $\tilde{b}(x,D)$. Hence a continuous map

$$\tilde{a}(x,D)\colon\ \dot{\mathscr{S}}'(\bar{\mathbb{R}}^n_+)\to\dot{\mathscr{S}}'(\bar{\mathbb{R}}^n_+)$$

is defined by (18.3.15) with $u\in\dot{\mathscr{S}}'(\bar{\mathbb{R}}^n_+)$ and $v\in\bar{\mathscr{S}}(\mathbb{R}^n_+)$. If $u\in\bar{\mathscr{S}}(\mathbb{R}^n_+)$ is extended to $u_0\in\dot{\mathscr{S}}'(\bar{\mathbb{R}}^n_+)$ by defining $u_0=u$ in $\bar{\mathbb{R}}^n_+$ and $u_0=0$ elsewhere, then $\tilde{a}(x,D)u_0$ is the function $\tilde{a}(x,D)u$ defined by (18.3.1). By the Hahn-Banach theorem the restriction map $\dot{\mathscr{S}}'(\bar{\mathbb{R}}^n_+)\to\bar{\mathscr{S}}'(\mathbb{R}^n_+)$ is surjective. The kernel is

$$\mathscr{S}'(\bar{\mathbb{R}}^n_+,\partial\mathbb{R}^n_+)=\{u\in\mathscr{S}'(\mathbb{R}^n);\ \operatorname{supp}u\subset\partial\mathbb{R}^n_+\}=\bigcup\mathscr{S}_k'(\bar{\mathbb{R}}^n_+,\partial\mathbb{R}^n_+), \tag{18.3.18}$$

$$\mathscr{S}_k'(\bar{\mathbb{R}}^n_+,\partial\mathbb{R}^n_+)=\{u\in\mathscr{S}'(\mathbb{R}^n);\ x_n^k u=0\}. \tag{18.3.18$'$}$$

Here we have used Theorem 2.3.5 and the fact that temperate distributions are of finite order. Now

$$(x_n^k\tilde{a}(x,D)u,v)=(u,\tilde{b}(x,D)x_n^k v)=0,\qquad v\in\bar{\mathscr{S}}(\mathbb{R}^n_+),\qquad u\in\mathscr{S}_k'(\bar{\mathbb{R}}^n_+,\partial\mathbb{R}^n_+)$$

for $\tilde{b}(x,D)x_n^k v$ vanishes of order k when $x_n=0$ by virtue of (18.3.7), so we can take out a factor x_n^k which annihilates u when moved to the left. Hence

$$\tilde{a}(x,D)\mathscr{S}_k'(\bar{\mathbb{R}}^n_+,\partial\mathbb{R}^n_+)\subset\mathscr{S}_k'(\bar{\mathbb{R}}^n_+,\partial\mathbb{R}^n_+)\qquad\text{for every } k. \tag{18.3.19}$$

Thus $\tilde{a}(x,D)$ induces a map $\bar{\mathscr{S}}'(\mathbb{R}^n_+)\to\bar{\mathscr{S}}'(\mathbb{R}^n_+)$ which is still defined by (18.3.15), now with $u\in\bar{\mathscr{S}}'(\mathbb{R}^n_+)$ and $v\in\dot{\mathscr{S}}(\bar{\mathbb{R}}^n_+)$. Its restriction to $C_0^\infty(\mathbb{R}^n_+)$ determines $\tilde{a}(x,D)$ on $\dot{\mathscr{S}}'(\bar{\mathbb{R}}^n_+)$ since $C_0^\infty(\mathbb{R}^n_+)$ is dense there.

In the standard calculus of pseudo-differential operators one ignores C^∞ functions since they are in the range of operators of order $-\infty$. A somewhat larger class of distributions is neglected in the totally characteristic calculus.

Theorem 18.3.9. *If $a\in S_{la}^{-\infty}$ and $u\in\dot{\mathscr{S}}'(\bar{\mathbb{R}}^n_+)$ then $\tilde{a}(x,D)u\in I^k(\mathbb{R}^n,\partial\bar{\mathbb{R}}^n_+)$ for some k, and $\operatorname{supp}\tilde{a}(x,D)u\subset\bar{\mathbb{R}}^n_+$.*

Proof. By Definition 18.2.6 the statement means that the order of the distribution $D^{\alpha'}(x_nD_n)^{\alpha_n}\tilde{a}(x,D)u$ has a bound independent of α. Now

$$(D^{\alpha'}(x_nD_n)^{\alpha_n}\tilde{a}(x,D)u,v)=(u,\tilde{b}(x,D)D^{\alpha'}(D_nx_n)^{\alpha_n}v)=(u,\tilde{b}_\alpha(x,D)v),$$

if $v\in\bar{\mathscr{S}}(\mathbb{R}^n_+)$ and

$$b_\alpha(x,\xi)=\xi^{\alpha'}(\xi_n-i-i\xi_n\partial/\partial\xi_n)^{\alpha_n}b(x,\xi)\in S_{la}^{-\infty}.$$

It suffices to verify this for the operator $\tilde{b}(x,D)D_nx_n=\tilde{b}(x,D)x_nD_n-i\tilde{b}(x,D)$. Then it follows from (18.3.1) since the Fourier transform of x_nD_nv is $-D_n\xi_n\hat{v}$ and

$$\xi_nD_{\xi_n}(e^{i\langle x,\xi\rangle}\tilde{b}(x,\xi))=e^{i\langle x,\xi\rangle}x_n\xi_n(\tilde{b}(x,\xi)-i\tilde{b}^{(n)}(x,\xi)).$$

Hence it follows from Theorem 18.3.5 that for suitable μ, μ' independent of α

$$|(D^{\alpha'}(x_n D_n)^{\alpha_n} \tilde{a}(x,D)u, v)| \leqq C \sup_{\mathbb{R}^n_+} \sum_{|\beta+\gamma|\leqq\mu} |x^\beta D^\gamma \tilde{b}_\alpha(x,D)v| \leqq C_\alpha \sup_{\mathbb{R}^n_+} \sum_{|\beta+\gamma|\leqq\mu'} |x^\beta D^\gamma v|.$$

Here we have used the following lemma:

Lemma 18.3.10. *The quotient topology in $\bar{\mathscr{S}}(\mathbb{R}^n_+)$ is defined by the semi-norms*

$$\bar{\mathscr{S}}(\mathbb{R}^n_+) \ni v \mapsto \sup_{\mathbb{R}^n_+} |x^\alpha D^\beta v|.$$

Proof. That these are continuous semi-norms in $\bar{\mathscr{S}}(\mathbb{R}^n_+)$ is obvious for they are continuous in $\mathscr{S}(\mathbb{R}^n)$ and constant in each equivalence class. On the other hand, let q be a continuous semi-norm on $\bar{\mathscr{S}}(\mathbb{R}^n_+)$. Then q is a continuous semi-norm on $\mathscr{S}(\mathbb{R}^n)$, so

$$q(v) \leqq C \sum_{|\alpha+\beta|\leqq k} \sup_{\mathbb{R}^n} |x^\alpha D^\beta v|$$

and $q(v)=0$ if $v=0$ in $\mathbb{R}^n_+$. If $v \in \mathscr{S}(\mathbb{R}^n)$ we now set

$$\tilde{v}(x) = v(x), \quad x_n \geqq 0, \quad \tilde{v}(x) = \chi(x_n) \sum_{j\leqq k} \partial_n^j v(x',0) x_n^j/j!, \quad x_n < 0,$$

where $\chi \in C_0^\infty$ and $\chi = 1$ in $(-1,1)$. Choose $\phi \in C_0^\infty(\mathbb{R}^n_-)$ with $\int \phi\, dx = 1$ and set $\phi_\varepsilon(x) = \phi(x/\varepsilon)\varepsilon^{-n}$. Then $\tilde{v} * \phi_\varepsilon \in \mathscr{S}$, $v * \phi_\varepsilon = \tilde{v} * \phi_\varepsilon$ in $\mathbb{R}^n_+$ and $v * \phi_\varepsilon \to v$ in $\mathscr{S}$ when $\varepsilon \to 0$. Hence

$$q(v) = \lim_{\varepsilon\to 0} q(v*\phi_\varepsilon) = \lim_{\varepsilon\to 0} q(\tilde{v}*\phi_\varepsilon) \leqq C' \sum_{|\alpha+\beta|\leqq 2k} \sup_{\mathbb{R}^n_+} |x^\alpha D^\beta v|,$$

which completes the proof. (Using the proof of Theorem 2.3.5 we could of course replace $2k$ by k here.)

Recall that for the distributions in the theorem the wave front set is contained in the conormal bundle of the boundary. In many contexts one can exclude such singularities for other reasons. However, we postpone the discussion of this and related matters until we have completed the remaining parts of the calculus, the product formula and the invariance under change of variables.

Theorem 18.3.11. *If $a_j \in S^{m_j}_{la}$, $j=1,2$, then $\tilde{a}_1(x,D)\tilde{a}_2(x,D) = \tilde{b}(x,D)$ where $b \in S^{m_1+m_2}_{la}$ is given by*

$$(18.3.20) \quad b(x,\xi) = e^{i\langle D_y, D_\eta\rangle} a_1(x,\eta) a_2(y', x_n y_n, \xi', \xi_n y_n)|_{y=(x',1),\eta=\xi}$$
$$\sim \sum a_1^{(\alpha)}(x,\xi) D_{x'}^{\alpha'} D_s^{\alpha_n} a_2(x', s x_n, \xi', s\xi_n)/\alpha!|_{s=1}.$$

Proof. Assume first that $a_1 \in S^{-\infty}_{la}$ and that $a_2(x,\xi)=0$ for large $|x|$. If $u \in \mathscr{S}(\mathbb{R}^n)$ it follows then that $\tilde{a}_2(x,D)u \in \mathscr{E}'$ has the Fourier transform

$$\eta \mapsto (2\pi)^{-n} \int e^{i\langle y, \xi-\eta\rangle} \tilde{a}_2(y,\xi)\hat{u}(\xi)\, dy\, d\xi.$$

When a is of order $-\infty$ then (18.3.1) remains valid for reasons of continuity when $u\in\dot{\mathscr{E}}'(\overline{\mathbb{R}}^n_+)$ and $x_n>0$, for $\xi\mapsto e^{i\langle x,\xi\rangle}\tilde{a}(x,\xi)$ is then in $\mathscr{S}$ and is a C^∞ function of x with values in $\mathscr{S}$. Hence we obtain as in the proof of Theorem 18.1.8 that $\tilde{a}_1(x,D)\tilde{a}_2(x,D)u=\tilde{b}(x,D)u$, $x_n>0$, if

$$\tilde{b}(x,\xi)=(2\pi)^{-n}\int e^{-i\langle x-y,\xi-\eta\rangle}\tilde{a}_1(x,\eta)\tilde{a}_2(y,\xi)\,dy\,d\eta,$$

that is, replacing ξ_n by ξ_n/x_n, η_n by η_n/x_n and y_n by $x_n y_n$,

$$(18.3.20)'\quad b(x,\xi)$$
$$=(2\pi)^{-n}\iint_{y_n>0} e^{-i\langle x'-y',\xi'-\eta'\rangle-i(1-y_n)(\xi_n-\eta_n)}a_1(x,\eta)\,a_2(y',x_n y_n,\xi',\xi_n y_n)\,dy\,d\eta.$$

We shall now show that (18.3.20)′ defines for arbitrary m_1, m_2 a continuous bilinear map $S^{m_1}_{la}\times S^{m_2}_{la}\ni(a_1,a_2)\mapsto b\in S^{m_1+m_2}_{la}$. It will then be easy to show that $\tilde{b}(x,D)=\tilde{a}_1(x,D)\tilde{a}_2(x,D)$ without the simplifying assumptions made above.

Choose $\chi\in C_0^\infty(\mathbb{R}_+)$ as in Lemma 18.3.7. To study (18.3.20)′ with a cutoff function $\chi(y_n)$ inserted in the integral we introduce

$$f_{x,\xi}(y,\eta)=a_1(x,\eta)\,a_2(y',x_n y_n,\xi',\xi_n y_n)\,\chi(y_n).$$

Then $(y,\eta)\mapsto(1+|\xi|)^{-m_2}(1+x_n)^\nu f_{x,\xi}(y,\eta)$ is uniformly bounded in S^{m_1} for any ν. In fact, y_n lies between fixed positive bounds in the support, a differentiation with respect to y_n bringing out a factor ξ_n is accompanied by a decrease in the order of a_2, and $|\xi_n|/(1+|\xi'|+|\xi_n y_n|)\le 1/|y_n|$. Hence it follows from Theorem 18.1.7 that

$$b_1(x,\xi)=e^{i\langle D_y,D_\eta\rangle}f_{x,\xi}(y,\eta)|_{y=(x',1),\eta=\xi},$$

is defined and that

$$|b_1(x,\xi)-\sum_{j<N}\langle iD_y,D_\eta\rangle^j a_1(x,\eta)\,a_2(y',y_n x_n,\xi',\xi_n y_n)|_{y=(x',1),\eta=\xi}/j!|$$
$$\le C_{N,\nu}(1+x_n)^{-\nu}(1+|\xi|)^{m_1+m_2-N}.$$

In view of Proposition 18.1.4 it follows that $b_1\in S^{m_1+m_2}_+$ and that b_1 has the stated asymptotic expansion.

Set

$$A_1(x,y)=(2\pi)^{-n}(1-\chi(1-y_n))\int e^{i\langle y,\xi\rangle}a_1(x,\xi)\,d\xi.$$

Since $\chi(1-y_n)=1$ in a neighborhood of 0, the proof of (18.3.11) gives

$$(18.3.11)'\qquad |D_x^\alpha D_y^\beta A_1(x,y)|\le C_{\alpha\beta N}(1+|y|)^{-N}(1+x_n)^{-N}$$

for any N; the constants $C_{\alpha\beta N}$ are semi-norms of a_1 in S^{m_1}. Since $A_1(x,y)=0$ when $y_n>1$ by the lacunary condition, we also have by Taylor's formula

$$(18.3.11)''\quad |D_x^\alpha D_y^\beta A_1(x,y)|\,|y_n-1|^{-N}\le C'_{\alpha\beta N}(1+|y|)^{-2N}(1+x_n)^{-N}.$$

Now we have $b=b_1+b_2$ where

$$b_2(x,\xi)=\int e^{-i\langle x'-y',\xi'\rangle-i(1-y_n)\xi_n}g(x,y,\xi)\,dy,$$
$$g(x,y,\xi)=A_1(x,x'-y',1-y_n)\,a_2(y',x_n y_n,\xi',\xi_n y_n).$$

If we observe again that a differentiation with respect to y_n which brings out a factor ξ_n is accompanied by a decrease of the order of a_2, we obtain

$$|D_x^\alpha D_y^\beta D_\xi^\gamma g(x,y,\xi)| \leq C_{\alpha\beta\gamma N}(1+|x-y|+x_n)^{-N}(1+|\xi|)^{m_2-|\gamma|}$$

in view of (18.3.11)″ and the fact that $1+|\xi'|+|\xi_n y_n| \geq |y_n|(1+|\xi|)/(1+|y_n|)$. Since for every β

$$\xi^\beta b_2(x,\xi) = \int e^{-i\langle x'-y',\xi'\rangle - i(1-y_n)\xi_n}(-D_y)^\beta g(x,y,\xi)\,dy$$

it follows at once that $b_2 \in S_+^{-\infty}$.

When $a_1 \in S_{la}^{-\infty}$ and $a_2 \in S_{la}^{m_2}$, $a_2(x,\xi)=0$ for large $|x|$, we have now shown that $a_1(x,D)a_2(x,D)u = b(x,D)u$ in $\mathbb{R}_+^n$, $u \in \mathscr{S}$, where $b \in S_+^{-\infty}$ is given by (18.3.20). Since this is 0 when $u=0$ in $\mathbb{R}_+^n$ we have $b \in S_{la}^{-\infty}$. If a_2 does not vanish for large $|x|$ we set $a_{2\nu}(x,\xi) = \chi(x/\nu)a_2(x,\xi)$ where $\chi \in C_0^\infty$ is equal to 1 in a neighborhood of 0. Then $\tilde{a}_1(x,D)\tilde{a}_{2\nu}(x,D) = \tilde{b}_\nu(x,D)$ where $b_\nu \in S_{la}^{-\infty}$ and $b_\nu \to b$ in $S_+^{-\infty}$ as $\nu \to \infty$ by the proof above, b always being defined by (18.3.20)′. Hence $b \in S_{la}^{-\infty}$ and $\tilde{a}_1(x,D)\tilde{a}_2(x,D) = \tilde{b}(x,D)$. The hypothesis that $m_1 = -\infty$ is now removed exactly as in the proof of Theorem 18.3.8 which completes the proof.

The proof of Theorem 18.1.11 is now easily modified to a proof of L^2 estimates for totally characteristic operators.

Theorem 18.3.12. *If $a \in S_{la}^0$ then $\tilde{a}(x,D)$ is continuous in $L^2(\mathbb{R}_+^n)$.*

Proof. If $a \in S_{la}^{-n-2}$ then the kernel K of $\tilde{a}(x,D)$ satisfies (18.3.14) and the L^2 continuity follows from Lemma 18.1.12. If $a \in S_{la}^{-k}$ we have

$$\|\tilde{a}(x,D)u\|^2 = (\tilde{b}(x,D)\tilde{a}(x,D)u,u) = (\tilde{c}(x,D)u,u),\quad u \in \mathscr{S},$$

where $\tilde{b}(x,D)$ is the adjoint of $\tilde{a}(x,D)$ and $c \in S_{la}^{-2k}$. If $\tilde{c}(x,D)$ is already known to be L^2 continuous we obtain the same result for $\tilde{a}(x,D)$, so the theorem follows for all $a \in S_{la}^{-k}$ if $k \leq -(n+2)/2$, then if $k \leq -(n+2)/4$ and so on. The proof is then completed by taking an approximate square root satisfying the lacunary condition just as in the proof of Theorem 18.1.11. (Note that $(M^2-|a|^2)^{\frac{1}{2}} - M \in S_+^0$ if $a \in S_+^0$ and $\sup|a| < M \in \mathbb{R}$.) The repetition is left as an exercise for the reader.

Since $\tilde{a}(x,D)$ commutes approximately with differential operators it is possible to extend Theorem 18.3.12 to $H_{(s)}$ spaces:

Theorem 18.3.13. *If $a \in S_{la}^0$ then $\tilde{a}(x,D)$ is continuous in $\dot{H}_{(s)}(\overline{\mathbb{R}}_+^n)$ and in $\bar{H}_{(s)}(\mathbb{R}_+^n)$ for every $s \in \mathbb{R}$.*

Proof. For $s=0$ this is just Theorem 18.3.12. If $u \in \mathscr{S}(\mathbb{R}^n)$ we have by Theorem 18.3.5

$$D_j\tilde{a}(x,D)u = \tilde{a}(x,D)D_ju - i\delta_{jn}\tilde{a}^{(n)}(x,D)D_nu - i\tilde{a}_{(j)}(x,D)u.$$

If s is a non-negative integer for which the theorem is proved we obtain if $u \in \dot{\mathscr{S}}(\overline{\mathbb{R}}^n_+)$

$$\|\tilde{a}(x,D)u\|^2_{(s+1)} = \|\tilde{a}(x,D)u\|^2_{(s)} + \sum_1^n \|D_j \tilde{a}(x,D)u\|^2_{(s)}$$
$$\leqq C(\|u\|^2_{(s)} + \sum_1^n \|D_j u\|^2_{(s)}) = C\|u\|^2_{(s+1)}.$$

Now $\dot{\mathscr{S}}(\overline{\mathbb{R}}^n_+)$ is dense in $\dot{H}_{(s+1)}(\overline{\mathbb{R}}^n_+)$ by Theorem B.2.1 so this proves continuity in $\dot{H}_{(s+1)}(\overline{\mathbb{R}}^n_+)$. Using (B.2.2) we obtain in the same way the continuity in $\bar{H}_{(s+1)}(\mathbb{R}^n_+)$. By Theorem 18.3.8 the adjoint of $\tilde{a}(x,D)$ in $\bar{H}_{(s)}(\mathbb{R}^n_+)$ is for every s the operator $\tilde{b}(x,D)$ in $\dot{H}_{(-s)}(\overline{\mathbb{R}}^n_+)$ where $b \in S^0_{la}$. By duality we therefore obtain the continuity for negative integers s also. The proof will be completed by an interpolation argument close to the proof of Corollary B.1.6, which establishes continuity in $\dot{H}_{(s+k)}(\overline{\mathbb{R}}^n_+)$ if k is an integer and $|s| \leqq \frac{1}{2}$.

Choose $\psi \in C_0^\infty(\mathbb{R}^n_+)$ with $\hat{\psi}(\xi) = 1 + O(|\xi|^2)$ as $\xi \to 0$. To do so we can first take $\phi \in C_0^\infty(\mathbb{R}^n_+)$ with $\hat{\phi}(0) = 1$ and set $\psi = 2\phi - \phi * \phi$, for $\hat{\psi} = 1 - (1-\hat{\phi})^2$ then. Writing $\psi_\varepsilon(x) = \varepsilon^{-n}\psi(x/\varepsilon)$ we have for $|s| \leqq \frac{1}{2}$

$$\int_0^1 \|\psi_\varepsilon * u\|^2_{(k+1)} \varepsilon^{1-2s} d\varepsilon + \int_0^1 \|u - \psi_\varepsilon * u\|^2_{(k-1)} \varepsilon^{-3-2s} d\varepsilon \leqq C\|u\|^2_{(s+k)}$$

when $u \in \dot{\mathscr{S}}$. This is equivalent to the elementary inequalities

$$\int_0^1 |\hat{\psi}(\varepsilon\xi)|^2 \varepsilon^{1-2s} d\varepsilon \leqq C(1+|\xi|^2)^{s-1}, \quad \int_0^1 |1 - \hat{\psi}(\varepsilon\xi)|^2 \varepsilon^{-3-2s} d\varepsilon \leqq C(1+|\xi|^2)^{s+1}$$

Both are obvious if $|\xi| < 1$, since $|1 - \hat{\psi}(\varepsilon\xi)|^2 \varepsilon^{-3-2s}$ is bounded then, and when $|\xi| > 1$ they follow if ε is replaced by $\varepsilon/|\xi|$ and the integrals are extended to $+\infty$. Set $v_\varepsilon = a(x,D)(\psi_\varepsilon * u)$, $w_\varepsilon = a(x,D)(u - \psi_\varepsilon * u)$. Then

$$\int_0^1 (\|v_\varepsilon\|^2_{(k+1)} \varepsilon^{1-2s} d\varepsilon + \|w_\varepsilon\|^2_{(k-1)} \varepsilon^{-3-2s}) d\varepsilon \leqq C_1 \|u\|^2_{(s+k)}$$

in view of the continuity in $\dot{H}_{(k\pm1)}(\overline{\mathbb{R}}^n_+)$, that is,

$$(2\pi)^{-n} \int \int_0^1 (|\hat{v}_\varepsilon(\xi)|^2 (\varepsilon^2 + |\varepsilon\xi|^2)^{k+1} + |\hat{w}_\varepsilon(\xi)|^2 (\varepsilon^2 + |\varepsilon\xi|^2)^{k-1}) \varepsilon^{-1-2(s+k)} d\xi\, d\varepsilon$$
$$\leqq C_1 \|u\|^2_{(s+k)}.$$

If $U = \tilde{a}(x,D)u$ we have $\hat{U} = \hat{v}_\varepsilon + \hat{w}_\varepsilon$, hence

$$|\hat{U}(\xi)|^2 \leqq C_2 (|\hat{v}_\varepsilon(\xi)|^2 (\varepsilon^2 + |\varepsilon\xi|^2)^{k+1} + |\hat{w}_\varepsilon(\xi)|^2 (\varepsilon^2 + |\varepsilon\xi|^2)^{k-1})$$

if $1 < 2\varepsilon(1+|\xi|) < 2$. Since

$$\int_{1 < 2\varepsilon(1+|\xi|) < 2} \varepsilon^{-1-2(s+k)} d\varepsilon \geqq C_3 (1+|\xi|^2)^{s+k}$$

we obtain the required estimate $\|U\|^2_{(s+k)} \leqq C_4 \|u\|^2_{(s+k)}$. The proof is complete.

The interpolation argument is also applicable to Besov spaces; we content ourselves with the case of interest here:

Theorem 18.3.14. *If $a \in S^0_{la}$ then $\tilde{a}(x, D)$ is continuous in ${}^\infty\dot{H}_{(s)}(\overline{\mathbb{R}}^n_+)$ for every s.*

Proof. First we show that for $u \in {}^\infty H_{(s+k)}(\overline{\mathbb{R}}^n_+)$ and $|s| \leqq \frac{1}{2}$, $0 < \varepsilon < 1$,

$$\|\psi_\varepsilon * u\|^2_{(k+1)} \varepsilon^{2-2s} + \|u - \psi_\varepsilon * u\|^2_{(k-1)} \varepsilon^{-2-2s} \leqq C^\infty \|u\|^2_{(k+s)}.$$

This follows if we show with the notation in (B.1.2) that

$$\sum_j \sup_{X_j} (|\hat{\psi}(\varepsilon\xi)|^2 (\varepsilon R_j)^{2-2s} + |1 - \hat{\psi}(\varepsilon\xi)|^2 (\varepsilon R_j)^{-2-2s}) < \infty.$$

The terms where $\varepsilon R_j < 1$ can all be estimated by $(\varepsilon R_j)^{2-2s}$ since $|1 - \hat{\psi}(\varepsilon\xi)|^2 \leqq C|\varepsilon\xi|^4$ then, and those with $\varepsilon R_j > 1$ can be estimated by $(\varepsilon R_j)^{-2-2s}$ since $|\hat{\psi}(\varepsilon\xi)| \leqq C/|\varepsilon\xi|^2$ then. The sum of a geometric series with ratio $\leqq \frac{1}{2}$ is at most twice the largest term which proves the statement. Now we obtain with the notation in the proof of Theorem 18.3.13

$$\varepsilon^{2-2s} \|v_\varepsilon\|^2_{(k+1)} + \varepsilon^{-2-2s} \|w_\varepsilon\|^2_{(k-1)} \leqq C^\infty \|u\|^2_{(k+s)},$$

that is,

$$\varepsilon^{-2(k+s)} \int (|\hat{v}_\varepsilon(\xi)|^2 (\varepsilon^2 + |\varepsilon\xi|^2)^{k+1} + |\hat{w}_\varepsilon(\xi)|^2 (\varepsilon^2 + |\varepsilon\xi|^2)^{k-1}) \, d\xi$$
$$\leqq C^\infty \|u\|^2_{(k+s)}.$$

If we restrict the integration to X_j and take $\varepsilon = 1/R_j$ it follows that

$$R_j^{k+s} \Big(\int_{X_j} |\hat{U}(\xi)|^2 \, d\xi\Big)^{\frac{1}{2}} \leqq C_1^\infty \|u\|^2_{(k+s)}$$

which completes the proof.

We can apply Theorem 18.3.14 to the spaces I^k in Definition 18.2.6:

Corollary 18.3.15. *If $a \in S^m_{la}$ and $u \in I^k(\mathbb{R}^n, \partial\overline{\mathbb{R}}^n_+) \cap \dot{\mathscr{E}}'(\overline{\mathbb{R}}^n_+)$, then*

$$\tilde{a}(x, D) u \in I^k(\mathbb{R}^n, \partial\overline{\mathbb{R}}^n_+) \cap \dot{\mathscr{D}}'(\overline{\mathbb{R}}^n_+).$$

Proof. We must show that $\tilde{P}(x, D)\tilde{a}(x, D) u \in {}^\infty H^{loc}_{(-k-n/4)}$ if $\tilde{P} \in \mathrm{Diff}^{m'}_b(\overline{\mathbb{R}}^n_+)$. Choose an even integer $M > m + m'$ and set $Q(\xi) = |\xi|^M$. By Theorem 18.3.11 we can use the argument in the proof of Theorem 18.1.9 to show that

$$\tilde{P}(x, D)\tilde{a}(x, D) = \tilde{F}(x, D)\tilde{Q}(x, D) + \tilde{G}(x, D)$$

for some $F, G \in S^0_{la}$, if the coefficients of P have compact support. Since $\tilde{Q}(x, D) u$ and u are in ${}^\infty H_{(-k-n/4)}$ by hypothesis, the statement follows now from Theorem 18.3.14.

We shall later on use a dual version of Corollary 18.3.15 to obtain the continuity of $\tilde{a}(x,D)$ in another space which occurs in the study of boundary problems. (See Proposition 18.3.23.)

All operators of order $-\infty$ do not map $\dot{H}_{(s)}$ into $\underline{\dot{H}_{(s')}}$ for any $s'>s$. In fact, let us consider $(\tilde{a}(x,D)u_\varepsilon, v_\varepsilon)=\iint K(x,y)\,u_\varepsilon(y)\,v_\varepsilon(x)\,dx\,dy$ with $u_\varepsilon(y)=\varepsilon^{-\frac{1}{2}}u(y',y_n/\varepsilon)$, $v_\varepsilon(x)=\varepsilon^{-\frac{1}{2}}v(x',x_n/\varepsilon)$ and suitable u, $v\in C_0^\infty$. As $\varepsilon\to 0$ it is easily seen that if $\tilde{a}(x,D)$ is continuous from $\dot{H}_{(s)}$ to $\dot{H}_{(s')}$ and $s'>s$ then the function F in Theorem 18.3.6 must vanish when $t=0$. However, operators of positive order cannot consume more differentiability than in the case of standard pseudo-differential operators:

Theorem 18.3.16. *If* $a\in S^m_{la}$ and $m\geqq 0$, then $\tilde{a}(x,D)$ *is continuous from* $\dot{H}_{(s)}(\overline{\mathbb{R}}^n_+)$ *to* $\dot{H}_{(s-m)}(\overline{\mathbb{R}}^n_+)$ *and from* $\bar{H}_{(s)}(\mathbb{R}^n_+)$ *to* $\bar{H}_{(s-m)}(\mathbb{R}^n_+)$ *for any* $s\in\mathbb{R}$.

Proof. The first statement implies the second one by duality. When proving it we first assume that m is a positive integer and that the statement has already been established for smaller values of m. We can write

$$a(x,\xi)=\sum_1^n \xi_j a_j(x,\xi)+a_0(x,\xi)$$

where $a_j\in S^{m-1}_{la}$. In fact, writing

$$b_j(x,\xi)=\xi_j a(x,\xi)/(1+|\xi|^2),\quad j\neq 0;\qquad b_0(x,\xi)=a(x,\xi)/(1+|\xi|^2);$$

we have $a(x,\xi)=\sum \xi_j b_j(x,\xi)+b_0(x,\xi)$ and can take

$$a_j=(b_j)_\rho,\quad j\neq 0;\qquad a_0=b_0+\sum_1^n \xi_j(b_j-a_j)=a-\sum_1^n \xi_j a_j,$$

with the notation in Lemma 18.3.4. Since

$$\tilde{a}(x,D)\,u=\sum_1^{n-1}\tilde{a}_j(x,D)\,D_j u+x_n\,\tilde{a}_n(x,D)\,D_n u+\tilde{a}_0(x,D)\,u$$

it follows from the inductive hypothesis that a has the stated continuity property. Moreover, $(a,u)\mapsto\tilde{a}(x,D)\,u$ is continuous from $S^m_{la}\times\dot{H}_{(s)}(\overline{\mathbb{R}}^n_+)$ to $\dot{H}_{(s-m)}(\overline{\mathbb{R}}^n_+)$.

A complex interpolation argument similar to that in the proof of Theorem 7.1.12 will now prove the general statement. Set

$$B_z(x,\xi)=a(x,\xi)(1+|\xi|^2)^{(z-m)/2},\qquad A_z=(B_z)_\rho+a-a_\rho.$$

Then A_z is analytic in z, $A_m=a$ and $A_z\in S^{\operatorname{Re} z}_{la}$; the semi-norms can be estimated by a power of $(1+|z|)$ when $\operatorname{Re} z$ is bounded. If M is an integer $>m$ it follows from the first part of the proof that when $\operatorname{Re} z=0$ or $\operatorname{Re} z=M$ we have for some C_s and μ

$$|(A_z(x,D)\,u,(1+|D|^2)^{-\bar{z}/2}\,v)|\leqq C_s(1+|z|)^\mu\,\|u\|_{(s)}\,\|v\|_{(-s)};\qquad u\in\dot{\mathscr{S}}(\overline{\mathbb{R}}^n_+),\ v\in\mathscr{S}(\mathbb{R}^n).$$

When $0 \leqq \operatorname{Re} z \leqq M$ we have a weaker estimate with $\|u\|_{(s)}$ replaced by $\|u\|_{(s+M)}$, for $A_z \in S_{la}^M$ and the semi-norms there can be estimated by a power of $(1+|z|)$. Since $1+|z| \leqq |2z+2|$, $\operatorname{Re} z \geqq 0$, it follows from the Phragmén-Lindelöf theorem that when $0 < \operatorname{Re} z < M$ we have for the same u and v

$$|(A_z(x,D)\,u, (1+|D|^2)^{-\bar{z}/2}\,v)| \leqq C_s |2z+2|^\mu \|u\|_{(s)} \|v\|_{(-s)}.$$

This means that

$$\|A_z(x,D)\,u\|_{(s-\operatorname{Re} z)} \leqq C_s |2z+2|^\mu \|u\|_{(s)}, \qquad u \in \mathscr{S}(\overline{\mathbb{R}}^n_+).$$

When $z=m$ this proves the first part of the theorem when $m>0$, hence the second part when $m<0$. The proof is completed by the argument at the beginning.

To discuss invariance under a change of variables we shall resume the study of the kernel K of $\tilde{a}(x,D)$ begun with Theorem 18.3.6, assuming only that $a \in S_{la}^m$ now. The inverse Fourier transform

$$A(x,y) = (2\pi)^{-n} \int e^{i\langle y,\eta\rangle} a(x,\eta)\, d\eta$$

is then a conormal distribution $\in I^m(\mathbb{R}^{2n}, \{y=0\})$ with principal symbol $a(x,\eta)\,|dx|^{\frac{1}{2}} |d\eta|^{\frac{1}{2}}$ on the normal bundle $\{(x,0,0,\eta)\}$ of the plane $y=0$. (To be quite precise we should observe that a and therefore A is only defined when $x_n \geqq 0$. However, this is obviously inessential since x_n is just a parameter on which A depends in a C^∞ fashion. Quite generally, if X is a manifold with boundary and Y a submanifold intersecting ∂X transversally, then the definition of $I^m(X,Y)$ in Section 18.2 can be applied if we just use local coordinate systems near the boundary such that X is defined by $x_n \geqq 0$, say, and Y by $x_1 = \ldots = x_k = 0$. We leave this slight extension for the reader.) The kernel K of $\tilde{a}(x,D)$ is given by (18.3.12),

$$K(x,y) = A(x, x'-y', (x_n - y_n)/x_n)/x_n, \quad x_n > 0.$$

By (18.3.1) K is a continuous function of x with values in $\mathscr{S}'$ when $x_n \geqq 0$, and $K=0$ when $x_n < 0$. To interpret this we introduce again the symmetric singular coordinates (18.3.8). Since $D(t,r)/D(x_n, y_n) = -1/t$, we obtain if K is transformed as a half density to $k(x',y',t,r) = K(x,y)\,t^{\frac{1}{2}}$

$$k(x',y',t,r) = A(x', t(1+r/2), x'-y', 2r/(2+r))(1+r/2)^{-1}\, t^{-\frac{1}{2}}. \tag{18.3.21}$$

Thus $t^{\frac{1}{2}} k(x',y',t,r)$ is a C^∞ function when $0<|r|<2$, $t \geqq 0$. All derivatives tend to 0 when $|r| \to 2$. In fact, this follows when $r \to -2$ from the fact that (18.3.11) is valid when $|y_n|$ is bounded away from 0. When $r \to 2$ the statement also follows from (18.3.11) and the fact that $A(x,y)=0$ when $y_n > 1$ by the lacunary condition. Defining $k=0$ when $|r| \geqq 2$ we obtain $t^{\frac{1}{2}} k \in I^m(\overline{\mathbb{R}}^{2n}_+, \hat{\Delta})$ where $\overline{\mathbb{R}}^{2n}_+$ is defined by $t \geqq 0$ and

$$\hat{\Delta} = \{(x',y',t,r);\ x'=y', r=0, t \geqq 0\},$$

which corresponds to the diagonal in the original coordinates. With the corresponding parametrization of the conormal bundle of $\hat{\Delta}$ as $\{(x',x',t,0,\xi',-\xi',0,\rho)\}$ the principal symbol of $t^{\frac{1}{2}}k$ is

$$a(x',t,\xi',\rho)\,|dx'|^{\frac{1}{2}}\,|dt|^{\frac{1}{2}}\,|d\xi'|^{\frac{1}{2}}\,|d\rho|^{\frac{1}{2}}. \tag{18.3.22}$$

In fact, $t^{\frac{1}{2}}(1+r/2)^{\frac{1}{2}}k$ is the pullback of A as a half density by the diffeomorphism $(x',y',t,r)\mapsto(x',t(1+r/2),x'-y',2r/(2+r))$. When $r=0$ the differential is $(dx',dy',dt,dr)\mapsto(dx',dt+t\,dr/2,dx'-dy',dr)$; the adjoint maps the element $(x,0,0,\xi)$ in the normal bundle of $\mathbb{R}^n\times\{0\}$ to $(x',x',x_n,0;\ \xi',-\xi',0,\xi_n)$. When $u\in C_0^\infty$ the substitution $y_n=s\,x_n$ gives, with all integrals representing the action of a distribution

$$\int K(x,y)\,u(y)\,dy=\iint k(x',y',x_n(1+s)/2,2(1-s)/(s+1))((1+s)/2)^{-\frac{1}{2}}$$
$$\cdot u(y',s\,x_n)\,x_n^{\frac{1}{2}}\,dy'\,ds.$$

If k_0 is the limit of $t^{\frac{1}{2}}k$ as $t\to 0$ then this converges to

$$\iint k_0(x',y',2(1-s)/(s+1))(2/(1+s))\,u(y',0)\,dy'\,ds$$

when $x_n\to 0$.

Conversely, given a distribution $t^{\frac{1}{2}}k\in I^m(\overline{\mathbb{R}}^{2n}_+,\hat{\Delta})$ of compact support, vanishing when $|r|>2$, we can reverse the argument by introducing for $x_n\geqq 0$

$$A(x,y)=k(x',x'-y',x_n(2-y_n)/2,2\,y_n/(2-y_n))\,x_n^{\frac{1}{2}}(1-y_n/2)^{-\frac{1}{2}} \tag{18.3.21$'$}$$

which vanishes for $y_n>1$, decreases rapidly as $y\to\infty$, and is conormal with respect to the plane $y=0$. From A we return to $a\in S^m_{la}$ by

$$a(x,\eta)=\int A(x,y)\,e^{-i\langle y,\eta\rangle}\,dy.$$

Hence we have an identification of the operators $\tilde{a}(x,D)$, $a\in S^m_{la}$, and the conormal distributions in $I^m(\overline{\mathbb{R}}^{2n}_+,\hat{\Delta})$, at least when the kernels have compact supports. Thus we recover locally the invariance under passage to the adjoint (Theorem 18.3.8). Invariance under coordinate transformations will follow when we have discussed the intrinsic meaning of the new coordinates (18.3.8).

If X is a C^∞ manifold and Y is a C^∞ submanifold, then a new manifold $\hat{X}$, the *blowup* of X along Y, can be defined as the union of $X\setminus Y$ and the projective normal bundle of Y, that is, the quotient by multiplication with real numbers $\neq 0$ of the normal bundle $(T(X)|_Y\setminus T(Y))/T(Y)$. If f and g are C^∞ functions in X vanishing on Y then f/g is well defined in

$$\{x\in X, g(x)\neq 0\}\cup\{\hat{t}\in\hat{X}\setminus X, \langle\hat{t},dg\rangle\neq 0\},$$

as a limit in the second set. We declare such sets to be open and f/g to be in C^∞ there. Together with the C^∞ functions in X lifted to $\hat{X}$ these quotients define a C^∞ structure on $\hat{X}$. To see this we choose local coordinates $x=(x_1,\dots,x_n)$ in X such that Y is defined by $x'=(x_1,\dots,x_k)=0$.

Then $\hat{X}$ can be identified with the quotient of $\mathbb{R}\times S^{k-1}\times\mathbb{R}^{n-k}$ by identification of (t,ω,x'') with

$$i(t,\omega,x'')=(-t,-\omega,x'').$$

Note that the projection π: $(t,\omega,x'')\mapsto(t\omega,x'')\in X$ is defined on $\hat{X}$ since $\pi i=\pi$. If $f,g\in C^\infty(X)$ and both vanish on Y then

$$(f/g)(0,\omega,x'')=\langle f'_{x'}(0,x''),\omega\rangle/\langle g'_{x'}(0,x''),\omega\rangle$$

when the denominator is not 0 and then we have $g(t\omega,x'')\neq 0$ for small $|t|$. Thus f/g is defined and C^∞ in an open subset of $\mathbb{R}\times S^{k-1}\times\mathbb{R}^{n-k}$ and even under the involution i. Where $\omega_k\neq 0$, for example, all such functions are C^∞ functions of $\omega_j/\omega_k (j<k)$, $x_k=t\omega_k$ and $x_{k+1},\dots,x_n$ so we have indeed a C^∞ structure. We could have taken these polar coordinates as the definition of the blowup of course, but then we would have been obliged to prove that the procedure is coordinate free.

If X and Y are two manifolds with boundary, we can blow up the *manifold with corner* $X\times Y$ with respect to the corner manifold $\partial X\times\partial Y$, of codimension 2. Without embedding X and Y in open manifolds we then obtain the *stretched product* $X\hat{\times}Y$ which is the union of $X\times Y\setminus(\partial X\times\partial Y)$ and the projective interior normal bundle of $\partial X\times\partial Y$ in $X\times Y$ defined as

$$(T_{\text{int}}(X)|_{\partial X}+T_{\text{int}}(Y)|_{\partial Y})\setminus T(\partial X\times\partial Y))/T(\partial X\times\partial Y)$$

modulo multiplication by positive reals. Choose local coordinates in X and in Y such that X is defined by $x_n\geq 0$ and Y by $y_m\geq 0$. Then $X\hat{\times}Y$ has local coordinates (x',y',t,r); $t\geq 0$, $-2\leq r\leq 2$; with the map into $X\times Y$ given by

$$(x',y',t,r)\mapsto(x',t(1+r/2),y',t(1-r/2)).$$

Thus the C^∞ structure defined by these coordinates is coordinate free and agrees with our earlier singular coordinates (see Fig. 2).

In the particular case where $X=Y$ the closure in $X\hat{\times}X$ of the diagonal in $(X\setminus\partial X)\times(X\setminus\partial X)$ is a C^∞ manifold $\hat{\Delta}\subset X\hat{\times}X$ which does not meet the corners of $X\hat{\times}X$, and it intersects the boundary transversally. In fact, with our local coordinates it is defined by $x'=y'$ and $r=0$. The restriction to $\hat{\Delta}$ of the C^∞ map $X\hat{\times}X\to X\times X$ is a diffeomorphism on the diagonal Δ in $X\times X$. Now recall that the normal and conormal bundles of Δ in $X\times X$ are naturally isomorphic to $T(X)$ and $T^*(X)$ lifted to Δ by the projection on one of the factors. More precisely, a cotangent vector $\gamma\in T^*_{x_0}(X)$ corresponds to $\pi_1^*\gamma-\pi_2^*\gamma\in N_{x_0,x_0}(\Delta)$ where π_1 and π_2 are the two projections, and if v is a tangent vector to $X\times X$ at (x_0,x_0) then $\pi_{1*}v-\pi_{2*}v$ is a tangent vector$\in T_{x_0}(X)$ depending only on the class of v modulo $T_{x_0,x_0}(\Delta)$. Now the C^∞ map $X\hat{\times}X\to X\times X$ defines maps

(18.3.23) $\quad T(X\hat{\times}X)\to T(X\times X),\qquad T^*(X\times X)|_\Delta\to T^*(X\hat{\times}X)|_{\hat{\Delta}}.$

Let $\tilde{T}(X)$ and $\tilde{T}^*(X)$ be the dual vector bundles obtained by pulling the normal and conormal bundles of $\hat{\Delta}$ in $X \hat{\times} X$ back to X by the inverse of the projection $\hat{\Delta} \to X$, which is a diffeomorphism since it is a product of two diffeomorphisms $\hat{\Delta} \to \Delta \xrightarrow{\pi_1} X$. Then we have natural maps

$$\tilde{T}(X) \to T(X), \quad T^*(X) \to \tilde{T}^*(X). \tag{18.3.24}$$

The first is the composition

$$\tilde{T}(X) \to T(X \hat{\times} X)|_{\hat{\Delta}}/T(\hat{\Delta}) \to T(X \times X)/T(\Delta) \to T(X)$$

where the maps in turn are obtained from the definition of $\tilde{T}$, the first part of (18.3.23) and the discussion of the normal bundle of Δ above. The second map is similarly the composition

$$T^*(X) \to N(\Delta) \to N(\hat{\Delta}) \to \tilde{T}^*(X)$$

where the second map comes from the second part of (18.3.23). With our local coordinates the cotangent vector $\langle \xi, dx \rangle$ at $x \in X$ corresponds to $\langle \xi, dx - dy \rangle$ at $(x,x) \in X \times X$ and is mapped to $\langle \xi', dx' - dy' \rangle + t\xi_n dr$. Thus we have in local coordinates the map

$$T^*(X) \ni (x, \xi) \mapsto (x', x', x_n, 0, \xi', -\xi', 0, x_n \xi_n) \in N(\hat{\Delta}) \cong \tilde{T}^*(X).$$

Every element in the normal bundle of $\hat{\Delta}$ at $(x', x', x_n, 0)$ has a unique representative of the form

$$\sum_1^{n-1} v_j \partial/\partial x_j + v_n \partial/\partial r.$$

By duality we see that (18.3.24) maps it to

$$\sum_1^{n-1} v_j \partial/\partial x_j + x_n v_n \partial/\partial x_n \in T(X).$$

In view of Lemma 18.2.5 it follows that the map $\tilde{T}(X) \to T(X)$ sends the sections of $\tilde{T}(X)$ to the vector fields which are tangent to ∂X. This closes a circle; it was on these vector fields that we set out to model an algebra of pseudo-differential operators in this section.

We sum up the preceding conventions and results in the following

Definition 18.3.17. Let X be a C^∞ manifold with boundary.

a) The stretched product $X \hat{\times} X$ is a C^∞ manifold with corner obtained by replacing $\partial X \times \partial X$ in $X \times X$ by the interior projective normal bundle. There is a natural C^∞ map $X \hat{\times} X \to X \times X$.

b) The diagonal Δ in $X \times X$ is the diffeomorphic image of a manifold $\hat{\Delta}$ in $X \hat{\times} X$ which only intersects the new smooth boundary and does so transversally.

c) The compressed cotangent bundle $\tilde{T}^*(X)$ is the pullback of the conormal bundle of $\hat{\Delta}$ in $X \hat{\times} X$ by the inverse of the diffeomorphic projection on

X. There is a natural map $T^*(X) \to \tilde{T}^*(X)$ which maps $T_x^*(X)$ linearly to $\tilde{T}_x^*(X)$, bijectively if $x \in X \setminus \partial X$ and with kernel $N_x(\partial X)$ and a hyperplane as range if $x \in \partial X$. The range can then be identified with $T_x^*(X)/N_x(\partial X) \cong T_x^*(\partial X)$ which makes $T^*(\partial X)$ a subbundle of $\tilde{T}^*(X)|_{\partial X}$. The sections of the dual bundle $\tilde{T}(X)$ are mapped in $T(X)$ to the vector fields tangent to ∂X.

The pullback of the symplectic form in $T^*(X)$ (see (6.4.8)) to $\tilde{T}^*(X)$, thus to the conormal bundle of $\hat{\Delta}$ in $X \hat{\times} X$, is a symplectic form with singularity at ∂X. In the local coordinates above it is given by

$$\sum_1^{n-1} d\xi_j \wedge dx_j + t^{-1} d\rho \wedge dt. \tag{18.3.25}$$

Thus the half density $|d\xi'|^{\frac{1}{2}} |dx'|^{\frac{1}{2}} t^{-\frac{1}{2}} |d\rho|^{\frac{1}{2}} |dt|^{\frac{1}{2}}$ is invariantly defined. The principal symbol of k, which is $t^{-\frac{1}{2}}$ times (18.3.22), can therefore invariantly be identified with the function a on $N(\hat{\Delta}) \setminus 0$, that is, on the compressed cotangent bundle $\tilde{T}^*(X) \setminus 0$.

We have now developed all that is needed for a global calculus:

Definition 18.3.18. If X is a C^∞ manifold with boundary then the space $\Psi_b^m(X; \Omega^{\frac{1}{2}}, \Omega^{\frac{1}{2}})$ of totally characteristic pseudo-differential operators A of order m on half densities in X is the set of all continuous linear maps $C_0^\infty(X, \Omega^{\frac{1}{2}}) \to C^\infty(X, \Omega^{\frac{1}{2}})$ with Schwartz kernel K obtained by pushforward from the stretched product $X \hat{\times} X$ to $X \times X$ of a distribution half density k such that $kt^{\frac{1}{2}} \in I^m(X \hat{\times} X, \hat{\Delta})$ and k vanishes of infinite order on $\partial(X \times X) \setminus (\partial X \times \partial X)$. Here t is a C^∞ function in $X \hat{\times} X$ which is positive in $X \times X \setminus (\partial X \times \partial X)$ and vanishes simply on the other part of the boundary.

The pushforward is defined as follows: If f: $X \hat{\times} X \to X \times X$ is the natural C^∞ map and ϕ is a half density of compact support in $X \times X$, then $f^* \phi$ is a half density of compact support in $X \hat{\times} X$ so the equation

$$K(\phi) = k(f^* \phi) \tag{18.3.26}$$

defines a distribution half density K in $X \times X$.

Locally A can be defined by (18.3.1). As we have just seen this leads to a principal symbol isomorphism

$$\Psi_b^m(X; \Omega^{\frac{1}{2}}, \Omega^{\frac{1}{2}}) / \Psi_b^{m-1}(X; \Omega^{\frac{1}{2}}, \Omega^{\frac{1}{2}}) \cong S^m(\tilde{T}^*(X)) / S^{m-1}(\tilde{T}^*(X)) \tag{18.3.27}$$

where $\tilde{T}^*(X)$ is the compressed cotangent bundle on X. The adjoint of an operator in Ψ_b^m is in Ψ_b^m, and the principal symbol is obtained by complex conjugation. If $A \in \Psi_b^m$ and $B \in \Psi_b^{m'}$ are properly supported then $AB \in \Psi_b^{m+m'}$ is properly supported and the principal symbol of AB is the product of those of A and of B; the isomorphism (18.3.27) also holds if one takes only properly supported operators in the left-hand side. All these basic facts of the calculus as well as extensions where vector bundles are present follow just as in Section 18.1. One just has to replace say Theorems 18.1.6, 18.1.7,

18.1.8, 18.1.17 by Theorems 18.3.5, 18.3.8, 18.3.11 and the connection with conormal distributions on $X \hat{\times} X$ established above. We have also proved that all operators in Ψ_b^m map $\dot{H}_{(s)}^{\text{comp}}(X)$ continuously into $\dot{H}_{(s-m)}^{\text{loc}}(X)$ if $m \geqq 0$.

The spaces $\dot{H}_{(s)}^{\text{comp}}(X), \dot{\mathscr{D}}'(X), \dot{\mathscr{E}}'(X), C_0^\infty(X), \ldots$ are all defined without any reference to an extension of X to an open manifold. So is the subspace

$$\dot{\mathscr{A}}^{(m)}(X) = I^m(X, \partial X) \subset \dot{\mathscr{D}}'(X) \tag{18.3.28}$$

of distributions u remaining in ${}^\infty\dot{H}_{(-m-n/4)}^{\text{loc}}$, $n = \dim X$, after any totally characteristic differential operator has been applied to u. (We tacitly allow values in a vector bundle but do not wish to burden the notation by making this explicit when it is not essential.) By Theorem 18.3.9 every operator in $\Psi_b^{-\infty}$ maps $\dot{\mathscr{E}}'(X)$ into $\dot{\mathscr{A}}(X) = \bigcup_m \dot{\mathscr{A}}^{(m)}(X)$, and by Corollary 18.3.15 $\dot{\mathscr{A}}_{\text{comp}}^{(m)}(X)$ is mapped into $\dot{\mathscr{A}}^{(m)}(X)$ by any operator in $\Psi_b^\infty = \bigcup_k \Psi_b^k$. In these respects the relation of $\dot{\mathscr{A}}(X)$ to totally characteristic operators is similar to the relation of $C^\infty(X)$ to pseudo-differential operators when X is an open manifold. It is therefore natural to expect that the dual space of $\dot{\mathscr{A}}$ will be a useful space of distributions, and we shall introduce it after a preparatory lemma.

Lemma 18.3.19. *There exist linear smoothing operators Q_ε: $\dot{\mathscr{D}}'(X) \to \dot{C}^\infty(X)$ such that $Q_\varepsilon u \to u$ in $\dot{\mathscr{A}}^{(m)}$, $\varepsilon \to 0$, whenever $u \in \dot{\mathscr{A}}^{(m')}$ for some $m' < m$. For every compact set $K \subset X$ there is another compact set $K' \subset X$ such that* $\operatorname{supp} Q_\varepsilon u \subset K'$ *if* $\operatorname{supp} u \subset K$ *and* $0 < \varepsilon < 1$.

Proof. By a partition of unity the proof is reduced to the case $X = \overline{\mathbb{R}}_+^n$. Choose $\chi \in C_0^\infty(\mathbb{R}_+^n)$ with $\hat{\chi}(0) = 1$ and set $Q_\varepsilon u = \chi_\varepsilon * u$, $\hat{\chi}_\varepsilon(\xi) = \hat{\chi}(\varepsilon\xi)$. Then the statement on the supports is obvious. Taking $\chi(x) = \phi(x')\psi(x_n)$, $x' = (x_1, \ldots, x_{n-1})$, we obtain if

$$u = \int a(x', \xi_n) e^{ix_n\xi_n} d\xi_n$$

that

$$Q_\varepsilon u = \int a_\varepsilon(x', \xi_n) e^{ix_n\xi_n} d\xi_n; \quad a_\varepsilon(x', \xi_n) = \hat{\psi}(\varepsilon\xi_n) \int a(x' - \varepsilon y', \xi_n) \phi(y') dy'.$$

If $a \in S^{\mu'}$ and $\mu > \mu'$ it follows in view of Proposition 18.1.2 that $a_\varepsilon \to a$ in S^μ when $\varepsilon \to 0$.

Definition 18.3.20. By $\mathscr{A}'(X)$ we shall denote the set of all $u \in \dot{\mathscr{D}}'(X)$ such that for every compact set $K \subset X$ and every $m \geqq -(n+2)/4$ the form $C_0^\infty(K) \ni \phi \mapsto u(\phi)$ is continuous in the topology of $\dot{\mathscr{A}}^{(m)}$, defined in (18.3.28).

We recall from the discussion after Definition 18.2.13 that $C_0^\infty(X) \subset \dot{\mathscr{A}}^{(m)}$ if $m \geqq -(n+2)/4$ which is the reason for this condition in the definition. Since the embedding $\dot{\mathscr{A}}^{(m')} \to \dot{\mathscr{A}}^{(m)}$ is continuous when $m' < m$ the continuity condition becomes stronger when m increases, and $\phi \mapsto u(\phi)$ is continuous in the $\dot{\mathscr{A}}^{(m)}$ topology for every m if $\phi \in C_0^\infty(K \cap X^\circ)$ where $X^\circ = X \setminus \partial X$ is the interior of the manifold. By Lemma 18.3.19 the restriction of $u(\phi)$ to $C_0^\infty(X^\circ)$ already determines a continuous extension to $\dot{\mathscr{A}}^{(m)} \cap \dot{\mathscr{E}}'$ uniquely. In particular, the restriction

$$\mathscr{A}'(X) \ni u \mapsto u|_{X^\circ} \in \bar{\mathscr{D}}'(X^\circ)$$

is injective. It is really the range of this map which is dual to $\dot{\mathscr{A}}$ which must be kept in mind when arguing by duality.

As topology in $\mathscr{A}'$ we shall use the weak topology defined by the seminorms $u \mapsto |u(\phi)|$ when $\phi \in \dot{\mathscr{A}} \cap \dot{\mathscr{E}}'$. Now $C^\infty(X)$ is a subset of $\mathscr{A}'$ since the topology in $\dot{\mathscr{A}}^{(m)}$ is stronger than that in ${}^\infty\dot{H}^{\text{loc}}_{(-m-n/4)}$, and $C_0^\infty(X)$ is weakly dense for if $\phi \in \dot{\mathscr{A}} \cap \dot{\mathscr{E}}'$ then $u(\phi) = 0$ for all $u \in C_0^\infty(X)$ implies $\phi = 0$.

Proposition 18.3.21. *There is a unique continuous restriction map $\mathscr{A}'(X) \to \mathscr{D}'(\partial X)$ which agrees with the standard restriction on $C^\infty(X)$.*

Proof. To underline the invariance we consider distributions with values in a vector bundle E. If $u \in C_0^\infty(X, E)$ then

$$\langle u|_{\partial X}, \phi \rangle = \langle u, T\phi \rangle, \quad \phi \in C_0^\infty(\partial X, E' \otimes \Omega(\partial X)) \tag{18.3.29}$$

where Ω is the density bundle and $T\phi = \phi \otimes \delta(x_n)$ if x_n as usual is a local coordinate vanishing on ∂X. This is independent of the choice of local coordinates since $\delta(x_n)$ is a distribution density on $\mathbb{R}$. Now the map

$$C_0^\infty(\partial X, E' \otimes \Omega(\partial X)) \ni \phi \mapsto T\phi \in \dot{\mathscr{A}}^{(2-n)/4}(X, E' \otimes \Omega(X))$$

is continuous so (18.3.29) defines a weakly continuous map $u \mapsto u|_{\partial X}$. Since the uniqueness is obvious this completes the proof.

Differentiation is continuous from $\dot{\mathscr{A}}^{(m)}$ to $\dot{\mathscr{A}}^{(m+1)}$. This implies that the space of restrictions to X° of elements in $\mathscr{A}'$ is invariant for differentiation, so using Proposition 18.3.21 we can define boundary values of arbitrary derivatives of u. (In view of Corollary 8.2.7 this means that the elements of $\mathscr{A}'$ have a certain regularity at $N(\partial X)$ although $\mathscr{E}'(X^\circ)$ is a subset of $\mathscr{A}'(X)$.) However, differentiation does not preserve the space $\mathscr{A}'$ unless one takes boundary terms into account as we already did in Theorem 3.1.9. To simplify the statement we take $X = \bar{\mathbb{R}}^n_+$:

Proposition 18.3.22. *If $u \in \mathscr{A}'(\bar{\mathbb{R}}^n_+)$ then $D_j u + i\delta_{jn} u|_{x_n=0} \otimes \delta(x_n) \in \mathscr{A}'(\bar{\mathbb{R}}^n_+)$; the restriction to $\mathbb{R}^n_+$ is equal to D_j applied to the restriction of u.*

Proof. $D_j u$ is defined by regarding u as an element in $\dot{\mathscr{D}}'(\bar{\mathbb{R}}^n_+)$, thus

$$\langle D_j u, \phi \rangle = -\langle u, D_j \phi \rangle, \quad \phi \in C_0^\infty(\bar{\mathbb{R}}^n_+).$$

If ϕ_0 is the element in $\dot{\mathscr{A}}$ defined by ϕ, that is, $\phi_0 = \phi$ when $x_n \geqq 0$ and $\phi_0 = 0$ when $x_n < 0$, then

$$D_j \phi_0 = (D_j \phi)_0 - i\delta_{jn} \phi(.,0) \otimes \delta(x_n).$$

The inclusion $C_0^\infty(\overline{\mathbb{R}}^n_+)\subset\dot{\mathscr{A}}^{(m)}$ identifies $D_j\phi$ with $(D_j\phi)_0$. Thus

$$\langle D_j u, \phi\rangle = -\langle u, D_j\phi_0\rangle - i\,\delta_{jn}\langle u, \phi(.,0)\otimes\delta(x_n)\rangle.$$

Since

$$\langle u, \phi(.,0)\otimes\delta(x_n)\rangle = \langle u|_{x_n=0}, \phi(.,0)\rangle = \langle u|_{x_n=0}\otimes\delta(x_n), \phi\rangle,$$

and

$$\dot{\mathscr{A}}^{(m)}\cap\mathscr{E}'(K)\ni\psi\mapsto\langle u, D_j\psi\rangle$$

is continuous for every m and every compact set K, the proposition is proved.

Proposition 18.3.23. *If $B\in\Psi_b^m(X)$ is properly supported and $u\in\mathscr{A}'(X)$ then $Bu\in\mathscr{A}'(X)$.*

Proof. $Bu\in\dot{\mathscr{D}}'(X)$ and if $\phi\in C_0^\infty(K)$, $K\Subset X$, then

$$(Bu, \phi) = (u, B^*\phi).$$

Here $B^*\phi\in C_0^\infty(X)$ has support in a fixed compact set, and the continuity in the $\dot{\mathscr{A}}^{(m)}$ topology follows since

$$\mathscr{E}'(K)\cap\dot{\mathscr{A}}^{(m)}\ni\phi\mapsto B^*\phi\in\dot{\mathscr{A}}^{(m)}$$

is continuous by Corollary 18.3.15.

Note that (18.3.7) can now be extended by continuity to $u\in\mathscr{A}'$.

The elements of $\dot{\mathscr{A}}(X)$ are smooth in the interior of X and have tangential smoothness at the boundary while those in $\mathscr{A}'(X)$ have normal derivatives on ∂X of all orders. This suggests the following

Proposition 18.3.24. *On any C^∞ manifold X with boundary one has*

$$\mathscr{A}'(X)\cap\dot{\mathscr{A}}(X) = C^\infty(X).$$

Proof. It is obvious that $C^\infty(X)\subset\mathscr{A}'(X)\cap\dot{\mathscr{A}}(X)$. To prove the opposite inclusion it suffices to show that if $u\in\mathscr{E}'(\overline{\mathbb{R}}^n_+)\cap\mathscr{A}'(\overline{\mathbb{R}}^n_+)\cap\dot{\mathscr{A}}(\overline{\mathbb{R}}^n_+)$ then $u\in C_0^\infty(\overline{\mathbb{R}}^n_+)$. Since $u\in\dot{\mathscr{A}}$ we have $u\in C^\infty(\mathbb{R}^n_+)$. From Proposition 18.3.22 it follows that for every α there is an element $u_\alpha\in\mathscr{A}'$ such that $D^\alpha u - u_\alpha$ has support in $\partial\mathbb{R}^n_+$.

Choose m so that $u\in\dot{\mathscr{A}}^{(m)}$. Then $v = x_n^N D^\alpha u\in\dot{\mathscr{A}}^{(m+\alpha_n-N)}$, that is,

$$D_{x'}^{\beta'}(x_n D_n)^{\beta_n} v\in{}^\infty H^{\mathrm{loc}}_{(N-\alpha_n-m-n/4)}\subset\dot{L}^2 \qquad \text{for all } \beta,$$

where the last inclusion holds if $N>\alpha_n+m+n/4$. Let $\chi\in C_0^\infty(\mathbb{R})$, $0\leqq\chi\leqq 1$, and $\chi=1$ in a neighborhood of 0. Set $\chi^\varepsilon(x)=\chi(x_n/\varepsilon)$. Then

$$D_{x'}^{\beta'}(x_n D_n)^{\beta_n}\chi^\varepsilon v \text{ is bounded in } \dot{L}^2\subset{}^\infty H_{(0)} \text{ for all } \beta, 0<\varepsilon\leqq 1.$$

In fact, $[x_n D_n, \chi^\varepsilon]=\chi_1^\varepsilon$ if $\chi_1(t)=t\,D_t\chi(t)$ so we can commute χ^ε through to the left, obtaining a sum of similar terms with a factor χ_j^ε to the left instead, $\chi_j(t)$

$=(tD_t)^j\chi(t)$ and the L^2 estimate is obvious for them. Thus $(1-\chi^\varepsilon)v$ is bounded in $\dot{\mathscr{A}}^{(-n/4)}$ for $0<\varepsilon\leqq 1$. Since $u_\alpha\in\mathscr{A}'$ and $D^\alpha u=u_\alpha$ when $x_n>0$, it follows that

$$(D^\alpha u,(1-\chi^\varepsilon)v)=\int |D^\alpha u|^2 x_n^N(1-\chi^\varepsilon(x_n))\,dx$$

is bounded when $\varepsilon\to 0$, hence

$$\int_{x_n>0} |D^\alpha u|^2 x_n^N\,dx<\infty \text{ if } N>\alpha_n+m+n/4.$$

Now we recall Hardy's inequality for functions $v\in C^\infty(\mathbb{R}_+)$ vanishing far away,

$$\int_0^\infty |v|^2 t^{2\mu}\,dt\leqq 4/(2\mu+1)^2\int_0^\infty |v'|^2 t^{2\mu+2}\,dt,\qquad \mu\geqq 0.$$

In the proof one can assume that v is smooth at 0 and integrate by parts, which gives

$$\|v t^\mu\|^2=-2/(2\mu+1)\int_0^\infty \operatorname{Re} v\bar{v}'\, t^{2\mu+1}\,dt\leqq 2/(2\mu+1)\,\|t^\mu v\|\,\|t^{\mu+1}v'\|.$$

Hardy's inequality follows after cancellation of a factor. Now we obtain

$$\int_{x_n>0} |D^\alpha u|^2\,dx\leqq 4\int_{x_n>0} |D^\alpha D_n^\nu u|^2 x_n^{2\nu}\,dx<\infty$$

if ν is a positive integer with $2\nu>\alpha_n+\nu+m+n/4$, that is, $\nu>\alpha_n+m+n/4$. In view of Theorem B.2.8 it follows that $u=U_1+U_2$ where $U_1\in C_0^\infty(\overline{\mathbb{R}}^n_+)$, considered as an element of $\mathscr{A}'$, and $x_n=0$ in $\operatorname{supp} U_2$. Thus $U_2\in\mathscr{A}'$ so $U_2=0$ in view of Lemma 18.3.19, which completes the proof.

For the wave front set of a distribution in a manifold with boundary we can now give a definition parallel to the characterization in Theorem 18.1.27:

Definition 18.3.25. If $u\in\dot{\mathscr{D}}'(X)$ then $WF_b(u)\subset\tilde{T}^*(X)$ is defined by

$$WF_b(u)=\bigcap \operatorname{Char} B \tag{18.3.30}$$

with intersection over all properly supported $B\in\Psi_b^0(X)$ such that $Bu\in\dot{\mathscr{A}}(X)$.

Here $\operatorname{Char} B\subset\tilde{T}^*(X)\setminus 0$ is the set of all $(x,\xi)\in\tilde{T}^*(X)\setminus 0$ such that the principal symbol b of B is not invertible at (x,ξ) (cf. Definition 18.1.25). We have used the space $\dot{\mathscr{A}}(X)$ and not $C^\infty(X)$ because it contains the residual terms in the calculus by Theorem 18.3.9. Thus it follows at once from the definition and Theorem 18.3.9 that $WF_b(Au)\subset\Gamma$ if the symbol of A is of order $-\infty$ outside the closed conic set $\Gamma\subset\tilde{T}^*(X)\setminus 0$. However, $C^\infty(X)$ may be used if we assume more about u:

Proposition 18.3.26. *If* $u\in\mathscr{A}'(X)$ *then* (18.3.30) *is valid with the intersection taken over all properly supported* $B\in\Psi_b^0(X)$ *such that* $Bu\in C^\infty(X)$.

Proof. If $u\in\mathscr{A}'(X)$ and $Bu\in\dot{\mathscr{A}}(X)$ as in Definition 18.3.25 then $Bu\in C^\infty(X)$ by Proposition 18.3.24, for $Bu\in\mathscr{A}'(X)$ by Proposition 18.3.23.

Theorem 18.3.27. *If X is a C^∞ manifold with boundary ∂X and interior X°, and $u\in\dot{\mathscr{D}}'(X)$ then*

(i) $WF_b(u)|_{X^\circ}=WF(u|_{X^\circ})$.

(ii) $WF_b(Bu)\subset WF_b(u)$ *if $B\in\Psi_b^m(X)$ is properly supported or if B is a differential operator with C^∞ coefficients.*

(iii) *If $WF_b(u)=\emptyset$ then $u\in\dot{\mathscr{A}}(X)$, hence $u\in C^\infty(X)$ if $u\in\mathscr{A}'(X)$.*

(iv) $WF(u|_{\partial X})\subset WF_b(u)\cap T^*(\partial X)$, $u\in\mathscr{A}'(X)$.

Note that $\tilde{T}^*(X)|_{X^\circ}$ is identified with $T^*(X^\circ)$ and that the map $T^*(X)|_{\partial X}\to\tilde{T}^*(X)|_{\partial X}$ defines an embedding $T^*(\partial X)\to\tilde{T}^*(X)|_{\partial X}$.

Proof of Theorem 18.3.27. (i) follows since every $B\in\Psi^0(X^\circ)$ with kernel of compact support in $X^\circ\times X^\circ$ is also in $\Psi_b^0(X)$ and vice versa. To prove (ii) assume that $\gamma\in\tilde{T}^*(X)\smallsetminus 0\smallsetminus WF_b(u)$, and choose $B_1\in\Psi_b^0(X)$ properly supported and non-characteristic at γ so that $B_1u\in\dot{\mathscr{A}}$. We can then find $C_1\in\Psi_b^0$ properly supported so that $C_1B_1=I+R_1$ where the symbol of R_1 (in local coordinates) is of order $-\infty$ in a conic neighborhood of γ. Thus

$$CBu=CBC_1B_1u-CBR_1u\in\dot{\mathscr{A}}$$

by Corollary 18.3.15 and Theorem 18.3.9 if the symbol of C is of order $-\infty$ outside a sufficiently small conic neighborhood of γ. Hence $\gamma\notin WF_b(Bu)$. It remains to prove (ii) when $X=\overline{\mathbb{R}}^n_+$, $u\in\dot{\mathscr{D}}'(\overline{\mathbb{R}}^n_+)$ and $B=D_j$. If $\tilde{a}(x,D)u\in\dot{\mathscr{A}}$ and a is non-characteristic at $(0,\xi_0)$ then we can choose $b(x,\xi)$ non-characteristic at $(0,\xi_0)$ so that $b(x,\xi)=b(0,\xi)$ when x is in a neighborhood U of 0, and the support of b is in a cone where a is non-characteristic. Then we have $\tilde{b}(x,D)=\tilde{c}(x,D)\,\tilde{a}(x,D)+\tilde{r}(x,D)$ where $c\in S^0_{la}$ and $r\in S^{-\infty}_{la}$, so $\tilde{b}(x,D)u\in\dot{\mathscr{A}}$ by Corollary 18.3.15 and Theorem 18.3.9. Choose $\chi\in C_0^\infty(U)$ with $\chi(0)\neq 0$. Then we have by (18.3.6)

$$\chi\tilde{b}(x,D)D_ju=\chi D_j\tilde{b}(x,D)u+\chi i\delta_{jn}\tilde{b}^{(n)}(x,D)D_nu.$$

Since χb and $\chi(b-ib^{(n)})$ are non-characteristic at $(0,\xi_0)$ the statement (ii) follows. To prove (iii) we choose for $\phi\in C_0^\infty(X)$ properly supported $B_j\in\Psi_b^0(X)$, $j=1,\dots,N$, with $B_ju\in\dot{\mathscr{A}}$ and $\bigcap\operatorname{Char}B_j$ empty over supp ϕ. Then we can choose $C_1,\dots,C_N\in\Psi_b^0(X)$ properly supported so that

$$C_1B_1+\dots+C_NB_N=\phi+R$$

where $R\in\Psi_b^{-\infty}$ and ϕ denotes multiplication by ϕ. Since $C_jB_ju\in\dot{\mathscr{A}}$ and $Ru\in\dot{\mathscr{A}}$, by Corollary 18.3.15 and Theorem 18.3.9, it follows that $\phi u\in\dot{\mathscr{A}}$. The last statement in (iii) follows from Proposition 18.3.24. To prove (iv), finally, we assume that $\gamma\in T^*(\partial X)\smallsetminus 0$ is not in $WF_b(u)$. Then we can find $B\in\Psi_b^0(X)$ properly supported with $\gamma\notin\operatorname{Char}B$ and $Bu\in C^\infty(X)$. We have

$$Bu|_{\partial X}=B_0(u|_{\partial X})$$

where $B_0 \in \Psi^0(\partial X)$; the principal symbol is equal to the restriction to $T^*(\partial X)$ of that of B by (18.3.7). Hence B_0 is non-characteristic at γ, so $\gamma \notin WF(u|_{\partial X})$. The proof is complete.

From (18.3.30) it follows at once that

$$WF_b(u) \subset WF_b(Bu) \cup \operatorname{Char} B, \quad u \in \dot{\mathscr{D}}'(X), \tag{18.3.31}$$

if $B \in \Psi_b^m(X)$ is properly supported. Assume now instead that P is any differential operator with coefficients in $C^\infty(X)$, of order m, and let $u \in \dot{\mathscr{D}}'(X)$, $Pu = f$. If ϕ is a C^∞ function vanishing simply on ∂X, then $\phi^m P \in \operatorname{Diff}_b^m$ and $\phi^m Pu = \phi^m f$. With the usual local coordinates at the boundary we can take $\phi = x_n$ and have

$$x_n^m P(x,D) = \sum_{|\alpha| \leq m} x_n^m a_\alpha(x) D^\alpha u = \sum x_n^{m-\alpha_n} a_\alpha(x) x_n^{\alpha_n} D^\alpha u$$

which shows that the principal symbol when $x_n = 0$ is $a_\alpha(x)\xi_n^m$, where $\alpha = (0, \dots, 0, m)$. Thus it is identically 0 in some fibers unless ∂X is non-characteristic, and then it vanishes precisely when $\xi_n = 0$, that is, in $T^*(\partial X)$. Thus we have

Proposition 18.3.28. *Let P be a differential operator of order m with coefficients in $C^\infty(X)$, let $\phi \in C^\infty(X)$ vanish simply on ∂X and assume that ∂X is non-characteristic for P. Then*

$$WF_b(u)|_{\partial X} \subset WF_b(\phi^m Pu)_{\partial X} \cup T^*(\partial X), \quad u \in \dot{\mathscr{D}}'(X). \tag{18.3.32}$$

We shall now prove a result which is closely related to Theorem 4.4.8′. As in that statement we just assume partial hypoellipticity at the boundary and not that the boundary is non-characteristic. Let X_0 be an open set in $\mathbb{R}^{n-1}$ and set $X = X_0 \times [0, c) \subset \mathbb{R}^n$, considered as a manifold with boundary $\partial X = X_0 \times \{0\}$ and interior $X^\circ = X_0 \times (0, c)$.

Proposition 18.3.29. *Let $u \in \mathscr{D}'(X^\circ)$ satisfy a differential equation of the form*

$$Pu = D_n^m u + a_{m-1} D_n^{m-1} u + \dots + a_0 u = f \quad \text{in } X^\circ$$

where a_j is a differential operator in $x' = (x_1, \dots, x_{n-1})$ with coefficients in $C^\infty(X)$, and $f \in \mathscr{A}'(X)$. Then there is a unique $U \in \mathscr{A}'(X)$ with restriction u to X°. We have $x_n^m(PU - f) = 0$ in $\dot{\mathscr{D}}'(X)$, and there is no other $U \in \dot{\mathscr{D}}'(X)$ with this property. Here u may have values in $\mathbb{C}^N$, the coefficients of a_j being $N \times N$ matrices then.

Proof. It suffices to prove the proposition when $m = 1$. In fact, if $u_j = D_n^j u$, $0 \leq j < m$, then

$$D_n u_{m-1} + \sum_{j \leq m-1} a_j u_j = f; \quad D_n u_j = u_{j+1}, \quad j < m-1.$$

The statement with $m=1$ will then give extensions $U_j \in \mathscr{A}'$ satisfying these equations multiplied by x_n. In particular, $x_n(D_n U_j - U_{j+1})=0$ for $j<m-1$, hence

$$x_n^j U_j = x_n^j D_n U_{j-1} = \ldots = x_n^j D_n^{j-1} U_1 = x_n^j D_n^j U_0, \quad j<m,$$

for if $k<j$ then $x_n^j D_n^k$ can be rewritten as a sum with a factor x_n to the right. Thus $x_n^m(PU_0 - f)=0$. If $V\in \dot{\mathscr{D}}'(X)$, $\operatorname{supp} V \subset \partial X$ and $x_n^m PV=0$ then $V=0$. In fact, if $Y \Subset X$ we can by Theorem 2.3.5 write in Y

$$V=\sum_0^{\mu} v_j \otimes \delta^{(j)}(x_n).$$

Now

$$x_n^m D_n^k \delta^{(j)}(x_n) = c_{mkj}\, \delta^{(j+k-m)}$$

which should be read as 0 if $m>k+j$; here $c_{mmj} \neq 0$. Thus the equation $x_n^m PV=0$ gives $v_\mu=0$, hence $v_{\mu-1}=0$ and so on.

From now on we assume $m=1$. Let $K_0 \Subset X_0$ and set $K=K_0\times[0,c/2]$. By Lemma 18.3.19 it suffices to show that $\phi \mapsto \langle u,\phi\rangle$ is continuous in the topology of $\mathscr{A}^{(k)}$ for every k if $\phi\in C_0^\infty(K)$. Recall that the semi-norms in this topology are

$$\phi \mapsto {}^\infty\| x_n^{\alpha_n} D^\alpha \phi\|_{(-k-n/4)}. \tag{18.3.33}$$

By hypothesis $|u(\phi)| \leqq C\|\phi\|_{(s)}$ for some s so the continuity is clear when $k< -s-n/4$. We must prove that continuity in $\dot{\mathscr{A}}^{(k)}$ follows from continuity in $\dot{\mathscr{A}}^{(k-1)}$. To do so we set $\psi = i\phi * (\delta\otimes H(x_n))$, that is,

$$\psi(x) = i\int_{-\infty}^{x_n} \phi(x',t)\,dt.$$

Then $D_n\psi=\phi$. Let $\chi\in C_0^\infty(-c,c)$ be equal to 1 in $(-c/2,c/2)$. With $\chi_n(x)=\chi(x_n)$ we have for $\phi\in C_0^\infty(K)$

$$\begin{aligned}\langle u,\phi\rangle &= \langle u, \chi_n D_n\psi\rangle = -\langle D_n u, \chi_n\psi\rangle - \langle u, \psi D_n\chi_n\rangle\\ &= -\langle f,\chi_n\psi\rangle + \langle u, {}^t a_0(\chi_n\psi) - \psi D_n \chi_n\rangle.\end{aligned}$$

Writing $H(x_n)=h_0(x_n)+h_2(x_n)$ where $h_2\in C^\infty$ and the support of h_0 is close to 0, we can estimate ψ in $C^\infty(X^\circ)$ by the semi-norms (18.3.33) on ϕ. We can also estimate

$${}^\infty\| D_j x_n^{\alpha_n} D^\alpha(\chi_n\psi)\|_{(-k-n/4)}$$

in this way for arbitrary j and α. It suffices to do so for $j=n$ for this implies an estimate of ${}^\infty\|x_n^{\alpha_n} D^\alpha(\chi_n\psi)\|_{(-k-n/4)}$. Since $\chi_n=1$ near 0 it suffices to estimate

$${}^\infty\|\chi_n D_n x_n^{\alpha_n} D^\alpha \psi\|_{(-k-n/4)}.$$

Now

$$D_n x_n^{\alpha_n} D_n^{\alpha_n}\psi = x_n^{\alpha_n} D_n^{\alpha_n}\phi - i\alpha_n x_n^{\alpha_n-1} D_n^{\alpha_n-1}\phi$$

where the last term should be dropped when $\alpha_n=0$, so this estimate is obvious. Thus the semi-norms ${}^{\infty}\|x_n^{\alpha_n} D^\alpha(\chi_n\psi)\|_{(1-k-n/4)}$ of $\chi_n\psi$ in $\dot{\mathscr{A}}^{(k-1)}$ can be estimated in terms of the semi-norms (18.3.33) of ϕ, which completes the proof.

In the non-characteristic case it follows from (18.3.32) that $WF_b(u)|_{\partial X} \subset T^*(\partial X)$ if $f\in C^\infty(X)$ We introduce a notation for such distributions:

Definition 18.3.30. If X is a C^∞ manifold with boundary ∂X then $\mathscr{N}(X)$ is the set of all $u\in\mathscr{A}'(X)$ with $WF_b(u)|_{\partial X}\subset T^*(\partial X)$.

From Propositions 18.3.28 and 18.3.29 we obtain

Corollary 18.3.31. *Let P be a differential operator with C^∞ coefficients in X such that ∂X is non-characteristic. If $f\in\mathscr{N}(X)$ and $u\in\bar{\mathscr{D}}'(X^\circ)$ satisfies the equation $Pu=f$ in the interior X° of X, then u has a unique extension $u_0\in\mathscr{N}(X)$.*

In Chapters XX and XXIV we shall study the wave front set when u also satisfies boundary conditions. It will then be technically simpler - at least in the present state of the art - not to use the nice invariant definitions of this section but rather work with pseudo-differential operators along the boundary. We have seen in Theorem 18.1.36 how they act on the wave front set in the interior and shall now study the behavior at the boundary.

Theorem 18.3.32. *Let X be an open subset of $\bar{\mathbb{R}}^n_+$ and set $\partial X=X\cap\partial\mathbb{R}^n_+$. If $u\in\mathscr{N}(X)$ and $b\in S^m(X\times\mathbb{R}^{n-1})$ defines a properly supported operator $b(x,D')$ in X, then $b(x,D')u\in\mathscr{N}(X)$ and*

$$WF_b(b(x,D')u)|_{\partial X}\subset WF_b(u)|_{\partial X}\cap\Gamma \tag{18.3.34}$$

if Γ is a closed cone $\subset\partial X\times\mathbb{R}^{n-1}$ such that b is of order $-\infty$ in a conic neighborhood of $(\partial X\times\mathbb{R}^{n-1})\setminus\Gamma$ in $X\times\mathbb{R}^{n-1}$. On the other hand,

$$WF_b(u)|_{\partial X}\subset WF_b(b(x,D')u)|_{\partial X}\cup \operatorname{Char} b_0 \tag{18.3.35}$$

where $b_0(x',\xi')=b(x',0,\xi')$.

Proof. Since $WF_b(u)$ is closed in $\dot{T}^*(X)\setminus 0$ and the restriction to ∂X is a subset of $T^*(\partial X)$, the set

$$Y=\{x\in X;\ x_n=0 \text{ or } 3x_n|\xi_n|<|\xi'| \text{ when } (x,\xi)\in WF(u)\}$$

is an open neighborhood of ∂X in X. If $\psi\in C_0^\infty(Y)$ we have $u_0=\psi u\in\mathscr{N}(X)\cap\mathscr{E}'(X)$, and the wave front set of ψu in $X\setminus\partial X$ has no element (x,ξ) with $\xi'=0$ causing trouble in Theorem 18.1.36. Choose $t\in S^0_+$

so that

$$t(x,\xi)=1 \quad \text{when } 2|\xi_n|<|\xi'|, |\xi'|>1, \quad x_n<1; \tag{18.3.36}$$
$$t(x,\xi)=0 \quad \text{when } |\xi_n|>|\xi'|.$$

We can take $t(x,\xi)$ independent of the x' variables. If $v=\tilde{t}_\rho(x,D)u_0$ we have

$$WF_b(u_0-v)=WF_b((1-\tilde{t}_\rho(x,D))u_0)=\emptyset$$

in view of Theorem 18.3.27 (ii) and the first part of (18.3.36). Hence $u_0-v\in\mathscr{A}'\cap\mathscr{A}=C^\infty(\overline{\mathbb{R}}^n_+)$; it follows from Theorem 18.3.5 that the difference is in $\mathscr{S}(\overline{\mathbb{R}}^n_+)$. If $b\in S^m(\mathbb{R}^n\times\mathbb{R}^{n-1})$ then

$$b(x,D')\tilde{t}_\rho(x,D)u_0=\tilde{a}(x,D)u_0 \tag{18.3.37}$$

where $a(x,\xi)=b(x,\xi')t_\rho(x,\xi)\in S^m_{la}$ by the second part of (18.3.36). In fact, if $w\in\mathscr{S}$ then the Fourier transform of $\tilde{t}_\rho(x,D)w$ with respect to x' is

$$(2\pi)^{-1}\int e^{ix_n\xi_n}\tilde{t}_\rho(x_n,\xi)\hat{w}(\xi)\,d\xi_n,$$

so (18.3.37) is valid with u_0 replaced by w. Letting $w\to u_0$ in $\mathscr{S}'$ we obtain (18.3.37). Since $\tilde{t}_\rho(x,D)u_0-u_0\in\mathscr{S}(\overline{\mathbb{R}}^n_+)$ we obtain $b(x,D')u_0-\tilde{a}(x,D)u_0\in\mathscr{S}(\overline{\mathbb{R}}^n_+)$, hence $b(x,D')u_0\in\mathscr{N}(X)$ and

$$WF_b(b(x,D')u_0)|_{\partial X}\subset WF_b(u_0)\cap\Gamma$$

since a is of order $-\infty$ in a conic neighborhood of $\{(x,\xi);\ x_n=\xi_n=0, (x',\xi')\notin\Gamma\}$. This proves (18.3.34).

Let $(y',\eta')\in T^*(\partial X)\setminus 0$ be a point not belonging to the right-hand side of (18.3.35). Then we can choose $\psi\in C_0^\infty(Y)$ equal to 1 in such a large subset that $b(x,D')(u_0-u)$ vanishes in a neighborhood of $(y',0)$ if $u_0=\psi u$ as above. Then $(y',\eta')\notin WF_b(b(x,D')u_0)=WF_b(\tilde{a}(x,D)u_0)$, and a is non-characteristic at $(y',0,\eta',0)$ since b_0 is non-characteristic at (y',η'). Hence $(y',\eta')\notin WF_b(u_0)$ as claimed.

Remark. In the open subset of X where $b(x,D')u$ is determined by the restriction of u to $\{x\in X^\circ;\ (x,\xi)\notin WF(u) \text{ when } \xi'=0\}$ we conclude using Theorem 18.1.36 that $WF_b(b(x,D')u)\subset WF_b(u)$. However, the set where this holds depends on u.

Corollary 18.3.33. *If $u\in\mathscr{N}(X)$ then $(y',\eta')\notin WF_b(u)$ if and only if $b(x,D')u\in C^\infty(X)$ for some properly supported tangential pseudo-differential operator $b(x,D')$ which is non-characteristic at $(y',0,\eta')$.*

Proof. In view of Proposition 18.3.24 this is an immediate consequence of Theorem 18.3.32.

Our definition of $WF_b(u)$ is obviously coordinate free. Hence this is also true for the alternative definition given by Corollary 18.3.33 which by Corollary 18.3.31 is applicable to solutions of differential equations with reasonably regular right-hand sides. The invariance is all that we shall actually use of the results of this section, but the general philosophy will also be helpful in discussions of regularity at a boundary.

18.4. Gauss Transforms Revisited

The reader will have noticed that the calculus in Sections 18.1 and 18.3 was mainly based on the results in Section 7.6 on the Gauss transform $\exp(i\langle D_x, D_\xi\rangle)$ in $\mathbb{R}^{2n}$. Indeed, the proof of Theorem 18.1.7 was based on Theorem 7.6.5 and a localization argument; the multiplicative properties were then reduced to Theorem 18.1.7. In this section we shall make a systematic study of the localization properties of $\exp(iA(D))$ when A is a real quadratic form. The results will be used in Section 18.5 to extend the calculus of Section 18.1 to more general symbols and to develop an alternative to it, the Weyl calculus, which in many respects has better properties.

To motivate the definitions we first recall that by Definition 18.1.1 the symbol class S^m is the set of all C^∞ functions a in $\mathbb{R}^{2n}$ such that

$$|a^{(\alpha)}_{(\beta)}(x,\xi)| \leqq C_{\alpha\beta}(1+|\xi|)^{m-|\alpha|}; \quad x,\xi\in\mathbb{R}^n.$$

To reinterpret this condition we introduce at (x,ξ) an orthonormal basis with respect to the metric

$$|dx|^2+|d\xi|^2/(1+|\xi|^2). \tag{18.4.1}$$

Then the derivatives of order k with respect to the new coordinates can be estimated by $C_k(1+|\xi|)^m$ for some C_k independent of ξ. Our generalization consists in considering in a finite dimensional vector space V any slowly varying metric in the sense of Section 1.4. By Lemma 1.4.3 it is no restriction to assume that it is Riemannian, that is, that for every $x\in V$ we have a positive definite quadratic form $g_x(y)$ in $y\in V$.

Definition 18.4.1. g is said to be slowly varying if there are positive constants c and C such that

$$g_x(y)\leqq c \Rightarrow g_{x+y}(t)\leqq C g_x(t). \tag{18.4.2}$$

This is precisely the condition in Definition 1.4.7 for the metric $|y|_x = (g_x(y)/c)^{\frac{1}{2}}$. Decreasing c if necessary we may therefore as observed in Section 1.4 give (18.4.2) a symmetric form:

$$g_x(y)\leqq c \Rightarrow g_x(t)/C\leqq g_{x+y}(t)\leqq C g_x(t). \tag{18.4.2$'$}$$

An example is the metric (18.4.1) or more generally

$$|dx|^2(1+|\xi|^2)^\delta+|d\xi|^2(1+|\xi|^2)^{-\rho} \tag{18.4.1$'$}$$

if $\rho \leqq 1$. Indeed, if $g_{x,\xi}(y,\eta) \leqq c$ then $|\eta|^2 \leqq c(1+|\xi|^2)$ which implies

$$(1+|\xi|)/2 \leqq 1+|\xi+\eta| \leqq 2(1+|\xi|)$$

if $c<\frac{1}{4}$. The slow variation follows at once.

If G is a fixed quadratic form and $u \in C^k$ in a neighborhood of $x \in V$ we shall norm the kth differential at x by

$$|u|_k^G(x) = \sup_{t_j \in V} |u^{(k)}(x; t_1, \ldots, t_k)| \Big/ \prod_1^k G(t_j)^{\frac{1}{2}}.$$

For fixed k an equivalent norm is of course the maximum of the derivatives of order k with respect to a G orthonormal coordinate system. Leibniz' rule gives

$$|u\,v|_k^G(x) \leqq \sum_0^k \binom{k}{j} |u|_j^G(x)\,|v|_{k-j}^G(x). \tag{18.4.3}$$

If $u(x)=1$ and we put $u=1-v$, then the kth differential at x of $1/u$ is equal to the kth differential of $\sum_{j \leqq k} v^j$ which can be estimated by a sum of products $|v|_{j_1} \ldots |v|_{j_\nu}$ with $j_\nu \geqq 1$ and $\sum j_\nu = k$. Thus

$$|1/u|_k^G(x) \leqq C_k(|u|_1^G(x) + \ldots + |u|_k^G(x)^{1/k})^k, \tag{18.4.4}$$

if $u(x)=1$. For general u we obtain a bound by homogenizing this estimate.

When g is a Riemannian metric we shall write $|u|_k^g(x)$ for $|u|_k^G(x)$ when $G = g_x$. Now we define corresponding symbol classes as follows:

Definition 18.4.2. If g is slowly varying then a positive real-valued function m in V is said to be g continuous if there are positive constants c and C such that

$$g_x(y) < c \Rightarrow m(x)/C \leqq m(x+y) \leqq C\,m(x). \tag{18.4.5}$$

We define $S(m,g)$ to be the set of all $u \in C^\infty(V)$ such that, for every integer $k \geqq 0$

$$\sup |u|_k^g(x)/m(x) < \infty. \tag{18.4.6}$$

It is obvious that $S(m,g)$ is a Fréchet space with the topology defined by the quantities in (18.4.6). It is important to note that the seminorms are indexed by the non-negative integers k so that it makes sense to talk about "the same seminorms" in the spaces $S(m,g)$ with different m and g.

If g is the metric (18.4.1)$'$ we can take $m=(1+|\xi|^2)^{\mu/2}$ for any real number μ. Then $S(m,g)$ becomes the symbol space $S^\mu_{\rho,\delta}$ introduced in Section 7.8 already.

The following lemma is an immediate consequence of (18.4.3) and (18.4.4):

Lemma 18.4.3. *If* $u \in S(m,g)$ *and* $v \in S(m',g)$ *then* $u\,v \in S(m\,m',g)$. *If* $1/|u| < C/m$ *for some* C, *then* $1/u \in S(1/m,g)$.

It is clear that $C_0^\infty(V)\subset S(m,g)$, for g_x and $m(x)$ are bounded from above and from below when x is in a compact set. Assuming always that (18.4.2)′ is fulfilled we can apply Lemma 1.4.9 and Theorem 1.4.10 to $|y|_x=(g_x(y)/c)^{\frac{1}{2}}$ and obtain

Lemma 18.4.4. *If $0<\varepsilon<1$ one can find a sequence $x_1,x_2,\ldots\in V$ such that the balls*

$$B_\nu=\{x;\,g_{x_\nu}(x_\nu-x)<R^2\}$$

cover V if $\varepsilon c<R^2$ and the intersection of more than N_ε balls B_ν is always empty if $R^2<c$. If $2\varepsilon c<R^2<c$ one can choose non-negative $\phi_\nu\in C_0^\infty(B_\nu)$ with $\sum\phi_\nu=1$ so that for all ν and k

$$|\phi_\nu|_k^g\leq C_{k,\varepsilon}. \tag{18.4.7}$$

The partition of unity can be used to regularize the metric and the weight function m. In fact, if we set

$$m_1(x)=\sum\phi_\nu(x)\,m(x_\nu)$$

and observe that $g_{x_\nu}(x-x_\nu)<c$ when $\phi_\nu(x)\neq 0$, it follows from (18.4.5) that

$$m(x_\nu)/C\leq m(x)\leq C\,m(x_\nu)\quad\text{ in supp }\phi_\nu$$

which implies that

$$m(x)/C\leq m_1(x)\leq C\,m(x).$$

In addition we obtain

$$|m_1|_k^g(x)\leq C'_k\,m(x),$$

which means that $m_1\in S(m,g)=S(m_1,g)$. In particular, we conclude that $S(m,g)\subset S(m',g)$ if and only if m/m' is bounded. In the same way we can of course regularize the metric g.

If $u\in S(m,g)$ it follows from (18.4.7), (18.4.3) and (18.4.5) that $u_\nu=\phi_\nu u$ has the bounds

$$|u_\nu|_k^g(x)\leq C'_k\,m(x_\nu)\quad\text{ for } x\in V,\text{ all } \nu \text{ and } k. \tag{18.4.7$'$}$$

Here g may be replaced by g_{x_ν}. Conversely, if we have any sequence of functions u_ν with supports in the balls B_ν with $R^2<c$ and satisfying (18.4.7)′, then $u=\sum u_\nu\in S(m,g)$. It is therefore clear that optimal estimates for linear functionals on symbol classes must be obtained by studying just functions with the properties of u_ν. We shall do so for the Gauss transforms.

Let A be a real-valued quadratic form in the dual space V' of V. Then $A(D)$ is a differential operator in V characterized by

$$A(D)\exp\langle ix,\xi\rangle=A(\xi)\exp\langle ix,\xi\rangle,\qquad x\in V,$$

for every fixed $\xi \in V'$. When u is in $\mathscr{S}$ or in $\mathscr{S}'$ we can define $\exp(iA(D))u$ as the inverse Fourier transform of $\exp(iA(\xi))\hat{u}(\xi)$ where $\hat{u}$ is the Fourier transform of u. Let g be a positive definite quadratic form in V, and let

$$K = \{x;\, g(x) < 1\}$$

be the corresponding unit ball. By (7.6.7) applied to a g orthogonal coordinate system we have for $u \in C_0^\infty(K)$

$$|\exp(iA(D))u - \sum_{j<k} (iA(D))^j u/j!| \leqq C \sup_{j \leqq s} \sup_{y \in K} |A(D)^k u|_j^g(y)/k!. \tag{18.4.8}$$

Here s is an integer $> \dim V/2$. Outside K the sum vanishes, and we shall then improve (18.4.8) by an argument equivalent to that used to prove Lemma 7.6.4. However, we phrase it differently to prepare for another analogous proof in Section 18.6.

Let L be a real affine linear (i.e. not necessarily homogeneous) function in V. Then

$$[\exp(iA(D)), L] = \exp(iA(D))\langle A'(D), L'\rangle$$

where L is regarded as a multiplication operator and $L' \in V'$. This is a case of (18.1.6) but of course perfectly elementary, for after Fourier transformation the identity becomes

$$[\exp(iA(\xi)), L(-D)] = \exp(iA(\xi))\langle A'(\xi), L'\rangle.$$

Set $L(y) = \langle y - x, \eta\rangle$ and assume that $L \neq 0$ in K. We have $\langle A'(D), \eta\rangle = 2A(D, \eta) = 2\langle A\eta, D\rangle$ where $A(\ ,\ )$ is the symmetric bilinear form defined by A in V' and A is the corresponding linear transformation $V' \to V$. Thus the preceding formula gives

$$\exp(iA(D))u(x) = 2\exp(iA(D))(\langle A\eta, D\rangle L^{-1}u)(x), \quad u \in C_0^\infty(K), \tag{18.4.9}$$

and this result may be iterated any number of times.

Lemma 18.4.5. *If L is linear and never* 0 *in* $\{y;\, g(y) < R^2\}$ *where* $R > 1$, *then*

$$|L(0)/L|_k^g(y) \leqq k!\, R/(R-1)^{k+1}, \quad y \in K. \tag{18.4.10}$$

Proof. It is no restriction to assume that $g(y) = \sum y_j^2$ and that $L(y) = 1 - a y_1$ where $0 \leqq aR \leqq 1$. Then (18.4.10) follows since $a^k(1 - a y_1)^{-k-1} \leqq R/(R-1)^{k+1}$ when $|y_1| < 1$.

If we iterate (18.4.9) and use (18.4.10), (18.4.3), and (18.4.8), with $k = 0$, it follows after k iterations that

$$|\exp(iA(D))u(x)| \leqq C_{k,R}(g(A\eta)^{\frac{1}{2}}/|L(0)|)^k \sup_{j \leqq s+k} \sup |u|_j^g, \quad u \in C_0^\infty(K).$$

To examine how small the new factor $g(A\eta)^{\frac{1}{2}}/|L(0)|=g(A\eta)^{\frac{1}{2}}/|\langle x,\eta\rangle|$ can be made we introduce the dual form of $\xi\mapsto g(A\xi)$

$$g^A(x)=\sup_{g(A\xi)<1}\langle x,\xi\rangle^2. \tag{18.4.11}$$

g^A is of course $+\infty$ except in the orthogonal space of the radical $\{\xi; A\xi=0\}$ $=\{\xi; A(\xi)=0\}$, that is, the range of A. If we set $x=A\eta$ and note that $\langle A\eta,\xi\rangle=\langle A\xi,\eta\rangle$ we obtain

$$g^A(A\eta)=\sup_{\operatorname{Im}A}|\langle y,\eta\rangle|^2/g(y)$$

so g^A is the composition of A^{-1} and the dual form of the restriction of g to the range of A. Note that $\sup_x\langle x,\xi\rangle^2/g^A(x)=g(A\xi)$.

Now assume that the g^A distance from x to RK is $\geqq a>0$. By the Hahn-Banach theorem we can then find η so that

$$\langle y,\eta\rangle<\langle x+z,\eta\rangle \text{ if } g(y)<R^2 \text{ and } g^A(z)<a^2.$$

This means that

$$\langle y,\eta\rangle<\langle x,\eta\rangle-a g(A\eta)^{\frac{1}{2}}, \quad y\in RK.$$

For $L(y)=\langle y-x,\eta\rangle$ it follows that

$$L(0)=-\langle x,\eta\rangle<-a g(A\eta)^{\frac{1}{2}}$$

We have therefore proved that for $k=0,1,\dots$

$$\begin{aligned}&|\exp(iA(D))u(x)|\\ &\leqq C_{k,R}(1+\inf_{y\in RK} g^A(x-y))^{-k/2}\sup_{j\leqq s+k}\sup|u|_j^g,\end{aligned} \tag{18.4.12}$$

if $u\in C_0^\infty(K)$. We sum up our results so far:

Proposition 18.4.6. *Let g be a positive definite quadratic form in V and A a real quadratic form in V'. Denote by K the unit ball with respect to g, and define g^A by* (18.4.11). *Then the estimates* (18.4.8), (18.4.12) *are valid for all $k\geqq 0$ and $R>1$ if $2s>\dim V$.*

Since $\exp(iA(D))$ commutes with differentiation it would have been easy to add estimates for the derivatives. The important point in (18.4.12) is that the right-hand side is very small at large g^A distance from a neighborhood of K. This localization property allows us to get estimates for $e^{iA(D)}$ in appropriate symbol classes; in a special case it was already used in the proof of Theorem 18.1.7. The following is the required condition.

Definition 18.4.7. The Riemannian metric g (and the positive function m) in V is said to be A temperate (resp. A, g temperate) with respect to $x\in V$ if g is slowly varying (and m is g continuous) and there exist constants C and N

such that for all $y, t \in V$

$$g_y(t) \leqq C g_x(t)(1+g_y^A(x-y))^N, \tag{18.4.13}$$

$$m(y) \leqq C m(x)(1+g_y^A(x-y))^N. \tag{18.4.14}$$

Note that (18.4.13) implies

$$g_x^A(t) \leqq C g_y^A(t)(1+g_y^A(x-y))^N. \tag{18.4.13}'$$

Conversely, (18.4.13)′ implies (18.4.13) if A is non-degenerate. When $t=x-y$ we obtain in particular

$$1+g_x^A(x-y) \leqq C(1+g_y^A(x-y))^{N+1}. \tag{18.4.13}''$$

To remove the condition on the support of u in Proposition 18.4.6 we shall use the partition of unity in Lemma 18.4.4. Choose R_0 with $R<R_0<c^{\frac{1}{2}}$ and introduce in addition to the balls B_ν containing $\operatorname{supp}\phi_\nu$ the neighborhoods

$$U_\nu=\{x; g_{x_\nu}(x-x_\nu) \leqq R_0^2\}, \quad U_\nu'=\{x; g_{x_\nu}(x-x_\nu) \leqq c\}.$$

When $u \in C_0^\infty$ we shall apply (18.4.12) to $u_\nu=\phi_\nu u$ with R and g replaced by R_0/R and g_{x_ν}/R^2. To add up the estimates for $\exp(iA(D))u_\nu$ we need the following lemma:

Lemma 18.4.8. *Assume that $g_x \leqq g_x^A$ and that g is A temperate with respect to x. Then there are constants C and N depending only on those in* (18.4.13) *such that*

$$\sum_\nu (1+d_\nu(x))^{-N} \leqq C, \quad d_\nu(x)=\inf_{y\in U_\nu} g_y^A(x-y). \tag{18.4.15}$$

Proof. We may assume that g_x is the square of the Euclidean norm $|\ |$, which is then a lower bound for g_x^A. Let $k \geqq 1$ and set

$$M_k=\{\nu; d_\nu(x) \leqq k\}.$$

When $\nu \in M_k$ we choose $y_\nu \in U_\nu$ with $g_{y_\nu}^A(x-y_\nu) \leqq k$. By (18.4.13)

$$g_{y_\nu}(t) \leqq C|t|^2 k^N.$$

Now $y_\nu+z \in U_\nu'$ if $g_{x_\nu}(z)<(c^{\frac{1}{2}}-R_0)^2$, and since g is slowly varying we have fixed upper and lower bounds for g_{x_ν}/g_{y_ν}. It follows that a Euclidean ball V_ν of radius $c_1 k^{-N/2}$ and center at y_ν is contained in U_ν'. In view of (18.4.13)″ we have

$$|x-y_\nu|^2=g_x(x-y_\nu) \leqq g_x^A(x-y_\nu) \leqq C k^{N+1};$$

hence the balls V_ν are contained in a Euclidean ball of radius $C' k^{(N+1)/2}$ with center at x. Since we have a bound for the number of U_ν' and therefore for the number of V_ν which can overlap, we obtain

$$c'|M_k| k^{-nN/2} \leqq \sum_{M_k} m(V_\nu) \leqq C m(\bigcup_{M_k} V_\nu) \leqq C' k^{n(N+1)/2},$$

if $|M_k|$ is the number of elements in M_k. Hence $|M_k| < C k^{N-1}$ for some new constants C and N. If we add successively the terms in (18.4.15) with $\nu \in M_1, M_2 \setminus M_1, \ldots, M_{2^k} \setminus M_{2^{k-1}}$, the estimate follows.

With $u_\nu = \phi_\nu u$ we obtain using (18.4.12)

$$|\exp(iA(D))\, u_\nu(x)| \leqq C_k(1+d_\nu(x))^{-k/2} \sup_{j \leqq s+k} \sup_{U_\nu} |u_\nu|_j^{g_{x_\nu}}. \tag{18.4.16}$$

Since g varies slowly we may replace g_{x_ν} by g here. With y_ν defined as above we have

$$m(y) \leqq C m(y_\nu) \leqq C' m(x)(1+d_\nu(x))^{N'}, \quad y \in U_\nu,$$

provided that m satisfies (18.4.14). If the hypotheses of Lemma 18.4.8 are also fulfilled then (18.4.16), (18.4.15), and (18.4.7) yield

$$\sum_\nu |\exp(iA(D))\, u_\nu(x)| \leqq C m(x) \sup_{j \leqq s+k} \sup |u|_j^g/m, \tag{18.4.17}$$

provided that k is large enough.

We shall use (18.4.17) to extend the definition of $\exp(iA(D))$ from C_0^∞ to $S(m,g)$. However, an arbitrary continuous linear form on $S(m,g)$ is not determined by its restriction to C_0^∞ for this is not a dense subset. We shall therefore be interested in a stronger continuity condition.

Definition 18.4.9. A continuous linear form on $S(m,g)$ will be called weakly continuous if the restriction to a bounded subset is continuous in the C^∞ topology.

A weakly continuous form is determined by its restriction to C_0^∞, since the partial sums of the partition $u = \sum u_\nu$ are bounded in $S(m,g)$ if $u \in S(m,g)$; they converge to u in the C^∞ topology since they are ultimately equal to u on any compact set.

The proof of (18.4.17) actually gave a convergent majorant series for the left-hand side of (18.4.17) which is valid for all u in a bounded set in $S(m,g)$. It follows that the sum

$$\exp(iA(D))\, u(x) = \sum_\nu \exp(iA(D))\, u_\nu(x)$$

defines a weakly continuous linear form on $S(m,g)$. Thus we have proved:

Theorem 18.4.10. *The map $C_0^\infty \ni u \mapsto \exp(iA(D))\, u(x) \in \mathbb{C}$ has a unique extension to a weakly continuous linear form on $S(m,g)$ for every x such that g is A temperate, $g_x \leqq g_x^A$, and m is A, g temperate with respect to x. We have*

$$|\exp(iA(D))\, u(x)| \leqq m(x)\, \|u\|, \tag{18.4.18}$$

where the seminorm $\|u\|$ in $S(m,g)$ only depends on the constants in (18.4.2)′, (18.4.5), (18.4.13) *and* (18.4.14).

The hypotheses in Theorem 18.4.10 also allow us to estimate the derivatives of $\exp(iA(D))u(x)$. Let us first assume that $u\in C_0^\infty$ so that we know that they exist. Then we have

$$(18.4.18)' \quad |\langle D,t_1\rangle\dots\langle D,t_k\rangle \exp(iA(D))u(x)| \leqq m(x)\prod_1^k g_x(t_j)^{\frac12}\,\|u\|,$$

where $\|u\|$ is a seminorm in $S(m,g)$ which depends only on k in addition to the constants in (18.4.2)', (18.4.5), (18.4.13), (18.4.14). To prove this we set

$$v=\langle D,t_1\rangle\dots\langle D,t_k\rangle u, \qquad m'(y)=m(y)\prod_1^k g_y(t_j)^{\frac12}.$$

m' is also a A, g temperate function with respect to x, with bounds depending only on the bounds for g and m in addition to k. Any seminorm of v in $S(m',g)$ is bounded by a fixed seminorm of u in $S(m,g)$. Hence (18.4.18) applied to v gives (18.4.18)'. (It is sometimes useful to observe that it is sufficient to have (18.4.6) for the differentials of order $\geqq k$.) The following statement is an immediate consequence:

Theorem 18.4.10'. *Assume that the hypotheses of Theorem 18.4.10 are fulfilled uniformly for all x in a linear subspace V_0 of V. Then the map*

$$S(m,g)\ni u\mapsto \exp(iA(D))\,u|_{V_0}$$

is weakly continuous with values in the space $S(m,g)|_{V_0}$ of symbols in V_0 corresponding to the restrictions of m and of g.

For the proof we just have to take $t_1,\dots,t_k\in V_0$ in (18.4.18)' and note that on bounded subsets of $S(m,g)$ the C^∞ topology is equal to the topology of pointwise convergence.

The preceding results can all be improved when

$$(18.4.19) \qquad h(x)^2=\sup_t g_x(t)/g_x^A(t)$$

is not only less than or equal to 1 but is small. (If the coordinates are chosen so that g_x is the Euclidean metric and $A(\xi)=\sum b_j\xi_j^2$, then $h(x)=\sup|b_j|$.) First of all we note that h is A, g temperate with the same constants as in (18.4.13). In fact, h is obviously g continuous. We can write (18.4.13) in the form $g_y\leqq M g_x$ which implies $g_y^A\geqq g_x^A/M$ and $g_y/g_y^A\leqq M^2 g_x/g_x^A$, hence $h(y)\leqq M h(x)$. Let us now return to the estimate (18.4.16). Recall that there is a bound for the number of overlapping balls U_ν' and that

$$g_{x_\nu}(x-y)\geqq (c^{\frac12}-R_0)^2=c_1>0 \quad \text{when } x\notin U_\nu' \text{ and } y\in U_\nu.$$

This implies that

$$c_2\leqq g_y(x-y)\leqq h(y)^2\,g_y^A(x-y)\leqq C\,h(x)^2(1+g_y^A(x-y))^{N'},$$

hence that

$$1 \leq C h(x)^2 (1 + d_\nu(x))^{N'} \quad \text{if } x \notin U'_\nu.$$

This means that (18.4.17) can be improved by any power of $h(x)$ in the right-hand side if we only sum over all ν with $x \notin U'_\nu$. The remaining boundedly many terms can be estimated by means of (18.4.8) with $A(D)$ as above which proves that the remainder term

$$R_N = \exp(iA(D))u - \sum_{j<N} (iA(D))^j u/j! \tag{18.4.20}$$

has the bound

$$|R_N^{(k)}(x; t_1, \dots, t_k)| \leq h(x)^N m(x) \prod_1^k g_x(t_j)^{\frac{1}{2}} \|u\|, \tag{18.4.21}$$

where $\|u\|$ is a fixed seminorm in $S(m, g)$. (The proof for $k \neq 0$ is reduced to the case $k=0$ as in the proof of Theorem 18.4.10'.) Thus we obtain

Theorem 18.4.11. *Assume that the hypotheses of Theorem* 18.4.10 *are satisfied uniformly for all x in a linear subspace* V_0 *of V, and define h by* (18.4.19). *With* R_N *defined by* (18.4.20) *it follows that*

$$S(m, g) \ni u \mapsto R_N \in S(m h^N, g)|_{V_0}$$

is weakly continuous. The seminorm in (18.4.21) *depends only on N, k and the constants in* (18.4.2)', (18.4.5), (18.4.13) *and* (18.4.14).

Thus it is justified to calculate $\exp(iA(D))u(x)$ by the formal expansion where $h(x)$ is small. An example is Theorem 18.1.7. There the metric is (18.4.1), and A is the quadratic form $(x, \xi) \mapsto \langle x, \xi \rangle$ in $\mathbb{R}^n \oplus \mathbb{R}^n$, so A is the map $(x, \xi) \mapsto (\xi, x)$ and $g^A = (1+|\xi|^2)|dx|^2 + |d\xi|^2$. Thus $h^2 = (1+|\xi|^2)^{-1}$, and (18.1.11) follows from Theorem 18.4.11 with the weight function $m(\xi) = (1+|\xi|^2)^{\mu/2}$, if m is replaced by μ in Theorem 18.1.7. In fact, (18.4.13) and (18.4.14) follow with $N=1$ and $N=\mu/2$ respectively since

$$(1+|\xi|^2)/(1+|\eta|^2) \leq (1+(|\eta|+|\xi-\eta|)^2)/(1+|\eta|^2) \leq (1+|\xi-\eta|)^2.$$

More generally, the metric g defined by (18.4.1)' is slowly varying if $\rho \leq 1$. We have

$$g^A = (1+|\xi|^2)^\rho |dx|^2 + (1+|\xi|^2)^{-\delta} |d\xi|^2,$$

hence $h^2 = (1+|\xi|^2)^{\delta-\rho} \leq 1$ if and only if $\delta \leq \rho$. The condition for g to be A temperate is

$$\begin{aligned}(1+|\xi|^2)^{-\delta}(1+|\eta|^2)^{\delta} + (1+|\xi|^2)^{\rho}(1+|\eta|^2)^{-\rho} \\ \leq C(1+|\xi-\eta|^2(1+|\eta|^2)^{-\delta})^N.\end{aligned}$$

If $|\xi| < |\eta|/2$ it is valid if and only if $\delta \leq N(1-\delta)$, that is, $\delta \leq N/(N+1)$. When $|\eta|/2 < |\xi| < 2|\eta|$ it is true for large C, and when $|\xi| > 2|\eta|$ it follows if $N \geq \rho$.

When the condition is fulfilled then $(1+|\xi|^2)^{\mu/2}$ is A, g temperate. Hence Theorem 18.1.7 remains valid if $0 \leqq \delta \leqq \rho \leqq 1$ and $\delta < 1$, as observed before Theorem 18.1.35.

18.5. The Weyl Calculus

Let V be an n dimensional vector space over $\mathbb{R}$ and V' its dual. In Section 18.1 we associated with any $a \in \mathscr{S}(W)$, $W = V \oplus V'$, the operator

$$a(x, D) u(x) = (2\pi)^{-n} \iint a(x, \xi) e^{i\langle x-y, \xi\rangle} u(y)\, dy\, d\xi, \quad u \in \mathscr{S}. \tag{18.5.1}$$

(Here dy is a Lebesgue (Haar) measure in V and $d\xi$ is the dual one in V' such that Fourier's inversion formula holds with the usual constant. Replacing dy by $c\,dy$ one must change $d\xi$ to $c^{-1}\, d\xi$ so $dy\, d\xi$ is invariantly defined.) The weak version of (18.5.1)

$$\begin{aligned} \langle a(x, D) u, v\rangle &= (2\pi)^{-n} \iiint a(x, \xi) e^{i\langle x-y, \xi\rangle} u(y) v(x)\, dy\, dx\, d\xi \\ &= (2\pi)^{-n} \iiint a(x, \xi) e^{i\langle t, \xi\rangle} u(x-t) v(x)\, dx\, dt\, d\xi \end{aligned}$$

makes sense for any $a \in \mathscr{S}'(W)$ and defines a continuous operator from $\mathscr{S}(V)$ to $\mathscr{S}'(V)$. The adjoint of $\bar{a}(x, D)$ is the operator

$$\tilde{a}(x, D) u(x) = (2\pi)^{-n} \iint a(y, \xi) e^{i\langle x-y, \xi\rangle} u(y)\, dy\, d\xi, \tag{18.5.2}$$

interpreted in the weak sense too. If a is a polynomial in ξ then $a(x, D)$ is obtained when ξ is replaced by $D = -i\partial/\partial x$ placed to the right of the coefficients. Putting the coefficients to the right instead we obtain the operator $\tilde{a}(x, D)$.

If $a \in S^m$ then Theorem 18.1.6 means that $a(x, D)$ maps $\mathscr{S}$ to $\mathscr{S}$. By Theorem 18.1.7 the class of operator (18.5.1) with $a \in S^m$ is the same as the class of operators (18.5.2) with $a \in S^m$ so they can be extended to continuous operators from $\mathscr{S}'$ to $\mathscr{S}'$ as well.

In the Weyl calculus one adopts the symmetric compromise

$$a^w(x, D) = (2\pi)^{-n} \iint a((x+y)/2, \xi) e^{i\langle x-y, \xi\rangle} u(y)\, dy\, d\xi, \tag{18.5.3}$$

again defined in the weak sense. The Schwartz kernel K is

$$K(x, y) = (2\pi)^{-n} \int a((x+y)/2, \xi) e^{i\langle x-y, \xi\rangle}\, d\xi, \tag{18.5.4}$$

that is,

$$K(x+t/2, x-t/2) = (2\pi)^{-n} \int a(x, \xi) e^{i\langle t, \xi\rangle}\, d\xi \tag{18.5.4$'$}$$

is the inverse Fourier transform of a with respect to ξ, so

$$a(x, \xi) = \int K(x+t/2, x-t/2) e^{-i\langle t, \xi\rangle}\, dt. \tag{18.5.4$''$}$$

(These formulas are analogous to (18.1.7) and (18.1.8).) If L is linear then $L(x, D) = \tilde{L}(x, D) = L^w(x, D)$. To explain the definition (18.5.3) we first compute

$L^w(x,D)\,a^w(x,D)$ for $a\in\mathscr{S}$ when L is linear. Since

$$\begin{aligned}(L(x,D_x)-L((x+y)/2,\xi))\,a((x+y)/2,\xi)\,e^{i\langle x-y,\xi\rangle}\\ =(L(0,D_x)\,a((x+y)/2,\xi)+a((x+y)/2,\xi)\,L(D_\xi/2,0))\,e^{i\langle x-y,\xi\rangle},\end{aligned}$$

we obtain after an integration by parts in ξ

$$L^w(x,D)\,a^w(x,D)=b^w(x,D),\qquad b=La+\{L,a\}/2i, \tag{18.5.5}$$

where

$$\{L,a\}=\langle\partial L/\partial\xi,\partial a/\partial x\rangle-\langle\partial L/\partial x,\partial a/\partial\xi\rangle$$

is the Poisson bracket introduced already in Section 6.4. Recall that it is the bilinear form in $W'\oplus W'$ which is dual to the symplectic form

$$\sigma(x,\xi;y,\eta)=\langle\xi,y\rangle-\langle x,\eta\rangle,\qquad (x,\xi)\in W,\quad (y,\eta)\in W.$$

This indicates already the symplectic invariance of the Weyl calculus which is an important property to which we shall return later on. At the moment we just observe that although the proof of (18.5.5) above assumed that $a\in\mathscr{S}$ the formula extends by continuity to all $a\in\mathscr{S}'$. Let L be a real linear form and set $a_t=\exp(itL)$. Then we obtain

$$iL(x,D)\,a_t^w(x,D)=\partial a_t^w(x,D)/\partial t$$

for $\{L,a_t\}=0$. A simple explicit computation of $a_t^w(x,D)u$ shows that $a_t^w(x,D)u\in\mathscr{S}$ if $u\in\mathscr{S}$. Now the closure in L^2 of $L(x,D)$ with domain $\mathscr{S}$ is self-adjoint. In fact, let $u\in L^2$ and $L(x,D)u=f\in L^2$ in the sense of distribution theory. Choose $\chi\in\mathscr{S}(W)$ with $\chi(0)=1$. Then $\mathscr{S}\ni\chi_\varepsilon(x,D)u\to u$ and $\chi_\varepsilon(x,D)f\to f$ in L^2 as $\varepsilon\to 0$, and

$$L(x,D)\,\chi_\varepsilon(x,D)\,u-\chi_\varepsilon(x,D)f=-\varepsilon i\{L,\chi\}(\varepsilon x,\varepsilon D)\,u\to 0$$

where $\chi_\varepsilon(x,\xi)=\chi(\varepsilon x,\varepsilon\xi)$. This proves the statement. It follows that

$$a_t^w(x,D)=\exp(it\,L(x,D))$$

in the operator theoretical sense. The general definition of a^w is deduced from this case by Fourier decomposition of a, so the preceding property characterizes the Weyl calculus.

From (18.5.3) it follows at once that the adjoint of a^w is equal to $\bar{a}^w$. In particular, a^w is its own adjoint when a is real valued, which is an essential advantage of the Weyl calculus and an important reason why it was introduced by Hermann Weyl for the purposes of quantum mechanics.

In order to motivate the conditions which will be placed on the symbol a we shall now derive a formula for the composition of $a_1^w(x,D)$ and $a_2^w(x,D)$ when a_1 and a_2 are in $\mathscr{S}(W)$. The kernel of $a_1^w(x,D)\,a_2^w(x,D)$ is equal to

$$(2\pi)^{-2n}\iiint a_1((x+z)/2,\zeta)\,a_2((z+y)/2,\tau)\,e^{i\langle x-z,\zeta\rangle+i\langle z-y,\tau\rangle}\,dz\,d\zeta\,d\tau,$$

by (18.5.4), so it follows from (18.5.4)″ that $a_1^w a_2^w = a^w$ if

$$a(x,\xi)=(2\pi)^{-2n}\iiiint a_1((x+z+t/2)/2,\zeta)\,a_2((x+z-t/2)/2,\tau)\,e^{iE}\,dz\,d\zeta\,dt\,d\tau,$$
$$\begin{aligned}E&=\langle x-z+t/2,\zeta\rangle+\langle z-x+t/2,\tau\rangle-\langle t,\xi\rangle\\&=\langle x-z+t/2,\zeta-\xi\rangle+\langle z-x+t/2,\tau-\xi\rangle.\end{aligned}$$

We introduce $\zeta-\xi$, $\tau-\xi$, $(z-x+t/2)/2$ and $(z-x-t/2)/2$ as new variables instead of ζ, τ, z and t. The Jacobian is 2^{2n}. Hence

$$a(x,\xi)=\pi^{-2n}\iiiint a_1(x+z,\xi+\zeta)\,a_2(x+t,\xi+\tau)\,e^{2i\sigma(t,\tau;z,\zeta)}\,dz\,d\zeta\,dt\,d\tau.$$

Here we regard the symplectic form as a quadratic form on $W\oplus W$.

For $f\in\mathscr{S}(\mathbb{R}^2)$ we have

$$\iint f(x,y)\,e^{2ixy}\,dx\,dy=(4\pi)^{-1}\iint \hat f(\xi,\eta)\,e^{-i\xi\eta/2}\,d\xi\,d\eta.$$

This follows from the Fourier inversion formula if $f(x,y)=g(x)h(y)$. Hence the formula above can be written in the form

(18.5.6) $$a(x,\xi)=\exp(i\sigma(D_x,D_\xi;D_y,D_\eta)/2)\,a_1(x,\xi)\,a_2(y,\eta)|_{(x,\xi)=(y,\eta)}.$$

We can therefore use the results of Section 18.4 to study a if a_1 and a_2 belong to suitable symbol classes. Since we have a product $a_1(x,\xi)\,a_2(y,\eta)$ in the right-hand side we shall encounter quadratic forms in $W\oplus W$ of the form

$$G(t_1,t_2)=g_1(t_1)+g_2(t_2)$$

where g_1 and g_2 are quadratic forms in W. If $(x,\xi,y,\eta)\in W\oplus W$ and $(\hat x,\hat\xi,\hat y,\hat\eta)$ denote dual variables, then the linear map associated in (18.4.11) with the quadratic form

$$A=2\sigma(\hat x,\hat\xi,\hat y,\hat\eta)=2\langle\hat\xi,\hat y\rangle-2\langle\hat x,\hat\eta\rangle$$

maps $(\hat x,\hat\xi,\hat y,\hat\eta)$ to $(-\hat\eta,\hat y,\hat\xi,-\hat x)$. Hence

$$G^A(x,\xi,y,\eta)=\sup|\langle x,\hat x\rangle+\langle\xi,\hat\xi\rangle+\langle y,\hat y\rangle+\langle\eta,\hat\eta\rangle|^2/(g_1(-\hat\eta,\hat y)+g_2(\hat\xi,-\hat x)).$$

If we write $(x,\xi)=w$ and $(\hat\xi,-\hat x)=w'\in W$ then $\langle x,\hat x\rangle+\langle\xi,\hat\xi\rangle=\sigma(w,w')$. Let

(18.5.7) $$g_j^\sigma(w)=\sup|\sigma(w,w')|^2/g_j(w')$$

be the dual quadratic form of g_j when W is identified with its dual by means of the duality defined by the symplectic form. Then we obtain

(18.5.8) $$G^A(w_1,w_2)=g_1^\sigma(w_2)+g_2^\sigma(w_1),$$

(18.5.9) $$G\leqq H^2G^A \;\Leftrightarrow\; g_2\leqq H^2 g_1^\sigma \;\Leftrightarrow\; g_1\leqq H^2 g_2^\sigma,$$

for the second and third conditions are both equivalent to

(18.5.9)′ $$|\sigma(w,w')|^2\leqq H^2 g_1^\sigma(w)\,g_2^\sigma(w'),$$

and the first is equivalent to the conjunction of the second and third.

If g_1 and g_2 are slowly varying Riemannian metrics in W, and G is the Riemannian metric $g_1 \oplus g_2$ in $W \oplus W$, it is obvious that G is slowly varying. To discuss (18.5.6) we must examine if G is uniformly A temperate with respect to the points on the diagonal, that is, if for arbitrary $w, w_1, w_2, t_1, t_2 \in W$

$$g^\sigma_{1w}(t_1)+g^\sigma_{2w}(t_2) \leqq C(g^\sigma_{1w_1}(t_1)+g^\sigma_{2w_2}(t_2))\, M^N, \tag{18.5.10}$$
$$M=1+g^\sigma_{1w_1}(w_2-w)+g^\sigma_{2w_2}(w_1-w).$$

If $g_1=g_2=g$ it follows in particular when $w_1=w_2$ that for $t \in W$

$$g^\sigma_w(t) \leqq C g^\sigma_{w_1}(t)(1+g^\sigma_{w_1}(w_1-w))^N; \quad w, w_1 \in W; \tag{18.5.11}$$

or equivalently

$$g_{w_1}(t) \leqq C g_w(t)(1+g^\sigma_{w_1}(w_1-w))^N; \quad w, w_1 \in W. \tag{18.5.11$'$}$$

We therefore introduce a definition parallel to Definition 18.4.7.

Definition 18.5.1. The metric g in $W=V\oplus V'$ is called σ temperate if it is slowly varying and (18.5.11)$'$ is valid. A positive function m in W is called σ, g temperate if m is g continuous and

$$m(w_1) \leqq C m(w)(1+g^\sigma_{w_1}(w-w_1))^N; \quad w, w_1 \subset W. \tag{18.5.12}$$

Note that by (18.5.11) $1/m$ is σ, g temperate if m is.

Proposition 18.5.2. *If g is σ temperate and m_1, m_2 are σ, g temperate in $W=V\oplus V'$ then the metric $G=g_1\oplus g_2$ in $W\oplus W$, where $g_1=g_2=g$, and the weight function $m=m_1\otimes m_2$ are uniformly A temperate and A, G temperate with respect to the diagonal. If $h(w)^2=\sup g_w/g^\sigma_w$ then $\sup G_{w,w}/G^A_{w,w}=h(w)^2$ too.*

Proof. The last statement follows from (18.5.9). To prove the others we must show that

$$g^\sigma_{w_1}(w_1-w)+g^\sigma_{w_2}(w_2-w) \leqq CM^N, \qquad M=1+g^\sigma_{w_1}(w_2-w)+g^\sigma_{w_2}(w_1-w).$$

Writing $w'=w_1+w_2-w$ we have $w'-w_1=w_2-w$, $w'-w_2=w_1-w$, hence

$$g^\sigma_{w'}(w_2-w) \leqq C g^\sigma_{w_1}(w_2-w)(1+g^\sigma_{w_1}(w_2-w))^N \leqq CM^{N+1},$$
$$g^\sigma_{w'}(w_1-w) \leqq CM^{N+1},$$
$$g^\sigma_{w_1}(w_1-w) \leqq C g^\sigma_{w'}(w_1-w)(1+g^\sigma_{w'}(w_2-w))^N \leqq C' M^{N'},$$
$$g^\sigma_{w_2}(w_2-w) \leqq C' M^{N'},$$

which proves the statement with the constants C' and N'.

Occasionally it is necessary to consider the general case where g_1 and g_2 are different. However, this is a rare situation so it might be best to skip the following result during a first reading.

Proposition 18.5.3. *Let g_1 and g_2 be σ temperate in W. Then $G=g_1\oplus g_2$ is uniformly A temperate with respect to the diagonal in $W\oplus W$ if and only if*

$$g_{1w}^\sigma(t)\leqq C g_{1w_1}^\sigma(t)(1+g_{2w}^\sigma(w_1-w))^N;\quad t,w,w_1\in W;\tag{18.5.13}$$
$$g_{2w}^\sigma(t)\leqq C g_{2w_2}^\sigma(t)(1+g_{1w}^\sigma(w_2-w))^N;\quad t,w,w_2\in W.$$

The metric $g=(g_1+g_2)/2$ is then σ temperate in W. If we set

$$h_j(w)^2=\sup g_{jw}/g_{jw}^\sigma;\quad H(w)^2=\sup g_{1w}/g_{2w}^\sigma=\sup g_{2w}/g_{1w}^\sigma,$$

then

$$\max(h_1(w)^2,h_2(w)^2,H(w)^2)\tag{18.5.14}$$
$$\leqq 4\sup g_w/g_w^\sigma\leqq h_1(w)^2+h_2(w)^2+2H(w)^2.$$

If m_j is σ, g_j temperate then $m=m_1\otimes m_2$ is uniformly A, G temperate with respect to the diagonal in $W\oplus W$ if and only if

$$m_1(w_1)\leqq C m_1(w)(1+g_{2w}^\sigma(w-w_1))^N;\quad w,w_1\in W;\tag{18.5.15}$$
$$m_2(w_2)\leqq C m_2(w)(1+g_{1w}^\sigma(w-w_2))^N;\quad w,w_2\in W.$$

These conditions are equivalent to m_j being σ, g temperate.

Proof. If (18.5.10) is valid we obtain (18.5.13) by taking $t_1=t$, $t_2=0$, $w_2=w$ or $t_1=0$, $t_2=t$, $w_1=w$. Assume now that (18.5.13) is valid. We shall then prove that for all $t,w,w_1\in W$

$$g_{jw_1}(t)\leqq C g_{jw}(t)(1+g_{w_1}^\sigma(w-w_1))^N.\tag{18.5.16}$$

This implies that g is σ temperate, so it follows from the proof of Proposition 18.5.2 that for some other C and N

$$g_{1w_1}(t_1)+g_{2w_2}(t_2)\leqq C(g_{1w}(t_1)+g_{2w}(t_2))(1+g_{w_1}^\sigma(w_2-w)+g_{w_2}^\sigma(w_1-w))^N$$

which is a stronger estimate than (18.5.10) since $g^\sigma\leqq 2g_j^\sigma$. Thus G is uniformly A temperate with respect to the diagonal.

To prove (18.5.16) we first observe that if F_1 and F_2 are positive definite quadratic forms on a vector space V, then the dual form of F_1+F_2 on V' is $\inf_{t\in V'}(F_1'(\cdot-t)+F_2'(t))$, if F_j' is the dual form of F_j on V'. If we diagonalize F_1 and F_2 simultaneously we find that it suffices to prove this for the forms $F_1(x)=x^2/a$ and $F_2(x)=ax^2$ on $\mathbb{R}$. Then $F_1'=F_2$ and $F_2'=F_1$ and the verification is elementary. Thus we have

$$g_w^\sigma(t)=\inf_{t_0}2(g_{1w}^\sigma(t-t_0)+g_{2w}^\sigma(t_0)).$$

The estimate (18.5.16) is therefore equivalent to

$$g_{jw_1}(t)\leqq C g_{jw}(t)M^N,\quad M=1+g_{1w_1}^\sigma(w-w_0)+g_{2w_1}^\sigma(w_0-w_1)\tag{18.5.16$'$}$$

for all $t, w, w_0, w_1 \in W$. Since g_j is σ temperate and (18.5.13) is assumed, we obtain for some C and N

$$g_{jw_1} \leqq C g_{jw_0}(1+g^\sigma_{2w_1}(w_0-w_1))^N \leqq C g_{jw_0} M^N,$$
$$g_{jw_0} \leqq C g_{jw}(1+g^\sigma_{1w_0}(w-w_0))^N.$$

Since again by (18.5.13) and the fact that g_2 is σ temperate

$$g^\sigma_{1w_0}(w-w_0) \leqq C g^\sigma_{1w_1}(w-w_0)(1+g^\sigma_{2w_1}(w_0-w_1))^N \leqq CM^{N+1},$$

we obtain (18.5.16)′. The same proof shows that m_j is σ, g temperate. Conversely, if m_j is σ, g temperate then (18.5.15) follows since $\frac{1}{2} g^\sigma_w \leqq g^\sigma_{jw}$; if m_j is g_k continuous it also follows that m_j is σ, g_k temperate.

Since $g_{jw} \leqq 2g_w$, hence $g^\sigma_w \leqq 2g^\sigma_{jw}$, we have for $j,k=1,2$

$$4g_w/g^\sigma_w \geqq g_{jw}/g^\sigma_{kw},$$

which proves the left-hand inequality in (18.5.14). To prove the right-hand one we use that

$$2g_w = g_{1w}+g_{2w} \leqq (h_j^2+H^2)\, g^\sigma_{jw},$$

which implies

$$2g_{jw} \leqq (h_j^2+H^2)\, g^\sigma_w, \qquad j=1,2,$$

and therefore

$$4g_w \leqq (h_1^2+h_2^2+2H^2)\, g^\sigma_w$$

which is the right-hand inequality in (18.5.14). The proof is complete.

Combining Theorem 18.4.11 and Proposition 18.5.2 we now obtain the main theorem of the Weyl calculus:

Theorem 18.5.4. *Let g be a σ temperate Riemannian metric in $W=V\oplus V'$ with $g\leqq g^\sigma$, and let m_1, m_2 be σ, g temperate weight functions in W. Then the composition formula (18.5.6) can be extended to a weakly continuous bilinear map $(a_1,a_2)\mapsto a=a_1 \# a_2$ from $S(m_1,g)\times S(m_2,g)$ to $S(m_1 m_2, g)$. If*

(18.5.17) $$h(x,\xi)^2 = \sup g_{x,\xi}/g^\sigma_{x,\xi},$$

then the map from $a_1\in S(m_1,g)$ and $a_2\in S(m_2,g)$ to the remainder term

$$a_1 \# a_2(x,\xi) - \sum_{j<N} (i\sigma(D_x,D_\xi;D_y,D_\eta)/2))^j a_1(x,\xi)\, a_2(y,\eta)/j!$$

evaluated for $(x,\xi)=(y,\eta)$, is continuous with values in $S(h^N m_1 m_2, g)$ for every integer N. It is zero if a_1 or a_2 is a polynomial of degree less than N.

The terms with j even (odd) are (skew) symmetric in a_1, a_2. This implies that

$$a_1 \# a_2 - a_2 \# a_1 - \{a_1,a_2\}/i \in S(h^3 m_1 m_2, g),$$
$$a_1 \# a_2 + a_2 \# a_1 - 2a_1 a_2 \in S(h^2 m_1 m_2, g).$$

In both cases we would have a factor h less in the calculus developed in Section 18.1.

Using Proposition 18.5.3 instead of Proposition 18.5.2 we obtain a more general result:

Theorem 18.5.5. *Let g_1 and g_2 be σ temperate metrics in $W=V\oplus V'$ satisfying (18.5.13), and assume that the function*

$$H(x,\xi)^2=\sup g_{1x,\xi}/g^\sigma_{2x,\xi}=\sup g_{2x,\xi}/g^\sigma_{1x,\xi} \tag{18.5.17$'$}$$

is $\leqq 1$. Set $g=(g_1+g_2)/2$, and let m_j be g_j continuous σ, g temperate weight functions. Then the composition formula (18.5.6) can be extended to a weakly continuous bilinear map $(a_1,a_2)\mapsto a=a_1 \# a_2$ from $S(m_1,g_1)\times S(m_2,g_2)$ to $S(m_1 m_2,g)$. The map to the Nth remainder term is continuous with values in $S(H^N m_1 m_2,g)$ for every integer N.

Note that the error terms in the calculus improve by powers of H which may be much smaller than the function h defined by (18.5.17).

To give some examples we first observe that if $B(x,\xi)=2\langle x,\xi\rangle$ then $g^B_{x,\xi}(y,\eta)=g^\sigma_{x,\xi}(y,-\eta)$ since

$$\langle y,\xi\rangle+\langle x,\eta\rangle=\sigma((y,-\eta),-(x,\xi)).$$

If the V and V' directions are g orthogonal at every point it follows that $g^B=g^\sigma$. From the discussion at the end of Section 18.4 it follows therefore that the metric (18.4.1)$'$ is σ temperate if $0\leqq\delta\leqq\rho\leqq 1$ and $\delta<1$. More general σ temperate metrics enter the theory in the following way:

Proposition 18.5.6. *Assume that g is σ temperate, that $G=mg$, where $m\geqq 1$, is slowly varying, and that $G\leqq G^\sigma$. Then it follows that G is σ temperate.*

Proof. We must show that for some C and N

$$G_{w_1}\leqq C G_w(1+G^\sigma_{w_1}(w-w_1))^N. \tag{18.5.18}$$

If $G_w(w-w_1)$ is sufficiently small this follows from the hypothesis that G is slowly varying, so we may assume that $G_w(w-w_1)\geqq c_1$. Next assume that $g_w(w-w_1)<c$ with c so small that this implies

$$C^{-1}\leqq g_w/g_{w_1}\leqq C.$$

Then (18.5.18) follows with $N=1$ if we show that

$$m(w_1)\leqq C' m(w)\, g^\sigma_{w_1}(w-w_1)/m(w_1).$$

Now

$$c_1\leqq m(w)\, g_w(w-w_1)\leqq C m(w)\, g_{w_1}(w-w_1)\leqq C m(w)\, h(w_1)^2\, g^\sigma_{w_1}(w-w_1)$$

if h is defined by (18.5.17), and $h(w_1)^2\, m(w_1)^2\leqq 1$ since $G\leqq G^\sigma$. This proves the assertion, and it only remains to study the case where $g_{w_1}(w-w_1)\geqq c_2$

for some fixed $c_2>0$. Then we have

$$g^\sigma_{w_1}(w-w_1)/m(w_1)=G^\sigma_{w_1}(w-w_1)\geqq G_{w_1}(w-w_1)\geqq c_2 m(w_1),$$

which implies

$$c_2 m(w_1)^2\leqq g^\sigma_{w_1}(w-w_1),\quad c_2 g^\sigma_{w_1}(w-w_1)\leqq G^\sigma_{w_1}(w-w_1)^2.$$

Since

$$g_{w_1}\leqq C' g_w(1+g^\sigma_{w_1}(w-w_1))^{N'}$$

and $m\geqq 1$, we obtain (18.5.18) with $N=2N'+1$. The proof is complete.

For conformal metrics the condition in Theorem 18.5.5 also simplifies:

Proposition 18.5.7. *Assume that g_1 and g_2 are conformal σ temperate metrics with $h_j(w)^2=\sup g_{jw}/g^\sigma_{jw}\leqq 1$. Then* (18.5.13) *is valid and the function H in* (18.5.17)′ *is $(h_1 h_2)^{\frac12}$.*

Proof. By hypothesis $g_2=m g_1$ for some m. Thus $h_2^2=m^2 h_1^2$ and

$$H(w)^2=\sup g_{2w}/g^\sigma_{1w}=m(w)\,h_1(w)^2=h_1(w)\,h_2(w).$$

The first estimate (18.5.13) follows from the slow variation of g_1 when $g_{1w}(w_1-w)\leqq c$. If $g_{1w}(w_1-w)\geqq c$ then

$$\begin{aligned} g^\sigma_{2w}(w-w_1)^2&=m(w)^{-2}\,g^\sigma_{1w}(w-w_1)^2\\ &\geqq m(w)^{-2}\,h_1(w)^{-2}\,g_{1w}(w-w_1)\,g^\sigma_{1w}(w-w_1)\\ &\geqq c\,g^\sigma_{1w}(w-w_1)\end{aligned}$$

since $m(w)\,h_1(w)=h_2(w)\leqq 1$. Since g_1 is σ temperate the first estimate (18.5.13) follows, and the other is proved in the same way.

Theorems 18.5.4 and 18.5.5 will give a farreaching generalization of Theorem 18.1.8 once we have an analogue of Theorem 18.1.6 allowing us to regard $a_1\#a_2$ as the symbol of a composition of operators in $\mathscr{S}$ or in $\mathscr{S}'$. Since this is not quite straight-forward we postpone the proof to Section 18.6. Instead we shall discuss here the invariance of the Weyl calculus under affine symplectic transformations χ, that is, affine maps χ in W with $\chi^*\sigma=\sigma$. In doing so we may assume that $V=\mathbb{R}^n$. Examples of such maps are

(a) The translation $x\mapsto x+x_0$ in V.
(b) The translation $\xi\mapsto\xi+\xi_0$ in V'.
(c) The map $\chi(x,\xi)$ replacing x_j,ξ_j by $\xi_j,-x_j$, leaving the other coordinates unchanged.
(d) The map $\chi(x,\xi)=(Tx,{}^tT^{-1}\xi)$ where T is a linear bijection in $\mathbb{R}^n$.
(e) The map $\chi(x,\xi)=(x,\xi-Ax)$ where A is a symmetric matrix.

Lemma 18.5.8. *Every affine symplectic map is a composition of maps of the preceding types* (a)–(e).

Proof. Since (a) and (b) supply all translations it is sufficient to consider linear maps χ. The group G generated by the maps of type (c)–(e) is transitive. In fact, the basis vector $e_1=(1,0,\dots,0)$ is mapped to $(1,0,\dots,0,\xi)$ for any desired ξ by the map (e) for an appropriate choice of A, hence to any (x,ξ) with $x\neq 0$ if we follow with a map of type (d). All elements of the form $(0,\xi)$, $\xi\neq 0$, are obtained if we use the maps (d) and (c). Any χ is therefore the product of an element in G and a symplectic linear map χ_1 with $\chi_1 e_1=e_1$. Thus

$$\sigma(\chi_1(x,\xi),e_1)=\sigma((x,\xi),e_1)=\xi_1$$

so χ_1 preserves the ξ_1 coordinate. If $x'=(x_2,\dots,x_n)$ and $\xi'=(\xi_2,\dots,\xi_n)$ it follows that the map $\chi_2(x',\xi')$ defined by $\chi_1(0,x',0,\xi')$ when the x_1 and ξ_1 coordinates are dropped is also symplectic. If $n>1$ and the statement is already proved for smaller values of n we can now write $\chi_1=\chi_3\chi_2$ where χ_2 is in the group G corresponding to the $x'\,\xi'$ variables and

$$\chi_3(x,\xi)=(x_1+a_1\xi_1,\dots,x_n+a_n\xi_1,\xi_1,\xi_2+b_2\xi_1,\dots,\xi_n+b_n\xi_1).$$

That χ_3 is symplectic means that

$$\sum_2^n(d\xi_j\wedge a_j\,d\xi_1+b_j\,d\xi_1\wedge dx_j)=0$$

so $a_2=\dots=a_n=b_2=\dots=b_n=0$. Thus χ_3 is the map (e) with $Ax=(a_1x_1,0,\dots,0)$ conjugated by (c) with $j=1$. This completes the proof.

Theorem 18.5.9. *For every affine symplectic transformation χ in $W=V\oplus V'$ there is a unitary transformation U in $L^2(V)$, uniquely determined apart from a constant factor of modulus* 1, *such that for all linear forms L in W*

$$U^{-1}L(x,D)\,U=(L\circ\chi)(x,D). \tag{18.5.19}$$

U is also an automorphism of $\mathscr{S}$ and of $\mathscr{S}'$, and

$$U^{-1}a^w(x,D)\,U=(a\circ\chi)^w(x,D) \tag{18.5.19$'$}$$

for every $a\in\mathscr{S}'(W)$.

Proof. It is sufficient to prove *uniqueness* when χ is the identity. So let U be a unitary transformation with

$$UL(x,D)=L(x,D)\,U$$

for all linear forms L. Then U commutes with the one parameter group generated by L. In particular, $U(fg)=f(Ug)$ if $g\in L^2(V)$ and f is a bounded exponential, hence if f is any function in $\mathscr{S}$, for it can be decomposed into exponentials by the Fourier inversion formula. It follows that $Ug=hg$ for some h of modulus 1 almost everywhere. Since U commutes with translations also, it is clear that h must be a constant. (Note that the result is extremely close to Lemma 7.1.4.)

To prove the *existence* of some U satisfying (18.5.19) it suffices to consider the cases (a)–(e) above. Then we can take

(a) $Uf(x)=f(x-x_0)$; (b) $Uf(x)=f(x)\exp\langle ix,\xi_0\rangle$. (c) $U=$ Fourier transformation with respect to x_j; (d) $Uf(x)=f(T^{-1}x)|\det T|^{-\frac{1}{2}}$; (e) $Uf(x)=\exp(-i\langle Ax,x\rangle/2)f(x)$. This completes the proof of (18.5.19). If $a(x,\xi)=\exp(iL(x,\xi))$ for a real linear form L then $a^w(x,D)=\exp(iL(x,D))$ in the operator theoretical sense so (18.5.19)′ is valid then. Since bounded exponentials are weakly dense in $\mathscr{S}'$ it follows that (18.5.19)′ is always true, which completes the proof.

We shall now study the relation between the Weyl calculus and the calculus established in Section 18.1. First assume that $a\in\mathscr{S}(W)$. Then $a(x,D)$ has a kernel $K\in\mathscr{S}$ given by (18.1.7), so we can write $a(x,D)=b^w(x,D)$ where $b\in\mathscr{S}$ is given by (18.5.4)″,

$$\begin{aligned} b(x,\xi)&=\int K(x+t/2,x-t/2)\,e^{-i\langle t,\xi\rangle}\,dt\\ &=(2\pi)^{-n}\iint a(x+t/2,\eta)\,e^{i\langle t,\eta-\xi\rangle}\,d\eta\,dt\\ &=\pi^{-n}\iint a(x+t,\xi+\eta)\,e^{2i\langle t,\eta\rangle}\,dt\,d\eta\\ &=e^{\langle D_x,D_\xi\rangle/2i}\,a(x,\xi).\end{aligned}$$

Here the last equality follows from the argument which led to (18.5.6). If $\tilde{c}(x,D)$ is defined by (18.5.2) then it follows from (18.1.9) that $a(x,D)=\tilde{c}(x,D)$ if

$$a(x,\xi)=e^{i\langle D_x,D_\xi\rangle}\,c(x,\xi).$$

For reasons of continuity these observations are also true if $a,b,c\in\mathscr{S}'$. Now we pointed out after Theorem 18.5.5 that a σ temperate metric g with $g_{x,\xi}(t,\tau)=g_{x,\xi}(t,-\tau)$ is also B temperate if $B(x,\xi)=2\langle x,\xi\rangle$, and m is then σ, g temperate if and only if m is B, g temperate. Hence we obtain:

Theorem 18.5.10. *Let g be σ temperate, $g\leqq g^\sigma$, and let m be σ, g temperate. If $g_{x,\xi}(t,\tau)\equiv g_{x,\xi}(t,-\tau)$ then $\exp\langle i\kappa D_x,D_\xi\rangle$ is a weakly continuous isomorphism of $S(m,g)$ for every $\kappa\in\mathbb{R}$,*

$$e^{i\kappa\langle D_x,D_\xi\rangle}\,a(x,\xi)-\sum_{j<N}\langle i\kappa D_x,D_\xi\rangle^j a(x,\xi)/j!\in S(h^N m,g)$$

for every integer N if h is defined by (18.5.17). If $a,b,c\in S(m,g)$ then $a(x,D)=b^w(x,D)=\tilde{c}(x,D)$ if and only if

$$\begin{aligned} b(x,\xi)&=e^{-i\langle D_x,D_\xi\rangle/2}\,a(x,\xi)=e^{i\langle D_x,D_\xi\rangle/2}\,c(x,\xi),\\ a(x,\xi)&=e^{i\langle D_x,D_\xi\rangle/2}\,b(x,\xi)=e^{i\langle D_x,D_\xi\rangle}\,c(x,\xi),\\ c(x,\xi)&=e^{-i\langle D_x,D_\xi\rangle}\,a(x,\xi)=e^{-i\langle D_x,D_\xi\rangle/2}\,b(x,\xi).\end{aligned}\tag{18.5.20}$$

If $g_{x,\xi}(0,\tau)\leqq|\tau|^2$ then the bilinear maps $(a,u)\mapsto a(x,D)u$, $(b,u)\mapsto b^w(x,D)u$ and $(c,u)\mapsto\tilde{c}(x,D)u$ are continuous from $S(m,g)\times\mathscr{S}$ to $\mathscr{S}$ and from $S(m,g)\times\mathscr{S}'$ to $\mathscr{S}'$.

Proof. It suffices to show that $(a,u)\mapsto a(x,D)u$ is continuous from $S(m,g) \times \mathscr{S}$ to $\mathscr{S}$. Since

$$m(x,\xi) \leqq C\,m(0,0)(1+g^{\sigma}_{0,0}(x,\xi))^N \leqq C'(1+|x|^2+|\xi|^2)^N$$

for some C, C' and N, the map is continuous with values in the space of continuous functions f with $\sup|f(x)|/(1+|x|^2)^N<\infty$. By (18.1.6) we have for some constants $c_{\alpha\beta}$

$$(1+|x|^2)^N a(x,D) = \sum_{|\alpha+\beta|\leqq N} c_{\alpha\beta}\, a^{(\alpha)}(x,D)\, x^{\beta},$$

and the assumption on the metric implies that $a\mapsto a^{(\alpha)}$ is a continuous map in $S(m,g)$ for every α. Hence $a(x,D)$ is continuous with values in the continuous bounded functions in $\mathbb{R}^n$. The continuity with values in $\mathscr{S}$ follows from the proof of Theorem 18.1.6 since for any $t\in W$

$$\langle D,t\rangle\, a\in S(m',g)$$

where $m'(x,\xi)=m(x,\xi)\,g_{x,\xi}(t)^{\frac{1}{2}}$ is also σ, g temperate. (See also the proof of Theorem 18.4.10′.) This completes the proof.

Remark. Theorem 18.6.2 below will show that the restriction imposed on the metric in the continuity statement is superfluous. However, the proof is much less elementary then, and the easy result proved above covers the metrics which are most frequent in the applications.

If $a_j\in S(m_j,g)$ and $u\in\mathscr{S}$ we can conclude under the assumptions made in the theorem that

$$a_1^w(x,D)\,a_2^w(x,D)\,u=(a_1 \# a_2)^w(x,D)\,u.$$

In fact, we can take sequences $a_{j\nu}\in\mathscr{S}$ which are bounded in $S(m_j,g)$ and equal to a_j on any compact set for large ν. Then $a_{2\nu}^w(x,D)u\to a_2^w(x,D)u$ in $\mathscr{S}$ as $\nu\to\infty$, so

$$(a_{1\nu}\# a_{2\nu})^w(x,D)\,u=a_{1\nu}^w(x,D)\,a_{2\nu}^w(x,D)\,u\to a_1^w(x,D)\,a_2^w(x,D)\,u \quad \text{in } \mathscr{S}.$$

Since $a_{1\nu}\# a_{2\nu}$ converges weakly to $a_1\# a_2$ the assertion follows.

Still under the hypotheses in Theorem 18.5.10 one can easily recover composition rules for standard pseudo-differential operators from the preceding results. Suppose for example that $a_j\in S(m_j,g)$ and set $a_1(x,D)\,a_2(x,D)=a_3(x,D)$. Then $a_j(x,D)=b_j^w(x,D)$ for

$$b_j(x,\xi)=e^{-i\langle D_x,D_\xi\rangle/2}\,a_j(x,\xi),$$

and

$$b_3(x,\xi)=e^{i(\langle D_\xi,D_y\rangle-\langle D_x,D_\eta\rangle)/2}\,b_1(x,\xi)\,b_2(y,\eta)|_{x,\xi=y,\eta}.$$

Since

$$\begin{aligned}(\langle D_x+D_y, D_\xi+D_\eta\rangle &+\langle D_\xi,D_y\rangle-\langle D_x,D_\eta\rangle \\ &-\langle D_x,D_\xi\rangle-\langle D_y,D_\eta\rangle)/2=\langle D_\xi,D_y\rangle\end{aligned}$$

it follows that

(18.5.21) $$a_3(x,\xi)=e^{i\langle D_\xi, D_y\rangle}\, a_1(x,\xi)\, a_2(y,\eta)|_{x,\xi=y,\eta},$$

which is precisely (18.1.14) which is therefore proved now in much greater generality than in Section 18.1.

If $a(x,D)$ is a pseudo-differential operator with polyhomogeneous symbol

$$a(x,\xi)\sim a_m(x,\xi)+a_{m-1}(x,\xi)+\dots,$$

where a_j is homogeneous in ξ of degree j, we can use Theorem 18.5.10 to write $a(x,D)=b^w(x,D)$ where $b(x,\xi)\sim\sum b_{m-j}(x,\xi)$. Then

$$b_m(x,\xi)=a_m(x,\xi),\qquad b_{m-1}(x,\xi)=a_{m-1}(x,\xi)+i\sum \partial^2 a_m(x,\xi)/\partial x_j\partial\xi_j/2,$$

and b_{m-1} is the subprincipal symbol introduced in (18.1.32). Thus the Weyl calculus explains the role of the subprincipal symbol, and the composition formula in the Weyl calculus gives immediately rules of computation for the subprincipal symbol.

18.6. Estimates of Pseudo-Differential Operators

In this section we shall first prove that $a^w(x,D)$ is continuous in $\mathscr{S}$ and in $\mathscr{S}'$ when $a\in S(m,g)$ if g is σ temperate and m is σ, g temperate. In a special but important case this was proved in Theorem 18.5.10. When the hypotheses of Theorem 18.5.4 or Theorem 18.5.5 are fulfilled the result gives as in the special case that $(a_1\# a_2)^w(x,D)$ is the composition of $a_1^w(x,D)$ and $a_2^w(x,D)$ in $\mathscr{S}$ as well as in $\mathscr{S}'$. When $g\leqq g^\sigma$ and m is bounded (resp. $m\to 0$ at ∞) we shall prove that $a^w(x,D)$ is bounded (resp. compact) in L^2. This result combined with the calculus in Section 18.5 will give efficient lower bounds for operators with non-negative symbol.

As usual we shall split the symbol by means of the partition of unity in Lemma 18.4.4. To handle an individual term it suffices to have the following elementary result:

Lemma 18.6.1. *For every $a\in\mathscr{S}(\mathbb{R}^{2n})$ we have with operator norm in L^2*

$$\|a^w(x,D)\|\leqq(2\pi)^{-2n}\|\hat a\|_{L^1}=\|a\|_{FL^1}.$$

Proof. The Schwartz kernel K of $a^w(x,D)$ is

$$K(x,y)=(2\pi)^{-n}\int a((x+y)/2,\xi)\,e^{i\langle x-y,\xi\rangle}\,d\xi$$
$$=(2\pi)^{-2n}\int \hat a(\theta, y-x)\,e^{i\langle x+y,\theta/2\rangle}\,d\theta.$$

Hence

$$\int|K(x,y)|\,dx\leqq\|a\|_{FL^1},\qquad \int|K(x,y)|\,dy\leqq\|a\|_{FL^1},$$

so the lemma follows from Lemma 18.1.12.

Remark. When using the lemma we shall need the obvious fact that $\|a\|_{FL^1}$ can be estimated by a finite number of seminorms in $\mathscr{S}$. It is also essential that this norm is invariant under composition with affine transformations, which is obvious since it is the total mass of characters in the Fourier decomposition of a.

Theorem 18.6.2. *If g is σ temperate and m is σ, g temperate, then $a^w(x,D)$ is a continuous map from $\mathscr{S}$ to $\mathscr{S}$ and from $\mathscr{S}'$ to $\mathscr{S}'$ for every $a\in S(m,g)$, and it is weakly continuous as a function of a.*

Proof. It suffices to prove the continuity in $\mathscr{S}$ since the continuity of $\bar{a}^w$ in $\mathscr{S}$ implies that of a^w in $\mathscr{S}'$. Let $\{\phi_\nu\}$ be the partition of unity introduced in Lemma 18.4.4, define the neighborhoods U_ν of $\operatorname{supp}\phi_\nu$ as before Lemma 18.4.8, and set

$$a=\sum a_\nu, \quad a_\nu=\phi_\nu a.$$

If w_ν is the center of U_ν, the seminorms of a_ν in $S(m(w_\nu), g_{w_\nu})$ have bounds independent of ν. By Lemma 18.6.1 and the remark following it we conclude that

$$\|a_\nu^w(x,D)\|\leqq C m(w_\nu),$$

for all positive definite quadratic forms are equivalent under arbitrary linear transformations. Thus we have with L^2 norms

$$\|a_\nu^w(x,D)u\|\leqq C m(w_\nu)\,\|u\|.$$

To obtain a better estimate we shall use an argument parallel to the proof of Proposition 18.4.6.

Let L be an arbitrary linear form on W which is positive in U_ν. By Lemma 18.4.5 we have a uniform bound for $L(w_\nu)/L$ in $S(1, g_{w_\nu})$ over the support of ϕ_ν. Now Theorem 18.5.4 or just (18.5.5) gives

$$a_\nu^w(x,D)u=(a_\nu/L)^w(x,D)\,L(x,D)u+i\{a_\nu/L,L\}^w(x,D)u/2. \tag{18.6.1}$$

Here the symbol of a_ν/L is uniformly bounded in $S(m(w_\nu)/L(w_\nu),\, g_{w_\nu})$. If we write $L(w)=\sigma(w,t)$, $t\in\mathbb{R}^{2n}$, then

$$\{a_\nu/L, L\}=-L^{-1}\langle t, da_\nu\rangle.$$

Thus we have uniform bounds for the seminorms of $\{a_\nu/L, L\}$ in

$$S(m(w_\nu)\,g_{w_\nu}(t)^{\frac{1}{2}}/L(w_\nu), g_{w_\nu}).$$

$L(x,D)u$ is a linear combination of $x_j u$ and $D_j u$ with coefficients bounded by the length of t in a fixed metric such as g_0. Iterated use of (18.6.1) gives

$$\|a_\nu^w(x,D)u\|\leqq C_N\, m(w_\nu)\, R_\nu^{-N} \sum_{|\alpha+\beta|\leqq N} \|x^\alpha D^\beta u\|, \quad u\in\mathscr{S}, \tag{18.6.2}$$

for every positive integer N, provided that $0<R_\nu\leqq L(w_\nu)$, $L\neq 0$ in U_ν, $g_0(t)+g_{w_\nu}(t)\leqq 1$, $L(w)=\sigma(w,t)$. By the Hahn-Banach theorem we can take for R_ν

the distance from 0 to U_ν in the norm dual to $(g_0+g_{w_\nu})^{\frac{1}{2}}$ with respect to σ. This norm is defined by

$$\|w\|_\nu=\inf(g^\sigma_{w_\nu}(w_1)+g^\sigma_0(w_2))^{\frac{1}{2}}, \quad w_1+w_2=w.$$

(See the proof of Proposition 18.5.3.). Thus we can take

$$R_\nu=\min_{w\in U_\nu}\|w\|_\nu.$$

In a moment we shall prove that for some N

$$\sum(1+R_\nu)^{-N}<\infty, \quad \text{hence } R_\nu\to\infty, \tag{18.6.3}$$

$$g_0(w_\nu)\leqq C(1+R_\nu)^N. \tag{18.6.4}$$

Admitting these estimates we conclude using (18.6.2) for large ν that $\|a^w(x,D)u\|$ can be estimated by a seminorm of u in $\mathscr{S}$ if

$$m(w_\nu)\leqq C(1+g_0(w_\nu))^N$$

for some N, which is always true when m is σ, g temperate. Let $M(x,D)$ be linear in (x,D). Then $M(x,D)\,a^w_\nu(x,D)$ has the symbol

$$M a_\nu+\{M,a_\nu\}/2i$$

which is bounded in $S(m'_\nu,g_{w_\nu})$ for some other m'_ν bounded by a power of $1+g_0(w_\nu)$. Hence the L^2 norm of $M(x,D)\,a^w_\nu(x,D)\,u$ can be estimated by a seminorm of u in $\mathscr{S}$. Repeating the argument we conclude that $a^w(x,D)$ is continuous in $\mathscr{S}$.

It remains to verify the estimates (18.6.3) and (18.6.4). The proof is parallel to that of Lemma 18.4.8. Let

$$M_k=\{\nu;\, R^2_\nu\leqq k\}.$$

When $\nu\in M_k$ we can choose $w'_\nu\in U_\nu$ and w''_ν so that

$$g^\sigma_{w_\nu}(w'_\nu-w''_\nu)\leqq k, \quad g^\sigma_0(w''_\nu)\leqq k.$$

In the first inequality we pass to the equivalent norm $g^\sigma_{w'_\nu}$ and conclude using (18.5.11) that

$$g^\sigma_{w''_\nu}(w'_\nu-w''_\nu)\leqq C k^N, \quad g^\sigma_0(w'_\nu-w''_\nu)\leqq C' k^{N'}.$$

Hence

$$g^\sigma_{w_\nu}\leqq C g^\sigma_{w'_\nu}\leqq C_1 g^\sigma_{w''_\nu} k^{N_1}\leqq C_2 g^\sigma_0 k^{N_2},$$

$$g_{w_\nu}\leqq C' g_{w'_\nu}\leqq C'_1 g_{w''_\nu} k^{N'_1}\leqq C'_2 g_0 k^{N'_2}.$$

$$g_0(w_\nu-w'_\nu)\leqq C_2 g_{w_\nu}(w_\nu-w'_\nu) k^{N_2}\leqq C_4 k^{N_2},$$

$$g_0(w'_\nu)/2\leqq g_0(w'_\nu-w''_\nu)+g_0(w''_\nu)\leqq C_5 k^{N_3}.$$

The estimate (18.6.4) is an immediate consequence, and (18.6.3) follows if we repeat the proof of Lemma 18.4.8. This is left for the reader.

Next we shall discuss continuity in L^2.

Theorem 18.6.3. *Assume that $g \leq g^\sigma$, that g is σ temperate, and that m is σ, g temperate. Then the operator $a^w(x,D)$ is L^2 continuous for every $a \in S(m,g)$ if and only if m is bounded. The L^2 norm of $a^w(x,D)$ is then a continuous semi-norm in $S(m,g)$.*

In the proof of necessity we shall need the following lemma:

Lemma 18.6.4. *If g is a positive definite quadratic form in $\mathbb{R}^{2n}$ then there is a linear symplectic map χ in $\mathbb{R}^{2n}$ such that*

$$g(\chi(x,\xi)) = \sum \lambda_j (x_j^2 + \xi_j^2).$$

Here λ_j are uniquely determined by g, and

$$\sup g/g^\sigma = \max \lambda_j^2.$$

Proof. Since $\sup g/g^\sigma$ is symplectically invariant it suffices to prove the last statement when $g(x,\xi) = \sum \lambda_j (x_j^2 + \xi_j^2)$. Then we have $g^\sigma(x,\xi) = \sum (x_j^2 + \xi_j^2)/\lambda_j$ so the statement is obvious. To prove the lemma we shall consider the eigenvalues and eigenvectors of the map

$$F\colon (x,\xi) \mapsto H_g(x,\xi)/2 = (\partial g/\partial \xi, -\partial g/\partial x)/2$$

which is defined in a symplectically invariant way. Note that if $g(x,\xi) = \sum \lambda_j (x_j^2 + \xi_j^2)$ then

$$F(x,\xi) = (\lambda_1 \xi_1, \dots, \lambda_n \xi_n, -\lambda_1 x_1, \dots, -\lambda_n x_n),$$

so $x_j = 1$, $\xi_j = \pm i$, all other coordinates 0, is an eigenvector with eigenvalue $\pm i \lambda_j$. Hence $\lambda_1, \dots, \lambda_n$ are uniquely defined by g apart from the order. For general g, if z is an eigenvalue with eigenvector $(y,\eta) \in \mathbb{C}^{2n}$, then

$$\partial g/\partial \eta = 2zy, \qquad \partial g/\partial y = -2z\eta.$$

Writing $(\theta, w) = \sum \theta_j \bar{w}_j$ if $\theta, w \in \mathbb{C}^n$, we obtain

$$(\partial g/\partial y, y) + (\partial g/\partial \eta, \eta) = 2z((y,\eta) - (\eta, y)).$$

In the left-hand side we have a positive definite Hermitian form, and

$$(y,\eta) - (\eta,y) = 2i(\langle \operatorname{Im} y, \operatorname{Re} \eta\rangle - \langle \operatorname{Re} y, \operatorname{Im} \eta\rangle) = -2i\sigma(\varepsilon_1, e_1)$$

if e_1 and ε_1 are the real and imaginary parts of (y,η). Hence $\lambda = z/i$ is real and not 0. Taking the complex conjugates of y, η, z if necessary we may assume that $\lambda > 0$. Then we have $\sigma(\varepsilon_1, e_1) > 0$, or $\sigma(\varepsilon_1, e_1) = 1$ if we multiply by an appropriate constant. Using the symplectic maps (c)–(e) in Lemma 18.5.8 it is easy to find a linear symplectic map χ mapping the unit vectors along the x_1 and ξ_1 axes to e_1 and ε_1 respectively. (See also Proposition 21.1.3 for a complete proof.) Replacing g by $g \circ \chi$ we have then

$$g(x,\xi) = \lambda(x_1^2 + \xi_1^2) + g_1(x,\xi)$$

where g_1 is independent of x_1 and ξ_1, for $\partial g/\partial x_j = \partial g/\partial \xi_j = 0$, $j \neq 1$, at the eigenvectors $(1,0,\dots,0,\pm i,0,\dots,0)$ of F. This proves the lemma by induction with respect to n.

Remark. In Theorem 21.5.3 we shall give a much more thorough discussion of the symplectic equivalence of quadratic forms using the tools developed in Section 21.1. Note that in Lemma 18.6.4 we can equally well reduce g to the quadratic form $\sum(x_j^2+\lambda_j^2\xi_j^2)$. Lemma 18.6.4 is well known in classical mechanics where λ_j occur as the fundamental frequencies of the small oscillations of a mechanical system around an equilibrium point.

In the proof of sufficiency in Theorem 18.6.3 we shall use the same decomposition of a as in the proof of Theorem 18.6.2. To control the sum we need the following lemma of Cotlar, Knapp, Stein, Calderón, Vaillancourt, Bony and Lerner on sums of almost orthogonal operators, with an additional remark on compactness which prepares for Theorem 18.6.6.

Lemma 18.6.5. *Let A_i, $i \in I$, be a countable number of bounded operators from one Hilbert space H_1 to another H_2 such that for $j \in I$*

$$\sum_{k\in I}\|A_j^*A_k\|^{\frac12}\le M,\quad \sum_{k\in I}\|A_jA_k^*\|^{\frac12}\le M. \tag{18.6.5}$$

Then $\sum_{j,k}|(A_ku,A_ju)|\le M^2\|u\|^2$, $u\in H_1$, the sum

$$Su=\sum_{j\in I}A_ju,\quad u\in H_1,$$

exists with norm convergence in H_2, and $\|S\|\le M$. If the sum

$$S_J=\sum_{j\in J}A_j$$

is compact for every $J\subset I$, then $\|A_j\|\to 0$ when $j\to\infty$ in I.

Proof. Set $T=\sum\alpha_{jk}A_j^*A_k$ where $\alpha_{jk}\in\mathbb{C}$, $|\alpha_{jk}|\leqq 1$ and only a finite number N are not zero. Then $T^*=\sum\overline{\alpha_{kj}}A_j^*A_k$. We have $\|T\|^2=\|T^*T\|$, and more generally $\|T\|^{2m}=\|(T^*T)^m\|$ by the spectral theorem, if m is a positive integer. For the terms in the expansion

$$(T^*T)^m=\sum\overline{\alpha_{j_2j_1}}A_{j_1}^*A_{j_2}\alpha_{j_3j_4}A_{j_3}^*A_{j_4}\dots\alpha_{j_{4m-1}j_{4m}}A_{j_{4m-1}}^*A_{j_{4m}}$$

we have the estimate

$$\begin{aligned}&\|A_{j_1}^*A_{j_2}\dots A_{j_{4m-1}}^*A_{j_{4m}}\|\\ &\leqq\min(\|A_{j_1}^*A_{j_2}\|\dots\|A_{j_{4m-1}}^*A_{j_{4m}}\|,\\ &\qquad\|A_{j_1}^*\|\,\|A_{j_2}A_{j_3}^*\|\dots\|A_{j_{4m-2}}A_{j_{4m-1}}^*\|\,\|A_{j_{2m}}\|).\end{aligned}$$

Taking the geometric mean of the two upper bounds, and noting that $\|A_j\|\leqq M$ by hypothesis, we obtain

$$\|T^{2m}\|\leqq M\sum|\alpha_{j_1j_2}|\,\|A_{j_1}^*A_{j_2}\|^{\frac12}\,\|A_{j_2}A_{j_3}^*\|^{\frac12}\dots\|A_{j_{4m-1}}^*A_{j_{4m}}\|^{\frac12}.$$

The sum is taken over $j_1,\dots,j_{4m}$. If we use (18.6.5) to estimate successively the sum over $j_{4m},\dots,j_2$, then only the sum over j_1 is left over and we see that

$$\|T\|^{2m}\leqq N\,M^{4m}.$$

Taking $2m^{\text{th}}$ roots and letting $m \to \infty$ yields $\|T\| \leqq M^2$, hence

$$|\sum \alpha_{jk}(A_k u, A_j u)| = |(Tu, u)| \leqq M^2 \|u\|^2, \quad u \in H_1,$$

and taking the supremum over all choices of (α_{jk}) we obtain

$$\sum_{j,k \in I} |(A_k u, A_j u)| \leqq M^2 \|u\|^2, \quad u \in H_1.$$

If $J \subset I$ is finite and $A(J) = \sum_{j \in J} A_j$ it follows that $\|A(J)u\| \leqq M\|u\|$, and if $J' \supset J$ is another finite subset then

$$\|(A(J') - A(J))u\|^2 \leqq \sum_{j,k \in I \setminus J} |(A_j u, A_k u)|.$$

Since $\sum |(A_j u, A_k u)|$ is convergent it follows that $Su = \sum_{j \in J} A_j u$ converges in H_2, and that $\|Su\| \leqq M\|u\|$.

To prove the last statement it suffices to show that if $\|A_j\| \geqq c > 0, j \in I$, then $\sum_{j \in J} A_j$ is not compact for some $J \subset I$. To do so we shall choose $j_1, j_2, \ldots \in I$ so that $\sum_{\nu \neq \mu} \|A_{j_\nu} A_{j_\mu}^*\| \leqq 1/\mu$ for every μ. Assume that $j_1, \ldots, j_{N-1}$ have already been chosen so that this is true with strict inequality for $\mu < N$ if $\nu < N$ in the sums. We must then choose $j_N = j$ so that

$$\sum_{\nu < N} \|A_{j_\nu} A_j^*\| < 1/N, \quad \sum_{\nu \neq \mu} \|A_{j_\nu} A_{j_\mu}^*\| + \|A_j A_{j_\mu}^*\| < 1/\mu, \quad \mu < N.$$

By the second part of (18.6.5) the first inequality holds except for finitely many j, and since $\|A_j A_{j_\mu}^*\| = \|A_{j_\mu} A_j^*\|$ the same is true for the other inequalities, by the inductive hypothesis. Set $J = \{j_1, j_2, \ldots\}$. For every j we choose $u_j \in H_2$ with $\|u_j\| = 1$ and $\|A_j^* u_j\| \geqq \|A_j^*\|/2 \geqq c/2$ if $j \in J$. Then $v_j = A_j^* u_j$ converges weakly to 0 in H_1. In fact,

$$(A_j^* u_j, A_k^* v) = (A_k A_j^* u_j, v) \to 0 \quad \text{as } j \to \infty$$

so $(A_j^* u_j, w) \to 0$ for every w in the closed linear hull of the ranges of the operators A_k^*. Since $A_k w = 0$ for every k if w is in the orthogonal complement, we have $(A_j^* u_j, w) = 0$ for all j then. Thus v_j converges weakly to 0. If S_J is compact, it follows that $\|S_J v_j\| \to 0$ as $j \to \infty$. We have with $j, k \in J$

$$S_J v_j = A_j A_j^* u_j + \sum_{k \neq j} A_k A_j^* u_j.$$

The norm of the sum is at most $1/\mu$ if $j = j_\mu$. Hence $\|A_j A_j^* u_j\| \to 0$, so

$$\|A_j^* u_j\|^2 = (A_j A_j^* u_j, u_j) \to 0.$$

This is a contradiction which completes the proof.

Proof of Theorem 18.6.3. First we prove the necessity so we assume that $a^w(x, D)$ is L^2 continuous for all $a \in S(m, g)$. Since the map

$$\mathcal{S}' \ni a \mapsto (a^w(x, D)u, v)$$

is continuous for all $u, v \in \mathscr{S}$, it follows that the map

$$S(m,g) \ni a \mapsto a^w(x,D) \in \mathscr{L}(L^2, L^2)$$

given by the hypothesis is closed. Hence it is continuous by the closed graph theorem. Let w_ν be a sequence going to ∞ in W. We shall prove that the sequence $m(w_\nu)$ has a bounded subsequence. By Lemma 18.6.4 we can take a linear symplectic transformation χ_ν from $\mathbb{R}^{2n}$, with the usual symplectic structure, to W such that $g_{w_\nu}(\chi_\nu(y,\eta))$ is reduced to the diagonal form

$$\sum (y_j^2 + \lambda_{j\nu}^2 \eta_j^2).$$

Since $g \leqq g^\sigma$, all $\lambda_{j\nu}$ are bounded by 1. Thus we may assume that their limits exist as $\nu \to \infty$. Let $\phi \in C_0^\infty(\mathbb{R}^{2n})$ and set

(18.6.6) $$e_\nu(x,\xi) = \phi(x, \lambda_\nu \xi) = b_\nu(w_\nu + \chi_\nu(x,\xi)),$$

where $\lambda_\nu \xi = (\lambda_{1\nu}\xi_1, \dots, \lambda_{n\nu}\xi_n)$. If the support of ϕ is small enough it follows that $m(w_\nu) b_\nu$ is a bounded sequence in $S(m,g)$. The norm of $b_\nu^w(x,D)$ is equal to the norm of $e_\nu^w(x,D)$ since these operators are unitarily equivalent by Theorem 18.5.9. Now

$$e_\nu^w(x,D)\, u = (2\pi)^{-n} \iint \phi((x+y)/2, \lambda_\nu \xi)\, e^{i\langle x-y, \xi\rangle} u(y)\, dy\, d\xi$$

has a non-zero limit as $\nu \to \infty$, for a suitable choice of $u \in C_0^\infty$. It follows that $\|e_\nu^w(x,D) u\|$ is bounded from below for large ν. Hence

$$\|m(w_\nu) b_\nu^w(x,D)\| > c\, m(w_\nu)$$

which proves that $m(w_\nu)$ must be bounded. This ends the proof of the necessity.

To prove the sufficiency we assume now that $m=1$. Writing $a = \sum a_\nu$ as in the proof of Theorem 18.6.2 we have, again by Lemma 18.6.1,

$$\|a_\nu^w(x,D)\| \leqq C.$$

To prove that (18.6.5) is valid for $A_\nu = a_\nu^w(x,D)$ we must consider the compositions

$$a_{\nu\mu}^w = \bar{a}_\nu^w a_\mu^w, \qquad b_{\nu\mu}^w = a_\nu^w \bar{a}_\mu^w.$$

Of course it suffices to discuss $a_{\nu\mu}$. By (18.5.6) we have

$$a_{\nu\mu}(x,\xi) = e^{i\sigma(D_x, D_\xi; D_y, D_\eta)/2}\, \bar{a}_\nu(x,\xi)\, a_\mu(y,\eta)|_{(x,\xi)=(y,\eta)}.$$

We apply (18.4.16) with d_ν replaced by d'_ν, defined by substituting g_x^A for g_y^A in the definition (18.4.15) of d_ν. This is legitimate since $1 + d'_\nu \leqq C'(1+d_\nu)^{N'}$ by (18.4.13)$''$. Hence we have for any positive integer k

$$|a_{\nu\mu}(w)| \leqq C_k (1 + M(w))^{-k},$$

where

$$M(w) = \min_{U_\nu} g_w^\sigma(w - w') + \min_{U_\mu} g_w^\sigma(w - w'').$$

It is clear that

$$g^\sigma_w(w'-w'')\leqq 2M(w)$$

if $w'\in U_\nu$ and $w''\in U_\mu$ are chosen so that the minimum in the definition of M is attained. We also have

$$(18.6.7)\qquad g^\sigma_{w_\nu}(w'-w'')\leqq C g^\sigma_{w'}(w'-w'')\leqq C_1 g^\sigma_w(w'-w'')(1+g^\sigma_w(w-w'))^N \leqq C_2(1+M(w))^{N+1},$$

$$(18.6.8)\qquad 1+g_{w_\nu}(w-w_\nu)\leqq C(1+g_{w'}(w-w'))\leqq C_1(1+g^\sigma_{w'}(w-w')) \leqq C_2(1+g^\sigma_w(w-w'))^{N+1}\leqq C_3(1+M(w))^{N+1}.$$

With the notation

$$d_{\nu\mu}=\min_{w'\in U_\nu,\, w''\in U_\mu} g^\sigma_{w_\nu}(w'-w'')$$

it follows from (18.6.7) with new constants C and N that for all w

$$1+d_{\nu\mu}\leqq C(1+M(w))^N.$$

Taking also (18.6.8) into account we conclude that for any k

$$(18.6.9)\qquad |a_{\nu\mu}(w)|\leqq C_k(1+d_{\nu\mu})^{-k}(1+g_{w_\nu}(w-w_\nu))^{-k}.$$

The same estimate is valid for any seminorm in $S(1,g_w)$. In fact, if $g_{w_\nu}(t)\leqq 1$ and we apply the differential operator $\langle D,t\rangle$ to $a_{\nu\mu}$, we obtain one term where the differentiation falls on a_ν, which does not affect the estimate, and one where it falls on a_μ. This may lead to a loss of a factor $(g_{w_\mu}(t)/g_{w_\nu}(t))^{\frac{1}{2}}$ in the estimates. Since it is bounded by some power of $1+d_{\nu\mu}$, our assertion is proved. Hence Lemma 18.6.1 combined with the remark following it gives

$$\|a^w_{\nu\mu}(x,D)\|\leqq C_N(1+d_{\nu\mu})^{-N}$$

for any N. We shall prove in a moment that for some C and N

$$(18.6.10)\qquad \sum_\nu(1+d_{\nu\mu})^{-N}\leqq C,\quad \sum_\mu(1+d_{\nu\mu})^{-N}\leqq C.$$

From Lemma 18.6.5 it follows then that $\|a^w(x,D)\|\leqq C'$.

It remains to prove (18.6.10), which is closely related to Lemma 18.4.8. Since for $w'\in U_\nu$ and $w''\in U_\mu$ we have

$$1+g^\sigma_{w_\nu}(w'-w'')\leqq C(1+g^\sigma_{w'}(w'-w''))\leqq C'(1+g^\sigma_{w''}(w'-w''))^{N'} \leqq C''(1+g^\sigma_{w_\mu}(w'-w''))^{N'},$$

it follows that

$$1+d_{\nu\mu}\leqq C''(1+d_{\mu\nu})^{N'}.$$

Hence it suffices to consider the sum in (18.6.10) with respect to μ. We choose g_{w_ν} orthonormal coordinates with 0 at w_ν. Then $g^\sigma_{w_\nu}$ is at least as large as the Euclidean metric g_{w_ν}. If $d_{\nu\mu}\leqq k$ then there is some point $w''\in U_\mu$ with Euclidean distance at most $Ck^{\frac{1}{2}}$ from 0. The metric at w'' can be estimated by a power of k times the Euclidean metric, so a ball of radius

k^{-N} with center at w'' is contained in U'_μ. As in the proof of Lemma 18.4.8 it follows at once that the number of indices μ with $d_{\nu\mu} \leqq k$ is bounded by a power of k, which proves (18.6.10).

Theorem 18.6.6. *Assume that $g \leqq g^\sigma$, that g is σ temperate, and that m is σ, g temperate. Then the operators $a^w(x, D)$ with $a \in S(m, g)$ are all compact in L^2 if and only if $m \to 0$ at ∞.*

Proof. With the notation used in the proof of Theorem 18.6.3 it is clear that a_ν^w is compact for every ν. If $m \to 0$ at ∞ we can for every $\varepsilon > 0$ choose a finite set $N(\varepsilon)$ such that

$$(a - \sum_{\nu \in N(\varepsilon)} a_\nu)/\varepsilon$$

is bounded in $S(1, g)$ when $\varepsilon \to 0$. Thus we have by Theorem 18.6.3

$$\|a^w(x, D) - \sum_{\nu \in N(\varepsilon)} a_\nu^w(x, D)\| \leqq C\varepsilon,$$

which proves that $a^w(x, D)$ is compact. Assume on the other hand that all $a^w(x, D)$ with $a \in S(m, g)$ are compact. Then it follows from Theorem 18.6.3 that m is bounded. If b_ν is defined as in the proof of necessity in Theorem 18.6.3 and $m_\nu = m(w_\nu)$ then

$$\sum_{\nu \in J} m_\nu b_\nu \in S(m, g)$$

for every subset J of the index set. Hencc

$$\sum_{\nu \in J} m_\nu b_\nu^w(x, D)$$

is compact by hypothesis, and b_ν^w satisfy (18.6.5) by the proof of sufficiency in Theorem 18.6.3, so it follows from Lemma 18.6.5 that $m_\nu \|b_\nu^w(x, D)\| \to 0$. Since $\|b_\nu^w(x, D)\|$ is bounded from below we conclude that $m_\nu \to 0$ as $\nu \to \infty$, which implies that $m(x, \xi) \to 0$ as $(x, \xi) \to \infty$. The proof is complete.

So far we have only considered scalar pseudo-differential operators. However, it is clear that the calculus developed in Section 18.5 is not changed at all if one allows the functions u to take their values in a Banach space B_1 and the symbol a to take its values in $\mathscr{L}(B_1, B_2)$, so that $a^w(x, D)u$ takes its values in B_2. However, the discussion of L^2 estimates here depends on Lemma 18.6.5 where the Hilbert space structure is vital. Theorems 18.6.3 and 18.6.6 are therefore applicable only when B_1 and B_2 are Hilbert spaces which in the second case must also be finite dimensional.

As a first application we shall prove a general version of Theorem 18.1.14. Note that with the notation there the Weyl symbol of $(a(x, D) + a(x, D)^*)/2$ is $\operatorname{Re}(a + i/2 \sum \partial^2 a/\partial x_j \partial \xi_j)$ modulo S^{m-2}. We only consider the case of Theorem 18.1.14 where $m = 0$, for spaces corresponding to $H_{(s)}$ have not been defined here in the general context.

Theorem 18.6.7. *Let g be σ temperate,*

$$h(x,\xi)^2=\sup g_{x,\xi}/g^\sigma_{x,\xi}\leqq 1, \tag{18.6.11}$$

and assume that $0\leqq a\in S(1/h,g)$. *Then it follows that*

$$(a^w(x,D)u,u)\geqq -C\|u\|^2,\quad u\in\mathscr{S}, \tag{18.6.12}$$

with scalar product and norm in $L^2(\mathbb{R}^n)$.

Proof. For the metric

$$G=(a+1)^{-1}h^{-1}g$$

we have

$$\sup G_w/G^\sigma_w=(a(w)+1)^{-2}h(w)^{-2}\sup g_w/g^\sigma_w=(a(w)+1)^{-2}\leqq 1.$$

If we show that G is slowly varying it will follow from Proposition 18.5.6 that G is σ temperate, for $h(a+1)$ may be replaced by $\min(1,h(a+1))$ since $h(a+1)$ is bounded. Thus $a+1$ is σ, G temperate. We shall also prove that

$$a+1\in S(a+1,G), \tag{18.6.13}$$

that is,

$$|a^{(k)}(w;t_1,\dots,t_k)|\leqq C_k(a(w)+1)^{1-k/2}h(w)^{-k/2}\prod_1^k g_w(t_j)^{\frac12}. \tag{18.6.13)'$$

When $k\geqq 2$ this follows from the hypothesis $a\in S(1/h,g)$, so we only have to show that

$$|h(w)a'(w)t|\leqq C(h(w)a(w))^{\frac12}g_w(t)^{\frac12}. \tag{18.6.13)''$$

If we introduce g_w orthonormal coordinates $z_1,\dots,z_{2n}$ with the origin at w and regard $h(w)a$ as a function $F(z)$, then

$$|D^\alpha_z F(z)|\leqq C_\alpha,\quad |z|<c,$$

for all α since $a\in S(1/h,g)$, and $F\geqq 0$. Thus it follows from Lemma 7.7.2 that

$$|F'(0)|\leqq CF(0)^{\frac12}$$

which proves (18.6.13)″. By Taylor's formula we obtain

$$F(z)+h(w)\geqq(F(0)+h(w))/2\quad \text{if } |z|^2<c_1(F(0)+h(w)),$$

that is,

$$2h(w)(a(w_1)+1)\geqq h(w)(a(w)+1)\quad \text{if } G_w(w-w_1)<c_1.$$

Thus $G_{w_1}\leqq CG_w$ then so G is slowly varying.

If F is a C^∞ function with

$$|t^jF^{(j)}(t)|<C_jF(t),\quad t>0,$$

and $m\in S(m,G)$, it is immediately verified that $F(m)\in S(F(m),G)$. (Cf. Lemma 18.4.3.) Taking $F(t)=t^{\frac12}$ and $m=a+1$ we obtain $b=(a+1)^{\frac12}\in S(b,G)$. Now

Theorem 18.5.4 gives

$$a^w(x,D)+1=b^w(x,D)^2+c^w(x,D)$$

where $c\in S((a+1)^{-1},G)\subset S(1,G)$. Hence $c^w(x,D)$ is bounded which proves the theorem since $b^w(x,D)^2$ is positive.

In the last part of the proof there was so much information given away that one might suspect that a better result is valid. Indeed, we shall now prove the stronger Fefferman-Phong inequality:

Theorem 18.6.8. *If g is σ temperate and* (18.6.11) *is fulfilled then* (18.6.12) *is valid for every* $a\in S(1/h^2,g)$ *with* $a\geqq 0$.

A crucial point in the proof of Theorem 18.6.7 was the application of Lemma 7.7.2. We shall now prove a similar result which takes derivatives of order $\leqq 4$ into account. By B_r we shall denote the ball $\{x\in\mathbb{R}^\nu; |x|<r\}$ where $|x|^2=e$ is the Euclidean metric form.

Lemma 18.6.9. *Let* $0\leqq f\in C^\infty(B_2)$ *and assume that*

$$|f|_4^e(x)\leqq 1,\quad x\in B_2, \tag{18.6.14}$$

$$\max(|f(0)|,|f|_2^e(0))=1. \tag{18.6.15}$$

Then we can find r>0 independent of f such that

$$\tfrac12<\max(|f(x)|,|f|_2^e(x))<2,\quad x\in B_r, \tag{18.6.16}$$

$$|f|_j^e(x)<8,\quad \text{if } j<4,\quad x\in B_r, \tag{18.6.17}$$

$$f(x)=f_1(x)+g(x)^2,\quad x\in B_r, \tag{18.6.18}$$

where $f_1,g\in C^\infty(B_r)$ *and* $f_1\geqq 0$, $\langle y,\partial\rangle f_1(x)=0$ *when* $x\in B_r$ *for some* $y\in\mathbb{R}^\nu\smallsetminus 0$. *The supremum of* $|D^\alpha f_1|$ *and of* $|D^\alpha g|$ *in* B_r *can be estimated by the supremum of* $|D^\beta f|$ *in* B_r *for* $|\beta|\leqq 2+|\alpha|$.

Proof. We shall first estimate $|f|_1^e(0)$ and $|f|_3^e(0)$. Let $f_j(x)$ be the Taylor polynomial of order j at 0. Then

$$0\leqq f(x)\leqq 1+f_1(x)+|x|^2/2+f_3(x)+|x|^4/24,\quad |x|<2,$$

by Taylor's formula. Hence

$$|f_1(x)+f_3(x)|\leqq 1+|x|^2/2+|x|^4/24,\quad |x|\leqq 2.$$

If $|x|=1$ we obtain if x is also replaced by $2x$

$$|f_1(x)+f_3(x)|\leqq 3/2+1/24,\quad |2f_1(x)+8f_3(x)|\leqq 3+2/3,$$

which implies

$$6|f_1(x)|\leqq 16,\quad 6|f_3(x)|\leqq 7.$$

Thus $|f|_1^e(0)\leqq 3$ and $|f|_3^e(0)\leqq 7$. From Taylor's formula it follows now that (8.6.16) and (8.6.17) hold if r is small enough.

First assume that $|f|_2^g(0)=1$ and that $f(0)$ is small. By Lemma 7.7.2 we have $|rf'(0)|^2 \leqq 2f(0)$ if $r<\frac{1}{2}$, so $f'(0)$ is also small. The quadratic form $f_2(x)$ has an eigenvector y with eigenvalue $\pm\frac{1}{2}$. After an orthogonal transformation we may assume that $y=e_1=(1,0,\dots,0)$. Thus $\partial^2 f(0)/\partial x_1^2 = \pm 1$ and $\partial^2 f(0)/\partial x_1\,\partial x_j = 0$ for $j \neq 1$. Since

$$0 \leqq f(x_1,0,\dots,0) \leqq f(0) + x_1\,\partial f(0)/\partial x_1 \pm x_1^2/2 + 7|x_1|^3/6 + x_1^4/24$$

we conclude by taking $x_1 = \pm\frac{1}{3}$ that $\partial^2 f(0)/\partial x_1^2 = +1$ if $f(0) < 1/100$, for $1/100 - 1/18 + 7/162 + 3^{-4}/24 < 0$. If $|x_1| = r$ and $|x'| < r$, $x' = (x_2,\dots,x_n)$, then the estimate

$$|\partial f(x)/\partial x_1 - \partial f(0)/\partial x_1 - x_1| < 4|x|^2 + |x|^3/6$$

implies that $\partial f/\partial x_1$ has the sign of x_1 if

$$r > |\partial f(0)/\partial x_1| + 4r^2 + r^3/6.$$

We fix r so small that $4r + r^2/6 < 1/2$ and $1/2 < \partial^2 f(x)/\partial x_1^2 < 2$ when $x \in B_{2r}$. If $f(0) < r^4/8$ then $|f'(0)| < r/2$ so we conclude that the equation

$$\partial_1 f(x_1, x') = 0 \tag{18.6.19}$$

has a unique solution $x_1 = X(x')$ with $|x_1| < r$ if $|x'| < r$. Since

$$\partial X/\partial x_j = -\partial_1\,\partial_j f(X,x')/\partial_1^2 f(X,x'),$$

we obtain successively bounds for all derivatives of X of order k from bounds for the derivatives of f of order $k+1$, for $\partial_1^2 f(X,x') \geqq 1/2$. By Taylor's formula and (18.6.19) we obtain $f = f_1 + g^2$ in B_r where

$$f_1(x) = f(X(x'), x')$$

is independent of x_1, and

$$g(x)^2 = f(x) - f(X(x'),x') = (x_1 - X(x'))^2\,Q(x),$$

$$Q(x) = \int_0^1 \partial_1^2 f(x_1 + t(X(x') - x_1), x')\,t\,dt.$$

We have $Q \geqq 1/4$, and the derivatives of Q of order k can be estimated in terms of the derivatives of f of order $k+2$, so $g = (x_1 - X(x'))\,Q^{\frac{1}{2}} \in C^\infty(B_r)$ has the required properties.

It remains to examine the case where $f(0) \geqq r^4/8$. Then we can find $r_0 < r$ so that

$$f(x) > r^4/9 \quad \text{if } |x| < r_0.$$

If we replace r by r_0 in the lemma then $g(x) = f(x)^{\frac{1}{2}}$ has the required properties. The proof is complete.

The following lemma is the special case of Theorem 18.6.8 where g is a constant metric; it is the essential step in the proof.

Lemma 18.6.10. *Let g be a positive definite quadratic form in $\mathbb{R}^{2n}$ with $g/g^\sigma \leqq \lambda^2 \leqq 1$. Let $0 \leqq a \in C^\infty(\mathbb{R}^{2n})$, and assume that*

(18.6.20) $$|a|_k^g(w) \leqq \lambda^{-2}, \quad \text{if } w \in \mathbb{R}^{2n} \text{ and } k \leqq N.$$

If N is large enough it follows that

(18.6.21) $$(a^w(x,D)u,u) \geqq -C\,\|u\|^2, \quad u \in \mathscr{S},$$

where C is independent of g and of a.

Proof. In the proof we may assume that $a \in \mathscr{S}(\mathbb{R}^{2n})$. By Lemma 18.6.4 and Theorem 18.5.9 it is no restriction to assume that

$$g(x,\xi) = \sum \lambda_j (x_j^2 + \xi_j^2).$$

Here $\lambda_j \leqq \lambda$. The hypotheses in the lemma therefore remain valid if we replace all λ_j by λ, and (18.6.20) can then be written

(18.6.20)′ $$|a|_k^e(w) \leqq \lambda^{(k-4)/2}, \quad k \leqq N, \quad w \in \mathbb{R}^{2n}.$$

Here e is again the Euclidean metric form. We may also assume that the lemma has already been proved for lower dimensions than $2n$. This implies that (18.6.21) is valid if a is independent of ξ_1, say, for $a^w(x,D)$ can then be regarded as an operator in the variables $x' = (x_2, \dots, x_n)$ depending on x_1 as a parameter, and it has a lower bound $-C$ for every x_1. Using Theorem 18.5.9 we conclude that this is also true if a is constant in some other direction. This will allow us to handle the term f_1 in Lemma 18.6.9.

As in the proof of Theorem 18.6.7 we shall change the metric λe to a metric G with $G_{x,\xi} = H(x,\xi)e$ such that $H \leqq 1$ and $a \in S(1/H^2, G)$. This requires in particular that

$$a \leqq H^{-2}, \quad |a|_2^e \leqq H^{-1}.$$

To make sure that these conditions are satisfied we define

(18.6.22) $$1/H(w) = \max(1, a(w)^{\frac{1}{2}}, |a|_2^e(w)).$$

Now we apply Lemma 18.6.9 to

$$f(z) = H(w)^2 a(w + z/H(w)^{\frac{1}{2}}), \quad z \in \mathbb{R}^{2n}.$$

From (18.6.20)′ we obtain $|f|_4^e \leqq 1$, and we have

$$f(0) = H(w)^2 a(w), \quad |f|_2^e(0) = H(w)\,|a|_2^e(w).$$

These quantities are $\leqq 1$ by (18.6.22). We can always apply Lemma 18.6.9 to $f(z) + 1 - f(0)$. From (18.6.16) we obtain

$$1/H(w) \leqq 2/H(w_1) \quad \text{if } |w - w_1|^2\, H(w) < r^2,$$

that is,

$$G_{w_1} \leqq 2G_w \quad \text{if } G_w(w - w_1) < r^2.$$

Hence G is slowly varying so it follows from Proposition 18.5.6 that G is σ temperate. The proof in the present situation is so simple that we repeat it to emphasize the uniformity. We must show that

$$G_{w_1} \leqq C G_w (1 + G^\sigma_{w_1}(w - w_1))$$

or equivalently that

$$H(w_1) \leqq C H(w)(1 + |w - w_1|^2 / H(w_1)).$$

In doing so we may assume that $G_w(w - w_1) \geqq c$, that is, that $H(w)|w - w_1|^2 \geqq c$, and then the estimate holds with $C = 1/c$.

Choose $\chi \in C_0^\infty\{z; |z| < r\}$ real valued and equal to 1 in $\{z; |z| < r/2\}$. Then we have uniform bounds for $\chi(z) f(z)$ and its derivatives of order $\leqq N$. This follows from Lemma 18.6.9 and the fact that (18.6.20)′ gives estimates for the derivatives of f of order $\geqq 4$. By an obvious modification of Lemma 18.4.4 we can find a sequence $w_\nu \in \mathbb{R}^{2n}$ such that there is a fixed bound for the number of overlapping balls $B_\nu = \{w; G_{w_\nu}(w - w_\nu) < r^2\}$ and for suitable real valued $\phi_\nu \in C_0^\infty(B'_\nu)$, $B'_\nu = \{w; G_{w_\nu}(w - w_\nu) < r^2/4\}$, we have $|\phi_\nu|^G_k \leqq C_k$ and

$$\sum \phi_\nu(w)^2 = 1.$$

Set

$$a_\nu(w) = \chi(H(w_\nu)^{\frac{1}{2}}(w - w_\nu))^2 \, a(w).$$

Then we have

$$\sum \phi_\nu(w)^2 \, a_\nu(w) = a(w); \quad a_\nu \in C_0^\infty(B_\nu); \quad |a_\nu|^G_k \leqq C_1 H^{-2}, \quad k \leqq N.$$

We shall prove in a moment that there is a constant C_2 such that

$$(a^w_\nu(x, D) u, u) \geqq -C_2 (u, u), \quad u \in \mathscr{S}. \tag{18.6.23}$$

Admitting this we replace u by $\phi^w_\nu(x, D) u$ and sum. We can consider $\{\phi_\nu(x, \xi)\}$ as a symbol in $S(1, G)$ with values in $l^2 \cong \mathscr{L}(\mathbb{C}, l^2)$, for at any point there are only a fixed number of terms to consider. Hence it follows from Theorem 18.6.3 that the operator

$$\Phi \colon u \mapsto \{\phi^w_\nu(x, D) u\} \in L^2(\mathbb{R}^n, l^2)$$

is L^2 continuous, that is,

$$\sum \|\phi^w_\nu(x, D) u\|^2 \leqq C_3 \|u\|^2. \tag{18.6.24}$$

The sum $\sum \phi^w_\nu(x, D) a^w_\nu(x, D) \phi^w_\nu(x, D)$ can be regarded as the composition of Φ, the operator A with the diagonal matrix symbol $\{a_\nu(x, \xi) \delta_{\nu\mu}\}$ in $L^2(\mathbb{R}^n, l^2)$, and the adjoint of Φ. We have control of as many seminorms of the symbol of A in $S(H^{-2}, G)$ as we like. Since $\phi_\nu \{a_\nu, \phi_\nu\} + \{\phi_\nu, a_\nu\} \phi_\nu = 0$ the first order terms in the composition formula in Theorem 18.5.4 cancel. Hence

$$\sum \phi^w_\nu(x, D) a^w_\nu(x, D) \phi^w_\nu(x, D) = a^w(x, D) + R^w(x, D),$$

where any desired number of seminorms of R in $S(1, G)$ are bounded. By Theorem 18.6.3 and (18.6.23), (18.6.24) it follows that

$$\begin{aligned}(a^w(x, D) u, u) &= \sum (a^w_\nu(x, D) \phi^w_\nu(x, D) u, \phi^w_\nu(x, D) u) - (R^w(x, D) u, u) \\ &\geqq -C_2 C_3 \|u\|^2 - C_4 \|u\|^2\end{aligned}$$

as claimed.

It remains to verify (18.6.23). In Lemma 18.6.9 we can write

$$\chi(z)^2 f(z)=\chi(z)^2 f_2(z)+(\chi(z)g(z))^2$$

where $f_2 \geqq 0$ is obtained by multiplying f_1 with a cutoff function which is constant in the direction y and 1 in supp χ. Then we have bounds for the derivatives of f_2 of order $\leqq N-2$. Going back to the original variables we set $H_\nu = H(w_\nu)$ and

$$\chi_\nu=\chi((.-w_\nu)H_\nu^{\frac{1}{2}}), \quad b_\nu=f_2((.-w_\nu)H_\nu^{\frac{1}{2}})/H_\nu^2,$$
$$c_\nu=(\chi g)((.-w_\nu)H_\nu^{\frac{1}{2}})/H_\nu$$

where f_2 and g are obtained from Lemma 18.6.9 applied to $H_\nu^2 a(w_\nu+./H_\nu^{\frac{1}{2}})$. By (18.6.18) we have

$$a_\nu=\chi_\nu^2 b_\nu+c_\nu^2,$$

and we have bounds for any desired number of seminorms of b_ν, χ_ν and c_ν in $S(H_\nu^{-2}, H_\nu e)$, $S(1, H_\nu e)$ and $S(H_\nu^{-1}, H_\nu e)$. By the inductive hypothesis

$$(b_\nu^w(x,D)u,u) \geqq -C_5 \|u\|^2.$$

From Theorem 18.5.4 we obtain as above a bound for any number of seminorms of the symbol of $a_\nu^w(x,D)-\chi_\nu^w(x,D)b_\nu^w(x,D)\chi_\nu^w(x,D) -c_\nu^w(x,D)c_\nu^w(x,D)$ in $S(1,H_\nu e)$ if N is large, so Theorem 18.6.3 gives a bound for the norm of this difference. Thus (18.6.23) and the lemma are proved.

Proof of Theorem 18.6.8. We just have to repeat a part of the proof of Lemma 18.6.10. Choose ϕ_ν as in Lemma 18.4.4 but so that $\sum \phi_\nu^2=1$. We can arrange so that $\phi_\nu\psi_\nu=\phi_\nu$ for some non-negative $\psi_\nu \in C_0^\infty(B_\nu)$ which are also bounded in $S(1,g)$. Set $a_\nu=\psi_\nu a$, which implies that $a=\sum \phi_\nu a_\nu \phi_\nu$. Hence we obtain as before that

$$\sum \phi_\nu^w(x,D)a_\nu^w(x,D)\phi_\nu^w(x,D)-a^w(x,D)$$

is a bounded operator. By Lemma 18.6.10 we have

$$(a_\nu^w(x,D)u,u) \geqq -C\|u\|^2$$

where C does not depend on ν, so it follows that

$$\sum (a_\nu^w(x,D)\phi_\nu^w(x,D)u, \phi_\nu^w(x,D)u) \geqq -C'\|u\|^2$$

since (18.6.24) remains valid for our new choice of ϕ_ν. This completes the proof.

Specializing to the metric (18.4.1)' we have proved:

Corollary 18.6.11. *If* $0 \leqq a \in S_{\rho,\delta}^{2(\rho-\delta)}(\mathbb{R}^n \times \mathbb{R}^n)$ *and* $0 \leqq \delta < \rho \leqq 1$ *then* $a^w(x,D)$ *is bounded from below and so is* $a(x,D)+a(x,D)^*$.

It is not possible to replace λ^{-2} in Lemma 18.6.10 by a larger power:

Example 18.6.12. Set $b(x,\xi)=x\xi$, $(x,\xi)\in\mathbb{R}^2$. Then

$$b^w(x,D)^2=a^w(x,D)+\tfrac{1}{4}$$

where $a=b^2$, for $-8^{-1}(\partial_\xi\partial_y-\partial_x\partial_\eta)^2 x\xi y\eta=\frac{1}{4}$. Thus

$$(a^w(x,D)u,u)=\|b^w(x,D)u\|^2-\|u/2\|^2,\quad u\in\mathscr{S}.$$

The equation $b^w(x,D)u=0$ can be written $2xu'+u=0$ so it is satisfied by $u(x)=x^{-\frac{1}{2}}$. Let $\chi\in C_0^\infty$ be equal to 1 in a neighborhood of 0 and set for small $\varepsilon>0$

$$u_\varepsilon(x)=(\chi(\varepsilon x)-\chi(x))/|x|^{\frac{1}{2}}.$$

Then it follows that

$$b^w(x,D)u_\varepsilon(x)=|x|^{-\frac{1}{2}}xD(\chi(\varepsilon x)-\chi(x)),$$

so $\|b^w(x,D)u_\varepsilon\|^2\leqq C$ but $\|u_\varepsilon\|^2/\log\varepsilon\to-2$ as $\varepsilon\to 0$. Thus $(a^w(x,D)u_\varepsilon,u_\varepsilon)<0$ if ε is small enough. Now choose $a_0\in C_0^\infty(\mathbb{R}^2)$ so that $a_0\geqq 0$ and $a_0(x,\xi)=x^2\xi^2$ in a neighborhood of 0. When $\lambda\to 0$ we obtain

$$(a_0^w(\sqrt{\lambda}x,\sqrt{\lambda}D)u_\varepsilon,u_\varepsilon)/\lambda^2\to(a^w(x,D)u_\varepsilon,u_\varepsilon),$$

for $a_0(\sqrt{\lambda}x,\sqrt{\lambda}\xi)/\lambda^2$ converges weakly to a in $S((1+x^2+\xi^2)^2,e)$ when $\lambda\to 0$. Since the right-hand side can be negative this proves that we cannot improve the exponent -2 in (18.6.20) or in Theorem 18.6.8.

However, the condition $a\geqq 0$ in Theorem 18.6.8 and Lemma 18.6.10 is too strong. The reason is that in the proof of Lemma 18.6.10 we discarded quadratic terms which can give essential contributions when combined with other such terms in a later stage of the inductive proof. We shall return to this matter in Section 22.3.

Theorem 18.6.7 remains valid in the vector valued case but cannot be improved then. The key to the proof of the positive result is the following rather weak analogue of Lemma 18.6.10.

Lemma 18.6.13. *Let g be a positive definite quadratic form in $\mathbb{R}^{2n}$ with $g/g^\sigma\leqq\lambda^2\leqq 1$. Let $0\leqq a\in C^\infty(\mathbb{R}^{2n},\mathscr{L}(H,H))$ where H is a Hilbert space, and assume that*

$$|a|_k^g(w)\leqq\lambda^{-1}\quad\text{if } w\in\mathbb{R}^{2n}\text{ and } k\leqq N. \tag{18.6.25}$$

If N is large enough it follows that

$$(a^w(x,D)u,u)\geqq -C\|u\|^2,\quad u\in\mathscr{S}(\mathbb{R}^n,H). \tag{18.6.26}$$

Proof. As in the proof of Lemma 18.6.10 we may assume that $g=\lambda e$ where e is the Euclidean metric form. Then

$$|a|_k^e(w)\leqq\lambda^{(k-2)/2}. \tag{18.6.25$'$}$$

Let $A_0+A_1(x,\xi)$ be the first order Taylor expansion of a at 0. By (18.6.25)′ we have for all $v\in H$

$$(A_0 v, v)+(A_1(x,0) v, v)+|x|^2 \|v\|^2/2\geqq 0,$$
$$(A_0 v, v)+(A_1(0,\xi) v, v)+|\xi|^2 \|v\|^2/2\geqq 0.$$

If $u\in\mathscr{S}(\mathbb{R}^n, H)$ we obtain by applying these inequalities with $v=u(x)$ or $v=\hat{u}(\xi)$ respectively, with x and ξ replaced by $2x$ and 2ξ,

$$(A_0 u, u)+(A_1(x,D) u, u)\geqq -\sum(\|x_j u\|^2+\|D_j u\|^2), \tag{18.6.27}$$

where the scalar products are now in $L^2(\mathbb{R}^n, H)$. (This is a substitute for Lemma 7.7.2.) Next we prove that

$$(a^w(x,D) u, u)\geqq -C\sum(\|x_j u\|^2+\|D_j u\|^2). \tag{18.6.28}$$

To do so we use Taylor's formula to write

$$a(x,\xi)=A_0+A_1(x,\xi)+\sum x_j x_k R_{jk}(x,\xi)+\sum \xi_j \xi_k S_{jk}(x,\xi)$$
$$+2\sum x_j \xi_k T_{jk}(x,\xi),$$

where R_{jk} and S_{jk} are symmetric in j and k and we control any number of seminorms of R_{jk}, S_{jk} and T_{jk} in $S(1,g)$. Then we have

$$a^w(x,D)=A_0+A_1(x,D)+\sum x_j R^w_{jk}(x,D) x_k+\sum D_j S^w_{jk}(x,D) D_k$$
$$+\sum x_j T^w_{jk}(x,D) D_k+\sum D_k T^w_{jk}(x,D) x_j-R^w(x,D)$$

where

$$4R=\sum \partial^2 R_{jk}/\partial\xi_j\partial\xi_k+\sum\partial^2 S_{jk}/\partial x_j\partial x_k-2\sum\partial^2 T_{jk}/\partial\xi_j\partial x_k$$

has bounded seminorms in $S(\lambda, g)$. This follows from the fact that left (right) multiplication by x_j or D_j means that the symbol is multiplied by $x_j \pm\frac{1}{2}i\partial/\partial\xi_j$ or $\xi_j\mp\frac{1}{2}i\partial/\partial x_j$, after a short calculation. The estimate (18.6.28) is an immediate consequence since $\|u\|^2\leqq\|x_j u\|^2+\|D_j u\|^2$.

From (18.6.28) it follows that more generally

$$(a^w(x,D) u, u)\geqq -C\sum(\|(x_j-y_j) u\|^2+\|(D_j-\eta_j) u\|^2) \tag{18.6.28′}$$

for arbitrary $(y,\eta)\in\mathbb{R}^{2n}$. We just have to apply (18.6.28) to $a(x+y,\xi+\eta)$ with u transformed according to Theorem 18.5.9 to verify this. Now choose $\phi\in C_0^\infty(\mathbb{R}^n)$ so that

$$\sum\phi(x-v)^2=1$$

when v runs over the lattice points. Set $\phi_v(x)=\phi(x-v)$ and apply (18.6.28)′ with $y=v$ and u replaced by $\phi_v u$. Then

$$\sum\|(x_j-v_j)\phi_v u\|^2\leqq C\|u\|^2.$$

We can use the calculus with the metric $|dx|^2+\lambda|d\xi|^2$ to compute

$$\sum\phi^w_v(x,D)\, a^w(x,D)\,\phi^w_v(x,D).$$

The main term in the sum is a and the first order terms cancel, so we conclude as in the proof of Lemma 18.6.10 that the symbol is $a+b$ where we have bounds for a large number of seminorms of b in $S(1,|dx|^2+\lambda|d\xi|^2)$, hence also a bound for the norm. It follows that

$$(18.6.28)'' \qquad (a^w(x,D)u,u) \geqq -C(\|u\|^2+\sum\|(D_j-\eta_j)u\|^2).$$

Here we replace u by $\phi_\nu^w(D)u$ and η by ν. Repetition of the preceding argument then gives (18.6.26). The proof of the lemma is complete.

From the lemma we obtain at once using the localization argument in the proof of Theorem 18.6.8:

Theorem 18.6.14. *Let g be a σ temperate metric and assume that* (18.6.11) *holds. If $a \in S(1/h,g)$ takes non-negative values in $\mathscr{L}(H,H)$ where H is a Hilbert space, it follows then that*

$$(a^w(x,D)u,u) \geqq -C\|u\|^2, \quad u \in \mathscr{S}(\mathbb{R}^n,H).$$

Notes

Pseudo-differential operators have developed from the theory of singular integral operators; these are essentially pseudo-differential operators with homogeneous symbol of order 0. In the theory of singular integral operators only the principal symbol is studied. Its multiplicative properties appear somewhat mysteriously since the Fourier representation is avoided, and this seems to be the historical reason for the term. (See Seeley [5].) Singular integral operators were introduced in the study of elliptic problems, but it was realized in the 1950's that they are not really essential then. The work by Calderón [1] on the uniqueness of the Cauchy problem gave another testimony to their importance, but in the predecessor of this book his results were proved and extended by direct methods based on partial integration, Fourier transforms and localization techniques. It seems likely that it was the solution by Atiyah and Singer [1] of the index problem for elliptic operators which led to the revitalization of the theory of singular integral operators. Anyway, shortly afterwards Kohn and Nirenberg [1] introduced pseudo-differential operators with general polyhomogeneous symbols. Thus they removed the artificial restriction to order 0 and gave rules of computation for terms of lower order, which made the new techniques highly competitive. Their proofs and definitions relied on the Fourier representation except for the change of variables which was based on a return to singular integral operators. This remnant was removed shortly afterwards by Hörmander [16]. The need to incorporate fundamental solutions of hypoelliptic operators of constant strength led to the introduction of symbols of type ρ, δ in Hörmander [18], by many considered as an excessive generalization. However, these were in fact inadequate for the study of differential operators of principal type, which led Beals and Fefferman [1], Beals [1] to such a great extension of the class of symbols allowed that one can adapt the symbols to the operator being studied. An extension of their techniques, adapted to the Weyl calculus which has a tradition in quantum mechanics going back to Weyl [5] is presented in Sections 18.4 through 18.6. We have mainly followed Hörmander [39] but some extensions are adopted from Dencker [2]. One of the applications of the general calculus given by Beals and Fefferman [1] was the "sharp Gårding inequality" first proved in Hörmander [17] and later extended to the vector valued case by Lax and Nirenberg [1]. In Section 18.1 we give an elementary proof of the original result which will be needed frequently in the later chapters. A generalization of the Lax-Nirenberg result as well as a much more precise statement in the scalar case due to Fefferman and Phong [1] are given in Section 18.6. For improvements in another direction we refer to Section 22.3 and the notes to Chapter XXII. The key to the estimates in Section 18.6 is a result on sums of almost orthogonal operators often referred to as Cotlar's lemma which was proved by Cotlar [1] and by Knapp and Stein [1]; the more general statement given here is due to Calderón and Vaillancourt [1, 2].

The conormal distributions discussed in Section 18.2 were defined in Hörmander [26]. They constitute the simplest case of Lagrangian (Fourier integral) distributions which will be discussed fully in Chapter XXV. The results on the transmission condition are due to Boutet de Monvel [1]; the extended version in Theorem 18.2.18 comes from old lecture notes inspired by the work of Višik and Eškin [1-5]. Section 18.3 is almost entirely due to Melrose [1]. We refer to his paper for further developments of the theme parallel to the theory of Fourier integral operators in Chapter XXV.

Chapter XIX. Elliptic Operators on a Compact Manifold Without Boundary

Summary

For an elliptic pseudo-differential operator on a compact manifold it follows immediately from the calculus of such operators that the kernel and the cokernel are both finite dimensional. Thus elliptic operators are Fredholm operators. The main topic of this chapter is the study of the index, that is, the difference between the dimensions of the kernel and the cokernel. This is an interesting quantity to study because it is very stable under perturbations; for many operators which occur in geometry the index gives important information on the topology. Thus the classical Riemann-Roch theorem, as well as some of its modern analogues for several complex variables, is a case of the index theorem.

We start in Section 19.1 by reviewing abstract Fredholm theory. In doing so we add some points which are not quite standard. These concern the stability of the index of strongly continuous families of operators, the expression of the index by means of traces in the case of operators in Hilbert space, and finally related results on invariance of Euler characteristics under passage to homology. The main analytical properties of the index of elliptic pseudo-differential operators are then established in Section 19.2. For pseudo-differential operators in $\mathbb{R}^n$ an explicit index formula is proved in Section 19.3 by means of evaluation of certain traces. As indicated at the end of Section 19.2 the results of Sections 19.2 and 19.3 can be used to derive the Atiyah-Singer index formulas. However, this is mainly a problem in differential geometry, and we do not wish to develop the necessary prerequisites on characteristic classes to discuss it so the remarks will be quite brief. Instead we pass in Section 19.4 to the Lefschetz formula of Atiyah and Bott which can be stated and proved without any extensive background in geometry. A discussion of the extent to which ellipticity is a necessary condition for the Fredholm property is given in Section 19.5; a more general notion of ellipticity is introduced in this context.

19.1. Abstract Fredholm Theory

If B_1 and B_2 are finite dimensional vector spaces (over $\mathbb{C}$) and T is a linear transformation $B_1 \to B_2$ then the rank of T is the codimension of the kernel

Ker T or equivalently the dimension of the range, that is,

$$\dim B_1 - \dim \operatorname{Ker} T = \dim B_2 - \dim \operatorname{Coker} T$$

where $\operatorname{Coker} T = B_2/TB_1$. This means that

$$\dim \operatorname{Ker} T - \dim \operatorname{Coker} T = \dim B_1 - \dim B_2$$

is independent of T. This is the basic reason for the stability properties of the left-hand side in the infinite dimensional case; it is called the index of T.

If B_1 and B_2 are Banach spaces and $T \in \mathscr{L}(B_1, B_2)$ then

$$\operatorname{Ker} T = \{f \in B_1; Tf = 0\}$$

is a closed subspace of B_1 but need not be finite dimensional. The range TB_1 need not be closed, but we have

Lemma 19.1.1. *If $T \in \mathscr{L}(B_1, B_2)$ and the range TB_1 has finite codimension in B_2 then TB_1 is closed.*

Proof. We may assume that T is injective for otherwise we can consider instead the map from $B_1/\operatorname{Ker} T$ to B_2 induced by T. If n is the codimension of TB_1 we can choose a lincar map

$$S: \mathbb{C}^n \to B_2$$

such that $S\mathbb{C}^n$ is a supplement of TB_1, that is, the map

$$T_1: B_1 \oplus \mathbb{C}^n \ni (x, y) \mapsto Tx + Sy \in B_2$$

is bijective. By Banach's theorem it is then a homeomorphism, which proves that $TB_1 = T_1(B_1 \oplus \{0\})$ is closed.

Definition 19.1.2. $T \in \mathscr{L}(B_1, B_2)$ is called a Fredholm operator if $\dim \operatorname{Ker} T$ is finite and TB_1 (is closed and) has finite codimension; one then defines

$$\operatorname{ind} T = \dim \operatorname{Ker} T - \dim \operatorname{Coker} T. \tag{19.1.1}$$

Sometimes we shall also use the definition (19.1.1) when only $\dim \operatorname{Ker} T$ is finite and TB_1 is closed. This situation can be characterized as follows:

Proposition 19.1.3. *If $T \in \mathscr{L}(B_1, B_2)$ the following conditions are equivalent:*

(i) $\dim \operatorname{Ker} T < \infty$ *and TB_1 is closed.*

(ii) *Every sequence $f_j \in B_1$ such that Tf_j is convergent and f_j is bounded has a convergent subsequence.*

For the proof we need a classical lemma of F. Riesz:

Lemma 19.1.4. *The unit ball in a Banach space B of infinite dimension is never compact (in the norm topology).*

Proof. We can choose a sequence $f_j \in B$ such that

$$\|f_j\| = 1 \leqq \|f_j + \sum_{k<j} a_k f_k\|$$

for all j and all $a_k \in \mathbb{C}$. In fact, if $f_1, \dots, f_{j-1}$ have already been chosen we can take g outside the linear space L spanned by these elements. There is a point $h \in L$ which minimizes $\|g-h\|$ for L is finite dimensional and the norm $\to \infty$ if $h \to \infty$. Then $f_j = (g-h)/\|g-h\|$ has the required properties. In particular we have $\|f_j - f_k\| \geqq 1$ for $j \neq k$ so there is no convergent subsequence.

Proof of Proposition 19.1.3. If (ii) is fulfilled it follows in particular that the unit ball in $\operatorname{Ker} T$ is compact, so $\operatorname{Ker} T$ is finite dimensional by Lemma 19.1.4. By the Hahn-Banach theorem the finite dimensional space $\operatorname{Ker} T$ has a topological supplement $B_0 \subset B_1$. It also follows from condition (ii) that

$$\|f\|_1 \leqq C \|Tf\|_2, \quad f \in B_0. \tag{19.1.2}$$

In fact, otherwise there is a sequence $f_j \in B_0$ with $\|f_j\|_1 = 1$ and $\|Tf_j\|_2 \leqq 1/j$. By condition (ii) a subsequence has a limit f such that $f \in B_0$, $\|f\| = 1$, $Tf = 0$ which is a contradiction proving (19.1.2). Conversely, assume that (19.1.2) is valid and that f_j is a bounded sequence in B_1 such that Tf_j is convergent. We can write $f_j = g_j + h_j$ where $g_j \in \operatorname{Ker} T$ and $h_j \in B_0$ are also bounded. Since $Th_j = Tf_j$ it follows from (19.1.2) that the sequence h_j is convergent, and the bounded sequence g_j in the finite dimensional space $\operatorname{Ker} T$ has a convergent subsequence. Thus (ii) holds. Now it follows from Banach's theorem that (19.1.2) holds if and only if T restricted to B_0 is injective with closed range. This proves the equivalence of (i) and (ii).

Our first stability result on the index is the following

Theorem 19.1.5. *If T satisfies the conditions in Proposition 19.1.3 and $S \in \mathscr{L}(B_1, B_2)$ has sufficiently small norm then* $\dim \operatorname{Ker}(T+S) \leqq \dim \operatorname{Ker} T$, *$T+S$ has closed range, and* $\operatorname{ind}(T+S) = \operatorname{ind} T$.

Proof. Assume first that T is bijective. Then

$$T + S = T(I + T^{-1} S)$$

is bijective if $\|T^{-1}\| \, \|S\| < 1$, for $I + T^{-1} S$ can then be inverted by the Neuman series. Thus the theorem is valid then. Next assume only that T is injective. Then we have by (19.1.2)

$$\|f\|_1 \leqq C \|Tf\|_2 \leqq C \|(T+S)f\|_2 + C \|S\| \, \|f\|_1.$$

Hence $\|f\|_1 \leqq 2C \|(T+S)f\|_2$ if $C\|S\| < \frac{1}{2}$, which means that $T+S$ is injective with closed range then. If $\dim \operatorname{Coker} T < \infty$ we can choose a finite dimensional supplementary space W of TB_1. Let q be the natural map

$B_2 \to B_2/W$. Then qT is bijective so $qT+qS$ is bijective if $\|S\|$ is small enough, thus

$$\operatorname{ind}(T+S) = -\dim W = \operatorname{ind} T$$

then. If we just know that dim Coker $T > \nu$ we can choose W of dimension $\nu+1$ with $W \cap TB_1 = \{0\}$ and obtain $\operatorname{ind}(T+S) < -\nu$ if $\|S\|$ is small enough. If we apply this with T replaced by $T+S$, we see that the sets

$$\{S;\, C\,\|S\| < \tfrac{1}{2}, \operatorname{ind}(T+S) = -\nu\}, \quad \{S;\, C\,\|S\| < \tfrac{1}{2}, \operatorname{ind}(T+S) < -\nu\}$$

are open for every ν. Because $\{S;\, C\,\|S\| < \tfrac{1}{2}\}$ is connected it follows that $\operatorname{ind}(T+S)$ must in fact be constant.

If $N = \operatorname{Ker} T \neq \{0\}$ we choose a topological supplement $B_0 \subset B_1$ of N, thus $B_1 = B_0 \oplus N$. Then it follows from what we have already proved that $T+S$ restricted to B_0 is injective with closed range and that

$$\operatorname{codim}(T+S)B_0 = \operatorname{codim} TB_0 = \operatorname{codim} TB_1$$

if $\|S\|$ is small enough. If $N' = \operatorname{Ker}(T+S)$ then $N' \cap B_0 = \{0\}$ and $B_1 = N' \oplus W \oplus B_0$ for some finite dimensional W. The codimension of $(T+S)B_1 = (T+S)W \oplus (T+S)B_0$ is equal to $\operatorname{codim} TB_1 - \dim W$, so

$$\begin{aligned}\operatorname{ind}(T+S) &= \dim N' + \dim W - \operatorname{codim} TB_1\\ &= \dim N - \dim \operatorname{Coker} T - \operatorname{ind} T\end{aligned}$$

which completes the proof.

Corollary 19.1.6. *The set of Fredholm operators in $\mathscr{L}(B_1, B_2)$ is open,* dim Ker T *is upper semi-continuous, and* ind T *is constant in each component.*

Corollary 19.1.7. *If $T_1 \in \mathscr{L}(B_1, B_2)$ and $T_2 \in \mathscr{L}(B_2, B_3)$ are Fredholm operators then $T_2 T_1 \in \mathscr{L}(B_1, B_3)$ is a Fredholm operator and we have "the logarithmic law"*

$$\operatorname{ind}(T_2 T_1) = \operatorname{ind} T_1 + \operatorname{ind} T_2.$$

Proof. Since T_1 maps Ker $T_2 T_1$ into Ker T_2 with kernel Ker T_1 we have $\dim \operatorname{Ker} T_2 T_1 \leqq \dim \operatorname{Ker} T_1 + \dim \operatorname{Ker} T_2$. Similarly, $\dim \operatorname{Coker} T_2 T_1 \leqq \dim \operatorname{Coker} T_1 + \dim \operatorname{Coker} T_2$ so it is clear that $T_2 T_1$ is a Fredholm operator. If I_2 is the identity operator in B_2 it follows in block matrix notation that

$$\begin{pmatrix} I_2 & 0 \\ 0 & T_2 \end{pmatrix} \begin{pmatrix} I_2 \cos t & I_2 \sin t \\ -I_2 \sin t & I_2 \cos t \end{pmatrix} \begin{pmatrix} T_1 & 0 \\ 0 & I_2 \end{pmatrix}$$

is for every t a Fredholm operator from $B_1 \oplus B_2$ to $B_2 \oplus B_3$. When $t=0$ it is the direct sum of the operators T_1 and T_2, and when $t = -\pi/2$ it is the operator $(f_1, f_2) \mapsto (-f_2, T_2 T_1 f_1)$. The index is clearly $\operatorname{ind} T_1 + \operatorname{ind} T_2$ and $\operatorname{ind}(T_2 T_1)$ in the two cases which proves the corollary.

Corollary 19.1.8. *If $T \in \mathscr{L}(B_1, B_2)$ is a Fredholm operator and $K \in \mathscr{L}(B_1, B_2)$ is compact, then $T+K$ is a Fredholm operator and* $\operatorname{ind}(T+K) = \operatorname{ind} T$.

Proof. If f_j is a bounded sequence in B_1 such that $(T+K)f_j$ is convergent, then the compactness allows us to choose a subsequence f_{j_k} such that Kf_{j_k} is convergent. Thus Tf_{j_k} is convergent, and it follows from condition (ii) in Propositon 19.1.3 that the sequence f_{j_k} has a convergent subsequence. Thus another application of Proposition 19.1.3, now to $T+K$, shows that $T+K$ has finite dimensional kernel and closed range. By Theorem 19.1.5 the index of $T+zK$ is therefore a locally constant function of $z\in\mathbb{C}$, hence independent of z. Thus $T+zK$ is always a Fredholm operator with index equal to ind T.

Corollary 19.1.9. *If $T\in\mathscr{L}(B_1,B_2)$ and for some $S_j\in\mathscr{L}(B_2,B_1)$ we have $TS_2=I_2+K_2$, $S_1T=I_1+K_1$ where K_j are compact, then T, S_1 and S_2 are Fredholm operators and* $\operatorname{ind} T=-\operatorname{ind} S_j$, $j=1,2$.

Proof. T is a Fredholm operator since

$$\dim\operatorname{Ker} T\leqq\dim\operatorname{Ker}(I_1+K_1),$$
$$\dim\operatorname{Coker} T\leqq\dim\operatorname{Coker}(I_2+K_2).$$

Since

$$S_1TS_2=S_1+S_1K_2=S_2+K_1S_2$$

it follows that S_2-S_1 is compact. Hence S_2T-I_1 and TS_1-I_2 are also compact so S_1 and S_2 are also Fredholm operators by the first part of the proof, and $\operatorname{ind} T+\operatorname{ind} S_j=\operatorname{ind}(I_j+K_j)=0$ by Corollary 19.1.7 and Corollary 19.1.8. The proof is complete.

The index has much stronger stability properties than stated in Corollary 19.1.6:

Theorem 19.1.10. *Let B_1 and B_2 be two Banach spaces and let I be a compact space. If $T_t\in\mathscr{L}(B_1,B_2)$ and $S_t\in\mathscr{L}(B_2,B_1)$ are strongly continuous as functions of $t\in I$, and if*

$$K_{1t}=S_tT_t-I_1,\qquad K_{2t}=T_tS_t-I_2$$

are uniformly compact in the sense that for $j=1,2$

$$M_j=\{K_{jt}f;\,t\in I, f\in B_j, \|f\|_j\leqq 1\}$$

is precompact in B_j, then T_t and S_t are Fredholm operators, $\dim\operatorname{Ker} T_t$ *and* $\dim\operatorname{Ker} S_t$ *are upper semi-continuous, and* $\operatorname{ind} T_t=-\operatorname{ind} S_t$ *is locally constant, hence constant if I is connected.*

Proof. T_t and S_t are Fredholm operators by Corollary 19.1.9. Let

$$N=\{(t,f)\in I\times B_1,\ \|f\|_1=1,\ T_tf=0\}.$$

This is a closed subset of $I\times\overline{M}_1$, for $T_tf=0$ implies $f=-K_{1t}f\in M_1$. Hence N is compact. The proof of Lemma 19.1.4 shows that

$$I_k=\{t\in I;\ \dim\operatorname{Ker} T_t\geqq k\}$$

is the set of all t such that there exist $f_1,\ldots,f_k$ with $(t,f_j)\in N$ and

$$\|f_j+\sum_{i<j} a_i f_i\| \geqq 1$$

for $j=1,\ldots,k$ and all $a_i\in\mathbb{C}$. Thus I_k is the projection in I of a compact set in N^k and therefore compact, which means that $\{t;\dim\operatorname{Ker} T_t<k\}$ is open. Hence $\dim\operatorname{Ker} T_t$ is upper semicontinuous and the same is of course true for $\dim\operatorname{Ker} S_t$.

Since $\operatorname{ind} T_t+\operatorname{ind} S_t=0$ the theorem will be proved if we show that $\operatorname{ind} T_t$ (and therefore $\operatorname{ind} S_t$) is upper semi-continuous. Let $t_0\in I$ and choose a topological supplement $B_0\subset B_1$ of $\operatorname{Ker} T_{t_0}$. Since the injection $B_0\to B_1$ has index $-\dim\operatorname{Ker} T_{t_0}$, the index of the restriction of T_t to B_0 is $\operatorname{ind} T_t-\dim\operatorname{Ker} T_{t_0}$. We must therefore show that $\operatorname{codim} T_t B_0\geqq \operatorname{codim} T_{t_0} B_1$ when t is in some neighborhood of t_0. Choose W with $\dim W=\dim\operatorname{Coker} T_{t_0}$ and $W\cap T_{t_0} B_1=\{0\}$. The set

$$N_W=\{(t,f)\in I\times B_0,\ \|f\|_1=1,\ T_t f\in W\}$$

is compact. In fact, N_W is closed. When $(t,f)\in N_W$ we have $f=S_t T_t f-K_{1t} f$ where $K_{1t} f\in M_1$, and

$$\|T_t f\|_2\leqq C\|f\|_1=C$$

since T_t is strongly continuous in t. The compactness of N_W follows now from the fact that

$$\{(t,g)\in I\times W;\ \|g\|_2\leqq C\}$$

is compact and that $(t,g)\mapsto S_t g$ is continuous there. We can now conclude that the projection in I

$$I^W=\{t;(t,f)\in N_W \text{ for some } f\}$$

is compact. The complement is therefore open and contains t_0 by the choice of W. For all t in a neighborhood of t_0 we have therefore proved that

$$\operatorname{codim} T_t B_0\geqq \dim W=\operatorname{codim} T_{t_0} B_1,$$

which completes the proof.

The hypotheses of Theorem 19.1.10 are very natural in the study of elliptic pseudo-differential operators T_t; the operators S_t are their parametrices. - We shall now leave the stability properties of the index and derive a trace formula for the index of Fredholm operators in Hilbert spaces. It will be used to prove explicit index formulas in Section 19.3. First we recall the basic facts on Hilbert-Schmidt and trace class operators.

Definition 19.1.11. If H_1 and H_2 are Hilbert spaces then the space $\mathscr{L}_2(H_1,H_2)$ of Hilbert-Schmidt operators from H_1 to H_2 consists of all $T\in\mathscr{L}(H_1,H_2)$ such that

$$\|T\|_2^2=\sum\|Te_i\|^2$$

is finite, if $\{e_i\}$ is a complete orthonormal system in H_1.

The notation already presupposes that the sum is independent of the choice of orthonormal basis, which is indeed the case since for any complete orthonormal system $\{f_j\}$ in H_2 we have

$$\sum \|Te_i\|^2 = \sum |(Te_i, f_j)|^2 = \sum |(e_i, T^* f_j)|^2 = \sum \|T^* f_j\|^2.$$

This shows also that $T^* \in \mathscr{L}_2(H_2, H_1)$ and that T^* has the same Hilbert-Schmidt norm as T. If $S_1 \in \mathscr{L}(H_1', H_1)$ and $S_2 \in \mathscr{L}(H_2, H_2')$ for some other Hilbert spaces H_1' and H_2' we also have $S_2 T S_1 \in \mathscr{L}_2(H_1', H_2')$ and

$$\|S_2 T S_1\|_2 \leqq \|S_2\| \, \|T\|_2 \, \|S_1\|,$$

the norms of S_1 and S_2 being operator norms. In fact,

$$\|S_2 T S_1\|_2 \leqq \|S_2\| \, \|T S_1\|_2 = \|S_2\| \, \|S_1^* T^*\|_2 \leqq \|S_2\| \, \|S_1^*\| \, \|T^*\|_2.$$

If $T \in \mathscr{L}(H_1, H_2)$ has finite rank, that is, the kernel N has finite codimension, then we can factor T as the canonical operator $H_1 \to H_1/N$ followed by an operator $H_1/N \to H_2$. Since H_1/N is finite dimensional the second operator is a Hilbert-Schmidt operator and we conclude that $T \in \mathscr{L}_2(H_1, H_2)$. Operators of finite rank form a dense subset of $\mathscr{L}_2(H_1, H_2)$ since we can approximate any $T \in \mathscr{L}_2(H_1, H_2)$ arbitrarily closely by changing the definition of Te_i to be 0 except for a large but finite subset of indices. In particular we conclude that all Hilbert-Schmidt operators are compact.

If $T_j \in \mathscr{L}_2(H_j, H_0)$, $j=1,2$, it follows that $T = T_2^* T_1 \in \mathscr{L}_2(H_1, H_2)$, but the product has additional properties. In fact, if $\{e_j\}$ and $\{f_j\}$ are orthonormal systems in H_1 and H_2 labelled by the same index set, then

$$\sum |(Te_j, f_j)| = \sum |(T_1 e_j, T_2 f_j)| \leqq \|T_1\|_2 \, \|T_2\|_2.$$

On the other hand, let $T \in \mathscr{L}(H_1, H_2)$ and assume that

$$\sum |(Te_j, f_j)| < \infty \tag{19.1.3}$$

for all orthonormal systems $\{e_j\}$ and $\{f_j\}$ in H_1 and H_2 labelled by the same index set. Let $A = (T^* T)^{\frac{1}{2}}$ which is a bounded self-adjoint operator in H_1. If the kernel is N, the range is dense in the orthogonal complement N' of N. Since

$$\|Tf\|^2 = (T^* Tf, f) = \|Af\|^2,$$

the map $Af \mapsto Tf$ is isometric, so its closure is an isometric operator $U: N' \to H_2$, and $T = UA$. If $\{e_j\}$ is a complete orthonormal system in N' it follows that $f_j = U e_j$ is an orthonormal system in H_2, and we have

$$\sum |(Te_j, f_j)| = \sum (A e_j, e_j).$$

Thus $B = A^{\frac{1}{2}}$ is a Hilbert-Schmidt operator and

$$\|B\|_2^2 = \sum (A e_j, e_j),$$

for B vanishes in N. Recalling that $\|UB\|_2 \leqq \|B\|_2$ we take $T_1 = B$ and $T_2^* = UB$, which gives $T = T_2^* T_1$ and

$$\|T_1\|_2 \, \|T_2\|_2 \leqq \|B\|_2^2 = \sum |(Te_j, f_j)|.$$

The opposite inequality is always true so we must have equality. Summing up, we introduce the following definition:

Definition 19.1.12. If H_1 and H_2 are Hilbert spaces then the space $\mathscr{L}_1(H_1,H_2)$ of trace class operators from H_1 to H_2 is the set of all $T\in\mathscr{L}(H_1,H_2)$ such that (19.1.3) is valid for all orthonormal systems $\{e_j\}$, $\{f_j\}$ in H_1 and H_2 respectively, or equivalently $T=T_2^* T_1$ where $T_j\in\mathscr{L}_2(H_j,H_0)$ for some other Hilbert space H_0. The norm in $\mathscr{L}_1(H_1,H_2)$ is defined by

$$\|T\|_1=\sup\sum|(Te_j,f_j)|=\inf\|T_1\|_2\,\|T_2\|_2, \tag{19.1.4}$$

the supremum (infimum) being taken over all such $\{e_j\}$, $\{f_j\}$ (resp. T_1, T_2).

The triangle inequality and the completeness follow from the first part of (19.1.4). From the second part we conclude that if $T\in\mathscr{L}_1(H_1,H_2)$ and $S_1\in\mathscr{L}(H_1',H_1)$, $S_2\in\mathscr{L}(H_2,H_2')$ then $S_2TS_1\in\mathscr{L}_1(H_1',H_2')$ and

$$\|S_2TS_1\|_1\leq\|S_2\|\,\|T\|_1\,\|S_1\|.$$

Hence the operators of finite rank are of trace class, and they form a dense subset since this is true for Hilbert-Schmidt operators.

If $T\in\mathscr{L}_1(H,H)$ we define the trace of T as

$$\operatorname{Tr}(T)=\operatorname{Tr}(T|H)=\sum(Te_j,e_j)$$

where $\{e_j\}$ is a complete orthonormal system in H. This is independent of the choice of $\{e_j\}$, for if we write $T=T_2^*T_1$ with $T_j\in\mathscr{L}_2(H,H_0)$ then

$$\|t_1T_1+t_2T_2\|_2^2=\sum_{\nu,\mu=1}^{2}\sum_j(T_\mu^*T_\nu e_j,e_j)\bar{t}_\mu t_\nu.$$

The coefficient of $t_1\bar{t}_2$ is independent of the choice of e_j and it is equal to $\operatorname{Tr}(T)$. Thus the trace is uniquely defined and it is a linear form of norm 1 on $\mathscr{L}_1(H,H)$.

Proposition 19.1.13. *If $T\in\mathscr{L}_1(H,H)$ and H' is a closed subspace of H such that $TH'\subset H'$ then the restriction T' of T to H' and the operator T'' in $H''=H/H'$ induced by T are both of trace class, and*

$$\operatorname{Tr}(T|H)=\operatorname{Tr}(T'|H')+\operatorname{Tr}(T''|H''). \tag{19.1.5}$$

If $T_1\in\mathscr{L}_1(H_1,H_1)$ and S is a continuous bijection $H_1\to H_2$, then

$$T_2=ST_1S^{-1}\in\mathscr{L}_1(H_2,H_2),\quad \operatorname{Tr}(T_1|H_1)=\operatorname{Tr}(T_2|H_2). \tag{19.1.6}$$

Proof. That T' is of trace class follows from the definition of trace class operators in terms of (19.1.3). We can identify H'' with $H\ominus H'$. If P is the orthogonal projection $H\to H''$ it is clear that T'' is the restriction of PT to H'', hence of trace class. If we take an orthonormal basis $\{e_i'\}$ in H' and one

$\{e_i''\}$ in H'' then they form together an orthonormal basis for H and (19.1.5) follows at once. If $T\in\mathscr{L}_1(H,H)$ is of finite rank and N is the kernel of T, it follows from (19.1.5) that $\operatorname{Tr}(T)$ is the trace of the operator induced by T in the finite dimensional vector space H/N. Hence it is independent of the norm in the Hilbert space, which proves (19.1.6) when T_1 is of finite rank. Since operators of finite rank are dense in $\mathscr{L}_1(H_1,H_1)$ and the map $\mathscr{L}_1(H_1,H_1)\ni T_1\mapsto ST_1S^{-1}\in\mathscr{L}_1(H_2,H_2)$ is continuous, we obtain (19.1.6).

We can now prove a well-known trace formula for the index:

Proposition 19.1.14. *Let $T\in\mathscr{L}(H_1,H_2)$, $S\in\mathscr{L}(H_2,H_1)$, and set*

$$R_1=I_1-ST,\qquad R_2=I_2-TS,$$

where I_j is the identity operator in H_j. Assume that R_1^N and R_2^N are of trace class for some $N>0$. Then it follows that T is a Fredholm operator and that

$$\operatorname{ind} T=\operatorname{Tr}(R_1^N)-\operatorname{Tr}(R_2^N). \tag{19.1.7}$$

Proof. Assume first that $N=1$. By Corollary 19.1.9 the compactness of R_j implies that T is a Fredholm operator. Note that

$$TR_1=R_2T$$

and write $H_j=N_j\oplus H_j'$ where N_1 is the kernel of T and H_2' is the range of T. Then T induces a bijection $H_1'\to H_2'$. If P_1 is the orthogonal projection $H_1\to H_1'$ then $TP_1R_1=TR_1=R_2T$. Hence R_2 maps H_2' to itself, and it follows from (19.1.6) that

$$\operatorname{Tr}(R_2|H_2')=\operatorname{Tr}(P_1R_1|H_1').$$

Now R_1 is the identity on N_1 so we obtain from (19.1.5)

$$\operatorname{Tr}(R_1)=\dim N_1+\operatorname{Tr}(P_1R_1|H_1').$$

Furthermore, $I_2-R_2=TS$ maps H_2 into H_2', so R_2 induces the identity on H_2/H_2'. Using (19.1.5) again we obtain

$$\operatorname{Tr}(R_2)=\dim N_2+\operatorname{Tr}(R_2|H_2')$$

which proves that

$$\operatorname{Tr}(R_1)-\operatorname{Tr}(R_2)=\dim N_1-\dim N_2=\operatorname{ind} T.$$

Thus Proposition 19.1.14 is proved when $N=1$. If $N>1$ we replace S by

$$S'=S(I_2+R_2+\ldots+R_2^{N-1}).$$

It is then obvious that $TS'=I_2-R_2^N$, and we have

$$S'T=S(I_2+(I_2-TS)+\ldots+(I_2-TS)^{N-1})\,T.$$

Since $S(TS)^kT=(ST)^{k+1}$ and

$$x(1+(1-x)+\ldots+(1-x)^{N-1})=1-(1-x)^N,$$

it follows that

$$S'T = I_1 - (I_1 - ST)^N = I_1 - R_1^N.$$

The proof is therefore reduced to the case $N=1$ and so complete.

The first part of the proof is a special case of the following result on invariance of Euler numbers which will be the basis for our work in Section 19.4.

Theorem 19.1.15. *Let*

$$(19.1.8) \quad H_{-1} = \{0\} \to H_0 \xrightarrow{T_0} H_1 \xrightarrow{T_1} H_2 \to \dots \xrightarrow{T_{N-1}} H_N \to \{0\} = H_{N+1}$$

be a complex of Hilbert spaces with all T_j continuous with closed range. Thus we assume $T_j T_{j-1} = 0$, that is,

$$T_{j-1} H_{j-1} \subset \{f \in H_j;\, T_j f = 0\} = N_j, \quad j = 1, \dots, N.$$

Let $R_j \in \mathscr{L}_1(H_j, H_j)$ commute with the complex, that is,

$$(19.1.9) \qquad R_j T_{j-1} = T_{j-1} R_{j-1}, \quad j = 1, \dots, N.$$

Then $R_j(T_{j-1} H_{j-1}) \subset T_{j-1} H_{j-1}$ and $R_j N_j \subset N_j$ so R_j induces an operator $\tilde{R}_j$ of trace class in $\tilde{H}_j = N_j / T_{j-1} H_{j-1}$. We have

$$(19.1.10) \qquad \sum_0^N (-1)^j \operatorname{Tr}(R_j | H_j) = \sum_0^N (-1)^j \operatorname{Tr}(\tilde{R}_j | \tilde{H}_j).$$

Proof. By (19.1.9) we have

$$R_j(T_{j-1} H_{j-1}) = T_{j-1} R_{j-1} H_{j-1} \subset T_{j-1} H_{j-1},$$

and $R_j N_j \subset N_j$ since $T_j R_j N_j = R_{j+1} T_j N_j = \{0\}$. From Proposition 19.1.13 it follows that $\tilde{R}_j$ as well as the operator R'_j induced by R_j in H_j/N_j are of trace class and that

$$\begin{aligned}\operatorname{Tr}(R_j | H_j) &= \operatorname{Tr}(R_j | N_j) + \operatorname{Tr}(R'_j | H_j/N_j) \\ &= \operatorname{Tr}(R_j | T_{j-1} H_{j-1}) + \operatorname{Tr}(\tilde{R}_j | \tilde{H}_j) + \operatorname{Tr}(R'_j | H_j/N_j).\end{aligned}$$

Now T_j defines an isomorphism $H_j/N_j \to T_j H_j$, and by (19.1.9) it transforms R'_j to R_{j+1} restricted to $T_j H_j$. Thus

$$\operatorname{Tr}(R_j | H_j) = \operatorname{Tr}(R_j | T_{j-1} H_{j-1}) + \operatorname{Tr}(\tilde{R}_j | \tilde{H}_j) + \operatorname{Tr}(R_{j+1} | T_j H_j).$$

The first and the third terms drop out when we form the alternating sum in (19.1.10) which proves this formula.

The complex (19.1.8) is called a *Fredholm complex* if all T_j have closed range and the spaces $\tilde{H}_j = \operatorname{Ker} T_j / T_{j-1} H_{j-1}$ are finite dimensional. A Fredholm operator defines a Fredholm complex with $N=1$. Conversely, a Fredholm complex can be split to a Fredholm operator:

Proposition 19.1.16. *The complex* (19.1.8) *is a Fredholm complex if and only if the operator A defined by*

$$H^{\text{even}}=\bigoplus H_{2j}\ni(f_0,f_2,\ldots)\mapsto(T_0f_0+T_1^*f_2,T_2f_2+T_3^*f_3,\ldots)\in\bigoplus H_{2j+1}=H^{\text{odd}}$$

is a Fredholm operator. The index of A is the Euler characteristic

$$\sum(-1)^j\dim\tilde H_j.$$

Proof. Since $T_{j-1}H_{j-1}\subset\operatorname{Ker}T_j$ the range of T_{j-1} is orthogonal to the range of T_j^*. Hence

$$\|A(f_0,f_2,\ldots)\|^2=\sum(\|T_{2j}f_{2j}\|^2+\|T_{2j-1}^*f_{2j}\|^2). \tag{19.1.11}$$

Assume first that we have a Fredholm complex. Then

$$\|f_{2j}\|^2\leqq C\|T_{2j}f_{2j}\|^2$$

when f_{2j} is orthogonal to $\operatorname{Ker}T_{2j}$, for T_{2j} has closed range. Since T_{2j-1} has closed range it follows that T_{2j-1}^* has closed range, hence that

$$\|f_{2j}\|^2\leqq C\|T_{2j-1}^*f_{2j}\|^2$$

when f_{2j} is orthogonal to $\operatorname{Ker}T_{2j-1}^*$. Now H_{2j} is the orthogonal sum of the orthogonal complement of $\operatorname{Ker}T_{2j}$, that of $\operatorname{Ker}T_{2j-1}^*$, and $\tilde H_{2j}$ which we identify with $\operatorname{Ker}T_{2j}\cap\operatorname{Ker}T_{2j-1}^*$. Hence

$$\|f_{2j}\|^2\leqq C(\|T_{2j}f_{2j}\|^2+\|T_{2j-1}^*f_{2j}\|^2)$$

if f_{2j} is orthogonal to $\tilde H_{2j}$, for we can then write $f_{2j}=g_{2j}+h_{2j}$ where g_{2j} is orthogonal to $\operatorname{Ker}T_{2j}$, hence in $\operatorname{Ker}T_{2j-1}^*$, and h_{2j} is orthogonal to $\operatorname{Ker}T_{2j-1}^*$, hence in $\operatorname{Ker}T_{2j}$. The inequality is therefore reduced to the two preceding ones applied to g_{2j} and to h_{2j}. Using (19.1.11) we can now conclude that A has closed range and kernel $\bigoplus\tilde H_{2j}$. In the same way it follows that the kernel of A^* is $\bigoplus\tilde H_{2j+1}$, which proves that A is a Fredholm operator with index equal to the Euler characteristic.

On the other hand, if A is assumed to be a Fredholm operator it follows from (19.1.11) that $W_j=\operatorname{Ker}T_{2j}\cap\operatorname{Ker}T_{2j-1}^*$ is finite dimensional and that

$$\|f_{2j}\|^2\leqq C(\|T_{2j}f_{2j}\|^2+\|T_{2j-1}^*f_{2j}\|^2)$$

when f_{2j} is orthogonal to W_j. When f_{2j} is orthogonal to $\operatorname{Ker}T_{2j}$ we have $T_{2j-1}^*f_{2j}=0$ so $\|f_{2j}\|^2\leqq C\|T_{2j}f_{2j}\|^2$ which proves that T_{2j} has closed range. When f_{2j} is orthogonal to $\operatorname{Ker}T_{2j-1}^*$ we obtain that T_{2j-1}^* has closed range, hence T_{2j-1} has closed range. The proof is complete.

We shall apply the preceding method to a complex defined as a product of the complexes associated with two Fredholm operators A_j: $H_0^j\to H_1^j$, $j=1,2$. First recall that the tensor product $H_1\otimes H_2$ of two Hilbert spaces H_1 and H_2 is defined as the set of continuous antilinear forms on $H_1\times H_2$ corresponding to antilinear maps $H_2\to H_1$ of Hilbert Schmidt class in the obvious sense. Thus $H_1 \otimes H_2 = \mathscr{L}_2(\overline{H}_2, H_1)$ where $\overline{H}_2$ is H_2 with multiplication by scalars changed by complex conjugation. (Equivalently, if we realize H_j as $L^2(X_j, \mu_j)$ where μ_j is a positive measure on X_j then $H_1 \otimes H_2$ is naturally isomorphic to $L^2(X_1 \times X_2)$.) Imitating the definition of the de Rham complex of differential forms in $\mathbb{R}^2$ we introduce the complex with the spaces

$$G_0 = H_0^1 \otimes H_0^2, \quad G_1 = H_1^1 \otimes H_0^2 \oplus H_0^1 \otimes H_1^2, \quad G_2 = H_1^1 \otimes H_1^2,$$

and the maps

$$T_0(f_0^1 \otimes f_0^2) = (A_1 f_0^1 \otimes f_0^2, f_0^1 \otimes A_2 f_0^2),$$
$$T_1(f_1^1 \otimes f_0^2, f_0^1 \otimes f_1^2) = A_1 f_0^1 \otimes f_1^2 - f_1^1 \otimes A_2 f_0^2.$$

Here $f_j^k \in H_j^k$. It is immediately clear that we obtain a complex. Splitting it as in Proposition 19.1.16 we obtain an operator $A: G_0 \oplus G_2 \to G_1$ defined by

$$A = \begin{pmatrix} A_1 \otimes I_0^2 & -I_1^1 \otimes A_2^* \\ I_0^1 \otimes A_2 & A_1^* \otimes I_1^2 \end{pmatrix} \tag{19.1.12}$$

in block matrix notation. If $F = (F_0, F_1) \in H_0^1 \otimes H_0^2 \oplus H_1^1 \otimes H_1^2$ then

$$AF = ((A_1 \otimes I_0^2) F_0 - (I_1^1 \otimes A_2^*) F_1, (I_0^1 \otimes A_2) F_0 + (A_1^* \otimes I_1^2) F_1)$$

and thus

$$\|AF\|^2 = \|(A_1 \otimes I_0^2) F_0\|^2 + \|(I_0^1 \otimes A_2) F_0\|^2 + \|(I_1^1 \otimes A_2^*) F_1\|^2 + \|(A_1^* \otimes I_1^2) F_1\|^2. \tag{19.1.13}$$

It follows immediately that

$$\operatorname{Ker} A = \operatorname{Ker} A_1 \otimes \operatorname{Ker} A_2 \oplus \operatorname{Ker} A_1^* \otimes \operatorname{Ker} A_2^*. \tag{19.1.14}$$

The estimate

$$\|F\|^2 \leq C \|AF\|^2$$

is valid for all F orthogonal to this space if

$$\|f\|^2 \leq C \|A_j f\|^2,$$

when $f \in H_0^j$ is orthogonal to $\operatorname{Ker} A_j$, and

$$\|g\|^2 \leq C \|A_j^* g\|^2$$

when $g \in H_1^j$ is orthogonal to $\operatorname{Ker} A_j^*$. These estimates are equivalent, and we see that the range of A is closed because A_j have closed range. Similarly we find that A^* has closed range and that

$$\operatorname{Ker} A^* = \operatorname{Ker} A_1^* \otimes \operatorname{Ker} A_2 \oplus \operatorname{Ker} A_1 \otimes \operatorname{Ker} A_2^*. \tag{19.1.15}$$

Thus A is a Fredholm operator and

$$\operatorname{ind} A = \dim \operatorname{Ker} A_1 \dim \operatorname{Ker} A_2 + \dim \operatorname{Coker} A_1 \dim \operatorname{Coker} A_2 - \dim \operatorname{Coker} A_1 \dim \operatorname{Ker} A_2 - \dim \operatorname{Ker} A_1 \dim \operatorname{Coker} A_2,$$

which completes the proof of the following theorem:

Theorem 19.1.17. *Let A_j: $H_0^j \to H_1^j$, $j=1,2$, be Fredholm operators. Then the operator A from $H_0^1 \otimes H_0^2 \oplus H_1^1 \otimes H_1^2$ to $H_1^1 \otimes H_0^2 \oplus H_0^1 \otimes H_1^2$ defined by* (19.1.12) *is a Fredholm operator, and*

$$\operatorname{ind} A = \operatorname{ind} A_1 \operatorname{ind} A_2. \tag{19.1.16}$$

We shall end this section by proving a more general version of Theorem 19.1.15 which is occasionally useful although we shall not need it in this book. The main point is to extend the last assertion in Proposition 19.1.13.

Proposition 19.1.18. *Let H_j be a Hilbert space and let $T_j \in \mathscr{L}_1(H_j, H_j)$, $j=1,2$. If there exists a closed linear operator S with domain dense in H_1 and range dense in H_2 such that S is injective and $T_2 S \subset S T_1$, then it follows that* $\operatorname{Tr}(T_1) = \operatorname{Tr}(T_2)$.

Proof. By a classical theorem of von Neumann $A = S^* S$ is self-adjoint and S is the closure of the restriction to the domain of A. Thus $B = A^{\frac{1}{2}}$ is a self-adjoint operator in H_1 with the same domain as S, and we have there $\|Bf\|_1 = \|Sf\|_2$. Let P_1^ε be the spectral projection for B corresponding to the interval $(\varepsilon, 1/\varepsilon)$. Since $Bf \neq 0$ when $f \neq 0$ it follows that $P_1^\varepsilon f \to f$ as $\varepsilon \to 0$ for every $f \in H_1$. Define $P_2^\varepsilon = S P_1^\varepsilon S^{-1}$ on SH_1 which by hypothesis is a dense subset of H_2. If $g = Sf$ then

$$\|P_2^\varepsilon g\| = \|S P_1^\varepsilon f\| = \|B P_1^\varepsilon f\| \leqq \|Bf\| = \|Sf\| = \|g\|.$$

Thus the closure Q of P_2^ε is defined in H_2 and has norm $\leqq 1$; $Q^2 = Q$ so Q is a projection. It is an orthogonal projection, for $\|h\| \leqq \|g+h\|$ for all h with $Qh = h$ and g with $Qg = 0$, so these spaces are mutually orthogonal. We shall denote the closure Q also by P_2^ε. If $g = Sf$ then $P_2^\varepsilon g = S P_1^\varepsilon f$,

$$B P_1^\varepsilon f = P_1^\varepsilon B f \to Bf, \quad \varepsilon \to 0.$$

Hence $S P_1^\varepsilon f$ is also convergent, necessarily to $Sf = g$, which proves that $P_2^\varepsilon g \to g$ for every fixed $g \in H_2$.

Since the range of P_1^ε is contained in the domain of S, we have by hypothesis

$$T_2 S P_1^\varepsilon = S T_1 P_1^\varepsilon,$$

hence in the domain of S

$$T_2 P_2^\varepsilon S = S T_1 P_1^\varepsilon.$$

If we multiply to the left by P_2^ε we obtain

$$P_2^\varepsilon T_2 P_2^\varepsilon S = S P_1^\varepsilon T_1 P_1^\varepsilon.$$

Since S is a continuous bijection of $P_1^\varepsilon H_1$ on $P_2^\varepsilon H_2$, with norm $\leqq \varepsilon^{-1}$ for the operator and its inverse, it follows from Proposition 19.1.13 that

$$\operatorname{Tr}(P_2^\varepsilon T_2 P_2^\varepsilon) = \operatorname{Tr}(P_1^\varepsilon T_1 P_1^\varepsilon).$$

The proof is now completed by the following lemma:

Lemma 19.1.19. *Let* $T\in\mathscr{L}_1(H,H)$ *and let* $S_j, R_j\in\mathscr{L}(H,H)$ *be sequences converging strongly to the identity. Then it follows that* $\operatorname{Tr}(S_j TR_j^*)\to\operatorname{Tr}(T)$.

Proof. Since

$$\|S_j TR_j^*\|_1 \leqq C^2 \|T\|_1,$$

if C is an upper bound for $\|S_j\|$ and $\|T_j\|$, it is sufficient to verify the statement when T is of finite rank. Then $N=\operatorname{Ker} T$ is of finite codimension and we can write T as a product $T=T'P$ where $P\colon H\to H/N$ and $T'\colon H/N\to H$ are bounded. Then $S_j T'\to T'$ in $\mathscr{L}_1(H/N,H)$ since H/N is finite dimensional, so

$$\|S_j T-T\|_1 \leqq \|P\|\,\|S_j T'-T'\|_1\to 0.$$

Similarly,

$$\|TR_j^*-T\|_1 = \|R_j T^*-T^*\|_1\to 0,$$

which implies that

$$\|S_j TR_j^*-T\|_1 \leqq \|S_j\|\,\|TR_j^*-T\|_1+\|S_j T-T\|_1\to 0.$$

Hence $\operatorname{Tr}(S_j TR_j^*)\to\operatorname{Tr}(T)$, which proves the lemma.

We can now extend Theorem 19.1.15 as follows:

Theorem 19.1.20. *Let* (19.1.8) *be a complex of Hilbert spaces and densely defined closed linear operators* T_j. *Thus we assume that the range* W_j *of* T_{j-1} *is contained in* $\operatorname{Ker} T_j$. *(We set* $W_0=\{0\}$ *and* $\operatorname{Ker} T_N=H_N$.*) Let* $R_j\in\mathscr{L}_1(H_j,H_j)$ *commute with the complex in the sense that*

$$R_{j+1}\,T_j\subset T_j R_j,\quad j=0,\dots,N-1.$$

Then $R_j W_j\subset W_j$, $R_j\operatorname{Ker} T_j\subset\operatorname{Ker} T_j$, *and* R_j *induces an operator* $\tilde R_j$ *of trace class in the Hilbert space* $\tilde H_j=\operatorname{Ker} T_j/\overline{W}_j$. *With this notation the invariance formula* (19.1.10) *remains true.*

The proof is identical to that of Theorem 19.1.15 with Proposition 19.1.13 improved by Proposition 19.1.18. We leave for the reader to verify that this is true.

19.2. The Index of Elliptic Operators

Let X be a compact C^∞ manifold without boundary, and let E, F be two C^∞ complex vector bundles over X with the same fiber dimension. Their antiduals will be denoted by E^* and F^*. We recall from Section 18.1 that a pseudo-differential operator $P\in\Psi^m(X;E\otimes\Omega^{\frac12},F\otimes\Omega^{\frac12})$, where Ω is the density line bundle on X, has a principal symbol

$$p\in S^m(T^*(X),\operatorname{Hom}(\pi^*E,\pi^*F))$$

uniquely determined modulo S^{m-1}. Here the fiber of $\operatorname{Hom}(\pi^* E, \pi^* F)$ at $\gamma \in T^*(X)$ consists of the linear maps $E_{\pi\gamma} \to F_{\pi\gamma}$ if π is the projection $T^*(X) \to X$. Also recall that P is elliptic if there exists some $q \in S^{-m}(T^*(X), \operatorname{Hom}(\pi^* F, \pi^* E))$ such that

$$qp - I \in S^{-1}(T^*(X), \operatorname{Hom}(\pi^* E, \pi^* E)),\quad pq - I \in S^{-1}(T^*(X), \operatorname{Hom}(\pi^* F, \pi^* F)); \tag{19.2.1}$$

the conditions are equivalent when E and F have the same fiber dimensions. By I we have denoted the identity maps in $\pi^* E$ and in $\pi^* F$ respectively.

Theorem 19.2.1. *If $P \in \Psi^m(X; E \otimes \Omega^{\frac12}, F \otimes \Omega^{\frac12})$ is elliptic then P defines a Fredholm operator from $H_{(s)}(X; E \otimes \Omega^{\frac12})$ to $H_{(s-m)}(X; F \otimes \Omega^{\frac12})$ with kernel $\subset C^\infty(X; E \otimes \Omega^{\frac12})$, thus independent of s; the range is the orthogonal space of the kernel $\subset C^\infty(X; F^* \otimes \Omega^{\frac12})$ of the adjoint $P^* \in \Psi^m(X; F^* \otimes \Omega^{\frac12}, E^* \otimes \Omega^{\frac12})$. Thus the index is independent of s and equal to the index of P as operator $C^\infty(X; E \otimes \Omega^{\frac12}) \to C^\infty(X; F \otimes \Omega^{\frac12})$ or $\mathscr{D}'(X; E \otimes \Omega^{\frac12}) \to \mathscr{D}'(X; F \otimes \Omega^{\frac12})$, and it depends only on the class of P modulo Ψ^{m-1}. If $E = F^*$ and $P - P^* \in \Psi^{m-1}$ then* $\operatorname{ind} P = 0$.

Proof. Choose $Q \in \Psi^{-m}(X; F \otimes \Omega^{\frac12}, E \otimes \Omega^{\frac12})$ so that (19.2.1) is valid for the principal symbols of P and of Q. Then

$$\begin{aligned} P&\colon H_{(s)}(X; E \otimes \Omega^{\frac12}) \to H_{(s-m)}(X; F \otimes \Omega^{\frac12}),\\ Q&\colon H_{(s-m)}(X; F \otimes \Omega^{\frac12}) \to H_{(s)}(X; E \otimes \Omega^{\frac12}),\\ K_1 = QP - I&\colon H_{(s)}(X; E \otimes \Omega^{\frac12}) \to H_{(s+1)}(X; E \otimes \Omega^{\frac12}),\\ K_2 = PQ - I&\colon H_{(s-m)}(X; F \otimes \Omega^{\frac12}) \to H_{(s-m+1)}(X; F \otimes \Omega^{\frac12}) \end{aligned}$$

are continuous for every s. The map $H_{(s+1)} \to H_{(s)}$ is compact for every s. (See Theorem 10.1.10 or else use the characterization of $H_{(s)}$ in Theorem 18.1.29 and note that symbols of order -1 can be approximated in S^0 by symbols of order $-\infty$. See also Theorem 18.6.6 for a general version of the statement.) Hence it follows from Corollary 19.1.9 that $P\colon H_{(s)} \to H_{(s-m)}$ and $Q\colon H_{(s-m)} \to H_{(s)}$ are Fredholm operators and that $\operatorname{ind} P + \operatorname{ind} Q = 0$. Since (19.2.1) is only a condition on the principal symbol, the choice of Q only depends on the equivalence class of P modulo Ψ^{m-1}, so $\operatorname{ind} P$ is independent of P within such a class. Now $u \in \operatorname{Ker} P$ implies

$$u = -K_1 u = \ldots = (-K_1)^N u \in H_{(s+N)}$$

for every positive integer N. Thus $u \in C^\infty$. The adjoint P^* of P is an elliptic pseudodifferential operator mapping $H_{(m-s)}(X, F^* \otimes \Omega^{\frac12})$ to $H_{(-s)}(X, E^* \otimes \Omega^{\frac12})$. If v is in the kernel then

$$v = -K_2^* v = \ldots = (-K_2^*)^N v \in H_{(m-s+N)}$$

for every N so $v \in C^\infty$. Since the range of P as operator from $H_{(s)}$ to $H_{(s-m)}$ is closed, it is the orthogonal space of this kernel, which completes the proof, for the last statement follows from the fact that $\operatorname{ind} P = -\operatorname{ind} P^*$.

From Corollary 19.1.7 it follows that for the product $P_1 P_2$ of two elliptic operators we have

$$\operatorname{ind}(P_1 P_2) = \operatorname{ind} P_1 + \operatorname{ind} P_2. \tag{19.2.2}$$

By Corollary 19.1.6 we have $\operatorname{ind} P_1 = \operatorname{ind} P_2$ if $P_j \in \Psi^m(X; E \otimes \Omega^{\frac{1}{2}}, F \otimes \Omega^{\frac{1}{2}})$ are elliptic and the principal symbol of $P_1 - P_2$ is sufficiently small in S^m. A stronger stability property follows from Theorem 19.1.10:

Theorem 19.2.2. *Let I be the interval $[0,1]$ on $\mathbb{R}$, and let $I \ni t \mapsto a(t) \in C^\infty(T^*(X); \operatorname{Hom}(\pi^* E, \pi^* F))$ and $I \ni t \mapsto b(t) \in C^\infty(T^*(X); \operatorname{Hom}(\pi^* F, \pi^* E))$ be continuous maps such that $a(t)$ is uniformly bounded in S^m, $b(t)$ is uniformly bounded in S^{-m}, and $a(t)b(t) - I$ is uniformly bounded in $S^{-1}(T^*(X); \operatorname{Hom}(\pi^* F, \pi^* F))$ while $b(t)a(t) - I$ is uniformly bounded in $S^{-1}(T^*(X); \operatorname{Hom}(\pi^* E, \pi^* E))$. If $A_0, A_1 \in \Psi^m(X; E \otimes \Omega^{\frac{1}{2}}, F \otimes \Omega^{\frac{1}{2}})$ have principal symbols $a(0)$ and $a(1)$ respectively, it follows then that* $\operatorname{ind} A_0 = \operatorname{ind} A_1$.

Proof. Let us first recall how one obtains an operator $A(t)$ with principal symbol $a(t)$. (See the discussion after Definition 18.1.20.) Choose a covering of X by coordinate patches X_j, with local coordinates $\kappa_j: X_j \to \tilde{X}_j \subset \mathbb{R}^n$, such that there are local trivializations

$$\phi_j: E|_{X_j} \to X_j \times \mathbb{C}^N, \qquad \psi_j: F|_{X_j} \to X_j \times \mathbb{C}^N.$$

We can choose the covering so fine that $X_j \cup X_k$ is also contained in such a coordinate patch whenever $X_j \cap X_k \neq \emptyset$. Set

$$a_j(t, \kappa_j(x), \xi) = \psi_j(x) a(t, x, {}^t\kappa_j'(x)\xi) \phi_j(x)^{-1}, \quad x \in X_j, \quad \xi \in \mathbb{R}^n,$$

which is then an $N \times N$ matrix with entries in $S^m(X_j \times \mathbb{R}^n)$, where n of course is the dimension of X and N is the fiber dimension of E and of F. Choose a partition of unity $1 = \sum \chi_j^2$ on X with $\chi_j \in C_0^\infty(X_j)$. Then we can define $A(t)$ for $u \in C_0^\infty(X, E)$ by

$$A(t) u = \sum \chi_j \psi_i^{-1} \kappa_j^* a_j(t, x, D)(\kappa_j^{-1})^* (\chi_j \phi_j u).$$

Similarly we define $B(t)$ with principal symbol $b(t)$. To compute $A(t)B(t)$ and $B(t)A(t)$ we can work in local coordinates using Theorems 18.1.8 and 18.1.17. It follows that

$$K_1(t) = B(t)A(t) - I, \quad K_2(t) = A(t)B(t) - I$$

are of order -1 and uniformly bounded as operators $H_{(s)} \to H_{(s+1)}$ for any s. Hence Theorem 19.1.10 gives at once that

$$\operatorname{ind} A_0 = \operatorname{ind} A(0) = \operatorname{ind} A(1) = \operatorname{ind} A_1,$$

which proves the theorem.

To determine the index of arbitrary elliptic operators it suffices to study operators with polyhomogeneous symbol of order 0:

Theorem 19.2.3. *Let $P\in\Psi^m(X;E\otimes\Omega^{\frac12},F\otimes\Omega^{\frac12})$ be elliptic, with principal symbol p, and let h be a homogeneous function of degree 1 on $T^*(X)$ which is positive and C^∞ outside the zero section. Then the operators with principal symbol $p(x,R\xi/h)$ are elliptic and have the same index as P when R is sufficiently large.*

Proof. First assume that $m=0$. Choose $q\in S^0$ so that $qp-I$ and $pq-I$ have compact support, and set for $0\leqq\varepsilon\leqq 1$

$$p_\varepsilon(x,\xi)=p(x,\xi\,\chi(\varepsilon h)),\qquad q_\varepsilon(x,\xi)=q(x,\xi\,\chi(\varepsilon h)),$$

where χ is a decreasing C^∞ function on $\mathbb{R}$ with $\chi(t)=1$ when $t<1$ and $\chi(t)=1/t$ when $t>2$. By hypothesis we can find C so that $pq=I$ and $qp=I$ when $h\geqq C$. This implies that

$$p_\varepsilon q_\varepsilon-I=pq-I,\qquad q_\varepsilon p_\varepsilon-I=qp-I$$

provided that $h\chi(\varepsilon h)\geqq C$ when $\varepsilon h\geqq 1$, that is, $\varepsilon C\leqq\min_{t\geqq 1} t\chi(t)$. We have $p_0=p$ so the statement follows from Theorem 19.2.2 if we show that p_ε and q_ε are uniformly bounded in S^0 for $0\leqq\varepsilon\leqq 1$. To do so it is sufficient to examine the set where $1\leqq\varepsilon h\leqq 2$, for $p_\varepsilon=p$ and $q_\varepsilon=q$ when $\varepsilon h\leqq 1$, and p_ε, q_ε are homogeneous of degree 0 when $\varepsilon h\geqq 2$. With the notation $\varepsilon\xi=\eta$ the statement for p_ε is that

$$p(x,\eta\,\chi(h(x,\eta))/\varepsilon)$$

is uniformly bounded in C^∞ in the compact subset of $T^*(X)\setminus 0$ where $1\leqq h(x,\eta)\leqq 2$. This follows at once from the fact that $p(x,\theta/\varepsilon)$ is uniformly bounded in C^∞ when $\frac12\leqq h(x,\theta)\leqq 2$, because p is a symbol of degree 0. The proof is now complete when $m=0$.

To handle the general case we choose a C^∞ positive hermitian metric in E^* and write it in the form

$$(a(x)v,v),\qquad v\in E_x^*.$$

Thus $a(x)\colon E_x^*\to E_x$ is bijective and its own adjoint. Choose now a pseudo-differential operator $A\in\Psi^{-m}(X;E^*\otimes\Omega^{\frac12},E\otimes\Omega^{\frac12})$ with principal symbol ah^{-m}. By Theorem 19.2.1 the index of the elliptic operator A is 0 so $PA\in\Psi^0(X;E^*\otimes\Omega^{\frac12},F\otimes\Omega^{\frac12})$ has the same index as P and principal symbol pah^{-m}. Thus it follows from the first part of the proof that P has the same index as $P_R a$ if a denotes multiplication by a and P_R has principal symbol

$$p(x,R\xi/h)\,h(x,R\xi/h)^{-m}=p(x,R\xi/h)\,R^{-m}$$

for some sufficiently large R. This completes the proof, for a is invertible.

If $p\in C^\infty(T^*(X)\setminus 0,\ \mathrm{Hom}(\pi^*E,\pi^*F))$ is homogeneous of degree 0 and elliptic, that is, $p(x,\xi)$ is invertible for every $(x,\xi)\in T^*(X)\setminus 0$, then we define **s-ind p as the index of the elliptic pseudo-differential** operators with principal symbol p. From Theorem 19.2.2 or just from Corollary 19.1.6 it follows that s-ind p_1 = s-ind p if

$$\sup_{x,\xi} \|p(x,\xi)^{-1} p_1(x,\xi) - I\| < 1, \tag{19.2.3}$$

the norm being taken with respect to some hermitian metric in E. In fact, $t p_1 + (1-t)p$ is then the principal symbol of an elliptic operator with index independent of t when $0 \leqq t \leqq 1$. This observation allows us to define s-ind p even if p is just continuous; we define s-ind p = s-ind p_1 if p_1 is any C^∞ homogeneous elliptic symbol of order 0 satisfying (19.2.3). The definition is independent of the choice of p_1 since (19.2.3) implies that

$$\sup_{x,\xi} \|p_2(x,\xi)^{-1} p_1(x,\xi) - I\| < 1$$

for some $p_2 \in C^\infty$ arbitrarily close to p, hence s-ind p_1 = s-ind p_2. For our extended definition we still have s-ind p = s-ind p_1 if p and p_1 are just continuous, elliptic and homogeneous symbols of degree 0 satisfying (19.2.3).

Theorem 19.2.3 suggests a further extension. Let $p \in C(T^*(X), \operatorname{Hom}(\pi^* E, \pi^* F))$ and assume that the set of (x,ξ) such that $p(x,\xi)$ is not invertible is compact. If we choose h as in Theorem 19.2.3 then

$$p_R(x,\xi) = p(x, R\xi/h)$$

is a homogeneous elliptic symbol when R is large, and we define

$$\text{s-ind}\, p = \text{s-ind}\, p_R$$

then; the definition is independent of the choice of R. It is also independent of the choice of h, for if h_1 is another positive continuous function which is homogeneous of degree 1 then

$$p(x, R\xi/(th + (1-t)h_1))$$

is for large R invertible when $\xi \neq 0$ and $0 \leqq t \leqq 1$. Thus s-ind p does not change if h is replaced by h_1.

Theorem 19.2.4. *There is a unique function* s-ind *defined on all* $p \in C(T^*(X), \operatorname{Hom}(\pi^* E, \pi^* F))$ *with invertible values outside a compact set, such that*

(i) s-ind p = ind P *if* P *is an elliptic pseudo-differential operator of order* 0 *with homogeneous principal symbol* p *of order* 0;

(ii) *if* p_t *for* $0 \leqq t \leqq 1$ *is a continuous function of* t *with values in* $C(T^*(X), \operatorname{Hom}(\pi^* E, \pi^* F))$ *and* $p_1(x,\xi)$ *is invertible for all such* t *except when* (x,ξ) *is in a compact set, then* s-ind p_t *is independent of* t.

Proof. We have already proved the existence, and the uniqueness follows from the homotopy used in the proof of Theorem 19.2.3. Note that Theorem 19.2.3 states that ind P = s-ind p for any $P \in \Psi^m(X; E \otimes \Omega^{\frac12}, F \otimes \Omega^{\frac12})$ with principal symbol p.

Before proceeding we shall discuss the elementary example where $X = \mathbb{R}/2\pi\mathbb{Z}$ is the circle and the bundles are just $X \times \mathbb{C}$. It is then enough to study the operator with principal symbol

$$a(x,\xi)=a_+(x), \quad \xi>0; \qquad a(x,\xi)=a_-(x), \quad \xi<0,$$

where a_+ and a_- have values in $\mathbb{C}\setminus\{0\}$. We can multiply by the (multiplication) operator a_-^{-1} which has index 0 to change the symbol to $a_+(x)/a_-(x)$ when $\xi>0$ and to 1 when $\xi<0$. If m is the winding number of $a_+(x)/a_-(x)$ then

$$a_+(x)/a_-(x)=e^{imx+\phi(x)}$$

where ϕ is periodic. Replacing ϕ by $t\phi$ for $0\leqq t\leqq 1$ gives a homotopy showing that it is enough to consider the case $\phi=0$.

The operator P defined by

$$P(\sum u_k e^{ikx})=\sum_{k\geqq 0} u_k e^{ikx}$$

which takes the analytic part of the Fourier series expansion of u is a pseudo-differential operator with symbol 1 for $\xi>0$ and 0 for $\xi<0$. In fact, if u is a trigonometrical polynomial then

$$\begin{aligned} Pu(x)&=\lim_{r\to 1-0}(2\pi i)^{-1}\oint \sum u_k z^k/(z-re^{ix})\,dz \\ &=\lim_{r\to 1-0}\int u(y)/(1-re^{i(x-y)})\,dy/2\pi. \end{aligned}$$

The limit of $1/(1-e^{-\varepsilon}e^{ix})$ as $\varepsilon\to +0$ is $i/(x+i0)$ plus a term which is in C^∞ in $(-\pi,\pi)$. Since $i/(2\pi(x+i0))$ is the inverse Fourier transform of the Heaviside function this proves the statement. What remains is to determine the index of the operator

$$u\mapsto e^{imx}Pu+u-Pu.$$

Let u_+ be the analytic function in the unit disc with boundary values Pu and let u_- be the analytic function outside the unit disc with boundary values $u-Pu$; it vanishes at ∞. If $u\in C^\infty(X)$ is in the kernel we have $z^m u_+(z)+u_-(z)=0$ when $|z|=1$, so the function which is $z^m u_+(z)$ when $|z|\leqq 1$ and $-u_-(z)$ when $|z|\geqq 1$ is an analytic function vanishing at ∞, hence equal to 0, if $m\geqq 0$. If $m<0$ then $q(z)=u_+(z)$ when $|z|\leqq 1$ and $q(z)=-z^{-m}u_-(z)$ when $|z|\geqq 1$ defines an analytic function which is $O(|z|^{|m|-1})$ at infinity. Thus q is a polynomial of degree $|m|-1$ and conversely every such polynomial defines an element in the kernel. Thus the dimension of the kernel is 0 when $m\geqq 0$ and $-m$ when $m<0$.

On the other hand, if $f\in C^\infty(X)$ and we write $f=f_+ + f_-$ as for u, then the equation

$$z^m u_+(z)+u_-(z)=f_+(z)+f_-(z)$$

is satisfied by $u_-=f_-$ and $u_+=z^{-m}f_+(z)$ if $m\leqq 0$. When $m>0$ it is equivalent to $u_-=f_-$ and $z^m u_+(z)=f_+(z)$ so it can be fulfilled if and only if $f_+(z)=O(|z|^m)$. Thus the dimension of the cokernel is 0 and m in the two cases. In both cases we find that the index is $-m$. Summing up, we obtain if $a(x,\xi)$ is a continuous function from $X\times\mathbb{R}$ to $\mathbb{C}$ without zero for large $|\xi|$ that s-ind a is the degree of mapping for the mapping from the boundary of $\{(x,\xi);\ x\in X,\ |\xi|<R\}$ to $a(x,\xi)/|a(x,\xi)|\in S^1$ if R is large enough. Here we have oriented the cotangent bundle of X by $dx\wedge d\xi>0$.

The extension of the definition of s-ind to continuous symbols was made above for two reasons. Firstly it underlines the topological character of this object, which was also seen in the explicit example just discussed. Secondly, the product operation which we shall discuss below leads to operators which are not quite pseudo-differential but are covered by the following theorem.

Theorem 19.2.5. *Let $P_j\in\Psi^m_{\mathrm{phg}}(X;E\otimes\Omega^{\frac12},F\otimes\Omega^{\frac12})$ and assume that $P_j\to P$ in $\mathscr{L}(H_{(s)}(X;E\otimes\Omega^{\frac12}),\ H_{(s-m)}(F\otimes\Omega^{\frac12}))$ for some $s\in\mathbb{R}$. Then the homogeneous principal symbol p_j of P_j converges uniformly on compact subsets of $T^*(X)\smallsetminus 0$ to $p\in C(T^*(X)\smallsetminus 0;\ \mathrm{Hom}(\pi^*E,\pi^*F))$. If p is everywhere an isomorphism then P is a Fredholm operator from $H_{(s)}(X;E\otimes\Omega^{\frac12})$ to $H_{(s-m)}(F\otimes\Omega^{\frac12})$ with index equal to* s-ind p *(if p is modified in a compact set to a continuous function). If the hypothesis is valid for all $s\in\mathbb{R}$ then* Ker P *is a subspace of $C^\infty(X;E\otimes\Omega^{\frac12})$ and the range is the orthogonal space in $H_{(s-m)}(X;F\otimes\Omega^{\frac12})$ of a subspace of $C^\infty(X;F^*\otimes\Omega^{\frac12})$.*

Proof. If $a\in S^m_{\mathrm{phg}}(\mathbb{R}^n\times\mathbb{R}^n)$ has principal symbol a_0, and $u,v\in C_0^\infty(\mathbb{R}^n)$, then

$$t^{-m}(a(x,D)(u\,e^{i\langle\cdot,t\xi\rangle}),v\,e^{i\langle\cdot,t\xi\rangle})=(t^{-m}a(x,D+t\xi)u,v)\to(a_0(\cdot,\xi)u,v).$$

We have

$$t^{-s}\|u\,e^{i\langle\cdot,t\xi\rangle}\|_{(s)}=((2\pi)^{-n}\int|\hat u(\eta)|^2(t^{-2}+|\xi+\eta/t|^2)^s\,d\eta)^{\frac12}\to\|u\|\,|\xi|^s,$$

where the norm is the L^2 norm. A similar estimate is true for v, so it follows that

$$|(a_0(.,\xi)u,v)|\leq M\,\|u\|\,\|v\|,\qquad |\xi|=1,$$

if M is the norm of $a(x,D)$ as operator from $H_{(s)}$ to $H_{(s-m)}$. Thus $\sup_{|\xi|=1}|a_0|\leq M$. If we apply this to $\phi(P_j-P_k)\psi$ with $\phi,\psi\in C_0^\infty(X)$ having support in a coordinate patch, it follows that $p_j-p_k\to 0$ uniformly on compact subsets of $T^*(X)\smallsetminus 0$ as j and $k\to\infty$, so $p=\lim p_j$ exists and is a continuous function, homogeneous of degree m.

Assume now that p is invertible. For a moment we also assume that $s=m=0$. Then we have a uniform bound for p_j^{-1} on $T^*(X)\smallsetminus 0$ when j is large. By the second part of Theorem 18.1.15 it follows that we can choose a pseudo-differential operator Q_j with principal symbol p_j^{-1} such that the norm of Q_j in $\mathscr{L}(H_2,H_1)$ is uniformly bounded if $H_1=L^2(X;E\otimes\Omega^{\frac12})$ and $H_2=L^2(X;F\otimes\Omega^{\frac12})$. Now

$$Q_jP=I+Q_j(P-P_j)+(Q_jP_j-I),$$

$I+Q_j(P-P_j)$ is invertible in $\mathscr{L}(H_1, H_1)$ for large j, and $Q_j P_j - I$ is compact, so $Q_j P$ is a Fredholm operator. Thus $\operatorname{Ker} P$ is finite dimensional. Since PQ_j is a Fredholm operator for similar reasons it follows that P has a closed range of finite codimension. Hence P is a Fredholm operator, and for large j

$$\operatorname{ind} P = \operatorname{ind} P_j = \text{s-ind}\, p_j = \text{s-ind}\, p.$$

For arbitrary s and m we choose as in the proof of Theorem 19.2.3 elliptic pseudo-differential operators

$$A \in \Psi^{-s}(X; E^* \otimes \Omega^{\frac{1}{2}}, E \otimes \Omega^{\frac{1}{2}}) \quad \text{and} \quad B \in \Psi^{s-m}(X; F \otimes \Omega^{\frac{1}{2}}, F^* \otimes \Omega^{\frac{1}{2}})$$

with index 0. Thus $B(P_j - P)A \to 0$ in $\mathscr{L}(L^2(X; E^* \otimes \Omega^{\frac{1}{2}}), L^2(X; F^* \otimes \Omega^{\frac{1}{2}}))$ so it follows from the case of the theorem already proved that BPA is a Fredholm operator. Let A' and B' be parametrices of A and of B. Then

$$P = B'(BPA)A' + (I - B'B)PAA' + P(I - AA')$$

is a compact perturbation of the Fredholm operator $B'(BPA)A'$, hence also a Fredholm operator. For large j the index is equal to that of P_j, thus equal to $\text{s-ind}\, p_j = \text{s-ind}\, p$.

If the hypothesis is fulfilled for all s then $\operatorname{Ker} P \subset H_{(s)}(X; E \otimes \Omega^{\frac{1}{2}})$ decreases with s. If P^*: $H_{(m-s)}(X; F^* \otimes \Omega^{\frac{1}{2}}) \to H_{(-s)}(X; E^* \otimes \Omega^{\frac{1}{2}})$ is the adjoint then $\dim \operatorname{Coker} P$ is equal to the dimension of $\operatorname{Ker} P^* \subset H_{(m-s)}(X; F^* \otimes \Omega^{\frac{1}{2}})$ which increases with s. Since the index is independent of s it follows that $\operatorname{Ker} P$ and $\operatorname{Ker} P^*$ are independent of s, which completes the proof.

In our application the approximating sequence of pseudodifferential operators will be obtained from the following lemma.

Lemma 19.2.6. *Let $a(x,y,\xi) \in S^m_{\text{phg}}(\mathbb{R}^{n+n'} \times \mathbb{R}^n)$ where $m > 0$, and let a_0 be the principal symbol. Choose $\chi \in C^\infty(\mathbb{R}^{n+n'})$ so that $\chi(\xi,\eta) = 1$ when $|\eta| \leqq \max(1, |\xi|)$, and $\chi(\xi,\eta) = 0$ when $|\eta| \geqq \max(2, 2|\xi|)$, while $\chi(\xi,\eta)$ is homogeneous of degree 0 when $|\eta| > 2$. Then*

$$a_\varepsilon(x,y,\xi,\eta) = a(x,y,\xi)\,\chi(\xi,\varepsilon\eta) \in S^m_{\text{phg}}(\mathbb{R}^{n+n'} \times \mathbb{R}^{n+n'}),$$

the principal symbol converges uniformly to a_0 when $|\xi|^2 + |\eta|^2 = 1$ as $\varepsilon \to 0$, and the norm of $a_\varepsilon(x,y,D_x,D_y) - a(x,y,D_x)$ as operator from $H_{(s)}$ to $H_{(s-m)}$ tends to 0 as $\varepsilon \to 0$ for every $s \in \mathbb{R}$.

Proof. The principal symbol of a_ε is $a_0(x,y,\xi)\,\chi_0(\xi,\varepsilon\eta)$ if χ_0 is the principal symbol of χ. Here $\chi_0(\xi,\varepsilon\eta) \to \chi_0(\xi,0) = 1$ if $\varepsilon \to 0$ and $\xi \neq 0$; since $a_0(x,y,0) = 0$ this proves the convergence of the principal symbols. We have

$$(a_\varepsilon(x,y,D_x,D_y) - a(x,y,D_x))\,u = a(x,y,D_x)\,v_\varepsilon,$$
$$v_\varepsilon = \chi(D_x, \varepsilon D_y)\,u - u.$$

For fixed y we can estimate the L^2 norm of $a(x,y,D_x)v$ with respect to x by the $H_{(m)}$ norm of v with respect to x. Thus

$$\begin{aligned}&\|(a_\varepsilon(x,y,D_x,D_y)-a(x,y,D_x))u\|^2_{(0)}\\&\leqq C\iint(1+|\xi|^2)^m|\chi(\xi,\varepsilon\eta)-1|^2\,|\hat u(\xi,\eta)|^2\,d\xi\,d\eta\\&\leqq\|u\|^2_{(m)}\sup C'(1+|\xi|^2)^m|\chi(\xi,\varepsilon\eta)-1|^2(1+|\xi|^2+|\eta|^2)^{-m}.\end{aligned}$$

When $\chi(\xi,\varepsilon\eta)\neq 1$ we have $1+|\xi|^2\leqq 2|\varepsilon\eta|^2$, hence

$$(1+|\xi|^2)/(1+|\xi|^2+|\eta|^2)\leqq 2\varepsilon^2|\eta|^2/(1+|\xi|^2+|\eta|^2)\leqq 2\varepsilon^2.$$

This proves that

$$\|(a_\varepsilon(x,y,D_x,D_y)-a(x,y,D_x))u\|_{(0)}\leqq C'\varepsilon^m\|u\|_{(m)}. \tag{19.2.4}$$

Next we shall prove that

$$\|(a_\varepsilon(x,y,D_x,D_y)-a(x,y,D_x))u\|_{(s)}\leqq C_s\varepsilon^m\|u\|_{(m+s)} \tag{19.2.4)'$$

if s is a positive integer. To do so we observe that $\chi(D_x,\varepsilon D_y)$ commutes with differentiations while the commutator of $a(x,y,D_x)$ and a differentiation is of the same form. Hence $D^\alpha(a_\varepsilon(x,y,D_x,D_y)-a(x,y,D_x))u$ is a sum of terms of the form

$$(b_\varepsilon(x,y,D_x,D_y)-b(x,y,D_x))D^\beta u$$

with $|\beta|\leqq|\alpha|$ and b of the same form as a. This gives (19.2.4)′ at once when s is a positive integer. If s is a negative integer and $u\in H_{(m+s)}$, we can write

$$u=\sum_{|\alpha|\leqq -s}D^\alpha u_\alpha,\qquad \sum_{|\alpha|\leqq -s}\|u_\alpha\|^2_{(m)}\leqq C\|u\|^2_{(m+s)}.$$

If we commute the derivatives D^α through $a(x,y,D_x)$ as above and apply (19.2.4), we obtain (19.2.4)′ also when s is a negative integer. By Corollary B.1.6 this proves the estimate (19.2.4)′ for arbitrary real s. The proof is complete.

If V_1 and V_2 are vector bundles on different spaces X_1 and X_2 then the vector bundle on $X_1\times X_2$ with fiber $V_{1x_1}\oplus V_{2x_2}$ at (x_1,x_2) is denoted by $V_1\boxtimes V_2$ in the following theorem.

Theorem 19.2.7. *Let X and Y be compact C^∞ manifolds, E_X and F_X hermitian vector bundles on X, E_Y and F_Y hermitian vector bundles on Y, and assume that*

$$p\in C(T^*(X),\operatorname{Hom}(\pi_X^*E_X,\pi_X^*F_X)),\qquad q\in C(T^*(Y),\operatorname{Hom}(\pi_Y^*E_Y,\pi_Y^*F_Y))$$

are isomorphisms outside compact subsets K_X and K_Y. Then

$$d(x,\xi,y,\eta)=\begin{pmatrix}p(x,\xi)\otimes I & -I\otimes q(y,\eta)^*\\ I\otimes q(y,\eta) & p(x,\xi)^*\otimes I\end{pmatrix}$$

is in

$$C(T^*(X\times Y),\ \mathrm{Hom}(\pi_X^*E_X\boxtimes\pi_Y^*E_Y\oplus\pi_X^*F_X\boxtimes\pi_Y^*F_Y,\ \pi_X^*F_X\boxtimes\pi_Y^*E_Y\oplus\pi_X^*E_X\boxtimes\pi_Y^*F_Y)).$$

d is an isomorphism outside $K_X\times K_Y$, and

$$\text{s-ind}\, d=(\text{s-ind}\, p)(\text{s-ind}\, q). \tag{19.2.5}$$

Proof. With the notation in Theorem 19.1.17 it follows from (19.1.14) and (19.1.15) that A is an isomorphism whenever A_1 or A_2 is an isomorphism. The finite dimensional case of this result shows that $d(x,\xi,y,\eta)$ is an isomorphism unless $(x,\xi)\in K_X$ and $(y,\eta)\in K_Y$. Thus d is an isomorphism outside $K_X\times K_Y$ so s-ind d is defined.

When proving (19.2.5) we may assume that p and q are homogeneous of degree 1 and C^∞ outside the zero section, for p and q are homotopic to such functions. Then d is also homogeneous of degree 1 but may not be C^∞ at (x,ξ,y,η) unless both ξ and η are $\neq 0$. If we choose pseudo-differential operators P and Q with principal symbols p and q this corresponds analytically to the fact that

$$D=\begin{pmatrix} P\otimes I & -I\otimes Q^*\\ I\otimes Q & P^*\otimes I\end{pmatrix}$$

is not a pseudo-differential operator. (Here P^* and Q^* are of course the pseudo-differential operators which are adjoint to P and Q with respect to the scalar products $\int_X(u,v)$ or $\int_Y(u,v)$ for sections u,v of $E\otimes\Omega^{\frac12}$ or of $F\otimes\Omega^{\frac12}$. Thus they are not the Hilbert space adjoints if P and Q are considered as operators in $H_{(s)}$ spaces.)

Let $1=\sum\phi_j^4$ and $1=\sum\psi_k^4$ be partitions of unity in X and in Y subordinate to coverings by coordinate patches where the bundles E and F are trivial. Let D_0 be defined as D with P and Q replaced by

$$P_0=\sum\phi_j^2P\phi_j^2,\quad Q_0=\sum\psi_j^2Q\psi_j^2$$

which have the same principal symbols. Using Lemma 19.2.6 we can find pseudo-differential approximations to $\phi_j\psi_k D\phi_j\psi_k$ for arbitrary j and k. If we multiply them left and right by ϕ_j and ψ_k and sum, we obtain pseudo-differential operators D_ε of order 1 converging to D_0 in $\mathscr{L}(H_{(s)},H_{(s-1)})$ for every s, and with principal symbols converging to d. Thus $\operatorname{ind} D_0=s-\operatorname{ind} d$ by Theorem 19.2.5. To compute $\operatorname{ind} D_0$ we only have to determine C^∞ elements in the kernel and cokernel. Now the calculation which gave (19.1.13) also gives

$$D_0^*D_0=\begin{pmatrix} P_0^*P_0\otimes I+I\otimes Q_0^*Q_0 & 0\\ 0 & P_0P_0^*\otimes I+I\otimes Q_0Q_0^*\end{pmatrix}$$

and a similar formula for $D_0D_0^*$ with $P_0^*P_0$ replaced by $P_0P_0^*$ and so on. It follows that (19.1.14) and (19.1.15) remain true, so $\operatorname{ind} D_0=\operatorname{ind} P_0\operatorname{ind} Q_0=$ (s-ind p)(s-ind q), which completes the proof.

We shall now discuss a method which in principle can be used to compute s-ind$p$ in two steps:

(i) Simplification of the underlying space X to a Euclidean space.

(ii) Explicit calculation of the index for elliptic operators in a Euclidean space.

In this context we shall only consider elliptic operators which are trivial at infinity, that is, are defined by multiplication with an invertible matrix function outside a compact set. Step (ii) is the subject of Section 19.3. Step (i) begins with embedding the manifold X in a Euclidean space:

Lemma 19.2.8. *For any compact C^∞ manifold X there is a C^∞ embedding $\Phi: X \to \mathbb{R}^\nu$ for some large enough ν.*

Proof. We can cover X by compact sets K_j, $j=1,\dots,J$, contained in coordinate patches X_j with local coordinates $\kappa_j: X_j \to \mathbb{R}^n$, if n is the dimension of X. Choose $\phi_j \in C_0^\infty(X_j)$ equal to 1 in a neighborhood of K_j. Then

$$\Phi: X \ni x \mapsto (\phi_1(x), \phi_1(x)\kappa_1(x), \dots, \phi_J(x), \phi_J(x)\kappa_J(x)) \in \mathbb{R}^{(n+1)J}$$

is a C^∞ map with injective differential. If $x \in K_j$ and $\Phi(x)=\Phi(y)$ then $\phi_j(y) = \phi_j(x) = 1$ so $y \in X_j$ and $\kappa_j(x) = \kappa_j(y)$. Thus Φ is injective which proves the lemma.

By composing Φ with a generic projection on a $2n+1$ dimensional subspace it is easy to reduce ν to $2n+1$, but this is not important here.

Let $N(X)$ be the normal bundle of the embedding,

$$N(X) = \{(x,y) \in X \times \mathbb{R}^\nu, {}^t\Phi'(x)\, y = 0\}.$$

Then the fibers of $N(X)$ inherit a Euclidean structure from the Euclidean metric in $\mathbb{R}^\nu$, and the implicit function theorem gives that

$$\{(x,y) \in N(X);\, |y| < \rho\} \ni (x,y) \mapsto \Phi(x) + y$$

is a diffeomorphism on a tubular neighborhood U_ρ of $\Phi(X)$ if ρ is sufficiently small. To accomplish step (i) in the plan above we shall generalize the product construction in Theorem 19.2.7 so that applied to an elliptic pseudo-differential operator in X it yields a pseudo-differential operator in U_ρ which can be regarded as an operator in $\mathbb{R}^\nu$ which is trivial outside U_ρ. To do this we need first of all an operator of index 1 which can be defined in the fibers of $N(X)$ (compactified at infinity to spheres) with reference only to the Euclidean structure but is independent of a choice of basis. Thus we must define in $\mathbb{R}^\mu$ a pseudo-differential operator invariant under the orthogonal group which has index 1 and, preferably, a very simple kernel and cokernel.

To construct such an operator we consider first in the case $\mu=1$ the operators

$$P = x + iD: H_1 \to H_0,$$

where $H_0 = L^2(\mathbb{R})$ with norm denoted by $\| \ \|$, and H_1 consists of all $u \in H_0$ with $xu \in H_0$ and $Du \in H_0$, the norm being defined by

$$(\|u\|^2 + \|xu\|^2 + \|Du\|^2)^{\frac{1}{2}}.$$

If $Pu = 0$ then $(x + \partial)u = 0$, that is, $u(x) = Ce^{-x^2/2}$ which is a function in H_1. The range of P is dense in H_0 for if $v \in H_0$ is orthogonal then $(x - \partial)v = 0$, so $v = C' e^{x^2/2}$ which cannot be in H_0 unless $C' = 0$. Since

$$\|Pu\|^2 = \|xu\|^2 + \|Du\|^2 + (xu, \partial u) + (\partial u, xu) = \|xu\|^2 + \|Du\|^2 - \|u\|^2$$

it follows from Proposition 19.1.3 that the range is closed, hence equal to H_0. In fact, if u_j is a bounded sequence in H_1 such that Pu_j is convergent, then $\|xu_j\|$ and $\|Du_j\|$ are bounded so there is a subsequence u_{j_k} which is L^2 convergent. If we apply the preceding identity to $u_{j_k} - u_{j_l}$ it follows that xu_{j_k} and Du_{j_k} are also L^2 convergent. Hence P is a Fredholm operator with index 1.

The analogue of the preceding example in $\mathbb{R}^n$ is obtained by considering the exterior algebra over $\mathbb{C}^n$ as a complex with exterior multiplication $\Lambda(w)$ by $w = x + i\xi$,

$$\Lambda(w) \colon 0 \to \mathbb{C} \to \Lambda^1 \mathbb{C}^n \to \ldots \to \Lambda^n \mathbb{C}^n \to 0.$$

This is exact when $w \neq 0$, that is, the range of each map is the kernel of the next. It suffices to prove this when w is a basis vector, and then it is clear that w only annihilates the forms where w can be factored out. We shall split the complex in the manner described in Proposition 19.1.16, after introducing the usual hermitian structure in $\Lambda^j \mathbb{C}^n$. Thus we write

$$\Lambda^e = \bigoplus \Lambda^{2j}(\mathbb{C}^n), \qquad \Lambda^o = \bigoplus \Lambda^{2j+1}(\mathbb{C}^n),$$

and when $u = (u_0, u_2, \ldots) \in \Lambda^e$ we define

$$p(w)u = (\Lambda(w)u_0 + \Lambda(w)^* u_2, \ldots) \in \Lambda^o;$$

then $p(w)^* \colon \Lambda^o \to \Lambda^e$ has a similar form. By the finite dimensional case of Proposition 19.1.16 we know that $p(w)$ is bijective for $0 \neq w \in \mathbb{C}^n$. (In particular, the dimensions of Λ^e and of Λ^o are equal, which follows already from the fact that $(1-1)^n = 0$.) Let H_1 now be the space of all forms of even degree

$$u = u_0 + u_2 + \ldots$$

such that $f, D_j f, x_j f \in L^2$ for all coefficients f and $j = 1, \ldots, n$, and let H_0 be the space of all forms of odd degree with coefficients in L^2.

Proposition 19.2.9. *The first order differential operator $p(x + iD)$ maps H_1 onto H_0 with one dimensional kernel generated by the scalar function $e^{-|x|^2/2}$.*

Proof. Let $q \in C^\infty(\mathbb{R}^{2n}, \mathscr{L}(\Lambda^o, \Lambda^e))$ be equal to $p(x + i\xi)^{-1}$ outside a compact set. Then $q \in S(R^{-1}, g)$ where

$$R(x, \xi) = (1 + |x|^2 + |\xi|^2)^{\frac{1}{2}}, \qquad g = (|dx|^2 + |d\xi|^2)/R^2.$$

This follows from the homogeneity. Hence $q(x,D)$ is a continuous map from H_0 to H_1, for $q(x,D)$, $x_j q(x,D)$ and $D_j q(x,D)$ are in Op $S(1,g)$ and therefore L^2 continuous by Theorem 18.6.3 or just Theorem 18.1.11. From the general calculus of pseudodifferential operators in Section 18.5 we obtain, with $p(x,\xi)=p(x+i\xi)$,

$$q(x,D)\,p(x,D)=I+K_1(x,D)$$

where $K_1\in S(1/R,g)$. If $u\in\mathscr{S}'$ and $p(x,D)u=0$ it follows that $u=(-K_1(x,D))^N u$ for every N, so $u\in\mathscr{S}$. We also have

$$p(x,D)\,q(x,D)=I+K_2(x,D)$$

where $K_2\in S(1/R,g)$. Thus $K_2(x,D)$, $x_j K_2(x,D)$ and $D_j K_2(x,D)$ are continuous from H_0 to H_0 so $K_2(x,D)$ is compact from H_0 to H_0. Hence the range of $p(x,D)\colon H_1\to H_0$ is closed and of finite codimension. If v is orthogonal to the range then $v=-K_2(x,D)^* v=(-K_2(x,D)^*)^N v$ for any N, so $v\in\mathscr{S}$. Thus it only remains to show that the multiples of $e^{-|x|^2/2}$ are the only forms (with coefficients) in $\mathscr{S}$ annihilated by $p(x,D)$ and that $p(x,D)^*$ is injective on $\mathscr{S}$. This requires a more explicit calculation.

If f is a q form in $\mathbb{R}^n$ with coefficients in $\mathscr{S}$, then

$$\Lambda(x+iD)f=\langle x,dx\rangle\wedge f+df=e^{-|x|^2/2}\,d(e^{|x|^2/2}f).$$

We can write

$$f=\sum{}' f_J\,dx^J$$

where $J=(j_1,\dots,j_q)$ is a q-tuple of indices between 1 and n, f_J is antisymmetric in the indices, and $\sum'$ denotes summation over increasing indices only. Then we have

$$\Lambda(x+iD)f=\sum_j\sum_J{}'(\partial_j+x_j)f_J\,dx_j\wedge dx^J,$$

$$\|\Lambda(x+iD)f\|^2=\sum_{j,l}\sum_{J,L}{}'((\partial_j+x_j)f_J,(\partial_l+x_l)f_L)\,\varepsilon^{jJ}_{lL}$$

where $\varepsilon^{jJ}_{lL}=0$ unless $j\notin J$, $l\notin L$ and $\{j\}\cup J=\{l\}\cup L$; in that case ε^{jJ}_{lL} is the sign of the permutation $\begin{pmatrix} jJ\\ lL\end{pmatrix}$. We shall rearrange the terms in the sum. First consider those with $j=l$. For these we must have $J=L$ and $j\notin J$ if $\varepsilon^{jJ}_{lL}\neq 0$, so the sum is

$$\sum_{j\notin J}\sum{}'\|(\partial_j+x_j)f_J\|^2.$$

Next consider the terms with $j\neq l$. If $\varepsilon^{jJ}_{lL}\neq 0$ we must then have $l\in J$ and $j\in L$, and deletion of l from J or j from L gives the same multi-index K. Since

$$\varepsilon^{jJ}_{lL}=\varepsilon^{jJ}_{jlK}\,\varepsilon^{jlK}_{ljK}\,\varepsilon^{ljK}_{lL}=-\varepsilon^{J}_{lK}\,\varepsilon^{jK}_{L},$$

the sum of these terms is

$$-\sum_{j\neq l}\sum_K{}'((\partial_j+x_j)f_{lK},(\partial_l+x_l)f_{jK}).$$

A simple calculation which we leave for the reader gives

$$\|\Lambda(x+iD)^* f\|^2 = \sum_{j,l}\sum_K{}' ((x_l-\partial_l) f_{lK}, (x_j-\partial_j) f_{jK}).$$

Since

$$[(x_j+\partial_j), (x_l-\partial_l)] = 2\delta_{jl}$$

we obtain after an integration by parts

$$\|\Lambda(x+iD) f\|^2 + \|\Lambda(x+iD)^* f\|^2 = \sum_j \sum_{|J|=q} \|(x_j+\partial_j) f_J\|^2 + 2q\|f\|^2.$$

Using (19.1.11) we conclude that if $u=\sum u_{2q}$ where u_{2q} is a $2q$ form with coefficients in $\mathscr{S}$ then

$$\|p(x+iD)u\|^2 = \sum_q 4q\|u_{2q}\|^2 + \sum_j \|(x_j+\partial_j)u\|^2.$$

Thus $p(x+iD)u=0$ implies that u is a multiple of $e^{-|x|^2/2}$. Similarly we have if $v=\sum v_{2q+1}$ where v_{2q+1} is a $2q+1$ form with coefficients in $\mathscr{S}$

$$\|p(x+iD)^* v\|^2 = \sum_q (4q+2)\|v_{2q+1}\|^2 + \sum_j \|(x_j+\partial_j)v\|^2.$$

Thus $p(x+iD)^*$ is injective which completes the proof.

Remark. The proof of Theorem 15.1.1 is essentially the case $q=1$ of the calculation above.

To continue our program we must modify $p(x+iD)$ to a pseudo-differential operator of order 0.

Lemma 19.2.10. *If $T_\varepsilon(x,\xi)=(1+|\varepsilon x|^2+|\varepsilon\xi|^2)^{-\frac{1}{2}}$ then $T_\varepsilon(x,D)$ is for small $\varepsilon>0$ an isomorphism of L^2 on the Hilbert space B of all $u\in L^2$ with $D_j u\in L^2$ and $x_j u\in L^2$ for $j=1,\dots,n$, and $T_\varepsilon(x,D)^{-1}f\to f$ in L^2 as $\varepsilon\to 0$ if $f\in B$.*

Proof. Let $R_\varepsilon = 1/T_\varepsilon$ and introduce the metrics

$$g_\varepsilon = \varepsilon^2(|dx|^2+|d\xi|^2)/R_\varepsilon(x,\xi)^2$$

which are uniformly σ temperate when $0<\varepsilon\leqq 1$. It is obvious that R_ε and T_ε are uniformly bounded in $S(R_\varepsilon, g_\varepsilon)$ and $S(T_\varepsilon, g_\varepsilon)$ respectively, for this is true when $\varepsilon=1$. Hence

$$R_\varepsilon(x,D)\,T_\varepsilon(x,D) = I + K_{1\varepsilon}(x,D)$$

where $K_{1\varepsilon}$ is uniformly bounded in $S(\varepsilon^2/R_\varepsilon^2, g_\varepsilon)$. Thus the operator norm in L^2 is $O(\varepsilon^2)$, so $I+K_{1\varepsilon}(x,D)$ is invertible in L^2 for small ε. Similarly

$$T_\varepsilon(x,D)\,R_\varepsilon(x,D) = I + K_{2\varepsilon}(x,D)$$

where $K_{2\varepsilon}$ is bounded in $S(\varepsilon^2/R_\varepsilon^2, g_\varepsilon)$. Hence the L^2 operator norm of $K_{2\varepsilon}, [x_j, K_{2\varepsilon}]$ and $[D_j, K_{2\varepsilon}]$ are $O(\varepsilon^2)$, so the operator norm of $K_{2\varepsilon}$ in B is $O(\varepsilon^2)$. Thus $I+K_{2\varepsilon}$ is invertible in B for small ε, so $T_\varepsilon(x,D)$ is an isomorphism $L^2 \to B$ and $R_\varepsilon(x,D)$ is an isomorphism $B \to L^2$. We have

$$T_\varepsilon(x,D)^{-1} f = R_\varepsilon(x,D)(I + K_{2\varepsilon}(x,D))^{-1} f, \quad f \in B,$$

and $(I + K_{2\varepsilon}(x,D))^{-1} f \to f$ in B as $\varepsilon \to 0$. Finally, when $g \in \mathscr{S}$ then

$$\begin{aligned}\|R_\varepsilon(x,D) g\|^2 &= (R_\varepsilon(x,D)^* R_\varepsilon(x,D) g, g)\\ &= \|g\|^2 + \varepsilon^2(\sum \|x_j g\|^2 + \|D_j g\|^2) + (K_{3\varepsilon}(x,D) g, g)\end{aligned}$$

where $K_{3\varepsilon}$ is uniformly bounded in $S(\varepsilon^2, g_\varepsilon)$ so the norm is $O(\varepsilon^2)$. Hence $R_\varepsilon(x,D)$ has uniformly bounded norm as operator from B to L^2. If $g \in \mathscr{S}$ then $R_\varepsilon(x,D) g \to g$ in $\mathscr{S}$ as $\varepsilon \to 0$ since $R_\varepsilon \to 1$ in C^∞ and R_ε is uniformly bounded in $S(R_1, g_1)$. Thus

$$T_\varepsilon(x,D)^{-1} f = R_\varepsilon(x,D) f + R_\varepsilon(x,D)((I + K_{2\varepsilon}(x,D))^{-1} f - f) \to f$$

in L^2 as $\varepsilon \to 0$, which completes the proof.

Fix ε now so small that $T_\varepsilon(x,D)$ is an isomorphism and

$$(T_\varepsilon(x,D)^{-1} e^{-|x|^2/2}, e^{-|x|^2/2}) > 0.$$

Then $p_\varepsilon(x,D) = p(x,D)\, T_\varepsilon(x,D)$ is a surjective Fredholm operator in $\mathscr{L}(L^2(\mathbb{R}^n, \Lambda^e), L^2(\mathbb{R}^n, \Lambda^o))$ with index 1 which is injective in the orthogonal plane of $e^{-|x|^2/2}$. Clearly

$$p_\varepsilon(x,\xi) = p(x,\xi)\, T_\varepsilon(x,\xi) - i \sum \partial p(x,\xi)/\partial \xi_j \, \partial T_\varepsilon(x,\xi)/\partial x_j.$$

To deform p_ε to a pseudo-differential operator which is trivial at infinity we choose $\phi \in C_0^\infty(\mathbb{R}^n)$ and $\psi \in C^\infty(\mathbb{R}^n)$ as decreasing functions of the radius so that

$$\begin{aligned}&\phi(x) = \psi(x) = 1 \quad \text{when } |x| < 1;\\ &\phi(x) = 0, \quad \psi(x) = 1/|x| \quad \text{when } |x| > 2.\end{aligned}$$

Lemma 19.2.11. *If $a \in S(1,g)$ then*

$$a_\delta(x,\xi) = a(x, \phi(\delta x)\, \psi(\delta \xi)\, \xi) \tag{19.2.6}$$

is uniformly bounded in $S(1,G)$ for $0 \leqq \delta \leqq 1$ if

$$G = |dx|^2/(1+|x|^2) + |d\xi|^2/(1+|\xi|^2).$$

We have $a_\delta(x,\xi) = a(x,0)$ if $|\delta x| > 2$, and $a_\delta(x,\xi) = a(x, \phi(\delta x)\, \xi/|\delta \xi|)$ if $|\delta \xi| > 2$.

Proof. The statement means that

$$|D_\xi^\alpha D_x^\beta a_\delta(x,\xi)| \leqq C_{\alpha\beta}(1+|\xi|)^{-|\alpha|}(1+|x|)^{-|\beta|}.$$

In view of the homogeneity in ξ when $|\delta \xi| \geqq 2$ we may assume that $|\delta \xi| \leqq 2$ in the proof. First we assume that $|\delta x| \leqq 1$. Then $a_\delta(x,\xi) = a(x, \psi(\delta \xi)\, \xi)$, and

$$|D_\xi^\alpha \psi(\delta \xi)\, \xi| \leqq C_\alpha \delta^{|\alpha|-1}(1+|\delta \xi|)^{-|\alpha|} \leqq C_\alpha (1+|\xi|)^{1-|\alpha|}, \quad \alpha \neq 0;$$

it suffices to prove the first estimate when $\delta=1$ and then it follows from the fact that $\psi(\xi)\,\xi$ is smooth and homogeneous of degree 0 for large $|\xi|$. Since $1+|x|+\psi(\delta\xi)\,|\xi|\geqq 1+|x|+c|\xi|$ the terms in $D_\xi^\alpha D_x^\beta a_\delta(x,\xi)$ where a has been differentiated $j+|\beta|$ times can be estimated by a constant times

$$(1+|x|+|\xi|)^{-j-|\beta|}(1+|\xi|)^{j-|\alpha|}\leqq C(1+|x|)^{-|\beta|}(1+|\xi|)^{-|\alpha|}$$

as claimed. When $|\delta x|>2$ we have $a_\delta(x,\xi)=a(x,0)$. The assertion is obvious then so we now assume that $1\leqq|\delta x|\leqq 2$. We can then estimate a derivative of a of order j by $C\delta^j$, and

$$|D_x^\beta D_\xi^\alpha \phi(\delta x)\,\psi(\delta\xi)\,\delta\xi|\leqq C_{\alpha\beta}(1+|x|)^{-|\beta|}(1+|\xi|)^{-|\alpha|}$$

since this is true when $\delta=1$ and $\delta/(1+|\delta x|)\leqq 1/(1+|x|)$. The required estimate is an immediate consequence.

We have now developed all the tools required to construct the operator to be used in the generalized product formula:

Theorem 19.2.12. *There exists a symbol $B\in S^0_{\mathrm{phg}}(\mathbb{R}^n\times\mathbb{R}^n)$ with values in $\mathscr{L}(\Lambda^e(\mathbb{R}^n),\Lambda^o(\mathbb{R}^n))$ such that*

(i) *$B(x,\xi)=\Lambda(x/|x|)+\Lambda(x/|x|)^*$ if $|x|$ is sufficiently large.*

(ii) *The kernel of $B(x,D)$ minus $B(x,0)\,\delta(x-y)$ has compact support.*

(iii) *$B(x,D)$ maps $L^2(\mathbb{R}^n,\Lambda^e(\mathbb{R}^n))$ onto $L^2(\mathbb{R}^n,\Lambda^o(\mathbb{R}^n))$ with one dimensional kernel $\subset C_0^\infty(\mathbb{R}^n,\Lambda^e(\mathbb{R}^n))$ consisting of rotationally symmetric scalar functions.*

(iv) *$O^*B(x,D)\,u=B(x,D)\,O^*u$ if O is an orthogonal transformation in $\mathbb{R}^n$ and $u\in L^2(\mathbb{R}^n,\Lambda^e(\mathbb{R}^n))$.*

Proof. Choose ε so small that the conclusions of Lemma 19.2.10 are valid and the kernel of $p_\varepsilon(x,D)$ is not orthogonal to $e^{-|x|^2/2}$. Next we apply Lemma 19.2.11 to $a=p_\varepsilon$ and to $b=p_\varepsilon^{-1}$ outside a compact subset $M=\{(x,\xi);\,|x|\leqq C,|\xi|\leqq C\}$ of $\mathbb{R}^{2n}$. Then $a_\delta(x,\xi)\,b_\delta(x,\xi)$ is the identity if $(x,\xi)\notin M$ and $2\delta C\leqq 1$. In fact, if $(x,\phi(\delta x)\,\psi(\delta\xi)\,\xi)\in M$ then $|\delta x|\leqq\delta C\leqq 1$, hence $\psi(\delta\xi)\,|\delta\xi|\leqq\delta C\leqq\frac12$ so $|\delta\xi|\leqq 1$ and $|\xi|\leqq C$. Now

$$b_\delta(x,D)\,a_\delta(x,D)=I+K_{1\delta}(x,D),$$
$$a_\delta(x,D)\,b_\delta(x,D)=I+K_{2\delta}(x,D)$$

where $K_{j\delta}$ are bounded in $S((1+|x|)^{-1}(1+|\xi|)^{-1},G)$ for small δ if G is defined as in Lemma 19.2.11. Hence it follows from Theorem 19.1.10 that $\operatorname{ind} a_\delta(x,D)=1$ and that $\dim\operatorname{Ker} a_\delta(x,D)=1$ if δ is small enough. If $u_\delta\in\operatorname{Ker} a_\delta(x,D)$ and $\|u_\delta\|=1$ then $u_\delta=-K_{1\delta}u_\delta$ belongs to a fixed compact subset of L^2. Any limit point u_0 when $\delta\to 0$ is in the kernel of $p_\varepsilon(x,D)$ and has norm 1, and is therefore not orthogonal to $e^{-|x|^2/2}$. If dO is the Haar measure in the orthogonal group and

$$U_\delta=\int O^*u_\delta\,dO$$

then U_δ is also in the kernel of $a_\delta(x,D)$ and has the same scalar product with $e^{-|x|^2/2}$ as u_δ, so $U_\delta \neq 0$ for small δ. Thus we have a generator for the kernel which is invariant under the orthogonal group, which means that it is a rotationally symmetric function.

The kernel K of $a_\delta(x,D)$ is $a_\delta(x,0)\,\delta(x-y)$ if $|\delta x|>2$. Now

$$K(x,y)=(2\pi)^{-n}\int e^{i\langle x-y,\xi\rangle}\,a_\delta(x,\xi)\,d\xi$$

is rapidly decreasing when $|\delta x|\leq 2$ and $y\to\infty$ (see the proof of Theorem 7.1.22). It follows that

$$(K(x,y)-a_\delta(x,0)\,\delta(x-y))(1-\phi(\gamma y))\to 0$$

in $\mathscr{S}$ as $\gamma\to 0$, if ϕ is defined as in Lemma 19.2.11. Hence the operator

$$B_0(x,D)=a_\delta(x,D)-(a_\delta(x,D)-a_\delta(x,0))(1-\phi(\gamma x))$$

which clearly has the property (ii) will also satisfy (iii) if γ is small enough. When $|\delta x|>2$ we have

$$\begin{aligned} B_0(x,\xi)&=p(x,0)\,T_\varepsilon(x,0)-i\sum \partial p(x,0)/\partial\xi_j\,\partial T_\varepsilon(x,0)/\partial x_j\\ &=p(x,0)/|\varepsilon x|+O(1/|x|)\quad \text{as } x\to\infty. \end{aligned}$$

If θ is a sufficiently small positive number then

$$\begin{aligned} B(x,\xi)-\phi(\theta x)\,(1+\varepsilon^2(|x|^2+\phi(\delta x)^2/\delta^2))^{\frac{1}{2}}\,(|x|^2+\phi(\delta x)^2)^{-\frac{1}{2}}\,B_0(x,\xi)\\ +(1-\phi(\theta x))\,p(x,0)/|x| \end{aligned}$$

will remain a function of x with invertible values if $|\delta x|>2$, so (i) is fulfilled and B inherits properties (ii), (iii) from B_0. Rotational symmetry has been respected in the entire construction so (iv) is valid, which completes the proof. Note that the factor of B_0 above was chosen so that the principal symbol is

(i)′ $$p(x+i\phi(\delta x)\,\xi/|\delta\xi|)\,(|x|^2+\phi(\delta x)^2)^{-\frac{1}{2}}.$$

The operator $B(x,D)$ is convenient to use analytically, and the symbol is homotopic to $p(x+i\xi)$ outside a compact set. We shall call B the *Bott operator* since the symbol is a generator in the Bott periodicity theorem which will not be used here though. For technical reasons it is preferable to regard B as an operator in $S^n=\mathbb{R}^n\cup\{\infty\}$, with the local coordinates $x_j/|x|^2$ at infinity. B maps sections of the trivial bundle $E_B=S^n\times\Lambda^e(\mathbb{C}^n)$ to sections of the bundle F_B which is equal to $\mathbb{R}^n\times\Lambda^o(\mathbb{C}^n)$ over $\mathbb{R}^n$ and to $(S^n\smallsetminus\{0\})\times\Lambda^e(\mathbb{C}^n)$ over $S^n\smallsetminus\{0\}$, with $(x,p(x/|x|)\,w)$ and (x,w) identified if $x\in\mathbb{R}^n\smallsetminus\{0\}$ and $w\in\Lambda^e(\mathbb{C}^n)$. With the first representation the definition of Bu is unchanged if $u\in\mathscr{E}'(\mathbb{R}^n)$, and with the second one the operator B acts as the identity if $\operatorname{supp} u$ is sufficiently close to ∞. The compatibility of these definitions follows from condition (ii) in Theorem 19.2.12.

We are now prepared for the proof of an index formula for the fiber products which occur in connection with an embedding. The statement is long but one should note the analogy with Theorem 19.2.7.

Theorem 19.2.13. *Let Y be a compact C^∞ manifold, E_Y and F_Y two complex C^∞ hermitian vector bundles over Y, and let*

$$q \in C(T^*(Y), \operatorname{Hom}(\pi_Y^* E_Y, \pi_Y^* F_Y))$$

be an isomorphism outside a compact set. Let V be a real vector bundle of fiber dimension n over Y, endowed with a Euclidean structure. Let $\tilde{V}_y$ be the sphere obtained by adding a point at infinity to V_y, and let $\tilde{V}$ be the sphere bundle over Y with these fibers. Thus $\tilde{V} = V \cup (Y \times \{\infty\})$ with the C^∞ structure at infinity defined so that the inversion $V_y \setminus \{0\} \ni v \mapsto v/|v|^2$ extends to a diffeomorphism $\tilde{V} \setminus V^0 \to V$ if $V^0 = Y \times \{0\}$ is the zero section of V. Denote by $V_{\mathbb{C}}$ the complexification of the vector bundle V, and let $\tilde{E}_B$ be the lifting from Y to $\tilde{V}$ of the vector bundle $\Lambda^e(V_{\mathbb{C}})$. Let $\tilde{F}_B$ be the vector bundle on $\tilde{V}$ obtained from the lifting of $\Lambda^o(V_{\mathbb{C}})$ to V and the lifting of $\Lambda^e(V_{\mathbb{C}})$ to $\tilde{V} \setminus V^0$ by identification of $(v, p(v/|v|)\, w)$ and (v, w) if $v \in V_y \setminus \{0\}$ and $w \in \Lambda^e(V_{y\mathbb{C}})$. Here $p(v) = \Lambda(v) + \Lambda(v)^*$ as above. Let $\tilde{E}_Y$, $\tilde{F}_Y$ be the bundles E_Y, F_Y lifted to $\tilde{V}$, and choose

$$\tilde{q} \in C(T^*(\tilde{V}), \operatorname{Hom}(\pi_{\tilde{V}}^* \tilde{E}_Y, \pi_{\tilde{V}}^* \tilde{F}_Y))$$

so that with π denoting the projection $\tilde{V} \to Y$

$$\tilde{q}(t, \pi^* \eta) = q(\pi t, \eta) \quad \text{if } t \in \tilde{V} \text{ and } \eta \in T_{\pi t}^*(Y). \tag{19.2.7}$$

Now introduce for $(t, \tau) \in T^*(\tilde{V})$, ξ denoting the restriction of τ to the fiber,

$$d(t,\tau) = \begin{pmatrix} \psi(x)\, p(x + i\xi\phi(x)) \otimes I & -I \otimes \tilde{q}(t,\tau)^* \\ I \otimes \tilde{q}(t,\tau) & \psi(x)\, p(x + i\xi\phi(x))^* \otimes I \end{pmatrix}$$

where x is t regarded as an element of $V_{\pi t}$ and $\psi(x)$, $\phi(x)$ are decreasing continuous functions of $|x|$ which are equal to 1 at 0 and equal to $1/|x|$ resp. 0 in a neighborhood of ∞. Then

$$d \in C(T^*(\tilde{V}), \operatorname{Hom}(\pi_{\tilde{V}}^* \tilde{E}_B \otimes \tilde{E}_Y \oplus \tilde{F}_B \otimes \tilde{F}_Y, \pi_{\tilde{V}}^* \tilde{F}_B \otimes \tilde{E}_Y \oplus \tilde{E}_B \otimes \tilde{F}_Y))$$

is an isomorphism outside a compact set, and

$$\text{s-ind}\, d = \text{s-ind}\, q. \tag{19.2.8}$$

Proof. First we recall that (19.1.12) is an isomorphism whenever A_1 or A_2 is an isomorphism. Now $\psi(x)\, p(x + i\xi\phi(x))$ is an isomorphism unless $x = \xi = 0$, thus $t \in V^0$ and τ is orthogonal to the fiber $V_{\pi t}$. But that means that $\tau = \pi^* \eta$ for some $\eta \in T_{\pi t}^*(Y)$, so it follows in view of (19.2.7) that d is an isomorphism outside a compact set. Thus s-ind d is defined and is independent of the choice of $\tilde{q}$ satisfying (19.2.7) and also of the choice of ϕ and of ψ. When proving (19.2.8) we can also assume that q is homogeneous of degree 1 and C^∞ outside the zero section of $T^*(Y)$. Choose $Q \in \Psi_{\text{phg}}^1(Y; E_Y \otimes \Omega^{\frac{1}{2}}, F_Y \otimes \Omega^{\frac{1}{2}})$ with principal symbol q. To change the Bott operator B to a first order operator we choose a pseudo-differential operator T in S^n such that the principal symbol at a cotangent vector (x, ξ) is its length with respect to the standard arc length in S^n. (This is $|dx|^2/(1+|x|^2)^2$ in $\mathbb{R}^n$ identified with

$S^n \setminus \{\infty\}$ so the principal symbol of T restricted to $T^*(\mathbb{R}^n)$ is $|\xi|(1+|x|^2)$.) Adding a large positive constant to T we can make T invertible, hence of index 0. The operator T also acts on sections of the trivial bundle E_B on S^n, and $B_1 = BT$ is now a first order elliptic pseudo-differential operator with kernel generated by a rotationally symmetric scalar function if T is chosen invariant under the orthogonal group. It is clear that B_1 also acts on sections of $E_B \otimes E_Y \otimes \Omega_Y^{\frac{1}{2}}$ for they can locally be considered as forms in $\mathbb{R}^n$ depending on the parameter y and with values in $E_y \otimes \Omega_y^{\frac{1}{2}}$, and B_1 is invariant under the orthogonal group. (In the spherical fibers we identify functions with densities.)

As in the proof of Theorem 19.2.7 we take a partition of unity $1 = \sum \psi_k^4$ subordinate to a covering of Y with coordinate patches Y_k where E_Y and F_Y are trivial and V with its Euclidean structure is equivalent to $Y_k \times \mathbb{R}^n$. In terms of these trivializations $\psi_k Q \psi_k$ defines an operator from sections of $\tilde{E}_B \otimes \tilde{E}_Y \otimes \Omega_Y^{\frac{1}{2}}$ to sections of $\tilde{E}_B \otimes \tilde{F}_Y \otimes \Omega_Y^{\frac{1}{2}}$ which can be approximated by means of Lemma 19.2.6. After multiplication left and right by ψ_k we can pull the operator back to the manifold. Summation over k gives an operator $\tilde{Q} = \sum \tilde{Q}_k$ from sections of $\tilde{E}_B \otimes \tilde{E}_Y \otimes \Omega_Y^{\frac{1}{2}}$ to sections of $\tilde{E}_B \otimes \tilde{F}_Y \otimes \Omega_Y^{\frac{1}{2}}$ such that

$$\tilde{Q}(1 \otimes \pi^* u) = 1 \otimes \pi^* Q_0 u, \tag{19.2.9}$$

if u is a section of $E_Y \otimes \Omega_Y^{\frac{1}{2}}$ on Y. Here $Q_0 = \sum \psi_k^2 Q \psi_k^2$ is another pseudo-differential operator in Y with principal symbol q. When we apply Lemma 19.2.6 to $\psi_k Q \psi_k$ we obtain pseudo-differential operators converging to $\tilde{Q}_k$ with principal symbols converging to $\psi_k^4 q(\pi\iota, \eta)$ at $(\iota, \pi^*\eta)$. (It is extended from there so that it is constant in the directions conormal to the plane $x = \text{constant}$ defined in terms of the local trivializations, but this information does not survive when we form $\tilde{Q}$.) Thus the principal symbols of the operators approximating $\tilde{Q}$ converge to some $\tilde{q}$ satisfying (19.2.7). In the same way we can also lift Q to an operator $\tilde{Q}_1$ from sections of $\tilde{F}_B \otimes \tilde{E}_Y \otimes \Omega_Y^{\frac{1}{2}}$ to sections of $\tilde{F}_B \otimes \tilde{F}_Y \otimes \Omega_Y^{\frac{1}{2}}$, and we have the intertwining properties

$$\tilde{Q}_1 B_1 = B_1 \tilde{Q}, \quad B_1^* \tilde{Q}_1 = \tilde{Q} B_1^*, \tag{19.2.10}$$

since they are true for the local constructions.

Now consider the operator

$$D = \begin{pmatrix} B_1 & -\tilde{Q}_1^* \\ \tilde{Q} & B_1^* \end{pmatrix}$$

from sections of $\tilde{E}_B \otimes \tilde{E}_Y \otimes \Omega_Y^{\frac{1}{2}} \oplus \tilde{F}_B \otimes \tilde{F}_Y \otimes \Omega_Y^{\frac{1}{2}}$ to sections of $\tilde{F}_B \otimes \tilde{E}_Y \otimes \Omega_Y^{\frac{1}{2}} \oplus \tilde{E}_B \otimes \tilde{F}_Y \otimes \Omega_Y^{\frac{1}{2}}$. Just as in the proof of Theorem 19.2.7 we can now apply Theorem 19.2.5 and conclude that D is a Fredholm operator whose index can be computed on smooth sections, for the limiting principal symbol

$$\begin{pmatrix} b_1 & -\tilde{q}_1^* \\ \tilde{q} & b_1 \end{pmatrix} \tag{19.2.11}$$

is an isomorphism outside the zero section of $T^*(\tilde{V})$. In fact, the principal symbol b_1 of B_1 is an isomorphism at (t,τ) except when τ is conormal to $\tilde{V}_{\pi t}$, and then it follows from (19.2.7) that $\tilde{q}(t,\tau)$ is an isomorphism, when $\tau \neq 0$. From the intertwining property (19.2.10) it follows that

$$D^* D = \begin{pmatrix} B_1^* B_1 + \tilde{Q}^* \tilde{Q} & 0 \\ 0 & B_1 B_1^* + \tilde{Q}_1 \tilde{Q}_1^* \end{pmatrix}.$$

Here $B_1 B_1^*$ is injective, and the elements in the kernel of $B_1^* B_1$ can be written in the form $u_0(|x|) \otimes v(\pi y)$ where u_0 is a scalar function of the Euclidean metric and v is a section of $E_Y \otimes \Omega_Y^{\frac{1}{2}}$. But the proof of (19.2.9) gives

$$\tilde{Q}(u_0 \otimes \pi^* v) = u_0 \otimes \pi^* Q_0 v,$$

so the kernel of D is isomorphic to that of Q_0. Similarly,

$$DD^* = \begin{pmatrix} B_1 B_1^* + \tilde{Q}_1^* \tilde{Q}_1 & 0 \\ 0 & B_1^* B_1 + \tilde{Q} \tilde{Q}^* \end{pmatrix}.$$

Here $B_1 B_1^*$ is injective so we see as above that the kernel consists of all $u_0 \otimes \pi^* v$, where v is a C^∞ section of $F_Y \otimes \Omega_Y^{\frac{1}{2}}$, annihilated by $\tilde{Q}^*$. Since $\tilde{Q}$ preserves the degree of forms along the fibers of $\tilde{V}$ this means that $u_0 \otimes \pi^* v$ must be orthogonal to $\tilde{Q} w$ for every section w of $\tilde{F}_Y \otimes \Omega_Y^{\frac{1}{2}}$ regarded as a subbundle of $\tilde{E}_B \otimes \tilde{F}_Y \otimes \Omega_Y^{\frac{1}{2}}$. If $\int$ denotes integration along the fibers then

$$\int u_0 \tilde{Q} w = Q_0 \int u_0 w$$

by the definition of $\tilde{Q}$. (Cf. (19.2.9).) Hence the condition on v is that v is orthogonal to the range of Q_0, that is, $Q_0^* v = 0$, so we have

$$\operatorname{ind} D = \dim \operatorname{Ker} Q_0 - \dim \operatorname{Coker} Q_0 = \operatorname{ind} Q_0 = \text{s-ind}\, q.$$

It remains to establish a homotopy between the symbol (19.2.11) of D and $d(t,\tau)$. The construction of B_1 gives

$$b_1 = |\xi|_x\, p(x + i\phi(\delta x)\, \xi/|\delta \xi|)/(|x|^2 + \phi(\delta x)^2)^{\frac{1}{2}},$$

wjere δ is obtained from Lemma 19.2.11 and $|\xi|_x = (1+|x|^2)|\xi|$ is the length of ξ with respect to the spherical Riemannian metric. By the homogeneity of p this is equal to

$$p(x|\xi|_x + i\phi(\delta x)(1+|x|^2)\, \xi/\delta)/(|x|^2 + \phi(\delta x)^2)^{\frac{1}{2}}.$$

In d we have at the corresponding position $\psi(x)\, p(x + i\phi(\delta x)\, \xi)$. For $0 \leqq \lambda \leqq 1$ a connecting homotopy of isomorphisms for $\xi \neq 0$ is given by

$$p(((1-\lambda)|\xi|_x + \lambda)\, x + i\phi(\delta x)\, \xi((1+|x|^2)/\delta)^{1-\lambda})(\lambda \psi(x) + (1-\lambda)/(|x|^2 + \phi(\delta x)^2)^{\frac{1}{2}}).$$

This completes the proof.

One can also make d quite trivial in a neighborhood of infinity by considering for $0\leqq\delta\leqq 1$ the homotopy

$$d_{1-\delta}(t,\tau)=\begin{pmatrix}\psi(x)\,p(x+i\xi\,\phi(x))\otimes I & -\phi(\delta x)\,I\otimes\tilde{q}(t,\tau)^*\\ \phi(\delta x)\,I\otimes\tilde{q}(t,\tau) & \psi(x)\,p(x+i\xi\,\phi(x))^*\otimes I\end{pmatrix}.$$

This is still an isomorphism when $x\neq 0$ or $\xi\neq 0$, and when $x=0$ we have changed nothing. Thus s-ind d = s-ind d_1 = s-ind d_0. Note that in the neighborhood of ∞ in $\tilde{V}$ where $\phi(x)=0$ we have

$$d_0(t,\tau)=\begin{pmatrix}p(x/|x|)\otimes I & 0\\ 0 & p(x/|x|)^*\otimes I\end{pmatrix}.$$

As in Theorem 19.2.3 we can choose a pseudo-differential operator of order 0 with a homotopic principal symbol such that the kernel of the operator differs from the kernel of a multiplication operator by a kernel of compact support in $V\times V$. This means that the kernel and cokernel of the operator can be computed on sections with compact support in V, so the extension by the section at infinity is no longer relevant. To simplify the operator further we shall trivialize the bundles:

Lemma 19.2.14. *If E is a C^∞ complex vector bundle over a compact C^∞ manifold X, then one can find another such bundle G such that $E\oplus G$ is isomorphic to $X\times\mathbb{C}^N$ for some N.*

Proof. This is very close to the proof of Lemma 19.2.8. Choose a covering of X by coordinate patches X_j, $j=1,\dots,J$, such that for every j there is a trivialization $\psi_j\colon E_{|X_j}\to X_j\times\mathbb{C}^\nu$. If $\phi_j\in C_0^\infty(X_j)$ and $\sum\phi_j=1$ then

$$E\ni e\mapsto(\pi e,\psi_1\phi_1 e,\dots,\psi_J\phi_J e)\in X\times\mathbb{C}^{\nu J},$$

where π is the projection $E\to X$, is an embedding of E as a subbundle of $X\times\mathbb{C}^N$, $N=J\nu$. Now we just have to take for G the bundle for which G_x is the orthogonal space of the image of E_x in $\mathbb{C}^N$. The lemma is proved.

If we apply the lemma with X replaced by $\tilde{V}$ and E replaced by $\tilde{E}_B\otimes\tilde{E}_Y\oplus\tilde{F}_B\otimes\tilde{F}_Y$ it follows that the direct sum of the identity in $C^\infty(\tilde{V},G)$ and the operator constructed just before Lemma 19.2.14 is an elliptic operator D_G on sections of the trivial bundle $\tilde{V}\times\mathbb{C}^N$ which is just multiplication by a vector bundle isomorphism in a neighborhood of ∞.

Suppose now that V is the normal bundle of an embedding of Y in $\mathbb{R}^\nu$ for some ν. As observed after Lemma 19.2.8 we can then identify V with a tubular neighborhood U_ρ of Y, and D_G defines an elliptic pseudo-differential operator on U_ρ with symbol d_G which is just a bundle isomorphism on $U_\rho\setminus\overline{U}_r$ for some $r<\rho$. For a fixed $\Xi\neq 0$ in $\mathbb{R}^\nu$ the symbol $d_G(X,\Xi)$ gives an isomorphism of $\mathbb{C}^N$ and the fiber of the range bundle at X, so this is also a trivial bundle. Thus D_G is a $N\times N$ matrix of pseudo-differential operators reducing to multiplication by an invertible matrix function in $U_\rho \setminus \overline{U}_r$ for some $r<\rho$. To extend it to all of $\mathbb{R}^\nu$ we need the following lemma:

Lemma 19.2.15. *Let $U \subset \mathbb{R}^\nu$ be open and bounded, let $K \subset U$ be compact and $a \in C^\infty(U \setminus K, GL(N,\mathbb{C}))$. Then one can find*

$$A_0 \in C^\infty(U, GL(N,\mathbb{C})), \quad A_\infty \in C^\infty(\mathbb{R}^\nu \setminus K, GL(N,\mathbb{C}))$$

such that $A_\infty = aA_0$ in $U \setminus K$ and A_∞ is homogeneous of degree 0 in a neighborhood of ∞.

Proof. Choose $\phi \in C_0^\infty(U)$ with $0 \leqq \phi \leqq 1$ and $\phi=1$ in a neighborhood of K. Then

$$E=\{(x,z_1,z_2)\in\mathbb{R}^\nu\times(\mathbb{C}^N\oplus\mathbb{C}^N);\, a(x)(1-\phi(x))\,z_1=\phi(x)\,z_2 \text{ if } x\in U,\ z_1=0 \text{ if } x\notin U\}$$

is a vector bundle over $\mathbb{R}^\nu$, hence trivial. The proof goes as follows. If $\frac{1}{2}<t<1$ then the orthogonal projection from the fiber $E_x\subset\mathbb{C}^{2N}$ to E_{tx} is the identity map for large x or $t=1$, hence an isomorphism for all x if $1-t$ is small. This defines an isomorphism $P(x)\colon E_x\to E_{tx}$ depending continuously on x. If we fix a large ball B containing $\operatorname{supp}\phi$ then

$$P(t^{k-1}x)\dots P(x)\colon E_x\to E_{t^k x}$$

trivializes E over B if k is so large that E is trivial over $t^k B$. Thus we can find C^∞ maps F and G from B to $N\times N$ matrices such that E restricted to B is

$$\{(x, F(x)\,w, G(x)\,w);\, x\in B,\, w\in\mathbb{C}^N\}.$$

Thus $a(x)(1-\phi(x))\,F(x)=\phi(x)\,G(x)$ if $0<\phi<1$. Set

$$A_0(x)=F(x)/\phi(x) \quad \text{when } \phi(x)>0;$$
$$A_\infty(x)=G(x)/(1-\phi(x)) \quad \text{when } \phi(x)<1.$$

Then $a(x)A_0(x)=A_\infty(x)$ when $0<\phi(x)<1$ so we can extend A_0 to U by defining $A_0(x)=a(x)^{-1}A_\infty(x)$ when $\phi(x)=0$ and extend A_∞ to $\complement K$ by defining $A_\infty(x)=a(x)A_0(x)$ when $\phi(x)=1$. Then $aA_0=A_\infty$ in $U\setminus K$. Since U is bounded we can take $B=\{x\in\mathbb{R}^\nu;\,|x|<2R\}$ with R so large that $|x|<R/2$ in U. With ψ as in Lemma 19.2.11 we may change the definition of $A_\infty(x)$ to $A_\infty(\psi(x/R)\,x)$ to achieve the desired homogeneity, and this completes the proof.

Now we just replace our operator D_G by $D_G A_0$ and obtain an operator acting on $\mathbb{C}^N$ valued functions which is just multiplication by A_∞ outside a compact subset of U_ρ. Thus it can be extended to all of $\mathbb{R}^\nu$. Summing up, starting from an elliptic symbol q in Y we have defined an elliptic $N\times N$ system of pseudodifferential operators in $\mathbb{R}^\nu$ which outside a compact subset of $\mathbb{R}^\nu$ is just multiplication by an invertible matrix $D(x)$, homogeneous of degree 0. For this operator acting on functions with compact support the index is equal to s-ind$q$. In Section 19.3 we shall determine the index of such operators explicitly. This will in principle yield a method for computing s-ind$q$. References to completely explicit formulas for s-ind$q$ will be given in the notes, but it would take us too far from the main path to develop these fully here.

19.3. The Index Theorem in $\mathbb{R}^n$

As in Section 19.2 we shall denote by g the σ temperate metric in $\mathbb{R}^{2n}$ defined by

$$g_{x,\xi}=h(x,\xi)(|dx|^2+|d\xi|^2), \quad h(x,\xi)=(1+|x|^2+|\xi|^2)^{-1}.$$

The symmetry in the x and ξ variables and the use of the Weyl calculus simplifies the proof of the following theorem:

Theorem 19.3.1. *If $a\in S(1,g)$ has values in $\mathscr{L}(\mathbb{C}^\nu,\mathbb{C}^\nu)$ such that a^{-1} exists and is bounded outside an open ball $B\subset\mathbb{R}^{2n}$ then $a^w(x,D)$ is a Fredholm operator in $L^2(\mathbb{R}^n,\mathbb{C}^\nu)$ with index*

$$\text{(19.3.1)}\quad \operatorname{ind} a^w(x,D)=-(-2\pi i)^{-n}(n-1)!/(2n-1)!\int_{\partial B}\operatorname{Tr}(a^{-1}da)^{2n-1},$$

if $\mathbb{R}^{2n}$ is oriented by $dx_1\wedge d\xi_1\wedge\ldots\wedge dx_n\wedge d\xi_n>0$.

Here $a^{-1}da$ is a one form with coefficients valued in $\nu\times\nu$ matrices, thus non-commuting. If $n=\nu=1$ then a is a scalar function and the right-hand side is

$$(2\pi i)^{-1}\int_{\partial B} da/a,$$

that is, the winding number of a considered as a map from ∂B to $\mathbb{C}\setminus\{0\}$. (See the example after Theorem 19.2.4.) To prove Theorem 19.3.1 we shall use the trace formula in Proposition 19.1.14. First recall that A is a Hilbert-Schmidt operator in $L^2(\mathbb{R}^n)$ if and only if the kernel $\mathscr{A}$ of A is square integrable, and that

$$\|A\|_2^2=\int|\mathscr{A}(x,y)|^2\,dx\,dy.$$

If $A=a^w(x,D)$ this means by Parseval's formula and (18.5.4) that $a\in L^2$ and that

$$\|a^w(x,D)\|_2^2=(2\pi)^{-n}\iint|a(x,\xi)|^2\,dx\,d\xi.$$

(This holds for $a(x,D)$ as well.) If $\mathscr{A}_1,\mathscr{A}_2\in L^2(\mathbb{R}^{2n})$ then the kernel $\mathscr{A}$ of the composition A of the corresponding operators is given by

$$\mathscr{A}(x,y)=\int\mathscr{A}_1(x,z)\,\mathscr{A}_2(z,y)\,dz,$$

and the trace of A is $\iint\mathscr{A}_1(x,z)\,\mathscr{A}_2(z,x)\,dx\,dz$. Now

$$\mathscr{A}(x,x+y)=\int\mathscr{A}_1(x,z)\,\mathscr{A}_2(z,x+y)\,dz$$

is by Fubini's theorem in L^1 as a function of x and is a continuous function of y with values in L^1 since

$$\int |\mathscr{A}_2(z,x+y)-\mathscr{A}_2(z,x+y')|^2\,dx\,dz$$

tends to 0 with $y-y'$. Hence the trace of A is $\int \mathscr{A}(x,x)\,dx$ if the kernel $\mathscr{A}$ of A is continuous. This is true if $A=a^w(x,D)$ is of trace class and the symbol is integrable, so then we have

$$\operatorname{Tr}(A)=(2\pi)^{-n}\iint a(x,\xi)\,dx\,d\xi.$$

If the values of a are $\nu\times\nu$ matrices then A is an operator in $L^2(\mathbb{R}^n,\mathbb{C}^\nu)$ and the formula remains valid with a replaced by the trace of a as a matrix in the right-hand side. We leave the verification as an exercise and pass to a very elementary sufficient condition for an operator to be of trace class.

Lemma 19.3.2. *$a^w(x,D)$ is of trace class in $L^2(\mathbb{R}^n)$ and*

$$\|a^w(x,D)\|_1 \leqq C \sum_{|\alpha|+\dots+|\beta'|\leqq n+1} \|x^\alpha\,\xi^\beta\,D_x^{\alpha'}\,D_\xi^{\beta'}\,a\|_{L^2} \tag{19.3.2}$$

if the right-hand side is finite.

Proof. From the operator calculus we obtain

$$h^w(x,D)\,(1+|x|^2+|D|^2)+c^w(x,D)=1$$

where $c=\Delta h/4\in S(h^2,g)$ is also a function of $|x|^2+|\xi|^2$. Hence $1+|x|^2+|D|^2$ commutes with h^w and c^w since the Poisson bracket of functions of $|x|^2+|\xi|^2$ is 0. Raising the identity to the power k where $2k\geqq n+1$ we obtain

$$\sum_0^k b_j^w(x,D)\,(1+|x|^2+|D|^2)^j=1$$

where $b_j^w(x,D)$ is a sum of products of j factors $h^w(x,D)$ and $k-j$ factors $c^w(x,D)$ in some order, hence $b_j\in S(h^{2k-j},g)$. If we multiply to the right by $a^w(x,D)$ and multiply the largest possible number of factors x and D less than or equal to $n+1$ into $a^w(x,D)$ then we obtain with summation for $|\alpha'|+\dots+|\beta''|\leqq n+1$

$$a^w(x,D)=\sum b^w_{\alpha',\alpha'',\beta',\beta''}(x,D)\,a^w_{\alpha',\alpha'',\beta',\beta''}(x,D),$$

where

$$a_{\alpha',\alpha'',\beta',\beta''}(x,\xi)=x^{\alpha'}\,\xi^{\beta'}\,D_\xi^{\alpha''}\,D_x^{\beta''}\,a(x,\xi)$$

while $b_{\alpha',\alpha'',\beta',\beta''}$ is independent of a and is in $S(h^{(n+1)/2},g)$ since at most $2k-(n+1)$ factors x and D will be left as right factors of b_j, and $b_j\in S(h^k,g)$. Thus $b^w_{\alpha',\alpha'',\beta',\beta''}$ is a Hilbert-Schmidt operator which completes the proof of the lemma.

Proof of Theorem 19.3.1. Choose $\psi\in C^\infty(\mathbb{R}^{2n})$ so that $1-\psi$ has compact support and a is invertible in $\operatorname{supp}\psi$. Set $b=\psi a^{-1}$, defined as 0 outside supp ψ. Then $b \in S(1,g)$ and $ba = ab = \psi I$ where I is the identity matrix. Hence

$$b^w(x,D)\,a^w(x,D) = I - R_1^w(x,D), \qquad a^w(x,D)\,b^w(x,D) = I - R_2^w(x,D)$$

where I is now the identity in $H = L^2(\mathbb{R}^n, \mathbb{C}^\nu)$ and $R_j \in S(h,g)$. The symbol of $R_j^w(x,D)^N$ is therefore in $S(h^N,g)$ so it follows from Lemma 19.3.2 that $R_j^w(x,D)^N$ is of trace class if $N \geqq n+1$. Using Proposition 19.1.14 we conclude that $a^w(x,D)$ is a Fredholm operator in H and that

$$\text{(19.3.3)} \qquad \operatorname{ind} a_\varepsilon^w(x,D) = \operatorname{Tr}(R_1^w(x,D)^N - R_2^w(x,D)^N)$$

To derive an explicit formula like (19.3.1) from (19.3.3) we must place ourselves in a situation where the asymptotic expansions of the calculus of pseudo-differential operators give a precise result in the whole space. This can be done by considering for $0 < \varepsilon \leqq 1$ the symbols

$$a_\varepsilon(x,\xi) = a(\varepsilon x, \varepsilon \xi), \qquad b_\varepsilon(x,\xi) = b(\varepsilon x, \varepsilon \xi)$$

which are uniformly bounded in $S(1, g_\varepsilon)$ if

$$g_\varepsilon = h_\varepsilon(x,\xi)(|dx|^2 + |d\xi|^2), \qquad h_\varepsilon(x,\xi) = \varepsilon^2 h(\varepsilon x, \varepsilon \xi).$$

Note that $g_\varepsilon \leqq g$ since $h_\varepsilon \leqq h$. If as before we define

$$R_{\varepsilon 1}^w(x,D) = I - b_\varepsilon^w(x,D)\,a_\varepsilon^w(x,D), \qquad R_{\varepsilon 2}^w(x,D) = I - a_\varepsilon^w(x,D)\,b_\varepsilon^w(x,D)$$

then $R_{\varepsilon j} - (1-\psi_\varepsilon)I$ is uniformly bounded in $S(h,g)$ when $0 < \varepsilon \leqq 1$ if $\psi_\varepsilon(x,\xi) = \psi(\varepsilon x, \varepsilon \xi)$. When ε is bounded away from 0 then $1 - \psi_\varepsilon$ also has such a bound so it follows that the symbol of $R_{\varepsilon j}^w(x,D)^N$ is uniformly bounded in $S(h^N, g)$ and a continuous function of ε with values in $C^\infty(\mathbb{R}^{2n})$. Since

$$\text{(19.3.3)}' \qquad \operatorname{ind} a_\varepsilon^w(x,D) = \operatorname{Tr}(R_{\varepsilon 1}^w(x,D)^N - R_{\varepsilon 2}^w(x,D)^N)$$

and the right-hand side is a continuous function of ε by Lemma 19.3.2 while the left-hand side is an integer, it follows that $a_\varepsilon^w(x,D)$ and $a^w(x,D)$ have the same index for every $\varepsilon \in (0,1)$. (This is also an easy consequence of Theorem 19.1.10.)

For $0 < \varepsilon \leqq 1$ we now introduce

$$C_\varepsilon^w(x,D) = R_{\varepsilon 1}^w(x,D)^N - R_{\varepsilon 2}^w(x,D)^N.$$

If we compute the right-hand side by means of the asymptotic formulas in Theorem 18.5.4, breaking off where the error term becomes bounded in $S(h_\varepsilon^N, g_\varepsilon)$, then the trace of the error term is bounded by

$$C \iint h_\varepsilon^N \,dx\,d\xi = C\varepsilon^{2(N-n)} \iint h^N \,dx\,d\xi \to 0 \quad \text{as } \varepsilon \to 0.$$

(Recall that $N > n$.) If S_ε is the finite sum of terms kept in the asymptotic expansion of C_ε it follows that

$$\text{(19.3.4)} \qquad \operatorname{ind} a^w(x,D) = \lim_{\varepsilon \to 0} (2\pi)^{-n} \iint \operatorname{Tr}(S_\varepsilon(x,\xi))\,dx\,d\xi,$$

where Tr is now just the trace of a matrix. Every term in S_ε is a product where every other factor is a derivative (possibly of order 0) of a_ε and every other factor is a derivative of b_ε. If we introduce εx and $\varepsilon\xi$ as new integration variables in (19.3.4) then every term containing μ differentiations in all becomes a constant times $\varepsilon^{\mu-2n}$. Hence the contributions to the integral when $\mu<2n$ must add up to 0 and

$$\operatorname{ind} a^w(x,D)=(2\pi)^{-n}\int \operatorname{Tr}(S_\varepsilon^0(x,\xi))\,dx\,d\xi, \tag{19.3.4$'$}$$

where S_ε^0 consists of the terms in S_ε where exactly $2n$ differentiations occur. The right-hand side is independent of ε so we can take $\varepsilon=1$ and write S^0 instead of S_ε^0 in what follows.

If $T\in GL(2n,\mathbb{R})$, the group of invertible linear transformations in $\mathbb{R}^{2n}$, and $A_T=a\circ T$, then $A_T\in S(1,g)$ so the preceding discussion can be applied also to A_T. It follows that $\operatorname{ind} A_T^w(x,D)$ is a continuous function of T, hence constant in each component of $GL(2n,\mathbb{R})$. Thus

$$\operatorname{ind} A_T^w(x,D)=\operatorname{ind} a^w(x,D) \quad \text{if } \det T>0.$$

In particular we can apply this to diagonal matrices with positive diagonal elements,

$$T(x,\xi)=(\varepsilon_1 x_1,\dots,\varepsilon_n x_n,\delta_1\xi_1,\dots,\delta_n\xi_n).$$

If we replace a and b by the composition with T and then introduce $\varepsilon_j x_j$ and $\delta_j\xi_j$ as new integration variables, we find that the contribution to $\operatorname{ind} a^w(x,D)$ from a term in S^0 containing differentiations of order $\alpha_1,\dots,\alpha_n$, $\beta_1,\dots,\beta_n$ with respect to $x_1,\dots,x_n$, $\xi_1,\dots,\xi_n$ is multiplied by $\varepsilon_1^{\alpha_1-1}\dots\delta_n^{\beta_n-1}$. Hence

$$\operatorname{ind} a^w(x,D)=(2\pi)^{-n}\iint \operatorname{Tr}(S^1(x,\xi))\,dx\,d\xi, \tag{19.3.5}$$

if S^1 denotes the terms in S^0 containing precisely one derivative with respect to each variable $x_1,\dots,x_n,\xi_1,\dots,\xi_n$.

An immediate consequence of (19.3.5) is that the index of the operator **$A_T^w(x,D)$ above is equal to $\operatorname{ind} a^w(x,D)$ times the sign of $\det T$ for every** $T\in GL(2n,\mathbb{R})$. In fact, (19.3.5) proves this for arbitrary diagonal matrices, and the assertion has already been proved when $\det T>0$. In particular we can apply the result when T is the linear transformation corresponding to an arbitrary permutation of the x,ξ variables.

Returning to the integration variables $T(x,\xi)$ we conclude that if $S^1(t,x,\xi)$ is obtained from $S^1(x,\xi)$ by making the permutation t of the differentiations $\partial/\partial x_j$ and $\partial/\partial\xi_k$ then

$$\operatorname{sgn} t \operatorname{ind} a^w(x,D)=(2\pi)^{-n}\iint \operatorname{Tr}(S^1(t,x,\xi))\,dx\,d\xi.$$

Hence it follows that

$$\operatorname{ind} a^w(x,D)=(2\pi)^{-n}\iint \operatorname{Tr}(S^2(x,\xi))\,dx\,d\xi/(2n)! \tag{19.3.6}$$

where

$$S^2(x,\xi)=\sum_t \operatorname{sgn} t\, S^1(t,x,\xi).$$

The alternation which occurs in S^2 eliminates all terms containing a factor differentiated more than once, for only sgn t is changed if we exchange the corresponding differentiations. Thus S^2 is a sum of products of derivatives of a and b of order less than or equal to 1, a derivative with respect to each variable occurring exactly once. To compute S^2 we can now return to the calculus formulas discarding all terms containing a derivative of order greater than 1. In the expansion of the symbol of $b^w(x,D)\,a^w(x,D)$ it is therefore sufficient to keep the terms

$$b\,a-i\{b,a\}/2=\psi I-i\{b,a\}/2,$$

and in the expansion of the symbol of $a^w(x,D)\,b^w(x,D)$ we just keep

$$a\,b-i\{a,b\}/2=\psi I-i\{a,b\}/2.$$

Note that since matrices do not commute, the Poisson bracket

(19.3.7) $$\{b,a\}=\sum(\partial b/\partial\xi_j\,\partial a/\partial x_j-\partial b/\partial x_j\,\partial a/\partial\xi_j)$$

does not only differ in sign from $\{a,b\}$. In the symbol series for C_1 we shall never keep terms where $\{b,a\}$ or $\{a,b\}$ is differentiated. The terms we keep will therefore be products of factors $\{b,a\}$ or $\{a,b\}$ and terms from the expansion of the symbol of a power of $\psi^w(x,D)$, for all factors involving ψ can be moved together since ψ is a scalar. However, since $\{\psi,\psi\}=0$ there are no such terms where ψ is differentiated only once. It follows that in C_1 we need only consider the terms

$$((1-\psi)I+i\{b,a\}/2)^N-((1-\psi)I-i\{a,b\}/2)^N.$$

The terms where $2n$ derivatives occur are

$$\binom{N}{n}(1-\psi)^{N-n}(i/2)^n(\{b,a\}^n-\{a,b\}^n).$$

If we introduce the explicit definition (19.3.7) of the Poisson bracket we find that $\{b,a\}^n$ contains $2^n n!$ terms where precisely one derivative occurs with respect to each variable. They occur with the same sign as coefficients of $d\xi_1\wedge dx_1\wedge\ldots\wedge d\xi_n\wedge dx_n$ in $(db\wedge da)^n$. If we perform the permutations t, multiply by sgn t and sum then all terms are obtained once. This proves that

$$S^2\,d\xi_1\wedge dx_1\wedge\ldots\wedge d\xi_n\wedge dx_n=n!\,i^n\binom{N}{n}(1-\psi)^{N-n}((db\wedge da)^n-(da\wedge db)^n).$$

Hence we obtain

(19.3.8) $$\operatorname{ind} a^w(x,D)=(n!/(2n)!)(i/2\pi)^n\int 2(1-\psi)^{N-n}\binom{N}{n}\operatorname{Tr}(db\wedge da)^n$$

where the integration is made over $\mathbb{R}^{2n}$ with the orientation given by

(19.3.9) $$d\xi_1 \wedge dx_1 \wedge \ldots \wedge d\xi_n \wedge dx_n > 0.$$

Here we have used the fact that

$$\operatorname{Tr}(da \wedge db)^n = -\operatorname{Tr}(db \wedge da)^n,$$

since the trace of a product of matrices is invariant under circular permutation whereas moving an exterior one form from the right to the left through a $2n-1$ form will change the sign.

Now we introduce the definition $b = \psi a^{-1}$ and

$$db = (d\psi) a^{-1} - \psi a^{-1}(da) a^{-1}, \quad db \wedge da = (d\psi) \wedge a^{-1} da - \psi (a^{-1} da)^2.$$

When forming the nth power we can only use one factor $d\psi$ since $d\psi$ (anti-) commutes with the other factors and $d\psi \wedge d\psi = 0$. Moreover

(19.3.10) $$\operatorname{Tr}(a^{-1} da)^{2n} = 0,$$

for moving a factor $a^{-1} da$ from the extreme right to the extreme left introduces a factor -1 without changing this quantity. By the invariance of the trace under circular permutations of factors of even exterior degree, we conclude that

$$\operatorname{Tr}(db \wedge da)^n = n \psi^{n-1} d\psi (-1)^{n-1} \wedge \operatorname{Tr}(a^{-1} da)^{2n-1}.$$

The form $\operatorname{Tr}(a^{-1} da)^{2n-1}$ is closed by (19.3.10). If we set

$$F(t) = \int_0^t \binom{N}{n} (1-s)^{N-n} n s^{n-1} ds$$

we obtain $F(1) = \binom{N}{n} n B(N-n+1, n) = 1$ (see (3.4.9)), hence by Stokes' formula

(19.3.1)′ $$\operatorname{ind} a^w(x, D) = -(2\pi i)^{-n} (n-1)!/(2n-1)! \int_{\partial B} \operatorname{Tr}(a^{-1} da)^{2n-1}$$

where B is a ball so large that $a(x, \xi)$ is invertible in its exterior. We recall that $\mathbb{R}^{2n}$ has here been oriented by (19.3.9). Choosing the standard orientation used in (19.3.1) introduces a factor $(-1)^n$ so (19.3.1) and (19.3.1)′ are equivalent. The proof is complete.

The proof of (19.3.10) shows that the differential form $\operatorname{Tr}(A^{-1} dA)^{2n}$ on $GL(v, \mathbb{C})$ is equal to 0, hence that $\operatorname{Tr}(A^{-1} dA)^{2n-1}$ is closed. This implies that the integral in (19.3.1) only depends on the homotopy class of the map $\partial B \to GL(v, \mathbb{C})$ defined by a. We shall use this homotopy invariance together with the stability properties of the index to extend Theorem 19.3.1 to symbols in $S(1, G)$ where

(19.3.11) $$G = |dx|^2/(1+|x|^2) + |d\xi|^2/(1+|\xi|^2).$$

Lemma 19.3.3. *Let* $\psi \in C_0^\infty(\mathbb{R}^{2n})$ *be a decreasing function of* $r = (|x|^2 + |\xi|^2)^{\frac{1}{2}}$ *which is* 1 *when* $r \leqq 1$ *and* $1/r$ *when* $r \geqq 2$, *and set* $\psi_\varepsilon(x, \xi) = \psi(\varepsilon x, \varepsilon \xi)$. *If*

$a\in S(1,G)$ then

$$a_\varepsilon(x,\xi)=a(\psi_\varepsilon(x,\xi)\,x,\psi_\varepsilon(x,\xi)\,\xi)$$

is uniformly bounded in $S(1,G)$ when $0\leqq\varepsilon\leqq 1$ and belongs to $S(1,g)$ if $0<\varepsilon\leqq 1$.

Proof. When $r\leqq 1/\varepsilon$ we have $a_\varepsilon(x,\xi)=a(x,\xi)$ so we may assume in the proof that $r>1/\varepsilon$. Then we have

$$(19.3.12)\quad (1+|x|/\varepsilon r)^{-1}\,|D_\xi^\alpha D_x^\beta \psi_\varepsilon(x,\xi)\,x_j|\leqq C_{\alpha\beta}(1+|x|)^{-|\beta|}(1+|\xi|)^{-|\alpha|}$$

and an analogous estimate with x and ξ interchanged. To prove these estimates we observe that ψ_ε is uniformly bounded in $S(1/(1+\varepsilon r),g)$. This gives the estimate

$$C_{\alpha\beta}(1+|x|/\varepsilon r)^{-1}(1+\varepsilon r)^{-1}((1+r)^{-|\alpha+\beta|}|x|+(1+r)^{1-|\alpha+\beta|})$$
$$\leqq C_{\alpha\beta}((1+r)^{-|\alpha+\beta|}+(1+|x|)^{-1}(1+r)^{1-|\alpha+\beta|}).$$

The second term can be omitted if $\beta=0$ so this proves (19.3.12). The lemma follows at once from (19.3.12) and its analogue with x and ξ interchanged since $\psi_\varepsilon\geqq 1/2\,\varepsilon r$ when $\varepsilon r\geqq 1$. The proof is complete since a_ε is homogeneous of degree 0 when $\varepsilon r\geqq 1$.

Theorem 19.3.1'. *Theorem* 19.3.1 *remains valid for every $a\in S(1,G)$ with values in $\mathscr{L}(\mathbb{C}^\nu,\mathbb{C}^\nu)$ such that a^{-1} exists and is bounded outside a compact set.*

Proof. Define a_ε as in Lemma 19.3.3 and set $b_\varepsilon=\psi\,a_\varepsilon^{-1}$ with ψ chosen as in the proof of Theorem 19.3.1. Then the symbols of $a_\varepsilon^w(x,D)\,b_\varepsilon^w(x,D)-I$ and $b_\varepsilon^w(x,D)\,a_\varepsilon^w(x,D)-I$ are bounded in $S((1+|x|)^{-1}(1+|\xi|)^{-1},G)$, so it follows from Theorem 19.1.10 in view of Theorem 18.6.6 that $a_\varepsilon^w(x,D)$ is a Fredholm operator with index independent of ε. Since (19.3.1) is valid for a_ε when $\varepsilon>0$ and both sides are independent of ε when $\varepsilon\geqq 0$ it follows that (19.3.1) is also valid for $\varepsilon=0$, which proves the theorem.

Theorem 19.3.1' is applicable in particular to the pseudodifferential operators in $\mathbb{R}^n$ which we encountered in Section 19.2. This is the reason why we have restricted ourselves to such a special metric in Theorem 19.3.1'. A moment's reflection shows that the proof of Theorem 19.3.1 is directly applicable to much more general metrics. It seems plausible that Theorem 19.3.1 is valid for every σ temperate metric such that $\sup g_{x,\xi}/g_{x,\xi}^\sigma\to 0$ as $(x,\xi)\to\infty$ but no proof seems to have been given in such generality.

In Theorem 19.3.1 and Theorem 19.3.1' one can replace $a^w(x,D)$ by $a(x,D)$. In fact, we have $a(x,D)=a_1^w(x,D)$ where

$$a_1-a\in S((1+|x|)^{-1}(1+|\xi|)^{-1},G).$$

If B is sufficiently large it follows that $a_t=t\,a_1+(1-t)\,a$, $0\leqq t\leqq 1$, is a homotopy between the maps $\partial B\to GL(\nu,\mathbb{C})$ defined by a and by a_1. Thus the right-hand side of (19.3.1) does not change if a is replaced by a_1, which proves the claim. This completes in principle the determination of the index of elliptic operators.

19.4. The Lefschetz Formula

In geometry elliptic complexes rather than elliptic operators occur naturally. To define this notion we assume that we have a compact C^∞ manifold X, C^∞ complex vector bundles $E_0, E_1, \ldots, E_N$ on X and polyhomogeneous pseudo-differential operators of the same order m

(19.4.1)
$$0 \to C^\infty(X, E_0 \otimes \Omega^{\frac{1}{2}}) \xrightarrow{D_0} C^\infty(X, E_1 \otimes \Omega^{\frac{1}{2}}) \xrightarrow{D_1} \ldots \xrightarrow{D_{N-1}} C^\infty(X, E_N \otimes \Omega^{\frac{1}{2}}) \to 0$$

forming a complex, that is, the compositions $D_j D_{j-1}$ vanish for $j=1, \ldots, N-1$. The complex is called *elliptic* if for all $(x,\xi) \in T^*(X) \setminus 0$ the principal symbol sequence

(19.4.2)
$$0 \to E_{0x} \xrightarrow{d_0(x,\xi)} E_{1x} \to \ldots \xrightarrow{d_{N-1}(x,\xi)} E_{Nx} \to 0$$

is exact, that is, the range of each map is exactly the kernel of the next. The basic example is the de Rham complex of differential forms on X with D_j defined by the exterior differentiation from j forms to $j+1$ forms.

After introducing a hermitian structure in each E_j we obtain adjoint operators

$$D_j^*: C^\infty(X, E_{j+1} \otimes \Omega^{\frac{1}{2}}) \to C^\infty(X, E_j \otimes \Omega^{\frac{1}{2}}).$$

The exactness of (19.4.2) means precisely that for every j

$$f \mapsto (D_{j-1} D_{j-1}^* + D_j^* D_j) f$$

is an elliptic operator on sections of $E_j \otimes \Omega^{\frac{1}{2}}$. The kernel

(19.4.3)
$$K_j = \{f \in \mathscr{D}'(X, E_j \otimes \Omega^{\frac{1}{2}});\, D_j f = 0,\, D_{j-1}^* f = 0\}$$

is therefore a finite dimensional subspace of $C^\infty(X, E_j \otimes \Omega^{\frac{1}{2}})$, and for every $f \in H_{(s)}$ orthogonal to K_j one can find $g \in H_{(s+2m)}$ with

$$(D_j^* D_j + D_{j-1} D_{j-1}^*) g = f.$$

If $D_j f = 0$ then f and $D_{j-1} D_{j-1}^* g$ are orthogonal to $D_j^* D_j g$ so it follows that $D_j^* D_j g = 0$ and that $f = D_{j-1} u$ where $u = D_{j-1}^* g \in H_{(s+m)}$. Thus D_{j-1} maps $H_{(s+m)}$ on the set of all $f \in H_{(s)}$ with $D_j f = 0$ orthogonal to K_j. Hence K_j can be identified with the quotient of the kernel of D_j by the range of D_{j-1} if D_j and D_{j-1} operate on C^∞ sections and also if they operate on distribution sections. The Euler characteristic

(19.4.4)
$$\sum (-1)^j \dim K_j$$

is equal to the index of the elliptic operator

$$C^\infty(X, \oplus E_{2j} \otimes \Omega^{\frac{1}{2}}) \ni (u_0, u_2, \ldots) \mapsto (D_0 u_0 + D_1^* u_2, \ldots) \in C^\infty(X, \oplus E_{2j+1} \otimes \Omega^{\frac{1}{2}})$$

obtained by splitting the complex, so it can be determined with the methods developed in Sections 19.2 and 19.3.

We shall now show that Theorem 19.1.15 is valid not only for trace class operators R_j but also for arbitrary R_j such that the calculus of wave front sets makes it possible to define a trace by restricting the Schwartz kernel to the diagonal.

Theorem 19.4.1. *Let* (19.4.1) *be an elliptic complex and let* K_j *be the finite dimensional cohomology spaces* $\operatorname{Ker} D_j/\operatorname{Im} D_{j-1}$ *isomorphic to* (19.4.3). *Assume given operators*

$$R_j\colon C^\infty(X, E_j\otimes\Omega_X^{\frac12})\to\mathscr{D}'(X, E_j\otimes\Omega_X^{\frac12})$$

commuting with the complex, that is, assume that

$$R_j D_{j-1} f = D_{j-1} R_{j-1} f, \quad j=1,\dots,N;\ f\in C^\infty(X, E_{j-1}\otimes\Omega_X^{\frac12}). \tag{19.4.5}$$

Then R_j *induces a linear map* $\tilde R_j\colon K_j\to K_j$. *Assume that the Schwartz kernel* $\mathscr{R}_j\in\mathscr{D}'(X\times X, E_j\boxtimes\overline{E}_j\otimes\Omega_{X\times X}^{\frac12})$ *has a wave front set disjoint with the conormal bundle of the diagonal in* $X\times X$, *and denote by* $\operatorname{Tr}(R_j)$ *the integral over* X *of the trace of the pullback* $\in\mathscr{D}'(X,\operatorname{Hom}(E_j,E_j)\otimes\Omega_X)$ *of* $\mathscr{R}_j$ *by the diagonal map* $X\to X\times X$. *Then we have*

$$\sum_0^N(-1)^j\operatorname{Tr}(R_j) - \sum_0^N(-1)^j\operatorname{Tr}(\tilde R_j|K_j). \tag{19.4.6}$$

Proof. Let F_j be a parametrix for the "Laplace operator" $D_{j-1}D_{j-1}^* + D_j^* D_j$. Thus the operators

$$(D_{j-1}D_{j-1}^* + D_j^* D_j)F_j - I_j, \quad F_j(D_{j-1}D_{j-1}^* + D_j^* D_j) - I_j \tag{19.4.7}$$

have C^∞ kernels if I_j is the identity operator on sections of $E_j\otimes\Omega^{\frac12}$. We define a *parametrix* of the complex by

$$G_j = D_j^* F_{j+1}, \quad j=0,\dots,N-1.$$

The essential property is that the operators

$$h_j = I_j - D_{j-1}G_{j-1} - G_j D_j \tag{19.4.8}$$

have C^∞ kernels. To prove this we first observe that

$$h_j = I_j - D_{j-1}D_{j-1}^* F_j - D_j^* F_{j+1} D_j.$$

Now we have

$$D_j(D_{j-1}D_{j-1}^* + D_j^* D_j) = D_j D_j^* D_j = (D_j D_j^* + D_{j+1}^* D_{j+1})D_j.$$

If we multiply to the left by F_{j+1} and to the right by F_j and use that the operators (19.4.7) have C^∞ kernels it follows that $F_{j+1}D_j - D_j F_j$ has a C^∞ kernel. Since

$$h_j = I_j - (D_{j-1}D_{j-1}^* + D_j^* D_j)F_j + D_j^*(D_j F_j - F_{j+1}D_j)$$

it follows that h_j has a C^∞ kernel as claimed.

h_j also commutes with the complex, that is, $D_j h_j = h_{j+1} D_j$, for both sides are equal to $D_j - D_j G_j D_j$. If $f \in \operatorname{Ker} D_j$ then $h_j f = f - D_{j-1} D_{j-1}^* F_j$ so h_j induces the identity in the quotient spaces K_j. Hence $h_j R_j h_j$ induces the map $\tilde{R}_j$ there, and $h_j R_j h_j$ is a continuous map from $\mathscr{D}'$ to C^∞ so it is a trace class operator. By Theorem 19.1.15 we have therefore proved that

$$\sum (-1)^j \operatorname{Tr}(\tilde{R}_j | K_j) = \sum (-1)^j \operatorname{Tr}(h_j R_j h_j).$$

Now $h_j R_j h_j - R_j$ is a sum of a number of terms each containing at least one factor F_j or F_{j+1}. We shall make these terms disappear by choosing a sequence of parametrices converging to 0 in a suitable sense. This requires a brief review of properties of the wave front set established in Section 8.2 and the construction of parametrices given in Section 18.1.

First recall from Theorem 8.1.9 (see also Theorem 18.1.16) that if $a \in S^m(\mathbb{R}^n \times \mathbb{R}^n)$ and $\mathscr{A}$ is the kernel of $a(x, D)$ then

$$WF(\mathscr{A}) \subset N(\Delta) = \{(x, x, \xi, -\xi); x \in \mathbb{R}^n, \xi \in \mathbb{R}^n \setminus 0\}.$$

If a^ν is a bounded sequence in S^m which converges to 0 in the C^∞ topology then the kernel of $a^\nu(x, D)$ converges to 0 in $\mathscr{D}'$, and it follows from the proof of Theorem 8.1.9 and the arguments following Definition 8.2.2 that it tends to 0 in $\mathscr{D}'_{N(\Delta)}(\mathbb{R}^n \times \mathbb{R}^n)$. For example, if $\chi \in C_0^\infty(\mathbb{R}^n)$ and $\chi = 1$ in the unit ball, then $a^\nu(x, \xi) = (1 - \chi(\xi/\nu))\, a(x, \xi)$ is a sequence to which this applies, and $a^\nu(x, D) - a(x, D) \in \operatorname{Op} S^{-\infty}$ has a C^∞ kernel. Writing a parametrix T in the form $\sum \phi_k T_k \psi_k$ where ϕ_k, ψ_k have support in a local coordinate patch and T_k is a pseudo-differential operator in the local coordinates we can apply this observation to each term. Thus we can choose sequences F_j^ν of parametrices above with kernels converging to 0 in $\mathscr{D}'_{N(\Delta)}$, and this implies that the kernel of $R_j - h_j^\nu R_j h_j^\nu$ converges to 0 in $\mathscr{D}'_M(X \times X)$ if $M = WF(R_j)$. Here we have used Theorem 8.2.14. But then it follows from Theorem 8.2.4 that the pullback by the diagonal map converges to 0 in $\mathscr{D}'(X)$, which completes the proof.

We shall now discuss the classical Lefschetz fixed point formula as an example of Theorem 19.4.1. Thus we consider the de Rham complex

$$0 \to C^\infty(X, \Lambda^0) \xrightarrow{d} C^\infty(X, \Lambda^1) \xrightarrow{d} C^\infty(X, \Lambda^2) \xrightarrow{d} \dots$$

where Λ^j are the exterior powers of the complexification of $T^*(X)$ and d the exterior differential operator. Then the spaces K_j are the de Rham cohomology groups, that is, the cohomology groups over $\mathbb{C}$ of the compact manifold X. If $\phi: X \to X$ is a C^∞ map, then the pullback of forms defines maps

$$\Phi_j: C^\infty(X, \Lambda^j) \to C^\infty(X, \Lambda^j)$$

commuting with the complex since $d\phi^* f = \phi^* df$ for every differential form f. The Lefschetz number $L(\phi)$ of ϕ is the alternating sum of the traces of the maps induced by ϕ^* in the de Rham cohomology spaces.

Now the kernel of Φ_j is a simple layer on the graph $\{(x, \phi(x)) \in X \times X\}$ of ϕ. (See Example 5.2.5.) By Example 8.2.5 the wave front set is the conormal bundle

$$\{(x, \phi(x), -{}^t\phi'(x)\eta, \eta); x \in X, \eta \in T^*_{\phi(x)}\}$$

of the graph which intersects the conormal bundle of the diagonal only when $x = \phi(x)$ and ${}^t\phi'(x)\eta = \eta$. Thus the wave front set and the conormal bundle of the diagonal are disjoint if and only if

$$\phi(x) = x \Rightarrow \det(I - \phi'(x)) \neq 0. \tag{19.4.9}$$

The fixed point x is then said to be *non-degenerate.* (Here $\phi'(x)$: $T_x(X) \to T_x(X)$, and I is the identity map in $T_x(X)$ so the determinant is well defined.) By the inverse function theorem non-degenerate fixed points are isolated so if all fixed points of ϕ are non-degenerate there are only a finite number of them. Let us choose local coordinates $x_1, \dots, x_n$ vanishing at a non-degenerate fixed point. In a neighborhood the kernel of Φ_0 is then $\delta(y - \phi(x))$ and the kernel of Φ_j is $A_j(x)\,\delta(y - \phi(x))$ where $A_j(x) = \Lambda^j\,{}^t\phi'(x)$ is the linear transformation in $\Lambda^j \mathbb{C}^n$ defined by composition of a j form with the differential $\phi'(x)$. The alternating sum of the traces of the pullbacks by the map $x \mapsto (x, x)$ is

$$\left(\sum (-1)^j \operatorname{Tr} A_j(x)\right) \delta(x - \phi(x)) = \det(I - \phi'(x))/|\det(I - \phi'(x))|\, \delta(x).$$

Here we have used Example 6.1.3 and made a standard expansion of $\det(I - {}^t\phi'(x))$. Theorem 19.4.1 now reduces to the Lefschetz fixed point formula: If all fixed points of ϕ are non-degenerate then the Lefschetz number $L(\phi)$ is

$$L(\phi) = \sum_{\phi(x) = x} \operatorname{sgn} \det(I - \phi'(x)), \tag{19.4.10}$$

that is, the number of fixed points counted with signs.

19.5. Miscellaneous Remarks on Ellipticity

In order not to complicate the exposition we did not discuss in Section 19.2 whether ellipticity is necessary for the Fredholm properties. We shall now show that this is true in some sense. By X we denote a compact C^∞ manifold and by E, F two C^∞ complex vector bundles on X not necessarily of the same fiber dimension.

Theorem 19.5.1. *If $P \in \Psi^m(X; E \otimes \Omega^{\frac{1}{2}}, F \otimes \Omega^{\frac{1}{2}})$ then the following conditions are equivalent:*

(i) *The operator P: $H_{(s)}(X; E \otimes \Omega^{\frac{1}{2}}) \to H_{(s-m)}(X; F \otimes \Omega^{\frac{1}{2}})$ has closed range and finite dimensional kernel for every s.*

(ii) *Condition* (i) *is fulfilled for some s.*

(iii) *The principal symbol p has a left inverse* $q\in S^{-m}(X;\operatorname{Hom}(\pi^* F,\pi^* E))$, *that is*, $q\,p-\mathrm{id}\in S^{-1}(X;\operatorname{Hom}(\pi^* E,\pi^* E))$.

(iv) *The operator P has a left parametrix* $Q\in\Psi^{-m}(X;F\otimes\Omega^{\frac12},\ E\otimes\Omega^{\frac12})$, *that is*, $QP=I+R$ *where R is an operator with kernel in* $C^\infty(\operatorname{Hom}(E,E)\otimes\Omega^{\frac12}_{X\times X})$.

Proof. (i)$\Rightarrow$(ii) is obvious. If (ii) is fulfilled then we can find a constant C such that

$$\|u\|_{(s)}\leqq C(\|Pu\|_{(s-m)}+\|u\|_{(s-1)}),\quad u\in H_{(s)}(X;E\otimes\Omega^{\frac12}). \tag{19.5.1}$$

Indeed, otherwise we could find a sequence u_j with $\|u_j\|_{(s)}=1$ and

$$\|Pu_j\|_{(s-m)}<1/j,\qquad \|u_j\|_{(s-1)}<1/j.$$

By Proposition 19.1.3 it follows that u_j has a subsequence converging in $H_{(s)}(X;E\otimes\Omega^{\frac12})$ which is absurd since the limit must have norm 1 and be equal to 0. It is no restriction to assume that $s=m$ for otherwise we can multiply P to the left by an elliptic operator of order $s-m$. Let Y be a local coordinate patch with local coordinates $x_1,\dots,x_n$, where E and F are identified with $Y\times\mathbb{C}^{N'}$ and $Y\times\mathbb{C}^{N''}$ respectively. Choose $\psi\in C_0^\infty(Y)$ equal to 1 in a large open subset Z, and set with $\phi\in C_0^\infty(Z)$, $\xi\in\mathbb{R}^n$ and $w\in\mathbb{C}^{N'}$, $|w|=1$,

$$u_\xi(x)=\phi(x)\,w\,e^{i\langle x,\xi\rangle}.$$

Then $(1-\psi)Pu_\xi$ is rapidly decreasing when $\xi\to\infty$ since the kernel of $(1-\psi)P\phi$ is in C^∞. We have $\psi P\phi u=a(x,D)u$ in terms of the local coordinates, and $a(x,\xi)-p(x,\xi)\phi\in S^{m-1}$. Applying (19.5.1) to u_ξ we obtain, since $\psi Pu_\xi=a(x,\xi)\,w\,e^{i\langle x,\xi\rangle}$,

$$(1+|\xi|)^s\,\|\phi\|\leqq C'\,\|p(x,\xi)\,w\,\phi(x)\|+C_\phi(1+|\xi|)^{s-1}$$

(see the proof of Theorem 19.2.5). Here $\|\ \|$ is the L^2 norm. This implies that

$$\varliminf_{\xi\to\infty}\ \inf_{|w|=1}\ \inf_{x\in Z}|p(x,\xi)\,w|/(1+|\xi|)^m\geqq 1/C'.$$

In fact, assume that there are sequences w_ν with $|w_\nu|=1$, $x_\nu\in Z$, $\xi_\nu\to\infty$ such that $|p(x_\nu,\xi_\nu)\,w_\nu|/(1+|\xi_\nu|)^m\leqq C_0<1/C'$. Passing to a subsequence we may assume that $w_\nu\to w$, $x_\nu\to x_0$. Then

$$|p(x,\xi_\nu)\,w|<\tfrac12(C_0+1/C')(1+|\xi_\nu|)^m$$

for all x in a neighborhood U of x_0 independent of ν. When $\phi\in C_0^\infty(Z\cap U)$ we obtain a contradiction since $s=m$. Thus

$$(p^*(x,\xi)\,p(x,\xi)\,w,w)\geqq(1+|\xi|)^{2m}|w|^2/2\,C'^2,\qquad x\in Z,$$

if $|\xi|$ is sufficiently large. The determinant $\in S^{2mN'}$ of $p^*(x,\xi)\,p(x,\xi)$ is therefore bounded from below by $(1+|\xi|)^{2mN'}/(2\,C'^2)^{N'}$ so it has an inverse in $S^{-2mN'}$ by Theorem 18.1.9. The product q of this inverse by the cofactor matrix of p^*p and with p^* is a left inverse of p belonging to S^{-m} over Z.

Piecing together such local constructions of q by a partition of unity in X we obtain (iii). If (iii) holds and Q_1 has principal symbol q then $Q_1 P = I + R_1$ where $R_1 \in \Psi^{-1}$. Thus $I + R_1$ has a parametrix, and (iv) is valid for the product of such a parametrix and Q_1. By Proposition 19.1.3 it follows at once from (iv) that (i) is valid, which completes the proof.

There is also an essentially dual result:

Theorem 19.5.2. *If $P \in \Psi^m(X; E \otimes \Omega^{\frac{1}{2}}, F \otimes \Omega^{\frac{1}{2}})$ then the following conditions are equivalent:*

(i) *The range of the operator $P: H_{(s)}(X; E \otimes \Omega^{\frac{1}{2}}) \to H_{(s-m)}(X; F \otimes \Omega^{\frac{1}{2}})$ has finite codimension for every s.*

(ii) *Condition (i) is fulfilled for some s.*

(iii) *The principal symbol p has a right inverse $q \in S^{-m}(X; \mathrm{Hom}(\pi^* F, \pi^* E))$, that is, $pq - \mathrm{id} \in S^{-1}(X; \mathrm{Hom}(\pi^* F, \pi^* F))$.*

(iv) *The operator P has a right parametrix $Q \in \Psi^{-m}(X; F \otimes \Omega^{\frac{1}{2}}, E \otimes \Omega^{\frac{1}{2}})$, that is, $PQ = I + R$ where R is an operator with kernel in $C^\infty(\mathrm{Hom}(F, F) \otimes \Omega^{\frac{1}{2}}_{X \times X})$.*

Proof. That (i)⇒(ii) is trivial. By Lemma 19.1.1 condition (ii) implies that the range is closed, hence

$$P^*: H_{(m-s)}(X; F^* \otimes \Omega^{\frac{1}{2}}) \to H_{(-s)}(X; E^* \otimes \Omega^{\frac{1}{2}})$$

has closed range and finite dimensional kernel. By Theorem 19.5.1 it follows that p^* has a left inverse $q^* \in S^{-m}$, and $q \in S^{-m}$ is then a right inverse of p. That (iii)⇒(iv) follows just as in the proof of Theorem 19.5.1 (see also the proof of Theorem 18.1.9), and (iv)⇒(i) is clear since $I + R$ is a Fredholm operator. The proof is complete.

Remark. When P is polyhomogeneous the conditions in Theorems 19.5.1 and 19.5.2 are of course equivalent to injectivity and surjectivity of the principal symbol.

Theorems 19.5.1 and 19.5.2 combined show that P defines a Fredholm operator $H_{(s)} \to H_{(s-m)}$ if and only if P is elliptic. This justifies the hypotheses made in Section 19.2. However, it should be understood that the class of Fredholm operators changes if the norms are changed. A particularly simple case of importance in Chapter XX occurs when we have decompositions

$$E = \bigoplus_1^K E_k, \quad F = \bigoplus_1^J F_j.$$

A section u of $E \otimes \Omega^{\frac{1}{2}}$ can then be written in the form $(u_1, \dots, u_K)$ where u_k is a section of $E_k \otimes \Omega^{\frac{1}{2}}$, and sections of $F \otimes \Omega^{\frac{1}{2}}$ can be written similarly as $(f_1, \dots, f_J)$. If $P \in \Psi^\infty(X; E \otimes \Omega^{\frac{1}{2}}, F \otimes \Omega^{\frac{1}{2}})$ the equation $Pu = f$ can then be written

$$\sum_1^K P_{jk} u_k = f_j, \quad j = 1, \dots, J,$$

where $P_{jk} \in \Psi^{\infty}(X; E_k \otimes \Omega^{\frac{1}{2}}, F_j \otimes \Omega^{\frac{1}{2}})$. Let $s_1, \ldots, s_J$ and $t_1, \ldots, t_K$ be real numbers. Then P defines a continuous operator

$$P: \oplus H_{(t_k)}(X; E_k \otimes \Omega^{\frac{1}{2}}) \to \oplus H_{(s_j)}(X; F_j \otimes \Omega^{\frac{1}{2}}) \tag{19.5.2}$$

if $P_{jk} \in \Psi^{t_k - s_j}(X; E_k \otimes \Omega^{\frac{1}{2}}, F_j \otimes \Omega^{\frac{1}{2}})$.

Theorem 19.5.3. *If $P_{jk} \in \Psi^{t_k - s_j}(X; E_k \otimes \Omega^{\frac{1}{2}}, F_j \otimes \Omega^{\frac{1}{2}})$ then $P = (P_{jk})$ defines a continuous operator* (19.5.2), *and the following conditions are equivalent:*

(i) (19.5.2) *is a Fredholm operator.*

(ii) *One can find $q_{kj} \in S^{s_j - t_k}(X; \mathrm{Hom}(\pi^* F_j, \pi^* E_k))$ such that $(p)(q) - \mathrm{id}$ and $(q)(p) - \mathrm{id}$ are in S^{-1} if p_{jk} is the principal symbol of P_{jk}.*

(iii) *One can find $Q_{kj} \in \Psi^{s_j - t_k}(X; F_j \otimes \Omega^{\frac{1}{2}}, E_k \otimes \Omega^{\frac{1}{2}})$ such that for the corresponding operator Q the kernels of $QP - \mathrm{id}$ and $PQ - \mathrm{id}$ are in C^{∞}.*

Proof. Let $A_k \in \Psi^{-t_k}(X; E_k \otimes \Omega^{\frac{1}{2}}, E_k \otimes \Omega^{\frac{1}{2}})$ and $B_j \in \Psi^{s_j}(X; F_j \otimes \Omega^{\frac{1}{2}}, F_j \otimes \Omega^{\frac{1}{2}})$ be elliptic operators. Then the diagonal operators $A = (A_k \delta_{kl})$ and $B = (B_j \delta_{ij})$ are Fredholm operators,

$$A: H_{(0)}(X; E \otimes \Omega^{\frac{1}{2}}) \to \oplus H_{(t_k)}(X; E_k \otimes \Omega^{\frac{1}{2}}),$$
$$B: \oplus H_{(s_j)}(X; F_j \otimes \Omega^{\frac{1}{2}}) \to H_{(0)}(X; F \otimes \Omega^{\frac{1}{2}}).$$

If (i) is fulfilled it follows that BPA is a Fredholm operator in L^2, hence elliptic by Theorems 19.5.1 and 19.5.2. If b_j and a_k are the principal symbols of B_j and A_k it follows that $(b_j p_{jk} a_k)$ has an inverse $(r_{kj}) \in S^0$, and (ii) is then satisfied by $q_{kj} = a_k r_{kj} b_j$. The proof that (ii)$\Rightarrow$(iii) is just like the proof of (iii) $\Rightarrow$(iv) in Theorem 19.5.1, and (iii)$\Rightarrow$(i) by Corollary 19.1.9. The proof is complete.

An operator satisfying the conditions in Theorem 19.5.3 is said to be elliptic in the sense of Douglis and Nirenberg, and (p_{jk}) is called its principal symbol. It is clear that all we have done for standard elliptic operators can be extended to such operators by the simple device of left and right multiplication by appropriate diagonal matrices as in the proof of Theorem 19.5.3. This is also true for the index if we take the principal symbols of A_k and B_j as positive multiples of the identity. It is clear that these operators have index 0 and we still obtain $\mathrm{ind}\, P = s - \mathrm{ind}\, p$. As a rule we shall therefore content ourselves with discussing only the standard elliptic case and leave the extension to Douglis-Nirenberg systems for the reader. Thus the reader is now asked to state and prove analogues of Theorems 19.5.1 and 19.5.2 for such systems.

Even in the scalar case there exist Fredholm operators defined by non-elliptic pseudo-differential operators provided that the spaces $H_{(s)}$ are suitably modified. Examples are hypoelliptic operators of constant strength (cf. Theorem 13.4.1) or more generally the operators studied in Section 22.1.

Notes

The first result on the index of pseudo-differential (or rather singular integral) operators seems to have been proved by Noether [1] to whom the example discussed after Theorem 19.2.4 is due. The extension of Fredholm theory from compact perturbations of the identity to compact perturbations of arbitrary "Fredholm operators" given in Atkinson [1] made it plain that the index was an interesting homotopy invariant of elliptic operators. It was determined for various special classes of elliptic operators by a number of authors at the end of the 1950's. However, the most natural and important examples arise in complex analysis and algebraic geometry, in the Riemann-Roch theorem (in higher dimensions) and the Hirzebruch index theorem. (See Hirzebruch [1].) These results suggested an expression for the index in terms of characteristic classes, and such a formula was found by Atiyah and Singer [1]. We have not given their explicit formula here but just the characteristic properties of the index established in part I of Atiyah-Singer [2]. (Their discussion of the group invariant situation has been omitted here.) For the purely geometric derivation of the index formula from these results we refer to part III of Atiyah-Singer [2]. The formula has the following character. If X is a compact C^∞ manifold then the principal symbol of an elliptic operator P defines a Chern character which is a cohomology class with compact support in $T^*(X)$. In the manifold X there is a cohomology class $\mathscr{I}(X)$, independent of P, defined in terms of the Pontrjagin classes. Then $\operatorname{ind} P$ is the integral over $T^*(X)$ of the product of the Chern character and $\mathscr{I}(X)$, lifted to $T^*(X)$. To verify it from the results proved here one must first show that the formula agrees with (19.3.1) for operators in $\mathbb{R}^n$ which are trivial at infinity. This is easy because $\mathscr{I}(X)=1$ then. One must further show that the product construction in Theorem 19.2.13 agrees with the formula at least when V is the normal bundle of an embedding in Euclidean space. It is here that $\mathscr{I}(X)$ appears as the integral of the factor of the Chern character coming from the Bott operator.

The indicated path to the index formula differs from that of Atiyah-Singer [2] only by relying on Theorem 19.3.1 and the construction of the Bott operator in Section 19.2 instead of the Bott periodicity theorem and some simpler normalization conditions. The proof of Theorem 19.3.1 is due to Fedosov [1], who has later on discussed more general manifolds in the same way. Richard Melrose has informed the author that in unpublished work he has given a proof of the full index formula for operators between trivial bundles by arguments of Fedosov's type based on a careful correspondence between symbols and operators on a Riemannian manifold. Thus his arguments give the quite complicated index class $\mathscr{I}(X)$. For another derivation of the index formula we refer to Atiyah-Bott-Patodi [1].

The Lefschetz formula of Section 19.4 is essentially due to Atiyah-Bott [2]. They only considered operators R_j related to composition with a diffeomorphism in X with non-degenerate fixed points but now that the notion of wave front set is available Theorem 19.4.1 seems the natural setting for their arguments. The reader should consult Atiyah-Bott [2] for less classical applications than the one given here.

Section 19.1 is largely a collection of classical results in Fredholm theory. We have added some strengthened results on the stability of the index taken from Hörmander [42]. (The reader consulting this reference is warned that misprints are abundant since the article was never proofread.) The Bott operator was also discussed there. The simpler treatment of it here is largely taken from Hörmander [39] and relies on later progress in the theory of pseudo-differential operators. We also refer to Hörmander [42] for a proof that the index formula remains valid for fairly large classes of hypoelliptic operators which will be studied from a local point of view in Section 22.1. They include the elliptic operators in the sense of Douglis and Nirenberg [1] introduced in Section 19.5. Lack of space has made it impossible to cover the closely related results concerning Toeplitz operators. The reader is referred to Boutet de Monvel [3] for a discussion including an index theorem.

Chapter XX. Boundary Problems for Elliptic Differential Operators

Summary

In Chapter XVII we discussed some of the classical methods for solving the Dirichlet problem for second order elliptic differential operators. The purpose of this chapter is to use the theory of pseudo-differential operators to study general boundary problems for elliptic systems of differential operators. (See also the end of Section 17.3.)

To illustrate the main idea in a simple case let us assume that we want to solve the boundary problem

$$\Delta u=0 \quad \text{in } X, \qquad b_0 u_0+b_1 u_1=f \quad \text{on } \partial X,$$

where X is an open set in $\mathbb{R}^n$ with smooth boundary ∂X, Δ is the Laplacean, and b_0, b_1 are differential operators in ∂X acting respectively on the boundary value u_0 and the normal derivative $u_1=du/dn$ of u. If E is the fundamental solution of the Laplacean (see Theorem 3.3.2) and u is smooth, we obtain from Green's formula

$$u(x)=\int dE(x-y)/dn\, u_0(y)\, dS(y)-\int E(x-y)\, u_1(y)\, dS(y), \tag{20.1}$$

if $x\in X$. Thus it suffices to determine u_0 and u_1. However, although (20.1) always defines a harmonic function u it may not have the boundary value and normal derivative u_0, u_1. To examine if that is true we let x approach ∂X and obtain

$$u_0=k_0 u_0+k_1 u_1, \tag{20.2}$$

where k_0 and k_1 are pseudo-differential operators by Theorem 18.2.17, for convolution by E is a pseudo-differential operator satisfying the transmission condition. Conversely, (20.2) implies if $n>2$ that there is a harmonic function u with u_0 and u_1 as boundary value and normal derivative. For if u is defined by (20.1) it follows from (20.2) that $u=u_0$ on ∂X. If we subtract Green's formula for u from (20.1) we obtain

$$\int E(x-y)(u_1(y)-du/dn)\, dS(y)=0, \qquad x\in X.$$

The integral is a continuous function of x which is harmonic outside $\overline{X}$ and vanishes on ∂X and at infinity. Hence it is identically 0 which implies that $u_1 = du/dn$. To solve the boundary problem is thus equivalent to solving the system of pseudo-differential equations

$$(1-k_0)u_0 - k_1 u_1 = 0, \quad b_0 u_0 + b_1 u_1 = f.$$

Here k_1 is an injective elliptic pseudo-differential operator, and it is easily seen that it has index 0, hence a pseudo-differential inverse k_1^{-1}. Thus the Dirichlet problem is solved by taking $u_1 = k_1^{-1}(1-k_0)u_0$. In general we obtain pseudo-differential equation

$$(b_0 + b_1 k_1^{-1}(1-k_0))u_0 = f.$$

When this is an elliptic pseudo-differential operator the boundary problem is called elliptic and then we can just apply the results of Chapter XIX.

To develop the preceding ideas in general we must examine the analogues of the operators k_0 and k_1 for general elliptic systems of differential equations in X. This requires only minor additions to the results proved in Section 18.2. They are given in Section 20.1 where we establish the Fredholm properties of general elliptic boundary problems.

Next we turn to the question of the index of an elliptic boundary problem. As a preliminary we discuss in Section 20.2 some elementary but important facts on ordinary systems of differential operators with constant coefficients. These are used in Section 20.3 to deform an elliptic boundary problem to another one with the same index which is just one half of the index of a natural extension of the operator to the double of the manifold, which no longer has a boundary. As in Chapter XIX we do not give an explicit formula for the index but only establish a procedure for obtaining it.

In Section 20.4, finally, we give some remarks on non-elliptic boundary problems. These are meant to motivate why we devote so much effort to existence theory for non-elliptic *pseudo-differential* operators in the subsequent chapters.

20.1. Elliptic Boundary Problems

Let X be a compact C^∞ manifold with boundary ∂X. Assume given an elliptic differential operator P of order m from $C^\infty(X,E)$ to $C^\infty(X,F)$ where E and F are complex C^∞ vector bundles on X with the same fiber dimension N. (We do not include half density bundles here since they would just complicate notation.) Thus the principal symbol $p(x,\xi)$ is a bijection $E_x \to F_x$ for every $\xi \in T_x^* \setminus \{0\}$. To set up a boundary problem we need in addition boundary differential operators B_j: $C^\infty(X,E) \to C^\infty(\partial X, G_j)$, $j=1,\dots,J$, where G_j are C^∞ vector bundles on ∂X. (This notion is discussed in the scalar case in Appendix B.2. The presence of bundles makes no difference locally and should be ignored by the reader who does not feel familiar with bundles.) The boundary problem we shall study is to find $u \in C^\infty(X,E)$ satisfying

(20.1.1) $$Pu=f \text{ in } X; \quad B_j u = g_j \text{ in } \partial X, \quad j=1,\dots,J;$$

where $f \in C^\infty(X,F)$ and $g_j \in C^\infty(\partial X, G_j)$ are given. Later on we shall state the problem under weaker smoothness assumptions also.

In a local coordinate patch X_κ such that X_κ is defined by $x_n \geqq 0$ and E, F are identified with $X_\kappa \times \mathbb{C}^N$, the equation $Pu=f$ has the form

$$\sum P_\alpha(x) D^\alpha u = f$$

where the coefficients P_α are C^∞ functions of x with values in $N \times N$ matrices, and the coefficient of D_n^m is invertible since P is elliptic. Thus we can write

$$D_n^m u = P_{0,\dots,n}^{-1}(f - \sum_{\alpha_n < m} P_\alpha(x) D^\alpha u).$$

If the transversal order of B_j exceeds $m-1$, that is, B_j contains terms $B_{j\alpha}(x) D^\alpha u$ with $\alpha_n \geqq m$, then we can replace $D_n^m u$ by this expression in all terms with $\alpha_n \geqq m$ to reduce the transversal order. Repeating the argument and using a partition of unity to combine the local constructions we conclude that

$$B_j = B_j' + C_j P$$

where B_j' has transversal order $<m$ and C_j is a boundary differential operator from $C^\infty(F)$ to $C^\infty(G_j)$. Then (20.1.1) is equivalent to

$$Pu=f \text{ in } X; \quad B_j' u = g_j - C_j f \text{ in } \partial X, \quad j=1,\dots,J.$$

It is therefore no restriction to assume that the transversal order of B_j is less than m, as we shall do from now on. However, we shall put no restriction on the total order m_j of B_j.

For any $s \geqq m$ we can extend P from C^∞ to a continuous map

$$P: \bar{H}_{(s)}(X^\circ, E) \to \bar{H}_{(s-m)}(X^\circ, F);$$

here X° is the interior of X and the spaces are defined in Appendix B.2. By Theorem B.2.10 we also have a continuous extension

$$B_j: \bar{H}_{(s)}(X^\circ, E) \to H_{(s-m_j-\frac{1}{2})}(\partial X, G_j)$$

since the transversal order of B_j is $\leqq m-1$. We want to examine when the operator

(20.1.2) $$\bar{H}_{(s)}(X^\circ, E) \ni u \mapsto (Pu, B_1 u, \dots, B_k u) \in \bar{H}_{(s-m)}(X^\circ, E) \oplus H_{(s-m_1-\frac{1}{2})}(\partial X, G_1) \oplus \dots \oplus H_{(s-m_J-\frac{1}{2})}(\partial X, G_J)$$

is a Fredholm operator.

Definition 20.1.1. The boundary problem (20.1.1) is called elliptic if

(i) $P(x,D)$ is elliptic.

(ii) The boundary conditions are elliptic in the sense that for every $x \in \partial X$ and $\xi \in T_x^*(X)$ not proportional to the interior conormal n_x of X the map

$$M_{x,\xi}^+ \ni u \mapsto (b_1(x, \xi + D_t n_x) u(0), \dots, b_J(x, \xi + D_t n_x) u(0)) \in G_{1x} \oplus \dots \oplus G_{Jx}$$

is bijective, if $M_{x,\xi}^+$ is the set of all $u \in C^\infty(\mathbb{R}, E_x)$ with $p(x, \xi + D_t n_x) u(t) = 0$ which are bounded on $\mathbb{R}_+$.

Here b_j is of course the principal symbol of B_j and p is that of P. In the definition it suffices to take one ξ in each equivalence class mod $\mathbb{R} n_x$ since translation by $s n_x$ is equivalent to conjugation by e^{ist}. By condition (i) we know that $p(x, \xi + \tau n_x)$ is non-singular when τ is real, for $\xi + \tau n_x \neq 0$ then by assumption, so all elements of $M_{x,\xi}^+$ are in fact exponentially decreasing on $\mathbb{R}_+$. (See also Section 20.2 for a discussion of systems of ordinary differential operators.) We have chosen the term elliptic since we shall see that the definition is equivalent to ellipticity of an associated system of pseudo-differential operators. However, the literature contains many other terms such as the Lopatinski-Schapiro condition, coerciveness and the covering conditions. The main result of this section is the following generalisation of the study of the Dirichlet problem in Section 17.3.

Theorem 20.1.2. *If the boundary problem* (20.1.1) *is elliptic and* $s \geqq m$, *then* (20.1.2) *is a Fredholm operator.*

The proof will be given after some preparations which allow a reduction along the lines indicated in the introduction to a system of pseudo-differential operators in ∂X. It is convenient to identify a neighborhood X_1 of ∂X in X with $\partial X \times [0,1)$ for example by introducing a Riemannian metric in X and taking geodesic normal coordinates with respect to ∂X (see Appendix C.5). We can also identify the bundles E and F restricted to X_1 with their liftings by the projection $\pi_1: X_1 \to \partial X$, for any bundle map $E \to \pi_1^* E$, for example, which is an isomorphism on ∂X must be an isomorphism on $\partial X \times [0, \varepsilon)$ for small ε. A change of scales will replace X_1 by $\partial X \times [0, \varepsilon)$. In what follows we denote points in X_1 by (x', x_n) where $x' \in \partial X$ and $x_n \in [0,1)$. Since $E_x = E_{(x',0)}$ the normal derivatives $D_n^j u$ of sections u of E are then well defined. We can also extend X to an open manifold $\hat{X} = X \cup (\partial X \times (-1, 0))$ with the obvious C^∞ structure. The bundles E and F can be extended to $\hat{X}$ so that the fibers are independent of x_n in $\partial K \times (-1,1)$. We can also extend the differential operator P to $\hat{X}$. Shrinking X_1 again if necessary we may assume that P is elliptic on $\hat{X}$ too. By T we denote a properly supported parametrix of P on $\hat{X}$. Thus $T \in \Psi^{-m}(\hat{X}; F, E)$ and

$$TP = I_E + R_E, \quad PT = I_F + R_F, \tag{20.1.3}$$

where I_E, I_F denote the identity operators on sections of E and F, and R_E, R_F are operators with C^∞ kernels on such sections.

If $u \in C^\infty(X, E)$ we write

$$\gamma u = (u, D_1 u, \ldots, D_n^{m-1} u)|_{\partial X} \in C^\infty(\partial X, E^m)$$

for the Cauchy data. In $\partial X \times [0,1)$ we have

$$P = \sum_0^m P_j D_n^j$$

where P_j is a differential operator of order $m-j$ in ∂X depending on the parameter x_n. We denote the principal symbol by p_j. With $u^0 = u$ in X and $u^0 = 0$ in $\hat{X} \setminus X$, we obtain in $\hat{X}$

$$Pu^0 = (Pu)^0 + P^c \gamma u \tag{20.1.4}$$

where for $U = (U_0, \ldots, U_{m-1}) \in C^\infty(\partial X, E^m)$

$$P^c U = i^{-1} \sum_{j<m} P_{j+1} \sum_{k \leqq j} U_{j-k} \otimes D_n^k \delta, \tag{20.1.5}$$

δ denoting the Dirac measure in x_n. In fact, induction with respect to j gives

$$D_n^{j+1} u^0 = (D_n^{j+1} u)^0 + i^{-1} \sum_0^j \gamma_{j-k} u \otimes D_n^k \delta.$$

If T is applied to (20.1.4) we obtain "Green's formula"

$$u^0 + R_E u^0 = T(Pu)^0 + TP^c \gamma u, \tag{20.1.6}$$

in view of (20.1.3). Note that if $Pu = 0$ in X and $R_E = 0$ it follows that $\gamma u = \gamma TP^c \gamma u$, where the boundary values on the right exist by Theorem 18.2.17 since every term in the symbol of T is a rational function. For $U \in C^\infty(\partial X, E^m)$ we therefore introduce

$$QU = \gamma TP^c U; \tag{20.1.7}$$

thus QU is the Cauchy data of the Green potential $TP^c U$. Explicitly,

$$(QU)_k = \sum_0^{m-1} Q_{kl} U_l, \quad k = 0, \ldots, m-1,$$

where

$$Q_{kl} U_l = \sum_{j=0}^{m-1-l} i^{-1} D_n^k TP_{j+l+1} U_l \otimes D_n^j \delta|_{\partial X},$$

the boundary values being taken from X of course. Thus Q_{kl} is a pseudo-differential operator in ∂X of order $k-l$ with principal symbol

$$q_{kl}(x', \xi') = (2\pi i)^{-1} \int^+ \sum_{j=0}^{m-l-1} \xi_n^{k+j} p(x', 0, \xi)^{-1} p_{j+l+1}(x', 0, \xi') \, d\xi_n, \tag{20.1.8}$$

that is, the sum of the residues of the integrand in the half plane $\operatorname{Im} \xi_n > 0$. This means that (Q_{kl}) is a Douglis-Nirenberg system as discussed in Section 19.5.

Theorem 20.1.3. *The system Q of pseudo-differential operators defined by (20.1.7) is an approximate projection in the space of Cauchy data in the sense that Q^2-Q has a C^∞ kernel. If we identify solutions of the ordinary differential equation $p(x',0,\xi',D_n)v=0$ with the Cauchy data $(v(0),\dots,D_n^{m-1}v(0))$ then the principal symbol $q(x',\xi')$ is for $\xi'\neq 0$ identified with the projection on the subspace M^+ of solutions exponentially decreasing on $\mathbb{R}_+$ along the subspace M^- of solutions exponentially decreasing on $\mathbb{R}_-$. One calls Q, and sometimes also q, the Calderón projector.*

Proof. If $u=TP^cU$ then the restriction of u to X is in C^∞ and by (20.1.3)

$$Pu=(I_F+R_F)P^cU=R_FP^cU \quad \text{in } X^\circ$$

where R_F has a C^∞ kernel. By definition $\gamma u=QU$, so if we apply (20.1.6) to u and take the Cauchy data from X, it follows that

$$QU+\gamma R_E(TP^cU)^0=\gamma T(R_FP^cU)^0+Q^2U.$$

Now the map $U\mapsto R_FP^cU$ is continuous from $\mathscr{D}'(\partial X,E^m)$ to $C^\infty(X,F)$, and since T satisfies the transmission condition it follows that the map $U\mapsto\gamma T(R_FP^cU)^0$ is continuous from $\mathscr{D}'$ to C^∞, hence has a C^∞ kernel. If $\phi\in C^\infty(X,E^*\otimes\Omega_X)$ and $U\in C^\infty$, then approximation of ϕ by functions vanishing near ∂X gives

$$((TP^cU)^0,\phi)=(P^cU,T^*\phi^0).$$

(Note that $T^*\phi^0\in C^{m-1}(\hat{X},E)$.) Since T^* also satisfies the transmission condition it follows that $T^*\phi^0\in C^\infty(X)$, which proves that $U\mapsto(TP^cU)^0$ can be extended to a continuous map from $\mathscr{D}'(\partial X,E^m)$ to $\mathscr{D}'(X,E)$. Hence $U\mapsto\gamma R_E(TP^cU)^0$ is also continuous from $\mathscr{D}'$ to C^∞ which proves that Q^2-Q has a C^∞ kernel.

To interpret the principal symbol we observe that if $U=(U_0,\dots,U_{m-1})\in E^m_{(x',0)}$ then the inverse Fourier transform

$$v(x_n)=(2\pi i)^{-1}\int p(x',0,\xi)^{-1}\sum_{j+l<m} p_{j+l+1}(x',0,\xi')\,\xi_n^j\,U_l\,e^{ix_n\xi_n}\,d\xi_n$$

is in $\mathscr{S}'$ and satisfies the equation

$$p(x',0,\xi',D_n)v=i^{-1}\sum_{j+l<m} p_{j+l+1}(x',0,\xi')\,U_l\,D_n^j\delta.$$

Thus v concides for $x_n>0$ with an element $v^+\in M^+$ and when $x_n<0$ with an element $v^-\in M^-$. For $x_n=0$ we have the jump conditions

$$D_n^k(v^+-v^-)=U_k,\quad k=0,\dots,m-1.$$

The equations (20.1.8) mean that

$$q(x',\xi')\,U=(v^+(0),\dots,D_n^{m-1}v^+(0)).$$

Hence $qU=U$ means that $v^-=0$, that is, that U is the Cauchy data of a solution in M^+. The equation $qU=0$ means that $v^+=0$, that is, that U is the Cauchy data of a solution in M^-. Hence q is the projection on M^+ along the complementary space M^-; that q is a projection is of course also a consequence of the fact that the symbol of Q^2-Q is of order $-\infty$. The proof is complete.

Corollary 20.1.4. *The map*

$$C^\infty(X,F)\ni f\mapsto Q\gamma Tf^0\in C^\infty(\partial X,E^m)$$

can be extended to a continuous map from $\bar{H}_{(0)}(X^\circ,F)$ *to* $C^\infty(\partial X,E^m)$.

Proof. If $f\in C^\infty(X,F)$ then $u=Tf^0|_X\in C^\infty(X,E)$ since T satisfies the transmission condition. It follows from (20.1.3) that $Pu=f+R_Ff^0$ in X. Thus we obtain from (20.1.6)

$$\gamma u+\gamma R_E u^0=\gamma Tf^0+\gamma T(R_F f^0)^0+Q\gamma u.$$

The map $f\mapsto R_Ff^0$ is continuous from $\bar{H}_{(0)}$ to C^∞, so $f\mapsto \gamma T(R_Ff^0)^0$ is continuous from $H_{(0)}$ to C^∞, again by the transmission condition. Now $f\mapsto Tf^0|_X$ is continuous from $\bar{H}_{(0)}$ to $\bar{H}_{(m)}(X^0)$, so $f\mapsto \gamma R_E u^0$ is continuous from $\bar{H}_{(0)}$ to C^∞. By Theorem B.2.10 the map $f\mapsto\gamma u$ is continuous from $\bar{H}_{(0)}(X^\circ,F)$ to $\bigoplus H_{(m-j-\frac12)}(\partial X,E)$, and since $Q-Q^2$ is continuous from $\bigoplus H_{(m-j-\frac12)}(\partial X,E)$ to $\bigoplus C^\infty(\partial X,E)$, the corollary is proved.

By hypothesis the operators B_j in (20.1.1) have transversal order $<m$, so they can be written in the form $B_j=B_j^c\gamma$, where B_j^c is a differential operator from sections of E^m on ∂X to sections of G_j on ∂X. If $U=\gamma u$ it follows from (20.1.6) that $U-QU$ is essentially determined by $f=Pu$, so we are led to replace (20.1.1) by equations of the form

$$U=\psi+QU;\quad B_j^cU=g_j,\quad j=1,\dots,J.$$

At first sight it might seem that there are too many equations here. For example, the Laplace equation with Dirichlet boundary conditions leads to 3 equations for 2 unknowns. However, ψ is not arbitrary since $Q\psi$ is essentially 0 by Corollary 20.1.4. Since Q is a projection this will effectively decrease the number of equations to be solved as shown by the following proposition which for the sake of simplicity is stated for standard elliptic operators rather than those of the Douglis-Nirenberg type.

Proposition 20.1.5. *Let H and G be C^∞ complex vector bundles on a compact manifold Y without boundary, let $Q\in\Psi^0_{\text{phg}}(Y;H,H)$, $B\in\Psi^\mu_{\text{phg}}(Y;H,G)$ and assume that $Q^2-Q\equiv 0$, that is, that $Q^2-Q\in\Psi^{-\infty}$. Denote the principal symbols by b and by q. Then*

(i) *if $b(y,\eta)$ restricted to $q(y,\eta)H_y$ is surjective for all $(y,\eta)\in T^*(Y)\setminus 0$, then one can find $S\in\Psi^{-\mu}_{\text{phg}}(Y;G,H)$ such that*

$$BS\equiv I_G,\quad QS\equiv S. \tag{20.1.9}$$

(ii) *if* $b(y,\eta)$ *restricted to* $q(y,\eta)H_y$ *is injective for all* $(y,\eta)\in T^*(Y)\setminus 0$ *then one can find* $S'\in\Psi^{-\mu}_{\text{phg}}(Y;G,H)$ *and* $S''\in\Psi^0_{\text{phg}}(Y;H,H)$ *such that*

$$S'B+S''\equiv I_H,\quad S''Q\equiv 0. \tag{20.1.10}$$

(iii) *if* $b(y,\eta)$ *restricted to* $q(y,\eta)H_y$ *is bijective, then* S, S', S'' *are uniquely determined* mod $\Psi^{-\infty}$, *and* $S'\equiv S$.

Proof. In case (i) we can by the polyhomogeneous form of Theorem 19.5.2 applied to BQ find $T\in S^{-\mu}_{\text{phg}}(Y;G,H)$ such that $BQT\equiv I_G$. Then $S=QT$ satisfies (20.1.9), for

$$BS=BQT\equiv I_G,\quad QS=(Q^2-Q)T+QT\equiv QT=S.$$

In case (ii) we can use the polyhomogeneous case of Theorem 19.5.1 for the Douglis-Nirenberg system $B\oplus(I_H-Q)$ to find $T'\in\Psi^{-\mu}_{\text{phg}}(Y;G,H)$ and $T''\in\Psi^0_{\text{phg}}(Y;H,H)$ such that

$$T'B+T''(I_H-Q)\equiv I_H.$$

Then $S'=T'$ and $S''=T''(I_H-Q)$ satisfy (20.1.10) since $S''Q=T''(Q-Q^2)\equiv 0$. In case (iii) we choose S, S' and S'' so that (20.1.9) and (20.1.10) hold. Then it follows that

$$S'\equiv S'BS\equiv S'BS+S''(S-QS)\equiv S'BS+S''S\equiv S.$$

Hence $S\equiv S'$ so these operators and therefore S'' are uniquely determined mod $\Psi^{-\infty}$. This completes the proof.

In the situation at hand $Q=(Q_{kl})$ where $Q_{kl}\in\Psi^{k-l}_{\text{phg}}(\partial X;E,E)$, and B is replaced by the matrix (B^c_{jk}) defined by

$$B^c_j U=\sum_0^{m-1} B^c_{jk}U_k,\quad B^c_{jk}\in\Psi^{m_j-k}_{\text{phg}}(\partial X;E,G_j).$$

By the arguments used to prove Theorem 19.5.3 it follows from Proposition 20.1.5, if the boundary problem is elliptic, that we can find operators

$$S_{kj}\in\Psi^{k-m_j}_{\text{phg}}(\partial X;G_j,E);\quad k=0,\dots,m-1;\ j=1,\dots,J;$$
$$S''_{kl}\in\Psi^{k-l}_{\text{phg}}(\partial X;E,E);\quad k,l=0,\dots,m-1;$$

such that (20.1.9), (20.1.10) are fulfilled for $S'=S=(S_{kj})$ and $S''=(S''_{kl})$. We can use these operators to construct a parametrix of the boundary problem.

First assume that $u\in C^\infty(X,E)$ and that (20.1.1) holds. With $U=\gamma u$ it follows from (20.1.6) that

$$U+\gamma R_E u^0=\gamma Tf^0+QU,$$

and the boundary conditions (20.1.1) give $B^c_j U=g_j$. By (20.1.10) we have with the obvious block matrix notation

$$SB^c+S''(I-Q)=I-R$$

where $R\in\Psi^{-\infty}(\partial X; E^m, E^m)$. Hence

$$U=Sg+S''(\gamma Tf^0-\gamma R_E u^0)+RU.$$

Using (20.1.6) again we obtain

$$u=(I+TP^c S''\gamma)\,Tf^0+TP^c Sg+Ku, \tag{20.1.11}$$

where

$$Ku=TP^c(R\gamma u-S''\gamma R_E u^0)-R_E u^0$$

is a continuous map from $\overline{H}_{(m)}(X^\circ, E)$ to $C^\infty(X,E)$. Thus the map

$$(f,g)\mapsto(I+TP^c S''\gamma)\,Tf^0+TP^c Sg \tag{20.1.12}$$

is an approximate left inverse of (20.1.2). It is also an approximate right inverse. In fact, with the notation u for the right-hand side of (20.1.12) we have by (20.1.3)

$$\begin{aligned}Pu&=(I_F+R_F)((I+P^c S''\gamma T)f^0+P^c Sg)\\&=f^0+R_F f^0+R_F P^c S''\gamma Tf^0+R_F P^c Sg\end{aligned}$$

where the second equality is valid in X°. Hence

$$P((I+TP^c S''\gamma)\,Tf^0+TP^c Sg)=f+K_1 f+K_2 g \tag{20.1.13}$$

where K_1 is a continuous map from $\overline{H}_{(0)}(X^\circ, F)$ to $C^\infty(X,F)$ and K_2 is a continuous map from $\mathscr{D}'(\partial X, \oplus G_j)$ to $C^\infty(X,F)$. We have

$$B^c\gamma u=B^c(I+QS'')\gamma Tf^0+B^c QSg.$$

Here $K_4=B^c QS-I$ is of order $-\infty$ by (20.1.9), and (20.1.9), (20.1.10) give

$$B^c(I+QS'')\equiv B^c(I+Q-QSB^c)\equiv B^c+B^c Q-B^c QSB^c\equiv B^c Q.$$

In view of Corollary 20.1.4 it follows that

$$B((I+TP^c S''\gamma)\,Tf^0+TP^c Sg)=g+K_3 f+K_4 g \tag{20.1.14}$$

where K_3 is a continuous map from $\overline{H}_{(0)}(X^\circ, F)$ to $C^\infty(X,\oplus G_j)$ and K_4 is a map in $C^\infty(X,\oplus G_j)$ with C^∞ kernel. Theorem 20.1.2 will follow from Corollary 19.1.9 and (20.1.11), (20.1.13), (20.1.14) when we have verified some continuity properties of (20.1.12).

In the following proposition the notation $\|u\|_{(s)}$ stands for the norm in $\overline{H}_{(s)}(X^\circ)$ if $u\in C^\infty(X,E)$, and similarly for sections of F in X°. For the sake of clarity we shall write $\tilde{T}v$ for the restriction of Tv to X°.

Proposition 20.1.6. *If $s\geqq m$ and $f\in C^\infty(X,F)$ then*

$$\|\tilde{T}f^0\|_{(s)}\leqq C\|f\|_{(s-m)}. \tag{20.1.15}$$

If $U=(U_0,\dots,U_{m-1})\in C^\infty(\partial X, E^m)$ we have for any s

$$\|\tilde{T}P^c U\|_{(s)}\leqq C\sum_0^{m-1}\|U_j\|_{(s-j-\frac{1}{2})}. \tag{20.1.16}$$

Proof. It suffices to prove (20.1.15) when f has support in a compact subset K of a local coordinate patch $Y\times[0,1)$ at the boundary, $Y\subset\mathbb{R}^{n-1}$. Choose $\phi\in C_0^\infty(Y\times[0,1))$ equal to 1 in a neighborhood of K, and let k be an integer $\geqq s-m$. We shall first estimate $\|\phi\tilde{T}f^0\|_{(s-k,k)}$, with the notation in Appendix B.2. To do so we note that

$$\|f^0\|_{(-m+s-k,k)}\leqq\|f^0\|_{(0,s-m)}=\|f\|_{(0,s-m)}\leqq\|f\|_{(s-m)}.$$

Here f^0 is considered as a distribution in $\mathbb{R}^n$. We can write

$$D^\alpha\phi T=\sum_{|\beta|\leqq|\alpha|}T_\beta D^\beta$$

where T_β is a pseudo-differential operator of order $-m$ and $\beta_n=0$ if $\alpha_n=0$. Since $\|D^\beta f^0\|_{(s-m-k)}\leqq\|f^0\|_{(s-m-k,k)}$ if $|\beta|\leqq k$ and $\beta_n=0$, it follows that

$$\|D^\alpha\phi Tf^0\|_{(s-k)}\leqq C\|f^0\|_{(s-m-k,k)}$$

if $\alpha_n=0$ and $|\alpha|\leqq k$. This gives an estimate of $\|\phi Tf^0\|_{(s-k,k)}$, so

$$\|\phi\tilde{T}f^0\|_{(s-k,k)}\leqq C\|f\|_{(s-m)}.$$

Now $P\tilde{T}f^0=f+R_Ff^0$ in X° by (20.1.3). Choose $\psi\in C_0^\infty(Y\times[0,1))$ so that $\psi=1$ in a neighborhood of K and $\phi=1$ in a neighborhood of $\operatorname{supp}\psi$. Theorem B.2.9 or rather its proof shows that

$$\|\psi\tilde{T}f^0\|_{(s)}\leqq C(\|f\|_{(s-m)}+\|\phi\tilde{T}f^0\|_{(s-k,k)}).$$

Since $(1-\psi)Tf^0$ is a continuous function of $f\in H_{(0)}$ with values in $C^\infty(X)$ when $\operatorname{supp}f\subset K$, we have proved (20.1.15).

We may also suppose that $\operatorname{supp}U\subset K$ when proving (20.1.16). By (20.1.5) we have

$$P^cU=\sum_0^{m-1}v_j\otimes D_n^j\delta,\quad v_j=\sum_{j+l<m}P_{j+l+1}U_l/i.$$

Since P_{j+l+1} is of order $m-j-l-1$ we have

$$\sum\|v_j\|_{(s-m+j+\frac12)}\leqq C\sum\|U_j\|_{(s-j-\frac12)}.$$

The Fourier transform of $v_j\otimes D_n^j\delta$ is $\hat{v}_j(\xi')\xi_n^j$, and

$$\int\xi_n^{2j}(1+|\xi'|^2+\xi_n^2)^{-m}d\xi_n\leqq C(1+|\xi'|^2)^{j-m+\frac12}$$

when $j<m$. Thus

$$\|v_j\otimes D_n^j\delta\|_{(-m+s-k,k)}\leqq\|v_j\otimes D_n^j\delta\|_{(-m,s)}\leqq C\|v_j\|_{(s-m+j+\frac12)},$$

if $j<m$ and k is an integer $\geqq\max(s,0)$. Hence

$$\|\tilde{T}P^cU\|_{(s-k,k)}\leqq C\sum_0^{m-1}\|U_j\|_{(s-j-\frac12)},$$

for commuting T with x' derivatives it is easily seen that T is continuous from $H_{(t-m,k)}$ to $H_{(t,k)}$ for every integer $k\geqq0$. (See also Lemma 21.1.9 below.)

Since $U \mapsto P\tilde{T}P^c U = R_F P^c U$ is continuous from $\mathscr{D}'(\partial X)$ to C^∞, we can improve the estimate to (20.1.16) in the same way as in the proof of (20.1.15). The proof is complete.

Remark. The estimates proved here by means of Theorem B.2.9 are in fact valid for every T satisfying the transmission conditions. We leave the proof for the reader.

Theorem 20.1.7. *The map* (20.1.12) *is continuous from*

$$\overline{H}_{(s-m)}(X^\circ, F) \oplus H_{(s-m_1-\frac{1}{2})}(\partial X, G_1) \oplus \ldots \oplus H_{(s-m_J-\frac{1}{2})}(\partial X, G_J)$$

to $\overline{H}_{(s)}(X^\circ, E)$ *for every* $s \geqq m$. *It is an approximate left inverse of* (20.1.2) *in the sense that* (20.1.11) *is valid with* K *continuous from* $\overline{H}_{(m)}(X^\circ, E)$ *to* $C^\infty(X, E)$, *and an approximate right inverse as well in the sense that* (20.1.13) *and* (20.1.14) *are valid with*

$$\begin{pmatrix} K_1 & K_2 \\ K_3 & K_4 \end{pmatrix}$$

continuous from $H_{(0)}(X, F) \oplus \mathscr{D}'(\partial X, \oplus G_j)$ *to* $C^\infty(X, F) \oplus C^\infty(\partial X, \oplus G_j)$.

Proof. By (20.1.15) we have

$$\sum \|\gamma_j T f^0\|_{(s-j-\frac{1}{2})} \leqq C \|f\|_{(s-m)},$$

so it follows from (20.1.16) and the fact that $S''_{kl} \in \Psi^{k-l}$ that

$$\|\tilde{T}P^c S'' \gamma T f^0\|_{(s)} \leqq C \|f\|_{(s-m)}.$$

The continuity of the other terms in (20.1.12) is an obvious consequence of Proposition 20.1.6, which completes the proof.

As already pointed out, Theorem 20.1.2 is an immediate consequence of Theorem 20.1.7 and Corollary 19.1.9. Using Theorems 19.5.1 and 19.5.2 it would also be easy to show that ellipticity is necessary for the conclusions to hold.

Theorem 20.1.8. *If the boundary problem* (20.1.1) *is elliptic then the kernel of the Fredholm operator* (20.1.2) *is in* $C^\infty(X, E)$ *and the range is the orthogonal space of a finite dimensional subspace of*

$$C^\infty(X, F^* \otimes \Omega_X) \oplus C^\infty(\partial X, G_1^* \otimes \Omega_{\partial X}) \oplus \ldots \oplus C^\infty(\partial X, G_J^* \otimes \Omega_{\partial X}).$$

Thus the index is independent of s. It is also independent of the lower order terms in P and in B_j, *and it is stable under arbitrary small perturbations of the coefficients.*

Proof. From the properties of the left inverse in Theorem 20.1.7 it follows that if $u \in \overline{H}_{(m)}(X^\circ, E)$ and (20.1.1) is valid with $f \in \overline{H}_{(s-m)}(X^\circ, F)$,

$g_j \in H_{(s-m_j-\frac{1}{2})}(\partial X, G_j)$ for some $s \geqq m$, then $u \in \bar{H}_{(s)}(X^\circ, E)$. In particular, the kernel of (20.1.2) is contained in $C^\infty(X, E)$, and it suffices to prove that the range is defined by C^∞ relations when $s=m$. Thus assume that

$$v \in H_{(0)}(X, F^* \otimes \Omega_X), \quad h_j \in H_{(m_j+\frac{1}{2}-m)}(\partial X, G_j^* \otimes \Omega_{\partial X}),$$

and that

$$(Pu, v) + \sum (B_j u, h_j) = 0 \quad \text{when } u \in C^\infty(X, E).$$

In particular we can define u by (20.1.12) with $f=0$,

$$u = TP^c Sg; \quad g_j \in C^\infty(\partial X, G_j).$$

Then $Pu = K_2 g$, $Bu = g + K_4 g$ where K_2 and K_4 are continuous maps from $\mathscr{D}'$ to C^∞. Thus

$$\sum (g_j, h_j) = -(K_2 g, v) - (K_4 g, h)$$

is a continuous linear form on g_j in the distribution topology, so $h_j \in C^\infty$. Since

$$(Pu, v) = -\sum (B_j u, h_j), \quad u \in C^\infty(X, E),$$

we obtain $P^* v = 0$ in X° by taking u equal to 0 near the boundary. Thus the Cauchy data of v are well defined (Theorems B.2.9 and B.2.8), and they are in C^∞ since they can be expressed in terms of h. Using Theorem 1.2.6 (Borel's theorem) we can define v in $\hat{X} \setminus X^\circ$ so that $v \in C_0^\infty(\hat{X} \setminus X^\circ)$ has these Cauchy data on ∂X and the higher order derivatives with respect to x_n on ∂X are chosen so that $P^* v$ vanishes of infinite order there. After this extension $v \in H_{(0)}^{\text{comp}}(\hat{X})$ and $P^* v \in C^\infty(\hat{X})$. Hence $v \in C^\infty(\hat{X})$ which completes the proof that the range is defined by smooth relations independent of s. Stability of the index under perturbations of lower order follows from Corollary 19.1.8 and stability for arbitrary small perturbations is a consequence of Corollary 19.1.6. The proof is complete.

Theorem 20.1.8 shows that the index of an elliptic boundary problem is determined by the topological data defined by the operators involved. In Section 20.3 we shall show that the determination of the index can be reduced to the same problem on a manifold with no boundary. The proof will require some elementary facts on ordinary differential operators with constant coefficients which will be discussed in Section 20.2. It will also be necessary to generalize the results of this section. First we note that there is no need to assume that B_j are differential operators. The proofs are unchanged if $B_j = B_j^c \gamma$ where B_j^c are pseudo-differential operators from sections of E^m to sections of G_j on ∂X satisfying the ellipticity condition. However, we must also allow more general operators P, and that requires more work. The most natural setting is to allow P to be any pseudo-differential operator satisfying the transmission condition. However, the boundary conditions are then far more difficult to state (see the references given in the notes) so we shall use another not quite invariant class of operators which has the advantage that its elements are close to being differential operators at the boundary.

We shall assume as above that a neighborhood of ∂X in X has been identified with $\partial X \times [0,1)$, and that E, F in this "collar" are identified with the liftings from ∂X. Write $X_\delta = \partial X \times [0,\delta)$ and define the extension $\hat{X}$ as before. We shall assume that $P = P^b + P^i$ where $P^i \in \Psi^m_{\text{phg}}(X^\circ)$ and the kernel has support in $(X \setminus X_{\frac{1}{2}}) \times (X \setminus X_{\frac{1}{2}})$, say, and

$$P^b = \sum_0^m P_j^b D_n^j.$$

Here m is a positive integer and P_j^b is a C^∞ function of $x_n \in (-1,1)$ with values in $\Psi^{m-j}_{\text{phg}}(\partial X)$ vanishing for $x_n > 2/3$ say. We assume that P_m^b is a vector bundle map $E \to F$. By definition the principal symbol of P is the sum of those of P^b and of P^i. As indicated above we allow B_j to be of the form

$$\sum_0^{m-1} B_{jk} D_n^k$$

where $B_{jk} \in \Psi^{m_j-k}_{\text{phg}}(\partial X; E, G_j)$. The principal symbol defined in the obvious way is homogeneous of degree m_j and is a polynomial in ξ_n of degree $< m$. It is clear that Definition 20.1.1 makes sense in this more general situation and that we still have a continuous map (20.1.2). To extend Theorem 20.1.2 we must reexamine the definition and the properties of the parametrix T, for we can no longer use a standard pseudo-differential operator. Since our present setup is not invariant we shall use a local variant of the arguments used to prove Theorem 20.1.2.

In terms of a local coordinate system on ∂X the symbol of P in the corresponding part of the collar is a sum of a standard symbol of order m and a polynomial of degree m in ξ_n for which the coefficient of ξ_n^j is a symbol of degree $m-j$ in the other variables. Thus the symbol will locally satisfy estimates of the form

$$|D_\xi^\alpha D_x^\beta a(x,\xi)| \leq C_{\alpha\beta}(1+|\xi|)^{m-\alpha_n}(1+|\xi'|)^{-|\alpha'|} \tag{20.1.17}$$

where $\xi' = (\xi_1, \ldots, \xi_{n-1})$. The first time we differentiate the symbol of P^b with respect to ξ' the term ξ_n^m drops out since P_m^b is just a multiplication operator. Thus $\partial a/\partial \xi_j$ satisfies (20.1.17) with m replaced by $m-1$ for every j and not only for $j=n$.

Before proceeding with our discussion of the boundary problem we must make some remarks on the calculus of pseudo-differential operators. Denote by $S^{m,m'}$ the set of all $a \in C^\infty(\mathbb{R}^n \times \mathbb{R}^n)$ such that

$$|D_\xi^\alpha D_x^\beta a(x,\xi)| \leq C_{\alpha\beta}(1+|\xi|)^{m-\alpha_n}(1+|\xi'|)^{m'-|\alpha'|}. \tag{20.1.17'}$$

Lemma 20.1.9. *If $a \in S^{m,m'}$ then $a(x,D)$ is continuous from $H_{(s+m,t+m')}(\mathbb{R}^n)$ to $H_{(s,t)}(\mathbb{R}^n)$ for all s, $t \in \mathbb{R}$. If $b \in S^{\mu,\mu'}$ then $a(x,D)\,b(x,D) = c(x,D)$ where*

$c\in S^{m+\mu,m'+\mu'}$, more precisely, $c-ab\in S^{m+\mu,m'+\mu'-1}$. If $a^{(j)}\in S^{m-1,m'}$ for every j then $c-ab\in S^{m+\mu-1,m'+\mu'}$.

Proof. $S^{m,m'}$ is the space of symbols of weight $(1+|\xi|)^m(1+|\xi'|)^{m'}$ with respect to the σ temperate metric

$$|dx|^2+|d\xi'|^2/(1+|\xi'|)^2+|d\xi_n|^2/(1+|\xi|^2).$$

Hence it follows from Theorems 18.5.10 and 18.5.4 that $c\in S^{m+\mu,m'+\mu'}$ and that $c-ab\in S^{m+\mu,m'+\mu'-1}$. Hence

$$a_{s,t}(x,D)=(1+|D|^2)^{s/2}(1+|D'|^2)^{t/2}\,a(x,D)\,(1+|D|^2)^{-(s+m)/2}(1+|D'|^2)^{-(t+m')/2}$$

is in Op $S^{0,0}$ and therefore L^2 continuous by Theorem 18.6.3. This proves the continuity statement.

In order to prove the last statement we write

$$b(x,\xi)-b(y,\xi)=\sum(x_j-y_j)\,b_j(x,y,\xi),$$

$$b_j(x,y,\xi)=\int_0^1 b_{(j)}(y+t(x-y),\xi)\,dt.$$

When $y\in\mathbb{R}^n$ is regarded as a parameter it is clear that b_j is uniformly bounded in $S^{\mu,\mu'}$, and so are the y derivatives. Now

$$a(x,D)\,b(x,D)=a(x,D)\,b(y,D)+\sum(x_j-y_j)\,a(x,D)\,b_j(x,y,D)+C(x,y,D);$$
$$C(x,y,D)=-i\sum a^{(j)}(x,D)\,b_j(x,y,D).$$

Thus $C(x,y,\xi)$ and its y derivatives are uniformly bounded in $S^{m+\mu-1,m'+\mu'}$ Taking $y=x$ we obtain

$$c(x,\xi)=a(x,\xi)\,b(x,\xi)+C(x,x,\xi)$$

which proves that $c-ab\in S^{m+\mu-1,m'+\mu'}$. The proof is complete.

The following lemma is essentially a case of Lemma 18.4.3:

Lemma 20.1.10. *Let $\chi\in C_0^\infty(\mathbb{R}^n)$, $a\in S^{m,0}$, and assume that*

$$|a(x,\xi)|\geqq c(1+|\xi|)^m \quad \text{if } \xi\in\operatorname{supp}(1-\chi).$$

Then $b(x,\xi)=(1-\chi(\xi))/a(x,\xi)$ is in $S^{-m,0}$. If $a^{(j)}\in S^{m-1,0}$ then $b^{(j)}\in S^{-m-1,0}$.

Proof. Assume that estimates of the form (20.1.17)′ have already been proved for derivatives of order $<|\alpha+\beta|$ of b. Then we obtain by differentiating the equation $ab=1-\chi$ that

$$|a(x,\xi)\,D_\xi^\alpha D_x^\beta b(x,\xi)|\leqq C_{\alpha\beta}(1+|\xi|)^{-\alpha_n}(1+|\xi'|)^{-|\alpha'|},$$

and this implies that

$$|D_\xi^\alpha D_x^\beta b(x,\xi)|\leqq C'_{\alpha\beta}(1+|\xi|)^{-m-\alpha_n}(1+|\xi'|)^{-|\alpha'|}.$$

Since

$$b^{(j)}(x,\xi) = -a^{(j)}(x,\xi)/a(x,\xi)^2$$

outside supp χ, the second statement follows from the first.

Remark. The lemma and the proof remain valid if a is $N \times N$ matrix valued and $a(x,\xi)^{-1}$ can be estimated by $(1+|\xi|)^{-m}$.

Assuming as we may that P is defined and elliptic on $\hat{X}$ we can now prove the interior regularity result we need:

Proposition 20.1.11. *If* $u \in H^{\text{comp}}_{(s+m-1)}(\hat{X})$ *and* $Pu \in H^{\text{comp}}_{(s)}(\hat{X})$ *then* $u \in H^{\text{comp}}_{(s+m)}(\hat{X})$. *For every compact subset K of X we have if* supp $u \subset K$

$$\|u\|_{(s+m)} \leqq C(\|Pu\|_{(s)} + \|u\|_{(s+m-1)}).$$

Proof. Since for every $\phi \in C_0^\infty$

$$\|P\phi u\|_{(s)} \leqq \|\phi Pu\|_{(s)} + C\|u\|_{(s+m-1)}$$

it is sufficient to prove the statement when u has compact support in a coordinate patch. We may assume that the coordinate patch corresponds to the set $\{x \in \mathbb{R}^n; |x|<2\}$ and that supp $u \subset \{x \in \mathbb{R}^n; |x| \leqq \frac{1}{2}\}$. Set $a(x,\xi) = p(\psi(x)x,\xi)$ where p is the symbol matrix of P in the local coordinates, for some trivializations of E and F, and ψ is chosen as in Lemma 19.2.11. Then $a \in S^{m,0}$, $a^{(j)} \in S^{m-1,0}$ for every j, $a(x,\xi) = p(x,\xi)$ in a neighborhood of supp u, and $|\xi|^m a(x,\xi)^{-1}$ is bounded for large $|\xi|$. Using Lemma 20.1.10 we can now choose a matrix symbol $b \in S^{-m,0}$ so that $b^{(j)}(x,\xi) \in S^{-m-1,0}$ for every j, and $b(x,\xi)a(x,\xi)$ is the identity for large $|\xi|$. By Lemma 20.1.9 it follows that $b(x,D)a(x,D) = I + R(x,D)$ where $R \in S^{-1,0}$, hence $R(x,D)$ is continuous from $H_{(s+m-1)}$ to $H_{(s+m)}$. Thus

$$\begin{aligned} \|u\|_{(s+m)} &\leqq \|b(x,D)a(x,D)u\|_{(s+m)} + \|R(x,D)u\|_{(s+m)} \\ &\leqq C(\|a(x,D)u\|_{(s)} + \|u\|_{(s+m-1)}). \end{aligned}$$

This proves the proposition.

Next we shall discuss regularity at the boundary. In doing so we consider a local coordinate patch where $P = P^b$; we trivialize E there first and then trivialize F using the identification of E and F given by multiplication with P^b_m; After extending the symbols of P and of B_j to the whole space as in the proof of Proposition 20.1.11 we obtain the following situation. The symbol of P is of the form

$$p(x,\xi) = \sum_0^m p_j(x,\xi')\,\xi_n^j$$

where $p_j \in S^{m-j}(\mathbb{R}^n \times \mathbb{R}^{n-1})$, p_m is the identity $N \times N$ matrix and $|\xi|^m p(x,\xi)^{-1}$ is bounded for large $|\xi|$. Similarly the symbol of B_j is

$$b_j(x',\xi) = \sum_0^{m-1} b_{jk}(x',\xi')\,\xi_n^k$$

where $b_{jk}\in S^{m-k}(\mathbb{R}^{n-1}\times\mathbb{R}^{n-1})$. As in Lemma 20.1.10 we define $\tau(x,\xi)=(1-\chi(\xi))\,p(x,\xi)^{-1}$ where $\chi\in C_0^\infty$ is equal to 1 in such a large set that $\tau\in S^{-m,0}$ and $\tau^{(j)}\in S^{-m-1,0}$ for every j. Corresponding to (20.1.3) we set

$$\tau(x,D)\,p(x,D)=I+r_E(x,D),\qquad p(x,D)\,\tau(x,D)=I+r_F(x,D)$$

where $r_E, r_F\in S^{-1,0}$. Thus the operators $r_E(x,D)$ and $r_F(x,D)$ are continuous from $H_{(s,t)}(\mathbb{R}^n)$ to $H_{(s+1,t)}(\mathbb{R}^n)$ for all s, t while $\tau(x,D)$ is continuous from $H_{(s,t)}(\mathbb{R}^n)$ to $H_{(s+m,t)}(\mathbb{R}^n)$. We must also establish estimates similar to those in Proposition 20.1.6. If $f\in\mathscr{S}(\mathbb{R}^n)$ we denote as usual by f^0 the function which is equal to f in $\overline{\mathbb{R}}^n_+$ and 0 elsewhere. Let $\tilde{\tau}(x,D)f^0$ be the restriction of $\tau(x,D)f^0$ to $\mathbb{R}^n_+$ and define $\tilde{r}_E(x,D)$, $\tilde{r}_F(x,D)$ similarly.

Lemma 20.1.12. *If $f\in\mathscr{S}$ then*

$$\|\tilde{\tau}(x,D)f^0\|_{(s+m,t)}+\|\tilde{r}_E(x,D)f^0\|_{(s+1,t)}+\|\tilde{r}_F(x,D)f^0\|_{(s+1,t)} \leqq C_{s,t}\|f\|_{(s,t)},\qquad s\geqq 0, \tag{20.1.18}$$

with the norms of $\bar{H}_{(.,.)}(\mathbb{R}^n_+)$. If $f\in H_{(-m,t)}(\mathbb{R}^n)$ and $x_n=0$ in supp *f, then we have for every v*

$$\|\tilde{\tau}(x,D)f\|_{(v,t-v)}+\|\tilde{r}_E(x,D)f\|_{(v+1-m,t-v)}+\|\tilde{r}_F(x,D)f\|_{(v+1-m,t-v)} \leqq C_{s,t,v}\|f\|_{(-m,t)} \tag{20.1.19}$$

where the norms on the left-hand side are in $\overline{H}_{(.,.)}(\mathbb{R}^n_+)$ and the norm of f on the right-hand side is in $H_{(-m,t)}(\mathbb{R}^n)$.

Proof. The essential point is that $p(x,\xi)$ has a nice asymptotic expansion as $\xi_n\to\infty$. (Thus it is really the transmission condition that we use.) With the notation

$$\Xi=\xi_n+i(|\xi'|^2+1)^{\frac{1}{2}}$$

we can write

$$p(x,\xi)=\sum_0^m P_j(x,\xi')\,\Xi^j$$

where $P_m=I$ and $P_j\in S^{m-j}(\mathbb{R}^n\times\mathbb{R}^{n-1})$. Then we can successively choose $\tau_j\in S^j(\mathbb{R}^n\times\mathbb{R}^{n-1})$ so that for $N=1,2,\ldots$

$$\Xi^N p(x,\xi)\sum_0^{N-1}\tau_j(x,\xi')\,\Xi^{-m-j}=\Xi^N I+\sum_0^{m-1}\rho_{N,j}(x,\xi')\,\Xi^{-j},$$

where $\rho_{N,j}$ is in S^{j+N}. In fact, for $N=1$ this means that $\tau_0=I$ and that $\rho_{1,j}=P_{m-1-j}$. If the identity is established for one value of N then

$$\begin{aligned}&\Xi^{N+1}p(x,\xi)\sum_0^N\tau_j(x,\xi')\,\Xi^{-m-j}\\ &\quad=\Xi^{N+1}I+\sum_0^{m-1}\rho_{N,j}(x,\xi')\,\Xi^{1-j}+\sum_0^m P_{m-j}(x,\xi')\,\Xi^{1-j}\tau_N(x,\xi')\end{aligned}$$

which is of the desired form if $\tau_N(x,\xi')=-\rho_{N,0}(x,\xi')$ and

$$\rho_{N+1,j}=\rho_{N,j+1}+P_{m-1-j}\tau_N(x,\xi')$$

for $j=0,\ldots,m-1$. If we multiply the identity by $\tau(x,\xi)=(1-\chi(\xi))\,p(x,\xi)^{-1}$ and by Ξ^{-N}, it follows that

$$\tau(x,\xi)=\sum_0^{N-1}\tau_j(x,\xi')\,\Xi^{-m-j}+\rho_N(x,\xi)$$

where

$$\rho_N(x,\xi)=-\chi(\xi)\sum_0^{N-1}\tau_j(x,\xi')\,\Xi^{-m-j}-(1-\chi(\xi))\,p(x,\xi)^{-1}\sum_0^{m-1}\rho_{N,j}(x,\xi')\,\Xi^{-j-N}.$$

Since $\|f\|_{(0,s+t)}\leqq\|f\|_{(s,t)}$, $s\geqq 0$, and the norm of f^0 in $H_{(0,s+t)}(\mathbb{R}^n)$ is equal to that of f in $\bar{H}_{(0,s+t)}(\mathbb{R}^n_+)$, the fact that $\rho_N\in S^{-m-N,N}$ gives when $N\geqq s$

$$\|\rho_N(x,D)f^0\|_{(s+m,t)}\leqq\|\rho_N(x,D)f^0\|_{(m+N,s+t-N)}\leqq C\|f^0\|_{(0,s+t)}\leqq C\|f\|_{(s,t)}.$$

The continuity of $\rho_N(x,D)$ from $H_{(-m,t)}$ to $H_{(\nu,t-\nu)}$ when $\nu\leqq N$ also gives an estimate for the term $\|\rho_N(x,D)f\|_{(\nu,t-\nu)}$ in (20.1.19).

For the operator

$$\Xi(D)^{-k}u(x)=i^{-k}\int_0^\infty \exp(-t(|D'|^2+1)^{\frac{1}{2}})\,t^{k-1}u(x',x_n+t)\,dt/(k-1)!$$

the restriction of the left-hand side to the half space $x_n>0$ is determined by the restriction of u to that half space. Since $\operatorname{Op}(\tau_j\Xi^{-m-j})$ is continuous from $H_{(s,t)}(\mathbb{R}^n)$ to $H_{(s+m,t)}(\mathbb{R}^n)$ we also have continuity in the corresponding restriction spaces in $\mathbb{R}^n_+$. This proves the estimate for $\tilde{\tau}(x,D)u$ in (20.1.18) and (20.1.19) since the terms just discussed give no contribution at all in (20.1.19).

To prove the estimates for $\tilde{r}_E(x,D)$ and $\tilde{r}_F(x,D)$ we shall examine $p(x,D)\,\tau(x,D)-(p\,\tau)(x,D)$ and $\tau(x,D)\,p(x,D)-(\tau\,p)(x,D)$ using the same decomposition of τ. First we observe that the symbols of $p(x,D)\,\rho_N(x,D)-(p\,\rho_N)(x,D)$ and $\rho_N(x,D)\,p(x,D)-(\rho_N\,p)(x,D)$ are in $S^{-N,N-1}$ for every N, and for large enough N we can therefore argue in the same way as when we estimated $\tau(x,D)u$. The symbols of

$$\tau_j(x,D')\,\Xi(D)^{-m-j}p(x,D)-(\tau_j\,p\,\Xi^{-m-j})(x,D)$$

and the corresponding difference with the order of the factors interchanged are in $S^{-1,0}$ for every j. Thus the operator is then continuous from $H_{(s,t)}(\mathbb{R}^n)$ to $H_{(s+1,t)}(\mathbb{R}^n)$ for all s,t which again implies continuity in the corresponding restriction spaces in $\mathbb{R}^n_+$. This completes the proof.

With P replaced by $p(x,D)$ and T replaced by $\tau(x,D)$ we now repeat the steps which led to the proof of Theorem 20.1.2. First we define the matrix Q of pseudo-differential operators by (20.1.7). Only the principal symbol of Q^2-Q is now equal to 0, which means that Q^2-Q improves differentiability by one unit. The proof of Corollary 20.1.4 therefore gives that

$$\overline{H}_{(s,t)}(\mathbb{R}^n_+)\ni f\mapsto Q\gamma\tau(x,D)f^0\in\bigoplus H_{(s+t+m-j+\frac{1}{2})}$$

is continuous for $s\geqq 0$. If the hypothesis on Q in Proposition 20.1.5 is only fulfilled on the principal symbol level, that is, $Q^2-Q\in\Psi^{-1}$, then the conclusions remain valid with $\equiv$ interpreted as equality of principal symbols. We can also obtain $BQS-I_G\in\Psi^{-\infty}$ by taking $S=QT$ where $BQ^2T-I_G\in\Psi^{-\infty}$. For our Douglis-Nirenberg systems we thus obtain $S_{kj}=S'_{kj}\in\Psi^{k-m_j}$ and $S''_{kl}\in\Psi^{k-l}$. The error term K in (20.1.11) is now continuous from $\overline{H}_{(s,t)}(\mathbb{R}^n_+)$ to $\overline{H}_{(s+1,t)}(\mathbb{R}^n_+)$ when $s\geqq m$. For the error terms $K_1,\dots,K_4$ in (20.1.13) and (20.1.14) we obtain:

K_1 is continuous from $\overline{H}_{(s,t)}(\mathbb{R}^n_+)$ to $\overline{H}_{(s+1,t)}(\mathbb{R}^n_+)$ when $s\geqq 0$;
K_2 is continuous from $H_{(t+m-m_j-\frac{1}{2})}(\mathbb{R}^{n-1})$ to $\overline{H}_{(s+1,t-s)}(\mathbb{R}^n_+)$ for arbitrary s and t;
K_3 is continuous from $\overline{H}_{(s,t)}(\mathbb{R}^n_+)$ to $\bigoplus H_{(s+t+m-m_j+\frac{1}{2})}(\mathbb{R}^{n-1})$ if $s\geqq 0$;
K_4 has a C^∞ kernel.

In the verification, which we leave for the reader, it is useful to recall from the proof of Proposition 20.1.6 that

$$\|P^c U\|_{(-m,s)}\leqq C\sum\|U_j\|_{(s-j-\frac{1}{2})}.$$

Suppose now that $u\in\overline{H}_{(m)}(\mathbb{R}^n_+)$ and that u vanishes outside some large ball. We shall prove that $u\in\overline{H}_{(s)}(\mathbb{R}^n_+)$ if

$$p(x,D)u\in\overline{H}_{(s-m)}(\mathbb{R}^n_+),\qquad b_j(x,D)u\in H_{(s-m_j-\frac{1}{2})}(\mathbb{R}^{n-1}).$$

In the proof we may assume that we know already that $u\in\overline{H}_{(s')}(\mathbb{R}^n_+)$ for some $s'\geqq m$ with $s\leqq s'+1$. Since

$$u=(I+\tau(x,D)P^cS''\gamma)\tau(x,D)f^0+\tau(x,D)P^cSg+Ku$$

the assertion follows from the continuity properties of K above and from the continuity of the other operators given by Lemma 20.1.12.

Next assume as in the proof of Theorem 20.1.8 that we have a relation

$$0=\int_{x_n>0}(Pu)v\,dx+\sum\int B_ju(x')h_j(x')\,dx',\qquad u\in C_0^\infty(\overline{\mathbb{R}}^n_+)$$

where $v\in\overline{H}^{\mathrm{loc}}_{(0,t)}(\mathbb{R}^n_+)$ and $h_j\in H^{\mathrm{loc}}_{(t+m_j+\frac{1}{2}-m)}(\mathbb{R}^{n-1})$.

We shall prove that this must remain true with t replaced by $t+1$, hence for all t. To do so we take $f\in C_0^\infty(\mathbb{R}^n)$ and set

$$u=\phi((I+\tau(x,D)P^cS''\gamma)\tau(x,D)f^0+\tau(x,D)P^cSg) \tag{20.1.20}$$

where $\phi\in C_0^\infty(\mathbb{R}^n)$ is fixed. Then

$$\begin{aligned}\|Pu-\phi f\|_{(0,-t)}&+\sum\|B_ju-\phi g_j\|_{(-t-m_j-\frac{1}{2}+m)}\\ &\leqq C(\|f\|_{(0,-t-1)}+\sum\|g_j\|_{(-t-m_j-\frac{3}{2}+m)})\end{aligned}$$

by the estimates for K_j given above and the continuity of the operators in the right-hand side of (20.1.20) which follows from Lemma 20.1.12. Hence it follows that

$$|\int_{x_n>0} \phi f v\,dx + \sum \int \phi g_j h_j\,dx'| \leq C'(\|f\|_{(0,-t-1)} + \sum \|g_j\|_{(-t-m_j-\frac{1}{2}+m)}).$$

This proves that $\phi v \in \bar{H}_{(0,t+1)}(\mathbb{R}^n_+)$ and $\phi h_j \in H_{(t+m_j+\frac{1}{2}-m)}$ as claimed. Thus we have $h_j \in C^\infty$. Since ${}^t P v \in C^\infty$ and the Cauchy data of v are in C^∞ it follows as in the proof of Theorem 20.1.8 that $v \in C^\infty(\bar{\mathbb{R}}^n_+)$.

Theorem 20.1.8′. *Let $P = P^b + P^i$ where P^i is a pseudo-differential operator of order m with kernel of compact support in $X^\circ \times X^\circ$ and P^b is a differential operator in x_n which is pseudo-differential along ∂X in the collar, as described above. Assume that P is elliptic and that the boundary operators B_j are elliptic with respect to P. Then the conclusions of Theorem 20.1.8 remain valid.*

Proof. First we prove that the operator

$$\bar{H}_{(m)}(X^\circ, E) \ni u \mapsto (Pu, B_1 u, \dots, B_J u) \in H_{(0)}(X^\circ, F) \oplus \bigoplus_j H_{(m-m_j-\frac{1}{2})}(\partial X, G_j) \tag{20.1.21}$$

is a Fredholm operator with range equal to the annihilator of a finite dimensional subspace of $C^\infty(X, F^* \otimes \Omega_X) \oplus \bigoplus_j C^\infty(\partial X, G_j^* \otimes \Omega_{\partial X})$. Let u_ν be a weakly convergent sequence in $\bar{H}_{(m)}(X^\circ, E)$ such that Pu_ν is norm convergent in $H_{(0)}(X^\circ, F)$ and $B_j u_\nu$ is norm convergent in $H_{(m-m_j-\frac{1}{2})}(\partial X, G_j)$ for every j. Then u_ν is norm convergent in $\bar{H}_{(m-1)}(X^\circ, E)$ so the sequence ϕu_ν has the same properties if $\phi \in C_0^\infty(X)$. When $\phi \in C_0^\infty(X^\circ)$ it follows from Proposition 20.1.11 that ϕu_ν is also norm convergent in $\bar{H}_{(m)}(X^\circ, E)$, and if $\operatorname{supp}\phi$ is sufficiently close to a boundary point the same conclusion follows from the fact that the operator K above is compact from $\bar{H}_{(m)}$ to $\bar{H}_{(m)}$. By Proposition 19.1.3 this proves that the operator (20.1.21) has finite dimensional kernel and closed range. Now we have proved that the orthogonal space of the range is in $C^\infty(X, F^* \otimes \Omega_X) \oplus \bigoplus_j C^\infty(\partial X, G_j^* \otimes \Omega_{\partial X})$. (The regularity in the interior follows from Proposition 20.1.11 applied to the adjoint of P.) By Lemma 19.1.4 and the closed graph theorem this implies finite dimensionality. If $s \geq 0$ and

$$(f, g_1, \dots, g_J) \in \bar{H}_{(s)}(X^\circ, F) \oplus \bigoplus_j H_{(s+m-m_j-\frac{1}{2})}(\partial X, G_j)$$

is orthogonal to this space we know that the boundary problem (20.1.1) has a solution $u \in \bar{H}_{(m)}(X^\circ, E)$. We have also proved that the solution is in fact in $\bar{H}_{(s+m)}(X^\circ, E)$ in a neighborhood of the boundary; the same conclusion follows in the interior from Proposition 20.1.11. The proof is complete.

Theorem 20.1.8′ contains a global regularity statement which it is useful to microlocalize to a result analogous to Theorem 8.3.1. In a preliminary step we keep the hypotheses on P and B_j.

Lemma 20.1.13. *Let P and B_j have the properties in Theorem* 20.1.8′, *let u and f be in $\mathcal{N}(X)$, and assume that*

$$Pu=f \quad in\ X, \qquad B_j u=g_j \quad in\ \partial X, \qquad j=1,\dots,J.$$

Then

$$WF_b(u)|_{\partial X}=WF_b(f)|_{\partial X}\cup\bigcup_{j=1}^{J} WF(g_j). \tag{20.1.22}$$

Proof. It follows from Theorems 18.3.32 and 18.3.27 that the left-hand side contains the right-hand side. To prove the opposite inclusion we first choose $\delta>0$ so that $WF(u)$ contains no element (x,ξ) with $\xi'=0$ and $0<x_n\leqq\delta$, which is possible since $u\in\mathcal{N}(X)$. (See the proof of Theorem 18.3.32.) Let W' be the union of the right-hand side of (20.1.22) and

$$\{(x,\xi');\ 0<x_n\leqq\delta, (x,\xi)\in WF(f) \text{ for some } \xi_n\}.$$

This is a closed set since $WF_b(f)$ is closed. We shall call a symbol $\chi\in S^\infty(\overline{\mathbb{R}}_+\times T^*(\partial X))$ admissible if $\chi(x,\xi')=0$ for $x_n\geqq\delta$ and χ is of order $-\infty$ in a conic neighborhood of W'. Thus $\chi(x,D')f\in C^\infty$ by Theorems 18.1.36 and 18.3.32 if χ is admissible.

We can choose s and t so that $u\in\overline{H}^{loc}_{(s,t)}(X_1^\circ)$, hence

$$\chi(x,D')u\in\overline{H}_{(s,t)} \quad \text{for every admissible } \chi\in S^0. \tag{20.1.23}$$

The same result follows with t replaced by $t-\mu$ if $\chi\in S^\mu$. From (20.1.23) it follows that for admissible $\chi\in S^0$

$$P\chi(x,D')u=\chi(x,D')f+[P,\chi(x,D')]u\in\overline{H}_{(s-m+1,t)}, \tag{20.1.24}$$

for $[P,\chi(x,D')]=\sum_{j<m} D_n^j\chi_j(x,D')$ where $\chi_j\in S^{m-1-j}$ is admissible. Hence

$$D_n^j\chi(x,D')u\in\overline{H}_{(s-j+1,t-1)} \tag{20.1.25}$$

if $j=m$. By induction we obtain the same result for $j<m$, for if it is already proved with j replaced by $j+1$ then

$$D_n^{j+1}\chi(x,D')u\in\overline{H}_{(s-j,t-1)}, \qquad D_n^j\chi(x,D')u\in\overline{H}_{(s-j,t)}$$

by the inductive hypothesis and (20.1.23). In view of Theorem B.2.3 we conclude that (20.1.25) holds. When $j=0$ we have proved (20.1.23) with (s,t) replaced by $(s+1,t-1)$. After a finite number of iterations we obtain (20.1.23) with some new (s,t) such that $s\geqq m$. Then

$$B_j\chi(x,D')u=\chi(x',0,D')g_j+[B_j,\chi(x,D')]u\in\overline{H}_{(s+t-m_j+\frac{1}{2})}$$

so recalling (20.1.24) we conclude using Theorem 20.1.8′ that $\chi(x,D')u\in\overline{H}_{(s+1,t)}$. Thus (20.1.23) is true with s replaced by $s+1$ and no change of t, which means that $\chi(x,D')u\in C^\infty(X)$ and proves that the left-hand side of (20.1.22) is contained in the right-hand side. The proof is complete.

Lemma 20.1.13 can be given a more general form:

Theorem 20.1.14. *Let X be a C^∞ manifold with boundary ∂X, and let P be a differential operator from $C^\infty(X,E)$ to $C^\infty(X,F)$ where E and F are complex C^∞ vector bundles with the same fiber dimension. We assume that ∂X is non-characteristic. Let $B_1,\dots,B_J$ be boundary differential operators, and define the characteristic set $\operatorname{Char}(P;B_1,\dots,B_J)\subset T^*(\partial X)\setminus 0$ as the set of all $(x',\xi')\in T^*(\partial X)\setminus 0$ such that either P is characteristic for some $(x,\xi)\in T^*(X)$ in the inverse image of (x',ξ') under the natural map, or else the map in condition* (ii) *of Definition* 20.1.1 *fails to be injective. If $u, f\in\mathcal{N}(X)$ and $Pu=f$ in X°, $B_j u=g_j$ in ∂X, $j=1,\dots,J$, then*

$$WF_b(u)|_{\partial X}\subset\operatorname{Char}(P;B_1,\dots,B_J)\cup WF_b(f)|_{\partial X}\cup\bigcup_{j=1}^{J} WF(g_j). \tag{20.1.26}$$

Proof. We may assume that $X=\mathbb{R}^n_+$ and have to show that $(0,\xi_0')\notin WF_b(u)$ if $(0,\xi_0')$ is not in the set on the right-hand side. Dropping some superfluous B_j's we may assume that the map in condition (ii) of Definition 20.1.1 is bijective at $(0,\xi_0')$. Write

$$P(x,\xi)=\sum_0^m P_j(x,\xi')\,\xi_n^j,$$

and set

$$\tilde{P}(x,\xi)=P_m(x)\,\xi_n^m+\sum_0^{m-1}\big(P_j(x,\xi')\,\psi(\xi')+P_j(0,\xi_0'\,|\xi'|/|\xi_0'|)(1-\psi(\xi'))\big)\,\xi_n^j$$

where $\psi\in C^\infty(\mathbb{R}^{n-1})$ is homogeneous of degree 0 in ξ' when $|\xi'|\geqq 1$ and equal to 1 in a conic neighborhood of ξ_0'. If the support of ψ is small enough then $\tilde{P}$ is elliptic in a neighborhood of 0, and the boundary problem $(\tilde{P};\tilde{B}_1,\dots,\tilde{B}_J)$ is elliptic too if $\tilde{B}_j$ are defined similarly. Now $\tilde{P}u=\tilde{f}$ in X°, $\tilde{B}_j u=\tilde{g}_j$ in ∂X where

$$(0,\xi_0')\notin WF_b(f-\tilde{f})\cup\bigcup WF(g_j-\tilde{g}_j)$$

so it follows from Lemma 20.1.13 that $(0,\xi_0')\notin WF_b(u)$. The proof is complete.

20.2. Preliminaries on Ordinary Differential Operators

Let E be a finite dimensional vector space over $\mathbb{C}$ and let $A\in\mathscr{L}(E,E)$. Then the differential equation

$$Du-Au=0 \tag{20.2.1}$$

where $u\in C^\infty(\mathbb{R},E)$ and $D=-i\,d/dt$ has $\dim E$ linearly independent solutions. The solution with $u(0)=u_0$ is given by

$$u(t)=(2\pi i)^{-1}\oint (zI-A)^{-1}u_0\,e^{itz}\,dz=\sum \operatorname{Res}(zI-A)^{-1}u_0\,e^{itz},$$

where the contour integral is taken along the boundary of a domain containing the eigenvalues of A. If A has no real eigenvalues then $u(t) = u_+(t) + u_-(t)$ where u_+ is the sum of the residues in the upper half plane and u_- is the sum of those in the lower half plane; u_+ is bounded on $\mathbb{R}_+$ and u_- is bounded on $\mathbb{R}_-$. Thus

$$q = \sum_{\operatorname{Im} z > 0} \operatorname{Res}(zI - A)^{-1} \tag{20.2.2}$$

is the projection on the space M^+ of solutions bounded on $\mathbb{R}_+$ annihilating the space M^- of solutions bounded on $\mathbb{R}_-$, if solutions are identified with their values at 0.

Without violating the "ellipticity condition" that A has no real eigenvalues one can deform A to a very special form:

Proposition 20.2.1. *If A has no real eigenvalues and q is defined by* (20.2.2), *then*

$$A_\tau = (1-\tau)A + \tau(i\lambda q - i\lambda(I-q)) \tag{20.2.3}$$

has no real eigenvalues for $0 \leqq \tau \leqq 1$ if $\lambda > 0$, and

$$q = \sum_{\operatorname{Im} z > 0} \operatorname{Res}(zI - A_\tau)^{-1}, \quad 0 \leqq \tau \leqq 1. \tag{20.2.2)'$$

Proof. Let q_z be the residue of $(zI - A)^{-1}$ at z. Then $q_z = 0$ unless z is an eigenvalue, and q_z is then the projection on the space of generalized eigenvectors belonging to the eigenvalue z along the space spanned by the other generalized eigenvectors. Thus the projections q_z commute and

$$q = \sum_{\operatorname{Im} z > 0} q_z.$$

The restriction of $A - zI$ to $q_z E$ is a nilpotent operator and the restriction of q is the identity (zero) if $\operatorname{Im} z > 0$ ($\operatorname{Im} z < 0$). The restriction of A_τ to $q_z E$ is thus the sum of a nilpotent operator and the identity times

$$(1-\tau)z + \tau i \lambda \operatorname{sgn} \operatorname{Im} z,$$

which has an imaginary part of the same sign as $\operatorname{Im} z$. Thus A_τ cannot have a real eigenvalue when $0 \leqq \tau \leqq 1$. The spaces of generalized eigenvectors do not depend on τ although the eigenvalues do. The proof is complete.

The interest of the proposition is that on one hand A_0 is an arbitrary operator with no real eigenvalues and on the other the homotopic A_1 is very special. We can write down the solutions of the equation $Du - A_1 u = 0$ explicitly,

$$u(t) = \exp(-\lambda t) q u_0 + \exp(\lambda t)(I-q) u_0.$$

Higher order operators are not as easy to deform as first order ones. We shall therefore discuss a reduction to first order systems. Let for some $m > 1$

$$p(z) = \sum_0^m p_j z^j \tag{20.2.4}$$

where $p_j \in \mathscr{L}(E,E)$ and $p_m = I$. Every solution of the equation $p(D)u=0$ can be written in the form

$$u(t) = (2\pi i)^{-1} \oint p(z)^{-1} U(z) e^{itz} dz = \sum \operatorname{Res}(p(z)^{-1} U(z) e^{itz})$$

with a unique polynomial $U(z)$ of degree $<m$ with values in E. In terms of the Cauchy data U is given by

$$U(z) = \sum_{k \leqq j < m} p_{j+1} D^{j-k} u(0) z^k.$$

(Cf. (20.1.5).)

In the standard reduction to first order systems one simply introduces $D^j u(t)$, $0 \leqq j < m$, as new unknowns. However, we must not introduce any real eigenvalues and must therefore as in Proposition 20.2.1 use the operators $D + i\lambda$ and $D - i\lambda$ instead for some $\lambda > 0$. First note that we can write

$$p(z) = \sum_0^m a_j (z - \lambda i)^j (z + \lambda i)^{m-j}. \tag{20.2.4$'$}$$

With $(z - \lambda i)/(z + \lambda i) = w$, that is, $z = \lambda i (w+1)/(1-w)$, this means that

$$\sum_0^m a_j w^j = ((1-w)/2\lambda i)^m p(\lambda i (1+w)/(1-w)).$$

Thus we conclude that for some constants c_{jk}

$$a_j = \sum_0^m c_{jk} p_k \lambda^{k-m}. \tag{20.2.5}$$

If $p(D)u = 0$ and we set

$$u_j = (D - \lambda i)^j (D + \lambda i)^{m-1-j} u, \qquad 0 \leqq j < m,$$

it follows that we obtain a system of first order equations

$$a_m (D - \lambda i) u_{m-1} + \sum_0^{m-1} a_j (D + \lambda i) u_j = 0,$$

$$(D + \lambda i) u_j - (D - \lambda i) u_{j-1} = 0, \quad 0 < j < m.$$

This is essentially the usual procedure modified by the Möbius transformation above.

In order to make a smooth transition to the original equation we define for $0 \leqq \tau \leqq 1$ a polynomial $P_\tau(z)$ with values in $\mathscr{L}(E^m, E^m)$ by the block matrix

(20.2.6)
$$\begin{pmatrix} p(z) & 0 & \dots & \\ 0 & M_0(z) & 0 & \dots \\ 0 & \dots & & \\ 0 & \dots & & M_0(z) \end{pmatrix} + \begin{pmatrix} R_0(\tau,z) & R_1(\tau,z) & \dots & \dots & R_{m-1}(\tau,z) \\ -\tau M_1(z) & 0 & \dots & \dots & \dots \\ 0 & -\tau M_1(z) & 0 & \dots & \dots \\ 0 & 0 & \dots & -\tau M_1(z) & 0 \end{pmatrix}.$$

Here $M_0(z)=(z+\lambda i)^m$, $M_1(z)=(z-\lambda i)(z+\lambda i)^{m-1}$, and

$$R_0(\tau,z)=-\tau^m p(z)+\tau^m a_0 M_0(z),$$
$$R_j(\tau,z)=\tau^{m-j} a_j M_0(z), \quad 0<j<m-1,$$
$$R_{m-1}(\tau,z)=\tau a_{m-1} M_0(z)+\tau a_m M_1(z).$$

The coefficient of z^m in $P_\tau(z)$ has determinant 1, and since $a_0+\ldots+a_m=I$ an easy calculation also gives that it has the factorization

$$\begin{pmatrix} I & \tau^{m-1}(a_1+\ldots+a_m) & \ldots & \tau(a_{m-1}+a_m) \\ 0 & I & 0 & \ldots \\ & \cdots & & \\ 0 & \ldots & 0 & I \end{pmatrix} \begin{pmatrix} I & 0 & \ldots & \\ -\tau I & I & 0 & \ldots \\ & \cdots & & \\ 0 & \ldots & -\tau I & I \end{pmatrix}.$$

Thus the coefficients of the inverse are just linear functions of $a_0,\ldots,a_m$, and this remains true if $a_0,\ldots,a_m$ are replaced by more general operators.

Note that the second term in (20.2.6) vanishes for $\tau=0$ and that $P_1(z)$ becomes of first order in z if one factors out $(z+\lambda i)^{m-1}$. Let us now determine the solutions $U=(U_0,\ldots,U_{m-1})$ of the equation $P_\tau(D)U=0$. Explicitly the equation is

$$(1-\tau^m)p(D)U_0+(D+\lambda i)^m(a_0\tau^m U_0+\ldots+a_{m-1}\tau U_{m-1})$$
$$+\tau(D+\lambda i)^{m-1}(D-\lambda i)a_m U_{m-1}=0,$$
$$(D+\lambda i)^m U_j=\tau(D+\lambda i)^{m-1}(D-\lambda i)U_{j-1}, \quad 0<j<m.$$

If U is a solution which is bounded on $\mathbb{R}_+$ then the derivatives are also bounded so we can cancel a factor $(D+\lambda i)^{m-1}$ because $(D+\lambda i)f=0$ has no solution bounded on $\mathbb{R}_+$ other than 0. Hence

$$(D+\lambda i)U_j=\tau(D-\lambda i)U_{j-1}; \quad (D+\lambda i)^j U_j=\tau^j(D-\lambda i)^j U_0.$$

The first equation can now be rewritten $p(D)U_0=0$.

If $\det p(z)\neq 0$ when $z\in\mathbb{R}$ then the space M_τ^+ of solutions of $P_\tau(D)U=0$ which are bounded on $\mathbb{R}_+$ is thus obtained by taking U_0 in the set M^+ of solutions of $p(D)u=0$ which are bounded on $\mathbb{R}_+$ and taking for U_j the exponentially decreasing solution U_j of

$$(D+\lambda i)^j U_j=\tau^j(D-\lambda i)^j U_0.$$

In particular, it follows that $\det P_\tau(z)\neq 0$ when $z\in\mathbb{R}$.

Suppose now that we have a set of boundary operators

$$B_j(D)=\sum_{k\leq m_j} b_{jk}D^k, \quad j=1,\ldots,J, \tag{20.2.7}$$

where $b_{jk}\in\mathscr{L}(E,G_j)$ for some other finite dimensional vector space G_j. Assume that

$$M^+\ni u\mapsto(B_1(D)u(0),\ldots,B_J(D)u(0))\in\bigoplus G_j \tag{20.2.8}$$

is bijective. Then it follows that

(20.2.9) $$M_\tau^+ \ni U \mapsto (B_1(D)\,U_0(0), \ldots, B_J(D)\,U_0(0)) \in \bigoplus G_j$$

is also bijective.

Assuming that $m_j < m$ for every j we can write

(20.2.7)′ $$B_j(z) = \sum_{k<m} \beta_{jk}(z - \lambda i)^k (z + \lambda i)^{m-1-k}.$$

Here β_{jk} can be computed from the identity

$$\sum \beta_{jk} w^k = ((1-w)/2\lambda i)^{m-1} B_j(\lambda i(1+w)/(1-w))$$

which shows that for some constants c'_{kl}

(20.2.10) $$\beta_{jk} = \sum_{l \leq m_j} c'_{kl}\, b_{jl}\, \lambda^{l+1-m}.$$

For $U \in M_\tau^+$ we have

$$\tau^k (D - \lambda i)^k (D + \lambda i)^{m-1-k} U_0 = (D + \lambda i)^{m-1} U_k.$$

If $\tau \neq 0$ it follows that

(20.2.11) $$B_j(D)\,U_0 = \sum_0^{m-1} \tau^{-k} \beta_{jk} (D + \lambda i)^{m-1} U_k.$$

For reference in Section 20.3 we sum up the preceding conclusions:

Proposition 20.2.2. *Let $p(z)$ be defined by (20.2.4) where $p_j \in \mathscr{L}(E,E)$ and $p_m = I$. Assume that $p(z)$ is invertible for real z. Define $a_j \in \mathscr{L}(E,E)$ by (20.2.4)′ or equivalently by (20.2.5), and define a polynomial $P_\tau(z)$ with coefficients in $\mathscr{L}(E^m, E^m)$ by (20.2.6). Then the coefficient of z^m has determinant 1 and the entries in the block matrix for the inverse are linear in $a_0, \ldots, a_m$; $P_\tau(z)$ is invertible for real z, and the projection of solutions of $P_\tau(D)\,U = 0$ on the first component is a bijection of the set M_τ^+ of solutions bounded on $\mathbb{R}_+$ on the set M^+ of solutions of $p(D)\,u = 0$ bounded on $\mathbb{R}_+$. If $B_j(D)$ is of order $m_j < m$ with coefficients in $\mathscr{L}(E, G_j)$, then (20.2.8) is bijective if and only if (20.2.9) is bijective. When $\tau \neq 0$ we have (20.2.11) if $U \in M_\tau^+$ and β_{jk} is defined by (20.2.7)′ or equivalently by (20.2.10).*

Remark. If $P_{\tau m}$ is the coefficient of z^m in P_τ then the preceding conclusions hold for $P_{\tau m}^{-1} P_\tau$ as well. For this operator the coefficient of D^m is also the identity. When $\tau = 0$ this does not affect the operator P_τ.

20.3. The Index for Elliptic Boundary Problems

As in Section 20.1 X will denote a compact C^∞ manifold with boundary ∂X, and E, F will be two C^∞ complex vector bundles on X of the same fiber dimension. We assume that a neighborhood of ∂X in X has been identified with $\partial X \times [0,1)$ and use the notation (x', x_n), $x' \in \partial X$ and $0 \leqq x_n < 1$, for points in this collar. There we also choose an identification of $E_{(x',x_n)}$, $F_{(x',x_n)}$ with $E_{(x',0)}$, $F_{(x',0)}$. We shall discuss the index of elliptic boundary problems for operators P of the form $P = P^b + P^i$ where $P^i \in \Psi^m_{\text{phg}}(X^\circ; E, F)$ has a kernel of compact support in $X^\circ \times X^\circ$, and

$$P^b = \sum_0^m P_j^b D_n^j.$$

Here m is a positive integer and P_j^b is a C^∞ function of $x_n \in (-1,1)$ with values in $\Psi^{m-j}_{\text{phg}}(\partial X)$ vanishing for $x_n > \frac{2}{3}$, say; P_m^b is a bundle map. We define the principal symbol and ellipticity as explained in Section 20.1. Next we determine the index in a very special situation:

Proposition 20.3.1. *Let $E = F$ and assume given a decomposition $E|_{\partial X} = E^+ \oplus E^-$. Choose $\phi \in C_0^\infty((-1,1))$ so that $0 \leqq \phi \leqq 1$ and $\phi = 1$ in a neighborhood of 0, and consider $\phi(x_n)$ as a function on X with support in the collar. Choose*

$$\Lambda^\pm \in \Psi^1_{\text{phg}}(\partial X, E^\pm, E^\pm), \quad \Lambda \in \Psi^1_{\text{phg}}(X^\circ; E, E)$$

with principal symbols equal to $\lambda^\pm$ and λ times the identity, where $\lambda^\pm > 0$ in $T^*(\partial X) \smallsetminus 0$ and $\lambda > 0$ in $T^*(X^\circ) \smallsetminus 0$. If $u \in C^\infty(X, E)$ the restriction to the collar can be written $u = (u^+, u^-)$ where $u^\pm$ is a section of $E^\pm$, and we define

$$P^b u = \phi(x_n)((D_n + i\Lambda^+)u^+, (-D_n + i\Lambda^-)u^-). \tag{20.3.1}$$

Set $P^i u = i(1-\phi(x_n))\Lambda(1-\phi(x_n))u$, $P = P^b + P^i$ and

$$Bu = u^-|_{\partial X} \in C^\infty(\partial X, E^-).$$

Then the boundary problem (P, B) is elliptic and the index of the corresponding operator $\bar{H}_{(s)}(X^\circ, E) \to \bar{H}_{(s-1)}(X^\circ, E) \oplus H_{(s-\frac{1}{2})}(\partial X, E)$ is equal to 0 when $s \geqq 1$.

Proof. If $(y, \eta) \in T^*(\partial X) \smallsetminus 0$ then the solutions of the equations

$$(D_n + i\lambda^+(y,\eta))u^+ = 0, \quad (-D_n + i\lambda^-(y,\eta))u^- = 0$$

which are bounded on $\mathbb{R}_+$ are $u^+ = 0$, $u^- = u_0^- \exp(-x_n \lambda^-(y,\eta))$, so the ellipticity is obvious. To determine the kernel, which we know is in C^∞, we choose a hermitian structure in E so that E^+ and E^- are orthogonal in the collar. We also choose a positive density in X which in the collar is the product of one in ∂X and the Lebesgue density dx_n. If $u \in C^\infty(X, E)$ then

$$\begin{aligned} 2\,\mathrm{Im}(P^b u, u) &= ((P^b u, u) - (u, P^b u))/i \\ &= 2\,\mathrm{Re}((\phi\Lambda^+ u^+, u^+) + (\phi\Lambda^- u^-, u^-)) + ((\phi D_n u^+, u^+) \\ &\quad - (u^+, \phi D_n u^+) - (\phi D_n u^-, u^-) + (u^-, \phi D_n u^-))/i. \end{aligned}$$

By a partial integration we move the operators D_n to the other side and obtain in view of the positivity of Λ^+ and Λ^-

$$2\,\mathrm{Im}(P^b u, u) \geqq \|u^+\|^2_{\partial X} - \|u^-\|^2_{\partial X} - C\|u\|^2_X.$$

Since $\operatorname{Re} P^i$ is also bounded from below it follows that

$$2\operatorname{Im}(Pu,u) \geqq -C'(u,u),$$

if $u \in C^\infty(X,E)$ and $Bu=0$. Thus the kernel of $(P+itI,B)$ is equal to $\{0\}$ if $2t>C'$. Replacing P by $P+itI$ we assume in what follows that the kernel is equal to $\{0\}$.

The hermitian structure and densities introduced above allow us to regard the adjoints of Λ and of $\Lambda^\pm$ as operators in the same bundles. Assume now that $v \in C^\infty(X,E)$ and $h \in C^\infty(\partial X, E^-)$ are orthogonal to the range of (P,B), that is,

$$(v,Pu)_X + (h,u^-)_{\partial X} = 0, \quad u \in C^\infty(X,E).$$

Then we have $P^*v=0$ and $v^+=0$ on ∂X since there is no boundary condition on u^+. Now $-P^*$ is of the same form as P but with the roles of E^- and E^+ interchanged. If the number t above is chosen large enough it follows that $v=0$, hence that $h=0$. Thus the index is equal to 0, and the proof is complete.

Note that the positivity of the imaginary part of the symbol makes all operators considered in Proposition 20.3.1 homotopic when E, E^+ and E^- are fixed. We shall now show that the index theorem on manifolds without boundary allows one to handle general symbols in the interior of X. Thus we keep the assumption that there is given a decomposition $E|_{\partial X} = E^+ \oplus E^-$. We also assume that E is identified with F in the collar $\partial X \times [0,1)$ but do not require that this is true in all of X. As before P^b is defined by (20.3.1), but P^i is now any operator in $\Psi^1_{\mathrm{phg}}(X^\circ; E, F)$ with kernel of compact support in $X^\circ \times X^\circ$ such that $P = P^b + P^i$ is elliptic. To determine the index we shall form the double $\tilde{X}$ of X consisting of two copies X_1 and X_2 of X identified in ∂X. The C^∞ structure is defined by identifying the collar in X_1 with $\partial X \times [0,1)$ and that in X_2 with $\partial X \times (-1,0]$ by changing the sign of the x_n coordinate. On $\tilde{X}$ we have a bundle $\tilde{E}$ obtained by gluing together the bundle E in X_1 and in X_2, and a bundle $\tilde{F}$ is obtained by gluing together the bundle F in X_1 with the bundle E in X_2. This is possible and the vector bundle structure is obvious since in the collars both bundles are obtained by lifting the same bundle from ∂X. We keep the definition (20.3.1) of P^b in the double collar $\partial X \times (-1,1) \subset \tilde{X}$ and choose $P_2^i \in \Psi^1_{\mathrm{phg}}(X_2^\circ; E, E)$ as in Proposition 20.3.1. If we regard P_2^i as an operator in $\tilde{X}$ then $\tilde{P} = P^b + P^i + P_2^i$ maps sections of $\tilde{E}$ on $\tilde{X}$ to sections of $\tilde{F}$, and $\tilde{P}$ has an elliptic principal symbol $\tilde{p}$ in our present sense although $\tilde{P}$ is not quite a pseudo-differential operator.

Proposition 20.3.2. *Under the preceding assumptions $\tilde{P}$ is a Fredholm operator $H_{(s)}(\tilde{X},E) \to H_{(s-1)}(\tilde{X},\tilde{F})$ for every $s \geqq 1$. The index of $\tilde{P}$ is equal to the index of the boundary problem (P,B) and also equal to $s-\operatorname{ind}\tilde{p}$.*

Proof. We can rewrite the equation

$$\tilde{P}\tilde{u} = \tilde{f} \tag{20.3.2}$$

where $\tilde{u}\in H_{(s)}(\tilde{X},\tilde{E})$ and $\tilde{f}\in H_{(s-1)}(\tilde{X},\tilde{F})$ by regarding the restrictions of $\tilde{u}$ and of $\tilde{f}$ to X_1 and to X_2 as sections $u_1,u_2\in\bar{H}_{(s)}(X,E)$ and $f_1\in\bar{H}_{(s-1)}(X,F)$, $f_2\in\bar{H}_{(s-1)}(X,E)$ such that

$$(20.3.2)' \quad Pu_1=f_1, \quad P_2u_2=f_2 \quad \text{in } X; \quad u_1^+=u_2^+, \quad u_1^-=u_2^- \quad \text{in } \partial X.$$

Here P_2 has the form in Proposition 20.3.1 but with the roles of E^+ and E^- interchanged since D_n is replaced by $-D_n$ when the sign of x_n is changed. The boundary operator

$$B_\tau(u_1,u_2)=(u_2^+-\tau u_1^+, u_1^- - \tau u_2^-)$$

is elliptic with respect to the system in (20.3.2)′ for any τ; the boundary conditions in (20.3.2)′ are obtained when $\tau=1$. In fact, for the bounded solutions of the ordinary system of differential equations which occurs in Definition 20.1.1 we find as at the beginning of the proof of Proposition 20.3.1 that $u_1^+=u_2^-=0$, and the boundary conditions then give $u_1^-=u_2^+=0$. The index is therefore independent of τ, and when $\tau=0$ it is equal to $\operatorname{ind}(P,B)$ by Proposition 20.3.1 for then there is no coupling between u_1 and u_2. Thus the index of the elliptic boundary problem $((P,P_2),B_1)$ is equal to the index of the boundary problem (P,B).

Now the index of the operator (20.1.2) is equal to the index of the operator

$$\{u\in\bar{H}_{(s)}(X^\circ,E);\, Bu=0\}\ni u\mapsto Pu\in\bar{H}_{(s-m)}(X^\circ,F)$$

provided that the equation $Bu=g$ alone can be solved for any $g\in C^\infty$, for this restriction of the domain does not change the kernel and gives an isomorphic cokernel. In our situation the map$(u_1,u_2)\mapsto B_1(u_1,u_2)$ is obviously surjective so this observation is applicable. Hence we conclude that $\tilde{P}$ is a Fredholm operator, and the index is equal to the index of (B,P) since we can return from (20.3.2)′ to (20.3.2).

By Lemma 19.2.6 we can choose a sequence $\Lambda_j^\pm\in\Psi_{\text{phg}}^1(\tilde{X};\tilde{E},\tilde{F})$ with principal symbols converging to those of $\Lambda^\pm$ so that $\Lambda^\pm-\Lambda_j^\pm\to 0$ in $\mathscr{L}(H_{(s)}(\tilde{X},\tilde{E}),\ H_{(s-1)}(\tilde{X},\tilde{F}))$. For $s\geqq 1$ we conclude for large j that the pseudo-differential operator $\tilde{P}_j$ obtained when $\Lambda^\pm$ is replaced by $\Lambda_j^\pm$ is a Fredholm operator with the same index as $\tilde{P}$. In view of Theorem 19.2.4 it follows that $\operatorname{ind}\tilde{P}=s-\operatorname{ind}\tilde{p}$, which completes the proof.

The boundary problems covered by Proposition 20.3.2 may seem very special but the homotopy invariance of the index gives it a surprisingly wide scope:

Proposition 20.3.3. *Let (P,B) be a first order elliptic boundary problem with B of order 0. Then the direct sum with a suitable operator of the form studied in Proposition 20.3.1, with $E=E^+$, is homotopic to a boundary problem for which the index is given by Proposition 20.3.2.*

Proof. As usual we may assume that $E_x = E_{(x',0)}$ in the collar. The coefficient P_1^b of D_n in P^b is an isomorphism of E on F in $\partial X \times [0,c]$ for some $c \in (0,1)$. We can use it to identify E and F then. Thus

$$P^b = D_n - A(x_n)$$

for $0 \leqq x_n \leqq c$, where $A(x_n) \in \Psi^1_{\text{phg}}(\partial X; E, E)$. Replacing $A(x_n)$ by

$$(1 - \tau \psi(x_n)) A(x_n) + \tau \psi(x_n) A(0), \quad 0 \leqq \tau \leqq 1,$$

we obtain a homotopy of elliptic boundary problems if $\psi = 1$ in a neighborhood of 0 and vanishes outside another small neighborhood, $0 \leqq \psi \leqq 1$. Changing c also we may therefore assume that A is independent of x_n for $0 \leqq x_n \leqq c$. We may also assume that the support of the kernel of P^i does not meet $X \times (\partial X \times [0,c])$ or $(\partial X \times [0,c]) \times X$.

Step I. Let a be the principal symbol of A and let

$$q(y,\eta) = (2\pi i)^{-1} \int^+ (zI - a(y,\eta))^{-1}\, dz$$

be the Calderón projector. It is a C^∞ projection map in E lifted to $T^*(\partial X) \setminus 0$ and is homogeneous of degree 0. Choose $Q \in \Psi^0_{\text{phg}}(\partial X; E, E)$ with principal symbol q and $\Lambda \in \Psi^1_{\text{phg}}(\partial X; E, E)$ with principal symbol equal to a positive multiple λ of the identity. With $\phi \in C_0^\infty((-c,c))$ equal to 1 in a neighborhood of 0 and $0 \leqq \phi \leqq 1$ everywhere we define $P_\tau^b = P^b$ when $x_n \geqq c$ and

$$P_\tau^b = D_n - (1 - \tau\phi) A - \tau\phi(i\Lambda Q - i\Lambda(I - Q)) \quad \text{when } x_n \leqq c.$$

By Proposition 20.2.1 $P_\tau = P_\tau^b + P^i$ is then elliptic and the Calderón projector is equal to q for all $\tau \in [0,1]$. If (P,B) is elliptic it follows that (P_1, B) is elliptic with the same index. Note that

$$P_1^b = \tilde{P}^b + (1-\phi) P^b; \quad \tilde{P}^b = \phi(D_n - i\Lambda Q + i\Lambda(I-Q)).$$

The kernel of $(1-\phi)P^b$ has compact support in $X^\circ \times X^\circ$ so using Lemma 19.2.6 we can find a genuine pseudo-differential operator $\tilde{P}^i$ with kernel of compact support in $X^\circ \times X^\circ$ and symbol so close to that of $(1-\phi)P^b + P^i$ that

$$(1-\tau) P_1 + \tau \tilde{P}$$

is elliptic for $0 \leqq \tau \leqq 1$ if $\tilde{P} = \tilde{P}^b + \tilde{P}^i$. At the boundary these operators do not depend on τ so $(\tilde{P}, B)$ is elliptic with the same index as (P,B).

Step II. We can now start with an operator P such that

$$P^b = \phi(D_n - i\Lambda Q + i\Lambda(I-Q)) = \phi(D_n + i\Lambda(I - 2Q)). \tag{20.3.3}$$

The difference between this situation and that in Proposition 20.3.2 is that the "projection operator" Q in (20.3.3) is a pseudo-differential operator and not as in Proposition 20.3.2 just a bundle map on ∂X. Let $B \in \Psi^0_{\text{phg}}(\partial X; E, G)$ be our boundary operator. (Since P is of first order the boundary operator only acts on the boundary values, that is, it does not involve D_n.) That (P,B)

is elliptic means that the principal symbol $b(y,\eta)$ is bijective from the range of $q(y,\eta)$ to G_y for all $(y,\eta)\in T^*(\partial X)\setminus 0$. Recall from the proof of Proposition 20.1.5 that this is equivalent to the existence of a (unique) map $s(y,\eta)$: $G_y\to E_{(y,0)}$ such that

$$b(y,\eta)\,s(y,\eta)=I_{G_y},\quad q(y,\eta)\,s(y,\eta)=s(y,\eta),\qquad s(y,\eta)\,b(y,\eta)\,q(y,\eta)=q(y,\eta). \tag{20.3.4}$$

The last equality follows since $sbs=s$ by the first one, the range of s is contained in that of q by the second one, and the equality of dimensions implies that they are equal. For $0\leqq\theta\leqq\pi/2$ we now define

$$b_\theta(y,\eta)=(\cos\theta\, b(y,\eta),\sin\theta\, I_{G_y}):\quad E_{(y,0)}\oplus G_y\to G_y,$$

$$s_\theta(y,\eta)=\begin{pmatrix}\cos\theta\, s(y,\eta)\\ \sin\theta\, I_{G_y}\end{pmatrix}:\quad G_y\to E_{(y,0)}\oplus G_y;$$

$$q_\theta(y,\eta)=\begin{pmatrix}(\cos\theta)^2\, q(y,\eta) & \sin\theta\cos\theta\, s(y,\eta)\\ \sin\theta\cos\theta\, b(y,\eta)\,q(y,\eta) & (\sin\theta)^2\, I_{G_y}\end{pmatrix}.$$

Direct calculation shows that $q_\theta(y,\eta)$ is a projection operator in $E_{y,0}\oplus G_y$ and that

$$b_\theta s_\theta=I_{G_y},\quad q_\theta s_\theta=s_\theta,\quad s_\theta b_\theta q_\theta=q_\theta; \tag{20.3.4)'}$$

the last equality follows as above since the range of the projection q_θ has dimension $\dim G_y$ which implies that s_θ and q_θ have the same range. When θ varies from 0 to $\pi/2$ we obtain a homotopy connecting the projection $q(y,\eta)\oplus 0$ to the projection on the second component.

To exploit this homotopy we must first choose a vector bundle H on X such that $H|_{\partial X}=G\oplus G'$ for some vector bundle G' on ∂X. This is possible by Lemma 19.2.14 which shows that we can take $H=X\times\mathbb{C}^N$ for sufficiently large N. The same decomposition of H is then valid in the collar. Now define P_H by Proposition 20.3.1 with E,F,E^+ replaced by H and E^- absent. Then P_H (with no boundary condition) has index 0. We define an operator P_0 from sections of $E\oplus H$ to sections of $F\oplus H$ by

$$P_0(u,v)=(Pu,P_H v), \tag{20.3.5}$$

if u is a section of E and v is a section of H, and we set

$$B_0(u,v)=Bu. \tag{20.3.6}$$

The boundary problem (P_0,B_0) is then elliptic and has the same index as (P,B). We shall prove that it is homotopic to a boundary problem of the form studied in Proposition 20.3.2.

Let q_G be the projection $H\to G$ which is well defined in the collar, and choose a decreasing function $\theta(x_n)$ such that $\phi(x_n)=1$ in $\operatorname{supp}\theta$ and $\theta(x_n)=1$ in a neighborhood of 0. For $0\leqq\tau\leqq\pi/2$ and $u\in C^\infty(X,E)$, $v\in C^\infty(X,H)$ we now define

$$P_\tau^b(u,v)=\phi((D_n u, D_n v)+i(\Lambda(u-2(\cos\tau\theta)^2 Q u-\sin 2\tau\theta\, S q_G v),$$
$$\Lambda_H(v-\sin 2\tau\theta\, BQ u-2(\sin\tau\theta)^2 q_G v))).$$

Here $S\in\Psi^0_{\text{phg}}(\partial X; G, E)$ has principal symbol s, and $\Lambda_H\in\Psi^1_{\text{phg}}(\partial X; H, H)$ has principal symbol λ times the identity and is assumed to be the operator which appears in the definition (20.3.1) of P_H^b. With P_0^i equal to the direct sum of the interior pseudodifferential operator terms in P and in P_H we set $P_\tau = P_\tau^b + P_0^i$ and

$$B_\tau(u,v)=\cos\tau\, Bu+\sin\tau\, q_G v.$$

When $\tau=0$ these definitions agree with (20.3.5), (20.3.6), and (P_τ, B_τ) is elliptic for every τ in view of (20.3.4)′. The index is therefore independent of τ. Now we have

$$P^b_{\pi/2}(u,v)=(D_n u+i\Lambda u, D_n v+i\Lambda_H(v-2q_G v))$$

in a neighborhood of ∂X, and

$$B_{\pi/2}(u,v)=q_G v.$$

This is of the form considered in Proposition 20.3.2. (When we identify E with F as here using the coefficient of D_n in P^b then there is a change of sign in the second component of (20.3.1) which makes it agree with $P^b_{\pi/2}$ for the decomposition $E\oplus H=(E\oplus G')\oplus G$ on ∂X.) As at the end of Step I we can cut $P^b_{\pi/2}$ off and add an approximation for the rest of this operator to the interior one, and this completes the proof.

Remark. It should be observed that the homotopy of P^b to the simple form obtained in Step II assumed that an elliptic boundary problem was known. There is a topological obstruction to the construction of such a homotopy, so elliptic boundary problems do not exist for every elliptic P. If B is of order $\mu\neq 0$ we can of course apply Proposition 20.3.3 after multiplying B to the left by an elliptic operator in ∂X of order $-\mu$.

We shall now study some operators P of order $m>1$ which are very close to first order operators. As usual we identify E_x with $E_{(x',0)}$ in the collar, and we use the coefficient of D_n^m in P^b to identify F with E in a neighborhood of ∂X. Let Λ^+ and Λ be as in Proposition 20.3.1 when $E^+ = E$, and assume that

$$P^b u=(D_n-A)(D_n+i\Lambda^+)^{m-1} \tag{20.3.7}$$

for some $A\in\Psi^1_{\text{phg}}(\partial X; E, E)$ when x_n is small. Also assume that P is elliptic and that each boundary operator B_j factors in a similar way

$$B_j u=\tilde B_j(D_n+i\Lambda^+)^{m-1}u \tag{20.3.8}$$

where $\tilde B_j\in\Psi^{m_j+1-m}_{\text{phg}}(\partial X; E, G_j)$.

Proposition 20.3.4. *Let $\tilde P^b=\phi(D_n-A)$ where $\phi\in C_0^\infty(\mathbb{R})$, $0\leq\phi\leq 1$, $\phi=1$ in a neighborhood of* 0, *and* (20.3.7) *is valid in* $\operatorname{supp}\phi$. *Choose $\psi\in C_0^\infty(-1,1)$ with*

$0 \leqq \psi \leqq 1$ and $\psi = 1$ in supp ϕ. Let $\tilde{P} = \tilde{P}^b + \tilde{P}^i$ where $\tilde{P}^i \in \Psi^1_{\text{phg}}(X^\circ; E, E)$ and the kernel has compact support in $X^\circ \times X^\circ$. If the principal symbol of $\tilde{P}$ multiplied by $(\psi(\xi_n + i\lambda^+) + i(1-\psi)^2 \lambda)^{m-1}$ is sufficiently close to that of P it follows that $(\tilde{P}, \tilde{B})$ is elliptic with the same index as (P, B).

Proof. The ellipticity of $\tilde{P}$ is obvious. To prove the ellipticity of the boundary conditions we observe that if a, $\tilde{b}_j$ are the principal symbols of A, $\tilde{B}_j$, then the solutions of the ordinary differential equation

$$(D_n - a(y,\eta))(D_n + i\lambda^+(y,\eta))^{m-1} u = 0; \quad (y,\eta) \in T^*(\partial X) \setminus 0;$$

which are bounded on $\mathbb{R}_+$ are by assumption mapped bijectively on $\oplus G_j$ by

$$u \mapsto (\tilde{b}_1(y,\eta) v(0), \dots, \tilde{b}_J(y,\eta) v(0))$$

where $v = (D_n + i\lambda^+(y,\eta))^{m-1} u$ satisfies $(D_n - a(y,\eta)) v = 0$. Conversely, if v is a solution of this equation which is bounded on $\mathbb{R}_+$ then $v = (D_n + i\lambda^+(y,\eta))^{m-1} u$ for a unique u bounded on $\mathbb{R}_+$. Thus $(\tilde{P}, \tilde{B})$ is elliptic. By Proposition 20.3.1 the operator Q,

$$Qu = \psi(D_n + i\Lambda^+) u + i(1-\psi) \Lambda (1-\psi) u,$$

has index 0 as operator from $\bar{H}_{(s)}$ to $\bar{H}_{(s-1)}$ for $s \geqq 1$. Hence it follows from Corollary 19.1.7 that for $s \geqq m$ the operator

$$\begin{aligned}\bar{H}_{(s)}(X^\circ, E) \ni u \mapsto (\tilde{P} Q^{m-1} u, \tilde{B}_1 Q^{m-1} u, \dots, \tilde{B}_J Q^{m-1} u) \in \bar{H}_{(s-m)}(X^\circ, F) \\ \oplus H_{(s-m_1-\frac{1}{2})}(\partial X, G_1) \oplus \dots \oplus H_{(s-m_J-\frac{1}{2})}(\partial X, G_J)\end{aligned}$$

has the same index as $(\tilde{P}, \tilde{B})$. The boundary operators are equivalent to B_j. By Lemma 20.1.9 $\tilde{P} Q^{m-1}$ is the sum of a compact operator and an operator in our standard class with symbol close to that of P. Hence the stability properties of the index show that the index is equal to that of (P, B). The proof is complete.

Using Proposition 20.2.2 we shall finally reduce an arbitrary elliptic boundary problem to the type handled by Proposition 20.3.4. In this final reduction we still assume that $E_x = E_{(x',0)}$ in the collar, and we identify E with F in a neighborhood of ∂X so that

$$P^b = \sum_0^m P_j^b D_n^j$$

with P_m^b equal to the identity for $0 \leqq x_n \leqq c$, say. Define Q as in the proof of Proposition 20.3.4 and set for sections $U = (U_0, \dots, U_{m-1})$ of E^m

$$\mathscr{P}_0 U = (P U_0, Q^m U_1, \dots, Q^m U_{m-1}); \quad \mathscr{B}_j^0 U = B_j U_0.$$

This defines a Fredholm operator in the usual spaces with the same index as that of (P, B). The operator Q^m is not quite of our standard class though. However, we can replace it by an operator Q_m which near the boundary keeps the form $(D_n+i\Lambda^+)^m$ and differs in principal symbol and in norm so little from Q^m that the index is not affected. We shall deform the operator $\mathscr{P}_0$ so modified, and the boundary operators as well, so that a new boundary problem is obtained which satisfies the hypotheses of Proposition 20.3.4.

When $0\leqq x_n\leqq c$ we define with c_{jk} as in (20.2.5)

$$A_j=\sum_0^m c_{jk}\, P_k^b\, F^{m-k}\in\Psi^0(\partial X;E,E)$$

where F is a parametrix of Λ^+. Since (20.2.4), (20.2.4)′ is equivalent to (20.2.5) we have

$$\sum_{j=0}^m c_{jk}=\delta_{mk},$$

so $A_0+\ldots+A_m=I$. We have for $0\leqq x_n\leqq c$

$$P^b=\sum_0^m A_j(D_n-i\Lambda^+)^j(D_n+i\Lambda^+)^{m-j}$$

apart from terms of order $-\infty$ in ∂X and order $<m$ with respect to x_n. Without changing the index of (P,B) we can change the definition of P so that this is an exact equality. Let $\chi\in C_0^\infty(-c,c)$ be equal to 1 in a neighborhood of 0, $0\leqq\chi\leqq 1$ everywhere. Now we define $\mathscr{P}_\tau^b$ by (20.2.6) with τ replaced by $\tau\chi$, $p(z)$ replaced by P^b, a_j replaced by A_j, M_0 replaced by $(D_n+i\Lambda^+)^m$ and M_1 replaced by $(D_n-i\Lambda^+)(D_n+i\Lambda^+)^{m-1}$. As pointed out in Section 20.2 the coefficient $\mathscr{P}_{\tau m}^b$ of D_n^m has an inverse whose block matrix entries are just linear combinations of the A_j. Without changing anything outside supp χ we can thus define

$$\hat{\mathscr{P}}_\tau^b=(\mathscr{P}_{\tau m}^b)^{-1}\,\mathscr{P}_\tau^b,\quad 0\leqq x_n\leqq c.$$

Since this is equal to $\mathscr{P}_0^b$ when $\chi(x_n)=0$, we can continue the definition by taking $\hat{\mathscr{P}}_\tau^b=\mathscr{P}_0^b$ for $x_n\geqq c$. Finally we define $\hat{\mathscr{P}}_\tau$ as the sum of $\hat{\mathscr{P}}_\tau^b$ and the direct sum of the interior pseudo-differential terms in P and in Q_m, taken $m-1$ times.

By Proposition 20.2.2 the boundary problem $(\mathscr{P}_\tau,\mathscr{B}^0)$ is elliptic for $0\leqq\tau\leqq 1$. When $\tau=1$ we can use (20.2.10) to choose $\beta_j\in\Psi^{m_j+1-m}(\partial X;E^m,G_j)$ such that $\mathscr{B}_j^0$ for $0\leqq\kappa\leqq 1$ may be replaced by

$$\mathscr{B}_j^\kappa U=(1-\kappa)B_jU_0+\kappa\beta_j(D_n+i\Lambda^+)^{m-1}U$$

because the principal parts of these operators all act in the same way on the functions occurring in Definition 20.1.1. Thus the index of (P,B) is equal to the index of $(\hat{\mathscr{P}}_1,\mathscr{B}^1)$, and this is a problem of the type discussed in Proposition 20.3.4, because $(D_n+i\Lambda^+)^{m-1}$ factors out on the right both in the operator $\hat{\mathscr{P}}_1$ and in the boundary operator $\mathscr{B}^1$.

We end this section by summing up the steps required to determine the index of an elliptic boundary problem:

(i) If $m>1$ one takes the direct sum with the mth power of the operator in Proposition 20.3.1 with $E^+=E$, repeated $m-1$ times, and then deforms to an operator and a boundary condition which can be factored to the right by such an operator raised to the power $m-1$. This factor is then cancelled (Proposition 20.3.4) so that one is left with a first order operator; the order of B is then reduced to 0.

(ii) When $m=1$ one can use Proposition 20.2.1 to find a homotopy to a system of the form (20.3.3) where Q is a pseudo-differential projector. (This is Step I in the proof of Proposition 20.3.3.)

(iii) After taking the direct sum with an operator provided by Proposition 20.3.1, with no boundary condition and containing the bundles where the boundary operators take their values, one can by a homotopy change the Calderón projector to a bundle map and change the boundary condition similarly. (Step II in the proof of Proposition 20.3.3.)

(iv) One can now double the manifold (Proposition 20.3.2) to determine the index.

We have developed the arguments in the opposite order since each step seems better motivated then.

20.4. Non-Elliptic Boundary Problems

In the introduction we indicated how the solution of the Dirichlet problem for the Laplacean can be used to reduce general boundary problems to pseudo-differential operators in the boundary. We shall here outline a general form of this procedure.

Let X be a C^∞ manifold with boundary, E and F two C^∞ complex vector bundles on X, and P: $C^\infty(X,E)\to C^\infty(X,F)$ an elliptic differential operator of order m. Assume that

$$B_j\colon C^\infty(X,E)\to C^\infty(\partial X,G_j),\quad j=1,\dots,J,$$

define an elliptic boundary problem for P as in Definition 20.1.1. Let now

$$C_k\colon C^\infty(X,E)\to C^\infty(\partial X,H_k),\qquad k=1,\dots,K,$$

be another set of boundary differential operators, of transversal order $<m$ of course. We want to study the boundary problem

$$Pu=f \ \text{ in } X;\quad C_k u=h_k \ \text{ in } \partial X;\quad k=1,\dots,K; \tag{20.4.1}$$

where f is a given section of F on X and h_k is a given section of H_k on ∂X. Suppose that u is a solution and set

$$B_j u=g_j,\qquad j=1,\dots,J.$$

Then we obtain using (20.1.11)

$$C_k u=C_k(I+TP^c S''\gamma)\,Tf^0+C_k\,TP^c Sg+C_k Ku.$$

We can write $C_k = C_k^c \gamma$ where C_k^c is a differential operator from $C^\infty(\partial X, E^m)$ to $C^\infty(\partial X, H_k)$. Thus

$$C_k^c Q S g = h_k - C_k(I + TP^c S'' \gamma) T f^0 - C_k K u \tag{20.4.2}$$

where $C_k K u \in C^\infty$ and the other terms in the right-hand side are known. Here we may replace QS by S if we add another C^∞ term on the right-hand side.

On the other hand, let us try to find a solution of (20.4.1) of the form (20.1.12)

$$u = (I + TP^c S'' \gamma) T\tilde{f}^0 + TP^c S g.$$

Then we have (20.1.13), and

$$C_k u = C_k(I + TP^c S'' \gamma) T\tilde{f}^0 + C_k^c Q S g.$$

To solve the boundary problem (20.4.1) is therefore equivalent to solving

$$\tilde{f} + K_1 \tilde{f} + K_2 g = f; \quad C_k^c Q S g = h_k - C_k(I + TP^c S'' \gamma) T\tilde{f}^0. \tag{20.4.3}$$

The first equation gives $\tilde{f} - f \in C^\infty(X, F)$, so the right-hand side in the second equation is $h_k - C_k(I + TP^c S'' \gamma) T f^0$ mod C^∞. Thus we can essentially reduce the study of the boundary problem to the study of the pseudo-differential operator $(C_1^c QS, \dots, C_K^c QS)$ from sections of $\oplus G_j$ to sections of $\oplus H_k$ on ∂X. We write down two cases of this principle, denoting by m_j and by μ_k the degrees of B_j and of C_k.

Theorem 20.4.1. *Let $m < s_0 \leq s_1$. Then the following regularity properties are equivalent:*

(i) *If $u \in \overline{H}_{(m)}(X^\circ, E)$, $Pu \in \overline{H}_{(s_1 - m)}(X^\circ, F)$ and*

$$C_k u \in H_{(s_1 - \mu_k - \frac{1}{2})}(\partial X, H_k), \quad k = 1, \dots, K,$$

then $u \in \overline{H}_{(s_0)}(X^\circ, E)$.

(ii) *If $g_j \in H_{(m - m_j - \frac{1}{2})}(\partial X, G_j)$ and*

$$C_k^c Q S g \in H_{(s_1 - \mu_k - \frac{1}{2})}(\partial X, H_k), \quad k = 1, \dots, K,$$

then $g_j \in H_{(s_0 - m_j - \frac{1}{2})}(\partial X, G_j)$.

Proof. (i) ⇒ (ii). With g_j as in (ii) we set $u = TP^c S g$. Then $u \in \overline{H}_{(m)}(X^\circ, E)$, $Pu \in C^\infty(X, F)$, and

$$C_k u = C_k^c Q S g \in H_{(s_1 - \mu_k - \frac{1}{2})}(\partial X, H_k).$$

By condition (i) this implies that $u \in \overline{H}_{(s_0)}(X^\circ, E)$, hence

$$B_j u \in H_{(s_0 - m_j - \frac{1}{2})}(\partial X, G_j).$$

Now $B_j u - g_j \in C^\infty$ by (20.1.14) so $g_j \in H_{(s_0 - m_j - \frac{1}{2})}$ which proves (ii).

(ii) ⇒ (i). With u as in (i) we set $g_j = B_j u$, $j = 1, \dots, J$. Then (20.4.2) gives

$$C_k^c Q S g \in H_{(s_1 - \mu_k - \frac{1}{2})}(\partial X, H_k), \quad k = 1, \dots, K,$$

so $g_j \in H_{(s_0-m_j-\frac{1}{2})}(\partial X, G_j)$ by (ii). From (20.1.11) it follows now that $u \in \bar{H}_{(s_0)}(X^\circ, F)$, which completes the proof.

In the same way one proves:

Theorem 20.4.2. *Let $m \leqq s_0 \leqq s_1$. Then the following existence statements are equivalent:*

(i) *For arbitrary $f \in \bar{H}_{(s_1-m)}(X^\circ, F)$ and $h_k \in H_{(s_1-\mu_k-\frac{1}{2})}(\partial X, H_k)$, $k=1, \dots, K$, one can find $u \in \bar{H}_{(s_0)}(X^\circ, E)$ with $Pu - f \in C^\infty(X, F)$ and $C_k u - h_k \in C^\infty(\partial X, H_k)$, $k=1, \dots, K$.*

(ii) *For arbitrary $h_k \in H_{(s_1-\mu_k-\frac{1}{2})}(\partial X, H_k)$ one can find*

$$g_j \in H_{(s_0-m_j-\frac{1}{2})}(\partial X, G_j)$$

such that $C_k^c QS\, g - h_k \in C^\infty$, $k=1, \dots, K$.

We leave the proof for the reader and refer to Theorem 26.4.2 for the relations between solvability mod C^∞ and solvability with finite dimensional cokernel.

Notes

The definition of elliptic boundary problems was probably first stated by Lopatinski [1], but a large class of boundary problems for second order equations had been studied earlier by Višik [1]. Elliptic boundary problems were discussed in great detail during the 1950's by many authors such as Agmon-Douglis-Nirenberg [1], Agmon [1], Browder [1], Peetre [1, 3], Schechter [1]. A thorough treatment was given in the predecessor of this book. It was based on methods of the type used here in Sections 17.1 and 17.3, and so is the exposition by Lions and Magenes [1].

Calderón [3] outlined another approach to boundary problems using singular integral operator theory. It was further developed by Seeley [1]. Similar ideas to handle also non-elliptic boundary problems for overdetermined elliptic systems by means of pseudo-differential operators were presented in Hörmander [17]. Section 20.1 follows that paper to a large extent. The discussion of the parametrix for the Calderón projector and the boundary operator given in Proposition 20.1.5 is closer to Grubb [1] though. In the proof of Theorem 20.1.8′ we have also used a case of the symbol classes studied in Grubb [2].

The index problem for elliptic boundary problems was solved by Atiyah and Bott [1]. By a sequence of homotopies of the principal symbol, using the given elliptic boundary operators, they defined a virtual vector bundle on the boundary of the unit ball bundle in $T^*(X)$ and outlined a proof that the index of the elliptic boundary problem is equal to the topological index of this extension. Thus they gave a reduction to the index problem on a manifold without boundary which was already solved. In Section 20.3 we have followed the arguments of Atiyah and Bott [1] closely in principle. (See also Luke [1].) However, they have been modified so that each step has a natural analytical interpretation. It is interesting to compare with the topologically motivated discussion of Atiyah and Bott, so we list an approximate correspondence between their five homotopies and the presentation here:

First homotopy: Change of degree in Proposition 20.3.4 using Proposition 20.3.1. (The latter comes mainly from Boutet de Monvel [2] though.)

Second homotopy: Proposition 20.2.2 and its application at the end of Section 20.3.

Third and fourth homotopies: Proposition 20.2.1 and its application in Step I of the proof of Proposition 20.3.3.

Fifth homotopy: Step II of the proof of Proposition 20.3.3.

Already Višik and Eškin [1–5] established Fredholm properties of boundary problems for elliptic singular integral operators. (See also Eškin [1].) Boutet de Monvel [1] introduced the transmission condition, and in [2] he developed a theory of elliptic boundary problems for such operators, including a determination of the index. More exactly, he considered operators of the form

$$\begin{pmatrix} A & K \\ T & Q \end{pmatrix}: C^\infty(X,E)\oplus C^\infty(\partial X,G)\to C^\infty(X,F)\oplus C^\infty(\partial X,H)$$

where Q is a pseudo-differential operator in ∂X, T is a "trace operator", K is a Poisson operator, that is, the adjoint of a trace operator, and A is essentially the sum of a pseudo-differential operator in X° and the product of a Poisson and a trace operator. (Here we have only considered a pair $\begin{pmatrix} A \\ T \end{pmatrix}$.) The more general setup has the advantage of a better multiplicative structure and a better symmetry under passage to adjoints. However, it would have taken us too far to develop this approach completely. A very detailed exposition has recently been given by Rempel and Schulze [1].

The study of non-elliptic boundary problems has largely been motivated by their occurrence in the theory of functions of several complex variables. (See Kohn-Nirenberg [2].) The study of boundary regularity for the $\bar\partial$ Neumann problem has been brought to great perfection but is beyond the scope of this book.

Chapter XXI. Symplectic Geometry

Summary

We have seen in Sections 6.4 and 18.1 that the principal symbol of a (pseudo-)differential operator in a C^∞ manifold X is invariantly defined on $T^*(X)\smallsetminus 0$, the cotangent bundle with the zero section removed. When discussing the integration theory for a first order differential equation $p(x, \partial\phi/\partial x)=0$, for example the characteristic equation of a (pseudo-) differential operator, we also saw in Section 6.4 the importance of certain geometrical constructions in $T^*(X)$ related to the canonical one form ω, given by $\sum \xi_j dx_j$ in local coordinates, and the symplectic form $\sigma = d\omega = \sum d\xi_j \wedge dx_j$. The integration theory was in fact reduced to the study of manifolds $\subset T^*(X)$ of dimension $\dim X$ where the restriction of σ is equal to 0. Maslov [1] has introduced the term *Lagrangian* for such manifolds. The conormal bundle of a submanifold of X is always Lagrangian. The extension of the theory of conormal distributions which we shall give in Chapter XXV will require a good description of arbitrary Lagrangian manifolds. We shall also need to know the structure of submanifolds of $T^*(X)$ where σ has a fixed rank or where the rank varies in a simple way. To develop such a study it is useful to start from an intrinsic point of view.

By a symplectic manifold S one means a manifold with a given non-degenerate two form σ such that $d\sigma = 0$. A classical theorem of Darboux, which we prove in Section 21.1, states that S is locally isomorphic to an open set in $T^*(\mathbb{R}^n)$. If a multiplication (free action) by positive reals is defined in S and σ is homogeneous of degree 1 with respect to it, then S is locally isomorphic to an open conic subset of $T^*(\mathbb{R}^n)\smallsetminus 0$ with multiplication by reals in S corresponding to multiplication in the fibers. The one form ω is invariantly defined in S and defines in $S/\mathbb{R}_+$ a contact structure. We shall not use this terminology here but rather talk about conic symplectic manifolds which is usually closer to the applications we have in mind.

A diffeomorphism preserving the symplectic form is called symplectic or canonical. In Section 21.2 we give a local symplectic classification of (homogeneous) Lagrangian manifolds as well as two other classes of submanifolds of a symplectic manifold, called involutive and isotropic, which can be thought of as larger resp. smaller than a Lagrangian manifold. A classification of pairs of Lagrangian manifolds is also given.

Section 21.3 is devoted to classification of certain functions under symplectic transformations and multiplication by non-vanishing functions. This is useful because we shall see in Chapter XXV that these changes of the principal symbol of a pseudo-differential operator can be achieved by means of a suitable conjugation and multiplication of the operator. Such arguments will be essential in Chapter XXVI when we study singularities of solutions.

The symplectic classification of geometrical objects is resumed in Section 21.4 with the study of symplectic transformations, or rather relations, having fold type of singularities so that they are in general 2 to 2 correspondences. We then continue with the study of intersecting transversal hypersurfaces when the restriction of the symplectic form to the intersection vanishes simply on a submanifold of codimension 1. The result, called equivalence of glancing hypersurfaces, is essential for a deeper understanding of mixed problems. However, in this book we shall never rely on any of the results proved in Section 21.4.

Already in Section 21.1 we present a small amount of symplectic linear algebra. The topic is pursued in Section 21.6 with a study of the Lagrangian Grassmannian, the space of Lagrangian planes in a symplectic vector space. The main purpose is the study of the density and Maslov bundles which will play a fundamental role in the global machinery to be established in Chapter XXV. Another aspect of symplectic linear algebra is discussed in Section 21.5 where we study the symplectic classification of quadratic forms. This is mainly a preparation for the study of hypoelliptic operators in Chapter XXII.

21.1. The Basic Structure

In the vector space $T^*(\mathbb{R}^n)=\{(x,\xi);\ x,\xi\in\mathbb{R}^n\}$ the symplectic form $\sigma=\sum d\xi_j\wedge dx_j$ is the bilinear form

$$\sigma((x,\xi),(x',\xi'))=\langle x',\xi\rangle-\langle x,\xi'\rangle$$

(see Section 6.4). If e_j and ε_j are the unit vectors along the x_j and ξ_j axes respectively, then we have for $j,k=1,\dots,n$

$$\sigma(e_j,e_k)=\sigma(\varepsilon_j,\varepsilon_k)=0;\quad \sigma(\varepsilon_j,e_k)=-\sigma(e_k,\varepsilon_j)=\delta_{jk} \tag{21.1.1}$$

where δ_{jk} is the Kronecker delta, equal to 1 when $j=k$ and 0 when $j\neq k$. We shall now put this situation in an abstract setting:

Definition 21.1.1. A vector space S over $\mathbb{R}$ (or $\mathbb{C}$) with a non-degenerate antisymmetric bilinear form σ is called a real (complex) symplectic vector space. If S_1, S_2 are symplectic vector spaces with symplectic forms σ_1,σ_2

and $T: S_1 \to S_2$ is a linear bijection with $T^*\sigma_2 = \sigma_1$, that is,

$$\sigma_1(\gamma, \gamma') = \sigma_2(T\gamma, T\gamma'); \quad \gamma, \gamma' \in S_1;$$

then T is called a symplectic isomorphism.

That σ is non-degenerate means of course that

$$\sigma(\gamma, \gamma') = 0 \,\forall\, \gamma' \in S \Rightarrow \gamma = 0.$$

We shall now show that every symplectic vector space is symplectically isomorphic to $T^*(\mathbb{R}^n)$ (resp. $T^*(\mathbb{C}^n)$; from now on we consider only the real case but the complex case is parallel).

Proposition 21.1.2. *Every finite dimensional symplectic vector space S is of even dimension $2n$ and admits a linear symplectic isomorphism $T: S \to T^*(\mathbb{R}^n)$.*

Proof. We just have to show that S has a *symplectic basis*, that is, a basis satisfying (21.1.1). Choose any $\varepsilon_1 \neq 0$ in S. Since σ is non-degenerate we can then choose $e_1 \in S$ with $\sigma(\varepsilon_1, e_1) = 1$. Let S_1 be the space spanned by e_1 and ε_1; it is two dimensional since $\sigma(\varepsilon_1, \varepsilon_1) = 0$. Then

$$S_0 = \{\gamma \in S; \sigma(\gamma, S_1) = 0\}$$

is of codimension 2 in S and supplementary to S_1. In fact, if $\gamma \in S_0 \cap S_1$ then $\gamma = x_1 e_1 + \xi_1 \varepsilon_1$ and

$$0 = \sigma(\gamma, \varepsilon_1) = -x_1, \quad 0 = \sigma(\gamma, e_1) = \xi_1.$$

S_0 is a symplectic vector space, for if $\gamma \in S_0$ and $\sigma(\gamma, S_0) = 0$ then $\sigma(\gamma, S) = 0$ because $\sigma(\gamma, S_1) = 0$ by the definition of S_0. Assuming that the proposition is already proved for symplectic vector spaces of dimension $< \dim S$ we can choose a symplectic basis $e_2, \varepsilon_2, \ldots, e_n, \varepsilon_n$ for S_0 which gives a symplectic basis $e_1, \varepsilon_1, \ldots, e_n, \varepsilon_n$ for S. The proof is complete.

A partial symplectic basis can always be extended to a full symplectic basis:

Proposition 21.1.3. *Let S be a symplectic vector space of dimension $2n$ and A, B two subsets of $\{1, \ldots, n\}$. If $\{e_j\}_{j \in A}$, $\{\varepsilon_k\}_{k \in B}$ are linearly independent vectors in S satisfying (21.1.1) then one can choose $\{e_j\}_{j \notin A}$, $\{\varepsilon_k\}_{k \notin B}$ so that a full symplectic basis for S is obtained.*

Proof. Assume for example that $J \in B \setminus A$. We choose $e_J = e$ so that

$$\sigma(e, e_j) = 0, \quad j \in A; \quad \sigma(e, \varepsilon_k) = -\delta_{J,k}, \quad k \in B,$$

which is possible since $\{e_j\}_{j \in A}$, $\{\varepsilon_k\}_{k \in B}$ are linearly independent and σ is non-degenerate. If

$$x e + \sum_A x_j e_j + \sum_B \xi_k \varepsilon_k = 0$$

then the σ scalar product with ε_j is 0, hence $x=0$. This implies $x_j=\xi_k=0$ so the extension of the system by e_j preserves the linear independence. We can argue in the same way if $A\setminus B$ is not empty. Finally, if $A=B\neq\{1,\dots,n\}$ and S_1 is the space spanned by $\{e_j\}_{j\in A}$ and $\{\varepsilon_j\}_{j\in A}$, then we find as in the proof of Proposition 21.1.2 that

$$S_0=\{\gamma\in S;\sigma(\gamma,S_1)=0\}$$

is supplementary to S_1 and symplectic. Choosing a symplectic basis in S_0 then completes the proof.

A coordinate system $(x,\xi)=(x_1,\dots,x_n,\xi_1,\dots,\xi_n)$ in S is called symplectic if it defines a symplectic map $S\to T^*(\mathbb{R}^n)$. There is an extension theorem for symplectic coordinates parallel to Proposition 21.1.3. To prove it we observe that if L is in the dual space S' of S, that is, a linear form on S, then there is a unique vector $H_L\in S$ such that

$$\langle t,L\rangle=\sigma(t,H_L),\quad t\in S.$$

One calls H_L the *Hamilton vector* of L. The map $S'\ni L\mapsto H_L\in S$ is an isomorphism, and we define a symplectic form in S' by

$$\{L,L'\}=\sigma(H_L,H_{L'})=\langle H_L,L'\rangle.$$

It is called the *Poisson bracket*. If $S=T^*(\mathbb{R}^n)$ then

$$H_L=\sum\partial L/\partial\xi_j\,\partial/\partial x_j-\sum\partial L/\partial x_j\,\partial/\partial\xi_j,$$
$$\{L,L'\}=\sum(\partial L/\partial\xi_j\,\partial L'/\partial x_j-\partial L/\partial x_j\,\partial L'/\partial\xi_j)$$

by an obvious calculation made already in Section 6.4. The coordinates x, ξ in $T^*(\mathbb{R}^n)$ satisfy the so called *commutation relations*

$$\{x_j,x_k\}=0,\quad\{\xi_j,\xi_k\}=0,\quad\{\xi_j,x_k\}=-\{x_k,\xi_j\}=\delta_{jk}. \tag{21.1.1$'$}$$

Application of Proposition 21.1.3 to S' now shows that any partial system of linearly independent forms on S satisfying (21.1.1)$'$ can be extended to a full coordinate system in S satisfying (21.1.1)$'$. But this means precisely that the coordinates define a symplectic isomorphism $S\to T^*(\mathbb{R}^n)$.

We shall continue this brief introduction to symplectic linear algebra in Section 21.2 but now we pass on to the analogue of the preceding discussion for manifolds.

Definition 21.1.4. A C^∞ manifold S with a non-degenerate closed C^∞ two form σ is called a symplectic manifold. If S_1, S_2 are symplectic manifolds with symplectic forms σ_1,σ_2 and $\chi\colon S_1\to S_2$ is a diffeomorphism with $\chi^*\sigma_2=\sigma_1$, then χ is called a symplectomorphism, or a canonical transformation.

An example of a symplectic manifold is $T^*(X)$ if X is any C^∞ manifold. If Y is another C^∞ manifold and $\kappa\colon X\to Y$ is a diffeomorphism, then

$$T^*(X)\ni(x,\xi)\mapsto(\kappa(x),{}^t\kappa'(x)^{-1}\xi)\in T^*(Y)$$

is a symplectomorphism. This is just another way of saying that the symplectic structure in $T^*(X)$ is invariantly defined.

Our next goal is to prove that every symplectic manifold is locally symplectomorphic to $T^*(\mathbb{R}^n)$. However, before doing so we must make some important preliminary observations on the symplectic structure. The symplectic form makes the tangent space $T_\gamma(S)$ of S at any $\gamma \in S$ a symplectic vector space, so S is of even dimension $2n$ by Proposition 21.1.2. If $f \in C^k(S)$, $k \geqq 1$, then df is a linear form on $T_\gamma(S)$ which defines a Hamiltonian vector $H_f(\gamma)$. Working in local coordinates we see at once that H_f is a C^{k-1} vector field on S. It is called the *Hamilton vector field* of f. The *Poisson bracket* of $f, g \in C^k$ is again defined by

$$\{f, g\} = H_f g \in C^{k-1}.$$

To express the condition $d\sigma = 0$ we need a lemma.

Lemma 21.1.5. *If ω is any C^1 two form on a C^2 manifold M and X, Y, Z are three C^1 vector fields on M, then with the notation $[X, Y] = XY - YX$*

$$\begin{aligned}\langle X \wedge Y \wedge Z, d\omega\rangle = {} & X\langle Y \wedge Z, \omega\rangle + Y\langle Z \wedge X, \omega\rangle \\ & + Z\langle X \wedge Y, \omega\rangle - \langle [X, Y] \wedge Z, \omega\rangle \\ & - \langle [Y, Z] \wedge X, \omega\rangle - \langle [Z, X] \wedge Y, \omega\rangle.\end{aligned} \tag{21.1.2}$$

Proof. Both sides are linear in X, Y, Z, ω. If X is replaced by fX, $f \in C^1$, then both sides are multiplied by f since

$$[fX, Y] = f[X, Y] - (Yf)X,$$
$$(Yf)\langle Z \wedge X, \omega\rangle + (Zf)\langle X \wedge Y, \omega\rangle + (Yf)\langle X \wedge Z, \omega\rangle - (Zf)\langle X \wedge Y, \omega\rangle = 0.$$

A similar remark is true for Y and Z. It is therefore sufficient to prove (21.1.2) when M is a neighborhood of 0 in $\mathbb{R}^N$, X, Y, Z are constant vector fields, and $\omega = f_1\, df_2 \wedge df_3$ with $f_1 \in C^1$ and f_2, f_3 equal to coordinates. Then the left-hand side is the determinant with the rows $(X(f_j), Y(f_j), Z(f_j))$, $j = 1, 2, 3$, and the right-hand side is the expansion by the first row.

Now we apply (21.1.2) to the symplectic manifold S with $\omega = \sigma$ and $X = H_f$, $Y = H_g$, $Z = H_h$, where $f, g, h \in C^2(S)$. We have

$$\langle H_g \wedge H_h, \sigma\rangle = \sigma(H_g, H_h) = \{g, h\} = H_g h,$$
$$\langle [H_f, H_g] \wedge H_h, \sigma\rangle = (H_f H_g - H_g H_f)h = \{f, \{g, h\}\} - \{g, \{f, h\}\}$$

so the right-hand side of (21.1.2) becomes

$$(1-2)(\{f, \{g, h\}\} + \{g, \{h, f\}\} + \{h, \{f, g\}\}).$$

Hence the fundamental *Jacobi identity*

$$\{f, \{g, h\}\} + \{g, \{h, f\}\} + \{h, \{f, g\}\} = 0; \quad f, g, h \in C^2(S); \tag{21.1.3}$$

follows from the hypothesis $d\sigma=0$ and is in fact equivalent to it. Replacing the last term by $-\{\{f,g\},h\}$ we can also write it in the equivalent form

$$H_{\{f,g\}}=[H_f,H_g]. \tag{21.1.3$'$}$$

This argument shows that (21.1.3) is in any case a first order differential operator on h and similarly on f or g. If $S=T^*(\mathbb{R}^n)$ we could therefore prove (21.1.3) by noting that it only needs to be verified when f,g,h are coordinate functions and then it is obviously true since any Poisson bracket of them is a constant so that the second Poisson bracket is zero. However, as already emphasized, in our general situation the Jacobi identity expresses the hypothesis $d\sigma=0$.

We are now ready to prove that a symplectic manifold is locally symplectomorphic to $T^*(\mathbb{R}^n)$. In doing so we observe that if $\phi\colon S_1\to S_2$ is a symplectomorphism then

$$\phi^*\{f,g\}=\{\phi^*f,\phi^*g\};\quad f,g\in C^2(S_2). \tag{21.1.4}$$

Indeed, if we use ϕ to identify S_1 and S_2 this formula just states that the Poisson bracket is determined by the symplectic form. Conversely, (21.1.4) implies that $\phi^*\sigma_2$ and σ_1 define the same Poisson bracket, hence that $\phi^*\sigma_2=\sigma_1$ so that ϕ is symplectic. It is of course sufficient that (21.1.4) holds for a system of coordinates. To define a local symplectomorphism from S to $T^*(\mathbb{R}^n)$ therefore means choosing local coordinates satisfying (21.1.1)$'$.

Theorem 21.1.6. *Let S be a symplectic manifold, of dimension $2n$, A and B two subsets of $\{1,\dots,n\}$. If q_j, $j\in A$, and p_k, $k\in B$, are C^∞ functions in a neighborhood of $\gamma_0\in S$ with linearly independent differentials satisfying the commutation relations*

$$\{q_i,q_j\}=0,\quad i,j\in A;\quad \{p_i,p_k\}=0,\quad i,k\in B; \tag{21.1.1$''$}$$
$$\{p_k,q_j\}=\delta_{jk},\quad k\in B,\ j\in A,$$

then one can find local coordinates x, ξ at γ_0 satisfying (21.1.1)$'$ *such that $x_j=q_j$, $j\in A$; $\xi_k=p_k$, $k\in B$.*

In particular, if we start from the empty sets A and B the theorem shows that a neighborhood of γ_0 in S is symplectomorphic to an open set in $T^*(\mathbb{R}^n)$. This is known as the *Darboux theorem.*

Proof. Set $x_j=q_j$, $j\in A$; $\xi_k=p_k$, $k\in B$. If we apply Proposition 21.1.3 to the cotangent space at γ_0 it follows that we can supplement these functions to a local coordinate system satisfying (21.1.1)$'$ at γ_0. Without restriction we may therefore assume that $S\subset\mathbb{R}^{2n}$ is a neighborhood of $\gamma_0=0$, that $q_j=x_j$ when $j\in A$, and that $p_k=\xi_k$ when $k\in B$; moreover (21.1.1)$'$ is valid at 0 which implies that $H_{q_j}=-\partial/\partial\xi_j$ and $H_{p_k}=\partial/\partial x_k$ at 0. Assume for example that $J\notin A$. Then we can find $q_J=q$ with $dq=dx_J$ at 0 satisfying the differential equations

(21.1.5) $\quad H_{q_j}q=\{q_j,q\}=0, \quad j\in A; \quad H_{p_k}q=\{p_k,q\}=\delta_{kJ}, \quad k\in B.$

Indeed, by the Jacobi identity (21.1.3)′ and (21.1.1)″ the vector fields H_{q_j}, H_{p_k} all commute, for the Poisson brackets $\{q_j,p_k\},\dots$, are constant. Hence it follows from the Frobenius theorem (see Appendix C, Corollary C.1.2) that (21.1.5) has a unique C^∞ solution q near 0 with

$$q(x,\xi)=x_J \quad \text{when } \xi_j=0, \quad j\in A, \quad x_k=0, \quad k\in B,$$

for we know that the Hamilton vectors H_{q_j}, H_{p_k} at 0 are linearly independent and span a supplementary plane. At 0 we have $dq=dx_J$ since (21.1.1)′ was already fulfilled there. Hence we can replace A by $A\cup\{J\}$ with $q_J=q$. Continuing in this way we obtain a complete set of symplectic coordinates.

Remark. The important role played by the Jacobi identity in the proof is very natural for we know that the theorem would be false if $d\sigma$ were not 0. Thus the full strength of the Jacobi identity is required.

With symplectic local coordinates we have $\sigma=\sum d\xi_j\wedge dx_j$, thus

$$\sigma^n=n!\,d\xi_1\wedge dx_1\wedge\dots\wedge d\xi_n\wedge dx_n$$

since two forms commute. Hence $\sigma^n/n!$ is a volume form which can be used to orient S and also identify the a-density bundle Ω^a on S with the trivial bundle. In fact, if $\psi\colon S_1\to S_2$ is symplectic, then $\psi^*\sigma_2^n/n!=\sigma_1^n/n!$ so the Jacobian of a symplectic change of variables is one. We can thus identify sections of Ω^a with the functions defining them when symplectic coordinates are used.

Let us also note that the non-degeneracy of σ at $\gamma\in S$ is equivalent to $\sigma^n\neq 0$ at γ, if $\dim S=2n$. In fact, the proof of Proposition 21.1.2 shows that we can always find local coordinates x, ξ such that

$$\sigma=\sum_1^k d\xi_j\wedge dx_j \quad \text{at } \gamma;$$

the construction just breaks off when we have chosen the beginning of a symplectic basis $e_1,\varepsilon_1,\dots,e_k,\varepsilon_k$ and σ is equal to 0 in the plane orthogonal to them with respect to σ. This is true in an odd dimensional space as well. At γ we therefore have $\sigma^k=k!\,d\xi_1\wedge dx_1\wedge\dots\wedge d\xi_k\wedge dx_k\neq 0$ while $\sigma^{k+1}=0$. Thus the rank of the form is $2k$ if k is the largest integer with $\sigma^k\neq 0$ at γ.

In Section 21.4 we shall also encounter degenerate symplectic structures for which an analogue of the Darboux theorem is valid. They will be of the form discussed in the following theorem, which will not be needed though since simplifying additional structure will be present.

Theorem 21.1.7. *Let S be a C^∞ manifold of dimension $2n$, let σ be a closed two form on S, and assume that $\sigma^n=m\,\omega$ where ω is a nowhere vanishing $2n$*

form and $m=0$, $dm \neq 0$ at a point $\gamma_0 \in S$. Assume also that the restriction of σ^{n-1} to the surface $S_0 = m^{-1}(0)$ does not vanish at γ_0. Then we can choose local coordinates x, ξ in a neighborhood of γ_0 such that

$$\sigma = \xi_1 \, d\xi_1 \wedge dx_1 + \sum_2^n d\xi_j \wedge dx_j.$$

Thus we would have the usual symplectic form if we could take $\xi_1^2/2$ as a local coordinate instead of ξ_1. This is of course not permissible since it would identify points with opposite values for ξ_1 and change the differentiable structure where $\xi_1 = 0$.

Proof. We make a preliminary choice of local coordinates such that γ_0 has coordinates 0 and $m = \xi_1$. If i is the inclusion of the surface S_0 where $\xi_1 = 0$ then $\sigma_1 = i^* \sigma$ has rank $2n-2$ by hypothesis, so the radical

$$\{t \in T_\gamma(S_0);\ \sigma_1(t, t') = 0 \ \forall\, t' \in T_\gamma(S_0)\}$$

is a line in $T_\gamma(S_0)$ for every $\gamma \in S_0$ near 0. Locally we can therefore choose a C^∞ vector field $\neq 0$ in the radical. Choosing new coordinates so that the integral curves of the vector field become parallel to the x_1 axis, we have

$$\sigma_1 = \sum_{1<j<k} a_{jk} \, dx_j \wedge dx_k + \sum_{1 \le j<k} b_{jk} \, d\xi_j \wedge d\xi_k + \sum_{j,k>1} c_{jk} \, dx_j \wedge d\xi_k.$$

Since $d\sigma_1 = 0$ the coefficients must all be independent of x_1 for no cancellation of terms in $d\sigma_1$ involving dx_1 can occur. By the Darboux theorem we can therefore choose new coordinates $x_2, \xi_2, \dots, x_n, \xi_n$ so that

$$\sigma_1 = \sum_2^n d\xi_j \wedge dx_j.$$

It follows that

$$\sigma = d\xi_1 \wedge \psi + \sum_2^n d\xi_j \wedge dx_j \quad \text{when } \xi_1 = 0,$$

$$\psi = \sum_1^n a_j(x, \xi') \, dx_j + \sum_2^n b_j(x, \xi') \, d\xi_j.$$

Here we have written $\xi' = (\xi_2, \dots, \xi_n)$. Thus

$$\sigma = \sum_2^n d(\xi_j + a_j \xi_1) \wedge d(x_j - b_j \xi_1) + a_1 \, d\xi_1 \wedge dx_1 \quad \text{when } \xi_1 = 0.$$

We replace the coordinates ξ_j and x_j by $\xi_j + a_j \xi_1$ and $x_j - b_j \xi_1$ when $j > 1$ and note that $a_1 = 0$ since $\sigma^n = 0$ when $\xi_1 = 0$.

The form σ has now been reduced to

$$\sigma = \xi_1 \theta + \sum_2^n d\xi_j \wedge dx_j, \quad \theta = \theta_0 + d\xi_1 \wedge \phi + O(\xi_1)$$

where θ_0 does not involve $d\xi_1$ or ξ_1. We have

$$\phi=\sum_1^n A_j(x,\xi')\,dx_j+\sum_2^n B_j(x,\xi')\,d\xi_j.$$

Hence

$$\sigma=\sum_2^n d(\xi_j+A_j\xi_1^2/2)\wedge d(x_j-B_j\xi_1^2/2)+\xi_1(A_1\,d\xi_1\wedge dx_1+\theta_0)+O(\xi_1^2).$$

Taking $\xi_j+A_j\xi_1^2/2$ and $x_j-B_j\xi_1^2/2$ as new variables we obtain

$$\sigma=\xi_1(A_1\,d\xi_1\wedge dx_1+\theta_0)+O(\xi_1^2)+\sum_2^n d\xi_j\wedge dx_j.$$

Since $d\sigma=0$ we have when $\xi_1=0$

$$d\xi_1\wedge\theta_0=0,$$

hence $\theta_0=0$ since θ_0 does not involve ξ_1 or $d\xi_1$. Now

$$\sigma^n=n!\,A_1\,\xi_1\,d\xi_1\wedge dx_1\wedge\ldots\wedge d\xi_n\wedge dx_n+O(\xi_1^2)$$

so $A_1\neq 0$ when $\xi_1=0$ by hypothesis. Changing the sign of x_1 if necessary we may assume $A_1>0$. If $\xi_1 A_1^{\frac{1}{2}}$ is then taken as new variable instead of ξ_1 we have with a smooth two form σ_2

$$\sigma=\sigma_0+\xi_1^2\sigma_2,\quad \sigma_0=\xi_1\,d\xi_1\wedge dx_1+\sum_2^n d\xi_j\wedge dx_j.$$

Repeating the argument once more, which we leave for the reader to do, we obtain with new coordinates $\sigma=\sigma_0+\xi_1^3\sigma_3$ where σ_3 is also smooth.

If $f\in C^\infty$ the Hamilton vector field H_f with respect to σ is defined when $\xi_1\neq 0$. With respect to σ_0 we get the Hamilton vector field

$$H_f^0=\xi_1^{-1}(\partial f/\partial\xi_1\,\partial/\partial x_1-\partial f/\partial x_1\,\partial/\partial\xi_1)+\sum_2^n(\partial f/\partial\xi_j\,\partial/\partial x_j-\partial f/\partial x_j\,\partial/\partial\xi_j)$$

as is obvious if $\xi_1^2/2$ is used as a new coordinate. We can write

$$\xi_1\sigma_3(t,H_f)=\sigma_0(t,TH_f)$$

where T is a linear transformation in $\mathbb{R}^{2n}$ with smooth coefficients. Hence

$$\sigma_0(t,(I+\xi_1^2T)H_f)=\sigma(t,H_f)=\langle t,df\rangle=\sigma_0(t,H_f^0)$$

which proves that $H_f=(I+\xi_1^2T)^{-1}H_f^0$. Thus $H_f-H_f^0=\xi_1R_f$ where R_f is smooth.

Following the proof of Theorem 21.1.6 closely we shall now choose local coordinate functions $q_1,\ldots,q_n,\,p_1,\ldots,p_n$ such that

$$\sigma=p_1\,dp_1\wedge dq_1+dp_2\wedge dq_2+\ldots+dp_n\wedge dq_n,$$

that is, so that we have the commutation relations

$$p_1 H_{p_1} p_j = p_1 H_{p_1} q_j = 0, \quad j>1;$$
$$H_{p_j} p_k = H_{q_j} q_k = 0, \quad H_{p_j} q_k = \delta_{jk}, \quad j,\ k>1;$$
$$p_1 H_{p_1} q_1 = 1, \quad H_{p_j} q_1 = H_{q_j} q_1 = 0, \quad j>1.$$

We set $p_1=\xi_1$ and then choose $p_2,\dots,p_n,\ q_2,\dots,q_n$ in arbitrary order, leaving the choice of q_1 to the end. We claim that this can be done uniquely with boundary conditions posed exactly as in the proof of Theorem 21.1.6, and that $p_j=\xi_j$, $q_j=x_j$ when $\xi_1=0$ for all j. Suppose that this is already shown until a certain step in the construction which is still incomplete. Then we have to find the next coordinate satisfying first of all the differential equation

$$p_1 H_{p_1} u = \partial u/\partial x_1 + \xi_1^2 R_{p_1} u = 0.$$

This equation is also satisfied by the other coordinate functions f already chosen so $\partial^2 f/\partial x_1 \partial\xi_1 = 0$ when $\xi_1=0$. Hence

$$f = f_0 + p_1 f_1 + p_1^2 f_2$$

where f_0 and f_1 are independent of x_1 and ξ_1, and

$$H_f = H_{f_0} + p_1(H_{f_1} + p_1 H_{f_2}) + (f_1 + 2p_1 f_2) H_{p_1}.$$

The equation $H_f u = c$ may now be replaced by the equation

$$(H_{f_0} + p_1(H_{f_1} + p_1 H_{f_2}))u = c$$

which means $H^0_{f_0} u = c$ when $\xi_1=0$. This will not affect the Frobenius condition when $\xi_1 \neq 0$ so it will be satisfied also when $\xi_1=0$ because of linear independence. Hence the construction of u is done as before, and since our old coordinates satisfy the required equations when $\xi_1=0$ we maintain the conditions $p_j=\xi_j$, $q_j=x_j$ when $\xi_1=0$ every time we choose a new coordinate function. Thus we can carry the construction on until it gives a system of local coordinates.

The definition of a symplectic manifold was motivated by the properties of a cotangent bundle $T^*(X)$ but omitted a great deal of its structure: $T^*(X)$ is also a vector bundle and the symplectic form vanishes in the fibers. We have even $\sigma=d\omega$ where ω is a one form which also vanishes in the fibers. We can recover ω from σ by noting that

$$\sigma(\rho,t)=\omega(t)$$

if t is a tangent vector field and ρ is the *radial vector field* which in local coordinates is given by $\rho=\sum \xi_j \partial/\partial\xi_j$. It is invariantly described by

$$\rho f = \frac{d}{dt} M_t^* f|_{t=1}$$

where $M_t(x,\xi)=(x,t\xi)$ is multiplication by t in the fibers of $T^*(X)$. Note that $M_t^*\sigma=t\sigma$. Although we shall continue to ignore much of the structure of $T^*(X)$ this multiplication will play an important role in our applications so we introduce the following terminology.

Definition 21.1.8. A C^∞ manifold S of dimension N is called conic if

(i) there is given a C^∞ map M: $\mathbb{R}^+\times S\to S$; we write $M_t(\gamma)=M(t,\gamma)$; $t\in\mathbb{R}^+$, $\gamma\in S$,

(ii) for every $\gamma\in S$ there is an open neighborhood V with $M(\mathbb{R}^+\times V)=V$ and a diffeomorphism ϕ: $V\to\Gamma$, where Γ is an open cone in $\mathbb{R}^N\smallsetminus 0$, such that

$$t\phi=\phi M_t,\quad t>0.$$

By a conic symplectic manifold we shall mean a conic manifold which is symplectic and for which $M_t^*\sigma=t\sigma$ where σ is the symplectic form.

Note that the definition implies $M_t M_{t'}=M_{tt'}$ since this is true in $\mathbb{R}^N\smallsetminus 0$. What we have assumed is therefore a free group action of $\mathbb{R}_+$. An example of a conic symplectic manifold is $T^*(X)\smallsetminus 0$, the cotangent bundle of X with the zero section removed. The required maps into $\mathbb{R}^{2n}\smallsetminus 0$ are given by $(x,\xi)\mapsto(x|\xi|,\xi)$ in terms of the usual local coordinates. We shall prove that for every conic symplectic manifold there is a local symplectomorphism ϕ, homogeneous of degree 1, on an open conic subset of $T^*(\mathbb{R}^n)\smallsetminus 0$; the homogeneity means of course that ϕ commutes with multiplication by positive reals as in part (ii) of Definition 21.1.8. However, we must first make some remarks on the definition.

First we note that in a conic manifold a radial vector field ρ is defined by

$$\rho f=\frac{d}{dt}M_t^*f|_{t=1};$$

with the notation in Definition 21.1.8 we have $\phi_*\rho=\sum y_j\partial/\partial y_j$ for if $F\in C^1(\mathbb{R}^N)$ then

$$\frac{d}{dt}M_t^*F(y)|_{t=1}=\frac{d}{dt}F(ty)|_{t=1}=\sum y_j\partial F/\partial y_j.$$

If a function f is homogeneous of degree μ, that is, $M_t^*f=t^\mu f$, then differentiation gives $\rho f=\mu f$ (Euler's homogeneity relation). Conversely a solution of this equation in a suitable neighborhood V of a point in S can be extended uniquely to a function homogeneous of degree μ in the conic neighborhood generated by V. This is perfectly obvious when $S\subset\mathbb{R}^N$ and V is convex.

Now assume that S is a conic symplectic manifold. The condition $M_t^*\sigma=t\sigma$ means that

$$tM_t^*\{f,u\}=\{M_t^*f,M_t^*u\};\quad f,\ u\in C^1(S). \tag{21.1.6}$$

To prove this we choose a positively homogeneous positive function T of degree 1 and set $\sigma_1 = \sigma/T$. Then $M_t^*\sigma_1 = t\sigma/tT = \sigma_1$, so for the Poisson bracket with respect to σ_1 we have (21.1.4). Thus

$$M_t^*(T\{f,u\}) = T\{M_t^* f, M_t^* u\}$$

which gives (21.1.6). Differentiation of (21.1.6) gives when $t=1$

$$\{f,u\} + \rho\{f,u\} = \{\rho f, u\} + \{f, \rho u\}, \tag{21.1.6$'$}$$

that is,

$$H_f + [\rho, H_f] = H_{\rho f}. \tag{21.1.6$''$}$$

In particular, if f is homogeneous of degree μ then

$$[H_f, \rho] = (1-\mu) H_f. \tag{21.1.6$'''$}$$

We shall now modify Theorem 21.1.6, taking into account that the standard local coordinates x, ξ in the cotangent bundle are homogeneous of degree 0 and 1 respectively.

Theorem 21.1.9. *Let S be a conic symplectic manifold of dimension $2n$, A and B two subsets of $\{1, \dots, n\}$, and q_j, $j \in A$; p_k, $k \in B$; C^∞ functions in a conic neighborhood of $\gamma \in S$ such that*

(i) *q_j and p_k are homogeneous of degree 0 and 1 respectively,*

(ii) *the commutation relations (21.1.1)$''$ are valid,*

(iii) *the Hamilton fields H_{q_j}, H_{p_k} with $j \in A$, $k \in B$, and the radial vector field ρ are linearly independent at γ.*

Let a_j, b_k be arbitrary real numbers such that

(iv) $$q_j(\gamma) = a_j, \quad j \in A; \quad p_k(\gamma) = b_k, \quad k \in B,$$

and assume that $b_J \neq 0$ for some $J \notin A$. In a conic neighborhood of γ one can then find C^∞ functions q_j, $j \notin A$, and p_k, $k \notin B$, so that (i), (ii), (iv) *remain valid for indices running from* 1 *to n and* (iii) *remains valid if $j \neq J$. In particular, (q,p) defines a homogeneous symplectomorphism of a conic neighborhood of γ in S on a conic neighborhood of $(a,b) \in T^*(\mathbb{R}^n) \setminus 0$.*

Note that starting from the empty sets A and B we conclude in particular that every conic symplectic manifold is locally isomorphic to $T^*(\mathbb{R}^n) \setminus 0$. The prescription of the values of $q_j(\gamma)$ is not of real interest since one can add constants to q_j without affecting (i), (ii), (iii). From (i) it follows that

$$\sigma(\rho, H_{q_j}) = 0, \qquad \sigma(\rho, H_{p_k}) = p_k.$$

Since $\sigma(H_{q_i}, H_{q_j}) = 0$ when $i, j \in A$ the symplectic form vanishes in the plane spanned by H_{q_j}, $j \in A$, and ρ, which must therefore be of dimension $\leqq n$ in view of Proposition 21.1.3. Hence (iii) implies that A has at most $n-1$ elements. For the extended system we must have

$$\rho = \sum_1^n c_j H_{q_j}$$

where $-c_j = \langle \rho, dp_j \rangle = p_j$. If (iii) is valid it follows that the conclusion in the theorem is false unless $b_j \neq 0$ for some $j \notin A$.

Proof of Theorem 21.1.9. It suffices to construct q_i, p_j in a neighborhood of γ satisfying the required conditions with (i) replaced by

(i)′ $$\rho q_j = 0, \quad \rho p_k = p_k,$$

for q_j, p_k can then be extended to homogeneous functions in a conic neighborhood of γ. By hypothesis the Hamilton vectors $\varepsilon_j = -H_{q_j}(\gamma)$, $e_k = H_{p_k}(\gamma)$ in the symplectic vector space $W = T_\gamma(S)$ satisfy the conditions

(i)′, (iv) $$\sigma(\rho, \varepsilon_j) = 0, \quad j \in A; \qquad \sigma(\rho, e_k) = b_k, \quad k \in B;$$

(ii) $$\sigma(\varepsilon_i, \varepsilon_j) = 0, \quad i, j \in A; \qquad \sigma(e_i, e_k) = 0, \quad i, k \in B;$$

$$\sigma(\varepsilon_j, e_k) = \delta_{jk}, \quad j \in A, \ k \in B;$$

(iii) $$\varepsilon_j, \ j \in A; \quad e_k, \ k \in B; \quad \text{and } \rho \text{ are linearly independent.}$$

The first step in the proof is to show that one can choose ε_j, $j \notin A \cup \{J\}$ and e_k, $k \notin B$ so that these conditions are preserved. This is just a problem in linear algebra, which we solve by looking at a number of cases:

a) If $k \in A \setminus B$ we choose $e = e_k$ satisfying the equations

$$\sigma(\rho, e) = b_k; \quad \sigma(e_i, e) = 0, \ i \in B; \quad \sigma(\varepsilon_j, e) = \delta_{jk}, \ j \in A.$$

This is possible since σ is non-degenerate and (iii) is fulfilled. Since $\sigma(\varepsilon_k, e) \neq 0$ but ε_k is σ orthogonal to ρ and to ε_j, $j \in A$, and to e_l, $l \in B$, the linear independence is preserved when we add e_k to our system.

b) If $k \in A \cap B$ then $W = W_0 \oplus W_1$ where W_1 is spanned by e_k and ε_k and W_0 is the symplectically orthogonal space of e_k and ε_k. Correspondingly we split $\rho = \rho_0 + \rho_1$ and obtain the same conditions for ρ_1 and the remaining vectors ε_i, e_i in W_0. The statement then follows from a lower dimensional case.

c) If A is empty and $k \notin B$ we have to choose $e = e_k$ so that

$$\sigma(\rho, e) = b_k, \quad \sigma(e_i, e) = 0, \quad i \in B.$$

The solutions form an affine space of dimension $2n - |B| - 1$ if $|B|$ is the number of elements in B. It must contain elements outside the span of ρ and e_i, $i \in B$, unless $|B| = n - 1$ and the equations are satisfied by all elements in this plane. This implies

$$0 = b_k = \sigma(\rho, e_i) = b_i$$

for all $i \in B$, which contradicts the hypothesis that some $b_j \neq 0$, which was preserved in the reduction made in case b).

d) If $B = \{1, \dots, n\}$ and A is empty, $n > 1$, we choose $\varepsilon = \varepsilon_j$ for some $j \neq J$ so that

$$\sigma(\rho, \varepsilon) = 0, \quad \sigma(e_k, \varepsilon) = -\delta_{jk}, \quad k = 1, \dots, n.$$

Suppose there is a linear relation

$$\varepsilon = c\rho + \sum_1^n c_k e_k.$$

Then

$$\delta_{jk} = \sigma(\varepsilon, e_k) = c\,\sigma(\rho, e_k) = c\,b_k.$$

Taking $k=j$ we find that $c \neq 0$ but this is a contradiction when $k=J$ since $b_J \neq 0$. Hence the linear independence is automatically preserved.

A combination of the preceding steps completes the construction of the vectors ε_j, $j \neq J$, and e_k, and we then choose ε_J so that (i)′, (iv), (ii) remain valid. It remains to modify this infinitesimal solution to a construction of actual coordinate functions. To do so we start by choosing local symplectic coordinates at γ so that $x_j = q_j$, $j \in A$, $\xi_k = p_k$, $k \in B$, and $H_{x_j}(\gamma) = -\varepsilon_j$, $H_{\xi_k}(\gamma) = e_k$ for all j, k. Then we have

$$\rho = \sum_A \xi_j \partial/\partial\xi_j + \rho_1; \qquad \rho_1 = \sum f_j \partial/\partial\xi_j + \sum g_k \partial/\partial x_k$$

where the coefficients of ρ_1 are independent of ξ_j, $j \in A$, and x_k, $k \in B$, because (21.1.6)‴ gives

$$[\partial/\partial\xi_j, \rho_1] = -[H_{q_j}, \rho] - \partial/\partial\xi_j = -H_{q_j} + H_{q_j} = 0,$$
$$[\partial/\partial x_k, \rho_1] = [H_{p_k}, \rho] = 0$$

To choose a new coordinate function, say $u = p_K$, $K \notin B$, means to find a solution of the equations

$$\partial u/\partial\xi_j = \delta_{jK}, \quad j \in A; \qquad \partial u/\partial x_k = 0, \quad k \in B; \qquad \rho u = u.$$

If we set $u = \xi_K + v$ this means that v is independent of ξ_j and x_k and that

$$\rho_2 v - v = \xi_K - \sum_{j \in A} \delta_{jK} \xi_j$$

where ρ_2 is obtained from ρ_1 by dropping $\partial/\partial\xi_j$, $j \in A$, and $\partial/\partial x_k$, $k \in B$. For v we therefore have a differential equation in the remaining variables, and $\rho_2 \neq 0$ by (iii). Hence the equation can be solved with prescribed initial data on a surface transversal to ρ_2 through γ in the plane of the variables ξ_j, $j \notin A$, x_k, $k \notin B$. We can then give du any value compatible with the equations at γ, in particular take $du = d\xi_K$ at γ. The construction of a coordinate function q_j, $j \notin A \cup \{J\}$ is exactly parallel. Finally, when all p_k and all q_j with $j \neq J$ are known, we obtain q_J by solving an ordinary differential equation in ξ_J. This completes the proof.

With the coordinates constructed in Theorem 21.1.9 we have a homogeneous local symplectomorphism to $T^*(\mathbb{R}^n)$. There ρ is the Hamilton field of the one form $\omega = \sum \xi_j dx_j$ which is therefore invariantly defined on any conic symplectic manifold. In fact, $\omega(t) = \sigma(\rho, t)$ if t is a tangent vector. We have $d\omega = \sigma$ since this is true in $T^*(\mathbb{R}^n) \setminus 0$, so the symplectic form is always exact on a conic symplectic manifold. The condition (iii) in Theorem 21.1.9 is often stated in the equivalent form that dq_j, dp_k and ω are linearly independent at γ if $j \in A$, $k \in B$.

There is also a homogeneous version of Theorem 21.1.7 which we give for the sake of completeness; it will never be used here.

Theorem 21.1.10. *Let S be a conic manifold of dimension $2n$ with a two form σ satisfying the hypothesis in Theorem 21.1.7 such that σ is homogeneous of degree 1 and the radial vector $\rho(\gamma_0)$ is not in the radical of σ restricted to S_0. In a conic neighborhood of γ_0 one can then choose coordinates ξ_1 homogeneous of degree $\frac{1}{2}$, $\xi_2, \dots, \xi_n$ homogeneous of degree 1 and x_j homogeneous of degree 0 such that the coordinates of γ_0 are $x=0$, $\xi=\varepsilon_n$ and*

$$\sigma = \xi_1\, d\xi_1 \wedge dx_1 + \sum_2^n d\xi_j \wedge dx_j.$$

Proof. Very little has to be changed in the proof of Theorem 21.1.7 so we shall only review it briefly. The surface S_0 defined by $m=0$ is obviously conic. By hypothesis the radical of the two form σ_1 in S_0 is not generated by the radial vector field, so we can take a conic hypersurface $S_1 \subset S_0$ through γ_0 which is transversal to the radical. The restriction of σ_1 to S_1 is then non-degenerate and makes S_1 a conic symplectic manifold. By Theorem 21.1.9 we can therefore choose homogeneous symplectic coordinates $x_2, \dots, x_n$, $\xi_2, \dots, \xi_n$ in S_1 with $\xi_n = 1$ and the others 0 at γ_0. We extend these functions to a conic neighborhood of S_1 in S_0 so that they are constant along the integral curves of the radical, which preserves the homogeneity properties, and we choose x_1 homogeneous of degree 0 and 0 at γ_0 with restriction to S_0 vanishing simply on S_1. Finally we choose ξ_1 homogeneous of degree $\frac{1}{2}$ vanishing simply on S_0. The restriction of σ to S_0 is then equal to

$$\sum_2^n d\xi_j \wedge dx_j.$$

Since the successive reductions made in the proof of Theorem 21.1.7 are all unique it is quite obvious that they preserve the homogeneities. The verification of this is left for the reader.

Example 21.1.11. The situation in Theorem 21.1.10 occurs if in $T^*(\mathbb{R}^{n+1})$ we form the intersection of the surfaces defined by $x_1 = 0$ and

$$\xi_1^2 = x_1\, \xi_n^2 + \xi_{n+1}\, \xi_n.$$

The intersection is transversal when $\xi_n > 0$, and the restriction of the symplectic form to the surface is

$$\sigma = d\xi_2 \wedge dx_2 + \dots + d\xi_n \wedge dx_n + d(\xi_1^2/\xi_n) \wedge dx_{n+1}.$$

Taking $x_{n+1}, x_2, \ldots, x_n, \xi_1(2/\xi_n)^{\frac{1}{2}}, \xi_2, \ldots, \xi_n$ as variables on the surface we have the situation described in Theorem 21.1.10. An equivalent formulation of Theorem 21.1.10 is therefore that a reduction to this example is always possible. We shall return to the example in Section 21.4.

21.2. Submanifolds of a Symplectic Manifold

As in Section 21.1 we begin with a brief discussion of linear algebra in a symplectic vector space S, with symplectic form σ. Just as in a Euclidean space a linear subspace V of S has an annihilator $V^\sigma \subset S$ with respect to the symplectic form,

$$V^\sigma = \{\gamma \in S; \sigma(\gamma, V) = 0\},$$

and $(V^\sigma)^\sigma = V$. However, in general V is not a supplementary space of V^σ. Since $\dim V + \dim V^\sigma = \dim S$ this is true if and only if $V \cap V^\sigma = \{0\}$, that is, V is *symplectic* in the sense that the restriction to V of the symplectic form is non-degenerate. In that case V^σ is of course also symplectic so S is the direct sum of the symplectic subspaces V and V^σ. This observation was used already in the proof of Proposition 21.1.2. In general there is a substitute for this construction:

Proposition 21.2.1. *If V is a linear subspace of the symplectic vector space S then*

$$S' = (V + V^\sigma)/(V \cap V^\sigma) \tag{21.2.1}$$

is a symplectic vector space, and

$$\dim S' = \dim S - 2\dim(V \cap V^\sigma) = 2\dim(V + V^\sigma) - \dim S. \tag{21.2.2}$$

Proof. The σ orthogonal space of $W = V + V^\sigma$ is $W^\sigma = V^\sigma \cap (V^\sigma)^\sigma = V \cap V^\sigma \subset W$. If $w \in W$ then $\sigma(w, w') = 0$ for all $w' \in W$ if and only if $w \in W^\sigma$, so σ induces a non-degenerate antisymmetric bilinear form σ' on S'. Since

$$\dim W + \dim W^\sigma = \dim S, \qquad \dim W - \dim W^\sigma = \dim S',$$

we obtain (21.2.2).

In particular, we have $\dim S' = \dim S - 2\dim V$ if $V^\sigma \supset V$ and $\dim S' = 2\dim V - \dim S$ if $V^\sigma \subset V$. These cases occur so frequently that they have special names:

Definition 21.2.2. A linear subspace V of a symplectic vector space is called isotropic, Lagrangian resp. involutive (or coisotropic) if $V \subset V^\sigma$, $V = V^\sigma$ and $V \supset V^\sigma$ respectively.

The dimension of V is thus $\leqq \dim S/2$, $=\dim S/2$, and $\geqq \dim S/2$ in these three cases. An isotropic (involutive) subspace V is Lagrangian if $\dim S = 2\dim V$.

Proposition 21.2.3. *V is isotropic (involutive) if and only if V is contained in (contains) a Lagrangian subspace.*

Proof. Assume first that V is isotropic and choose a basis $e_1, \dots, e_k$ for V. Then $\sigma(e_i, e_j) = 0$, $i,j = 1, \dots, k$ so by Proposition 21.1.3 we can extend this to a symplectic basis $e_1, \dots, e_n$, $\varepsilon_1, \dots, \varepsilon_n$ for S. Then $e_1, \dots, e_n$ span a Lagrangian subspace containing V. If V is involutive then V^σ is isotropic, so $V^\sigma \subset \lambda$ for some Lagrangian plane, hence $\lambda \subset V$.

S' always splits in a natural way into the direct sum of two orthogonal symplectic subspaces $V/(V \cap V^\sigma)$ and $V^\sigma/(V \cap V^\sigma)$. The symplectic form of S vanishes on $V \cap V^\sigma$. Now start by choosing a basis $e_1, \dots, e_k$ for $V \cap V^\sigma$ as in the proof of Proposition 21.2.3 and extend it by elements $e_{k+1}, \dots, e_{k+l}$, $\varepsilon_{k+1}, \dots, \varepsilon_{k+l}$ whose residue classes form a symplectic basis for $V/(V \cap V^\sigma)$. Next add elements $e_{k+l+1}, \dots, e_n$, $\varepsilon_{k+l+1}, \dots, e_n$ representing a symplectic basis in $V^\sigma/(V \cap V^\sigma)$. We have then obtained a basis for $V + V^\sigma$ satisfying (21.1.1). Since $\dim S = \dim(V + V^\sigma) + \dim(V \cap V^\sigma) = 2n$ we can now use Proposition 21.1.3 to add elements $\varepsilon_1, \dots, \varepsilon_k$ to get a symplectic basis for S. With corresponding symplectic coordinates V is then defined by

$$\xi_1 = \dots = \xi_k = 0, \quad x_{k+l+1} = \dots = x_n = \xi_{k+l+1} = \dots = \xi_n = 0.$$

The dimension is $k + 2l$ while the dimension of S' is $2(n-k)$ so it is not in general determined by the dimension of V.

The preceding discussion can be extended to manifolds:

Theorem 21.2.4. *Let S be a conic symplectic manifold, V a conic submanifold such that the restriction of the symplectic form to V has constant rank $2l$ in a neighborhood of a point γ where the canonical one form ω does not vanish identically on $T_\gamma(V)$. Then there are homogeneous symplectic coordinates x, ξ in a conic neighborhood of γ in S such that V is defined by*

$$\xi_1 = \dots = \xi_k = 0, \quad x_{k+l+1} = \dots = x_n = \xi_{k+l+1} = \dots = \xi_n = 0 \tag{21.2.3}$$

where $k + 2l = \dim V$.

Note that $l \neq 0$ for V cannot be isotropic since the tangential radial vector $\rho(\gamma)$ by hypothesis is not orthogonal to $T_\gamma(V)$. There is an obvious non-homogeneous version of the theorem where ω does not occur so that this condition drops out. We shall usually leave such variants for the reader.

Proof. We can choose $f_1, \dots, f_r \in C^\infty$ homogeneous of degree 1 and vanishing on V so that $df_1, \dots, df_r$ are linearly independent at γ and define $T_\gamma(V)$. Here $r = 2n - \dim V$ where $2n = \dim S$. Set $k = \dim V - 2l$.

a) If $k+l<n$ it follows from the discussion of the linear case that some Poisson bracket, say $\{f_1,f_2\}$, does not vanish at γ. Thus H_{f_1} cannot be proportional to the radial vector ρ, since $\rho f_2=0$. By Theorem 21.1.9 we can therefore choose local homogeneous symplectic coordinates such that $\xi_n=f_1$ and γ corresponds to $(0,\varepsilon_1)$, $\varepsilon_1=(1,0,\dots,0)$, say. Then the equation $f_2=0$ can be solved for x_n since

$$\partial f_2/\partial x_n=\{f_1,f_2\}\neq 0 \quad \text{at } \gamma.$$

In a sufficiently small conic neighborhood of γ the equation $f_2=0$ can therefore be written in the form

$$x_n=g(x_1,\xi_1,\dots,\xi_{n-1},\xi_n)$$

where g is homogeneous of degree 0. If $X_n=x_n-g(x_1,\xi_1,\dots,\xi_{n-1},\xi_n)$ we have $\xi_n=0$, $X_n=0$ on V, and $\{\xi_n,X_n\}=1$ in a neighborhood of γ. Using Theorem 21.1.9 again we can find a new homogeneous symplectic coordinate system where X_n and ξ_n are the last coordinates. Changing notation we may therefore assume that $x_n=\xi_n=0$ on V. This means that V can be considered as a submanifold of $T^*(\mathbb{R}^{n-1})$, and the theorem follows if it has already been proved for symplectic manifolds S of lower dimension.

b) If $k+l=n$ but $k\neq 0$, that is, $V\neq S$, we have $\{f_i,f_j\}=0$ on V for $i,j=1,\dots,r$ by the discussion of the linear case, and by hypothesis ρ is not a linear combination of $H_{f_1},\dots,H_{f_r}$. We choose homogeneous symplectic coordinates such that $f_1=\xi_1$ and γ has coordinates $(0,\varepsilon_n)$, say. Then $\xi_1=0$ on V and $\partial f_j/\partial x_1=\{\xi_1,f_j\}=0$ on V, $j=2,\dots,r$. Hence V is cylindrical in the x_1 direction and defined near γ by $\xi_1=0$, $f_j(0,x_2,\dots,x_n,0,\xi_2,\dots,\xi_n)=0$, $j=2,\dots,r$. The differentials of these functions and ω are linearly independent. If the theorem is already proved for lower dimensional S we can choose new coordinates $x_2,\dots,x_n,\xi_2,\dots,\xi_n$ such that $V\cap\{x_1=\xi_1=0\}$ is defined by $\xi_2=\dots=\xi_k=0$, and this completes the proof.

In part b) of the proof we found that V was generated by a family of curves. These are manifestations of an important foliation which we shall now discuss. (See Appendix C.1 for terminology.)

Definition 21.2.5. A submanifold V of a symplectic manifold is called symplectic, isotropic, Lagrangian resp. involutive if at every point of V the tangent space of V has this property.

Remark. Note that the constant rank condition of Theorem 21.2.4 is automatically fulfilled in these cases.

Theorem 21.2.6. *If V satisfies the hypothesis in Theorem 21.2.4 then the subspace $T_\gamma(V)\cap T_\gamma^\sigma(V)$ of $T_\gamma(V)$ defines a foliation with isotropic leaves transversal to the radial vector field, of dimension* $\dim V-\operatorname{rank}\sigma|_V$.

Proof. In the local coordinates of Theorem 21.2.4 we have the foliation of V by the parallels of the $x_1 \dots x_k$ plane.

The reader is advised not to ignore this important result because of the brevity of the proof; it really depends on all we have done so far in this chapter, and the foliation will be used very frequently.

As pointed out after the statement of Theorem 21.2.4 there is a non-homogeneous version of that result so we have a foliation of any submanifold V of a symplectic manifold S such that the symplectic form restricted to V has constant rank. If S and V are conic and the radial vector field ρ is tangent to a leaf B of the foliation at some point γ, then B is conic. In fact, the ray through γ must stay in the leaf since the homogeneity implies that ρ is in the plane of the foliation along it. The homogeneity also implies that $M_t B$ is a leaf of the foliation for any t, if M_t denotes multiplication by t; since it has points on the ray through γ in common with B it must agree with B. Hence we have the following supplement to Theorem 21.1.6.

Theorem 21.2.7. *If S is a conic symplectic manifold, V a conic submanifold such that the symplectic form restricted to V has constant rank, then $T_\gamma(V) \cap T_\gamma^\sigma(V)$ defines a foliation of V with isotropic leaves which are either transversal to the radial vector field or else conic.*

When V is involutive the constant rank condition is automatically fulfilled, as already pointed out, and the leaves of the foliation are then generated by the integral curves of the Hamilton fields H_f of functions vanishing on V. Indeed, we have $H_f(\gamma) \in T_\gamma^\sigma(V) = T_\gamma(V) \cap T_\gamma^\sigma(V)$ when $\gamma \in V$ then. The foliation is therefore usually called the *Hamilton foliation.*

If V is a submanifold of a symplectic manifold S, such that the symplectic form restricted to V has constant rank, then we have a natural vector bundle S_V on V with fiber

$$(T_\gamma(V) + T_\gamma^\sigma(V))/(T_\gamma(V) \cap T_\gamma^\sigma(V))$$

at $\gamma \in V$. The fiber is a symplectic vector space so S_V is called a symplectic vector bundle. In the local coordinates of Theorem 21.2.4 S_V is the product of V and the symplectic vector space $T^*(\mathbb{R}^{n-k})$. It is easy to see that we have in fact a vector bundle with these local trivializations. Note that the restriction of S_V to a leaf B of the foliation of V is equal to the bundle S_B, so the construction is mainly of interest in the isotropic case where the foliation is trivial (the leaves become all of V). For a manifold which is neither isotropic nor involutive the symplectic bundle S_V is in a non-trivial way the direct symplectically orthogonal sum of two symplectic bundles S'_V and S''_V with fibers

$$T_\gamma(V)/(T_\gamma(V) \cap T_\gamma^\sigma(V)) \quad \text{and} \quad T_\gamma^\sigma(V)/(T_\gamma(V) \cap T_\gamma^\sigma(V)).$$

There is also an analogue of Theorem 21.2.4 when the restriction of ω to V is identically 0. This implies that the restriction of $\sigma=d\omega$ to V vanishes, so V is isotropic. Conversely, if V is isotropic and conic then the restriction of ω to V is 0, for $\omega(t)=\sigma(\rho,t)=0$ if t is a tangent vector of V, since ρ is also a tangent vector of V.

Theorem 21.2.8. *Let S be a conic symplectic manifold, V a conic isotropic submanifold. In a conic neighborhood of any $\gamma\in V$ one can then find homogeneous symplectic coordinates such that V is defined by*

$$x_1=\ldots=x_n=0, \quad \xi_{k+1}=\ldots=\xi_n=0 \tag{21.2.4}$$

where k is the dimension of V.

Proof. We shall first prove that the coordinates can be chosen so that V is defined by

$$x_1=0, \quad \xi_2=\ldots=\xi_k=0, \quad x_{k+1}=\xi_{k+1}=\ldots=x_n=\xi_n=0. \tag{21.2.5}$$

Part a) of the proof of Theorem 21.2.4 is applicable with no change and reduces the proof to the case where $\dim V=n$, that is, V is Lagrangian. If $n>1$ we can still choose one of the functions f_j vanishing on V so that H_{f_j} is not proportional to ρ, so the reduction in part b) of the proof of Theorem 21.2.4 is applicable. Finally, if $n=1$ the fact that the one form $\xi_1\,dx_1$ vanishes on V implies that x_1 is a constant which we choose as 0. Since there is a local homogeneous symplectomorphism mapping any conic isotropic V of dimension k, and in particular (21.2.4), to the same model given by (21.2.5), we can also choose (21.2.4) as our model.

In particular any conic Lagrangian manifold can be defined locally by $x=0$, that is, as the fiber $T_0^*(\mathbb{R}^n)\setminus 0$ in a suitable system of homogeneous symplectic coordinates. Note that if X is any C^∞ manifold and $Y\subset X$ is a submanifold, then the conormal bundle $N(Y)$ of Y is Lagrangian since $\langle dx,\xi\rangle=0$ on $T_x(Y)$ if $\xi\in N_x(Y)$. The theorem is therefore applicable in particular to any conormal bundle. The following theorem shows that any conic Lagrangian submanifold of a cotangent bundle contains large open subsets of this type.

Theorem 21.2.9. *Let X be a C^∞ manifold and V a C^∞ conic Lagrangian $\subset T^*(X)\setminus 0$. If $(x_0,\xi_0)\in V$ and the restriction to V of the projection π: $T^*(X)\to X$ has constant rank r in a neighborhood of (x_0,ξ_0), then there is a C^∞ submanifold $Y\subset X$ of dimension r in a neighborhood of x_0 with conormal bundle equal to V in a neighborhood of (x_0,ξ_0).*

Note that $r<n$ since $\pi_*\rho=0$ for the radial vector field ρ which is a tangent of V.

Proof. By Proposition C.3.3 a neighborhood of (x_0, ξ_0) in V is mapped by π on an r dimensional C^∞ manifold $Y \subset X$. Since $\sum \xi_j dx_j = 0$ on V and π_* maps the tangent space of V onto that of Y, it follows that ξ must be a normal to Y, that is, $V \subset N^*(Y)$ in a neighborhood of (x_0, ξ_0). Equality must hold since the dimensions are the same.

The hypothesis in Theorem 21.2.9 is always fulfilled in an open dense subset of V where the rank is equal to the maximum in any neighborhood. However, in general the projection of V can be quite singular. A simple example is the Lagrangian in $T^*(\mathbb{R}^2) \setminus 0$ defined by

$$x_1 = 3(\xi_1/\xi_2)^2, \quad x_2 = -2(\xi_1/\xi_2)^3, \quad \xi_2 \neq 0. \tag{21.2.6}$$

The projection is the curve $(x_1/3)^3 - (x_2/2)^2 = 0$, and V is the closure of the conormal bundle of the regular part.

We shall now derive a normal form for intersecting conic Lagrangians. The intersection is never transversal, for the canonical one form vanishes on both tangent planes at the intersection. We must therefore use the notion of clean intersection discussed in Appendix C.3.

Theorem 21.2.10. *Let S be a conic symplectic manifold of dimension $2n$, V_1 and V_2 two conic Lagrangian submanifolds intersecting cleanly at $\gamma \in S$. Then one can choose homogeneous symplectic coordinates x, ξ at γ such that the coordinates of γ are $(0, \varepsilon_1)$, $\varepsilon_1 = (1, 0, \dots, 0)$ and near $(0, \varepsilon_1)$*

$$V_1 = \{(0, \xi)\}, \; V_2 = \{(0, x'', \xi', 0)\} \tag{21.2.7}$$

where $x' = (x_1, \dots, x_k)$, $x'' = (x_{k+1}, \dots, x_n)$ and k is the dimension of $V_1 \cap V_2$ (or equivalently the excess of the intersection).

Thus V_1 is locally the conormal bundle of a point and V_2 the conormal bundle of a manifold of codimension k through it. Note that the conormal bundles of manifolds intersecting cleanly must also intersect cleanly since they are linear spaces in coordinates where the intersecting manifolds are linear spaces.

Proof. By Theorem 21.2.8 we may assume that S is a conic neighborhood of $(0, \varepsilon_1)$ in $T^*(\mathbb{R}^n) \setminus 0$ and that V_1 is defined by $x = 0$. The intersections V_1^c, V_2^c and $V_1^c \cap V_2^c$ of $V_1, V_2, V_1 \cap V_2$ with the sphere bundle $S_1 = \{(x, \xi);\; |\xi| = 1\}$ have dimensions decreased by one, and the direct sums of their tangent spaces and ρ are equal to those of $V_1, V_2, V_1 \cap V_2$ at S_1. Hence V_1^c and V_2^c intersect cleanly with $k-1$ dimensional intersection.

a) Assume first that $k = 1$. Then the intersection is 0 dimensional which means that the tangent spaces of V_1^c and V_2^c have only 0 in common. Since V_1^c is the sphere defined by $x = 0$ it follows that the differential of π: $V_2^c \to \mathbb{R}^n$ is injective, hence of rank $n-1$. By Theorem 21.2.9 it follows that V_2 in a neighborhood of $(0, \varepsilon_1)$ is the conormal bundle of a hypersurface which we can take as the surface $x_1 = 0$ by a change of variables in $\mathbb{R}^n$.

b) Assume now that $k > 1$. Then V_1^c and V_2^c are of dimension $n-1$ and have a $k-1$ dimensional clean intersection in a $2n-1$ dimensional manifold, so the excess is k by (C.3.3). Hence we can find $k \geqq 2$ functions $f_1, \dots, f_k$ homogeneous of degree 1 which vanish on $V_1 \cup V_2$ and have linearly independent differentials at γ. Then H_{f_j} and ρ must be linearly independent for some j, say $j = 1$. As in part b) of the proof of Theorem 21.2.4 we can choose new homogeneous symplectic coordinates such that $f_1 = \xi_n$ and conclude that V_1 and V_2 are both locally equal to the product of the x_n axis and Lagrangians in $x_n = \xi_n = 0$ intersecting cleanly of dimension $k-1$. If the theorem is already proved for smaller values of k and n we can choose new coordinates $x_1, \xi_1, \dots, x_{n-1}, \xi_{n-1}$ so that V_1 and V_2 are defined near $(0, \varepsilon_1)$ by

$$V_1\colon\ x_1 = \dots = x_{n-1} = 0, \quad \xi_n = 0;$$
$$V_2\colon\ x_1 = \dots = x_{k-1} = 0, \quad \xi_k = \dots = \xi_n = 0,$$

which intersect in the plane where all variables except $x_n, \xi_1, \dots, \xi_{k-1}$ are 0. In particular, the pair in (21.2.7) can be presented in this form which completes the proof just as in the case of Theorem 21.2.8.

Theorem 21.2.10 is of course also valid in the non homogeneous case. The linear case is very elementary but instructive: If V_1 and V_2 are Lagrangian subspaces of a symplectic vector space S then we can choose a symplectic basis $e_1, \dots, e_n, \varepsilon_1, \dots, \varepsilon_n$ such that V_1 is spanned by $\varepsilon_1, \dots, \varepsilon_n$ and V_2 is spanned by $e_{k+1}, \dots, e_n, \varepsilon_1, \dots, \varepsilon_k$. To do so we first choose a basis $\varepsilon_1, \dots, \varepsilon_k$ for $V_1 \cap V_2$ and extend it to a basis $\varepsilon_1, \dots, \varepsilon_n$ for V_1. The symplectic form restricted to $V_1 \times V_2$ gives a duality between $V_1/(V_1 \cap V_2)$ and $V_2/(V_1 \cap V_2)$. Hence we can choose $e_{k+1}, \dots, e_n \in V_2$ forming a dual basis to $\varepsilon_{k+1}, \dots, \varepsilon_n$, and then we just have to extend to a full symplectic basis using Proposition 21.1.3. The following consequence will be useful below:

Corollary 21.2.11. *Let S be a symplectic vector space and V_1, V_2 two Lagrangian subspaces. Then we can find a Lagrangian subspace V transversal to both.*

Proof. If V_1 and V_2 have the form (21.2.7) we can take

$$V = \{(x, \xi);\ x'' = \xi'', \xi' = 0\}.$$

If S_1 and S_2 are symplectic manifolds then a symplectomorphism or canonical transformation $\chi\colon S_2 \to S_1$ is a diffeomorphism with $\chi^* \sigma_1 = \sigma_2$ if σ_j is the symplectic form in S_j (Definition 21.1.4). This implies that S_1 and S_2 have equal dimensions. The graph of χ

$$G = \{(\chi(\gamma), \gamma);\ \gamma \in S_2\} \subset S_1 \times S_2$$

is a Lagrangian submanifold of $S_1 \times S_2$ with the symplectic form

$$\sigma_{12} = \pi_1^* \sigma_1 - \pi_2^* \sigma_2$$

where π_1: $S_1 \times S_2 \to S_1$ and π_2: $S_1 \times S_2 \to S_2$ are the two projections. (In what follows we shall usually write σ_j instead of $\pi_j^* \sigma_j$ when no misunderstanding seems possible.) In addition π_j restricted to G is a diffeomorphism on S_j for $j=1,2$. By dropping this condition we can extend the notion of canonical transformation and in particular drop the condition that S_1 and S_2 have equal dimensions.

Definition 21.2.12. Let S_1, S_2 be symplectic manifolds with symplectic forms σ_1, σ_2. A Lagrangian manifold $G \subset S_1 \times S_2$ with respect to the symplectic form $\sigma_1 - \sigma_2$ is then called a canonical relation from S_2 to S_1. If S_j are conic and G is conic we call the relation homogeneous.

As in Section 5.2 a relation maps a set $E \subset S_2$ to a set $G(E) \subset S_1$,

$$G(E) = \{\gamma_1 \in S_1; (\gamma_1, \gamma_2) \in G \text{ for some } \gamma_2 \in E\} = \pi_1(G \cap \pi_2^{-1} E).$$

Relations can be composed as functions: If $G_1 \subset S_1 \times S_2$ and $G_2 \subset S_2 \times S_3$ then

$$G_1 \circ G_2 = \{(\gamma_1, \gamma_3); (\gamma_1, \gamma_2) \in G_1 \text{ and } (\gamma_2, \gamma_3) \in G_2 \text{ for some } \gamma_2\}.$$

We can also interpret this as the projection on $S_1 \times S_3$ of the intersection of $G_1 \times G_2$ with $S_1 \times \Delta(S_2) \times S_3$ where $\Delta(S_2) = \{(\gamma, \gamma); \gamma \in S_2\}$ is the diagonal in $S_2 \times S_2$. Note that $\Delta(S_2)$ is isotropic with respect to the symplectic form $-\pi_{21}^* \sigma_2 + \pi_{22}^* \sigma_2$ where π_{2j} is the projection $S_2 \times S_2 \to S_2$ on the jth factor. There is of course no reason in general why $G_1 \circ G_2$ should be smooth but we shall prove that for canonical relations G_1 and G_2 we always obtain a smooth canonical relation if the intersection is clean and suitable global restrictions are made. First we study the construction in the linear case, in a somewhat more general context.

Proposition 21.2.13. *Let S be a symplectic vector space with symplectic form σ, let Δ be an isotropic subspace, and denote by S' the symplectic vector space Δ^σ/Δ. If $\lambda \subset S$ is any Lagrangian subspace of S then $\lambda' = (\lambda \cap \Delta^\sigma)/(\lambda \cap \Delta)$ is a Lagrangian subspace of S'.*

Proof. That λ' is isotropic follows from the fact that λ is isotropic and from the definition of the symplectic form in S'. Now

$$\begin{aligned}\dim \lambda + \dim \Delta^\sigma &= \dim(\lambda \cap \Delta^\sigma) + \dim(\lambda + \Delta^\sigma)\\ &= \dim(\lambda \cap \Delta^\sigma) + \dim S - \dim(\lambda \cap \Delta),\end{aligned}$$

hence by (21.2.2)

$$\dim \lambda' = \dim \lambda - \dim \Delta = (\dim S')/2.$$

Remark. Although the map $\lambda \mapsto \lambda'$ is always defined it is not continuous when $\lambda \cap \Delta^\sigma$ changes dimension. As an example consider $S = T^*(\mathbb{R}^2)$,

$$\Delta = \{(0, x_2, 0, 0)\}, \quad \Delta^\sigma = \{(x_1, x_2, \xi_1, 0)\}.$$

Then $\lambda = \{(a x_2, x_2, \xi_1, -a\xi_1)\}$ is Lagrangian for every a but $\lambda' = \{(x_1, 0)\}$ if $a \neq 0$, $\lambda' = \{(0, \xi_1)\}$ if $a = 0$.

We are now prepared to discuss the composition of general canonical relations.

Theorem 21.2.14. *Let S_j be a symplectic manifold with symplectic form σ_j, $j = 1, 2, 3$. If $G_1 \subset S_1 \times S_2$ and $G_2 \subset S_2 \times S_3$ are Lagrangians for the symplectic forms $\sigma_1 - \sigma_2$ and $\sigma_2 - \sigma_3$, and $G_1 \times G_2$ intersects $S_1 \times \Delta(S_2) \times S_3$ cleanly in G, then the projection π from G to $S_1 \times S_3$ has rank* $(\dim S_1 + \dim S_3)/2$ *and the range $G_1 \circ G_2$ of π is locally a Lagrangian manifold with respect to $\sigma_1 - \sigma_3$.*

Under these hypotheses we shall say that the composition is clean.

Proof. Let $(\gamma_1, \gamma_2, \gamma_2, \gamma_3) \in G$. By the definition of clean intersection the tangent space of G there is

$$\lambda \cap (T_{\gamma_1}(S_1) \times \Delta(T_{\gamma_2}(S_2)) \times T_{\gamma_3}(S_3)), \quad \lambda = T_{\gamma_1, \gamma_2}(G_1) \times T_{\gamma_2, \gamma_3}(G_2).$$

λ is a Lagrangian subspace of the symplectic vector space

$$T_{\gamma_1}(S_1) \oplus T_{\gamma_2}(S_2) \oplus T_{\gamma_2}(S_2) \oplus T_{\gamma_3}(S_3)$$

with the symplectic form $\sigma_1 - \sigma_{21} + \sigma_{22} - \sigma_3$. The range of the differential of π is thus obtained by the construction in Proposition 21.2.13 applied to the isotropic subspace $\{0\} \times \Delta(T_{\gamma_2}(S_2)) \times \{0\}$ which proves the statement on the rank. Hence it follows from Proposition C.3.3 that the range of π is locally a Lagrangian submanifold of $S_1 \times S_3$.

Remark. If Ω is an open set in $S_1 \times S_3$ and the map π: $\pi^{-1}(\Omega) \to \Omega$ is proper, it follows that $\pi(\pi^{-1}(\Omega))$ is an immersed closed Lagrangian submanifold of Ω and that $\pi^{-1}(\Omega)$ is a fiber space with compact fibers over it.

The construction in Theorem 21.2.14 is applicable in particular when S_3 is reduced to a point. Then it produces a Lagrangian in S_1 from one in S_2 and one in $S_1 \times S_2$. This idea leads to an important method of parametrizing C^∞ Lagrangian submanifolds Λ of the cotangent bundle $T^*(X)$ of a manifold. Recall from Section 6.4 that if Λ is a *section* of $T^*(X)$ then Λ can locally be represented in the form $\{(x, \phi'(x))\}$ where $\phi \in C^\infty$ in an open subset of X. This representation breaks down if Λ is not a section. In particular it can never be used if Λ is conic since the radial vector field ρ is then a tangent of Λ. However, assume that $\phi(x, \theta)$ is a C^∞ function in the product of an open set in X and another C^∞ manifold Θ. Then

$$G = \{(x, \phi'_x(x, \theta); \theta, -\phi'_\theta(x, \theta))\}$$

is a Lagrangian submanifold of $T^*(X) \times T^*(\Theta)$ (the minus sign is caused by the conventions above about the symplectic form in the product of two symplectic manifolds). The zero section G_0 in $T^*(\Theta)$ is also Lagrangian. If $T^*(X)\times\Delta(T^*(\Theta))$ and $G\times G_0$ intersect cleanly, that is,

$$\{(x,\theta);\, \phi'_\theta(x,\theta)=0\}$$

is a submanifold of $X\times\Theta$ with tangent space everywhere defined by the equations $d\phi'_\theta(x,\theta)=0$ then

$$\{(x,\phi'_x(x,\theta));\, \phi'_\theta(x,\theta)=0\} \tag{21.2.8}$$

is locally a Lagrangian submanifold of $T^*(X)$. We recall that this construction occurred already in Theorem 8.1.9 in its natural analytical context. It is clear that (21.2.8) is conic if ϕ is homogeneous of degree 1.

Definition 21.2.15. If X is a C^∞ manifold and $\phi(x,\theta)$ a C^∞ real valued function in an open conic set $\Gamma\subset X\times(\mathbb{R}^N\setminus 0)$ which is homogeneous of degree 1 in θ, then ϕ is called a clean phase function if $d\phi\neq 0$ and

$$C=\{(x,\theta)\in\Gamma;\, \phi'_\theta(x,\theta)=0\}$$

is a C^∞ manifold with tangent plane defined by the equations $d\phi'_\theta=0$. The number of linearly independent differentials $d(\partial\phi/\partial\theta_j)$, $j=1,\dots,N$, is equal to $N-e$ on C, where $e=\dim C-\dim X$ is called the excess. The map $C\ni(x,\theta)\mapsto(x,\phi'_x)$ is locally a fibration with fibers of dimension e and a conic Lagrangian Λ as base; Λ is said to be parametrized by ϕ. If $e=0$ the phase function is said to be non-degenerate.

Theorem 21.2.16. *If X is a C^∞ manifold of dimension n and $\Lambda\subset T^*(X)\setminus 0$ is a C^∞ conic Lagrangian submanifold, then Λ can always be locally parametrized by a non-degenerate phase function ϕ in an open cone in $X\times(\mathbb{R}^n\setminus 0)$. If $\gamma_0\in\Lambda$ and x_0 is the projection of γ_0 in X, the local coordinates $x_1,\dots,x_n$ at x_0 can be chosen so that with the corresponding coordinates $x_1,\dots,x_n$, $\xi_1,\dots,\xi_n$ in $T^*(X)$ the Lagrangian plane $\xi=$constant through γ_0 is transversal to Λ. Then there is a unique phase function of the form*

$$\phi(x,\theta)=\sum_1^n x_j\theta_j-H(\theta)$$

parametrizing Λ; $H\in C^\infty$, H is homogeneous of degree 1, and near γ_0

$$\Lambda=\{(H'(\xi),\xi)\}.$$

Proof. Choose local coordinates $y_1,\dots,y_n$ at x_0 such that the coordinates of x_0 are all 0 and $\gamma_0=(0,dy_1(0))$. Using these coordinates we consider Λ as a subset of $T^*(\mathbb{R}^n)\setminus 0$. The tangent plane λ_0 of Λ at $(0,\varepsilon_1)$, $\varepsilon_1=(1,0,\dots,0)$ is a Lagrangian plane. It may not be transversal to the plane $\xi=\varepsilon_1$, which is the graph of the differential of the y_1 coordinate. However, we know from Corollary 21.2.11 that there is a Lagrangian plane through $(0,\varepsilon_1)$ which is transversal both to λ_0 and to the fiber $T_0^*(\mathbb{R}^n)$. Hence it is a section of

$T^*(\mathbb{R}^n)$ of the form

$$\eta = d(y_1 + A(y))$$

where $A(y)$ is a quadratic form. Now take $x_1 = y_1 + A(y)$, $x_2 = y_2, \dots, x_n = y_n$ as new local coordinates. The tangent plane of Λ at γ_0 is then transversal to the graph of dx_1 which passes through γ_0. The transversality means in standard local coordinates that the map

$$\Lambda \ni (x, \xi) \mapsto \xi$$

has bijective differential at γ_0. Hence we can locally take ξ as a parameter on Λ which is then defined by $x = \phi(\xi)$ where ϕ is a C^∞ function from a neighborhood of ε_1 to $\mathbb{R}^n$. Since ϕ is homogeneous of degree 0, we can take a homogeneous extension to a conic neighborhood still defining Λ. Now Λ is a conic Lagrangian so we have $\omega = \sum \xi_j dx_j = 0$ on Λ, thus

$$0 = \sum \xi_j d\phi_j(\xi) = dH(\xi) - \sum \phi_j(\xi) d\xi_j, \quad H(\xi) = \sum \xi_j \phi_j(\xi).$$

This means that $\phi_j(\xi) = \partial H(\xi)/\partial \xi_j$ and completes the proof.

It is often but not always possible to use a smaller number of θ variables when parametrizing a Lagrangian:

Theorem 21.2.17. *Let the hypotheses be as in Theorem 21.2.16 and let k be the dimension of $T_{\gamma_0}(\Lambda) \cap T^*_{x_0}(X)$. Then Λ can be parametrized near γ_0 by a non-degenerate phase function in $X \times (\mathbb{R}^k \smallsetminus 0)$ but not by one involving fewer parameters.*

Proof. Assume first that Λ is parametrized by the non-degenerate phase function $\phi(x, \theta)$ where $\theta \in \mathbb{R}^\kappa$. Then $\mu = T_{\gamma_0}(\Lambda) \cap T^*_{x_0}(X)$ is the image of the set in $T_{x_0}(X) \times \mathbb{R}^\kappa$ defined by $d\phi'_\theta = 0$, $dx = 0$, so it is at most of dimension κ. Hence $\kappa \geq k$.

To prove the first statement we choose local coordinates in X at x_0 such that $\gamma_0 = (0, \varepsilon_1)$ and Λ is defined by $x = H'(\xi)$ nearby. By a linear change of variables preserving x_1 and therefore γ_0 we can make μ defined by $dx = 0$, $d\xi'' = 0$ where $\xi'' = (\xi_{k+1}, \dots, \xi_n)$. Set $\xi' = (\xi_1, \dots, \xi_k)$. The Lagrangian plane $T_{\gamma_0}(\Lambda)$ contains μ so it is contained in the orthogonal plane μ^σ defined by $dx' = 0$. Hence the map

$$\Lambda \ni (x, \xi) \mapsto (x'', \xi') \in \mathbb{R}^n$$

has bijective differential at γ_0. In fact, $dx' = dx'' = 0$ in the kernel so it is contained in μ and satisfies $d\xi' = 0$. The statement is now a consequence of the following

Theorem 21.2.18. *Let $\Lambda \subset T^*(\mathbb{R}^n) \smallsetminus 0$ be a conic Lagrangian manifold in a neighborhood of a point γ_0 where*

$$\Lambda \ni (x, \xi) \mapsto (x'', \xi') \in \mathbb{R}^n$$

is a bijection for a splitting $x=(x',x'')$, $\xi=(\xi',\xi'')$ *of the coordinates. Then there is a function* $S(x'',\xi')$ *such that near* γ_0

$$\Lambda=\{(\partial S/\partial\xi',x'';\xi',-\partial S/\partial x'')\}.$$

Λ *is defined by the phase function* $\phi(x,\xi')=\langle x',\xi'\rangle-S(x'',\xi')$.

Proof. On Λ we can regard x' and ξ'' as functions of (x'',ξ'), and

$$0=\langle\xi,dx\rangle=d(\langle x',\xi'\rangle)-\langle x',d\xi'\rangle+\langle\xi'',dx''\rangle.$$

If we take $S(x'',\xi')=\langle x',\xi'\rangle$ it follows that $\partial S/\partial x''=-\xi''$ and that $\partial S/\partial\xi'=x'$. The last statement is obvious since $\partial\phi/\partial\xi'=0$ means that $x'=\partial S/\partial\xi'$, and $\partial\phi/\partial x=(\xi',-\partial S/\partial x'')$.

The reader may by now have raised – and answered – the question why we bother to introduce the notion of clean phase function when there is always a non-degenerate one. The answer is that the composition studied in Theorem 21.2.14 leads naturally to a clean but degenerate phase function when the intersection taken there is just clean and not transversal. However, before elaborating this point we must make an important remark on the definition of a canonical relation G in the product $(T^*(X)\setminus 0)\times(T^*(Y)\setminus 0)$ by means of a phase function. Recall that by definition G is Lagrangian with respect to the difference $\sigma_X-\sigma_Y$ of the symplectic forms of $T^*(X)$ and $T^*(Y)$ lifted to $T^*(X)\times T^*(Y)=T^*(X\times Y)$. This differs in sign from the symplectic form $\sigma_X+\sigma_Y$ of $T^*(X\times Y)$, so it is the *twisted canonical relation*

$$G'=\{(x,\xi,y,-\eta);(x,\xi,y,\eta)\in G\}$$

which is Lagrangian with respect to the standard symplectic structure in $T^*(X\times Y)$. (Recall that this sign change occurred already in the calculus of wave front sets in Section 8.2, particularly in Theorem 8.2.14.) If we take a clean phase function ϕ defining G' it follows that it defines G through

$$G=\{(x,\phi'_x(x,y,\theta),y,-\phi'_y(x,y,\theta));\phi'_\theta(x,y,\theta)=0\}. \tag{21.2.9}$$

When we say that a canonical relation is defined by a phase function we shall always refer to the formula (21.2.9) with a sign change in the second fiber variable.

Proposition 21.2.19. *Let* X,Y,Z *be* C^∞ *manifolds and* G_1,G_2 *homogeneous canonical relations* $\subset(T^*(X)\setminus 0)\times(T^*(Y)\setminus 0)$ *and* $(T^*(Y)\setminus 0)\times(T^*(Z)\setminus 0)$, *parametrized locally by non-degenerate phase functions* $\phi(x,y,\theta)$, $\theta\in\mathbb{R}^\nu$, *and* $\psi(y,z,\tau)$, $\tau\in\mathbb{R}^\mu$, *defined in conic neighborhoods of* (x_0,y_0,θ_0) *and* (y_0,z_0,τ_0) *where* $\phi'_\theta=0$, $\psi'_\tau=0$, $\phi'_y+\psi'_y=0$. *If the composition* $G_1\circ G_2$ *is clean at the corresponding point then*

$$\Phi(x,z,y,\theta,\tau)=\phi(x,y,\theta)+\psi(y,z,\tau)$$

is a clean phase function defining the composition, if (y, θ, τ) *are now regarded as the parameters. The excess of* Φ *is equal to the excess in the clean intersection of* $G_1 \times G_2$ *and* $T^*(X) \times \Delta(T^*(Y)) \times T^*(Z)$.

Note that (y, θ, τ) belongs to a conic manifold so there is an obvious interpretation of the homogeneity condition on phase functions in this situation. We might also take

$$(y(|\theta|^2 + |\tau|^2)^{\frac{1}{4}}, \theta, \tau) \in \mathbb{R}^{n_Y + \nu + \mu}$$

as parameter to agree literally with Definition 21.2.15. Here $n_Y = \dim Y$.

Proof. That ϕ and ψ are non-degenerate means that (locally)

$$M = \{(x, y, \theta); \phi'_\theta = 0\}, \qquad N = \{(y, z, \tau); \psi'_\tau = 0\}$$

are manifolds, with tangent planes defined by $d\phi'_\theta = 0$ resp. $d\psi'_\tau = 0$ and that the maps

$$M \ni (x, y, \theta) \mapsto (x, \phi'_x, y, -\phi'_y) \in G_1,$$
$$N \ni (y, z, \tau) \mapsto (y, \psi'_y, z, -\psi'_z) \in G_2$$

are local diffeomorphisms. That $G_1 \circ G_2$ is clean means first of all that

$$G = \{(x, \xi, y', \eta', y'', \eta'', z, \zeta) \in G_1 \times G_2; y' = y'', \eta' = \eta''\}$$

is a manifold, of dimension $n_X + n_Y + n_Y + n_Z - 2n_Y + e = n_X + n_Z + e$ where e is the excess of the intersection. Taking $(x, y', \theta, y'', z, \tau) \in M \times N$ as a parametrization of $G_1 \times G_2$ we know that G is defined by the equations

$$\xi = \phi'_x(x, y', \theta), \quad y' = y'', \eta' = -\phi'_y(x, y', \theta), \quad \eta' = \eta'',$$
$$\eta'' = \psi'_y(y'', z, \tau), \quad \zeta = -\psi'_z(y'', z, \tau)$$

and that the tangent plane of G is defined by the vanishing of all differentials in addition to those of ϕ'_θ, ψ'_τ defining the tangent plane of $M \times N$. All this just expresses the hypothesis that the intersection is clean. The claim is that

$$\{(x, z, y, \theta, \tau); \phi'_\theta(x, y, \theta) = 0, \psi'_\tau(y, z, \tau) = 0, \phi'_y(x, y, \theta) + \psi'_y(y, z, \tau) = 0\}$$

is a manifold with tangent plane defined by the equations

$$d\phi'_\theta = 0, \quad d\psi'_\tau = 0, \quad d(\phi'_y + \psi'_y) = 0.$$

Putting $y' = y'' = y$ and defining $\eta' = -\phi'_y(x, y', \theta)$, $\eta'' = \psi'_y(y'', z, \tau)$ identifies the sets and the tangent planes. Since the excess of the phase function is by definition $\dim G - \dim(X \times Z)$, the proposition is proved.

In Section 21.6 we shall return to the composition and in particular discuss operations defined on half densities in the canonical relations. As in Section 18.2 these will occur as symbols for the distributions associated with a Lagrangian in Chapter XXV.

21.3. Normal Forms of Functions

In questions concerning regularity and existence of solutions of a pseudo-differential equation

$$Pu=f$$

the theory of pseudo-differential operators developed in Section 18.1 allows one to replace the principal symbol p of P by the product with any non-vanishing homogeneous function q. In fact, q can be taken as the symbol of an elliptic and therefore essentially invertible operator; the equation $Pu=f$ is then nearly equivalent to $QPu=Qf$. The microlocal point of view where one first looks only for equations modulo functions with no wave front set at a given point (x_0, ξ_0) in the cotangent bundle allows one to localize the argument so that only invertibility of q at a point is needed. In particular, the points where p is non-characteristic are then of no further interest.

In Chapter XXV we shall extend the theory of pseudo-differential operators with a machinery which also allows one to change the principal symbol by composition with a local homogeneous canonical transformation. To benefit from this one must know simple normal forms, which can be studied quite explicitly, such that general principal symbols can be reduced to one of them by the two basic operations

(i) multiplication by a factor $\neq 0$;

(ii) composition with a local homogeneous canonical transformation.

This section is devoted to the geometrical investigation of normal forms. The material will not be used until Chapters XXVI, XXVII so the reader may well postpone studying it without loss of continuity.

Let p be a C^∞ complex valued function homogeneous of degree m in a conic neighborhood of $(0, \xi_0) \in T^*(\mathbb{R}^n) \setminus 0$. By multiplication with the non-vanishing function $|\xi|^{s-m}$ we can always change the degree to any convenient value s; it often turns out that $s=1$ is a good choice. We first observe an immediate consequence of Theorem 21.1.9 where this is so.

Theorem 21.3.1. *Let p be a real valued C^∞ function in a conic neighborhood of $(0, \xi_0) \in T^*(\mathbb{R}^n) \setminus 0$, with $p(0, \xi_0)=0$, such that p is homogeneous of degree 1 and the Hamilton field H_p is not in the radial direction at $(0, \xi_0)$. Then there is a homogeneous canonical transformation χ from a conic neighborhood of $(0, \varepsilon_n)$, $\varepsilon_n=(0, 0, \dots, 1)$ to $(0, \xi_0)$ such that $\chi^* p=\xi_1$.*

Proof. By hypothesis the conditions in Theorem 21.1.9 are satisfied by $p_1=p$, $a=0$, $b=\varepsilon_n$, so we can complete to a homogeneous system of symplectic coordinates q_j, p_k near $(0, \xi_0)$ with $q_j=0$, $p_k=\delta_{kn}$ at $(0, \xi_0)$.

When p is complex valued our equivalence problem becomes much more difficult but we shall discuss a number of important cases. In doing so we write $p=p_1+ip_2$ where p_1 and p_2 are real valued. If dp_1 and dp_2 are linearly independent at a point, then the equation $p=0$ defines a manifold of codimension 2 in a neighborhood. We shall first study the case where it is *involutive*.

Theorem 21.3.2. *Assume that p is a C^∞ homogeneous function in a conic neighborhood of a zero $(0,\xi_0)\in T^*(\mathbb{R}^n)\setminus 0$ and that*

(21.3.1) $\operatorname{Re} H_p$, $\operatorname{Im} H_p$ *and the radial direction ρ are linearly independent at* $(0,\xi_0)$;

(21.3.2) $\{p,\bar{p}\}=-2i\{\operatorname{Re} p, \operatorname{Im} p\}=0$ *in a neighborhood of $(0,\xi_0)$ when $p=0$.*

Then there is a C^∞ homogeneous function $a(x,\xi)\neq 0$ and a homogeneous canonical transformation χ from a conic neighborhood of $(0,\varepsilon_n)$ to $(0,\xi_0)$ such that $\chi^*(a\,p)=\xi_1+i\xi_2$.

Proof. The hypothesis (21.3.1) implies that $V=p^{-1}(0)$ near $(0,\xi_0)$ is a C^∞ conic manifold of codimension 2 such that the canonical one form does not vanish in the tangent plane at $(0,\xi_0)$. Condition (21.3.2) shows that we have $\sigma(\operatorname{Re} H_p, \operatorname{Im} H_p)=0$ on V that is, V is involutive. By Theorem 21.2.4 we can therefore by a homogeneous canonical transformation bring V and ξ_0 to the desired position, so we assume from now on that $\xi_0=\varepsilon_n$ and that V is defined by $\xi_1=\xi_2=0$. We may also assume that p is of degree 1. Now

$$H_p=a(x,\xi)\,\partial/\partial x_1+b(x,\xi)\,\partial/\partial x_2 \quad \text{on } V$$

where $a=\partial p/\partial\xi_1$ and $b=\partial p/\partial \xi_2$ span the complex plane. This is an elliptic operator in x_1, x_2 - essentially the Cauchy-Riemann operator - depending smoothly on the parameters $(x_3,\xi_3,\dots,\xi_n)$, and a, b are homogeneous of degree 0. (Outside V we might have a very complicated equation but it will be avoided.) By Theorem 13.3.3 we can therefore find local C^∞ solutions of the equation $H_p u=f$ on V if $f\in C^\infty$; if f is homogeneous of degree μ we can take u homogeneous of degree μ. We shall now show that it is possible to make $\{p,\bar{p}\}$ equal to 0 identically by multiplying p with a non-vanishing function. This requires several steps.

a) By hypothesis we know that $\{p,\bar{p}\}=0$ on V, so by Taylor's formula (Theorem 1.1.9) we have near $(0,\varepsilon_n)$

$$\{p,\bar{p}\}=f_0\bar{p}+f_1 p$$

with $f_j\in C^\infty$. Now set $q=e^{w}p$ where w is homogeneous of degree 0. Then

$$\{q,\bar{q}\}=e^{2\operatorname{Re} w}(\{p,\bar{p}\}+\{p,w\}\,\bar{p}+p\{\bar{w},\bar{p}\}+\{\bar{w},w\}\,p\,\bar{p}) \tag{21.3.3}$$

is $O(p^2)$ locally at $(0,\xi_0)$ if on V

$$f_0+\{p,w\}=0, \qquad f_1+\{\bar{w},\bar{p}\}=0.$$

Since $\overline{\{p,\bar{p}\}}=\{\bar{p},p\}=-\{p,\bar{p}\}$ we can always replace f_0 by $(f_0-\bar{f}_1)/2$ and f_1 by $(f_1-\bar{f}_0)/2$, hence arrange that $\bar{f}_0=-f_1$. The two equations are then identical so choosing w with $H_p w=-f_0$ on V, we obtain $\{q,\bar{q}\}=O(p^2)$.

b) Assume that $\{p,\bar{p}\}=O(p^k)$ at $(0,\xi_0)$ for some $k>1$. As in step a)

$$\{p,\bar{p}\}=\sum_0^k f_j p^j \bar{p}^{k-j}, \quad f_{k-j}=-\bar{f}_j.$$

Choose now $q=e^{\bar{w}}p$ where

$$w=\sum_0^{k-1} w_j p^j \bar{p}^{k-j-1}$$

Then the last term in (21.3.3) is $O(p^{2k-1})=O(p^{k+1})$ since we can only loose two of our $2k$ factors $p,\bar{p}$ and a Poisson bracket $\{p,\bar{p}\}=O(p^k)$ appears at the same time. In the preceding two terms the contributions where a Poisson bracket of p and $\bar{p}$ occurs are $O(p^{2k-1})=O(p^{k+1})$. Hence (21.3.3) is $O(p^{k+1})$ if

$$f_j+H_p w_j-\overline{H_p w_{k-j}}=0 \quad \text{on } V.$$

Since $f_{k-j}=-\bar{f}_j$ the equation with j replaced by $k-j$ is equivalent so it suffices to solve these equations when $j\leqq k/2$. We can then take $w_j=0$ when $j>k/2$ and have the equations

$$H_p w_j=-f_j, \quad j<k/2, \quad H_p w_j=-f_j/2 \quad \text{if } j=k/2,$$

for f_j is purely imaginary if $j=k/2$ is an integer. Now f_j can be chosen homogeneous of degree $1-k$, and extending a solution of the preceding equation by homogeneity from the case $|\xi|=1$ we obtain w_j homogeneous of the same degree, hence w homogeneous of degree 0.

c) By repeating step b) we obtain w^j vanishing of order j on V and homogeneous of degree 0 such that $\{q_k,\bar{q}_k\}$ vanishes of order $k+2$ in a fixed neighborhood of $(0,\varepsilon_n)$ on V if

$$q_k=p\exp(\bar{w}^0+\ldots+\bar{w}^k).$$

If we choose w with the Taylor expansion $w^0+w^1+\ldots$ at V and homogeneous of degree 0, the problem is solved to infinite order.

d) We have now reduced the proof to the case where

$$\{p_1,p_2\}=c_1p_1+c_2p_2$$

with c_j homogeneous of degree 0 and vanishing of infinite order on V. Choose f_1 homogeneous of degree 0 so that

$$\{e^{f_1}p_1,p_2\}=e^{f_1}c_2p_2, \quad \text{that is, } H_{p_2}f_1=c_1$$

and so that $f_1=0$ when $x_2=0$, if $\partial p_2/\partial\xi_2\neq 0$ as we may assume. The solution of this first order differential equation is unique, hence homogeneous of degree 0, and it vanishes of infinite order on V since H_{p_2} is tangential to V; we just have to differentiate the equation to prove this successively. Next we observe that

$$\{e^{f_1}p_1,e^{f_2}p_2\}=0 \quad \text{if } \{e^{f_1}p_1,f_2\}+e^{f_1}c_2=0$$

which again has a solution f_2 vanishing of infinite order on V. Thus

$$a=(e^{f_1}p_1+ie^{f_2}p_2)/(p_1+ip_2)=1+((e^{f_1}-1)p_1+i(e^{f_2}-1)p_2)/(p_1+ip_2)$$

is infinitely differentiable, and differs from 1 by a function vanishing of infinite order on V. Now $q=ap$ has the desired property $\{\operatorname{Re} q, \operatorname{Im} q\}=0$. In view of Theorem 21.1.9 we can make $\operatorname{Re} q=\xi_1$, $\operatorname{Im} q=\xi_2$ in a new homogeneous symplectic coordinate system, which completes the proof.

Remark. Note that after the reduction the Hamilton field is exactly the Cauchy-Riemann operator in x_1+ix_2. This simplicity is essential for it allows us to connect with classical function theory and also gives an operator which we know how to handle even outside V.

We shall now study the opposite situation where $\{p,\bar{p}\}\neq 0$ at a zero γ of p. This implies that $\operatorname{Re} H_p$, $\operatorname{Im} H_p$ and the radial vector field ρ are linearly independent at γ, for a linear relation

$$a \operatorname{Re} H_p+b \operatorname{Im} H_p+c\rho=0$$

implies $a\{\operatorname{Re} H_p, \operatorname{Im} H_p\}=0$ when applied to $\operatorname{Im} H_p$ and $b\{\operatorname{Im} H_p, \operatorname{Re} H_p\}=0$ when applied to $\operatorname{Re} H_p$, in view of the homogeneity. In particular, $p^{-1}(0)$ is near γ a manifold of codimension 2. Note that when $p=0$ we have

$$\{ap, \overline{ap}\}=|a|^2\{p,\bar{p}\}$$

so the sign of the Poisson bracket $\{\operatorname{Re} p, \operatorname{Im} p\}$ cannot be changed by the operations we allow. The following theorem shows that it is the only invariant:

Theorem 21.3.3. *Let p be C^∞ and homogeneous in a conic neighborhood of $(0,\xi_0)$ where $p=0$, $\{\operatorname{Re} p, \operatorname{Im} p\}\gtrless 0$. Then there is a homogeneous canonical transformation χ of a conic neighborhood of $(0,\pm\varepsilon_n)$ on a conic neighborhood of $(0,\xi_0)$ and a C^∞ function a with $a(0,\xi_0)\neq 0$ such that*

$$\chi^*(ap)=\xi_1+ix_1\xi_n.$$

For the proof we need a lemma which we state in sufficient generality for another application later on.

Lemma 21.3.4. *Let p and q be real valued C^∞ functions near a point γ in a symplectic manifold S where $p=q=0$ and $\{p,q\}>0$; and let a, b be positive numbers with $a+b=1$. Then there is a unique positive $u\in C^\infty$ in a neighborhood of γ such that*

$$\{u^a p, u^b q\}=1. \tag{21.3.4}$$

If S is conic and p, q are homogeneous of degree m and m' respectively, then $u^a p$ and $u^b q$ are homogeneous of degree $a(1-m')+bm$ and $am'+b(1-m)$.

Proof. The equation (21.3.4) can be written in the form

$$u\{p,q\}+(bq\,H_p-ap\,H_q)u=1. \tag{21.3.4)$'$}$$

On the manifold V of codimension 2 where $p=q=0$ it reduces to $u\{p,q\}=1$. If we set $u_0=1/\{p,q\}$, $p_0=p\,u_0^a$, $q_0=q\,u_0^b$ we have therefore $\{p_0,q_0\}=1$ on V. We simplify notation by assuming that p and q already have this property. By the Darboux theorem (Theorem 21.1.6) we can choose symplectic coordinates so that $\xi_1=p$, $\gamma=(0,0)$. Thus $\partial q/\partial x_1\neq 0$ at $(0,0)$ so the equation $q=0$ can be solved for x_1 to give $x_1=X(x',\xi)$, $x'=(x_2,\dots,x_n)$. Since we have $\{\xi_1,x_1-X(x',\xi)\}=1$ another application of Theorem 21.1.6 shows that we can take $x_1-X(x',\xi)$ as a new symplectic coordinate instead of x_1. With such coordinates we have $p=\xi_1$ and $q=Qx_1$, $Q\in C^\infty$. The equation (21.3.4)′ is then of the form

$$u\,\partial q/\partial x_1+(b\,\partial q/\partial x_1\,x_1\,\partial/\partial x_1+a\,\xi_1\,\partial q/\partial x_1\,\partial/\partial\xi_1)u+\dots=1$$

where dots indicate a differential operator vanishing of second order when $x_1=\xi_1=0$. Division by $\partial q/\partial x_1$ reduces the equation to the form discussed in Theorem C.2.1 in the appendix and proves the first part of the lemma.

Assume now that we have homogeneous p and q. Choose u so that (21.3.4) is valid in a neighborhood of γ and set $P=u^a p$, $Q=u^b q$. Then $\{P,Q\}=1$. Set $M_t^*P=P_t$, $M_t^*Q=Q_t$ and $M_t^*u=u_t$ where M_t is multiplication by t in S. Then

$$\{P_t,Q_t\}=t$$

by (21.1.6) and since $P_t=u_t^a\,t^m\,p$, $Q_t=u_t^b\,t^{m'}\,q$ we obtain

$$\{u_t^a\,p,u_t^b\,q\}=t^{1-m-m'}.$$

In view of the uniqueness of the solution of (21.3.4) it follows that

$$u_t=t^{1-m-m'}\,u$$

in a fixed neighborhood of γ when t is near 1. Hence u can be extended to a homogeneous function of degree $1-m-m'$, and (21.3.4) remains valid in a conic set. This proves that P and Q have the stated homogeneity properties.

Proof of Theorem 21.3.3. Assume that $\{\operatorname{Re}p,\operatorname{Im}p\}>0$. If we apply Lemma 21.3.4 to $\operatorname{Re}p$, $\operatorname{Im}p$ with $a=b=\frac12$ and $\gamma=(0,\xi_0)$ we obtain a function u such that $u^{\frac12}p=\tilde p$ is homogeneous of degree $\frac12$ and $\{\operatorname{Re}\tilde p,\operatorname{Im}\tilde p\}=1$. Now we choose a real valued homogeneous function h of degree 1 such that

$$\{\operatorname{Re}\tilde p,h\}=\{\operatorname{Im}\tilde p,h\}=0$$

and $h(\gamma)=1$. To do so we just solve these differential equations with homogeneous C^∞ values prescribed on $V=p^{-1}(0)$, equal to 1 at γ. This is possible since $[H_{\operatorname{Re}\tilde p},H_{\operatorname{Im}\tilde p}]=0$ by the Jacobi identity, which maked the Frobenius theorem applicable (see Corollary C.1.2 in the appendix). Note that the plane spanned by $H_{\operatorname{Re}\tilde p}$ and $H_{\operatorname{Im}\tilde p}$ is transversal to V. The uniqueness of the solution proves its homogeneity. Now Theorem 21.1.9 shows that we can find homogeneous symplectic coordinates near $(0,\xi_0)$ such that

$$\eta_1=h^{\frac{1}{2}}\operatorname{Re}\tilde{p}, \quad y_1=h^{-\frac{1}{2}}\operatorname{Im}\tilde{p}, \quad \eta_n=h,$$

and so that $y_1=\ldots=y_n=\eta_1=\ldots=\eta_{n-1}=0$ at γ. Then

$$h^{\frac{1}{2}}\tilde{p}=\eta_1+i\,y_1\,\eta_n$$

which proves the theorem when $\{\operatorname{Re}p,\operatorname{Im}p\}>0$. For the opposite sign we apply this result to $\bar{p}$ and conclude that there are symplectic coordinates such that $(0,\xi_0)$ corresponds to $(0,\varepsilon_n)$ and the product of p by a non-vanishing function is equal to $\eta_1-i\,y_1\,\eta_n$ Replacing the coordinates y_n,η_n by $-y_n,-\eta_n$ then proves the statement.

The zero set of p may be a symplectic manifold of codimension 2 even if $\{\operatorname{Re}p,\operatorname{Im}p\}$ is identically 0 in it. An example is

$$p=p_1+i\,p_2=\xi_1+i\,x_1^k\,\xi_n, \qquad k>1.$$

Then $\{p_1,p_2\}=k\,x_1^{k-1}\,\xi_n$, $H_{p_1}=\partial/\partial x_1$, $H_{p_2}=-k\,x_1^{k-1}\,\xi_n\,\partial/\partial\xi_1+x_1^k\,\partial/\partial x_n$ so all Poisson brackets formed with at most k factors vanish when $\xi_1=x_1=0$ but $H_{p_1}^k p_2=\{p_1,\{p_1,\ldots,\{p_1,p_2\}\ldots\}\}=k!\,\xi_n\neq 0$ if $\xi_n\neq 0$. This situation is characterized invariantly in the following theorem.

Theorem 21.3.5. *Let $p=p_1+i\,p_2$ be a homogeneous C^∞ function in a conic neighborhood of $\gamma=(0,\xi_0)$ where $p=0$, $H_{p_1}\neq 0$, and assume that p_2 has a zero of fixed order $k>1$ near γ on each integral curve of H_{p_1} in $p_1^{-1}(0)$ (that is, the leaves of the Hamilton foliation of the involutive surface $p_1^{-1}(0)$). Then the zeros of p form a manifold V near γ. On V we have $H_{p_2}=b\,H_{p_1}$, all Poisson brackets of p_1 and p_2 with at most k factors vanish, and for every $a\in C^\infty$*

$$H_{\operatorname{Re}ap}^k\operatorname{Im}ap=|a|^2(\operatorname{Re}(a+i\,b\,a))^{k-1}\,H_{\operatorname{Re}p}^k\operatorname{Im}p. \tag{21.3.5}$$

There is a homogeneous canonical transformation χ from a conic neighborhood of $(0,\varepsilon_n)$ to a conic neighborhood of $(0,\xi_0)$ and a C^∞ function a with $a(0,\xi_0)\neq 0$ such that

$$\chi^*(a\,p)=\xi_1+i\,x_1^k\,\xi_n$$

unless the sign on the right-hand side of (21.3.5) is always <0, that is, k is odd and $H_{\operatorname{Re}p}^k\operatorname{Im}p<0$; then $\operatorname{Im}p$ changes sign from $+$ to $-$ along the integral curves of H_{p_1} and the statement is true with ε_n replaced by $-\varepsilon_n$.

Before the proof we give a general reduction which follows from the Malgrange preparation theorem. It does not really provide a normal form but it will be of fundamental importance in Chapters XXVI and XXVII.

Theorem 21.3.6. *Let $p=p_1+i\,p_2$ be a C^∞ homogeneous function in a conic neighborhood of $(0,\xi_0)$ where $p=0$ and H_p is linearly independent of the radial vector. Then there is a homogeneous symplectic transformation χ of a* *conic neighborhood of* $(0,\varepsilon_n)$ *on a conic neighborhood of* $(0,\xi_0)$ *and a* C^∞ *homogeneous function a with* $a(0,\xi_0)\neq 0$ *such that*

$$\chi^*(ap)=\xi_1+if(x,\xi')$$

where $\xi'=(\xi_2,\dots,\xi_n)$. *Thus the imaginary part is independent of* ξ_1.

Proof. We may assume that p is of degree 1 and that $H_{\operatorname{Re}p}$ does not have the radial direction. By Theorem 21.1.9 homogeneous symplectic coordinates can then be chosen so that $p_1=\xi_1$ and $\xi_0=\varepsilon_n$. In particular $\partial p/\partial\xi_1\neq 0$ so by Malgrange's preparation theorem (Theorem 7.5.6) we can find $q, r\in C^\infty$ near $(0,\varepsilon_n)$ so that

$$\xi_1=qp+r$$

and r is independent of ξ_1. Restricting to $\xi_n=1$ and extending by homogeneity from there we can make q and r homogeneous of degree 0 and 1 respectively. Thus writing $r=r_1+ir_2$ we have

$$qp=\xi_1-r_1-ir_2$$

and we can apply Theorem 21.1.9 again to choose new homogeneous symplectic coordinates with $y_1=x_1$ and $\eta_1=\xi_1-r_1$. Then $H_{y_1}=H_{x_1}=-\partial/\partial\xi_1$ so $H_{y_1}r_2=0$. In the new coordinates we therefore have $-r_2=f(y,\eta')$ so $qp=\eta_1+if(y,\eta')$ with f independent of η_1. Hence $a=q$ and χ: $(y,\eta)\mapsto(x,\xi)$ have the required properties.

Proof of Theorem 21.3.5. In the hypersurface V_1 defined by $p_1=0$ the zeros of $q=H_{p_1}^{k-1}p_2$ form a hypersurface V_2 where $\{p_1,q\}\neq 0$ and $H_{p_1}^j p_2=0$, $j<k$. Thus the restriction of p_2 to V_1 is divisible by q^k, so

$$p_2=bp_1+cq^k,$$
$$\{p_1,p_2\}=p_1\{p_1,b\}+\{p_1,c\}q^k+ck\{p_1,q\}q^{k-1}.$$

where b, $c\in C^\infty$. Thus the zero set V of p is locally equal to V_2 and $H_{p_2}=bH_{p_1}$ there. All Poisson brackets with at most k factors p_1, p_2 vanish on V. In fact, taking a Poisson bracket of $\{p_1,p_2\}$ with cq^k gives a zero of order $k-1$ which cannot be removed by $k-2$ subsequent Poisson brackets, and the Poisson bracket of functions divisible by p_1 is again divisible by p_1. Hence it suffices to prove (21.3.5) when a is a constant. Then we have

$$\{\operatorname{Re}ap,\operatorname{Im}ap\}=i\{ap,\overline{ap}\}/2=|a|^2\{\operatorname{Re}p,\operatorname{Im}p\},$$

and $H_{\operatorname{Re}ap}^{k-1}$ may be replaced by $(\operatorname{Re}(a+iab))^{k-1}(H_{\operatorname{Re}p})^{k-1}$.

Now choose a and new symplectic coordinates such that ap has the form

$$q=\xi_1+if(x,\xi')$$

and $\gamma=(0,\varepsilon_n)$ (Theorem 21.3.6). Since $H_{\operatorname{Re}q}=\partial/\partial x_1\neq 0$, it follows from the part of the theorem already proved that the zeros of q near $(0,\varepsilon_n)$ form a manifold of codimension 2 where

$$\xi_1=0, \quad \partial^j f(x,\xi')/\partial x_1^j=0 \text{ when } j<k \text{ and } \partial^k f(x,\xi')/\partial x_1^k \neq 0.$$

The equation $\partial^{k-1} f(x,\xi')/\partial x_1^{k-1}=0$ defines x_1 as a C^∞ function $X(x',\xi')$ and $\partial^j f(x,\xi')/\partial x_1^j$ must vanish when $x_1=X(x',\xi')$ since the zero set of q is a manifold of codimension 2. Thus the quotient $f(x,\xi')/(x_1-X(x',\xi'))^k$ is smooth and positive, if $\partial^k f(x,\xi')/\partial x_1^k>0$ at $(0,\varepsilon_n)$ as we now assume. Hence it has a positive k^{th} root, so $f(x,\xi')=g(x,\xi')^k$ where g is C^∞ and homogeneous of degree $1/k$, and $\{\xi_1,g(x,\xi')\}>0$ at $(0,\varepsilon_n)$. We can now apply Lemma 21.3.4 with p replaced by ξ_1 and q replaced by g, with $a=k/(k+1)$ and $b=1/(k+1)$. This gives a function u such that

$$\{\phi,\psi\}=1 \quad \text{if } \phi=u^{k/(k+1)}\xi_1, \quad \psi=u^{1/(k+1)}g.$$

Here ϕ is homogeneous of degree $k/(k+1)$ and ψ is homogeneous of degree $1/(k+1)$. As in the proof of Theorem 21.3.3 we can now choose h homogeneous of degree 1 with $h(0,\varepsilon_n)=1$ so that

$$\{\phi,h\}=\{\psi,h\}=0.$$

By Theorem 21.1.9 there are homogeneous symplectic coordinates (y,η) at $(0,\varepsilon_n)$ with $y=0$, $\eta=\varepsilon_n$ at $(0,\varepsilon_n)$ and

$$\eta_1=h^{1/(k+1)}\phi, \quad y_1=h^{-1/(k+1)}\psi, \quad \eta_n=h$$

Then

$$u^{k/(k+1)}h^{1/(k+1)}(\xi_1+if)=\eta_1+iy_1^k\eta_n$$

so the theorem is proved when $\partial^k f/\partial x_1^k>0$. If the opposite sign holds we can apply this conclusion to $\bar{p}$, hence reduce p to the form $\eta_1-iy_1^k\eta_n$ at $(0,\varepsilon_n)$. If k is even we multiply by -1 and take $-y_1$, $-\eta_1$ as new coordinates instead of y_1,η_1 to obtain the desired form. If k is odd we replace y_n, η_n by $-y_n$, $-\eta_n$ and obtain the equivalence with $\eta_1+iy_1^k\eta_n$ in a neighborhood of $(0,-\varepsilon_n)$. The sign of (21.3.5) is then invariant so it is not possible to have equivalence with $\eta_1+iy_1^k\eta_n$ at $(0,\varepsilon_n)$.

21.4. Folds and Glancing Hypersurfaces

In the preceding sections we have studied a number of symplectic classification problems in non-degenerate situations. We shall now consider some of the simplest singular cases, starting with Lagrangian submanifolds Λ of a cotangent bundle $T^*(X)$. If Λ is a section, that is, the projection $\pi\colon \Lambda\to X$ is a diffeomorphism, we know that Λ is locally the graph of the differential of a function in X; indeed, it was this observation which first brought us in contact with Lagrangian manifolds in Section 6.4. We shall now examine the case where π is a folding map. (For definitions and basic facts see Appendix C 4.)

Theorem 21.4.1. *Let X be a C^∞ manifold and Λ a Lagrangian submanifold of $T^*(X)$ such that the projection $\Lambda \to X$ has a fold at $(x_0, \xi_0) \in \Lambda$. Then there exist local coordinates x in X, vanishing at x_0, and a C^∞ function $\psi(x)$ near 0 such that at (x_0, ξ_0) the manifold Λ is parametrized by*

$$\Lambda = \{(x, \partial\psi/\partial x', \partial\psi/\partial x_n \pm x_n^{\frac{1}{2}});\ x_n \geqq 0\}. \tag{21.4.1}$$

Here $x' = (x_1, \ldots, x_{n-1})$.

That (21.4.1) defines a manifold with a folding projection is clear if we put $x_n = t_n^2$ for then the set is of the form

$$\{(t', t_n^2, \partial_1 \psi(t', t_n^2), \ldots, \partial_n \psi(t', t_n^2) + t_n)\}.$$

When $t_n \neq 0$ we have the graph of the differentials of the two functions $\psi(x) \pm \frac{2}{3} x_n^{\frac{3}{2}}$ so it is clear that Λ is Lagrangian.

Proof of Theorem 21.4.1. By Theorem C.4.2 we can choose local coordinates $t_1, \ldots, t_n$ in Λ and $x_1, \ldots, x_n$ in X so that the projection of the point in Λ with coordinates (t', t_n) has the coordinates (t', t_n^2) in X. Then Λ is defined by

$$\Lambda = \{(t', t_n^2, \xi(t))\}$$

where $\xi(t)$ is a C^∞ function of t with values in $\mathbb{R}^n$ and we are using standard coordinates in the cotangent bundle. That Λ is Lagrangian means that

$$\sum_1^{n-1} \xi_j dt_j + \xi_n d(t_n^2)$$

is closed, hence the differential of a function $\phi(t)$ with

$$\xi_j = \partial\phi/\partial t_j, \quad j < n; \quad 2t_n \xi_n = \partial\phi/\partial t_n.$$

Hence Λ is the graph of the differentials of the functions $\phi(x', \pm x_n^{\frac{1}{2}})$ when $x_n > 0$. To put it in the desired form we write $\phi = \phi_e + \phi_o$ where $\phi_e(t) = (\phi(t', t_n) + \phi(t', -t_n))/2$ is even under the involution in Λ and ϕ_o is odd. Thus we know by Theorem C.4.4 that $\phi_e(t) = \psi(t', t_n^2)$ for some $\psi \in C^\infty$ at 0. Since $t_1, \ldots, t_n$ are parameters in Λ we know that $\partial\xi_j/\partial t_n \neq 0$ at 0 for some j. Now $\partial\xi_j/\partial t_n = \partial^2\phi/\partial t_j \partial t_n = 0$ at 0 if $j < n$, so we must have $\partial\xi_n/\partial t_n \neq 0$ at 0, that is,

$$\partial^3 \phi_o(t)/\partial t_n^3 = \partial^3 \phi(0)/\partial t_n^3 \neq 0.$$

Changing the sign of t_n if necessary we may assume that the sign is positive. Since $\partial^j \phi_o/\partial t_n^j = 0$ when $t_n = 0$ if $j < 3$ we can now write $\phi_o(t) = t_n^3 F(t)$ where $F \in C^\infty$, $F(0) > 0$, and F is an even function of t_n. Hence Theorem C.4.4 shows that there is a C^∞ function G near 0 such that

$$(\tfrac{3}{2}\phi_o(t))^{\frac{2}{3}} = (\tfrac{3}{2}F(t))^{\frac{2}{3}} t_n^2 = G(t', t_n^2).$$

We have $G = 0$ and $\partial G/\partial x_n > 0$ when $x_n = 0$, and $\phi_o(x', \pm x_n^{\frac{1}{2}}) = \pm \frac{2}{3} G(x)^{\frac{3}{2}}$, $x_n > 0$. If we take $G(x)$ as a new variable instead of x_n, the theorem is proved.

Remark. ψ and the coordinate function t_n are uniquely determined where $t_n>0$ but otherwise arbitrary apart from the smoothness. The other variables are any variables which define a coordinate system. Another form of the theorem is therefore that there exist two C^∞ functions ϕ and ψ at x_0 with $\phi(x_0)=0$, $\phi'(x_0)\neq 0$, such that the projection of Λ is locally defined by $\phi\geqq 0$ and Λ is the graph of the differentials of the functions $\psi\pm\frac{2}{3}\phi^{\frac{3}{2}}$. Lagrangians of the form just considered occur frequently in diffraction theory.

Next we shall study folds over a symplectic manifold. This will lead to the degenerate symplectic structure discussed already in Section 21.1.

Theorem 21.4.2. *Let S be a C^∞ symplectic manifold, T a C^∞ manifold of the same dimension $2n$ and $f\colon T\to S$ a C^∞ map with a fold at t_0. Then there exist a local symplectic coordinate system (x,ξ) vanishing at $f(t_0)$ in S, and local coordinates (y,η) vanishing at t_0 in T such that*

$$f(y,\eta)=(y_1,\dots,y_n,\eta_1^2/2,\eta_2,\dots,\eta_n). \tag{21.4.2}$$

Thus the pullback $f^*\sigma$ of the symplectic form in S is given by

$$\eta_1\,d\eta_1\wedge dy_1+\sum_2^n d\eta_j\wedge dy_j. \tag{21.4.3}$$

Conversely, let η, y be local coordinates vanishing at t_0 which are even with respect to the involution of the fold apart from η_1 which is odd. If $f^*\sigma$ is equal to (21.4.3) *then there exist local symplectic coordinates x, ξ at $f(t_0)$ such that* (21.4.2) *holds.*

Proof. By Theorem C.4.2 we can choose local coordinates $t_1,\dots,t_{2n}$ in T vanishing at t_0 and $s_1,\dots,s_{2n}$ in S vanishing at $f(t_0)$ such that

$$f(t)=(t_1,\dots,t_{2n-1},t_{2n}^2/2).$$

By the Darboux Theorem 21.1.6 we can choose local symplectic coordinates x, ξ in S vanishing at $f(t_0)$ such that $\xi_1=s_{2n}$. Then

$$y_j=f^*x_j,\quad j\geqq 1;\quad \eta_j=f^*\xi_j,\quad j>1,\ \eta_1=t_{2n},$$

are local coordinates at t_0. In fact, if the differentials of y_j, η_j at t_0 in the direction γ vanish for all j, then $\langle\gamma,dt_{2n}\rangle=0$ and

$$\langle f'(t_0)\gamma,dx_j\rangle=\langle f'(t_0)\gamma,d\xi_j\rangle=0,\quad j=1,\dots,n,$$

since $\langle f'(t_0)\gamma,d\xi_1\rangle=\langle\gamma,dt_{2n}^2/2\rangle=0$; hence $\gamma=0$. Since $\eta_1^2/2=f^*\xi_1$ we have (21.4.2), which proves the first part of the theorem.

To prove the second part we choose, using Theorem C.4.4, C^∞ functions q_j, p_k in a neighborhood of $f(t_0)$ such that $f^*q_j=y_j$ and $f^*p_k=\eta_k$ if $k\neq 1$, $f^*p_1=\eta_1^2/2$. From (21.4.3) it follows that $\sigma=\sum dp_j\wedge dq_j$ in the range of f; in particular we have local coordinates at $f(t_0)$. They are symplectic when

$p_1>0$ but fairly arbitrary when $p_1<0$. Now define $\xi_j=p_j$, $x_j=q_j$ in a neighborhood of 0 when $p_1>0$, and extend the definition to a full neighborhood of 0 so that

$$H_{q_1}x_j=0, \quad H_{q_1}\xi_j=-\delta_{1j}.$$

These equations are valid when $p_1>0$, and since $H_{q_1}=-\partial/\partial\xi_1$ when $p_1=0$ they define x, ξ uniquely in a neighborhood of 0. For the new coordinates the commutation relations are also valid when $\xi_1<0$, for

$$H_{q_1}\{x_j,x_k\}=\{q_1,\{x_j,x_k\}\}=-\{x_k,\{q_1,x_j\}\}+\{x_j,\{q_1,x_k\}\}=0$$

by the Jacobi identity and similarly for the other Poisson brackets. With these symplectic coordinates (21.4.2) remains valid, and the proof is complete.

The pullback $\sigma_T=f^*\sigma$ of σ to T is of course a closed two form. If i is the involution of the fold we have $i^*\sigma_T=\sigma_T$, because $f\circ i=f$. This is also obvious from (21.4.3) since i just changes the sign of η_1. The following version of Theorem 21.1.7 for manifolds with involution gives a characterization of this situation.

Theorem 21.4.3. *Assume that S and σ satisfy the hypotheses of Theorem 21.1.7 and that i is a C^∞ involution with $i^*\sigma=\sigma$ and S_0 as fixed point set. Then there exist local coordinates y, η such that σ is equal to (21.4.3) and the involution only changes the sign of the η_1 coordinate.*

Proof. We shall give a proof which is completely independent of Theorem 21.1.7 since the existence of the involution actually simplifies matters. Let $s_1,\dots,s_{2n}$ be local coordinates at a point $s_0\in S_0$ such that the involution just changes the sign of s_{2n}. Write

$$\sigma=\sum_{i<j}a_{ij}\,ds_i\wedge ds_j.$$

Since $i^*\sigma=\sigma$ we have $i^*a_{ij}=a_{ij}$ if $j\neq 2n$, and $i^*a_{ij}=-a_{ij}$ if $j=2n$. By Theorem C.4.4 it follows that for suitable $A_{ij}\in C^\infty$

$$\sigma=\sum(f^*A_{ij})\,df_i\wedge df_j$$

where $f(s)=(s_1,\dots,s_{2n-1},s_{2n}^2/2)$, for this means only that $f^*A_{ij}=a_{ij}$ when $j<2n$ and that $s_{2n}f^*A_{ij}=a_{ij}$ when $j=2n$. The form

$$\tilde\sigma=\sum A_{ij}(t)\,dt_i\wedge dt_j$$

is closed when $t_{2n}\geqq 0$ since the pullback by f is closed, and it is non-degenerate at 0. By the Poincaré lemma

$$\tilde\sigma=d(\sum B_j(t)\,dt_j)$$

when $t_{2n}\geqq 0$ and $|t|<\delta$ say, and $B_j\in C^\infty$ then. Extending B_j to a C^∞ function in $\{t\in\mathbb{R}^{2n};\ |t|<\delta\}$ we find that the definition of A_{ij} can be chosen so that $\tilde\sigma$

is a non-degenerate closed two form in a full neighborhood of 0. Since $\sigma = f^*\tilde{\sigma}$ Theorem 21.4.3 is now an immediate consequence of Theorem 21.4.2.

Theorem 21.4.4. *Assume that S and σ satisfy the hypotheses of Theorem 21.1.7 and that f, g are two involutions with $f^*\sigma=\sigma$, $g^*\sigma=\sigma$ having S_0 as fixed point set and with linearly independent reflection bundles. Then there exist local coordinates x, ξ such that*

$$\sigma=\xi_1\, d\xi_1\wedge dx_1+\sum_2^n d\xi_j\wedge dx_j, \tag{21.4.4}$$

$$f(x,\xi)=(x,-\xi_1,\xi'); \quad g(x,\xi)=(x_1+\xi_1,x',-\xi_1,\xi'). \tag{21.4.5}$$

Proof. Using Theorem 21.4.3 we choose coordinates x, ξ such that σ and f have the desired form, and using Theorem C.4.6 in the appendix we choose another set of local coordinates y, η, vanishing at the same point, such that f and g have the desired form. We have to reconcile the two coordinate systems. Note that in the $x\xi$ system the vector field $\partial/\partial x_1$ when $\xi_1=0$ is determined as the radical of the symplectic form while in the $y\eta$ system the vector field $\partial/\partial y_1$ when $\eta_1=0$ is uniquely given by the span of the reflection bundles. A first step in the proof is therefore to show that these vector fields do agree. To do so we shall prove that

$$\eta_1\{\eta_1,y_j\}=\eta_1\{\eta_1,\eta_j\}=0,\ j>1,\quad \text{when } \eta_1=0, \text{ that is, } \xi_1=0. \tag{21.4.6}$$

(Note that $\eta_1 H_{\eta_1}$ is a smooth vector field.) To prove this we set $u=y_j$ or η_j, for some $j>1$, and observe that $f^*u=u$, $g^*u=u$ but that $-\eta_1=f^*g^*y_1-y_1$. This means that when $\eta_1\neq 0$

$$-\{\eta_1,u\}=\{f^*g^*y_1,f^*g^*u\}-\{y_1,u\}=f^*g^*\{y_1,u\}-\{y_1,u\}.$$

Hence

$$-\eta_1\{\eta_1,u\}=f^*g^*(\eta_1\{y_1,u\})-\eta_1\{y_1,u\}=0 \quad \text{when } \eta_1=0,$$

which proves (21.4.6). Now

$$\eta_1 H_{\eta_1}=\eta_1/\xi_1(\partial\eta_1/\partial\xi_1\,\partial/\partial x_1-\partial\eta_1/\partial x_1\,\partial/\partial\xi_1)+\ldots$$

where the dots indicate a smooth vector field vanishing when $\xi_1=0$. We have $\eta_1=0$ when $\xi_1=0$ so $\partial\eta_1/\partial x_1=0$, $\partial\eta_1/\partial\xi_1\neq 0$ and

$$\eta_1 H_{\eta_1}=(\partial\eta_1/\partial\xi_1)^2\,\partial/\partial x_1 \quad \text{when } \xi_1=0.$$

Hence (21.4.6) gives

$$\partial y_j/\partial x_1=\partial\eta_j/\partial x_1=0, \quad j=2,\ldots,n; \qquad \partial\eta_1/\partial x_1=0 \quad \text{when } \xi_1=0; \tag{21.4.7}$$

proving the statement about agreement of $\partial/\partial x_1$ and $\partial/\partial y_1$ made above.

We can express the $x\xi$ coordinates in terms of the $y\eta$ coordinates,

$$x_j=\phi_j(y,\eta_1^2,\eta'),\quad \text{all } j;$$

$$\xi_1=\eta_1\psi_1(y,\eta_1^2,\eta'),\quad \xi_j=\psi_j(y,\eta_1^2,\eta'),\quad j>1.$$

Here we have used that all coordinates except ξ_1 and η_1 are even under f and that these are odd. Differentiation with respect to x_1 gives when $\xi_1=0$, if we use (21.4.7) and note that it implies $\partial y_1/\partial x_1 \neq 0$,

$$\partial\phi_j/\partial y_1=0,\quad \partial\psi_j/\partial y_1=0;\quad j\neq 1,\quad \xi_1=0.$$

Thus

$$\tilde{x}_j=\phi_j(0,y',0,\eta'),\quad \tilde{\xi}_j=\psi_j(0,y',0,\eta'),\quad j\neq 1,$$

are equal to x_j resp. ξ_j when $\xi_1=0$ and are even under f and g. Set $\tilde{x}_1=x_1$, which is an even function with respect to f.

If $v\in C^\infty$, $f^*v=v$ and $\partial v/\partial x_1=0$ when $\xi_1=0$, then H_v is a smooth vector field equal to

$$\partial^2 v/\partial\xi_1^2\,\partial/\partial x_1+\sum_2^n(\partial v/\partial\xi_j\,\partial/\partial x_j-\partial v/\partial x_j\,\partial/\partial\xi_j) \tag{21.4.8}$$

when $\xi_1=0$. In fact, when $\xi_1\neq 0$ we have

$$H_v=\frac{1}{\xi_1}(\partial v/\partial\xi_1\,\partial/\partial x_1-\partial v/\partial x_1\,\partial/\partial\xi_1)+\sum_2^n(\partial v/\partial\xi_j\,\partial/\partial x_j-\partial v/\partial x_j\,\partial/\partial\xi_j).$$

Since v and $\partial v/\partial x_1$ are even functions of ξ_1 we have $\partial v/\partial x_1=O(\xi_1^2)$ and $\partial v/\partial\xi_1=O(\xi_1)$, hence $\partial v/\partial\xi_1=\xi_1\partial^2 v/\partial\xi_1^2+O(\xi_1^2)$.

We can now complete the proof by following that of the Darboux Theorem 21.1.6 rather closely. To be able to use (21.4.8) we shall choose symplectic coordinates q, p with the required properties in the order p_2, $q_2,\dots,p_n,q_n,p_1,q_1$. Set $p_2=\tilde{\xi}_2$. Then $p_2=\xi_2$ when $\xi_1=0$, and $f^*p_2=g^*p_2=p_2$. Now we determine q_2 by solving the differential equation

$$\{p_2,q_2\}=H_{p_2}q_2=1;\quad q_2=0\quad\text{when } \tilde{x}_2=0.$$

The solution is unique near 0 since $d\tilde{x}_2=dx_2$ there. Now the Poisson bracket condition and the boundary condition are both invariant under f and under g, so f^*q_2 and g^*q_2 are also solutions, hence $f^*q_2=g^*q_2=q_2$. When $\xi_1=0$ the equation is satisfied by $q_2=x_2$ since (21.4.8) shows that

$$H_{p_2}=\partial/\partial x_2+\partial^2 p_2/\partial\xi_1^2\,\partial/\partial x_1$$

then, so $q_2=x_2$ when $\xi_1=0$. The equation (21.4.8) is therefore applicable to q_2, and we can now choose p_3 with

$$H_{p_2}p_3=H_{q_2}p_3=0,\quad p_3=\tilde{\xi}_3\quad\text{when } p_2=q_2=0.$$

Continuing in this way we obtain $p_2,q_2,\dots,p_n,q_n$ satisfying the commutation relations, all invariant under f and g, so that these functions are equal to $\xi_2, x_2, \ldots, \xi_n, x_n$ when $\xi_1=0$. Now we determine p_1 from the differential equations

$$H_{p_j} p_1 = H_{q_j} p_1 = 0, \quad j>1; \quad p_1=\eta_1 \quad \text{when } p_2=\ldots=q_n=0.$$

Since η_1 is odd with respect to f and g, this is also true for p_1. Finally the solution q_1 of the equations

$$H_{p_j} q_1 = H_{q_j} q_1 = 0, \quad j>1; \quad p_1 H_{p_1} q_1 = 1;$$
$$q_1=0 \quad \text{when } p_2=\ldots=q_n=\tilde{x}_1=0$$

is even with respect to f since $\tilde{x}_1$ is, but q_1 has no simple relation to g.

Now introduce q and p as new coordinates z, ζ. In terms of these coordinates the symplectic form is given by (21.4.4), f is given by (21.4.5), and

$$g(z,\zeta) = (z_1 + G(z,\zeta), z', -\zeta_1, \zeta')$$

for some so far unknown G. Now the fact that g is symplectic gives

$$-\zeta_1 d(-\zeta_1) \wedge d(z_1 + G(z,\zeta)) = \zeta_1 d\zeta_1 \wedge dz_1$$

so $d\zeta_1 \wedge dG = 0$ which means that G is a function of ζ_1 only. That g is an involution means that G is odd, and $G'(0) \neq 0$ since the reflection bundles of f and g are not the same. We shall now make a final change of the $z_1 \zeta_1$ variables to obtain the desired G. In this step the other variables are no longer involved so we drop them and write $t=z_1$, $\tau=\zeta_1$ to avoid subscripts.

Put $\kappa(t,\tau) = (tA(\tau), B(\tau))$. Then $\kappa \circ f = f \circ \kappa$ if B is odd and A is even, and κ preserves the symplectic form $\tau\, d\tau \wedge dt$ if

$$B(\tau)\, B'(\tau)\, A(\tau) = \tau. \tag{21.4.9}$$

Finally we have $\kappa^{-1} \circ g \circ \kappa = g_0$, $g_0(t,\tau) = (t+\tau, -\tau)$ if

$$\begin{aligned} g \circ \kappa(t,\tau) &= (tA(\tau) + G(B(\tau)), -B(\tau)) = \kappa \circ g_0(t,\tau) \\ &= ((t+\tau)\, A(-\tau), B(-\tau)). \end{aligned}$$

This means that

$$G(B(\tau)) = \tau A(\tau) = \tau^2 / B(\tau)\, B'(\tau),$$

which is easily integrated and gives

$$\tau^3/3 = \int_0^{B(\tau)} s\, G(s)\, ds.$$

Since G is odd and $G'(0) \neq 0$ the right-hand side can be written $B^3 G_1(B)/3$ where G_1 is an even function with $G_1(0) \neq 0$, so the equation becomes

$$\tau = B\, G_1(B)^{\frac{1}{3}}$$

which has a unique solution by the implicit function theorem. We can determine A from (21.4.9) with A even and smooth at 0 since B has a simple zero. This completes the proof of Theorem 21.4.4. if we take $\kappa^{-1}(t,\tau)$ as new variables.

As a first application of Theorem 21.4.4 we shall give a normal form for canonical relations with folds.

Theorem 21.4.5. *Let S_1 and S_2 be symplectic manifolds of the same dimension and with symplectic forms σ_1, σ_2, and let G be a canonical relation from S_2 to S_1 near $(s_1^0, s_2^0) \in S_1 \times S_2$, that is, a Lagrangian submanifold with respect to $\sigma_1 - \sigma_2$. Assume that the projections π_j: $G \to S_j$ have folds at (s_1^0, s_2^0). Then one can choose local symplectic coordinates (x, ξ) in S_1 vanishing at s_1^0 and (y, η) in S_2 vanishing at s_2^0 such that*

$$G = \{(x, \xi; y, \eta); \xi_1 = \eta_1 = 2(x_1 - y_1)^2, x' = y', \xi' = \eta'\}$$

Proof. Set $\sigma = \pi_1^* \sigma_1 = \pi_2^* \sigma_2$. By Theorem 21.4.2 this is a degenerate symplectic form in G of the type (21.4.3). The reflection bundles for the involutions f_1 and f_2 in G defined by π_1 and π_2 are different, for they are the kernels of $d\pi_j$ and (π_1, π_2) is the injection of G in $S_1 \times S_2$. Hence it follows from Theorem 21.4.4 that we can choose local coordinates $t_1, \dots, t_n, \tau_1, \dots, \tau_n$ in G vanishing at (s_1^0, s_2^0) such that

$$\sigma = \tau_1 \, d\tau_1 \wedge dt_1 + \sum_2^n d\tau_j \wedge dt_j;$$

$$f_1(t, \tau) = (t_1 + \tau_1, t', -\tau_1, \tau'); \quad f_2(t, \tau) = (t, -\tau_1, \tau')$$

where $t' = (t_2, \dots, t_n)$. Hence $t_1, \dots, t_n, \tau_1^2, \tau_2, \dots, \tau_n$ are invariant under f_2 so by Theorem 21.4.2 there are local symplectic coordinates (y, η) at s_2^0 such that

$$\pi_2(t, \tau) = (t, \tau_1^2/2, \tau').$$

Similarly $t_1 + \tau_1/2, t_2, \dots, t_n, \tau_1^2, \tau_2, \dots, \tau_n$ are invariant under f_1 since $t_1 + \tau_1 + (-\tau_1)/2 = t_1 + \tau_1/2$. Since

$$\sigma = \tau_1 \, d\tau_1 \wedge d(t_1 + \tau_1/2) + \sum_2^n d\tau_j \wedge dt_j$$

we can also choose symplectic coordinates (x, ξ) in S_1 such that

$$\pi_1(t, \tau) = (t_1 + \tau_1/2, t', \tau_1^2/2, \tau').$$

This means that

$$G = \{(t_1 + \tau_1/2, t', \tau_1^2/2, \tau'; t_1, t', \tau_1^2/2, \tau')\}$$

which is clearly a Lagrangian with the required properties. Note that the canonical relation affects only one pair of symplectic coordinates.

To motivate the next application of Theorem 21.4.4 we shall sketch the situation where it occurs; this gives an important clue to the constructions used in the proof. We consider a (second order) hyperbolic equation in X, with principal symbol $p(x, \xi)$. The bicharacteristics of p, that is, the curves in the Hamilton foliation of $p^{-1}(0)$, are the light rays of geometrical optics.

When they arrive at a surface S, defined by $s(x)=0$, the reflection law is applied. This means that the position $(x, \xi_-)\in T^*(X)$ where the incoming ray hits S, or rather $T^*(X)|_S$, is replaced by the starting point (x, ξ_+) for the outgoing ray in such a way that $\xi_+ - \xi_-$ is conormal to S and $p(x, \xi_+)=0$ of course. Another way of describing this is that we go from ξ_- to ξ_+ along the Hamilton foliation of $T^*(X)|_S$, that is, the zeros of s. We then follow the Hamilton foliation of p until the next encounter with $T^*(X)|_S$ and so on. With this formulation the functions p and s take on surprisingly symmetrical roles. Some conditions must of course be imposed for the procedure to make sense for all rays near a ray tangent at (x_0, ξ_0) say - which is the really interesting case. The first is that we can find precisely one other zero of $p(x, \xi)$ near ξ_- in the normal direction (possibly coinciding with ξ_-). This is guaranteed if we know that the second order ξ derivative in the normal direction is not 0, that is, $\{s, \{s, p\}\}\neq 0$. The second is that there is some good control of how the bicharacteristics of p meet the surface S. The simplest situation is where the given ray does not have higher order contact, that is, $H_p^2 s = \{p, \{p, s\}\}\neq 0$ at (x_0, ξ_0). Together with the obvious condition that dp and ds are linearly independent these are the conditions which we shall impose:

Definition 21.4.6. Let S be a C^∞ symplectic manifold and F, G two C^∞ hypersurfaces intersecting transversally at $s_0\in S$. Then F and G are said to be glancing at s_0 if the Hamilton foliation of F (of G) is simply tangent to G (to F) at s_0.

If we choose functions f and g with $f=0$ on F, $g=0$ on G and $df\neq 0$, $dg\neq 0$ at s_0, the transversal intersection means that df, dg are linearly independent at s_0, and the other conditions can be written

(21.4.10) $\quad \{f, g\}=0, \quad \{f, \{f, g\}\}\neq 0$ and $\{g, \{g, f\}\}\neq 0$ at s_0.

Thus we have precisely the situation in the preceding motivating discussion. Set $J=F\cap G$, which is a manifold of codimension 2, and

$$K=\{s\in J; \{f, g\}=0\}.$$

This is the set where the intersection is not symplectic. K is a manifold of codimension 3 for df, dg and $d\{f, g\}$ are linearly independent at s_0 since $H_f\{f, g\}\neq 0$ but $H_f g = H_f f = 0$, and H_f, H_g are linearly independent. To clarify the geometrical picture we shall introduce local coordinates such that the surfaces F, J and K are conveniently located:

Lemma 21.4.7. *There is a neighborhood U of s_0 in S and local symplectic coordinates y, η vanishing at s_0 such that*

$$F\cap U=\{(y, \eta)\in U; y_1=0\}, \quad J\cap U=\{(y, \eta)\in U; y_1=0, \eta_n=\eta_1^2\},$$
$$K\cap U=\{(y, \eta)\in U; y_1=\eta_1=\eta_n=0\}.$$

Proof. Choose local symplectic coordinates x, ξ vanishing at s_0 such that F is defined by $x_1=0$ and G by $g(x,\xi)=0$ near $(0,0)$. From (21.4.10) it follows that

$$\partial g/\partial \xi_1=0, \qquad \partial^2 g/\partial \xi_1^2 \neq 0 \quad \text{at } (0,0).$$

Since $g(0,0)=0$ this means that g has a double root at 0, as a function of ξ_1. The Malgrange preparation theorem (Theorem 7.5.6) implies the existence of C^∞ functions a, b vanishing at 0 and independent of ξ_1 and a C^∞ function ρ with $\rho(0,0)\neq 0$ such that

$$g=\rho(\xi_1^2+a\xi_1+b)=\rho((\xi_1+a/2)^2+b-a^2/4).$$

Now, using Theorem 21.1.6, we can choose new symplectic coordinates (y,η) vanishing at s_0 such that

$$\eta_1=\xi_1+a/2, \qquad y_1=x_1.$$

Since $\partial/\partial\eta_1=-H_{y_1}=-H_{x_1}=\partial/\partial\xi_1$ we have $b(x,\xi')-a(x,\xi')^2/4=-c(y,\eta')$ where c is independent of η_1. In the new coordinates the surface G is defined by $g_1(y,\eta)=\eta_1^2-c(y,\eta')=0$, where $c(0,0)=0$. The linear independence of dy_1 and dg_1 gives

$$d_{y'\eta'}c(0,0)\neq 0.$$

By a symplectic change of the (y',η') coordinates we can make $c(0,y',\eta')$ one of them, so we may assume that $c(0,y',\eta')=\eta_n$. Then F,J,K have the form described in the lemma.

Let us now list some simple but important consequences of the lemma. Choose U as a ball in the y,η coordinates, and set

$$i_F(y,\eta)=(y,-\eta_1,\eta'), \qquad (y,\eta)\in J\cap U$$

with the notation in the lemma. Since the curves in the Hamilton foliation of F are just the parallels of the ξ_1 axis it follows from the lemma that (y,η) and $i_F(y,\eta)$ are the only points in J on the same curve in this foliation; they are different if $(y,\eta)\notin K$. Hence the C^∞ involution i_F of J with fixed point set K is invariantly defined, without reference to the local coordinates. It expresses the construction in the introductory motivation above. The reflection bundle of i_F is generated by H_f since this is $\partial/\partial\eta_1$ in the local coordinates. The restriction of the symplectic form to J is

$$\sum_2^{n-1} d\eta_j\wedge dy_j+2\eta_1\,d\eta_1\wedge dy_n$$

so it satisfies the hypotheses in Theorem 21.1.7, with n replaced by $n-1$. Finally, if $\psi\in C^\infty(U)$ then H_ψ is tangential to J at s_0 if and only if $d\psi$ vanishes in the direction H_f and in the direction of the radical of the symplectic form restricted to K. In fact, these conditions mean respectively that $\partial\psi/\partial\eta_1=0$ and that $\partial\psi/\partial y_n=0$, and the tangent plane of J is defined by $dy_1=d\eta_n=0$ at $(0,0)$, so H_ψ is tangential when $\partial\psi/\partial\eta_1=\partial\psi/\partial y_n=0$.

We are now prepared for the proof that all pairs of glancing hypersurfaces are equivalent:

Theorem 21.4.8. *Let S be a C^∞ symplectic manifold and F, G two C^∞ hypersurfaces in S with a transversal glancing intersection at $s_0 \in S$. Then there are local symplectic coordinates (x, ξ) vanishing at s_0 such that F is defined by $x_1 = 0$ and G is defined by $\xi_1^2 - x_1 - \xi_n = 0$.*

In particular, the dimension $2n$ of S is at least 4. That $n > 1$ is immediately clear since $\{f, g\} = \sigma(H_f, H_g) = 0$ means that H_f and H_g are proportional if $n = 1$.

Proof of Theorem 21.4.8. The manifold J with the restriction $\sigma|_J$ of the symplectic form and the involutions i_F, i_G - defined as i_F - satisfy the hypotheses of Theorem 21.4.4 by the preceding remarks. Hence we can choose local coordinates $t_1, \dots, t_{n-1}, \tau_1, \dots, \tau_{n-1}$ in J at s_0 so that

$$\sigma|_J = 2\tau_1\, d\tau_1 \wedge dt_1 + \sum_2^{n-1} d\tau_j \wedge dt_j,$$
$$i_F(t, \tau) = (t, -\tau_1, \tau_2, \dots, \tau_{n-1}),$$
$$i_G(t, \tau) = (t_1 + 2\tau_1, t_2, \dots, t_{n-1}, -\tau_1, \tau_2, \dots, \tau_{n-1}).$$

We have inserted factors 2 here to get simpler formulas below. This is permissible since Theorem 21.4.4 shows the equivalence with the old model which is also immediately verified by a change of variables. The theorem states that there are local coordinates (x, ξ) in S such that

$$J = \{(0, x_2, \dots, x_n, \xi_1, \dots, \xi_{n-1}, \xi_1^2)\}, \quad \sigma|_J = 2\xi_1\, d\xi_1 \wedge dx_n + \sum_2^{n-1} d\xi_j \wedge dx_j,$$
$$i_F(0, x_2, \dots, x_n, \xi_1, \dots, \xi_{n-1}, \xi_1^2) = (0, x_2, \dots, x_n, -\xi_1, \xi_2, \dots, \xi_{n-1}, \xi_1^2),$$
$$i_G(0, x_2, \dots, x_n, \xi_1, \dots, \xi_{n-1}, \xi_1^2) = (0, x_2, \dots, x_n + 2\xi_1, -\xi_1, \xi_2, \dots, \xi_{n-1}, \xi_1^2).$$

In fact, the Hamilton vector field defined by x_1 is $\partial/\partial\xi_1$ and that of $\xi_1^2 - x_1 - \xi_n$ is $2\xi_1 \partial/\partial x_1 - \partial/\partial x_n + \partial/\partial \xi_1$, so $x_2, \dots, x_{n-1}, \xi_2, \dots, \xi_{n-1}$, $\xi_1^2 - x_1 = \xi_n$ and $\xi_1 + x_n$ are constant on the curves of the Hamilton foliation of G.

Introduce now the local coordinates y, η of Lemma 21.4.7 which make F, J, i_F and K have the desired form. Thus J is regarded as a subset of $T^*(\mathbb{R}^n)$,

$$J = \{(0, y_2, \dots, y_n, \eta_1, \dots, \eta_{n-1}, \eta_1^2)\} \cap U$$

parametrized by $y_2, \dots, y_n, \eta_1, \dots, \eta_{n-1}$. We define our new coordinates x, ξ first on J by

(21.4.11) $\quad x_j = t_j, \quad \xi_j = \tau_j, \quad 1 < j < n; \quad x_1 = 0, \quad \xi_1 = \tau_1, \quad x_n = t_1, \quad \xi_n = \tau_1^2.$

where t, τ are regarded as functions of $y_2, \ldots, y_n, \eta_1, \ldots, \eta_n$. Then we have

(21.4.12)
$$\sum_2^n d\xi_j \wedge dx_j = \sigma|_J \quad \text{on } J.$$

We must now extend the functions ξ_j and x_j from J to a neighborhood of $(0,0)$ so that we obtain a symplectic coordinate system with $x_1=0$ on F and $\xi_1^2 - x_1 - \xi_n = 0$ on G.

The map
$$J \ni (0, y_2, \ldots, y_n, \eta_1, \ldots, \eta_{n-1}, \eta_1^2) \mapsto (y_2, \ldots, y_n, \eta_2, \ldots, \eta_{n-1}, \eta_1^2)$$
is a fold map with involution i_F, and since $x_2, \ldots, x_n$, $\xi_2, \ldots, \xi_n$ are even under i_F there exist functions $\tilde{x}_2, \ldots, \tilde{x}_n$, $\tilde{\xi}_2, \ldots, \tilde{\xi}_n$ of (y', η) in a neighborhood of 0 which restrict to $x_2, \ldots, x_n$, $\xi_2, \ldots, \xi_n$ on J if we regard them as functions in $F = \{(y, \eta);\, y_1 = 0\}$ independent of η_1. From (21.4.12) we obtain

(21.4.12)′
$$\sum_2^n d\tilde{\xi}_j \wedge d\tilde{x}_j = \sum_2^n d\eta_j \wedge dy_j \quad \text{in } F$$

in the set $\eta_n > 0$ where $\tilde{\xi}_j$ and $\tilde{x}_j$ are uniquely determined. As in the proof of Theorem 21.4.2 we can then change the definition when $\eta_n < 0$ so that $\tilde{x}_2, \ldots, \tilde{x}_n$, $\tilde{\xi}_2, \ldots, \tilde{\xi}_n$ become symplectic coordinates in $T^*(\mathbb{R}^{n-1})$ in a full neighborhood of 0. Set $\tilde{x}_1 = 0$, still when $y_1 = 0$.

To extend the functions $\tilde{x}_j$, $\tilde{\xi}_j$ to a full neighborhood of 0 we observe that $x_n + \xi_1$ is a function on J which is invariant under i_G. Hence the preceding argument with F and G interchanged shows that there is a C^∞ extension Q to a neighborhood of 0,0 which is constant along the foliation of G. Thus

(21.4.13)
$$Q = x_n + \xi_1 \quad \text{on } J.$$

The derivative of Q along the radical of $\sigma|_K$ is not 0 because of the term x_n so H_Q is tangential to G but not to J, and G is near $(0,0)$ the H_Q flowout of J. It follows also that H_Q is transversal to F, for $J = F \cap G$. We can therefore define $x_1, \ldots, x_n$, $\xi_2, \ldots, \xi_n$ uniquely near 0 by

(21.4.14)
$$H_Q x_j = 0, \quad j = 2, \ldots, n, \quad H_Q \xi_k = 0, \quad k = 2, \ldots, n-1,$$
$$H_Q x_1 = 1, \quad H_Q \xi_n = -1,$$

with initial values given by $\tilde{x}_j, \tilde{\xi}_j$ on F. Then we obtain

(21.4.15)
$$\{\xi_j, x_k\} = \delta_{jk}, \quad \{\xi_j, \xi_{j'}\} = 0,$$
$$\{x_k, x_{k'}\} = 0; \quad j, j' = 2, \ldots, n; \quad k, k' = 1, \ldots, n.$$

By the Jacobi identity and (21.4.14) the Poisson brackets of these quantities and Q are equal to 0 so it is sufficient to verify (21.4.15) on F. There the derivatives with respect to η_1 all vanish so the Poisson brackets reduce to Poisson brackets in the variables $y_2, \ldots, y_n$, $\eta_2, \ldots, \eta_n$ which satisfy the relations (21.4.15) in view of (21.4.12)′ and the fact that $\tilde{x}_1 = 0$.

Now define the extension of ξ_1 by

(21.4.13)′ $$\xi_1 = Q - x_n$$

which is legitimate by (21.4.13). Then the remaining commutation relations

$$\{\xi_1, \xi_j\} = 0, \quad \{\xi_1, x_k\} = \delta_{1k}, \quad j, k = 1, \ldots, n$$

follow from (21.4.14). It remains to verify that $\xi_1^2 - x_1 - \xi_n$ vanishes on G, which so far we know on J by (21.4.11). Since G is the H_Q flowout of J it suffices to verify that

$$H_Q(\xi_1^2 - x_1 - \xi_n) = \{\xi_1 + x_n, \xi_1^2 - x_1 - \xi_n\} = 0$$

which follows from the commutation relations. The proof is complete.

In some approaches to the study of diffractive boundary problems one needs the analogues of Theorems 21.4.5 and 21.4.8 in the homogeneous case. (See Section 24.4 for other methods.) To prove them one must reexamine the preceding theorems from Theorem 21.4.3 on.

Theorem 21.4.9. *Assume that S and σ satisfy the hypotheses of Theorem 21.1.10 and that i is a homogeneous C^∞ involution with $i^*\sigma = \sigma$ and fixed point set S_0. Then there exist local coordinates with the properties stated in Theorem 21.1.10 such that the involution only changes the sign of the ξ_1 coordinate.*

Proof. We can follow the proof of Theorem 21.4.3 closely. By the proof of Theorem C.4.5 there are local coordinates $s_1, \ldots, s_{2n}$ in a conic neighborhood of any $s_0 \in S_0$ such that s_j is homogeneous of degree 1 and even with respect to i for $j \neq 2n$ while s_{2n} is homogeneous of degree $\frac{1}{2}$ and odd. By Theorem 21.4.3 the Hamilton field of $s_{2n}^2/2$ is smooth and is a non-zero element in the radical of σ at s_0, hence non-radial. As in the proof of Theorem 21.4.3 we have $\sigma = f^*\tilde{\sigma}$ where $f = (s_1, \ldots, s_{2n-1}, s_{2n}^2/2)$ and $\tilde{\sigma}$ is a closed non-degenerate homogeneous two form in $U_+ = \{t \in U; t_{2n} \geqq 0\}$ where U is a conic neighborhood of $f(s_0)$. Thus $\tilde{\sigma} = d\tilde{\omega}$ if $\tilde{\omega}(.) = \tilde{\sigma}(\partial/\partial t, .)$, for $\partial/\partial t$ is the radical vector field. A homogeneous extension of the one form $\tilde{\omega}$ gives a homogeneous closed non-degenerate extension of $\tilde{\sigma}$ to a full conic neighborhood U' of $f(s_0)$. Since

$$f^*\{t_{2n}, u\}_{\tilde{\sigma}} = \{s_{2n}^2/2, f^*u\}_\sigma$$

is not 0 at s_0 for all homogeneous u vanishing at $f(s_0)$, the Hamilton field of t_{2n} is not radial. By the homogeneous Darboux Theorem 21.1.9 we can therefore choose homogeneous symplectic coordinates y, η in U' such that $\eta_1 = t_{2n}$. Then $x_j = f^*y_j$, $\xi_k = f^*\eta_k$ for $k \neq 1$, $\xi_1 = s_{2n}$, have the required properties since $f^*\eta_1 = s_{2n}^2/2 = \xi_1^2/2$. This completes the proof, which is independent of Theorem 21.1.10.

The analogue of Theorem 21.4.4 in the homogeneous case is more complicated both to state and to prove:

Theorem 21.4.10. *Let S be a conic manifold of dimension $2n$ with a two form σ which is homogeneous of degree 1 and satisfies the hypotheses in Theorem 21.1.7. Let f and g be two homogeneous C^∞ involutions of S with $f^*\sigma=\sigma$, $g^*\sigma=\sigma$, having S_0 as fixed point set, and assume that their reflection bundles at γ_0 and the radial vector $\rho(\gamma_0)$ are linearly independent. Then $n\geqq 2$ and in a conic neighborhood of γ_0 one can choose coordinates (x,ξ) homogeneous of degree 0 and 1 respectively and equal to $(0,\varepsilon_n)$ at γ_0, such that*

$$\sigma=d(\xi_1^2/\xi_n)\wedge dx_1+\sum_2^n d\xi_j\wedge dx_j \tag{21.4.16}$$

$$f(x,\xi)=(x,-\xi_1,\xi_2,\dots,\xi_n), \tag{21.4.17}$$
$$g(x,\xi)=(x_1+2\xi_1/\xi_n, x_2,\dots,x_n-\tfrac{2}{3}(\xi_1/\xi_n)^3, -\xi_1,\xi_2,\dots,\xi_n).$$

The preceding normal form really comes from the intersection of the glancing hypersurfaces at the end of Section 21.1; we shall return to this point in Theorem 21.4.12.

Proof. As a first step we show that σ, f, g as given in (21.4.16), (21.4.17) satisfy all the hypotheses. To do so we observe that the reflection bundles are generated by $(0,\varepsilon_1)$ and $(-e_1,\xi_n\varepsilon_1)$, hence independent of $(0,\varepsilon_n)$. That $f^*\sigma=\sigma$ is clear and also that we have involutions. Finally $g^*\sigma=\sigma$ because

$$d(\xi_1^2/\xi_n)\wedge d(2\xi_1/\xi_n)-d\xi_n\wedge d(\tfrac{2}{3}(\xi_1/\xi_n)^3)$$
$$=2\xi_1^2/\xi_n^2\, d\xi_n\wedge d(\xi_1/\xi_n)-2(\xi_1/\xi_n)^2\, d\xi_n\wedge d(\xi_1/\xi_n)=0.$$

Theorem C.4.8 in the appendix really says that all conic manifolds with homogeneous involutions satisfying the hypotheses are equivalent. By Theorem 21.4.4 the radical of the symplectic form σ restricted to S_0 is in the linear span of the reflection bundles, so the hypotheses of Theorem 21.1.10 are also fulfilled. By Theorems 21.4.9 and Theorem C.4.8 in the appendix we can choose homogeneous coordinates x, ξ such that σ and f have the desired form, and also homogeneous coordinates y, η such that f and g have the desired form; both sets of coordinates are equal to $0,\varepsilon_n$ at γ_0. If we think of $\xi_1(2/\xi_n)^{\frac{1}{2}}$ as a variable instead of ξ_1 we see that the situation is essentially the same as in the proof of Theorem 21.4.4. One difference is that y_n is not invariant with respect to g, but $y_n+\eta_1^2 y_1/3\eta_n^2$ is. With ϕ_j, ψ_j as in the proof of Theorem 21.4.4 we therefore set

$$\tilde{x}_j=\phi_j(0,y^*,0,\eta'),\quad \tilde{\xi}_j=\psi_j(0,y^*,0,\eta'),\quad j\neq 1,$$

where $y^*=(y_2,\dots,y_{n-1},y_n+\eta_1^2 y_1/3\eta_n^2)$. Then $\tilde{x}_j$ and $\tilde{\xi}_j$ are invariant under f and g. We define $p_2, q_2,\dots,p_n$ as before but postpone the choice of q_n since the Hamilton field would become radial (see also the proof of Theorem 21.1.9). Instead we determine p_1 using the commutation relations with $p_2, \dots, p_n$ and the boundary condition $p_1 = \eta_1$ when $p_2 = \dots = q_{n-1} = \tilde{x}_n = 0$. Next q_1 is determined with $q_1 = 0$ when $p_2 = \dots = q_{n-1} = \tilde{x}_n = x_1 = 0$. Finally q_n is obtained by integrating the commutation relations with boundary conditions where $p_1 = q_1 = \dots = q_{n-1} = \tilde{x}_n = 0$. The coordinates p_j, q_j automatically get the required homogeneity because of the homogeneity of the equations and the boundary conditions posed. However, in the choice of the last two coordinates q_1 and q_n the boundary conditions were not invariant under g so we have no information how they are related to g.

If we take q and p as new local coordinates, changing notation so that they become x, ξ, then (21.4.16) holds, f is given by (21.4.17), and

$$g(x,\xi) = (x_1 + v(x,\xi), x_2, \dots, x_{n-1}, x_n + w(x,\xi), -\xi_1, \xi_2, \dots, \xi_n)$$

for some v, w homogeneous of degree 0. Since g preserves σ we have

$$d(\xi_1^2/\xi_n) \wedge dv + d\xi_n \wedge dw = 0 \tag{21.4.18}$$

which implies that v and w can only depend on the variables ξ_1 and ξ_n. The others no longer play any role so we simplify notation by assuming $n=2$. We can write $v(\xi) = V(t)$, $w(\xi) = W(t)$ where $\xi_1 = t\xi_2$, which simplifies (21.4.18) to $t^2\, d\xi_2 \wedge dV + d\xi_2 \wedge dW = 0$, that is,

$$t^2 V'(t) + W'(t) = 0. \tag{21.4.18$'$}$$

That g is an involution means that V and W are odd functions of t. Note that $V(t) = 2t$ implies $W(t) = -2t^3/3$ by (21.4.18)$'$ since $W(0) = 0$, which explains the look of (21.4.17) and shows that it suffices to make $V(t) = 2t$.

As at the end of the proof of Theorem 21.4.4 we make a simple change of variables by introducing

$$\kappa(x,\xi) = (x_1 S(t), x_2 + x_1 T(t), \xi_2 R(t), \xi_2), \quad t = \xi_1/\xi_2.$$

Then $\kappa^*\sigma = \sigma$ if

$$d(\xi_2 R^2) \wedge d(x_1 S) + d\xi_2 \wedge d(x_1 T) = d(\xi_2 t^2) \wedge dx_1,$$

which means explicitly, if we write $U = R^2$, that

$$US + T = t^2, \quad US' + T' = 0, \quad U'S = 2t. \tag{21.4.19}$$

The middle equation follows from the other two; it is clear that S and T can be determined as even functions with $S(0) \neq 0$ satisfying (21.4.19) by

$$S = 2t/U', \quad T = t^2 - 2tU/U'$$

if R is chosen odd with a simple zero at 0. We then have a diffeomorphism κ. Now κ commutes with f under these conditions and

$$g \circ \kappa(x,\xi) = (x_1 S + V(R), x_2 + x_1 T + W(R), -\xi_2 R, \xi_2).$$

If g_1 is defined as g with V and W replaced by V_1 and W_1 then

$$\kappa \circ g_1(x,\xi) = ((x_1 + V_1)S, x_2 + W_1 + (x_1 + V_1)T, -\xi_2 R, \xi_2),$$

so $g\circ\kappa=\kappa\circ g_1$ if
$$V_1 S=V(R), \quad W_1+V_1 T=W(R).$$
In particular, $V_1(t)=2t$ if
$$V(R)=2tS(t)=2t^2/R(t)\,R'(t),$$
that is,
$$2t^3/3=\int_0^R V(s)s\,ds.$$
Since $V'(0)\neq 0$ because the reflection bundles are different, this determines R as a function of t with a simple zero at 0 (see the proof of Theorem 21.4.4), so we have obtained a change of variables giving σ, f and g the desired form. The proof is complete.

In the applications of Theorem 21.4.10 which follow the important point is not the explicit normal form given but that all S, σ, f, g satisfying the hypothesis are equivalent. We can therefore use any other more convenient model.

We come now to the homogeneous version of Theorem 21.4.5. The normal form for a homogeneous canonical relation from $T^*(\mathbb{R}^n)\smallsetminus 0$ to $T^*(\mathbb{R}^n)\smallsetminus 0$ which we shall use is perhaps best described by means of the phase function

(21.4.20)
$$\phi(x,y,s,\xi)=\langle x-y,\xi\rangle+s\xi_1-s^3\xi_n/3;\quad x,y,\xi\in\mathbb{R}^n,\ \xi_n>0,\ s\in\mathbb{R}.$$

The equation $\phi'_\xi=\phi'_s=0$ means that
$$x_1-y_1+s=0;\quad x_n-y_n-s^3/3=0;\quad x_j=y_j,\quad 1<j<n;\quad \xi_1=s^2\xi_n.$$
The corresponding canonical relation is

(21.4.21)
$$G=\{(x,\xi,y,\eta);\ (x_1-y_1)^2\xi_n=\xi_1,\ 3(y_n-x_n)+(y_1-x_1)^3=0,\ \xi=\eta,\ x_j=y_j,\ 1<j<n,\ \xi_n>0\}.$$

With the parameters $x,s,\xi'=(\xi_2,\dots,\xi_n)$ we have

(21.4.21)′
$$G=\{(x,s^2\xi_n,\xi',x_1+s,x_2,\dots,x_{n-1},x_n-s^3/3,s^2\xi_n,\xi');\ \xi_n>0\}$$

which shows that G is a $2n$ dimensional manifold. In view of the symmetry between x and y it follows that we have fold maps to each factor $T^*(\mathbb{R}^n)\smallsetminus 0$, the range being defined by $\xi_1\geqq 0$ and $\eta_1\geqq 0$ respectively. The kernel of the differential of the map when $s=0$ is the y_1 direction and x_1 direction respectively. With the coordinates in (21.4.21)′ the corresponding involutions are $s\mapsto -s$ and $s\mapsto -s$, $x_1\mapsto x_1+2s$, $x_n\mapsto x_n-2s^3/3$ respectively, and the pullback of the symplectic form to G from either copy of $T^*(\mathbb{R}^n)$ is
$$d(s^2\xi_n)\wedge dx_1+\sum_2^n d\xi_j\wedge dx_j.$$

Theorem 21.4.11. *Let S_1 and S_2 be $2n$ dimensional conic symplectic manifolds with symplectic forms σ_1, σ_2, and let G be a homogeneous canonical relation from S_2 to S_1 near $(s_1^0, s_2^0) \in S_1 \times S_2$. Assume that the projections $\pi_j: G \to S_j$ have folds at (s_1^0, s_2^0) and that the canonical one forms ω_1 and ω_2 in S_1 and S_2 lifted to G do not both vanish on the tangent plane of G at (s_1^0, s_2^0). Then $n \geqq 2$ and we can choose local homogeneous symplectic coordinates (x, ξ) in S_1 and (y, η) in S_2, equal to $(0, \varepsilon_n)$ at s_1^0 and s_2^0, so that G is locally given by (21.4.21).*

Proof. The radial vector ρ at (s_1^0, s_2^0) is not the sum $\rho_1 + \rho_2$ of tangent vectors to G with zero component in S_2 and S_1 respectively, for by hypothesis at least one of the components of ρ in $T_{s_1^0}(S_1)$ and $T_{s_2^0}(S_2)$ is not in the Lagrangian tangent plane of G. The hypotheses of Theorem 21.4.10 are therefore fulfilled by G, $\sigma_G = \pi_1^* \sigma_1 = \pi_2^* \sigma_2$, and the involutions f_1, f_2 defined by π_1, π_2. In particular, this is true for the model we are aiming for, so we can choose local coordinates $t_1, \dots, t_n, s, \tau_2, \dots, \tau_n$ in G, with $t_1, \dots, t_n, s$ homogeneous of degree 0 and $\tau_2, \dots, \tau_n$ homogeneous of degree 1, $\tau_n = 1$ and the other coordinates 0 at (s_1^0, s_2^0) so that

$$\sigma_G = d(s^2 \tau_n) \wedge dt_1 + \sum_2^n d\tau_j \wedge dt_j,$$

$$f_1(t, s, \tau') = (t, -s, \tau'), \quad f_2(t, s, \tau') = (t_1 + 2s, t_2, \dots, t_{n-1}, t_n - 2s^3/3, -s, \tau').$$

Then there are C^∞ functions x, ξ in S_1 and y, η in S_2, homogeneous of degree 0 and 1 as usual, such that

$$\pi_1^* x_j = t_j, \quad \pi_1^* \xi_j = \pi_2^* \eta_j = \tau_j, \quad j \neq 1, \quad \pi_1^* \xi_1 = \pi_2^* \eta_1 = s^2 \tau_n,$$
$$\pi_2^* y_j = t_j, \quad j = 2, \dots, n-1, \quad \pi_2^* y_1 = t_1 + s, \quad \pi_2^* y_n = t_n - s^3/3.$$

They form local homogeneous symplectic coordinates in a neighborhood of $(0, \varepsilon_n)$ when $\xi_1 \geqq 0$ and $\eta_1 \geqq 0$ respectively. As in the proof of Theorem 21.4.2 they have unique modifications when $\xi_1 < 0$, $\eta_1 < 0$ to symplectic coordinates in a full neighborhood of $(0, \varepsilon_n)$ with x_1 and y_1 unchanged. These are automatically homogeneous because of the uniqueness. The proof is complete.

Finally we shall prove the homogeneous form of the equivalence of glancing hypersurfaces. Here we take the model surfaces in Example 21.1.11,

$$F = \{(x, \xi) \in T^*(\mathbb{R}^n) \setminus 0, x_1 = 0\},$$
$$G = \{(x, \xi) \in T^*(\mathbb{R}^n), \xi_{n-1} > 0, \xi_1^2 = x_1 \xi_{n-1}^2 + \xi_{n-1} \xi_n\}, \qquad n > 2.$$

The intersection is transversal,

$$F \cap G = J = \{(0, x', \xi_1, \dots, \xi_{n-1}, \xi_1^2/\xi_{n-1})\}$$

is parametrized by $x_2, \dots, x_n, \xi_1, \dots, \xi_{n-1}$ and the restriction of the symplectic form to J is

$$\sigma|_J = d(\xi_1^2/\xi_{n-1}) \wedge dx_n + \sum_2^{n-1} d\xi_j \wedge dx_j.$$

The Hamilton foliation of F consists of the lines parallel to the ξ_1 axis so the corresponding involution i_F just changes the sign of ξ_1. Since

$$\xi_2, \dots, \xi_n, x_2, \dots, x_{n-2}, x_1-(\xi_1/\xi_{n-1})^2,$$
$$x_n+\xi_1/\xi_{n-1} \quad \text{and } x_{n-1}-(\xi_1/\xi_{n-1})^3/3+x_1\xi_1/\xi_{n-1}$$

are constant on the Hamilton foliation of G we find that the corresponding involution i_G changes ξ_1 to $-\xi_1$, x_{n-1} to $x_{n-1}-2(\xi_1/\xi_{n-1})^3/3$ and x_n to $x_n+2\xi_1/\xi_{n-1}$ while leaving the other coordinates unchanged. The reflection bundles when $\xi_1=0$ are given by $(0,\varepsilon_1)$ and $(-e_n, \xi_{n-1}\varepsilon_1)$, and since $\xi_{n-1}\neq 0$ they are linearly independent of the radial vector $(0,\xi)$.

Theorem 21.4.12. *Let S be a conic symplectic manifold of dimension $2n\geqq 6$ and symplectic form σ, and let F, G be two conic C^∞ hypersurfaces in S with a transversal glancing intersection at $s_0\in S$. Assume that the tangents of the foliations of F and G at s_0 and the radial direction are linearly independent. Then there are local homogeneous symplectic coordinates (x,ξ) equal to $(0,\varepsilon_{n-1})$ at s_0, such that F is defined by $x_1=0$ and G is defined by $\xi_1^2=x_1\xi_{n-1}^2+\xi_n\xi_{n-1}$.*

Proof. All the hypotheses of Theorem 21.4.10 are fulfilled by $J=F\cap G$ with the form $\sigma|_J$ and the involutions i_F, i_G defined by the Hamilton foliations of F and G. This proves that $n-1\geqq 2$ and that we can find local coordinates $t_1,\dots,t_{n-1}$, $\tau_1,\dots,\tau_{n-1}$ in J, homogeneous of degree 0 and 1 respectively, with $\tau_{n-1}=1$ and the others 0 at s_0, so that

$$\sigma|_J = d(\tau_1^2/\tau_{n-1})\wedge dt_1+\sum_2^{n-1} d\tau_j\wedge dt_j;$$

i_F changes the sign of τ_1, and i_G changes the sign of τ_1 and replaces t_1 by $t_1+2\tau_1/\tau_{n-1}$, t_{n-1} by $t_{n-1}-2(\tau_1/\tau_{n-1})^3/3$. The proof of Lemma 21.4.7 shows that we can find homogeneous symplectic coordinates y, η in S, equal to 0, ε_{n-1} at s_0, such that F is defined by $y_1=0$ and

$$J=\{(0,y',\eta_1,\dots,\eta_{n-1},\eta_1^2/\eta_{n-1}\}; \tag{21.4.22}$$

hence i_F has the desired form. In fact, the proof of Lemma 21.4.7 works with no change until we have obtained a coordinate system where F is defined by $y_1=0$ and G by $\eta_1^2-c(y,\eta')=0$ where c is homogeneous of degree 2 and $d_{y'\eta'}c$ is not proportional to dy_{n-1} at $(0,\varepsilon_{n-1})$. Set $c_0(y',\eta')=c(0,y',\eta')$. We can choose $\phi(y',\eta')$ homogeneous of degree 1 with $\{c_0,\phi\}=0$ and $\phi=1$ at $(0,\varepsilon_{n-1})$, for H_{c_0} does not have the radial direction so we can put homogeneous initial data on a conic surface transversal to H_{c_0}. Then c_0/ϕ is homogeneous of degree 1 and $\{c_0/\phi,\phi\}=0$, so we can use Theorem 21.1.9 to find a homogeneous symplectic coordinate system $z_2,\dots,z_n$, $\zeta_2,\dots,\zeta_n$ with $\zeta_{n-1}=\phi$ and $\zeta_n=c_0/\phi$. Changing notation from z',ζ' to y',η' we have achieved (21.4.22).

We can now define our new coordinates x, ξ on J by (21.4.11) except that $\xi_n = \tau_1^2/\tau_{n-1}$. This makes (21.4.12) valid and we can find extensions $\tilde{x}_1, \dots, \tilde{x}_n$, $\xi_2, \dots, \xi_n$ to F as before. The definition of Q in (21.4.13) must be changed; the terms have different homogeneities. However, $x_n \xi_{n-1} + \xi_1$ is now invariant under i_G, homogeneous of degree 1, and has the same differential as $x_n + \xi_1$ at $(0, \varepsilon_{n-1})$. We can therefore find a C^∞ extension Q which is homogeneous of degree 1 and constant on the G foliation so that

(21.4.13)′ $$Q = x_n \xi_{n-1} + \xi_1 \quad \text{on } J.$$

H_Q is still transversal to F, so we can extend $\tilde{x}_1, \dots, \tilde{x}_n$, $\tilde{\xi}_2, \dots, \tilde{\xi}_n$ by solving the differential equations

(21.4.14)′ $$H_Q x_j = 0, \quad j \neq 1, n-1; \quad H_Q x_1 = 1, \quad H_Q x_{n-1} = x_n,$$
$$H_Q \xi_j = 0, \quad j \neq 1, n, \quad H_Q \xi_n = -\xi_{n-1},$$

with initial values $\tilde{x}_j$, $\tilde{\xi}_j$ on F. With $\xi_1 = Q - x_n \xi_{n-1}$ we then have the desired coordinate system as before.

21.5. Symplectic Equivalence of Quadratic Forms

The general purpose in this section is to study the simplest singularities of a Hamiltonian vector field. Recall that a vector field v in a manifold X is said to have a singularity at $x \in X$ if $v = 0$ at x; then

$$\phi \mapsto d(v\phi)(x), \quad \phi \in C^2(X),$$

defines a linear map $T_x^* \to T_x^*$ with adjoint V: $T_x \to T_x$ defined by

$$\langle Vt, d\phi \rangle = \langle t, d(v\phi) \rangle \quad \text{if } \phi \in C^2(X),\ t \in T_x.$$

In local coordinates the matrix of V is of course $(\partial v_j / \partial x_k)_{j,k=1,\dots,n}$. We shall use this invariant concept for Hamiltonian vector fields:

Definition 21.5.1. If S is a symplectic manifold and $f \in C^2(S)$ is real valued, $df = 0$ at $s_0 \in S$, then the linear map F in $T_{s_0}(S)$ defined by $H_f/2$ will be called the Hamilton map of f.

Some authors use the term "fundamental matrix" which seems awkward when one wants to have an invariant setup. The factor $\frac{1}{2}$ is just a traditional convention which will be seen to be convenient. The Hamilton map is of course determined by the Hessian f'' of f at s_0, so no generality is lost if we assume that S is a real symplectic vector space and that f is a real quadratic form Q. The definition of F (at 0) is then that for $X \in S$ and linear functions ϕ

$$\langle FX, \phi \rangle = \langle X, H_Q \phi/2 \rangle = -\langle X, H_\phi Q/2 \rangle = -\langle X, Q(.,H_\phi) \rangle = -Q(X, H_\phi)$$

where $Q(X,Y)$ denotes the symmetric bilinear polarized form of Q, that is, $Q(X,Y)=Q(Y,X)$ and $Q(X,X)=Q(X)$. The left-hand side is $\sigma(FX,H_\phi)$ so the definition of F is equivalent to

(21.5.1) $$\sigma(Y,FX)=Q(Y,X); \qquad X,Y\in S.$$

(There is no factor 2 here thanks to the factor $\frac{1}{2}$ in the definition.) In symplectic coordinates (x,ξ) the matrix of F becomes "the fundamental matrix"

(21.5.1)′ $$2^{-1}\begin{pmatrix} Q''_{\xi x} & Q''_{\xi\xi} \\ -Q''_{xx} & -Q''_{x\xi}\end{pmatrix}.$$

The symmetry of Q means that

(21.5.2) $$\sigma(FX,Y)=-\sigma(X,FY),$$

that is, F is skew symmetric with respect to σ. Note that the preceding definition is still applicable if Q is complex valued provided we replace S by its complexification

$$S_{\mathbb{C}}=\{X+iY; X,Y\in S\}$$

with the obvious complex symplectic structure. To study the symplectic classification of quadratic forms by means of the spectral decomposition of the Hamilton map F we must anyway make this extension even if Q is real. Let V_λ denote the space of generalized eigenvectors of F in $S_{\mathbb{C}}$ belonging to the eigenvalue $\lambda\in\mathbb{C}$.

Lemma 21.5.2. *$\sigma(V_\lambda,V_\mu)=0$ if $\lambda+\mu\neq 0$.*

Proof. $F+\mu$ is a bijection on V_λ when $\lambda+\mu\neq 0$, and

$$\sigma((F+\mu)^N V_\lambda, V_\mu)=\sigma(V_\lambda,(-F+\mu)^N V_\mu)=0$$

if N is large enough.

From the lemma it follows that V_λ is the dual of $V_{-\lambda}$ with respect to the symplectic form. Thus V_0 and $V_\lambda\oplus V_{-\lambda}$, $\lambda\neq 0$ are symplectic vector spaces, orthogonal with respect to σ and invariant under F, hence Q orthogonal too. We shall use these observations to determine the structure of F and Q in the cases of interest for us, namely when Q is real and positive semi-definite or hyperbolic in the sense that the negative index of inertia is 1. In both cases $Q\geqq 0$ in a hyperplane, so Q cannot be $\leqq 0$ in a two dimensional plane which only meets

$$\operatorname{Rad} Q=\{X; Q(X,Y)=0 \text{ for all } Y\}=\operatorname{Ker} F\subset V_0$$

at the origin. If $\operatorname{Re}\lambda\neq 0$ it follows from Lemma 21.5.2 that Q vanishes on $\operatorname{Re} V_\lambda\subset V_\lambda+V_{\bar\lambda}$ so the dimension of this space must be at most 1. Hence λ must be real and V_λ one dimensional if V_λ is non-trivial. Choose a real basis vector e_1 for V_λ then and a dual one ε_1 for $V_{-\lambda}$ with $\sigma(\varepsilon_1, e_1)=1$. Then $Fe_1 = \lambda e_1$, $F\varepsilon_1 = -\lambda\varepsilon_1$, and $Q(e_1,e_1)=Q(\varepsilon_1,\varepsilon_1)=0$, $Q(e_1,\varepsilon_1)=\lambda$, so

$$Q(x_1 e_1 + \xi_1 \varepsilon_1) = 2\lambda x_1 \xi_1.$$

Assume now that $\lambda = i\mu$, $\mu>0$ and that $FX = i\mu X$, $X = X_1 + iX_2 \neq 0$. Then

$$Q(\bar{X}, X) = \sigma(\bar{X}, FX) = \sigma(\bar{X}, i\mu X)$$

so

$$-2\mu\sigma(X_1, X_2) = Q(\bar{X}, X) = Q(X_1) + Q(X_2),$$

$$\begin{aligned} Q(t_1 X_1 + t_2 X_2) &= \sigma(t_1 X_1 + t_2 X_2, F(t_1 X_1 + t_2 X_2)) \\ &= \sigma(t_1 X_1 + t_2 X_2, -t_1 \mu X_2 + t_2 \mu X_1) \\ &= -\mu\sigma(X_1, X_2)(t_1^2 + t_2^2). \end{aligned}$$

Since Q cannot be ≤ 0 in a two dimensional plane not meeting the radical we conclude that $\sigma(X_1, X_2) < 0$ so we have an invariant symplectic two plane. Normalizing so that $\sigma(X_1, X_2) = -1$ we can take X_1, X_2 as basis vectors there in a symplectic coordinate system, which makes Q equal to μ times the Euclidean form, and can then pass to examining the σ and Q orthogonal complement. Repeating this argument we find that S is the direct sum of symplectically and Q orthogonal subspaces where Q is given by $\mu(x_j^2 + \xi_j^2)$, $\mu>0$, or $2\lambda x_j \xi_j$, $\lambda>0$, and the real part V_0' of V_0. It only remains to study V_0', so we assume from now on that $S = V_0'$, that is, that F is *nilpotent*.

If one can find $X \in S$ with

$$F^2 X = 0, \quad Q(X) = \sigma(X, FX) \neq 0,$$

then X, FX span a two dimensional F invariant symplectic space where

$$Q(t_1 X + t_2 FX) = \sigma(t_1 X + t_2 FX, t_1 FX) = t_1^2 Q(X),$$

and we can pass to studying the σ orthogonal complement. Such a reduction is also possible if $\operatorname{Ker} F$ contains two elements X and Y with $\sigma(X, Y) \neq 0$. What remains is therefore to study the situation where

a) $\operatorname{Ker} F$ is isotropic

b) $F^2 X = 0 \Rightarrow Q(X) = 0$.

By our hypothesis b) implies that $\operatorname{Ker} F^2/\operatorname{Ker} F$ has dimension at most one. In the exact sequence

$$0 \longrightarrow \operatorname{Ker} F \longrightarrow \operatorname{Ker} F^2 \xrightarrow{F} \operatorname{Ker} F \cap \operatorname{Im} F \longrightarrow 0$$

we have $\operatorname{Im} F = (\operatorname{Ker} F)^\sigma \supset \operatorname{Ker} F$ by a), so

$$0 \longrightarrow \operatorname{Ker} F \longrightarrow \operatorname{Ker} F^2 \xrightarrow{F} \operatorname{Ker} F \longrightarrow 0$$

is exact and $\operatorname{Ker} F$ has dimension 1. It follows now from the Jordan canonical form of F that there is a number N and an element X such that S has the basis $X, FX, \dots, F^{N-1}X$ while $F^N X = 0$. The dimension N must be even and >2. Now

$$Q(F^jX, F^kX) = (-1)^j \sigma(X, F^{j+k+1}X)$$

vanishes if $j+k+1 \geqq N$, so Q vanishes in the two dimensional space spanned by $F^{N-3}X$, $F^{N-2}X$ if $2(N-3)+1 \geqq N$, and it does not meet $\operatorname{Ker} F$. Hence $2(N-2) \leqq N$, that is, $N=4$. We must also have $\sigma(X, F^3X)<0$ since $Q \leqq 0$ in the two dimensional space spanned by FX and F^2X otherwise. We normalize so that $\sigma(X, F^3X) = -1$ and set $Y = X + tF^2X$. Then $\sigma(Y, FY) = \sigma(X, FX) - 2t = 0$ if $t = \sigma(X, FX)/2$. A symplectic basis is thus given by

$$e_1 = -F^3Y, \quad e_2 = -FY, \quad \varepsilon_1 = Y, \quad \varepsilon_2 = F^2Y.$$

With the corresponding symplectic coordinates we have $Q = x_2^2 - 2\xi_1\xi_2$. We sum up our conclusions as follows:

Theorem 21.5.3. *If Q is a positive semi-definite quadratic form in the symplectic vector space S of dimension $2n$ then one can choose symplectic coordinates x, ξ such that*

$$Q(x,\xi) = \sum_1^k \mu_j(x_j^2 + \xi_j^2) + \sum_{k+1}^{k+l} x_j^2, \tag{21.5.3}$$

where $\mu_j > 0$. If Q is hyperbolic, that is, has negative index of inertia equal to 1, then the symplectic coordinates can be chosen so that

$$Q(x,\xi) = \sum_1^k \mu_j(x_j^2 + \xi_j^2) + \sum_{k+1}^{k+l} x_j^2 + q(x,\xi), \tag{21.5.3$'$}$$

where $\mu_j > 0$ and either $k+l < n-1$ and

$$q(x,\xi) = x_n^2 - 2\xi_{n-1}\xi_n \tag{21.5.4}$$

or $k+l<n$ and

$$q(x,\xi) = -x_n^2 \quad \text{or} \quad q(x,\xi) = 2\lambda x_n \xi_n, \quad \lambda > 0. \tag{21.5.5}$$

The cases listed are of course all inequivalent; $i\mu_j$ are the eigenvalues of F on the positive imaginary axis, counted with multiplicity, (21.5.4) occurs when there is a 4×4 box in the Jordan decomposition of F, and the second case (21.5.5) means that λ is an eigenvalue of F.

We shall now give a modified form of the result when Q is complex valued and $\operatorname{Re} Q$ is positive definite. More generally the results in the semi-definite case have analogues when

$$|\operatorname{Im} Q(X)| \leqq C \operatorname{Re} Q(X), \quad X \in S, \tag{21.5.6}$$

as we shall now assume. Set

$$\Gamma = \{w \in \mathbb{C};\ |\operatorname{Im} w| \leqq C \operatorname{Re} w\}, \tag{21.5.7}$$

which is then a convex angular region containing the values of Q. We could use any such angle but (21.5.7) is convenient in the proof.

Theorem 21.5.4. *If* (21.5.6) *is valid, then*

a) $Q(\bar{X},X)=0 \Leftrightarrow FX=0 \Leftrightarrow F\,\mathrm{Re}\,X=F\,\mathrm{Im}\,X=0 \Leftrightarrow \bar{F}X=0$ *so* $\mathrm{Ker}\,F$ *is spanned by its real elements.*

b) $F^2 V_0=0$, *that is,*

$$X\in V_0 \Leftrightarrow Q(X,Y)=0 \quad \textit{when } \sigma(\mathrm{Ker}\,F,Y)=0.$$

c) $\lambda/i\in\Gamma$ *or* $\lambda/i\in-\Gamma$ *if* λ *is an eigenvalue of* F.

Proof. Write $X=X_1+iX_2$ with real X_j. Then

$$Q(\bar{X},X)=Q(X_1)+Q(X_2)$$

so (21.5.6) shows that $Q(\bar{X},X)=0$ implies $\mathrm{Re}\,Q(X_j)=0$, hence $\mathrm{Im}\,Q(X_j)=0$. If $z\in\mathbb{C}$ is so close to 1 that $\mathrm{Re}\,zQ$ is positive semidefinite, it follows that

$$\mathrm{Re}\,zQ(Y,X_j)=0, \qquad Y\in S,$$

hence $Q(Y,X_j)=\sigma(Y,FX_j)=0$ so $FX_j=0$. The other implications in a) are obvious. To prove b) assume that $X\in S_{\mathbb{C}}$ and that $F^3X=0$. Then $\bar{F}F^2X=0$ by a) and

$$Q(\overline{FX},FX)=\sigma(\overline{FX},F^2X)=-\sigma(\bar{X},\bar{F}F^2X)=0$$

so $F^2X=0$ by a) again. Since $\sigma(\mathrm{Ker}\,F,Y)=0 \Leftrightarrow Y=FZ$ for some Z, we have $Q(X,Y)=\sigma(X,FY)=0$ for all such Y if and only if $F^2X=0$. Finally c) follows since $FX=\lambda X$, $X=X_1+iX_2$, implies

$$Q(X_1)+Q(X_2)=Q(\bar{X},X)=\sigma(\bar{X},FX)=\lambda\sigma(\bar{X},X)=2i\lambda\,\sigma(X_1,X_2).$$

Here $\sigma(X_1,X_2)$ is real and not 0 unless $FX=0$ which proves c).

Let $W=S\cap\mathrm{Ker}\,F$ and set $S'=W^\sigma/(W\cap W^\sigma)$ which is a new symplectic space. Then

$$S'_{\mathbb{C}}=\mathrm{Im}\,F/(\mathrm{Ker}\,F\cap\mathrm{Im}\,F)\cong\bigoplus_{\lambda\neq 0}V_\lambda$$

by Theorem 21.5.4 a) and b), and for the quadratic form Q' induced by Q in S' the Hamilton map is isomorphic to the restriction of F to the eigenspaces with eigenvalues $\neq 0$. When studying these further we may therefore assume that $\mathrm{Re}\,Q$ is positive definite.

Definition 21.5.5. A complex Lagrangian plane $\lambda\subset S_{\mathbb{C}}$ is called positive if

$$i\sigma(\bar{X},X)\geq 0, \qquad X\in\lambda. \tag{21.5.8}$$

If $i\sigma(\bar{X},X)>0$ when $0\neq X\in\lambda$ we say that λ is strictly positive.

Note that $i\sigma(\bar{Y},X)$ is a hermitian symmetric form since

$$-i\sigma(Y,\bar{X})=i\sigma(\bar{X},Y).$$

Theorem 21.5.6. *If* Re Q *is positive definite, then*

$$S^+ = \bigoplus_{\lambda \in i\Gamma} V_\lambda$$

is a strictly positive Lagrangian plane, invariant under F.

Proof. That S^+ is Lagrangian follows from Lemma 21.5.2, and the invariance under F is clear. To prove that S^+ is positive we first assume that Q is real. Then we can choose symplectic coordinates so that

$$Q(x,\xi) = \sum \mu_j(x_j^2 + \xi_j^2).$$

Thus $F(x,\xi) = (\mu_1 \xi_1, \ldots, -\mu_1 x_1, \ldots)$ so S^+ is defined by $\xi_j = i x_j$ and

$$\sigma(\bar{x}, \bar{\xi}; x, \xi) = -2i \sum |x_j|^2$$

which proves the positivity. Returning to the general case we first observe that if S^+ is positive and $i\sigma(\overline{X}, X) = 0$ for some $X \in S^+$, then $i\sigma(\overline{X}, Y) = 0$, $Y \in S^+$, by Schwarz' inequality. Taking $Y = FX$ we obtain $Q(\overline{X}, X) = 0$, hence $X = 0$ so S^+ is strictly positive. The general proof follows now by a continuity argument. Set

$$Q_t(X) = \operatorname{Re} Q(X) + it \operatorname{Im} Q(X), \quad X \in S,\ 0 \leqq t \leqq 1.$$

The Lagrangian plane S_t^+ defined by Q_t varies continuously with t since the projection of $S_{\mathbb{C}}$ on S_t^+ is given by

$$(2\pi i)^{-1} \int_\gamma (F_t - z)^{-1}\, dz$$

where F_t is the Hamilton map of Q_t and γ is a circle in the upper half plane containing the eigenvalues of F_t there in its interior. Let T be the supremum of all $t \in [0,1]$ for which S_t^+ is strictly positive. Then S_T^+ is positive for reasons of continuity, hence strictly positive as we proved above. This proves that $T = 1$ and completes the proof.

Remark. Since

$$i\sigma(\overline{X}, FX) = iQ(\overline{X}, X) \in i\Gamma$$

the numerical range of the restriction of F to S^+, with respect to the positive definite hermitian form there, is contained in $i\Gamma$. This implies that all eigenvalues are in $i\Gamma$ but is a much stronger statement.

A strictly positive Lagrangian plane gives S additional structure:

Proposition 21.5.7. *If* $S^+ \subset S_{\mathbb{C}}$ *is a strictly positive Lagrangian plane then* $S^+ \ni X \mapsto \operatorname{Re} X \in S$ *is a bijection giving* S *a complex structure. There is a unique positive definite hermitian form* H *in* S *with respect to this structure such that*

$$\operatorname{Im} H(X,Y) = -\sigma(X,Y); \quad X, Y \in S;$$

and Re H *gives a natural Euclidean structure in* S.

Proof. If $X \in S^+$ then

$$0 < i\sigma(\overline{X}, X) = 2\sigma(\operatorname{Im} X, \operatorname{Re} X), \qquad X \neq 0,$$

which proves that $S^+ \ni X \mapsto \operatorname{Re} X \in S$ is injective, hence bijective since the real dimensions are equal. The hermitian form H we look for may be regarded as one on S^+, and then we must have if $X \in S^+$

$$H(X, iX) = i \operatorname{Im} H(X, iX) = -i\sigma(\operatorname{Re} X, \operatorname{Re} iX) = i\sigma(\operatorname{Re} X, \operatorname{Im} X),$$

hence

$$H(X, X) = \sigma(\operatorname{Im} X, \operatorname{Re} X) = i\sigma(\overline{X}, X)/2.$$

Polarization gives

$$H(X, Y) = i\sigma(\overline{Y}, X)/2 \tag{21.5.9}$$

and proves the theorem since the argument can be reversed.

One can use Proposition 21.5.7 to choose symplectic coordinate systems which are well adapted to a strictly positive Lagrangian plane:

Corollary 21.5.8. *Let S^+ be a strictly positive Lagrangian plane $\subset S_{\mathbb{C}}$ and let $\lambda \subset S$ be Lagrangian. Then one can choose symplectic coordinates x, ξ in S such that λ is defined by $x = 0$ and $S^+ = \{(x, ix);\ x \in \mathbb{C}^n\}$.*

Proof. The hermitian form H in Theorem 21.5.7 is real in λ. We can therefore choose a real orthonormal basis $\varepsilon_1, \dots, \varepsilon_n$ in λ. Choose $X_j \in S$ so that $\varepsilon_j - iX_j \in S^+$. The definition (21.5.9) of H means that then

$$i\sigma(\varepsilon_j + iX_j, \varepsilon_k - iX_k)/2 = \delta_{jk},$$

that is, $\sigma(X_j, X_k) = 0$ and $\sigma(\varepsilon_j, X_k) + \sigma(\varepsilon_k, X_j) = 2\delta_{jk}$. Since S^+ is Lagrangian we also have $\sigma(\varepsilon_j - iX_j, \varepsilon_k - iX_k) = 0$, hence $\sigma(\varepsilon_j, X_k) = \sigma(\varepsilon_k, X_j)$ which proves that $X_1, \dots, X_n, \varepsilon_1, \dots, \varepsilon_n$ is a symplectic basis for S and proves the corollary.

When we apply the corollary to the space S^+ in Theorem 21.5.6 we conclude that $Q(x, ix) = 0$ with the coordinates chosen. Thus Q is in the ideal generated by $x_j + i\xi_j$, $j = 1, \dots, n$.

A strictly positive Lagrangian plane has a very simple representation also for arbitrary symplectic coordinates:

Proposition 21.5.9. *Let $\lambda \subset T^*(\mathbb{C}^n)$ be a strictly positive Lagrangian plane. Then there is a symmetric matrix $A = A_1 + iA_2$ where A_1 is real and A_2 is positive definite, such that $\lambda = \{(z, Az);\ z \in \mathbb{C}^n\}$. Conversely, every such plane is strictly positive.*

Proof. From Definition 21.5.5 it follows at once that $\lambda \cap T_0^*(\mathbb{C}^n) = \{0\}$. Thus λ is the graph of the differential of a quadratic form, that is, $\lambda = \{(z, Az), z \in \mathbb{C}^n\}$ where A is a symmetric matrix. If $A = A_1 + iA_2$ with A_1 and A_2 real, we have

$$i\sigma((\bar{z}, \overline{Az}), (z, Az)) = 2\langle A_2 z, \bar{z}\rangle$$

which proves the proposition.

The structure of an arbitrary positive Lagrangian plane is clarified by the following result.

Proposition 21.5.10. *If* $\lambda \subset S_{\mathbb{C}}$ *is a positive Lagrangian plane and* $X \in \lambda$, $\sigma(X, \bar{X}) = 0$, *then* $\bar{X} \in \lambda$, *hence* $\operatorname{Re} X \in \lambda$ *and* $\operatorname{Im} X \in \lambda$.

Proof. By Cauchy-Schwarz' inequality we have $i\sigma(\bar{X}, Y) = 0$ for all $Y \in \lambda$. Since λ is Lagrangian this implies that $\bar{X} \in \lambda$.

Corollary 21.5.11. *Let* $\lambda \subset S_{\mathbb{C}}$ *be a positive Lagrangian plane and let* $\lambda_{\mathbb{R}}$ *be the isotropic plane* $\lambda \cap S$. *Set* $S' = \lambda_{\mathbb{R}}^{\sigma}/\lambda_{\mathbb{R}}$. *Then the image* λ' *of* λ *in* $S'_{\mathbb{C}}$ *is a strictly positive Lagrangian plane.*

Proof. That λ' is Lagrangian follows from the complex analogue of Proposition 21.2.13. Since λ' contains no real elements it follows from Proposition 21.5.10 that λ' is strictly positive.

Alternatively we can write $S = S_1 \oplus S_2$ where S_1, S_2 are symplectically orthogonal symplectic subspaces and $\lambda_{\mathbb{R}}$ is Lagrangian in S_1. Then λ is the direct sum of the complexification of $\lambda_{\mathbb{R}}$ in $S_{1\mathbb{C}}$ and a strictly positive Lagrangian plane in $S_{2\mathbb{C}}$.

21.6. The Lagrangian Grassmannian

Lagrangian manifolds were discussed at some length in Section 21.2 in preparation for the theory of Lagrangian distributions in Chapter XXV. To make that theory global we shall need a careful invariant analysis of the infinitesimal case involving Lagrangian planes in symplectic vector spaces. This will be provided in the present section.

Definition 21.6.1. If S is a finite dimensional symplectic vector space with symplectic form σ, then the Lagrangian Grassmannian $\Lambda(S)$ of S is the set of all Lagrangian planes in S.

We shall first show that $\Lambda(S)$ has a natural structure as an analytic manifold of dimension $n(n+1)/2$ where $2n = \dim S$. First we consider the subset

$$\Lambda_\mu(S) = \{\lambda \in \Lambda(S);\ \lambda \cap \mu = \{0\}\}, \qquad \mu \in \Lambda(S),$$

of Lagrangian planes transversal to a given one.

Lemma 21.6.2. *$\Lambda_\mu(S)$ has a natural structure as an affine space with the symmetric tensor product $\mu \underset{s}{\otimes} \mu$ as underlying vector space.*

Proof. The bilinear form $\sigma(X,Y)$ with $X\in\mu$ and $Y\in S$ is zero for all $X\in\mu$ precisely when $Y\in\mu$ so it gives an identification $S/\mu\to\mu'$, the dual space of μ. If $\lambda\in\Lambda_\mu(S)$, then the composed map

$$\lambda\to S/\mu\to\mu'$$

is an isomorphism, so it has an inverse T which we regard as a map into S with range λ. It has the properties

$$\langle X,\phi\rangle=\sigma(X,T\phi);\quad X\in\mu,\ \phi\in\mu', \tag{21.6.1}$$

$$\sigma(T\phi,T\psi)=0;\quad \phi,\psi\in\mu'. \tag{21.6.2}$$

Conversely, every such $T\colon \mu'\to S$ defines an element of $\Lambda_\mu(S)$. If T is one solution of (21.6.1), (21.6.2) and $T+R$ is any other, then by (21.6.1)

$$\sigma(X,R\phi)=0;\quad X\in\mu,\ \phi\in\mu',$$

that is, R is a map $\mu'\to\mu$, and (21.6.2) gives

$$\sigma(T\phi,R\psi)+\sigma(R\phi,T\psi)=0;\quad \phi,\psi\in\mu',$$

since $\sigma(R\phi,R\psi)=0$. Thus R corresponds to a bilinear form

$$A(\phi,\psi)=\sigma(R\phi,T\psi)=\sigma(R\psi,T\phi);\quad \phi,\psi\in\mu';$$

which is symmetric. Conversely, every such A defines a solution R of (21.6.1), (21.6.2) as claimed.

It is useful later to have seen the proof in terms of coordinates also: If we choose symplectic coordinates (x,ξ) so that μ is defined by $x=0$, then λ, as a Lagrangian section of $T^*(\mathbb{R}^n)$, is defined by the graph of the differential of a quadratic form $A/2$ in $\mathbb{R}^n$,

$$\lambda=\{(x,Ax);\,x\in\mathbb{R}^n\}$$

where A is the symmetric matrix of the quadratic form. This also identifies $\Lambda_\mu(S)$ with $\mathbb{R}^{n(n+1)/2}$ since $n(n+1)/2$ is the number of elements on or above the diagonal in an $n\times n$ matrix.

If μ_1 and μ_2 are two different Lagrangians we can choose symplectic coordinates (x,ξ) such that μ_1 is defined by $x=0$, and symplectic coordinates $y=L(x,\xi)$, $\eta=M(x,\xi)$ such that μ_2 is defined by $y=0$. A Lagrangian plane $\lambda\in\Lambda_{\mu_1}(S)\cap\Lambda_{\mu_2}(S)$ is defined by $\eta=Ay$ and by $\xi=Bx$ in the two coordinate systems. Thus the equations $M(x,\xi)=AL(x,\xi)$ can be solved for ξ and give $\xi=Bx$, so the coefficients. of B are rational functions with non-zero denominators of those of A. Hence the local charts we have given make $\Lambda(S)$ an analytic manifold of dimension $n(n+1)/2$. (Recall that the Grassmannian $G(N,k)$ of k planes in $\mathbb{R}^N$ is a manifold of dimension $k(N-k)$, the number of coefficients required to express $N-k$ coordinates in terms of the remaining k. In particular, $G(2n,n)=n^2$ is much larger than $\dim \Lambda(S)$ except when $n=1$.)

Our coordinate patches leave out a rather small subset of $\Lambda(S)$:

Lemma 21.6.3. *If* $\mu\in\Lambda(S)$ *then*

$$\{\lambda\in\Lambda(S);\ \dim\lambda\cap\mu=k\} \tag{21.6.3}$$

is a submanifold of dimension $n(n+1)/2-k(k+1)/2$.

Proof. The intersection $V=\lambda\cap\mu$ is an element in the Grassmannian of k planes in μ, which has dimension $k(n-k)$. For given V there is a one to one correspondence between Lagrangians $\lambda\subset S$ with $\lambda\cap\mu=V$ and Lagrangians λ' in $S'=V^\sigma/V$ transversal to μ/V, given by the construction in Proposition 21.2.13, $\lambda'=\lambda/V$. This is easily seen to be a fibration so the dimension of the space in (21.6.3) is $(n-k)(n-k+1)/2+k(n-k)$ as claimed.

In particular, $\Lambda(S)\setminus\Lambda_\mu(S)$ is a null set, the union of manifolds of codimension $k(k+1)/2$ for $k=1,2,\dots,n$. Any countable intersection $\bigcap\Lambda_{\mu_j}(S)$ is therefore non-empty which gives a much stronger version of Corollary 21.2.11.

In the analytical context in Chapter XXV S will occur as the tangent space of a cotangent bundle at some point. Then there is a distinguished Lagrangian plane given, the tangent space of the fiber. In what follows we shall therefore assume that a distinguished element $\lambda_0\in\Lambda(S)$ is given.

For a linear subspace V of $\mathbb{R}^n$ the simplest prototype of the conormal distributions studied in Section 18.2 is a translation invariant simple layer on V. This is a (half density) solution of the equation

$$L(x,D)u=0$$

for all linear functions $L(x,\xi)$ vanishing on the conormal bundle $V\times V^\circ$ of V, where V° is the annihilator of V. In fact, $L_1(x)u=0$ for all L_1 vanishing on V means that u is a simple layer on V, and $L_2(D)u=0$ for all L_2 vanishing on V° means that u is translation invariant along V. We shall now extend this construction to general Lagrangian planes.

To do so we choose any $\lambda_1\in\Lambda_{\lambda_0}(S)$, that is, λ_1 is Lagrangian and transversal to the distinguished Lagrangian plane λ_0. Recall that the bilinear map

$$\lambda_0\times\lambda_1\ni(X,Y)\mapsto\sigma(X,Y)$$

is then non-degenerate so it defines an isomorphism $\lambda_1\to\lambda_0'$ which we can use to identify half densities in λ_1 and in λ_0'. Introducing symplectic coordinates x,ξ in S *adapted* to λ_1 in the sense that $x=0$ in λ_0 and $\xi=0$ in λ_1, we define for any $\lambda \in \Lambda(S)$

(21.6.4) $$I(\lambda, \lambda_1) = \{u \in \mathscr{S}'(\lambda_0', \Omega^{\frac{1}{2}});\ L(x, D)u = 0 \text{ for all linear forms } L(x, \xi) \text{ vanishing on } \lambda\}.$$

Here $\Omega^{\frac{1}{2}}$ is the half density bundle. Note that the definition is independent of the choice of basis in λ_0, thus independent of the choice of dual basis in λ_1. We shall prove that the dependence on λ_1 is very mild and that $I(\lambda, \lambda_1)$ can be used to define the fiber of λ in a line bundle on $\Lambda(S)$. – Note that if $0 \neq u \in I(\lambda, \lambda_1)$ and $L(x, D)u = 0$ for some linear L, then $\{L, L'\}u = i[L(x, D), L'(x, D)]u = 0$ for all linear L' vanishing on λ. Thus L vanishes on λ, so the system of equations in (21.6.4) is maximal, and λ is determined by u.

With the symplectic coordinates x, ξ adapted to λ_1 any other $\lambda_2 \in \Lambda_{\lambda_0}(S)$ can be written in the form

$$\lambda_2 = \{(x, \xi); \xi = Ax\}$$

where A is a symmetric matrix. Thus the coordinates $y = x$ and $\eta = \xi - Ax$ are symplectic and adapted to λ_2. If $L(x, \xi) = 0$ on λ we obtain $L(y, \eta + Ay) = 0$ on λ in the y, η coordinates. Now

$$L(x, D)u_1 = 0 \Leftrightarrow L(y, D + Ay)u_2 = 0$$

if $u_2(y) = u_1(y) e^{-i\langle Ay, y\rangle/2}$. The map

(21.6.5) $$I(\lambda, \lambda_1) \ni u_{\lambda_1} \mapsto u_{\lambda_1} e^{-i\langle Ax, x\rangle/2} = u_{\lambda_2} \in I(\lambda, \lambda_2)$$

is therefore an isomorphism allowing us to identify $I(\lambda, \lambda_1)$ with $I(\lambda, \lambda_2)$. It is coordinate independent since it can be described as in the proof of Lemma 21.6.2, and it is transitive. The set $I(\lambda)$ of equivalence classes can be thought of as the set of maps

$$\Lambda_{\lambda_0}(S) \ni \lambda_1 \mapsto u_{\lambda_1} \in I(\lambda, \lambda_1)$$

satisfying (21.6.5). For every $\lambda_1 \in \Lambda_{\lambda_0}(S)$ we have of course an isomorphism $I(\lambda, \lambda_1) \to I(\lambda)$.

For every $\lambda \in \Lambda(S)$ there is some $\lambda_1 \in \Lambda_{\lambda_0}(S)$ transversal to λ. It is then easy to describe $I(\lambda, \lambda_1)$:

Lemma 21.6.4. *Let λ_1 be transversal to λ and λ_0, and choose symplectic coordinates x, ξ such that λ_0 and λ_1 are defined by $x = 0$ and $\xi = 0$ respectively. Every $u \in I(\lambda, \lambda_1)$ is then an oscillatory integral*

(21.6.6) $$u(x) = c(2\pi)^{-3n/4} \int e^{i(\langle x, \xi\rangle - \langle B\xi, \xi\rangle/2)}\, d\xi$$

where B is the symmetric matrix defining λ through

(21.6.7) $$\lambda = \{(B\xi, \xi); \xi \in \mathbb{R}^n\},$$

and c is a constant.

Proof. Since λ is transversal to λ_1 we can define λ by an equation $x = B\xi$, and B is symmetric since λ is Lagrangian. The equations defining $I(\lambda, \lambda_1)$ are then

$$x_j u - (BD)_j u = 0; \quad j = 1, \dots, n;$$

or after Fourier transformation

$$D\hat{u} + B\xi\,\hat{u} = 0.$$

The solution is $\hat{u}(\xi) = c\exp(-i\langle B\xi, \xi\rangle/2)$ for some constant c, which gives (21.6.6) with another constant. (We have inserted a factor $(2\pi)^{-3n/4}$ in (21.6.6) to get agreement with (18.2.11)′ when $k=n$. A better justification will follow shortly.)

Remark. It is easy to evaluate the integral (21.6.6) using Theorem 7.6.1; if $\langle B\xi, \xi\rangle = \langle B'\xi', \xi'\rangle$ is a non-singular quadratic form in $\xi' = (\xi_1, \dots, \xi_k)$ then

$$u = c(2\pi)^{(n-2k)/4} e^{i\langle B'^{-1}x', x'\rangle/2}\,\delta(x'')\,e^{-\pi i(\operatorname{sgn} B')/4}\,|\det B'|^{-\frac{1}{2}}. \tag{21.6.6}'$$

In view of (21.6.5) it follows for arbitrary λ and for $\lambda_1 \in \Lambda_{\lambda_0}(S)$ that $I(\lambda, \lambda_1)$ is a translation invariant simple layer on a subspace multiplied by a Gaussian of constant absolute value. However, such representations will be avoided since they hide how nicely (21.6.6) depends on B, that is, on λ.

For fixed $\lambda_1 \in \Lambda_{\lambda_0}(S)$ and arbitrary $\lambda \in \Lambda_{\lambda_1}(S)$ we can identify $I(\lambda)$ with $\mathbb{C}$ by first identifying $I(\lambda)$ with $I(\lambda, \lambda_1)$ and then identifying $u \in I(\lambda, \lambda_1)$ with $c \in \mathbb{C}$ when (21.6.6) holds. We shall prove that this makes

$$I = \{(u, \lambda); u \in I(\lambda)\}$$

a line bundle over $\Lambda(S)$. To do so we must determine the transition function which the coefficient c in (21.6.6) is multiplied by when λ_1 is replaced by another Lagrangian plane $\lambda_2 \in \Lambda_{\lambda_0}(S)$ such that $\lambda \in \Lambda_{\lambda_2}(S)$. First note that it follows from (21.6.6) that the oscillatory integral of u is well defined; we have

$$\int u(x)\,dx = (2\pi)^{n/4} c. \tag{21.6.8}$$

Since u is a half density, c transforms as a $-1/2$ density in λ_1, hence $c|d\xi|^{\frac{1}{2}}$ is a half density in the dual space λ_0 depending only on the choice of λ_1 and not on the choice of x coordinates. Now the projection

$$\pi_{\lambda_1}\colon \lambda \to \lambda_0$$

of λ on λ_0 along λ_1 is bijective since $\lambda_1 \cap \lambda = \{0\}$, so the half density $c|d\xi|^{\frac{1}{2}}$ can be lifted to a half density

$$u^{\#}_{\lambda_1} = \pi^{*}_{\lambda_1} c|d\xi|^{\frac{1}{2}}$$

in λ. When $\lambda_2 \in \Lambda_{\lambda_0}(S)$ the element $u \in I(\lambda, \lambda_1)$ defined by (21.6.6) is according to (21.6.5) identified with $v \in I(\lambda, \lambda_2)$ defined by

$$v(y) = c(2\pi)^{-3n/4} e^{-i\langle Ay, y\rangle/2} \int e^{i(\langle y, \xi\rangle - \langle B\xi, \xi\rangle/2)}\,d\xi.$$

If $\lambda \in \Lambda_{\lambda_2}(S)$ we can define $u^\#_{\lambda_2}$ as we defined $u^\#_{\lambda_1}$, noting that when v is written in the form (21.6.6) then the coefficient c is replaced by

$$c_2 = (2\pi)^{-n/4} \int u_2(y)\, dy = c(2\pi)^{-n} \iint e^{i(\langle y,\xi\rangle - \langle B\xi,\xi\rangle/2 - \langle Ay,y\rangle/2)}\, d\xi\, dy$$
$$= c e^{\pi i s/4} \left| \det \begin{pmatrix} -B & I \\ I & -A \end{pmatrix} \right|^{-\frac{1}{2}},$$

where s is the signature of the matrix $\begin{pmatrix} -A & I \\ I & -B \end{pmatrix}$. It is non-singular because λ and λ_2 are transversal, which means that the equations $x - B\xi = 0$ and $\xi - Ax = 0$ only have the solution $x = \xi = 0$. The matrix has $2n$ rows and columns, so the signature is an even number, hence $e^{\pi i s/4}$ is a power of the imaginary unit.

Next we compare the densities $\pi^*_{\lambda_1}|d\xi|$ and $\pi^*_{\lambda_2}|d\eta|$ on λ. In the x, ξ coordinate system ξ parametrizes the point $(B\xi, \xi) \in \lambda$, and in the y, η system η parametrizes the same point if $\eta = \xi - Ay = \xi - AB\xi$, so $\pi_{\lambda_2} \pi^{-1}_{\lambda_1} \xi = (I - AB)\xi$ and

$$\pi^*_{\lambda_2}|d\eta| = |\det(I - AB)|\, \pi^*_{\lambda_1}|d\xi|.$$

Since $|\det(I - AB)| = \left| \det \begin{pmatrix} -B & I \\ I & -A \end{pmatrix} \right|$ we obtain

$$u^\#_{\lambda_2} = e^{\pi i s/4} u^\#_{\lambda_1}, \quad s = \operatorname{sgn} \begin{pmatrix} -A & I \\ I & -B \end{pmatrix}. \tag{21.6.9}$$

Apart from the powers of i here we have thus been able to identify $I(\lambda)$ with the translation invariant half densities $\Omega(\lambda)^{\frac{1}{2}}$ on λ.

The integer

$$\sigma(\lambda_0, \lambda; \lambda_2, \lambda_1) = \tfrac{1}{2} \operatorname{sgn} \begin{pmatrix} -A & I \\ I & -B \end{pmatrix} \tag{21.6.10}$$

has been defined for any four Lagrangian planes λ_0, λ, λ_2, λ_1 such that each of the first two is transversal to each of the last two, by means of a symplectic coordinate system such that

$$\lambda_0\colon x = 0; \quad \lambda\colon x = B\xi; \quad \lambda_2\colon \xi = Ax; \quad \lambda_1\colon \xi = 0. \tag{21.6.11}$$

For reasons of continuity σ is constant when λ_0, λ, λ_2, λ_1 belongs to a connected open set in $\Lambda(S)^4$ where these transversality conditions are valid. When $u \in I(\lambda)$ we have assigned to every Lagrangian λ_1 transversal to λ_0 and to λ a translation invariant half density $u^\#_{\lambda_1}$ on λ such that, if λ_2 is another such Lagrangian,

$$u^\#_{\lambda_2} = i^{\sigma(\lambda_0, \lambda; \lambda_2, \lambda_1)} u^\#_{\lambda_1}. \tag{21.6.9$'$}$$

Thus we have modulo 4

$$\sigma(\lambda_0, \lambda; \lambda_2, \lambda_1) = -\sigma(\lambda_0, \lambda; \lambda_1, \lambda_2), \tag{21.6.12}$$

$$\sigma(\lambda_0, \lambda; \lambda_3, \lambda_1) = \sigma(\lambda_0, \lambda; \lambda_3, \lambda_2) + \sigma(\lambda_0, \lambda; \lambda_2, \lambda_1) \tag{21.6.13}$$

when all terms are defined. These equalities are in fact valid exactly. It suffices to prove (21.6.13) which gives (21.6.12) when $\lambda_3=\lambda_1$. Let λ_0, λ, λ_2, λ_1 be defined by (21.6.11) and λ_3 by $\xi=A'x$. With the coordinates $y=x$, $\eta=\xi-Ax$ adapted to λ_2 we have the equation $\eta=(A'-A)y$ for λ_3 and $y=B'\eta$ for λ if $B\xi=B'(\xi-AB\xi)$, that is, $B'=B+B'AB$ (the resolvent equation!). Thus we must prove that

$$\operatorname{sgn}\begin{pmatrix}-A' & I\\ I & -B\end{pmatrix}=\operatorname{sgn}\begin{pmatrix}A-A' & I\\ I & -B'\end{pmatrix}+\operatorname{sgn}\begin{pmatrix}-A & I\\ I & -B\end{pmatrix}.$$

In doing so we may assume that B and B' are invertible, for the equality is stable for small perturbations. Now

$$\begin{pmatrix}I & B^{-1}\\ 0 & I\end{pmatrix}\begin{pmatrix}-A & I\\ I & -B\end{pmatrix}\begin{pmatrix}I & 0\\ B^{-1} & I\end{pmatrix}=\begin{pmatrix}-A+B^{-1} & 0\\ 0 & -B\end{pmatrix}$$

so the equality becomes

$$\begin{aligned}\operatorname{sgn}(-A'+B^{-1})-\operatorname{sgn}B&=\operatorname{sgn}(A-A'+B'^{-1})-\operatorname{sgn}B'^{-1}\\&\quad+\operatorname{sgn}(-A+B^{-1})-\operatorname{sgn}B\end{aligned}$$

which is true since $B^{-1}=B'^{-1}+A$. This proves (21.6.12), (21.6.13). We also have the anti-symmetry

$$\sigma(\lambda_0,\lambda;\lambda_2,\lambda_1)=-\sigma(\lambda_1,\lambda_2;\lambda,\lambda_0). \tag{21.6.14}$$

It follows from the symplectic change of variables $y=-\xi$, $\eta=x$, which makes λ_0 defined by $\eta=0$, λ by $\eta=-Bx$, λ_2 by $y=-A\eta$ and λ_1 by $y=0$, and

$$\operatorname{sgn}\begin{pmatrix}-A & I\\ I & -B\end{pmatrix}=-\operatorname{sgn}\begin{pmatrix}B & I\\ I & A\end{pmatrix}.$$

Definition 21.6.5. The complex line bundle M defined on $\Lambda(S)$ by the covering with the open sets $\Lambda_{\lambda_1}(S)$, $\lambda_1\in\Lambda_{\lambda_0}(S)$, and the transition functions

$$g_{\lambda_2\lambda_1}=i^{\sigma(\lambda_0,\lambda;\lambda_2,\lambda_1)},\qquad \lambda\in\Lambda_{\lambda_1}(S)\cap\Lambda_{\lambda_2}(S), \tag{21.6.15}$$

is called the Maslov bundle on $\Lambda(S)$.

Thus we have identified I with the tensor product $M\otimes\Omega^{\frac{1}{2}}$ of the Maslov bundle and the bundle with fiber at λ equal to the translation invariant half densities in λ. One should think of the elements of $M\otimes\Omega^{\frac{1}{2}}$ as the symbols of the elements of I.

If we apply (21.6.5) with some λ_2 transversal to λ and λ_0, and use Lemma 21.6.4 to represent u_{λ_2}, it follows that every $u\in I(\lambda,\lambda_1)$ can be written in the form

$$u(x)=(2\pi)^{-3n/4}\int e^{iQ(x,\xi)}\,d\xi,\qquad x\in\lambda_1,$$

where $Q(x,\xi)=\langle x,\xi\rangle+(\langle Ax,x\rangle-\langle B\xi,\xi\rangle)/2$ for some symmetric A and B. Thus we have such a representation even if λ is not transversal to λ_1. We shall now study even more general representations of the form

$$u(x)=a(2\pi)^{-(n+2N)/4}\int e^{iQ(x,\theta)}\,d\theta, \quad x\in\lambda_1, \tag{21.6.16}$$

where Q is a real quadratic form in $\lambda_1\times\mathbb{R}^N$ and a transforms as a half density if the coordinates x are changed. At first we assume that Q is non-degenerate as a phase function in the sense that $\partial Q/\partial\theta_1,\dots,\partial Q/\partial\theta_N$ have linearly independent differentials. Since $\sum t_j\partial Q/\partial\theta_j$ is identically 0 precisely when $(0,t)$ is in the radical of Q, an equivalent condition is that $(0,t)\in\operatorname{Rad}Q$ implies $t=0$. Thus

$$|\theta|\le C(|\partial Q(x,\theta)/\partial x|+|\partial Q(x,\theta)/\partial\theta|) \quad \text{if } C|x|<|\theta|,$$

which makes (21.6.16) defined as an oscillatory integral. (The homogeneities are not the same as in Section 7.8 but the arguments used there apply. Regarded as a phase function Q does not have the homogeneity assumed in Section 21.2 either.) We have

$$L(x,D)u(x)=a(2\pi)^{-(n+2N)/4}\int L(x,\partial Q/\partial x)\,e^{iQ(x,\theta)}\,d\theta=0$$

if $L(x,\partial Q/\partial x)=\sum t_j\partial Q/\partial\theta_j$ for some t_j, that is, if L vanishes on the Lagrangian

$$\lambda=\{(x,\partial Q/\partial x);\ \partial Q/\partial\theta=0\}$$

parametrized by Q. Thus $u\in I(\lambda,\lambda_1)$.

If $\lambda\in\Lambda_{\lambda_1}(S)$ we can define λ by (21.6.7), and u must be of the form (21.6.6) with

$$c=(2\pi)^{-n/4}\int u(x)\,dx=a(2\pi)^{-(n+N)/2}\iint e^{iQ(x,\theta)}\,dx\,d\theta.$$

Here Q is non-singular since $\partial Q/\partial\theta=\partial Q/\partial x=0$ implies $(x,0)\in\lambda$, so $x=0$ by (21.6.7). The integral can thus be evaluated by Theorem 7.6.1 which gives

$$c=a\,e^{\pi i/4\operatorname{sgn}Q}\,|\det Q''|^{-\frac{1}{2}}; \tag{21.6.17}$$

the power of 2π in (21.6.16) was chosen so that we get cancellation here. There is an important geometrical interpretation of (21.6.17). The map

$$\mathbb{R}^{n+N}\ni(x,\theta)\mapsto Q'_\theta(x,\theta)\in\mathbb{R}^N$$

is surjective, so the pullback $\delta(Q'_\theta)$ of the δ function in $\mathbb{R}^N$ is a density d_C on $C=\{(x,\theta)\in\mathbb{R}^{n+N};\ Q'_\theta(x,\theta)=0\}$. By (6.1.1) we have

$$d_C=|D(t,\partial Q/\partial\theta)/D(x,\theta)|^{-1}\,|dt|$$

if $t_1,\dots,t_n$ are linear functions in $\mathbb{R}^{n+N}$ which restrict to coordinates on C, and $|dt|$ is the Lebesgue density on C defined by these coordinates. Since

$$C\ni(x,\theta)\mapsto(x,\partial Q/\partial x)\in\lambda$$

is a bijection for any λ, we can always regard d_C as a density on λ. When λ is transversal to λ_1, we can take $t=\partial Q/\partial x$, for the restriction to C agrees with ξ. Thus we have

$$d_C^{\frac{1}{2}} = |\det Q''|^{-\frac{1}{2}} |d\xi|^{\frac{1}{2}}$$

where both sides are considered as half densities on λ. If we recall that $u_{\lambda_1}^{\#} = c|d\xi|^{\frac{1}{2}}$, it follows from (21.6.17) that

$$u_{\lambda_1}^{\#} = a\, d_C^{\frac{1}{2}}\, e^{\pi i/4 \operatorname{sgn} Q}. \tag{21.6.17$'$}$$

If we have another representation

$$\tilde{u}(x) = \tilde{a}(2\pi)^{-(n+2\tilde{N})/4} \int e^{i\tilde{Q}(x,\tilde{\theta})}\, d\tilde{\theta} \tag{21.6.16$'$}$$

of the same element as in (21.6.16), then $\tilde{Q}$ must parametrize the same Lagrangian as Q, and it follows from (21.6.17)$'$ that

$$a\, d_C^{\frac{1}{2}}\, e^{\pi i(\operatorname{sgn} Q - \operatorname{sgn} \tilde{Q})/4} = \tilde{a}\, d_{\tilde{C}}^{\frac{1}{2}}. \tag{21.6.17$''$}$$

The signature difference is not influenced by the x dependence, for we have

$$\operatorname{sgn} Q - \operatorname{sgn} \tilde{Q} = \operatorname{sgn} Q(0,.) - \operatorname{sgn} \tilde{Q}(0,.). \tag{21.6.18}$$

(The signature of a singular quadratic form is also defined by the convention $\operatorname{sgn} t = t/|t|$ when $t \in \mathbb{R} \setminus 0$ and $\operatorname{sgn} 0 = 0$.) To prove (21.6.18) we first change the θ coordinates so that $Q(0,\theta)$ is a non-singular quadratic form $Q_0(\theta'')$ for a splitting $\theta = (\theta', \theta'')$ of the variables. Then we can write

$$Q(x,\theta) = Q_0(\theta'' + L(x,\theta')) + Q_1(x,\theta') \tag{21.6.19}$$

where L is a linear map, $Q_1(0,.) = 0$ and

$$\operatorname{sgn} Q = \operatorname{sgn} Q_0 + \operatorname{sgn} Q_1 = \operatorname{sgn} Q(0,.) + \operatorname{sgn} Q_1.$$

It is therefore sufficient to prove (21.6.18) when Q and $\tilde{Q}$ are linear in θ and $\tilde{\theta}$, which implies that the right-hand side is 0. Thus assume that

$$Q(x,\theta) = q(x) + \langle x, T\theta \rangle$$

where $T: \mathbb{R}^N \to \mathbb{R}^n$ is injective since Q is a non-degenerate phase function. We have

$$\lambda = \{(x, q'(x) + T\theta);\ {}^tTx = 0\}$$

where $q(x) = \langle q'(x), x \rangle/2 = \langle \xi, x \rangle/2$ if ${}^tTx = 0$, $(x,\xi) \in \lambda$. The kernel $\mathcal{N}$ of ${}^tT$ and the restriction $q|_{\mathcal{N}}$ are therefore determined by λ. Conversely, $\mathcal{N}$ and $q|_{\mathcal{N}}$ determine λ since the range of T is the orthogonal space of $\mathcal{N}$. If we choose x coordinates (x', x'') such that $\mathcal{N}$ is defined by $x' = 0$ and make a linear change of θ variables, we have $\langle x, T\theta \rangle = \langle x', \theta \rangle$ with no change of the signature. Now the signature of $q(x', x'') + \langle x', \theta \rangle$ is equal to that of the restriction to the plane where the (x', θ) derivatives vanish, that is, $x' = 0$ and $\theta + \partial q(0, x'')/\partial x' = 0$. Thus the signature is equal to that of $q|_{\mathcal{N}}$, which only depends on λ. This proves (21.6.18).

From (21.6.17)$''$ and (21.6.18) it follows that

$$\tilde{a}\, d_{\tilde{C}}^{\frac{1}{2}} = a\, d_C^{\frac{1}{2}}\, e^{\pi i(\operatorname{sgn} Q(0,.) - \operatorname{sgn} \tilde{Q}(0,.))/4} \tag{21.6.17$'''$}$$

when (21.6.16) and (21.6.16)′ are valid. So far we have proved this only when λ is transversal to λ_1. Then it follows from (21.6.18) that the Maslov factor in the right-hand side does not change if we make small perturbations of Q and of $\tilde{Q}$ so that they still parametrize the same Lagrangian. This conclusion remains true even if λ is not transversal to λ_1, and so does (21.6.17)‴. In fact, we can always find λ_2 transversal to λ_0 and to λ. Replacing λ_1 by λ_2 means that $-\langle Ax, x\rangle/2$ is added to Q and $\tilde{Q}$ which then become non-singular. This does not affect the signature difference in (21.6.17)‴ or the definition of d_C, $d_{\tilde{C}}$ on λ.

We can now give an alternative definition of the Maslov bundle as follows. Fix λ_1 and cover $\Lambda(S)$ by open sets ω such that λ is defined by a non-degenerate Q_λ depending continuously on λ when $\lambda \in \omega$. For another continuous function $\tilde{Q}_\lambda$, $\lambda \in \tilde{\omega}$, we obtain a locally constant transition function

$$e^{\pi i(\operatorname{sgn} Q_\lambda(0,.) - \operatorname{sgn} \tilde{Q}_\lambda(0,.))/4}, \quad \lambda \in \omega \cap \tilde{\omega},$$

defining the Maslov bundle. (If we always take the same number of fiber variables N, say $N = n$, then the transition function is a power of the imaginary unit.)

Every real quadratic form $Q(x, \theta)$ in $\lambda_1 \times \mathbb{R}^N$ defines a Lagrangian plane

$$\lambda = \{(x, Q_x'); Q_\theta' = 0\}.$$

Let

$$R = \{\theta; Q_x'(0, \theta) = Q_\theta'(0, \theta) = 0\}$$

be the intersection of the radical and the plane $x = 0$. We choose a splitting $\theta = (\theta', \theta'')$ of the θ variables such that $R \ni \theta \mapsto \theta''$ is bijective. This means that R has an equation of the form $\theta' = \psi(\theta'')$, so

$$Q(x, \theta', \theta'') = Q(x, \theta' - \psi(\theta''), 0).$$

The quadratic form $Q(x, \theta', 0)$ is a non-degenerate phase function defining λ, so

$$U = a(2\pi)^{-(n+2N)/4} \int e^{iQ(x,\theta)}\, d\theta', \quad x \in \lambda_1,$$

is a well defined element in $I(\lambda, \lambda_1)$ independent of θ''. We can interpret (21.6.16) invariantly as a translation invariant density on R with values in $I(\lambda, \lambda_1)$. To do so we take a projection $T: \mathbb{R}^N \to R$ and consider

(21.6.16)″ $$C_0(R) \ni \chi \mapsto a(2\pi)^{-(n+2N)/4} \int e^{iQ(x,\theta)} \chi(T\theta)\, d\theta.$$

Since $Q(x, \theta) = Q(x, \theta' - \psi(\theta''), 0)$ we replace θ' by $\theta' + \psi(\theta'')$ and note that

$$\int \chi(T(\theta' + \psi(\theta''), \theta''))\, d\theta'' = \int \chi(\psi(\theta''), \theta'')\, d\theta''$$

because T is linear and $T(\psi(\theta''), \theta'') = (\psi(\theta''), \theta'')$. Hence the right-hand side of (21.6.16)″ is equal to $U \int \chi(\psi(\theta''), \theta'')\, d\theta''$, thus an element of $I(\lambda, \lambda_1)$ independent of the choice of T. Thus (21.6.16)″ interprets (21.6.16) as an element of $I(\lambda, \lambda_1) \otimes \Omega(R)$.

If Δ is an isotropic subspace of S then $S_\Delta = \Delta^\sigma/\Delta$ is a symplectic vector space (Proposition 21.2.1). By Proposition 21.2.13 every $\lambda \in \Lambda(S)$ defines another Lagrangian

$$\lambda_\Delta = (\lambda \cap \Delta^\sigma)/(\lambda \cap \Delta) \in \Lambda(S_\Delta).$$

In particular $\lambda_{0\Delta}$ gives a distinguished Lagrangian in S_Δ. We shall now show that there is a natural map from the fiber of $M \otimes \Omega^{1/2}$ at λ to the fiber at λ_Δ of the corresponding line bundle $M_\Delta \otimes \Omega_\Delta^{\frac{1}{2}}$ on $\Lambda(S_\Delta)$, tensored with the density bundle $\Omega(\lambda \cap \Delta)$ and another factor depending only on Δ.

In the proof we use symplectic coordinates putting Δ in a convenient position. Since Δ is isotropic we have

$$\{0\} \subset \Delta \cap \lambda_0 \subset \Delta^\sigma \cap \lambda_0 \subset \lambda_0.$$

We choose a basis $\varepsilon_1, \dots, \varepsilon_k$ for $\Delta \cap \lambda_0$ and extend it first to a basis $\varepsilon_1, \dots, \varepsilon_{k+l}$ for $\Delta^\sigma \cap \lambda_0$ and then to a basis $\varepsilon_1, \dots, \varepsilon_n$ for λ_0. The symplectic form makes $\Delta/(\Delta \cap \lambda_0)$ dual to $\lambda_0/(\Delta^\sigma \cap \lambda_0)$ so we can extend $\varepsilon_1, \dots, \varepsilon_k$ with $e_{k+l+1}, \dots, e_n$ to a basis for Δ so that the commutation relations are fulfilled. Completing to a full symplectic basis we then obtain symplectic coordinates (x, ξ), $x = (x', x'', x''')$, $\xi = (\xi', \xi'', \xi''')$ such that Δ is the $\xi' x'''$ plane and $\lambda_0 = \{(0, \xi)\}$. We shall write $n' = k$, $n'' = l$, $n''' = n - k - l$ for the various dimensions.

Define λ_1 by $\xi = 0$ and recall that $I(\lambda)$ is isomorphic to $I(\lambda, \lambda_1)$ where every element u can be written in the form (21.6.16) with Q nondegenerate as a phase function defining λ. Now consider the integral of u with respect to x''' restricted to $x' = 0$,

$$u_\Delta(x'') = a(2\pi)^{-(n+2N)/4} \iint e^{iQ(0, x'', x''', \theta)} \, dx''' \, d\theta. \tag{21.6.20}$$

Here we have to take into account the radical in the parameter direction

$$R = \{(x''', \theta);\ \partial Q/\partial x'' = \partial Q/\partial x''' = \partial Q/\partial \theta = 0 \text{ at } (0, 0, x''', \theta)\}.$$

The bijection $\{(x, \theta);\ \partial Q/\partial \theta = 0\} \ni (x, \theta) \mapsto (x, \partial Q/\partial x) \in \lambda$ sends R to

$$\{(x, \xi) \in \lambda;\ x' = x'' = \xi'' = \xi''' = 0\} = \lambda \cap \Delta.$$

The Lagrangian parametrized by $Q(0, x'', x''', \theta)$, where (x''', θ) is now the parameter, is given by

$$\begin{aligned}&\{(x'', \partial Q(0, x'', x''', \theta)/\partial x'');\ \partial Q/\partial x''' = \partial Q/\partial \theta = 0\}\\ &= \{(x'', \xi'');\ (0, x'', x''', \xi', \xi'', 0) \in \lambda \text{ for some } x''', \xi'\}.\end{aligned}$$

Since Δ^σ is defined by $x' = \xi''' = 0$ and Δ is the $\xi' x'''$ plane, the coordinates x'', ξ'' induce symplectic coordinates in S_Δ and the Lagrangian obtained is precisely λ_Δ. Thus (21.6.20) defines an element in $I(\lambda_\Delta, \lambda_{1\Delta}) \otimes \Omega(\lambda \cap \Delta)$, hence one in $I(\lambda_\Delta) \otimes \Omega(\lambda \cap \Delta)$. Passing to "symbols" we have a map from the fiber of $M \otimes \Omega^{\frac{1}{2}}$ at λ to the fiber of $M_\Delta \otimes \Omega_\Delta^{\frac{1}{2}}$ at λ_Δ tensored with $\Omega(\lambda \cap \Delta)$, but so far it depends on a number of choices made.

First we show the independence of the choice of Q. If Q is of the form (21.6.19) then integration with respect to the θ'' variables yields the same factor in (21.6.16) as in (21.6.20). Hence Q and Q_1 define the same map. We may therefore assume that Q is linear in θ,

$$Q(x,\theta)=q(x)+\langle x, T\theta\rangle.$$

Here T is injective, $\operatorname{Ker}{}^tT$ is uniquely determined, and so is the restriction of q to $\operatorname{Ker}{}^tT$. Adding a polynomial vanishing in $\operatorname{Ker}{}^tT$ to q changes neither (21.6.16) nor (21.6.20), which proves the independence of the choice of Q.

Suppose now that we keep the x variables but replace ξ by $\xi - Ax$ where A is a symmetric matrix. We require that Δ is still the $\xi' x'''$ plane. Then $A(0,0,x''')$ must be in the ξ' direction so the quadratic form $\langle A(0,x'',x'''),(0,x'',x''')\rangle$ is independent of x'''. Passing to the new coordinates means by (21.6.5) that u is multiplied by $e^{-i\langle Ax,x\rangle/2}$, and this will only lead to multiplication of u_Δ by $e^{-i\langle A(0,x'',0),(0,x'',0)\rangle/2}$ which agrees with the change of coordinates induced in S_Δ.

Finally, if we keep the plane λ_1 fixed we can only change the x variables and the dual ξ variables so that the plane $x'=0$ and the plane $x'=x''=0$ are preserved, for these are the orthogonal spaces of $\Delta\cap\lambda_0$ and of $\Delta^\sigma\cap\lambda_0$. Since u transforms as $|dx|^{\frac12}=|dx'|^{\frac12}|dx''|^{\frac12}|dx'''|^{\frac12}$ when we change coordinates, it follows that u_Δ transforms as

$$|dx'|^{\frac12}|dx''|^{\frac12}|dx'''|^{-\frac12},$$

that is, as

$$|d\xi'|^{-\frac12}|dx'''|^{-\frac12}|dx''|^{\frac12}.$$

Since (ξ',x''') are coordinates in Δ, this means that we have actually obtained a map

$$I(\lambda)\to I(\lambda_\Delta)\otimes\Omega(\lambda\cap\Delta)\otimes\Omega(\Delta)^{-\frac12}.$$

Passing to "symbols" we obtain a map

$$M(\lambda)\otimes\Omega(\lambda)^{\frac12}\to M_\Delta(\lambda_\Delta)\otimes\Omega(\lambda_\Delta)^{\frac12}\otimes\Omega(\lambda\cap\Delta)\otimes\Omega(\Delta)^{-\frac12}.$$

It is clear that we have a bundle map if λ is allowed to vary only in a manifold where $\lambda\cap\Delta$ has a fixed dimension. Since the transition functions in M and M_Δ have absolute value one and those in the density bundles are positive, the map is the tensor product of two uniquely defined maps

$$M(\lambda)\to M(\lambda_\Delta) \quad\text{and}\quad \Omega(\lambda)^{\frac12}\to\Omega(\lambda_\Delta)^{\frac12}\otimes\Omega(\lambda\cap\Delta)\otimes\Omega(\Delta)^{-\frac12}$$

where the second map preserves positivity. It can also be described in a much more direct and geometrical way. In fact, since λ_Δ is the quotient space of $\lambda\cap\Delta^\sigma$ by $\lambda\cap\Delta$ we have

$$\Omega(\lambda_\Delta)^{\frac12}\cong\Omega(\lambda\cap\Delta^\sigma)^{\frac12}\otimes\Omega(\lambda\cap\Delta)^{-\frac12}. \tag{21.6.21}$$

The symplectic form makes $\lambda/(\lambda\cap\Delta^\sigma)$ and $\Delta/(\lambda\cap\Delta)$ dual, so

$$\Omega(\lambda)^{\frac12}\otimes\Omega(\lambda\cap\Delta^\sigma)^{-\frac12}\cong\Omega(\lambda\cap\Delta)^{\frac12}\otimes\Omega(\Delta)^{-\frac12}. \tag{21.6.22}$$

Combination of (21.6.21) and (21.6.22) gives

$$\begin{aligned}\Omega(\lambda)^{\frac12} &\cong \Omega(\lambda\cap\Delta^\sigma)^{\frac12}\otimes\Omega(\lambda\cap\Delta)^{\frac12}\otimes\Omega(\Delta)^{-\frac12}\\ &\cong \Omega(\lambda_\Delta)^{\frac12}\otimes\Omega(\lambda\cap\Delta)\otimes\Omega(\Delta)^{-\frac12}.\end{aligned}$$

We shall prove that apart from a power of 2π this is identical to the isomorphism defined above by means of (21.6.16), (21.6.20). When doing so we observe that the passage from S to S_Δ can be made in two steps, first from S to $S_{\Delta'}$ where $\Delta'=\Delta\cap\lambda$, and then from $S_{\Delta'}$ to S_Δ; in that case we have $\lambda_{\Delta'}\cap\Delta_{\Delta'}=\{0\}$. It is therefore sufficient to examine these two situations separately.

a) Assume that $\Delta\subset\lambda$ and choose the coordinates so that Δ is the $\xi' x'''$ plane as above. Then $x'=\xi'''=0$ in λ. Changing the ξ'' variables if necessary we may assume that λ_Δ is of the form $x''=B\xi''$. This means that λ_Δ is parametrized by $\langle x'',\xi''\rangle-\langle B\xi'',\xi''\rangle/2$ and that λ is parametrized by

$$Q(x,\xi)=\langle x',\xi'\rangle+\langle x'',\xi''\rangle-\langle B\xi'',\xi''\rangle/2.$$

Thus (21.6.16) and (21.6.20) become apart from the volume factor $|d\xi'|\,|dx'''|$

$$u(x)=a(2\pi)^{-(n+2n'+2n'')/4}\int e^{iQ(x,\xi)}\,d\xi'\,d\xi'',$$
$$u_\Delta(x'')=a(2\pi)^{-(n+2n'+2n'')/4}\int e^{i(\langle x'',\xi''\rangle-\langle B\xi'',\xi''\rangle/2)}\,d\xi''$$

with the coordinates ξ', x''' in $\Delta=\lambda\cap\Delta$. Since $\partial Q/\partial(\xi',\xi'')=(x',x''-B\xi'')$ the set $C=\{(x,\xi',\xi''),\partial Q/\partial(\xi',\xi'')=0\}$ is parametrized by ξ', ξ'', x''' and $d_C=|d\xi'|\,|d\xi''|\,|dx'''|$. The positive half density on λ associated with u is then by (21.6.17)'

$$|a|\,|d\xi'|^{\frac12}\,|d\xi''|^{\frac12}\,|dx'''|^{\frac12}.$$

That associated with u_Δ on λ_Δ is $|a|\,(2\pi)^{-(3n'+n''')/4}\,|d\xi''|^{\frac12}$, which corresponds with the coordinates used to

$$|a|\,(2\pi)^{-(3n'+n''')/4}\,|d\xi''|^{\frac12}\,|d\xi'|\,|dx'''|\,|d\xi'|^{-\frac12}\,|dx'''|^{-\frac12}\in\Omega(\lambda_\Delta)^{\frac12}\otimes\Omega(\lambda\cap\Delta)\otimes\Omega(\Delta)^{-\frac12}.$$

This differs by the factor

$$(2\pi)^{-(3n'+n''')/4}=(2\pi)^{(\dim\Delta-2\dim\Delta\cap\lambda_0-2\dim\Delta\cap\lambda)/4}$$

from the identification of $\Omega(\lambda)^{\frac12}$ with $\Omega(\lambda_\Delta)^{\frac12}\otimes\Omega(\Delta)^{\frac12}$ given by (21.6.21) and (21.6.22).

b) If $\lambda\cap\Delta=\{0\}$ we can choose the coordinates so that Δ is the $\xi' x'''$ plane and λ is transversal to the plane $\xi=0$. To do so we observe again that

$$\{0\}\subset\Delta\cap\lambda_0\subset\Delta^\sigma\cap\lambda_0\subset\lambda_0.$$

Choose a basis $\varepsilon_1,\dots,\varepsilon_k$ for $\Delta\cap\lambda_0$ and extend it to a basis $\varepsilon_1,\dots,\varepsilon_k$, $e_{k+l+1},\dots,e_n$ for Δ, where $2l=\dim S_\Delta$. In S_Δ we can find $e^\Delta_{k+1},\dots,e^\Delta_{k+l}$ spanning a plane transversal both to $\lambda_{0\Delta}$ and to λ_Δ. Let $\varepsilon^\Delta_{k+1},\dots,\varepsilon^\Delta_{k+l}$ be a biorthogonal basis for $\lambda_{0\Delta}$, and choose $e_{k+1},\dots,e_{k+l}\in\Delta^\sigma$ and $\varepsilon_{k+1},\dots,\varepsilon_{k+l}\in\Delta^\sigma\cap\lambda_0$ in the classes of $e^\Delta_{k+1},\dots,\varepsilon^\Delta_{k+l}$. Then $\varepsilon_1,\dots,\varepsilon_{k+l}$,

$e_{k+1},\dots,e_n$ form a partial symplectic basis in S, and $\varepsilon_1,\dots,\varepsilon_{k+l}$ span $\Delta^\sigma\cap\lambda_0$. If μ is the isotropic plane spanned by $e_{k+1},\dots,e_n$ then $\mu\cap\lambda\subset\Delta\cap\lambda=\{0\}$ and $\mu\cap\lambda_0=\{0\}$ since $\mu\cap\lambda_0\subset\Delta\cap\lambda_0$ and $\varepsilon_1,\dots,\varepsilon_k$, $e_{k+1},\dots,e_n$ are linearly independent. To choose $e_1,\dots,e_k$ we observe that $\varepsilon_1,\dots,\varepsilon_k$ are in μ^σ and linearly independent modulo μ. Thus we can find $e_1^\mu,\dots,e_k^\mu\in S_\mu$ biorthogonal to the classes of $\varepsilon_1,\dots,\varepsilon_k$ and spanning a Lagrangian plane in S_μ transversal to λ_μ and to $\lambda_{0\mu}$. Now we pick $e_1,\dots,e_k\in\mu^\sigma$ in the classes of $e_1^\mu,\dots,e_k^\mu$ and orthogonal to $\varepsilon_{k+1},\dots,\varepsilon_{k+l}$, which is possible since $e_{k+1},\dots,e_{k+l}\in\mu$. This extends the partial symplectic basis, and the plane spanned by $e_1,\dots,e_n$ is transversal to λ and to λ_0 since the intersections are contained in μ. Because the Lagrangian plane spanned by $e_1,\dots,e_n$ is transversal to λ_0 the remaining elements $\varepsilon_{k+l+1},\dots,\varepsilon_n$ in a symplectic basis can be chosen uniquely in λ_0.

With the coordinates chosen λ is of the form $x=B\xi$ and (21.6.6) becomes

$$u(x)=c(2\pi)^{-3n/4}\int e^{i(\langle x,\xi\rangle-\langle B\xi,\xi\rangle/2)}\,d\xi.$$

We have

$$u_\Delta(x'')=c(2\pi)^{-3n/4}\iint e^{i(\langle x'',\xi''\rangle+\langle x''',\xi'''\rangle-\langle B\xi,\xi\rangle/2)}\,dx'''\,d\xi.$$

The phase function $Q(x'',x''',\xi)=\langle x'',\xi''\rangle+\langle x''',\xi'''\rangle-\langle B\xi,\xi\rangle/2$ is non-degenerate, for

$$\begin{aligned}C&=\{(x'',x''',\xi);\ \partial Q/\partial x'''=0,\ \partial Q/\partial\xi=0\}\\&\cong\{(B\xi,\xi);\ \xi'''=0,\ (B\xi)'=0\}=\lambda\cap\Delta^\sigma\end{aligned}$$

has dimension n''. We can take ξ'' as coordinates there, hence

$$\begin{aligned}d_C&=|d\xi''|\,|D(\partial Q/\partial x'',\partial Q/\partial x''',\partial Q/\partial\xi)/D(x'',x''',\xi)|^{-1}\\&=|d\xi''|\,|\det\partial^2Q/\partial\xi'\partial\xi'|^{-1}.\end{aligned}$$

The absolute value of the half density in λ_Δ associated with u_Δ is

$$(2\pi)^{(n'''-n')/4}|d\xi''|^{\frac12}|\det\partial^2Q/\partial\xi'\partial\xi'|^{-\frac12}$$

for $-3n/4+(n''+2n+2n''')/4=(n'''-n')/4$. With u we have associated the half density

$$c|d\xi|^{\frac12}=c|d\xi'|^{\frac12}|d\xi''|^{\frac12}|d\xi'''|^{\frac12}$$

on λ parametrized by ξ as $\{(B\xi,\xi)\}$. If we take x',ξ'',ξ''' as parameters instead, noting that $|Dx'/D\xi'|=|\det\partial^2Q/\partial\xi'\partial\xi'|$ then

$$c|d\xi|^{\frac12}=c|\det\partial^2Q/\partial\xi'\partial\xi'|^{-\frac12}|dx'|^{\frac12}|d\xi''|^{\frac12}|d\xi'''|^{\frac12}.$$

These variables are useful since Δ^σ is defined by $x'=\xi'''=0$; the quotient $\lambda/(\lambda\cap\Delta^\sigma)$ is thus parametrized by x' and ξ''' which in (21.6.22) are paired with ξ' and x''' by means of the symplectic form, identifying $|dx'|^{\frac12}|d\xi'''|^{\frac12}$ with $|d\xi'|^{-\frac12}|dx'''|^{-\frac12}$. According to (21.6.21), (21.6.22) one thus identifies $c|d\xi|^{\frac12}$ with

$$c|\det\partial^2Q/\partial\xi'\partial\xi'|^{-\frac12}|d\xi'|^{-\frac12}|dx'''|^{-\frac12}|d\xi''|^{\frac12}$$

where ξ'' is the parameter on $\lambda\cap\Delta^\sigma\approx\lambda_\Delta$. Apart from the factor

$$(2\pi)^{(n'''-n')/4}=(2\pi)^{(\dim\Delta-2\dim\Delta\cap\lambda_0)/4}$$

this agrees with the density obtained from u_Δ. Summing up, we have proved:

Theorem 21.6.6. *Let Δ be an isotropic subspace of the symplectic vector space S and let $S_\Delta=\Delta^\sigma/\Delta$ with the induced symplectic structure. If $\lambda\in\Lambda(S)$ then $\lambda_\Delta=(\lambda\cap\Delta^\sigma)/(\lambda\cap\Delta)\in\Lambda(S_\Delta)$. There is a natural map*

$$M(\lambda)\otimes\Omega(\lambda)^{\frac12}\to M(\lambda_\Delta)\otimes\Omega(\lambda_\Delta)^{\frac12}\otimes\Omega(\lambda\cap\Delta)\otimes\Omega(\Delta)^{-\frac12}$$

which in symplectic coordinates $x', x'', x''', \xi', \xi'', \xi'''$ such that Δ is the $x'\xi'''$ plane maps the symbol of (21.6.16) *to that of* (21.6.20) *multiplied by $|d\xi'|^{-\frac12}|dx'''|^{-\frac12}$. It is a bundle map when the dimension of $\lambda\cap\Delta$ is constant, and it induces maps*

$$M(\lambda)\to M(\lambda_\Delta),\qquad \Omega(\lambda)^{\frac12}\to\Omega(\lambda_\Delta)^{\frac12}\otimes\Omega(\Delta)^{-\frac12}\otimes\Omega(\lambda\cap\Delta).$$

The latter is equal to the isomorphism defined by (21.6.21), (21.6.22) *multiplied by*

$$(2\pi)^{(\dim\Delta-2\dim\Delta\cap\lambda_0-2\dim\Delta\cap\lambda)/4}. \tag{21.6.23}$$

Our main application of Theorem 21.6.6 concerns the composition of linear canonical relations defined in Theorem 21.2.14.

Theorem 21.6.7. *Let S_j be a symplectic vector space with symplectic form σ_j, $j=1,2,3$, and distinguished Lagrangians λ_{j0}, and let $G_1\subset S_1\times S_2$, $G_2\subset S_2\times S_3$ be linear canonical relations from S_2 to S_1 and S_3 to S_2 respectively, that is, Lagrangians for $\sigma_1-\sigma_2$ and for $\sigma_2-\sigma_3$. Set $G=G_1\circ G_2\subset S_1\times S_3$ and $N=\{\gamma\in S_2;\ (0,\gamma)\in G_1,\ (\gamma,0)\in G_2\}$. Then* (21.6.16), (21.6.20) *give rise to natural bilinear maps*

$$M(G_1)\times M(G_2)\to M(G),\qquad \Omega(G_1)^{\frac12}\times\Omega(G_2)^{\frac12}\to\Omega(G)^{\frac12}\otimes\Omega(N).$$

The latter map is $(2\pi)^{-e/2}$ times the map derived from (21.6.21), (21.6.22) *where $e=\dim N$ is the excess of the (clean) composition, and $\Omega(\Delta(S_2))^{\frac12}$ has been trivialized by means of the symplectic form in S_2.*

Proof. This is the special case of Theorem 21.6.6 where Δ is the diagonal in S_2 times 0 in S_1 and in S_3. We have identified $\Omega(\Delta)$ with $\mathbb{C}$ by means of the natural density defined by $\sigma_2^{n_2}/n_2!$ in $S_2\cong\Delta$. Note that (21.6.23) simplifies because $\dim\Delta=2\dim\Delta\cap\lambda_0$.

The map of densities in Theorem 21.6.7 simplifies a great deal if one of the relations, say G_1, is the graph of a linear canonical transformation. Indeed, with $\lambda=G_1\times G_2$ and $\Delta=\Delta(S_2)$ we have $\lambda\cap\Delta=\{0\}$ so (21.6.21) is just the identification of λ_Δ with $\lambda \cap \Delta^\sigma = G_2$. (21.6.22) expresses the duality between Δ and $\lambda/(\lambda \cap \Delta^\sigma)$ given by the symplectic form. Now

$$\lambda = (G_1 \times \{0\}) \oplus (\lambda \cap \Delta^\sigma)$$

because $(\gamma_1, \gamma_2, \gamma_2', \gamma_3) \in \lambda$ implies $(G_1 \gamma_2', \gamma_2', \gamma_2', \gamma_3) \in \lambda \cap \Delta^\sigma$. The duality is therefore the self duality of S_2 when Δ and G_1 are both identified with S_2. Thus the identification of $\Omega(\Delta)^{\frac{1}{2}}$ with $\mathbb{C}$ means that we identify $\Omega(G_1)^{\frac{1}{2}} \cong \mathbb{C}$ using the natural density defined by σ_2 in S_2 (or σ_1 in S_1), lifted to G_1. Multiplication of $d_1 \in \Omega^{\frac{1}{2}}(G_1)$ and $d_2 \in \Omega^{\frac{1}{2}}(G_2)$ therefore means that d_1 is identified with a constant c_1 and that one forms $c_1 d_2 \in \Omega(G_2)^{\frac{1}{2}} \cong \Omega(G)^{\frac{1}{2}}$.

As in the discussion preceding Definition 21.2.15 one can consider the general representation (21.6.16) of elements in $M(\lambda) \otimes \Omega(\lambda)^{\frac{1}{2}}$ as a special case of the map in Theorem 21.6.6. However, this argument would now be circular since (21.6.16) was used when we defined the map in Theorem 21.6.6.

We have insisted on invariance throughout this section in order to have an extension to symplectic vector bundles immediately available. Thus assume that Y is a manifold and $E \to Y$ a symplectic vector bundle with fiber dimension $2n$ over Y. This means that every point in Y is assumed to have a neighborhood U where there is a vector bundle isomorphism

$$U \times T^*(\mathbb{R}^n) \to E|_U$$

respecting the symplectic structure. It is clear that the set $\Lambda(E)$ of Lagrangian planes in the fibers of E forms another fiber bundle on Y with fiber $\Lambda(T^*(\mathbb{R}^n))$. Assume now that we have a distinguished section λ_0 of $\Lambda(E)$. Then the Maslov line bundles M defined on the fibers $\Lambda_y(E)$ define a line bundle $M(E) \to \Lambda(E)$. In fact, we can trivialize E locally so that λ_0 corresponds to the fiber of $T^*(\mathbb{R}^n)$ at 0, and this gives a local trivialization of $M(E)$. Different trivializations are obviously compatible. If now λ is any other section of $\Lambda(E)$ we can pull $M(E)$ back to a line bundle $\lambda^* M(E)$ on Y which will also be called the Maslov bundle.

The typical situation we have in mind is a Lagrangian submanifold $Y \subset T^*(X)$ where X is a C^∞ manifold. Then the tangent spaces of $T^*(X)$ define a symplectic vector bundle E on Y and the tangents of the fibers in $T^*(X)$ give a distinguished Lagrangian section. We also have the Lagrangian section of E given by the tangent planes of Y. This defines on Y a well defined Maslov line bundle, with structure group $\mathbb{Z}_4$ of course. It is needed for the invariant (and global) statements in Chapter XXV.

So far in this section we have emphasized the analytical origin of the Maslov bundle. To clarify its meaning we shall now supplement this with a brief purely geometrical discussion although this will not be required in the further developments here. In doing so we shall restrict ourselves to a symplectic vector space S again. We can take $S=T^*(\mathbb{R}^n)$ with the usual symplectic coordinates x, ξ.

i) Let $n=1$. Every line in $T^*(\mathbb{R})$ is Lagrangian so $\Lambda(S)$ is the projective line $P_1(\mathbb{R})$. A generator for $\pi_1(\Lambda(S))$ is therefore obtained by turning the x axis in the positive direction an angle of π. Let U_j, $j=1,2$, be the set of lines transversal to λ_j where λ_1 is defined by $\xi=0$ and λ_2 is defined by $\xi=x$. It is clear that U_1 and U_2 cover $\Lambda(S)$. When $\lambda \in U_1 \cap U_2$ we can define λ by $x=B\xi$ where $B \neq 1$. The integer (21.6.10)

$$\sigma(\lambda_0, \lambda; \lambda_2, \lambda_1)=\tfrac{1}{2}\operatorname{sgn}\begin{pmatrix} -1 & 1 \\ 1 & -B \end{pmatrix}$$

is half the signature of the quadratic form $-x^2+2x\xi-B\xi^2$, that is, -1 if $B>1$ and 0 if $B<1$. Thus the Maslov 1 cocycle is defined by the cocycle $c_{21}(B)=-1$ if $B>1$, $c_{21}(B)=0$ if $B<1$. We can now compute the change of an element in the Maslov bundle as we go around $P_1(\mathbb{R})$ once, that is, follow the generator for $\pi_1(P_1(\mathbb{R}))$ described above. If we have the value c in the trivialization corresponding to U_1 then the value in U_2 on the line λ defined by $x=B\xi$ is c/i if B is large positive and c when B is large negative; when B goes from $+\infty$ to $-\infty$ the element in the Maslov bundle will therefore be multiplied by i. We could make the same calculation with the cohomology class in $H^1(\Lambda(S))$ defined by σ and conclude that its intersection with the generator of $\pi_1(\Lambda(S))$ is equal to 1.

ii) When $n>1$ we shall prove that every closed C^1 curve in $\Lambda(S)$ is homotopic to a sum of closed curves of the form $\lambda_1(t)\times\lambda_2$ where $\lambda_1(t) \subset \Lambda(T^*(\mathbb{R}))$ is the path described in i) and λ_2 is a fixed Lagrangian plane in $T^*(\mathbb{R}^{n-1})$. The choice of λ_2 is unimportant from the point of view of homotopy, for it follows from Lemma 21.6.2 that $\Lambda(T^*(\mathbb{R}^{n-1}))$ is connected. Let $t\mapsto\lambda(t)$, $\lambda(t+1)=\lambda(t)$, be the given C^1 curve in $\Lambda(S)$. We shall first prove that it is homotopic to a curve intersecting λ_0 in a very simple way. For any t_0 we can choose $\varepsilon>0$ and suitable symplectic coordinates such that λ_0 is still defined by $x=0$ and

$$\lambda(t)=\{(A(t)\xi, \xi)\}, \quad |t-t_0|\leqq\varepsilon,$$

where A is a symmetric matrix which is a C^1 function of t. Let us examine the intersection with a Lagrangian plane $x=B\xi$ where B is symmetric. The intersections are given by $(B-A(t))\xi=0$, so we must ask if $B=A(t)+B_k$ where B_k is of rank $k<n$. The dimension of the manifold M_k of all symmetric B_k of rank k was essentially determined in Lemma 21.6.3; to choose B_k one must first choose a subspace of codimension k and then a non-degenerate quadratic form in the orthogonal space, so the dimension is

$$k(n-k)+k(k+1)/2=n(n+1)/2-(n-k)(n-k+1)/2.$$

From the Morse-Sard theorem it follows that the range of the map $[t_0-\delta, t_0+\delta]\times M_k \ni (t, B_k)\mapsto A(t)+B_k$ is of measure 0 if $k<n-1$, and so is the set of critical values when $k=n-1$. If B is not in these sets and $\det(B-A(t))=0$ for some t with $|t-t_0|\leqq\delta$ it follows that the rank of $B-A(t)$ is $n-1$

and that $A'(t)$ is not a tangent to M_{n-1} at $B-A(t)$, that is, the derivative of $\det(B-A(t))$ with respect to t is not equal to 0. Now the maps

$$(x,\xi)\mapsto(x-\varepsilon B\xi,\xi),\quad 0\leqq\varepsilon\leqq 1,$$

are symplectic. If we use them to deform $\lambda(t)$ we obtain the desired simple intersection with λ_0 when $|t-t_0|\leqq\delta$, and choosing B small enough we do not disturb such a property already achieved in another closed set. After a finite number of such deformations we find that every homotopy class contains a curve such that

a) $\lambda(t)\cap\lambda_0=\{0\}$ except for finitely many values $t_1,\dots,t_N$ of $t \bmod 1$, and in a neighborhood of each of these points $\lambda(t)$ can be written in the form

$$x=A(t)\xi.$$

b) If $A(t)$ is singular then $A(t)$ has rank $n-1$ and $\det A(t)$ has a simple zero.

Indeed, in order to obtain the second property in a) we just have to choose as the plane $\xi=0$ a Lagrangian plane transversal to λ_0 and a finite number of other Lagrangian planes.

Except in a neighborhood of $t_1,\dots,t_N \bmod 1$ we can represent $\lambda(t)$ in the form $\xi=B(t)x$. Choose $\chi_j\in C^\infty$ with $0\leqq\chi_j\leqq 1$, periodic with period 1, disjoint supports and $\chi_j=1$ in a neighboorhood of t_j. Set $\chi=\Sigma\chi_j$. Then the curves λ^ε defined by

$$\lambda^\varepsilon(t)=\{(x,(\varepsilon\chi(t)+(1-\varepsilon))B(t)x)\},\quad 0\leqq\varepsilon\leqq 1,$$

where $\chi\neq 1$, and $\lambda^\varepsilon(t)=\lambda(t)$ where $\chi(t)=1$, give a homotopy which connects the curve $t\mapsto\lambda(t)$ to a finite number of loops starting and ending at $\lambda_1=\{(x,0)\}$, each containing only one Lagrangian not transversal to λ_0. It remains to examine a curve with only one such point.

Assume as in b) that $\det A(t)$ has a simple zero when $t=0$, and let $A(0)\xi_0=0$. We can choose coordinates so that ξ_0 is the first unit vector, that is,

$$A(0)=\begin{pmatrix}0 & 0\\ 0 & A_0\end{pmatrix},\quad \det A_0\neq 0,$$

where A_0 is a symmetric $(n-1)\times(n-1)$ matrix. The derivative of $\det A$ is $A'_{11}(0)\det A_0$ so $A'_{11}(0)\neq 0$. The sign has a symplectically invariant interpretation. In fact, if $\xi(t)\in C^1$ and $\xi(0)=(1,\dots,0)$ then

$$\sigma(0,\xi(0),A(t)\xi(t),\xi(t))=\langle A(t)\xi(t),\xi(0)\rangle$$

has the derivative $A'_{11}(0)$ at 0. Hence

$$\operatorname{sgn}A'_{11}(0)=\operatorname{sgn}\frac{d}{dt}\sigma(\gamma(0),\gamma(t))|_{t=0} \tag{21.6.24}$$

if $t\mapsto\gamma(t)$ is a curve with $\gamma(t)\in\lambda(t)$ and $0\neq\gamma(0)\in\lambda_0$ also.

We may make a further homotopy of A near 0 so that in a neighborhood we obtain

$$A(t) = \begin{pmatrix} A'_{11}(0)t & 0 \\ 0 & A_0 \end{pmatrix}.$$

The deformation above can then be made in the variables $\xi_2, \dots, \xi_n$ for all t and gives a homotopy to $\lambda(t)$ defined by

$$\lambda(t) = \{(x, \xi_1, 0);\ \xi_1 t A'_{11}(0) = \chi(t) x_1\}$$

when $|t| < 1/2$ and periodically continued to $\mathbb{R}$. This is the example in i), perhaps with a change of sign. We have therefore proved:

Theorem 21.6.8. *The first homotopy group $\pi_1(\Lambda(S))$ is isomorphic to $\mathbb{Z}$. Every homotopy class contains a curve $t \mapsto \lambda(t)$ such that $\lambda(t) \cap \lambda_0 = \{0\}$ except for a finite number of values of t where the intersection is simple and has a well defined sign given by* (21.6.24). *The sum of the signs gives the isomorphism with $\mathbb{Z}$. The intersection of the corresponding homology class with the Maslov cohomology class defined by* (21.6.10) *is equal to this integer ν; going around the curve multiplies the elements in the Maslov line bundle by i^ν.*

Notes

Most of Section 21.1 comes from classical mechanics and can be found in another form in, for example, Carathéodory [1]. The theory of Fourier integral operators created new interest in these matters. A special case of Theorem 21.1.9 for example can be found in Duistermaat-Hörmander [1], and a more general version was given by Melrose [2]. There is a great deal in the literature about degenerate symplectic structures, but we have only included what is directly related to the equivalence theorem of Melrose [2]. Also Section 21.2 begins with classical results. The importance of allowing clean intersections was stressed by Duistermaat-Guillemin [1] and Weinstein [3]. The discussion of clean phase functions is taken from these papers. The non-degenerate case was discussed in Hörmander [26] and goes back to classical results on generating functions of canonical transformations.

The normal form in Theorem 21.3.2 was established by Duistermaat-Hörmander [1]. That in Theorem 21.3.3 is due to Sato-Kawai-Kashiwara [1] in the analytic case and Duistermaat-Sjöstrand [1] in the C^∞ case. For the extension in Theorem 21.3.5 see also Yamamoto [1]. Theorem 21.3.6 is a key result in Nirenberg-Trêves [2].

The equivalence of glancing hypersurfaces was conjectured by Sato in the analytic case but Oshima [1] gave a counterexample. In the C^∞ case the result was proved true by Melrose [2], and Section 21.4 mainly follows his paper. The results on canonical relations with folds are due to Melrose and

Taylor (see Taylor [3]); they are also important in the study of mixed problems.

The symplectic reduction of quadratic forms is an old topic in mechanics, for it occurs in the study of Hamiltonian systems at an equilibrium point. The presentation here depends on Melin [1], Boutet de Monvel [4], Sjöstrand [3], Ivrii-Petkov [1] and Hörmander [36].

The Maslov index studied in Section 21.6 was defined by Maslov [1] in connection with his canonical operators. It expresses the classical observation in geometrical optics that a phase shift of $\pi/2$ takes place at a caustic (see Section 12.2), and it was formulated quite explicitly already by Keller [1]. Arnold [1] gave a clear presentation of the Maslov index which was rephrased somewhat in Hörmander [26]. The presentation here is modified in order to fit the case of Lagrangians with clean intersections. For the case of complex Lagrangians we refer to Melin-Sjöstrand [1] and Wang-Tsui [1].

Chapter XXII. Some Classes of (Micro-)Hypoelliptic Operators

Summary

In Chapter XI we have proved that a differential operator $P(D)$ with constant coefficients is hypoelliptic if and only if for some $\rho>0$

$$P^{(\alpha)}(\xi)/P(\xi)=O(|\xi|^{-\rho|\alpha|}) \quad \text{for every } \alpha \text{ when } \xi\to\infty \text{ in } \mathbb{R}^n. \tag{22.1}$$

Recall that $P(D)$ is called hypoelliptic if

$$\operatorname{sing\,supp} u=\operatorname{sing\,supp} P(D)\,u, \qquad u\in\mathscr{D}'. \tag{22.2}$$

For operators with constant coefficients (22.2) is equivalent to microhypoellipticity,

$$WF(u)=WF(P(D)\,u), \qquad u\in\mathscr{D}'. \tag{22.3}$$

Microhypoellipticity implies hypoellipticity but the converse is not always true. We shall mainly consider microhypoellipticity here.

In Chapter XIII we extended the results just mentioned to operators of constant strength. These are differential operators $P(x, D)$ such that

$$|D_\xi^\alpha D_x^\beta P(x,\xi)|\leqq C_{\alpha\beta}|P(x,\xi)|\,|\xi|^{-\rho|\alpha|} \quad \text{for large } |\xi|. \tag{22.1$'$}$$

However, the perturbation methods based on pseudo-differential operators are much more powerful than those in Chapter XIII. This will be seen in Section 22.1 where we improve the results of Section 13.4 by studying pseudo-differential operators (of type ρ, δ) with a pseudo-differential paramatrix. These results still have serious flaws though. One is that pseudo-differential operators of type ρ, δ are not always invariant under changes of variables so the classes of hypoelliptic operators obtained are not invariantly defined either. A striking example is given how the hypoellipticity of the heat equation may be masked by a change of variables from the point of view of the criteria in Section 22.1. Another flaw is that the results proved in Section 22.1 do not cover some simple types of hypoelliptic operators occurring in probability theory such as the Kolmogorov equation (7.6.13) for which we have already given an elementary proof of hypoellipticity.

Section 22.2 is therefore devoted to a study of "generalized Kolmogorov operators" of the form

$$\sum X_j^2 + X_0$$

where X_j are real vector fields. It is proved that such operators are hypoelliptic if the Lie algebra generated by all X_j gives a basis for all vector fields at any point. The methods used in the proof can also be applied to pseudo-differential operators with non-negative principal symbol p_m under similar symplectically invariant conditions on p_m and on the subprincipal symbol p_{m-1}^s. It is shown that if $Pu \in H_{(s)}$ at some point then $u \in H_{(s+m-2+\varepsilon)}$ there for an $\varepsilon \in (0, 1]$; it is customary to say that one has hypoellipticity with loss of $2-\varepsilon$ derivatives then, since this is the deficit compared to the elliptic case. In the proof we use the Fefferman-Phong lower bound of pseudo-differential operators proved in Section 18.6 but otherwise we shall only rely on facts concerning pseudo-differential operators contained in Section 18.1 until we reach Section 22.4. There the general theory of Sections 18.5 and 18.6 will be used freely.

The lower bound of pseudo-differential operators due to Fefferman and Phong is very precise in some respects and very weak in others. In Section 22.3 we therefore prove another lower bound due to Melin which gives necessary and sufficient conditions for a first order operator to majorize a positive constant operator. This estimate improves the theorems on hypoellipticity in Section 22.2 for the case of loss of exactly one derivative. Complete results on such operators with principal symbol taking values in a proper convex angle in $\mathbb{C}$ are proved in Section 22.4. As already mentioned, the proofs there are less elementary and Section 22.4 will not be referred to in later chapters. However, the methods are interesting and will reappear later in Chapters XXVI and XXVII in related contexts.

22.1. Operators With Pseudo-Differential Parametrix

Let X be an open set in $\mathbb{R}^n$ and $P=(P_{jk})_{j,k=1,\dots,N}$ a $N\times N$ matrix of properly supported pseudo-differential operators $P_{jk}\in\Psi^{\infty}_{\rho,\delta}$ for some ρ, δ with $0\leqq\delta<\rho\leqq 1$. (See the end of Section 18.1.) Under hypotheses similar to (22.1) we want to prove that P is hypoelliptic, that is, that solutions u of the system

$$\sum_{k=1}^{N} P_{jk} u_k = f_j, \quad j=1,\dots,N,$$

are smooth where f is smooth. To state the hypotheses on the symbol $p(x,\xi)\in S^{\infty}_{\rho,\delta}$ of P we introduce a notion of temperate norm on $\mathbb{C}^N$, parametrized by $T^*(X)$: A norm $\|\ \|_{x,\xi}$ on $\mathbb{C}^N$ defined for every $(x,\xi)\in T^*(X)$ will be called temperate if for every compact set $K\subset X$ there are constants C

and M such that with $\|z\|=(\sum|z_j|^2)^{\frac{1}{2}}$

(22.1.1) $C^{-1}(1+|\xi|)^{-M}\|z\| \leqq \|z\|_{x,\xi} \leqq C(1+|\xi|)^M\|z\|; \quad x\in K,\ \xi\in\mathbb{R}^n,\ z\in\mathbb{C}^N.$

Definition 22.1.1. A symbol $p(x,\xi)\in S^{\infty}_{\rho,\delta}(X\times\mathbb{R}^n)$ with values in $L(\mathbb{C}^N,\mathbb{C}^N)$ is said to be hypoelliptic if there exist two temperate $\mathbb{C}^N$ norms $\|\ \|'_{x,\xi}$ and $\|\ \|''_{x,\xi}$ parametrized by $T^*(X)$ such that for every compact set $K\subset X$ there are constants C, $C_{\alpha\beta}$ with

(22.1.2) $\|z\|'_{x,\xi} \leqq C\|p(x,\xi)z\|''_{x,\xi}; \quad x\in K,\ |\xi|>C,\ z\in\mathbb{C}^N;$

(22.1.3) $\|p^{(\alpha)}_{(\beta)}(x,\xi)z\|''_{x,\xi} \leqq C_{\alpha\beta}(1+|\xi|)^{-\rho|\alpha|+\delta|\beta|}\|z\|'_{x,\xi}; \quad x\in K.$

It would not really have been necessary to assume $p\in S^{\infty}_{\rho,\delta}$ for this is a consequence of (22.1.3). To clarify the meaning of these conditions we assume first that $N=1$. Then $\|z\|'_{x,\xi}=M(x,\xi)\|z\|''_{x,\xi}$ and the conditions are

$$M(x,\xi)\leqq C|p(x,\xi)|, \quad |p^{(\alpha)}_{(\beta)}(x,\xi)|\leqq C_{\alpha\beta}M(x,\xi)(1+|\xi|)^{-\rho|\alpha|+\delta|\beta|}$$

so $M(x,\xi)$ must be equivalent to $|p(x,\xi)|$ for large $|\xi|$, and

$$|p^{(\alpha)}_{(\beta)}(x,\xi)|\leqq C_{\alpha\beta}|p(x,\xi)|(1+|\xi|)^{-\rho|\alpha|+\delta|\beta|}, \quad x\in K,\ |\xi|>C.$$

This weakened form of (22.1)′ together with polynomial bounds for p and $1/p$ for large $|\xi|$ is equivalent to the hypoellipticity in the definition.

If we choose

$$\|z\|'_{x,\xi}=(1+|\xi|)^t\|z\|, \quad \|z\|''_{x,\xi}=(1+|\xi|)^s\|z\|,$$

the condition (22.1.3) is that $p\in S^{t-s}_{\rho,\delta}$ and (22.1.2) means that $1/p\in S^{s-t}_{\rho,\delta}$ for large $|\xi|$. More generally, we could split the z variables into several groups for which we use different exponents s and t. This grading leads to the Agmon-Douglis-Nirenberg elliptic systems where the notion of principal symbol depends on the location in the matrix symbol. (See Section 19.5.) – Here we have tacitly assumed the following simple but important lemma.

Lemma 22.1.2. *If the hypotheses of Definition* 22.1.1 *are fulfilled, then*

(22.1.4) $\|D^\alpha_\xi D^\beta_x p(x,\xi)^{-1}z\|'_{x,\xi}\leqq C'_{\alpha\beta}(1+|\xi|)^{-\rho|\alpha|+\delta|\beta|}\|z\|''_{x,\xi}; \quad x\in K,\ |\xi|>C.$

Proof. (22.1.4) concides with (22.1.2) when $\alpha=\beta=0$ and will be proved in general by induction for increasing $|\alpha+\beta|$. Differentiation of the equation

$$p(x,\xi)\,p(x,\xi)^{-1}=I$$

gives when $|\xi|>C$

$$p(x,\xi)\,D^\alpha_\xi D^\beta_x p(x,\xi)^{-1}=-\sum c_{\alpha'\beta'}D^{\alpha'}_\xi D^{\beta'}_x p(x,\xi)\,D^{\alpha''}_\xi D^{\beta''}_x p(x,\xi)^{-1}$$

where $\alpha'+\alpha''=\alpha$, $\beta'+\beta''=\beta$ and $|\alpha''+\beta''|<|\alpha+\beta|$ in the sum. By the inductive hypothesis and (22.1.3) we obtain

$$\|p(x,\xi)\,D^\alpha_\xi D^\beta_x p(x,\xi)^{-1}z\|''_{x,\xi}\leqq C''_{\alpha\beta}(1+|\xi|)^{-\rho|\alpha|+\delta|\beta|}\|z\|''_{x,\xi}$$

so (22.1.4) follows from (22.1.2).

We can now prove an extension of Theorem 18.1.9.

Theorem 22.1.3. *Assume that $P \in \Psi^{\infty}_{\rho,\delta}(X; \mathbb{C}^N, \mathbb{C}^N)$ is properly supported and that the symbol satisfies the hypotheses in Definition 22.1.1 with $0 \leqq \delta < \rho \leqq 1$. Then one can find Q satisfying the same hypotheses with the norms in Definition 22.1.1 interchanged and*

$$PQ = I + R_1, \qquad QP = I + R_2$$

where R_1 and R_2 are of order $-\infty$.

Proof. As in the proof of Theorem 18.1.9 it is sufficient to construct Q_1, Q_2 with $PQ_1 = I + R_1$, $Q_2 P = I + R_2$ and R_1, R_2 of order $-\infty$, for $Q_1 - Q_2$ is then automatically of order $-\infty$. By Lemma 22.1.2 we can choose Q_0 properly supported with symbol $p(x,\xi)^{-1}$ for large $|\xi|$. The symbol of

$$R_0 = PQ_0 - I$$

has the asymptotic expansion

$$r_0(x,\xi) \sim \sum_{\alpha \neq 0} (iD_\xi)^\alpha p(x,\xi)\, D_x^\alpha p(x,\xi)^{-1}/\alpha!$$

so using (22.1.3), (22.1.4) we obtain

$$\|r^{(\alpha)}_{0(\beta)}(x,\xi)\, z\|''_{x,\xi} \leqq C_{\alpha\beta}(1+|\xi|)^{\delta-\rho-\rho|\alpha|+\delta|\beta|}\, \|z\|''_{x,\xi}.$$

The symbol of R_0^k satisfies the same condition with $\delta - \rho$ replaced by $k(\delta - \rho)$ in the exponent. Since $\| \ \|''_{x,\xi}$ is a temperate norm it follows that $R_0^k \in \Psi^{M+k(\delta-\rho)}_{\rho,\delta}$ for some fixed M, at least on a compact set, so there is an operator T with

$$T - \sum_{j<N} (-R_0)^j \in \Psi^{M+N(\delta-\rho)}_{\rho,\delta}$$

for every N. Thus $PQ_0 T = I + R_1$ with R_1 of order $-\infty$. The symbol of $Q_1 = Q_0 T$ satisfies estimates of the form (22.1.4). Similarly we construct Q_2 noting that $Q_0 P = I + R_0'$ where R_0' has the same properties as R_0 with $\| \ \|''$ replaced by $\| \ \|'$. This completes the proof.

The construction of Q_2 could also have been obtained from the construction of Q_1 by passage to the adjoint. In fact, if P satisfies the hypotheses of Theorem 22.1.3 then P^* also satisfies them with respect to the duals of the norms $\| \ \|''_{x,\xi}$ and $\| \ \|'_{x,\xi}$. We leave the simple verification for the reader. Instead we give the main application of Theorem 22.1.3.

Theorem 22.1.4. *Assume that $P \in \Psi^{\infty}_{\rho,\delta}(X; \mathbb{C}^N, \mathbb{C}^N)$ is properly supported and that the symbol is hypoelliptic in the sense of Definition 22.1.1. Then P is microhypoelliptic, that is,*

(22.1.5) $$WF(u) = WF(Pu), \qquad u \in \mathscr{D}'(X, \mathbb{C}^N).$$

Proof. The inclusion

$$WF(Pu) \subset WF(u)$$

is the global ρ, δ version of (18.1.25), which also gives

$$WF(QPu) \subset WF(Pu)$$

if Q is the parametrix in Theorem 22.1.3. Since $QPu - u \in C^\infty$, this proves (22.1.5).

The result in Theorem 22.1.4 can of course be micro-localized further; if (22.1.2), (22.1.3) are only satisfied in an open cone $\Gamma \subset T^*(X) \setminus 0$ then (22.1.5) is valid in Γ. On the other hand, when $1-\rho \leqq \delta < \rho \leqq 1$ the results in this section can be given a global form where X is a manifold and $P \in \Psi^\infty_{\rho,\delta}(X; E, F)$ for two vector bundles E, F on X with equal fiber dimension. The norms $\| \; \|'_{x,\xi}$ and $\| \; \|''_{x,\xi}$ are then interpreted as hermitian norms on the bundles E and F lifted to $T^*(X)$. One just has to show that the conditions in Theorem 22.1.1 are invariant under a change of variables. We leave for the reader to verify that this follows from the ρ, δ version of Theorem 18.1.17. Note that such invariant classes of hypoelliptic operators are only obtained when $\rho > \frac{1}{2}$. The heat operator $\partial^2/\partial x_1^2 - \partial/\partial x_2$ satisfies (22.1) with $\rho = \frac{1}{2}$. If we introduce

$$y_1 = x_1, \quad y_2 = x_2 + x_1^2/2$$

then

$$\begin{aligned}\partial^2/\partial x_1^2 - \partial/\partial x_2 &= (\partial/\partial y_1 + y_1\, \partial/\partial y_2)^2 - \partial/\partial y_2 \\ &= \partial^2/\partial y_1^2 + y_1^2\, \partial^2/\partial y_2^2 + 2 y_1\, \partial^2/\partial y_1\, \partial y_2.\end{aligned}$$

The symbol $-(\eta_1 + y_1 \eta_2)^2$ does not satisfy the conditions in Definition 22.1.1. This shows that the restriction $\rho > \frac{1}{2}$ is essential and also that just comparing the symbol and its derivatives as we have done here is not always a satisfactory method for determining if an operator is hypoelliptic.

On the other hand, Theorem 22.1.4 proves the hypoellipticity of many operators which are very far from the operators of constant strength discussed in Chapter XIII. An example is the symbol

$$p(x, \xi) = c + |x|^{2\nu} |\xi|^{2\mu}$$

where $c > 0$. The hypotheses of Theorem 22.1.4 are fulfilled with $\rho = 1$ and $\delta = \mu/\nu$ if $\mu < \nu$. Since

$$\langle |x|^2 |D|^2 \delta, \phi \rangle = -\Delta |x|^2 \phi(0) = -2n\, \phi(0)$$

we see that $2n + |x|^2 |D|^2$ is not hypoelliptic, so the condition $\mu < \nu$ cannot be omitted.

22.2. Generalized Kolmogorov Equations

As an example of Fourier transforms of Gaussian functions we constructed in Section 7.6 a fundamental solution of the Kolmogorov equation

$$\partial^2 u/\partial x^2 + x\,\partial u/\partial y - \partial u/\partial t = f,$$

which represents Brownian motion on $\mathbb{R}$, x denoting the velocity and y the position of a particle. The fundamental solution is smooth outside the diagonal and one can easily deduce that the Kolmogorov equation is hypoelliptic. It is clear that it does not satisfy the conditions in Definition 22.1.1 for any choice of variables.

In this section we shall study more general equations of the form

$$\left(\sum_1^r L_j^2 + L_0 + c\right) u = f \tag{22.2.1}$$

where L_j are real C^∞ vector fields in the n dimensional manifold X and $c \in C^\infty(X)$. It is easy to see that (22.2.1) cannot be hypoelliptic if $c=0$ and the rank of the Lie algebra generated by $L_0, \dots, L_r$ is $<n$ in an open set Y, that is, the vector fields

$$L_j, [L_{j_1}, L_{j_2}], [L_{j_1}, [L_{j_2}, L_{j_3}]], \dots \tag{22.2.2}$$

have rank $<n$ at every point in Y. In fact, in a neighborhood of a point where the rank k is maximal we can by the Frobenius theorem (Theorem C.1.1 in the appendix) choose local coordinates such that

$$L_j = \sum_1^k a_{j\nu}\,\partial/\partial x_\nu, \quad j=1, \dots, r.$$

All functions of x_n are then solutions of the homogeneous equation (22.2.1). On the other hand, we shall prove that the equation is microhypoelliptic if the rank is n everywhere:

Theorem 22.2.1. *Let L_j, $j=0, \dots, r$, be real C^∞ vector fields in the manifold X generating a Lie algebra of rank* dim X *at every point. Then the equation* (22.2.1) *is microhypoelliptic.*

Example. The operator $\partial^2/\partial x_1^2 + \sum_2^n x_1^{m_j}\,\partial/\partial x_j$ is microhypoelliptic if $0 \leqq m_2 < m_3 < \dots < m_n$ are integers.

The statement is local so we may assume in the proof that $X \subset \mathbb{R}^n$. Set

$$P = -\sum_1^r L_j^2 - L_0 - c.$$

If $L_j(x,\xi)$ is the symbol of L_j/i, which is a real linear form in ξ, then the principal symbol of P is

$$p_2(x,\xi) = \sum L_j(x,\xi)^2.$$

We can write

$$P=\sum_1^r L_j^* L_j+T+c$$

where $L_j^*=-L_j-c_j$ is the adjoint of L_j and $T=-L_0+\sum c_j L_j$. Then T and $L_1, \dots, L_r$ also generate a Lie algebra of rank n. Denote the L^2 norm by $\| \ \|$.

Lemma 22.2.2. *If K is a compact subset of X then*

$$\sum_1^r \|L_j u\|^2 \leqq \operatorname{Re}(Pu, u)+C_K\|u\|^2, \qquad u\in C_0^\infty(K), \tag{22.2.3}$$

$$\sum_1^n \|p_2^{(\nu)}(x,D)u\|_{(0)}^2+\sum_1^n \|p_{2(\nu)}(x,D)u\|_{(-1)}^2 \leqq C_K' \operatorname{Re}(Pu,u)+C_K''\|u\|^2, \quad u\in C_0^\infty(K). \tag{22.2.4}$$

Proof. (22.2.3) follows from the identity

$$\operatorname{Re}(Pu,u)=\sum_1^r \|L_j u\|^2+(((T+T^*)/2+c)u,u)$$

since $T+T^*$ is just multiplication by a function. Now

$$p_2^{(\nu)}=2\sum_1^r \partial L_j/\partial \xi_\nu L_j, \qquad p_{2(\nu)}=2\sum_1^r \partial L_j/\partial x_\nu L_j$$

so $p_2^{(\nu)}(x,D)$ is a linear combination of $L_j(x,D)$ with C^∞ coefficients while

$$p_{2(\nu)}(x,D)-2\sum L_{j(\nu)}(x,D)L_j(x,D)$$

is of order 1. This proves (22.2.4).

We shall prove Theorem 22.2.1 using only the estimates in Lemma 22.2.2 and that p_2 is real valued. This will allow us to generalize the theorem at the end of the section without repeating the proof. Thus we now denote by $P=p(x,D)$ a properly supported pseudo-differential operator of order 2 with real principal symbol p_2 such that (22.2.4) is fulfilled. We denote by Q_1 the set of all properly supported first order operators $q(x,D)$ with real principal symbol such that for every K

$$\|q(x,D)u\|^2 \leqq C_K' \operatorname{Re}(Pu,u)+C_K''\|u\|^2, \quad u\in C_0^\infty(K). \tag{22.2.3)'$$

It is clear that Q_1 contains all operators of order 0 so this is just a condition on the principal symbol. Let E_s be a properly supported self adjoint pseudo-differential operator with symbol $(1+|\xi|^2)^{s/2}$. Then *the assumption* (22.2.4) *means that $P^\nu=p_2^{(\nu)}(x,D)$ and $P_\nu=E_{-1}\,p_{2(\nu)}(x,D)$ are in Q_1.* Let Q_2 be the set consisting of $(P-P^*)/i$ and all commutators $[q,q']/i$ with $q,q'\in Q_1$, and define Q_k successively for $k>2$ as the set of all $[q,q']/i$ with $q\in Q_{k-1}$ and $q'\in Q_1$ or $q\in Q_{k-2}$ and $q'\in Q_2$. In view of the Jacobi identity

$$[q,[q',q'']]=[[q,q'],q'']+[[q'',q],q']$$

we just have to use the element $(P-P^*)/i$ from Q_2 here. The main step in the proof of Theorem 22.2.1 is now the following

Lemma 22.2.3. *If* $q_k \in Q_k$ *and* $\varepsilon \leqq 2^{1-k}$ *we have for every* $K \Subset X$

$$\|q_k u\|_{(\varepsilon-1)} \leqq C(\|Pu\| + \|u\|), \quad u \in C_0^\infty(K). \tag{22.2.5}$$

Proof. (22.2.5) follows from (22.2.3)′ when $k=1$. In the proof for $k>1$ it is convenient to have $\operatorname{Re}(Pu, u) \geqq 0$ which we can always achieve by adding a large constant to P, since we work in a compact set. Write

$$P' = (P+P^*)/2, \quad P'' = (P-P^*)/2i$$

for the real and imaginary parts of P. Thus P' is assumed to be positive. To prove (22.2.5) for $k=2$ and $q_k = P''$ we have to estimate

$$\|E_{-\frac{1}{2}} P'' u\|^2 = (P'' u, Au) = ((P-P')u, Au)/i$$

where $A = E_{-\frac{1}{2}}^2 P''$ is of order 0. It is clear that

$$|(Pu, Au)| \leqq C\|Pu\|\,\|u\|,$$

and Cauchy-Schwarz' inequality gives in view of the positivity of P'

$$|(P' u, Au)| \leqq (P' u, u)^{\frac{1}{2}} (P' Au, Au)^{\frac{1}{2}} \leqq (\operatorname{Re}(Pu, u) + \operatorname{Re}(PAu, Au))/2.$$

The calculus gives

$$[P, A] = \sum (A_j' P_j + A_j'' P^j) + A_0 \tag{22.2.6}$$

where A_j', A_j'' and A_0 are of order 0. Hence

$$\begin{aligned}\operatorname{Re}(PAu, Au) &= \operatorname{Re}(APu, Au) + \operatorname{Re}([P, A]u, Au)\\ &\leqq C(\|Pu\| + \sum \|P_j u\| + \sum \|P^j u\| + \|u\|)\,\|u\|\end{aligned}$$

which proves (22.2.5) for P''.

Before the general inductive proof we observe that (22.2.5) implies

$$\|Bq_k Au\|_{(s)} \leqq C'(\|Pu\| + \|u\|), \quad u \in C_0^\infty(K), \tag{22.2.5$'$}$$

$s + \operatorname{order} A + \operatorname{order} B \leqq \varepsilon - 1$. In fact, the order of $B[q_k, A]$ is at most equal to $\varepsilon - 1 - s \leqq -s$, and by (22.2.5) the desired estimate is obviously valid for $\|BAq_k u\|_{(s)}$. It is also useful to note that

$$\|BPAu\|_{(s)} + \|B[P, A]u\|_{(s)} \leqq C(\|Pu\| + \|u\|), \quad u \in C_0^\infty(K), \tag{22.2.7}$$

if $s + \operatorname{order} A + \operatorname{order} B \leqq 0$. For the commutator this is clear since we have an identity of the form (22.2.6) with A_j', A_j'', A_0 of the same order as A. Since $\|BAPu\|_{(s)}$ has an estimate of the desired form, this proves (22.2.7). In the commutator term we may replace P by P' since $B\,[P'', A]$ is of order $\leqq -s$.

Assume now that (22.2.5) is valid for a certain value of k. To prove (22.2.5) for higher values of k we must show that if $q \in Q_1$ then

$$\|[q, q_k]u\|_{(\varepsilon/2-1)} \leqq C(\|Pu\| + \|u\|), \quad u \in C_0^\infty(K).$$

This means that we must estimate

$$([q,q_k]u,Au)=(q_k u,qAu)-(qu,q_k Au)$$

where $A=E^2_{\varepsilon/2-1}[q,q_k]$ is of order $\varepsilon-1$. (We may assume q,q_k self-adjoint since only the principal symbol is important.) But

$$\|q_k u\|_{(\varepsilon-1)}\|qAu\|_{(1-\varepsilon)}+\|qu\|_{(0)}\|q_k Au\|_{(0)}\leqq C(\|Pu\|+\|u\|)^2$$

by (22.2.5)′, so $[q,q_k]u$ has the desired estimate.

Finally we shall prove that (22.2.5) implies that

$$\|[P'',q_k]u\|_{(\varepsilon/4-1)}\leqq C(\|Pu\|+\|u\|),\quad u\in C_0^\infty(K).$$

This means that we must estimate

$$([P'',q_k]u,Au)=i((P^*-P')q_k u,Au)+i(q_k(P-P')u,Au)$$

where A is of order $\varepsilon/2-1$. The term

$$|(q_k u,PAu)|\leqq\|q_k u\|_{(\varepsilon-1)}\|PAu\|_{(1-\varepsilon)}$$

has an estimate of the required form by the assumed estimate (22.2.5) and (22.2.7), and so has the term $(Pu,q_k Au)$. What remains is to estimate $(q_k u,P'Au)$ and $(P'u,q_k Au)$. We may commute P' and A in the first term, for

$$\|q_k u\|_{(\varepsilon-1)}\|[P',A]u\|_{(1-\varepsilon)}$$

has the appropriate estimate since $1-\varepsilon+\operatorname{order}A<0$. Now an application of the Cauchy-Schwarz inequality as in the estimate of P'' above shows that we only have to establish an estimate by $(\|Pu\|+\|u\|)^2$ for

$$(P'q_k Au,q_k Au)=\operatorname{Re}(Pq_k Au,q_k Au),$$
$$(P'A^*q_k u,A^*q_k u)=\operatorname{Re}(PA^*q_k u,A^*q_k u).$$

These estimates follow since by the inductive hypothesis (22.2.5)′ and by (22.2.7) we have

$$\|Pq_k Au\|_{(-\varepsilon/2)}+\|PA^*q_k u\|_{(-\varepsilon/2)}+\|q_k Au\|_{(\varepsilon/2)}+\|A^*q_k u\|_{(\varepsilon/2)}$$
$$\leqq C(\|Pu\|+\|u\|),\quad u\in C_0^\infty(K).$$

The proof is complete.

The hypothesis that the vector fields (22.2.2) have rank n at every point means that the operators L_j, $[L_{j_1},L_{j_2}]$, $[L_{j_1},[L_{j_2},L_{j_3}]]$, ... have no characteristic point in common. In the following lemma we use a generalized version of this assumption:

Lemma 22.2.4. *Assume that for some integer N the operators in $Q_1\cup\ldots\cup Q_N$ have no characteristic point in common. If $\varepsilon\leqq 2^{-N}$ it follows then that for every compact set $K\subset X$ and every $s\in\mathbb{R}$*

(22.2.8)
$$\|u\|_{(s+2\varepsilon)}+\sum\|P_j u\|_{(s+\varepsilon)}+\sum\|P^j u\|_{(s+\varepsilon)}$$
$$\leqq C_{s,K}(\|Pu\|_{(s)}+\|u\|_{(s)}),\quad u\in C_0^\infty(K).$$

Proof. Choose $q_1, \dots, q_J \in Q_1 \cup \dots \cup Q_N$ with no common characteristic point over K. By Lemma 22.2.3

$$\sum_1^J \|q_j(x,D)u\|_{(2\varepsilon-1)} \leqq C(\|Pu\|+\|u\|), \quad u \in C_0^\infty(K),$$

hence

(22.2.9) $$\|u\|_{(2\varepsilon)} \leqq C(\|Pu\|+\|u\|), \quad u \in C_0^\infty(K).$$

By hypothesis P_j and P^j are in Q_1; hence we have by (22.2.3)′ for every $B>0$

(22.2.10) $$\sum(\|P_j u\|+\|P^j u\|) \leqq C(B\|Pu\|_{(-\varepsilon)}+B^{-1}\|u\|_{(\varepsilon)}+\|u\|_{(0)}), \quad u \in C_0^\infty(K).$$

Since we have such estimates for every K we may replace u by $E_s u$ in (22.2.9) and by $E_{s+\varepsilon}u$ in (22.2.10). Since the symbol of E_s mod $S^{-\infty}$ is $(1+|\xi|^2)^{s/2}$, we obtain

$$\|u\|_{(s+2\varepsilon)} \leqq \|E_s u\|_{(2\varepsilon)} + C\|u\|_{(s)},$$
$$\|P_j u\|_{(s+\varepsilon)} \leqq \|P_j E_{s+\varepsilon} u\|_{(0)} + C\|u\|_{(s+\varepsilon)}$$

and similarly for P^j, for $[P_j, E_{s+\varepsilon}]$ and $[P^j, E_{s+\varepsilon}]$ are of order $s+\varepsilon$. The calculus gives

$$E_{-\varepsilon}PE_{s+\varepsilon} = E_s P + i(s+\varepsilon)\sum D_j E_{s-1} P_j + R,$$

where R is of order s. Hence

$$\|PE_{s+\varepsilon}u\|_{(-\varepsilon)} \leqq \|Pu\|_{(s)} + C\sum\|P_j u\|_{(s)} + C\|u\|_{(s)}.$$

Combining these estimates we now obtain for sufficiently large B

(22.2.11) $$\|u\|_{(s+2\varepsilon)} + \sum\|P_j u\|_{(s+\varepsilon)} + \sum\|P^j u\|_{(s+\varepsilon)} \leqq C(\|Pu\|_{(s)} + \sum\|P_j u\|_{(s)} + \sum\|P^j u\|_{(s)} + \|u\|_{(s+\varepsilon)}).$$

Now it follows from Hölder's inequality that the $H_{(s)}$ norms are logarithmically convex functions of s, so we have for example

$$\|f\|_{(s)} \leqq \|f\|_{(s+\varepsilon)}^{1/(1+\varepsilon)} \|f\|_{(s-1)}^{\varepsilon/(1+\varepsilon)} \leqq \delta\|f\|_{(s+\varepsilon)} + \delta^{-1/\varepsilon}\|f\|_{(s-1)}, \quad \delta>0.$$

If we estimate $C\|P_j u\|_{(s)}$ in (22.2.11) by $\|P_j u\|_{(s+\varepsilon)}/2 + C'\|u\|_{(s)}$ and $C\|u\|_{(s+\varepsilon)}$ by $\|u\|_{(s+2\varepsilon)}/2 + C'\|u\|_{(s)}$ in (22.2.11), we obtain (22.2.8).

Lemma 22.2.5. *Let the assumptions in Lemma 22.2.4 be fulfilled. If* $Pu = f \in H_{(s)}^{\mathrm{loc}}$ *at* $(x_0, \xi_0) \in T^*(X) \smallsetminus 0$, *it follows then that*

(22.2.12) $$u \in H_{(s+2\varepsilon)}^{\mathrm{loc}}, \quad P_j u \in H_{(s+\varepsilon)}^{\mathrm{loc}}, \quad P^j u \in H_{(s+\varepsilon)}^{\mathrm{loc}} \quad \text{at } (x_0, \xi_0).$$

Proof. We may assume in the proof that

(22.2.13) $$u \in H_{(s+\varepsilon)}^{\mathrm{loc}}, \quad P_j u \in H_{(s)}^{\mathrm{loc}}, \quad P^j u \in H_{(s)}^{\mathrm{loc}} \quad \text{at } (x_0, \xi_0).$$

In fact, this hypothesis is certainly fulfilled with s replaced by $s - N\varepsilon$ if N is large enough. If the statement is proved under this additional hypothesis we can therefore successively conclude that (22.2.13) is valid with s replaced by

$s-k\varepsilon$, $k=N, N-1, \dots, 0$. We may also assume that $u\in\mathscr{E}'(X)$ and can then modify P outside a neighborhood K of $\operatorname{supp} u$, where (22.2.8) remains valid, so that $P\in\operatorname{Op} S^2(\mathbb{R}^n\times\mathbb{R}^n)$. Choose $\psi\in S^0$ with $1/\psi\in S^0$ in a conic neighborhood of (x_0,ξ_0) so that ψ is of order $-\infty$ outside a conic neighborhood of (x_0,ξ_0) where (22.2.13) is fulfilled. We can choose $\psi(x,D)$ properly supported, in fact so that $\operatorname{supp}\psi(x,D)v\subset K$ for all v with support near $\operatorname{supp} u$. Choose $\chi\in C_0^\infty(\mathbb{R}^n)$ with $\hat\chi(0)=1$ and set

$$\psi_\delta(x,\xi)=\psi(x,\xi)\,\hat\chi(\delta\xi),\qquad \Psi_\delta=\psi_\delta(x,D).$$

Then ψ_δ is bounded in S^0 and $\Psi_\delta u\in C_0^\infty(K)$ for small δ. It is clear that $\|\Psi_\delta u\|_{(s)}$ is uniformly bounded when $\delta\to 0$. Since

$$P\Psi_\delta u=\Psi_\delta Pu-i\sum\psi_{j\delta}(x,D)P^j u+i\sum\psi_\delta^j(x,D)P_j u-R_\delta(x,D)u$$

where R_δ, $\psi_{j\delta}$ and ψ_δ^j are bounded in S^0 and of order $-\infty$ outside a cone where (22.2.13) holds, we conclude when (22.2.8) is applied to $\Psi_\delta u$ that

$$\|\Psi_\delta u\|_{(s+2\varepsilon)}+\sum\|P_j\Psi_\delta u\|_{(s+\varepsilon)}+\sum\|P^j\Psi_\delta u\|_{(s+\varepsilon)}$$

is bounded when $\delta\to 0$. This proves (22.2.12).

With Lemma 22.2.5 we have completed the proof of Theorem 22.2.1 and even proved that $Pu\in H_{(s)}$ at (x,ξ) implies $u\in H_{(s+2\varepsilon)}$ at (x,ξ) for some $\varepsilon>0$. We shall now examine the more general results which are actually contained in the preceding lemmas. Let X be a C^∞ manifold and P a properly supported pseudo-differential operator in $\Psi^m(X)$ such that the refined principal symbol $\sigma(P)$ (cf. Theorem 18.1.33) satisfies

$$\operatorname{Im}\sigma(P)\in S^{m-1},\qquad \operatorname{Re}\sigma(P)+r\geqq 0 \tag{22.2.14}$$

for some $r\in S^{m-2}$. If $X\subset\mathbb{R}^n$ and $p(x,\xi)$ is the full symbol of P this means that

$$\operatorname{Im} p\in S^{m-1},\qquad \operatorname{Re} p+r\geqq 0 \tag{22.2.14$'$}$$

for some $r\in S^{m-2}$. For the sake of simplicity we assume $X\subset\mathbb{R}^n$ in what follows, but since the conclusions will be local they are valid for a manifold. As before we denote by E_s a properly supported pseudo-differential operator with symbol $(1+|\xi|^2)^{s/2}$. Then $u\in H_{(\mu)}$ is equivalent to $E_s u\in H_{(\mu-s)}$, and $E_s P$ satisfies (22.2.14)$'$ with m replaced by $m+s$ since the real part of the symbol is $(1+|\xi|^2)^{s/2}\operatorname{Re} p(x,\xi) \bmod S^{m+s-2}$. Taking $s=2-m$ we reduce the study to the case of operators with $m=2$ which we assume in what follows. By the Fefferman-Phong inequality (Corollary 18.6.11) we have then for every $K\subset X$

$$\operatorname{Re}(Pu,u)\geqq -C_K\|u\|^2,\qquad u\in C_0^\infty(K).$$

More generally, let Q_1 be the set of all properly supported $q\in\Psi^1(X)$ with real principal symbol such that for every $K\Subset X$

$$|q(x,\xi)|^2\leqq C_K'\operatorname{Re} p(x,\xi)+C_K'',\qquad x\in K. \tag{22.2.15}$$

If we apply the Fefferman-Phong inequality to $P-\delta\, q(x,D)^*\, q(x,D)$ with $\delta\, C'_K<1$ it follows that (22.2.3)′ is valid for every $q\in Q_1$, for the real part of the symbol of $q(x,D)^*\, q(x,D)$ is $|q(x,\xi)|^2 \bmod S^0$, and the imaginary part is in S^1.

The inequality (22.2.15) is satisfied by the operators P_j and P^j with principal symbol $(1+|\xi|^2)^{-\frac{1}{2}}\, p_{(j)}(x,\xi)$ and $p^{(j)}(x,\xi)$ respectively. In fact, if $R\geqq 1$ then the derivatives of order $\leqq 2$ of

$$f_R(x,\xi)=\operatorname{Re} p(x,R\,\xi)/R^2$$

have a bound independent of R when $\frac{1}{2}<|\xi|<2$ and x is in a neighborhood of K. Adding a constant to p we may assume that $f_R\geqq 0$. Then Lemma 7.7.2 gives

$$|\partial f_R(x,\xi)/\partial(x,\xi)|^2\leqq C f_R(x,\xi), \qquad |\xi|=1.$$

Since $\operatorname{Im} p\in S^1$ this proves (22.2.15) for the principal symbols of P_j and P^j.

From Lemmas 22.2.3 to 22.2.5 we now obtain:

Theorem 22.2.6. *Let $P\in\Psi^m(X)$ be properly supported and assume that the refined principal symbol satisfies (22.2.14) for some $r\in S^{m-2}$, in a conic neighborhood of $(x_0,\xi_0)\in T^*(X)\setminus 0$. Let Q_1 be the set of all $Q\in\Psi^{m-1}$ with real principal symbol q such that in a conic neighborhood of (x_0,ξ_0)*

$$|q(x,\xi)|^2\leqq C' \operatorname{Re} p(x,\xi)(1+|\xi|^2)^{(m-2)/2}+C''(1+|\xi|^2)^{m-2},$$

and assume that (x_0,ξ_0) is non-characteristic with respect to some commutator of ν_1 factors in Q_1 and ν_2 factors $P-P^*$ with $\nu_1+2\nu_2\leqq N$. If $\varepsilon=2^{-N}$ it follows that

(22.2.16) $\quad u\in\mathscr{D}'(X), Pu\in H^{\mathrm{loc}}_{(s)}$ *at* (x_0,ξ_0) *implies* $u\in H^{\mathrm{loc}}_{(s+m-2+2\varepsilon)}$ *at* (x_0,ξ_0).

Note that a commutator $[q_1(x,D),[q_2(x,D),[\dots,q_k(x,D)]\dots]]$ of k operators of order $m-1$ is an operator of order $k(m-1)-k+1$ with principal symbol

$$\{q_1(x,\xi),\{q_2(x,\xi),\{\dots,q_k(x,\xi)\}\dots\}\}\, i^{-k}.$$

The hypothesis is therefore that some of these repeated Poisson brackets of q_j satisfying (22.2.15) or $q_j=\operatorname{Im} p^s_{m-1}$ is invertible in a conic neighborhood of (x_0,ξ_0). We have stated the result with local conditions on the symbol of P. This improvement is immediately obtained by adding to P an operator with symbol $|\xi|^m a(x,\xi)$ where $0\leqq a\in S^0$ is 0 in a conic neighborhood of (x_0,ξ_0) and is positive outside another such neighborhood where the hypotheses are applicable.

22.3. Melin's Inequality

The inequality of Fefferman and Phong played an important role in the proof of Theorem 22.2.6. It is a very precise result in the sense that for self-adjoint operators $p(x,D)$ of order >2 it may happen that $p(x,D)$ is not bounded from below in L^2 although $\operatorname{Re} p(x,\xi)$ is bounded from below. However, the result is weak in the sense that an operator $p(x,D)$ of order two may be bounded from below in L^2 although the symbol tends to $-\infty$ as fast as $|\xi|$ in some directions. The reason for this is that the harmonic oscillator D^2+x^2 on $\mathbb{R}$ has the lower bound 1 because

$$((D^2+x^2)u,u)=\|(D-ix)u\|^2+\|u\|^2, \quad u\in C_0^\infty(\mathbb{R}).$$

We shall prove some lower bounds for pseudo-differential operators with polyhomogeneous symbols by imitating this identity. As an application this will give a supplement to Theorem 22.2.6. We begin with an algebraic lemma.

Lemma 22.3.1. *Let Q be a positive semi-definite quadratic form in $\mathbb{R}^{2n}$ and V_λ the space of generalized eigenvectors in $\mathbb{C}^{2n}$ with eigenvalue λ of the Hamilton map F of Q. Choose a unitary basis $v_1,\dots,v_k$ for $V^+=\bigoplus_{\mu>0} V_{i\mu}$ with the hermitian form $Q(\bar v,v)/2$, and an orthogonal basis $v_{k+1},\dots,v_{k+l}$ for the real part of $V_0/\operatorname{Ker} F$ with the quadratic form induced by Q there. If $L_j(x,\xi)=Q(v_j,(x,\xi))$ it follows that*

$$(Q(x,D)+Q(x,D)^*)/2=\sum L_j(x,D)^*L_j(x,D)+\operatorname{Tr}^+Q, \tag{22.3.1}$$

or equivalently

$$\sum|L_j(x,\xi)|^2=Q(x,\xi), \tag{22.3.2}$$

$$\sum\{\operatorname{Re}L_j,\operatorname{Im}L_j\}+\operatorname{Tr}^+Q=0, \quad (x,\xi)\in\mathbb{R}^{2n}.$$

Here $\operatorname{Tr}^+Q=\sum_1^k\mu_j$ *where $i\mu_j$ are the eigenvalues of F restricted to V^+.*

Proof. The symbol of $L_j(x,D)^*$ is $\overline{L_j(x,\xi)}$ so that of the self-adjoint operator $\sum L_j(x,D)^*L_j(x,D)$ is

$$\sum|L_j(x,\xi)|^2+i^{-1}\sum\partial\overline{L_j(x,\xi)}/\partial\xi_\nu\,\partial L_j(x,\xi)/\partial x_\nu.$$

The symbol of $(Q(x,D)+Q(x,D)^*)/2$ is

$$Q(x,\xi)+\sum\partial^2Q/\partial x_\nu\partial\xi_\nu/2i.$$

Taking the real part of the symbols we find that (22.3.1) implies (22.3.2); the converse is clear since two self-adjoint operators cannot differ by a purely imaginary constant $\neq 0$.

The condition (22.3.2) is clearly symplectically invariant, and (22.3.1) is obviously invariant under unitary transformations of $v_1,\dots,v_k$ and orthogonal transformations of $v_{k+1},\dots,v_n$. By Theorem 21.5.3 it is therefore sufficient to verify one of the conditions (22.3.1), (22.3.2) when

$$Q(x,\xi)=\sum_1^k \mu_j(x_j^2+\xi_j^2)+\sum_{k+1}^{k+l} x_j^2;$$

$$v_j=\mu_j^{-\frac{1}{2}}(e_j+i\varepsilon_j),\ 1\leqq j\leqq k; \quad v_j=e_j,\ k<j\leqq k+l.$$

Here e_j and ε_j are unit vectors along the x_j and ξ_j axes. (22.3.2) is then obvious for $Q(v_j,(x,\xi))=\mu_j^{\frac{1}{2}}(x_j+i\xi_j),\ j\leqq k;\ Q(v_j,(x,\xi))=x_j,\ k<j\leqq k+l.$

For operators with polyhomogeneous symbol we can now give a substantial improvement of the sharp Gårding inequality (Theorem 18.1.14).

Theorem 22.3.2. *Assume that* $P\in\Psi_{\mathrm{phg}}^{2m}(X,\Omega^{\frac{1}{2}})$ *is properly supported and self-adjoint. Let* p_{2m} *and* p_{2m-1}^s *be the principal and subprincipal symbols and assume that*

$$p_{2m}(x,\xi)\geqq 0 \quad \textit{in } T^*(X)\smallsetminus 0, \tag{22.3.3}$$

$$p_{2m-1}^s(x,\xi)+\mathrm{Tr}^+ Q_{x,\xi}\geqq 0 \quad \textit{if } (x,\xi)\in T^*(X)\smallsetminus 0,\ p_{2m}(x,\xi)=0. \tag{22.3.4}$$

Here $Q_{x,\xi}$ *is the Hessian of* $p_{2m}/2$ *at* (x,ξ). *Finally we assume that the characteristic set* $\Sigma=\{(x,\xi)\in T^*(X)\smallsetminus 0,\ p_{2m}(x,\xi)=0\}$ *is a* C^∞ *manifold, that* $T_{x,\xi}(\Sigma)$ *is the radical of* $Q_{x,\xi}$ *when* $(x,\xi)\in\Sigma$ *and that the symplectic form has constant rank on* Σ. *For every compact set* $K\subset X$ *one can then find a constant* C_K *such that*

$$(Pu,u)\geqq -C_K\|u\|_{(m-1)}^2, \quad u\in C_0^\infty(K). \tag{22.3.5}$$

Before the proof we make some observations on the statement. That P is self-adjoint implies that p_{2m} and p_{2m-1}^s are real valued. The condition (22.3.3) implies that $dp_{2m}=0$ on Σ, so the Hessian is invariantly defined on Σ. It is automatically semi-definite and vanishes on $T_{x,\xi}(\Sigma)$. The assumption on the radical is therefore that $Q_{x,\xi}$ is positive definite in a plane transversal to $T_{x,\xi}(\Sigma)$; the condition is often referred to as *transversal ellipticity*.

Proof of Theorem 22.3.2. Let E_s be a properly supported operator with principal symbol $(1+|\xi|^2)^{s/2}$ (defined with respect to some Riemannian metric). Then it is sufficient to prove that

$$(PE_{1-m}u, E_{1-m}u)\geqq -C_K\|u\|_{(0)}^2, \quad u\in C_0^\infty(K),$$

for this gives (22.3.5) when u is replaced by Au where A is a parametrix of E_{1-m}. The principal symbol of $E_{1-m}^*PE_{1-m}$ is $|\xi|^{2-2m}p_{2m}(x,\xi)$ and the subprincipal symbol is $|\xi|^{2-2m}p_{2m-1}^s(x,\xi)$ since it must be real valued. The hypotheses in the theorem are therefore fulfilled by $E_{1-m}^*PE_{1-m}$, so we may assume that $m=1$ in the proof.

We can apply Lemma 22.3.1 to the quadratic form $Q_{x,\xi}$ in $T_{x,\xi}(T^*(X))$, $(x,\xi)\in\Sigma$. The radical is $T_{x,\xi}(\Sigma)$ and the rank of the symplectic form there is constant. In the notation of Lemma 22.3.1 this means that $2n-2k-l$ and $2(n-k-l)$ are constant, so k and l are constant. It follows that the spaces

V^+ and V_0 vary smoothly with $(x, \xi) \in \Sigma$, although this is of course not true for the individual spaces $V_{i\mu}$. For all $(x, \xi) \in \Sigma$ in a neighborhood of any given $(x_0, \xi_0) \in \Sigma$ we can then choose C^∞ complex tangent vectors $v_1(x, \xi), \ldots, v_k(x, \xi)$ forming a unitary basis for V^+ and real tangent vectors $v_{k+1}(x, \xi), \ldots, v_{k+l}(x, \xi)$ forming an orthogonal basis for the real part of $V_0 \bmod T_{x,\xi}(\Sigma)$. Then

$$L_j(v) = Q_{x,\xi}(v_j, v), \quad v \in T(T^*(X))$$

is a (complex) cotangent vector of $T^*(X)$ at (x, ξ) which is conormal to Σ. With $\Lambda_{x,\xi} = (\operatorname{Re} L_1, \operatorname{Im} L_1, \ldots, \operatorname{Re} L_k, \operatorname{Im} L_k, L_{k+1}, \ldots, L_{k+l})$ we have by (22.3.2)

$$Q_{x,\xi}(v) = \sum_1^{2k+l} \Lambda_{x,\xi,j}(v)^2, \quad v \in T_{x,\xi}(T^*(X)). \tag{22.3.6}$$

Now the Morse lemma (Lemma C.6.1) shows that near (x_0, ξ_0) we can write

$$p_2(x, \xi) = \sum_1^{2k+l} b_j(x, \xi)^2 \tag{22.3.7}$$

where $b_j \in C^\infty$. Indeed, if we choose local coordinates y, z so that Σ is defined by $z = 0$ then $\partial^2 p_2 / \partial z^2$ is non-singular by hypothesis so the lemma shows that p_2 is a quadratic form in z for some other local coordinates. If we restrict b_j to the surface $|\xi| = |\xi_0|$ and then extend to a homogeneous function of degree 1 we still have (22.3.7), $b_j \in C^\infty$, and b_j is homogeneous of degree 1.

On Σ we have $Q_{x,\xi} = \sum db_j^2$, hence

$$\Lambda_{x,\xi,i} = \sum_j O_{ij}(x, \xi)\, db_j(x, \xi)$$

where O is a C^∞ function and the matrix (O_{ij}) is orthogonal. We restrict O to $|\xi| = |\xi_0|$ and then extend O to a C^∞ function of (x, ξ), homogeneous of degree 0, in a full conic neighborhood of (x_0, ξ_0). Set

$$c_i(x, \xi) = \sum O_{ij}(x, \xi)\, b_j(x, \xi).$$

Then

$$p_2(x, \xi) = \sum c_i(x, \xi)^2, \tag{22.3.7$'$}$$

and $dc_i(x, \xi) = \Lambda_{x,\xi,i}$ in a neighborhood of (x_0, ξ_0) on Σ when $|\xi| = |\xi_0|$. Set

$$X_j(x, \xi) = c_{2j-1}(x, \xi) + i c_{2j}(x, \xi), \quad j = 1, \ldots, k,$$
$$X_j(x, \xi) = c_{k+j}(x, \xi), \quad j = k+1, \ldots, k+l,$$

and let $X_j \in \Psi^1_{\mathrm{phg}}(X)$ be properly supported with this principal symbol in a conic neighborhood of (x_0, ξ_0). Then

$$Q = P - \sum X_j^* X_j$$

is self-adjoint and of first order in a conic neighborhood of (x_0, ξ_0). The principal symbol q_1 is real there. On Σ we have

$$q_1(x,\xi)=p_1(x,\xi)-i^{-1}\sum \overline{\partial X_j(x,\xi)/\partial\xi_\nu}\, \partial X_j(x,\xi)/\partial x_\nu,$$

and since q_1 is real we obtain by taking the real part

$$q_1(x,\xi)=\operatorname{Re} p_1(x,\xi)+\sum_1^k \{\operatorname{Im} X_j, \operatorname{Re} X_j\}=p_1^s(x,\xi)+\operatorname{Tr}^+ Q_{x,\xi}$$

when $(x,\xi)\in\Sigma$ is close to (x_0,ξ_0). The equality follows from the second part of (22.3.2).

The hypothesis (22.3.4) is that $q_1\geqq 0$ on Σ. We can therefore choose a homogeneous function $\tilde{q}_1$, of degree 1, which is non-negative everywhere and is equal to q_1 on Σ in a neighborhood of (x_0,ξ_0). Then

$$q_1(x,\xi)=\tilde{q}_1(x,\xi)+\sum \overline{r_j(x,\xi)}\, X_j(x,\xi)+\sum r_j(x,\xi)\,\overline{X_j(x,\xi)}$$

in a neighborhood of (x_0,ξ_0), where r_j is homogeneous of degree 0. Choose $\tilde{Q}$ and R_j with principal symbols $\tilde{q}_1$ and r_j. Then the symbol of

$$P-\sum X_j^* X_j-\tilde{Q}-\sum R_j^* X_j-\sum X_j^* R_j$$

is of order 0 in a conic neighborhood of (x_0,ξ_0) so this is also true for that of

$$P-\sum (X_j+R_j)^*(X_j+R_j)-\tilde{Q}.$$

If $\psi\in S^0$ has support in a sufficiently small conic neighborhood of (x_0,ξ_0) it follows that we have a microlocal form of (22.3.5)

$$(P\psi(x,D)u, \psi(x,D)u)\geqq -C\|u\|^2,\ u\in C_0^\infty(K), \tag{22.3.5$'$}$$

for the operator $\tilde{Q}$ is bounded from below by Theorem 18.1.14. The same result is quite obvious if ψ is supported in a small neighborhood of a non-characteristic point (x_0,ξ_0). In fact, if $Q\in\Psi^1_{\text{phg}}$ has symbol $q_1+q_0+\dots$ then $P-Q^*Q$ is of order 0 in a conic neighborhood of (x_0,ξ_0) if $q_1^2=p_2$ and $2q_1 q_0=p_1^s$ there, that is, $q_1=p_2^{\frac{1}{2}}$ and $q_0=p_1^s p_2^{-1/2}/2$.

To complete the proof of (22.3.5) we choose real valued $\psi_j(x,\xi)\in S^0$ with so small support that (22.3.5)′ is valid for each ψ_j and

$$\sum_1^J \psi_j(x,\xi)^2=1 \qquad \text{for } x \text{ near } K.$$

Then we have, with $\Psi_j=\psi_j(x,D)$ assumed properly supported,

$$(Pu,u)=\sum(P\Psi_j u, \Psi_j u)+[((1-\Psi_j^*\Psi_j)Pu,u)+\sum([\Psi_j,P]u,\Psi_j u)].$$

Here $[\Psi_j, P]$ is a first order operator with purely imaginary principal symbol so the self adjoint part of $\Psi_j^*[\Psi_j,P]$ is of order 0. Furthermore, $\sum\Psi_j^*\Psi_j-1$ is of order -1 in a neighborhood of K, self-adjoint and with purely imaginary symbol of order -1. Hence this difference is actually of order -2, so (22.3.5) follows from (22.3.5)′. The proof is complete.

If the inequality (22.3.4) is assumed strict, the other conditions in Theorem 22.3.2 can be relaxed and we obtain Melin's inequality:

Theorem 22.3.3. *Assume that $P \in \Psi^{2m}_{\mathrm{phg}}(X)$ is properly supported and self-adjoint. Let p_{2m} and p^s_{2m-1} be the principal and subprincipal symbols, and assume that* (22.3.3) *is fulfilled and that*

$$p^s_{2m-1}(x,\xi)+\operatorname{Tr}^+ Q_{x,\xi}>0 \text{ if } (x,\xi)\in T^*(X)\setminus 0,\quad p_{2m}(x,\xi)=0. \tag{22.3.4$'$}$$

For every compact set $K\subset X$ one can then find $c_K>0$ and C_K such that

$$(Pu,u)\geqq c_K\|u\|^2_{(m-1/2)}-C_K\|u\|^2_{(m-1)},\quad u\in C_0^\infty(K). \tag{22.3.8}$$

Proof. Exactly as in the proof of Theorem 22.3.2 we can reduce to the case $m=1$ and also show that (22.3.8) follows from analogous microlocal estimates similar to (22.3.5)$'$. Let (x_0,ξ_0) be a characteristic point. We can choose local coordinates y, z there such that Q_{x_0,ξ_0} is equal to $|z|^2$. Then the Morse lemma (Lemma C.6.2) shows that in a conic neighborhood of (x_0,ξ_0)

$$p_2(x,\xi)\geqq \sum b_j(x,\xi)^2.$$

Here b_j is homogeneous of degree 1 and $\sum db_j^2=Q_{x_0,\xi_0}$ at (x_0,ξ_0). We can also choose

$$c_i(x,\xi)=\sum O_{ij}b_j(x,\xi),$$

where O is an orthogonal matrix, so that $dc_i(x,\xi)=\Lambda_{x,\xi,i}$ at (x_0,ξ_0),

$$\operatorname{Tr}^+ Q_{x_0,\xi_0}=\sum_1^k\{c_{2j},c_{2j-1}\}(x_0,\xi_0).$$

Here we have used the notation in the proof of Theorem 22.3.2 and also the fact that the arguments there are valid for the fixed point (x_0,ξ_0).

Now we can split the symbol of P as follows,

$$p_2(x,\xi)+p_1(x,\xi)=\sum_1^3 q^j(x,\xi),$$

$$q^1(x,\xi)=p_2(x,\xi)-\sum_{j\leqq 2k+l}c_j(x,\xi)^2,$$

$$q^2(x,\xi)=\sum_{j\leqq 2k+l}c_j(x,\xi)^2-\sum_{j=1}^k\{c_{2j},c_{2j-1}\},$$

$$q^3(x,\xi)=\sum_{j=1}^k\{c_{2j},c_{2j-1}\}+p_1(x,\xi).$$

We have $q^1\geqq 0$ in a conic neighborhood of (x_0,ξ_0) and q^1 vanishes of third order at (x_0,ξ_0). Hence it follows from Theorem 18.1.14 and the first remark following its proof that we can find $q^4\in S^1$ with principal symbol vanishing at (x_0,ξ_0) such that if $\psi\in S^0$ has support in a small conic neighborhood of

(x_0, ξ_0) and $\Psi = \psi(x, D)$ then

$$\operatorname{Re}(q^1(x, D)\Psi u, \Psi u) \geqq \operatorname{Re}(q^4(x, D)\Psi u, \Psi u) - C_K\|u\|^2, \quad u \in C_0^\infty(K).$$

(We could have taken $q^4 \in S^0_{\text{phg}}$ by the Fefferman-Phong inequality but this precision is not required here.) With $X_j \in S^1_{\text{phg}}$ properly supported and with principal symbol $c_{2j-1} + ic_{2j}$, $j \leqq k$, and c_{k+j}, $j > k$, we have

$$\operatorname{Re}(q^2(x, D)u, u) - \sum (X_j^* X_j u, u) \geqq -C_K\|u\|^2.$$

In fact, the self-adjoint operator $(q^2(x, D) + q^2(x, D)^*)/2 - \sum X_j^* X_j$ is of order 0 since the principal symbol is 0 and the first order symbol is purely imaginary. (These calculations are much more transparent if one uses the Weyl calculus but we have not wanted to require this background here.) Finally the real part of the principal symbol of $q^4 + q^3$ is positive at (x_0, ξ_0), hence

$$\operatorname{Re}((q^4 + q^3)(x, D)\Psi u, \Psi u) \geqq \varepsilon_K \|\Psi u\|^2_{(1)} - C_K\|u\|^2_{(0)}, \quad u \in C_0^\infty(K)$$

if ε_K is sufficiently small. Summing up the preceding estimates we have proved the desired local version of (22.3.8).

Remark. In Section 22.4 we shall prove that conversely (22.3.8) implies $p_{2m} \geqq 0$ and (22.3.4)$'$.

As an application we shall now give an improvement of Theorem 22.2.6 at points where $\operatorname{Tr}' Q_{x,\xi} > 0$. (See Proposition 22.4.1.)

Theorem 22.3.4. *Assume that $P \in \Psi^m_{\text{phg}}(X)$ is properly supported, and that the principal symbol p_m is non-negative in a conic neighborhood of $(x_0, \xi_0) \in T^*(X) \setminus 0$ but vanishes at (x_0, ξ_0). Let Q be the Hessian of $p_m/2$ at (x_0, ξ_0), and assume that the subprincipal symbol p^s_{m-1} satisfies*

(22.3.9) $$p^s_{m-1}(x_0, \xi_0) + \operatorname{Tr}^+ Q \notin \overline{\mathbb{R}}_- = \{t \in \mathbb{R}, t \leqq 0\}.$$

Then $u \in \mathscr{D}'(X)$, $Pu \in H_{(s)}$ at (x_0, ξ_0) implies that $u \in H_{(s+m-1)}$ at (x_0, ξ_0).

Proof. As usual we may assume $m = 2$. The hypothesis (22.3.9) means that we can choose a complex number z with

(22.3.10) $$\operatorname{Re} z > 0, \quad \operatorname{Re} z(p^s_1(x_0, \xi_0) + \operatorname{Tr}^+ Q) > 0.$$

Then $P_z = (zP + \bar{z}P^*)/2$ has the principal symbol $\operatorname{Re} z p_m \geqq 0$ and the subprincipal symbol $\operatorname{Re} z p^s_1$. If $\psi \in S^0$ has support in a small conic neighborhood of (x_0, ξ_0) and $\Psi = \psi(x, D)$ we have then by Theorem 22.3.3 if $K \Subset X$

$$\|\Psi u\|^2_{(\frac{1}{2})} \leqq C' \operatorname{Re}(zP\Psi u, \Psi u) + C''_K\|u\|^2, \quad u \in C_0^\infty(K).$$

Hence

(22.3.11) $$\|\Psi u\|^2_{(\frac{1}{2})} \leqq C'^2 \|zP\Psi u\|^2_{(-\frac{1}{2})} + 2C''_K\|u\|^2, \quad u \in C_0^\infty(K).$$

Let P^j and P_j be properly supported of order 1 with principal symbols $\partial p_2/\partial\xi_j$ and $|\xi|^{-1}\partial p_2/\partial x_j$ respectively. Then the principal symbol of

$$P_z-\delta\sum P^{j*}P^j-\delta\sum P_j^*P_j$$

is non-negative near (x_0,ξ_0) for small δ, as observed in the proof of Theorem 22.2.6. For small δ we can apply Theorem 22.3.3 again and obtain

$$\begin{aligned}(22.3.12)\quad \sum\|P^j\Psi u\|^2+\sum\|P_j\Psi u\|^2 &\leqq C'\operatorname{Re}(zP\Psi u,\Psi u)+C''\|u\|^2\\ &\leqq C'\|zP\Psi u\|_{(-\frac12)}\|\Psi u\|_{(\frac12)}+C''\|u\|^2,\quad u\in C_0^\infty(K).\end{aligned}$$

(22.3.11), (22.3.12) are a microlocal version of (22.2.9), (22.2.10) with $\varepsilon=\frac12$ so the proof of the theorem is completed by inspection of the proofs of Lemmas 22.2.4 and 22.2.5.

Not even the condition (22.3.9) is quite optimal. In Section 22.4 we shall determine the precise set which $p_{m-1}^s(x_0,\xi_0)$ must avoid for the conclusion of the theorem to be valid.

22.4. Hypoellipticity with Loss of One Derivative

In this section we shall determine necessary and sufficient conditions on P for the conclusion of Theorem 22.3.4 to be valid when the condition $\operatorname{Re} p_m\geqq 0$ is relaxed to

$$(22.4.1)\qquad p_m(x,\xi)\in\Gamma=\{z\in\mathbb{C};\ |\operatorname{Im} z|\leqq\gamma\operatorname{Re} z\}.$$

(Any other convex angle with opening $<\pi$ could be used, but this is a convenient choice.) However, we shall first prove the converse of Theorem 22.3.3 which we postponed since it is technically closer to the arguments in this section.

Proposition 22.4.1. *Assume that $P\in\Psi_{\text{phg}}^m(X)$ is self-adjoint and that for every compact set $K\subset X$*

$$(22.4.2)\qquad (Pu,u)\geqq -C_K\|u\|^2_{(m/2-1)},\quad u\in C_0^\infty(K).$$

Then it follows that the principal symbol p_m is non-negative and that

$$(22.4.3)\qquad p_{m-1}^s(x,\xi)+\operatorname{Tr}^+Q_{x,\xi}\geqq 0\ \textit{ if } (x,\xi)\in T^*(X)\setminus 0 \textit{ and } p_m(x,\xi)=0.$$

Here $p_{m-1}^s(x,\xi)$ is the subprincipal symbol and $Q_{x,\xi}$ is the Hessian of $p_m/2$ at (x,ξ).

Proof. It suffices to prove (22.4.3) when $X\subset\mathbb{R}^n$, $x=0$ and 0 is an interior point of K. Choose $\chi\in C_0^\infty(X)$ equal to 1 in a neighborhood of K and supp Pu for every $u\in C_0^\infty(K)$. Then $Pu=\chi P\chi u=p(x,D)u$, $u\in C_0^\infty(K)$, where $p\in S^m_{\text{phg}}$ and $p\sim p_m+p_{m-1}+\dots$ in K. We shall apply (22.4.2) to

$$u(y)=e^{i\langle y,\xi\rangle}\psi(y|\xi|^{1/2}), \tag{22.4.4}$$

where $\psi\in C_0^\infty(\mathbb{R}^n)$. Then we have $u\in C_0^\infty(K)$ when $|\xi|$ is large. Since $\hat u(\eta)=\hat\psi((\eta-\xi)/|\xi|^{\frac12})|\xi|^{-n/2}$ we have

$$\begin{aligned}|\xi|^{\frac12 n-2s}\|u\|^2_{(s)}&=|\xi|^{-\frac12 n-2s}(2\pi)^{-n}\int(1+|\eta|^2)^s|\hat\psi((\eta-\xi)/|\xi|^{\frac12})|^2\,d\eta\\ &=(2\pi)^{-n}\int(|\xi|^{-2}+|\xi/|\xi|+\eta/|\xi|^{\frac12}|^2)^s|\hat\psi(\eta)|^2\,d\eta\to\|\psi\|^2\end{aligned} \tag{22.4.5}$$

when $\xi\to\infty$. An elementary calculation gives

$$p(y,D)u(y)=e^{i\langle y,\xi\rangle}\phi_\xi(y|\xi|^{\frac12}), \tag{22.4.6}$$

$$\phi_\xi(y)=(2\pi)^{-n}\int e^{i\langle y,\eta\rangle}p(y/|\xi|^{\frac12},\xi+\eta|\xi|^{\frac12})\hat\psi(\eta)\,d\eta. \tag{22.4.7}$$

By Taylor's formula we have

$$p(y/|\xi|^{\frac12},\xi+\eta|\xi|^{\frac12})=\sum_{|\alpha+\beta|\le 2}|\xi|^{(|\alpha|-|\beta|)/2}p^{(\alpha)}_{(\beta)}(0,\xi)\,y^\beta\eta^\alpha/\alpha!\,\beta!+R_\xi(y,\eta),$$

$$|R_\xi(y,\eta)|\le C(|y|+|\eta|)^3|\xi|^{m-3/2},\ |\eta|<|\xi|^{\frac12}/2,$$

$$|R_\xi(y,\eta)|\le C(1+|y|+|\eta|)^2(1+|\eta|)^{2|m|},\ |\eta|>|\xi|^{\frac12}/2.$$

In the first case this follows since $|\xi+\eta|\xi|^{\frac12}|>|\xi|/2$ if $|\eta|<|\xi|^{\frac12}/2$ and in the second case we have estimated each term separately and used that $|\xi+\eta|\xi|^{\frac12}|<6|\eta|^2$. Hence

$$\begin{aligned}&|\phi_\xi(y)-\sum_{|\alpha+\beta|\le 2}|\xi|^{(|\alpha|-|\beta|)/2}p^{(\alpha)}_{(\beta)}(0,\xi)\,y^\beta D^\alpha\psi/\alpha!\,\beta!|\\ &\qquad\le C(1+|y|)^3|\xi|^{m-\frac32},\end{aligned} \tag{22.4.8}$$

which by dominated convergence implies that

$$|\xi|^{\frac12 n-m}(Pu,u)=|\xi|^{-m}(\phi_\xi,\psi)=p_m(0,\xi)|\xi|^{-m}\|\psi\|^2+O(|\xi|^{-\frac12}).$$

In view of (22.4.5) it follows from (22.4.2) that $p_m(0,\xi)\ge 0$. If $p_m(0,\xi_0)=0$ it follows that $p^{(\alpha)}_{m(\beta)}(0,\xi_0)=0$, $|\alpha+\beta|=1$, and taking $\xi=t\xi_0$ we obtain

$$\begin{aligned}|\xi|^{\frac12 n+1-m}(Pu,u)&=|\xi|^{1-m}(\phi_\xi,\psi)\\ &=((Q(y,D)+p_{m-1}(0,\xi_0))\psi,\psi)+O(|\xi|^{-\frac12})\end{aligned}$$

where $Q=Q_{0,\xi_0}$. Hence (22.4.2) implies

$$((Q(y,D)+p_{m-1}(0,\xi_0))\psi,\psi)\ge 0,\qquad \psi\in C_0^\infty.$$

It is now convenient to switch to the Weyl calculus, noting that

$$Q(y,D)+p_{m-1}(0,\xi_0)=Q^w(y,D)+p^s_{m-1}(0,\xi_0)$$

since $Q^w(y,D)$ is obtained from $Q(y,D)$ when y_jD_j is replaced by $(y_jD_j+D_jy_j)/2=y_jD_j-i/2$. With the notation $c=p^s_{m-1}(0,\xi_0)$ we have

$$(Q^w(y,D)\psi,\psi)+c(\psi,\psi)\ge 0,\qquad \psi\in\mathcal{S}, \tag{22.4.9}$$

for C_0^∞ is dense in $\mathscr{S}$. We now use the important property of the Weyl calculus that such inequalities are invariant under composition of Q with linear symplectic maps (Theorem 18.5.9). To analyze (22.4.9) we may therefore assume that Q has the form (21.5.3),

$$Q(y,\eta)=\sum_1^k \mu_j(y_j^2+\eta_j^2)+\sum_{k+1}^{k+l} y_j^2.$$

The inequality (22.4.9) can then be written in the form

$$\sum_1^k \mu_j\|(y_j+iD_j)\psi\|^2+\sum_{k+1}^{k+l}\|y_j\psi\|^2+\Big(c+\sum_1^k\mu_j\Big)\|\psi\|^2\geqq 0,$$

as in Lemma 22.3.1. If we recall that $\operatorname{Tr}^+ Q=\sum_1^k \mu_j$ and take

$$\psi(y)=\sum_1^k e^{-y_j^2/2}\chi(y_{k+1}/\varepsilon,\dots,y_n/\varepsilon)\,\varepsilon^{(k-n)/2},$$

it follows when $\varepsilon\to 0$ that

$$(c+\operatorname{Tr}^+ Q)\,\|\chi\|^2\geqq 0.$$

Hence $c+\operatorname{Tr}^+ Q\geqq 0$ as claimed.

The proof of Proposition 22.4.1 also leads to necessary conditions for an operator to be hypoelliptic with loss of one derivative as in Theorem 22.3.4. First we must prove a lemma:

Lemma 22.4.2. *Let $P\in\Psi^m(X)$ be properly supported, $X\subset\mathbb{R}^n$, and assume that $u\in\mathscr{E}'(X)$ and $Pu\in H_{(s)}$ implies $u\in H_{(s+m-1)}$. For every compact set $K\subset X$ one can then find C_K so that*

$$\|u\|_{(s+m-1)}\leqq C_K(\|Pu\|_{(s)}+\|u\|_{(s+m-2)}),\quad u\in C_0^\infty(K). \tag{22.4.10}$$

Proof. The set of all $u\in H_{(s+m-2)}\cap\mathscr{E}'(K)$ with $Pu\in H_{(s)}$ is a Banach space with the norm in the right-hand side of (22.4.10). Since it is embedded in $\mathscr{E}'(K)\cap H_{(s+m-1)}$ and the embedding is closed, the statement follows from the closed graph theorem.

Let E_s be a properly supported pseudo-differential operator with symbol $(1+|\xi|^2)^{s/2}$. If we replace u by $E_{-s}u$ in (22.4.10) we conclude that (22.4.10) is equivalent to the same estimate with P replaced by E_sPE_{-s}, and s replaced by 0. If the symbol of P is $p_m+p_{m-1}+\dots$ then that of E_sPE_{-s} is

$$p_m+p_{m-1}-is|\xi|^{-2}\sum\xi_j\partial p_m/\partial x_j+\dots$$

so the first two terms are not changed where $\partial p_m/\partial x=0$. Keeping this in mind we assume $s=0$ in the following proposition.

Proposition 22.4.3. *Let $P \in \Psi^m_{\text{phg}}(X)$ be properly supported and assume that for every compact set $K \subset X \subset \mathbb{R}^n$*

$$\|u\|_{(m-1)} \leqq C_K(\|Pu\|_{(0)} + \|u\|_{(m-2)}), \quad u \in C_0^\infty(K). \tag{22.4.10$'$}$$

Let $Q(y, \eta)$ be any limit of $(y, \eta) \mapsto p_m(x + y|\xi|^{-\frac{1}{2}}, \xi + \eta|\xi|^{\frac{1}{2}})|\xi|^{1-m}$ as $x \to x^0 \in X$, $\xi \to \infty$ and $\xi/|\xi| \to \xi^0$. Then we have $p^{(\alpha)}_{m(\beta)}(x^0, \xi^0) = 0$, $|\alpha + \beta| < 2$, Q is a quadratic polynomial with principal part

$$\sum_{|\alpha+\beta|=2} p^{(\alpha)}_{m(\beta)}(x^0, \xi^0)\, y^\beta \eta^\alpha / \alpha!\, \beta!$$

and it follows that

$$\int |\psi|^2\, dy \leqq C \int |Q^w(y, D)\psi + p^s_{m-1}(x^0, \xi^0)\psi|^2\, dy, \quad \psi \in C_0^\infty(\mathbb{R}^n). \tag{22.4.11}$$

Here C is bounded by the constant C_K in (22.4.10)$'$ if x^0 is in the interior of K. One can replace Q by the homogeneous polynomial $y_0^2 Q(y/y_0, \eta/y_0)$.

Proof. Let $x_\nu \to x^0$, $\xi_\nu/|\xi_\nu| \to \xi^0$ be a sequence giving the limit Q, and assume that x^0 is in the interior of K. We shall apply (22.4.10)$'$ to

$$u_\nu(y + x_\nu) = e^{i\langle y, \xi_\nu\rangle} \psi(y|\xi_\nu|^{\frac{1}{2}}) \tag{22.4.4$'$}$$

where $\psi \in C_0^\infty(\mathbb{R}^n)$, which is just a translation of (22.4.4). Then we have

$$|\xi_\nu|^{n/2 - 2s} \|u_\nu\|^2_{(s)} \to \|\psi\|^2, \tag{22.4.5$'$}$$

$$Pu_\nu(y + x_\nu) = e^{i\langle y, \xi_\nu\rangle} \phi_\nu(y|\xi_\nu|^{\frac{1}{2}}), \tag{22.4.6$'$}$$

$$\phi_\nu(y) = (2\pi)^{-n} \int e^{i\langle y, \eta\rangle} p(x_\nu + y/|\xi_\nu|^{\frac{1}{2}}, \xi_\nu + \eta|\xi_\nu|^{\frac{1}{2}}) \hat\psi(\eta)\, d\eta. \tag{22.4.7$'$}$$

We claim that for any N

$$\begin{aligned} \Big\|\phi_\nu(y) - \sum_{|\alpha+\beta| \leqq 2} |\xi_\nu|^{(|\alpha| - |\beta|)/2} p^{(\alpha)}_{(\beta)}(x_\nu, \xi_\nu)\, y^\beta D^\alpha \psi / \alpha!\, \beta!\Big\| \\ \leqq C(1 + |y|)^{-N} |\xi_\nu|^{m - \frac{3}{2}}. \end{aligned} \tag{22.4.8$'$}$$

When $N = -3$ this is precisely the estimate (22.4.8). Since

$$y^\gamma \phi_\nu(y) = (2\pi)^{-n} \int e^{i\langle y, \eta\rangle} (-D_\eta)^\gamma p(x_\nu + y/|\xi_\nu|^{\frac{1}{2}}, \xi_\nu + \eta|\xi_\nu|^{\frac{1}{2}}) \hat\psi(\eta)\, d\eta,$$

the proof of (22.4.8) also gives (22.4.8)$'$ with $N = -3$ and any power y^γ in the left-hand side. This is equivalent to (22.4.8)$'$. By hypothesis and Taylor's formula

$$|\xi_\nu|^{1-m} \sum_{|\alpha+\beta| \leqq 2} |\xi_\nu|^{(|\alpha| - |\beta|)/2} p^{(\alpha)}_{m(\beta)}(x_\nu, \xi_\nu)\, y^\beta \eta^\alpha \to Q(y, \eta)$$

so $p^{(\alpha)}_{m(\beta)}(x_\nu, \xi_\nu/|\xi_\nu|) = O(|\xi_\nu|^{|\beta+\alpha|/2 - 1}) \to 0$ if $|\alpha| + |\beta| < 2$, and

$$\begin{aligned} |\xi_\nu|^{n/2 + 2(1-m)} \|Pu_\nu\|^2_{(0)} &= \| |\xi_\nu|^{1-m} \phi_\nu\|^2_{(0)} \\ &\to \|Q(y, D)\psi + p_{m-1}(x^0, \xi^0)\psi\|^2, \quad \nu \to \infty. \end{aligned}$$

Hence it follows from (22.4.10)$'$ and (22.4.5)$'$ that

$$\|\psi\| \leqq C_K \|Q(y, D)\psi + p_{m-1}(x^0, \xi^0)\psi\|.$$

As in the proof of Proposition 22.4.1 we have

$$Q(y,D)+p_{m-1}(x^0,\xi^0)=Q^w(y,D)+p^s_{m-1}(x^0,\xi^0)$$

so this proves (22.4.11). If we replace ξ_ν by $t^2\xi_\nu$ where $t>0$, then $Q(y,\eta)$ is replaced by $t^2Q(y/t,\eta/t)$. Substitution of $\psi(-y)$ for $\psi(y)$ changes the sign of t, so this proves the last statement.

Unfortunately the theorem gives no information on the lower order terms of Q. However, they can be determined easily if p_m vanishes precisely to the second order on the characteristic set Σ of p_m, that is, if for some positive constants C_1 and C_2

(22.4.12) $$C_2\,d(x,\xi)^2\leqq|p_m(x,\xi)|\leqq C_1\,d(x,\xi)^2;\quad x\in K,\ |\xi|=1,$$

where d denotes the distance to Σ. Indeed, let

$$|\xi_\nu|^{1-m}p_m(x_\nu+y|\xi_\nu|^{-\frac12},\ \xi_\nu+\eta|\xi_\nu|^{\frac12})\to Q(y,\eta).$$

Then $|\xi_\nu|^{1-m}p_m(x_\nu,\xi_\nu)=|\xi_\nu|\,p_m(x_\nu,\xi_\nu/|\xi_\nu|)$ is bounded, so (22.4.12) gives

$$d(x_\nu,\xi_\nu/|\xi_\nu|)=O(|\xi_\nu|^{-\frac12}).$$

Choose $(x'_\nu,\xi'_\nu)\in\Sigma$ with $|\xi'_\nu|=|\xi_\nu|$ so that $|x_\nu-x'_\nu|=O(|\xi_\nu|^{-\frac12})$ and $|\xi_\nu-\xi'_\nu|=O(|\xi_\nu|^{\frac12})$. Passing to a subsequence we have

$$|\xi_\nu|^{1-m}p_m(x'_\nu+y|\xi'_\nu|^{-\frac12},\ \xi'_\nu+\eta|\xi'_\nu|^{\frac12})\to Q_0(y,\eta)$$

where Q_0 is a homogeneous polynomial. If $(x_\nu-x'_\nu)|\xi_\nu|^{\frac12}\to y_1$ and $(\xi_\nu-\xi'_\nu)|\xi_\nu|^{-\frac12}\to\eta_1$, it follows that $Q(y,\eta)=Q_0(y+y_1,\eta+\eta_1)$.

Now a translation of ψ or the Fourier transform $\hat\psi$ of ψ shows that a real translation of the polynomial Q does not affect the validity of (22.4.11). When (22.4.12) is valid it follows that the condition in Proposition 22.4.3 is not changed if we take the lower order terms of Q to be 0.

Our next goal is to determine when an estimate of the form (22.4.11) is valid for some C and then to determine when the same constant can be used for a family of (Q,κ), $\kappa=p^s_{m-1}(x^0,\xi^0)$. In doing so we shall keep in mind that Q is obtained from p_m as described in Proposition 22.4.3, which implies that

(22.4.1)′ $$Q(x,\xi)\in\Gamma,\ (x,\xi)\in\mathbb{R}^{2n},$$

if p_m satisfies (22.4.1). At first we even assume that $\operatorname{Re}Q$ has a positive definite principal part.

Lemma 22.4.4. *Let Q be a complex valued quadratic polynomial in $\mathbb{R}^{2n}$ with principal part Q_2 such that $\operatorname{Re}Q_2$ is positive definite. Let $\mu_1,\dots,\mu_n$ be the eigenvalues of the Hamilton map of Q_2/i in the half plane $\operatorname{Re}\lambda>0$, and let c be the value of Q at the unique point $(z,\zeta)\in\mathbb{C}^{2n}$ where $dQ(z,\zeta)=0$. Then*

(22.4.13) $$\|\psi\|\leqq C\,\|Q^w(x,D)\psi\|,\ \psi\in\mathscr{S},$$

holds for some C if and only if

(22.4.14)
$$c+\sum_1^n (2\alpha_j+1)\mu_j \neq 0, \quad 0 \leqq \alpha_j \in \mathbb{Z},$$

and then we also have for some other constant C′

(22.4.13)′
$$\sum_{|\alpha+\beta|\leqq 2} \|y^\beta D^\alpha \psi\| \leqq C' \|Q^w(x,D)\psi\|, \psi \in \mathscr{S}.$$

Proof. Let B be the Hilbert space of all $\psi \in L^2$ with $y^\beta D^\alpha \psi \in L^2$, $|\alpha+\beta| \leqq 2$, and the norm

$$\Big(\sum_{|\alpha+\beta|\leqq 2} \|y^\beta D^\alpha \psi\|^2\Big)^{\frac{1}{2}}.$$

The operator Q^w is a Fredholm operator from B to $L^2(\mathbb{R}^n)$. In fact, $e(x,\xi) = \overline{Q(x,\xi)}/(1+|Q(x,\xi)|^2)$ is a symbol of weight $(1+|x|+|\xi|)^{-2}$ with respect to the metric

$$(|dx|^2+|d\xi|^2)/(1+|x|^2+|\xi|^2)$$

so $e^w(x,D)$ is a continuous map from L^2 to B. We have

$$e^w Q^w = I + R_1^w, \; Q^w e^w = I + R_2^w,$$

where R_1 and R_2 also have weight $(1+|x|+|\xi|)^{-2}$, so R_1^w and R_2^w are compact in L^2 and in B, which proves the Fredholm property. Also note that $Q^w u = 0$ implies $u = -R_1^w u = (-R_1^w)^N u$ for every N, hence $u \in \mathscr{S}$. If $v \in \mathscr{S}'$ is orthogonal to $Q^w \mathscr{S}$ then $v = -\bar{R}_2^w v$, hence $v \in \mathscr{S}$. Since

$$(x,\xi) \mapsto tQ(x,\xi) + (1-t)(1+|x|^2+|\xi|^2)$$

satisfies the hypotheses in the lemma for $0 \leqq t \leqq 1$ it follows that the index is the same as for the operator $|x|^2+|D|^2+1$ which is injective on $\mathscr{S}$ with range dense in L^2 hence of index 0 as operator from B to L^2. The index is therefore always 0, so (22.4.13)′ is valid if $Q^w(x,D)\mathscr{S}$ is dense in L^2.

In order to study the kernel and cokernel of Q^w we use Theorem 21.5.6 and Corollary 21.5.8 to choose symplectic coordinates such that $Q_2(x, ix) = 0$. With $t \in \mathbb{C}^n$ to be chosen later we set

$$E = e^{i\langle x,t\rangle - |x|^2/2}$$

and note that $E \in \mathscr{S}$ and that

$$E^{-1} Q^w E v = \tilde{Q}^w v$$

where

$$\tilde{Q}(x,\xi) = Q(x, \xi + ix + t)$$

is linear in x. If s is another vector in $\mathbb{C}^n$ we have if v is an entire function

$$\delta_{-s} * \tilde{Q}^w(v * \delta_s) = R^w v$$

where $v * \delta_s(x) = v(x-s)$ and

$$R(x,\xi) = Q(x+s, \xi + i(x+s) + t).$$

Now the radical of Q_2 is $\{0\}$ (cf. Theorem 21.5.4) so the equation $dQ(z,\zeta)=0$ has a unique solution, characterized by

$$Q(x+z,\xi+\zeta)=Q_2(x,\xi)+Q(z,\zeta).$$

With $s=z$, $t=\zeta-iz$ and the notation $c=Q(z,\zeta)$ we obtain

$$R^w v=Q_2^w(x,D+ix)v+cv.$$

The fact that $Q_2(x,ix)=0$ means that R^w maps polynomials of degree $\leqq\mu$ to polynomials of degree $\leqq\mu$, for every μ. Two cases can occur:

a) $R^w v=0$ for some polynomial $v\neq 0$. Then

$$Q^w(E(v*\delta_s))=0$$

so (22.4.13) is not valid.

b) $R^w v\neq 0$ for every polynomial $v\neq 0$. Then R is surjective on the space of polynomials of degree $\leqq\mu$. Now the products of $E_s=E*\delta_s$ by polynomials are dense in L^2. In fact, if $f\in L^2$ and

$$\int fE_s v\,dx=0$$

for every polynomial v, then the derivatives of the entire analytic function $(\widehat{fE_s})$ are 0 at 0, so $\widehat{fE_s}=0$, hence $f=0$. Then (22.4.13)′ must be valid.

What remains is just to show that (22.4.14) is equivalent to injectivity of R^w on the space of polynomials. Using the polarized form of Q_2 and the Hamilton map F defined by (21.5.1) we have

$$\begin{aligned}Q_2(x,\xi+ix;x,\xi+ix)&=2Q_2(0,\xi;x,ix)+Q_2(0,\xi)\\&=2\sigma((0,\xi),F(x,ix))+Q_2(0,\xi).\end{aligned}$$

The space $\{(x,ix)\}$ is spanned by the generalized eigenvactors of F with eigenvalues $i\mu_j$, so $F(x,ix)=(Mx,iMx)$ where M has the same eigenvalues. Hence

$$R^w=2\langle Mx,D\rangle+\operatorname{Tr}M/i+c+Q_2(0,D).$$

The first three terms preserve the order of a homogeneous polynomial while the third lowers it by two units. Since the eigenvalues of a triangular matrix are determined by the diagonal elements it follows that $Q_2(0,D)$ does not affect the eigenvalues of R^w as an operator on the space of polynomials. By Lemma C.2.2 in the appendix the eigenvalues of $2\langle Mx,D\rangle+\operatorname{Tr}M/i+c$ as an operator on homogeneous polynomials of degree μ are precisely the sums

$$2\sum\alpha_j\mu_j+\operatorname{Tr}M/i+c,\qquad |\alpha|=\mu.$$

Hence (22.4.14) means precisely that no eigenvalue is equal to 0, which completes the proof.

Proposition 22.4.5. *Let Q be a quadratic form in $\mathbb{R}^{2n}$ satisfying* (22.4.1)′, *denote by μ_j the eigenvalues in $\Gamma\setminus 0$ of the Hamilton map of Q/i, and let V_0 be the space of generalized eigenvectors belonging to the eigenvalue* 0. *Then* *the estimate*

$$\|\psi\| \leqq C\|(Q^w(x,D)+\kappa)\psi\|, \quad \psi\in\mathscr{S}(\mathbb{R}^n), \tag{22.4.15}$$

is valid for some C if and only if

$$\kappa+Q(\bar v,v)+\sum(2\alpha_j+1)\mu_j\neq 0 \quad \textit{when } v\in V_0 \textit{ and } 0\leqq\alpha_j\in\mathbb{Z}. \tag{22.4.16}$$

Proof. Since $\operatorname{Re} Q$ is semi-definite we can use Theorem 21.5.3 to choose symplectic coordinates such that

$$\operatorname{Re} Q=\sum_1^k \lambda_j^2(x_j^2+\xi_j^2)+\sum_{k+1}^{k+l} x_j^2; \quad \lambda_j>0. \tag{22.4.17}$$

In view of Theorem 21.5.4 it follows that Q is then a quadratic form in $x'=(x_1,\dots,x_k)$, $\xi'=(\xi_1,\dots,\xi_k)$ and $x''=(x_{k+1},\dots,x_{k+l})$. Thus $Q^w(x,D)$ only involves derivatives with respect to x' and can be regarded as a differential operator in x' depending on the parameters x''. The estimate (22.4.15) is therefore valid if and only if for every fixed x''

$$\|\psi\| \leqq C\|(Q^w(x',x'',D')+\kappa)\psi\|, \quad \psi\in\mathscr{S}(\mathbb{R}^k). \tag{22.4.15)$'$}$$

As observed after Theorem 21.5.4

$$S'_{\mathbb{C}}=\operatorname{Im} F/(\operatorname{Ker} F\cap \operatorname{Im} F)\cong \bigoplus_{\lambda\neq 0} V_\lambda;$$

here F is the Hamilton map of Q and $\operatorname{Im} F=(\operatorname{Ker} F)^\sigma$ is the complex $x'\xi'\xi''$ plane, $\operatorname{Ker} F\cap\operatorname{Im} F$ is the ξ'' plane, so (x',ξ') are symplectic coordinates in $S'_{\mathbb{C}}$. The map F' induced by F in $S'_{\mathbb{C}}$ is the Hamilton map of $Q(x',0,\xi')$ which is therefore a non-degenerate quadratic form. By b) in Theorem 21.5.4 V_0 is the Q orthogonal space of $\operatorname{Im} F$, that is, V_0 is defined by $d_{x'\xi'}Q(x',x'',\xi')=0$. Hence it follows from Lemma 22.4.4 that

$$\|\psi\| \leqq C(x'')\|(Q^w(x',x'',D')+\kappa)\psi\|, \quad \psi\in\mathscr{S}(\mathbb{R}^k), \tag{22.4.15)$''$}$$

if and only if

$$\kappa+Q(z',x'',\zeta')+\sum(2\alpha_j+1)\mu_j\neq 0, \quad 0\leqq\alpha_j\in\mathbb{Z}$$

when $(z',x'',\zeta')\in V_0$.

Any $v\in V_0$ can be written in the form $v=v_1+iv_2$ where $v_1,v_2\in V_0$ and the x'' coordinates are real, for (x',ξ') is a linear function of x''. Hence $\bar v-v_1+iv_2$ has x'' coordinates 0, so

$$Q(\bar v,v)=Q(v_1-iv_2,v_1+iv_2)=Q(v_1)+Q(v_2).$$

The range of $Q(\bar v,v)$ when $v\in V_0$ is therefore the same convex angle as the range of $Q(v)$ when $v\in V_0$ and v is required to have real x'' coordinates. This proves that (22.4.16) is necessary and sufficient for (22.4.15)$''$ to be valid with $C(x'')<\infty$ for every x''. However, the fact that (22.4.13) is equivalent to (22.4.13)$'$ shows that the best constant is locally bounded. We can also take

$C(x'')$ bounded for large x''. In fact, we have

$$\operatorname{Re}(Q(x,\xi)+\kappa)\geqq 1 \quad \text{if } |x''|>M.$$

This implies

$$\|\psi\|^2 \leqq \operatorname{Re}((Q^w(x,D)+\kappa)\psi,\psi), \qquad \psi\in\mathscr{S}(\mathbb{R}^k),\ |x''|>M,$$

as follows immediately from Lemma 22.3.1 if we remove the first order terms in x', D' by a translation of ψ and of $\hat\psi$. Hence

$$\|\psi\| \leqq \|(Q^w(x,D)+\kappa)\psi\|, \qquad \psi\in\mathscr{S}(\mathbb{R}^k),\ |x''|>M,$$

which completes the proof.

It is easy to see that the estimate (22.4.15) remains valid with a small change of C if κ and Q are replaced by κ', Q' and $\kappa-\kappa'$, $Q-Q'$ are sufficiently small, rank $Q=$ rank Q', and the dimension of the zero eigenspace remains fixed. However, small perturbations of Q may increase the rank and decrease the zero eigenspace. To obtain explicit conditions for (22.4.11) to hold we must therefore study families of estimates (22.4.15).

Proposition 22.4.6. *Let M be a set of pairs (Q,κ) where $\kappa\in\mathbb{C}$ and Q is a quadratic form in $\mathbb{R}^{2n}$ satisfying* (22.4.1)′. *Assume that κ and the eigenvalues of the Hamilton map of* $\operatorname{Re} Q$ *are uniformly bounded when $(Q,\kappa)\in M$. Then there is a constant C such that* (22.4.15) *is valid for all $(Q,\kappa)\in M$, if and only if there exist positive constants ε, δ such that*

$$\delta\leqq|\kappa+Q(\bar v,v)+\sum(2\alpha_j+1)\mu_j| \tag{22.4.18}$$

if $(Q,\kappa)\in M$, $0\leqq\alpha_j\in\mathbb{Z}$, μ_j are the eigenvalues in $\Gamma\setminus 0$ of the Hamilton map of Q/i and v is in the space spanned by the generalized eigenvectors of the Hamilton map of Q corresponding to eigenvalues of modulus $<\varepsilon$.

In view of the symplectic invariance and Theorem 21.5.3 we may assume in the proof that for every $(Q,\kappa)\in M$ the real part of Q is of the form (22.4.17). After splitting M into a finite number of sets we can assume that k and l are fixed and that $k+l=n$. By hypothesis we have a uniform bound for λ_j. It will be sufficient to show that every sequence $(Q_\nu,\kappa_\nu)\in M$ satisfying one of the conditions in the proposition has a subsequence satisfying both.

Write

$$\operatorname{Re} Q_\nu=\sum_1^k \lambda_{j\nu}^2(x_j^2+\xi_j^2)+\sum_{k+1}^n x_j^2. \tag{22.4.17$'$}$$

Since there is a uniform bound for λ_{j_ν} we may assume that $\lim_{\nu\to\infty}\lambda_{j_\nu}=\lambda_j$ exists for every j. The coordinates can be labelled so that $\lambda_j\neq 0$ when $j\leqq n'$ and $\lambda_j=0$ when $j>n'$. We shall write

$$x'=(x_1,\dots,x_{n'}),\quad x''=(x_{n'+1},\dots,x_k),\quad x'''=(x_{k+1},\dots,x_n).$$

By hypothesis $|\operatorname{Im} Q_\nu| \leqq \gamma \operatorname{Re} Q_\nu$, so if we write x''/λ_ν for the vector with coordinates $x_j/\lambda_{j\nu}, n'<j\leqq k$, then

$$|\operatorname{Im} Q_\nu(x', x''/\lambda_\nu, x''', \xi', \xi''/\lambda_\nu)| \leqq \gamma \operatorname{Re} Q_\nu(x', x''/\lambda_\nu, x''', \xi', \xi''/\lambda_\nu).$$

The right-hand side converges to

$$\gamma\left(\sum_1^{n'} \lambda_j^2(x_j^2+\xi_j^2)+|x''|^2+|\xi''|^2+|x'''|^2\right)$$

so the coefficients of $\operatorname{Im} Q_\nu$ are uniformly bounded after this dilation. Passing to a subsequence if necessary we obtain

$$Q_\nu(x', x''/\lambda_\nu, x''', \xi', \xi''/\lambda_\nu) \to Q(x', x'', x''', \xi', \xi'')$$

where Q is elliptic in the indicated variables. Note that we have made a non-symplectic change of scales here. The justification for this will be given by the following lemmas, which also involve the limit

$$Q_\infty(x,\xi)=Q(x',0,x''',\xi',0)$$

when the scale is not changed. The Hamilton map of Q_∞ has n' non-zero eigenvalues in $i\Gamma$, and the generalized eigenspace belonging to 0 is given by $(x',\xi')=H_\infty(x''')$ where H_∞ is a linear map. For any convergent sequence of linear maps in a finite dimensional vector space, the linear hull of the generalized eigenvectors belonging to eigenvalues converging to 0 converges to the zero eigenspace of the limit. It follows that the space N_ν spanned by generalized eigenvectors of the Hamilton map of Q_ν belonging to eigenvalues converging to 0 can for large ν be written in the form $(x',\xi') = H_\nu(x'', x''', \xi'')$.

Lemma 22.4.7. *$H_\nu(x''/\lambda_\nu, x''', \xi''/\lambda_\nu)$ has a limit $H(x'', x''', \xi'')$ as $\nu\to\infty$, and $(x',\xi')=H(x'',x''',\xi'')$ is equivalent to*

$$Q(x,\xi;\, y,\eta)=0 \quad \textit{when } y''=y'''=\eta''=\eta'''=0, \tag{22.4.19}$$
$$\textit{that is, } \partial Q(x,\xi)/\partial(x',\xi')=0.$$

(If F and F_∞ are the Hamilton maps of Q and Q_∞ this means that $F_\infty F(x,\xi) = 0$.)

Proof. $Q_\nu(x', x''/\lambda_\nu, x''', \xi', \lambda_\nu \xi'')$ differs from Q_ν by a symplectic dilation so the space spanned by the generalized eigenvectors with small eigenvalues of the corresponding Hamilton map is defined by

$$(x',\xi')=H_\nu(x''/\lambda_\nu, x''', \lambda_\nu \xi'').$$

This equation must therefore converge to the equation of the zero generalized eigenspace of the Hamilton map of $Q(x,\xi',0)$. Here we may interchange the roles of x'' and ξ'', which proves that the asserted limit H exists. Moreover, from b) in Theorem 21.5.4 it follows that (22.4.19) is valid if $(x',\xi') = H(x'',x''',\xi'')$ and either ξ'' or x'' is 0. In view of the linearity the last condition can immediately be removed. Since (22.4.19) determines (x', ξ') uniquely in terms of the other variables, the lemma is proved.

Lemma 22.4.8. *The limit of the range of* $Q_\nu(\bar{v}, v)$ *when* $v \in N_\nu$ *is equal to the range of* $Q(\bar{v}, v)$ *when* $v = (x, \xi)$ *and* $(x', \xi') = H(x'', x''', \xi'')$. *This does not change if we require* x'', ξ'', x''' *to be real.*

Proof. The range of Q_ν in question is the range of

$$Q_\nu(\bar{x}', \bar{x}''/\lambda_\nu, \bar{x}''', \bar{\xi}', \bar{\xi}''/\lambda_\nu; x', x''/\lambda_\nu, x''', \xi', \xi''/\lambda_\nu)$$

when $(x', \xi') = H_\nu(x''/\lambda_\nu, x''', \xi''/\lambda_\nu)$. The quadratic form converges to Q, the restriction on (x, ξ) converges to $(x', \xi') = H(x'', x''', \xi'')$, and Q is elliptic in the variables $(x', x'', x''', \xi', \xi'')$ which proves the first statement. To prove the last one we just have to repeat an argument in the proof of Proposition 22.4.5, this time using (22.4.19). This is left for the reader.

Proposition 22.4.6 is proved in one direction by Lemma 22.4.8 and

Lemma 22.4.9. *Assume that* (22.4.15) *is fulfilled by the sequence* (Q_ν, κ_ν) *and that* $\kappa_\nu \to \kappa$. *With the preceding notation it follows that*

$$\kappa + Q(\bar{v}, v) + \sum (2\alpha_j + 1)\mu_j \neq 0 \tag{22.4.16$'$}$$

if μ_j *are the eigenvalues in* Γ *of the Hamilton map of* Q_∞/i, $0 \leqq \alpha_j \in \mathbb{Z}$, *and* $v = (x, \xi)$ *satisfies* $(x', \xi') = H(x'', x''', \xi'')$.

Proof. The last statement in Lemma 22.4.8 shows that we may assume that x'', x''', ξ'' are real. By a rotation in the $x_j \xi_j$ plane for $n' < j \leqq k$ we can arrange that $\xi'' = 0$. A symplectic change of scales shows that (22.4.15) is equivalent to

$$\|\psi\| \leqq C \|Q_\nu^w(x', x''/\lambda_\nu, x''', D', \lambda_\nu D'')\psi + \kappa_\nu \psi\|, \quad \psi \in \mathscr{S}(\mathbb{R}^n),$$

where C is independent of ν. Letting $\nu \to \infty$ we conclude that

$$\|\psi\| \leqq C \|Q^w(x, D', 0)\psi + \kappa\psi\|, \quad \psi \in \mathscr{S}(\mathbb{R}^n).$$

The zero eigenspace of the corresponding Hamilton map is defined by $(x', \xi') = H(x'', x''', 0)$, so (22.4.16)$'$ follows from (22.4.16) when $\xi'' = 0$.

Remark. All positive multiples of the small eigenvalues which occur in (22.4.16) are contained in the range of Q on the corresponding eigenspace. Hence (22.4.16)$'$ implies that (22.4.16) is satisfied uniformly by κ_ν, Q_ν when ν is large. However, simple examples show that (22.4.16)$'$ is a good deal stronger than that.

When proving the sufficiency of (22.4.16)$'$ we shall regard Q^w as a (pseudo-)differential operator in x'', x''' with values in operators on functions of x'. To shorten notation we write $t=(x'',x''',\xi'')$. From the proof of Lemma 22.4.4 we know that (22.4.16)′ implies that $Q^w(x',D',t)+\kappa$ for every real t has an inverse $E(t)$ which is a continuous operator from L^2 to the Hilbert space B of all $u\in L^2(\mathbb{R}^{n'})$ with

$$\|u\|_B=\Big(\sum_{|\alpha+\beta|\leq 2}\|x'^\alpha D'^\beta u\|^2\Big)^{\frac12}<\infty.$$

It follows that $E(t)$ also exists and is an analytic function of t when t is in some small complex neighborhood of $\mathbb{R}^{n+k-2n'}$. In particular, $E\in C^\infty$.

Lemma 22.4.10. *For all multi-indices α, β, γ with $|\alpha+\beta|\leq 2$ we have*

$$(1+|t|)^{|\gamma|+2-|\alpha|-|\beta|}\|D_t^\gamma x'^\alpha D'^\beta Eu\|\leq C_\gamma\|u\|,\quad u\in C_0^\infty(\mathbb{R}^{n'}). \tag{22.4.20}$$

Proof. It suffices to prove (22.4.20) when $|t|$ is large. Assume first that $\gamma=0$. There is a constant $c>0$ such that for large t

$$c(|x'|^2+|\xi'|^2+|t|^2)\leq \operatorname{Re}(Q(x',\xi',t)+\kappa).$$

Hence the end of the proof of Proposition 22.4.5 shows that

$$\begin{aligned} c\int((|x'|^2+|t|^2)|v|^2+|D'v|^2)\,dx' &\leq \operatorname{Re}((Q^w(x',D',t)+\kappa)v,v)\\ &\leq |t|^{-2}\|(Q^w+\kappa)v\|^2/2c+c|t|^2\|v\|^2/2,\quad v\in\mathscr{S}(\mathbb{R}^{n'}).\end{aligned}$$

Hence

$$c^2|t|^2\int((|x'|^2+|t|^2)|v|^2+|D'v|^2)\,dx'\leq\|(Q^w(x',D',t)+\kappa)v\|^2,$$

so we have

$$\sum_{|\alpha+\beta|<2} t^{2-|\alpha|-|\beta|}\|x'^\alpha D'^\beta v\|\leq C\|(Q^w(x',D',t)+\kappa)v\|. \tag{22.4.20$'$}$$

(22.4.20) follows when $\gamma=0$ if we combine this result with the estimate

$$\sum_{|\alpha+\beta|=2}\|x'^\alpha D'^\beta v\|\leq C(\|Q^w(x',D',0)v\|+\|v\|)$$

which follows from Lemma 22.4.4. Since $D_tE=-E(D_tQ)E$, we obtain (22.4.20) inductively by repeated differentiations.

It is clear that we may replace $Q+\kappa$ in (22.4.20)′ and therefore in (22.4.20) by any sufficiently small perturbation which is independent of ξ'''. In particular we can for large ν replace $Q+\kappa$ by

$$Q_\nu(x',x''/\lambda_\nu,x''',\xi',\xi''/\lambda_\nu)+\kappa_\nu$$

and obtain an operator $E_\nu(x'',x''',\xi'')$ such that (22.4.20) is valid uniformly and

$$(Q_\nu^w(x',x''/\lambda_\nu,x''',D',\xi''/\lambda_\nu)+\kappa_\nu)E_\nu(x'',x''',\xi'')=I.$$

Set

$$E_\nu'(x'',x''',\xi'')=E_\nu(\lambda_\nu x'',x''',\lambda_\nu\xi'').$$

For $u \in C_0^\infty(\mathbb{R}^n)$ we denote by $\hat{u}(x', \xi'', x''')$ the Fourier transform with respect to the x'' variables and set

$$\mathscr{E}_\nu u(x) = (2\pi)^{-n''} \int e^{i\langle x'', \xi''\rangle} E'_\nu(x'', x''', \xi'') \hat{u}(x', \xi'', x''') \, d\xi''.$$

It follows from (22.4.20) that the norm of $D^\gamma_{x'',\xi''} E'_\nu$ as operator in $L^2(\mathbb{R}^{n'})$ is bounded by $C'(1+|x''|+|\xi''|)^{-|\gamma|}$, so the vector valued version of Theorem 18.1.11 proves that $\mathscr{E}_\nu$ is uniformly bounded in $L^2(\mathbb{R}^n)$. A direct computation gives

$$(Q^w_\nu(x, D) + \kappa_\nu) \mathscr{E}_\nu = I + \mathscr{R}_\nu,$$

where $\mathscr{R}_\nu$ is defined as $\mathscr{E}_\nu$ but with E'_ν replaced by

$$\begin{aligned} R'_\nu(x'', x''', \xi'') &= (Q^w_\nu(x, D', D'' + \xi'') - Q^w_\nu(x, D', \xi''))\, E'_\nu(x'', x''', \xi'') \\ &= R_\nu(\lambda_\nu x'', x''', \lambda_\nu \xi''), \\ R_\nu(x'', x''', \xi'') &= (Q^w_\nu(x', x''/\lambda_\nu, x''', D', (\lambda^2_\nu D'' + \xi'')/\lambda_\nu) \\ &\quad - Q^w_\nu(x', x''/\lambda_\nu, x''', D', \xi''/\lambda_\nu))\, E_\nu. \end{aligned}$$

It follows immediately from (22.4.20) that

$$\|D^\gamma_t R_\nu u\| \leq C \sum_{n'+1}^{k} |\lambda_{j\nu}|^2 (1+|t|)^{-|\gamma|} \|u\|, \quad u \in C_0^\infty(\mathbb{R}^{n'}).$$

Hence we conclude that the norm of $\mathscr{R}_\nu$ as operator in $L^2(\mathbb{R}^n)$ tends to 0 as $\nu \to \infty$. Since we know that $(Q^w_\nu + \kappa_\nu)$ is invertible, the inverse is $\mathscr{E}_\nu (I + \mathscr{R}_\nu)^{-1}$ which has uniformly bounded norm as $\nu \to \infty$. The proof of Proposition 22.4.6 is now completed in view of Lemmas 22.4.8, 22.4.9 and the remarks which followed the statement.

Proposition 22.4.6 assumes a much simpler form if one has a compact family where Q has constant rank. This is the only case which occurs in our application when the characteristic set is a manifold and p_m is transversally elliptic.

Corollary 22.4.11. *Let M be a compact set of pairs (Q, κ) of complex numbers κ and quadratic forms Q in $\mathbb{R}^{2n}$ satisfying* (22.4.1)'. *Assume that all forms Q occurring have the same rank r. Then there is a uniform estimate* (22.4.15) *when $(Q, \kappa) \in M$ if and only if* (22.4.16) *is valid for each $(Q, \kappa) \in M$.*

Proof. Let $Q_\nu \to Q$ and assume that all Q_ν have the same rank as Q. Denote by N_0 the space of generalized eigenvectors of the Hamilton map of Q with eigenvalue 0 and by N_ν the space spanned by generalized eigenvectors of the Hamilton map of Q_ν belonging to eigenvalues $\to 0$. By Proposition 22.4.6 it suffices to show that the range of $Q(\bar{v}, v)$ when $v \in N_0$ is the limit of the range of $Q_\nu(\bar{v}, v)$ when $v \in N_\nu$, $\nu \to \infty$. It is obvious that $\dim N_\nu = \dim N_0$ for large ν, and N_ν resp. N_0 contains the null spaces N'_ν and N'_0 of the Hamilton maps of Q_ν and Q. Since N'_0 has the same dimension as N'_ν by hypothesis, it follows that $N'_\nu \to N'_0$. We can choose a linear bijection $T_\nu: N_0 \to N_\nu$ converging to the identity as $\nu \to \infty$ so that $T_\nu N'_0 = N'_\nu$. Then $Q_\nu \circ T_\nu$ induces a hermitian form Q'_ν on N_0/N'_0 converging to the elliptic form Q' induced there by Q. This implies that the range of Q'_ν converges to that of Q', which proves the corollary.

In Proposition 22.4.6 we have given necessary and sufficient conditions for the uniform validity of the test estimate (22.4.11). Our next goal is to prove a converse of Proposition 22.4.3. In doing so it is convenient to assume $m=1$, which is no essential restriction.

Lemma 22.4.12. *Let $P \in \Psi^1_{\text{phg}}(X)$ be properly supported with principal symbol p_1 satisfying* (22.4.1). *Suppose that for every compact set $K \subset X$ there is a constant C such that* (22.4.11) *is valid for all limits defined in Proposition* 22.4.3 *with $x^0 \in K$. Then there is a function $\varepsilon(\xi) \to 0$, $\xi \to \infty$, such that for every $x \in K$*

$$(22.4.21) \quad \int |\psi|^2\,dy \leqq (C+1) \int \Big| \sum_{|\alpha+\beta| \leqq 2} p^{(\alpha)}_{1(\beta)}(x,\xi) |\xi|^{(|\alpha|-|\beta|)/2} y^\beta D^\alpha \psi/\alpha!\,\beta! + p_0(x,\xi)\psi\Big|^2 dy + \varepsilon(\xi) \int \sum_{|\alpha+\beta| \leqq 3} |y^\beta D^\alpha \psi|^2\,dy, \quad \psi \in C_0^\infty(\mathbb{R}^n).$$

Proof. We just have to show that (22.4.21) is valid for large $|\xi|$ if $\varepsilon(\xi)$ is replaced by a fixed positive number ε. Assume that this is false. Then there exists for some $\varepsilon > 0$ sequences $x_j \in K$ and $\xi_j \to \infty$, $\psi_j \in C_0^\infty$, such that

$$(22.4.22) \quad 1 = \int |\psi_j|^2\,dy > (C+1) \int \Big| \sum_{|\alpha+\beta| \leqq 2} p^{(\alpha)}_{1(\beta)}(x_j,\xi_j) |\xi_j|^{(|\alpha|-|\beta|)/2} y^\beta D^\alpha \psi_j/\alpha!\,\beta! + p_0(x_j,\xi_j)\psi_j\Big|^2 dy + \varepsilon \int \sum_{|\alpha+\beta| \leqq 3} |y^\beta D^\alpha \psi_j|^2\,dy.$$

It follows that the sequences $x^\alpha D^\beta \psi_j$ are precompact in L^2 when $|\alpha+\beta| \leqq 2$, so we may assume that they converge to $x^\alpha D^\beta \psi$ in L^2 for some $\psi \in L^2$. Since (22.4.1) implies by Lemma 7.7.2 that for $|\alpha+\beta|=1$

$$|p^{(\alpha)}_{1(\beta)}(x,\xi)|\,|\xi|^{(|\alpha|-|\beta|)/2} \leqq C(\operatorname{Re} p_1(x,\xi))^{\frac12}, \qquad x \in K,$$

the sequence $p_1(x_j,\xi_j)$ must be uniformly bounded, for the right-hand side of (22.4.22) would otherwise be asymptotically equal to $|p_1(x_j,\xi_j)|^2 (C+1)$. Hence we obtain a limit Q as in Proposition 22.4.3 and a function ψ with $x^\alpha D^\beta \psi \in L^2$, $|\alpha|+|\beta| \leqq 3$, for which (22.4.11) is not valid. Approximation of ψ with a C_0^∞ function gives a contradiction which proves the lemma.

When we use (22.4.21) it is convenient to change scales by substituting $y|\xi|^{\frac12}$ for y and then $y-x$ for y, $D-\xi$ for D (which amounts to a translation of ψ or $\hat{\psi}$). This gives the equivalent estimate

$$(22.4.21)' \quad \int |\psi|^2\,dy \le (C+1)\int \Big| \sum_{|\alpha+\beta|\le 2} p_{1(\beta)}^{(\alpha)}(x,\xi)(y-x)^\beta (D-\xi)^\alpha \psi/\alpha!\,\beta! + p_0(x,\xi)\psi\Big|^2 dy + \varepsilon(\xi)\int \sum_{|\alpha+\beta|\le 3} |\xi|^{(|\beta|-|\alpha|)} |(y-x)^\beta (D-\xi)^\alpha \psi|^2\,dy, \quad \psi\in\mathscr{S}(\mathbb{R}^n),\ x\in K.$$

Lemma 22.4.13. *Assume that the hypotheses in Lemma 22.4.12 are fulfilled, and let P^j, P_j be properly supported operators with principal symbols $\partial p_1/\partial\xi_j$ and $|\xi|^{-1}\partial p_1/\partial x_j$ respectively. Then there is for every compact set $K\subset X$ a constant C_K such that for arbitrary $\delta>0$ and $u\in C_0^\infty(K)$*

$$(22.4.23) \quad \|u\|_{(0)} \le C_K \|Pu\|_{(0)} + \delta\Big(\|u\|_{(0)} + \sum_1^n \|P^j u\|_{(\frac12)} + \sum_1^n \|P_j u\|_{(\frac12)}\Big) + C_\delta \|u\|_{(-\frac12)}.$$

Proof. If $\chi\in C_0^\infty(X)$ is equal to 1 in a neighborhood K_1 of K then $Pu=P\chi u = p(x,D)u$, $u\in C_0^\infty(K)$, where $p\in S^1_{\mathrm{phg}}(\mathbb{R}^n)$ has compact support in x and p is equal to the symbol of $P \bmod S^{-\infty}$ in K_1. We choose $\chi_1\in C_0^\infty$ equal to 1 in K and with support in the interior of K_1. The estimate (22.4.21)′ is valid for some $\varepsilon(\xi)\to 0$, $\xi\to\infty$, for all $x\in K_1$.

We shall now localize with respect to the metric

$$g_\delta = |dx|^2(\delta^2|\xi|+1) + \delta^4 |d\xi|^2/(\delta^2|\xi|+1), \qquad 0<\delta<1.$$

The quotient by the dual metric is δ^4; since the metric is slowly varying and σ temperate when $\delta=1$ it follows that this is true uniformly when $0<\delta<1$. Choose $\psi_j\in C_0^\infty(\mathbb{R}^{2n})$ and $(x_j,\xi_j)\in\operatorname{supp}\psi_j$ so that ψ_j is real valued,

$$\sum \psi_j^2 = 1,$$

ψ_j is uniformly bounded in $S(1,g_\delta)$ and there is a fixed bound for the number of overlapping supports and for the g_δ distance from (x_j,ξ_j) to points in $\operatorname{supp}\psi_j$. Then $\{\psi_j\}$ is a symbol in $S(1,g_\delta)$ taking its values in $\ell^2 = \mathscr{L}(\mathbb{C},\ell^2) = \mathscr{L}(\ell^2,\mathbb{C})$, so the calculus gives

$$\sum \|\psi_j(x,D)u\|^2 = (u,u) + (r(x,D)u,u)$$

where r has a uniform bound in $S(\delta^2,g_\delta)$. It follows that

$$(22.4.24) \qquad \|u\|^2 \le \sum \|\psi_j(x,D)u\|^2 + C\delta^2\|u\|^2, \qquad u\in C_0^\infty.$$

The proof of (22.4.23) will be complete if we show that the sum in the right-hand side can be estimated by the right-hand side of (22.4.23).

If $\operatorname{supp}\psi_j$ and $\operatorname{supp}\chi_1$ are disjoint then Theorem 18.5.4 gives

$$\psi_j(x,D)u = \psi_j(x,D)\chi_1 u = \phi_j(x,D)u, \qquad u\in C_0^\infty(K)$$

where $\{\phi_j\}$ is bounded in $S(\delta^2,g_\delta)$. Hence

$$(22.4.25) \qquad \sum{}' \|\psi_j(x,D)u\|^2 \le C\delta^2\|u\|^2, \qquad u\in C_0^\infty(K)$$

if $\sum'$ denotes the sum for such indices. For any finite sum we have

(22.4.25)′
$$\sum'' \|\psi_j(x,D)u\|^2 \leqq C\|u\|^2_{(-1)}.$$

From now on we therefore assume that $x_j \in K_1$ and that $|\xi_j|$ is larger than some suitable number depending on δ. Then we have by (22.4.21)′

(22.4.26)
$$\|\psi_j(x,D)u\|^2 \leqq (C+1)\|T_j(x,D)\psi_j(x,D)u\|^2 + \varepsilon(\xi_j) \sum_{|\alpha+\beta|\leqq 3} \|\psi_{\alpha\beta j}(x,D)u\|^2,$$

where

$$T_j(x,D) = \sum_{|\alpha+\beta|\leqq 2} p^{(\alpha)}_{1(\beta)}(x_j,\xi_j)(x-x_j)^\beta (D-\xi_j)^\alpha/\alpha!\,\beta! + p_0(x_j,\xi_j),$$

$$\psi_{\alpha\beta j}(x,\xi) = (|\xi_j|^{\frac12}(x-x_j))^\beta (|\xi_j|^{-\frac12}(D_x+\xi-\xi_j))^\alpha \psi_j(x,\xi).$$

Now $(|\xi_j|^{-\frac12}D_x)^\alpha\psi_j$ is for large j bounded in $S(\delta^{|\alpha|}, g_\delta)$ and the components of $|\xi_j|^{\frac12}(x-x_j)$ and $|\xi_j|^{-\frac12}(\xi-\xi_j)$ are bounded in $S(\delta^{-1}, g_\delta)$ in $\operatorname{supp}\psi_j$, so $\psi_{\alpha\beta j}$ is in a bounded set in $S(\delta^{-|\alpha+\beta|}, g_\delta)$. Hence

(22.4.27)
$$\sum_{|\alpha+\beta|\leqq 3}\sum_{j>J} \varepsilon(\xi_j)\|\psi_{\alpha\beta j}(x,D)u\|^2 = (Ru,u) \leqq C\delta^2\|u\|^2$$

since the symbol of R is bounded in $S(\delta^2, g_\delta)$ if J is large enough.

To estimate the first term in the right-hand side of (22.4.26) we observe first that the symbol of $T_j(x,D)\psi_j(x,D)$ is

$$\sum_{|\alpha|\leqq 2} T_j^{(\alpha)}(x,\xi) D_x^\alpha \psi_j(x,\xi)/\alpha! = R_j(x,\xi) + S_j(x,\xi),$$

$$R_j(x,\xi) = \sum_{|\alpha|\leqq 1} (T_j^{(\alpha)}(x,\xi) - p^{(\alpha)}(x,\xi)) D_x^\alpha \psi_j(x,\xi)/\alpha!,$$

$$S_j(x,\xi) = \sum_{|\alpha|\leqq 1} p^{(\alpha)}(x,\xi) D_x^\alpha \psi_j(x,\xi)/\alpha! \; - \sum_{|\alpha|=2} p_1^{(\alpha)}(x_j,\xi_j)\psi_{j(\alpha)}(x,\xi)/\alpha!.$$

Here $T_j(x,\xi) - p(x,\xi)$ is bounded in $S(\delta^{-2}(1+\delta^2|\xi|)^{-\frac12}, g_\delta)$ in $\operatorname{supp}\psi_j$ by Taylor's formula, for the norm of the kth differential of p with respect to g_δ is bounded by $(1+|\xi|)((\delta^2|\xi|+1)/\delta^4(1+|\xi|)^2)^{k/2} C_k$. Hence $\{R_j\}$ is bounded in $S(\delta^{-2}(1+\delta^2|\xi|)^{-\frac12}, g_\delta)$ so it follows that

(22.4.28)
$$\sum \|R_j u\|^2 \leqq C\delta^{-4}\|(1+\delta^2|D|)^{-\frac12}u\|^2 \leqq C_\delta\|u\|^2_{(-\frac12)}.$$

Finally

$$S_j(x,D) = \psi_j(x,D)p(x,D) + i\sum \psi_j^{(v)}(x,D)p_{(v)}(x,D) - i\sum \psi_{j(v)}(x,D)p^{(v)}(x,D) + V_j(x,D)$$

where $\{V_j\}$ is bounded in $S(\delta^2 + (1+\delta^2|\xi|)^{-1}, g_\delta)$. In fact, in the asymptotic expansion of the symbol of say $\psi_j(x,D)p(x,D)$ the terms improve successively by at least a factor $(1+\delta^2|\xi|)^{\frac12}(1+|\xi|)^{-1}$, and $(1+\delta^2|\xi|)(1+|\xi|)^{-1}$

$\leq \delta^2+1/(1+\delta^2|\xi|)$, so the error when one breaks off after two terms has this weight. The last term in S_j has a similar bound. It follows that

$$\sum \|V_j(x,D)u\|^2 \leq C\delta^4\|u\|^2 + C_\delta \|u\|^2_{(-1)}.$$

If we note that $\psi_j^{(\nu)}$ and $\psi_{j(\nu)}$ are bounded in $S(\delta^2(\delta^2|\xi|+1)^{-\frac{1}{2}}, g_\delta)$ and in $S((\delta^2|\xi|+1)^{\frac{1}{2}}, g_\delta)$ respectively, we conclude that

$$(22.4.29)\quad \sum \|S_j u\|^2 \leq C(\|p(x,D)u\|^2 + \delta^2 \sum \|p^{(\nu)}(x,D)u\|^2_{(\frac{1}{2})} + \delta^2 \sum \|p_{(\nu)}(x,D)u\|^2_{(-\frac{1}{2})} + \delta^4\|u\|^2) + C_\delta \|u\|^2_{(-1)}.$$

This estimate combined with (22.4.28), (22.4.27), (22.4.26), (22.4.25) and (22.4.24) yields (22.4.23).

From (22.4.1) it follows that $\operatorname{Re} p_1 \geq 0$ and that $\gamma \operatorname{Re} p_1 - \operatorname{Im} p_1 \geq 0$, hence

$$(22.4.30)\quad |\xi| \sum |p_1^{(\nu)}(x,\xi)|^2 + |\xi|^{-1} \sum |p_{1(\nu)}(x,\xi)|^2 \leq C_K \operatorname{Re} p_1(x,\xi), \quad x\in K,$$

by Lemma 7.7.2, as already observed. Theorem 18.1.14 gives

$$\sum \|p^{(\nu)}(x,D)u\|^2_{(\frac{1}{2})} + \sum \|p_{(\nu)}(x,D)u\|^2_{(-\frac{1}{2})} \leq C_K \operatorname{Re}(p(x,D)u,u) + C'_K \|u\|^2, \quad u\in C_0^\infty(K).$$

This means that with new constants

$$\sum \|P^\nu u\|_{(\frac{1}{2})} + \sum \|P_\nu u\|_{(\frac{1}{2})} \leq C_K(\|Pu\| + \|u\|),$$

which combined with (22.4.23) gives for small δ

$$\|u\|_{(0)} \leq C_K(\|Pu\| + \|u\|_{(-\frac{1}{2})}), \quad u\in C_0^\infty(K).$$

The logarithmic convexity of $\|u\|_{(s)}$ in s shows that $C_K\|u\|_{(-\frac{1}{2})}$ can be estimated by $\|u\|_{(0)}/2 + C'_K\|u\|_{(-1)}$, so we can replace $\|u\|_{(-\frac{1}{2})}$ by $\|u\|_{(-1)}$ in the right hand side. Summing up, we have

$$(22.4.31)\quad \|u\|_{(0)} + \sum \|P^j u\|_{(\frac{1}{2})} + \sum \|P_j u\|_{(\frac{1}{2})} \leq C_K(\|Pu\|_{(0)} + \|u\|_{(-1)}), \quad u\in C_0^\infty(K).$$

This is a perfect substitute for (22.2.9), (22.2.10) with $\varepsilon=\frac{1}{2}$ although there is a slight discrepancy in the subscripts caused by the fact that P is now a first order operator. However, nothing has to be changed in proving that (22.4.31) implies that P is microhypoelliptic with loss of one derivative. Summing up the results of this section we have therefore proved

Theorem 22.4.14. *Let $P\in\Psi^m_{\mathrm{phg}}(X)$ be properly supported and assume that the principal symbol satisfies* (22.4.1). *If for some $s\in\mathbb{R}$ we have*

$$u\in\mathscr{E}'(X),\ Pu\in H_{(s)} \Rightarrow u\in H_{(s+m-1)}$$

then there is for every compact set $K\subset X$ a constant $\delta>0$ such that

$$\delta \leq |p^s_{m-1}(x^0,\xi^0) + Q(\bar v, v) + \sum (2\alpha_j+1)\mu_j|, \quad 0\leq\alpha_j\in\mathbb{Z},$$

if $x^0 \in K$, $|\xi^0|=1$, $Q(y, y_0, \eta, \eta_0)$ *is any limit of*

$$y_0^2\, p_m(x+y|\xi|^{-\frac{1}{2}}/y_0,\ \xi+\eta|\xi|^{\frac{1}{2}}/y_0)\,|\xi|^{1-m}$$

as $x \to x^0$, $\xi \to \infty$, $\xi/|\xi| \to \xi^0$. *The numbers* μ_j *are the eigenvalues in* Γ *of the Hamilton map of* Q/i, *and* v *is in the space spanned by its generalized eigenvectors belonging to eigenvalues* $< \delta$ *in absolute value. Conversely, if this condition is fulfilled then*

$$u \in \mathscr{D}'(X),\ Pu \in H_{(s)} \ \text{ at } (x^0,\xi^0) \ \text{ implies } u \in H_{(s+m-1)} \ \text{ at } (x^0,\xi^0)$$

for arbitrary $s \in \mathbb{R}$ *and* $(x^0,\xi^0) \in T^*(X) \setminus 0$, *so* P *is microhypoelliptic.*

The condition in the theorem is deplorably complicated, but it simplifies if one makes stronger hypotheses on the geometry:

Theorem 22.4.15. *Assume in addition to* (22.4.1) *that the zeros of* p_m *form a* C^∞ *manifold where* p_m *is transversally elliptic. Then the conditions in Theorem* 22.4.14 *are equivalent to*

$$P^s_{m-1}(x^0,\xi^0) + Q(\bar{v},v) + \sum(2\alpha_j+1)\mu_j \neq 0, \quad 0 \leqq \alpha_j \in \mathbb{Z}, \ |\xi^0| = 1$$

if $p_m(x^0,\xi^0) = 0$, Q *is the Hessian of* $p_m/2$ *at* (x^0,ξ^0), μ_j *are the eigenvalues in* Γ *of the Hamilton map of* Q/i, *and* v *is a generalized eigenvector belonging to the eigenvalue* 0.

Proof. By a remark after the proof of Proposition 22.4.3 we only have to consider the limit Q which occurs here, and Corollary 22.4.11 shows that the uniformity of the test estimate (22.4.11) is automatic.

Notes

In the 1950's the results on hypoelliptic operators of constant strength proved in Section 13.4 seemed quite general, but Trèves [4] constructed an example of a hypoelliptic operator not satisfying these conditions for any choice of coordinates. Such examples can now be found in the class of hypoelliptic operators discussed in Section 22.1. It was introduced in Hörmander [18] and reformulated in Hörmander [42] which we have followed here. For second order differential operators no such class of operators is invariant under coordinate changes. This motivated the study in Hörmander [20] of hypoelliptic operators related to the equation of Kolmogorov [1]. (See Section 7.6 for the original treatment.) Simpler proofs of slightly less precise but more general results were later given by Kohn [2] and Olejnik-Radkevič [1], and we have followed their arguments in Section 22.2.

The precise lower bound in Theorem 22.3.3 is due to Melin [1]. It was observed in Hörmander [36] that the stronger Theorem 22.3.2 is valid under additional assumptions. The proof of Theorem 22.3.3 given here exploits that fact. See also Hörmander [37, 39] and particularly Fefferman-Phong [1, 2] for strong versions without such supplementary conditions, some of which were discussed in Section 18.6 above.

Melin [1] proved that his inequality implies some cases of Theorem 22.4.14 due to Radkevič [1]. Sjöstrand [3] went much further, and a great deal of the methods in Section 22.4 come from his paper; some special model cases had been discussed before by Grušin [2]. The generality was increased further in Boutet de Monvel [4], and Theorem 22.4.14 was proved in full generality by Hörmander [35]. An independent proof of Theorem 22.4.15 was given by Boutet de Monvel-Grigis-Helffer [1].

From the study of hypoelliptic operators with double characteristics there has developed an extensive study of hypoelliptic operators on nilpotent Lie groups. We shall not discuss these interesting developments here but refer the reader to Helffer-Nourrigat [1], Melin [2], Rothschild [1] and the references given in these papers.

Chapter XXIII. The Strictly Hyperbolic Cauchy Problem

Summary

The Cauchy problem for a constant coefficient differential operator P with data on a non-characteristic plane is correctly posed for arbitrary lower order terms if and only if P is strictly hyperbolic (Corollary 12.4.10 and Definition 12.4.11). If the plane is defined by $x_n=0$ this means that the principal symbol $P_m(\xi',\xi_n)$ has m distinct real zeros ξ_n for every $\xi'=(\xi_1,\dots,\xi_{n-1})\in\mathbb{R}^{n-1}\setminus 0$. In Section 23.2 we shall prove that this condition also guarantees the correctness of the Cauchy problem in a very strong sense when the coefficients are variable. The converse will be discussed in Section 23.3 and in the notes.

The proofs in Section 23.2 depend on factorization of $P(x,D)$ into first order factors of the form $D_n-a(x,D')$ where the principal symbol of a is one of the zeros of $P_m(x,\xi',\xi_n)$. We shall therefore study such first order operators in Section 23.1. The basic tool is the energy integral method. It could be applied directly to the operator $P(x,D)$ but it is more transparent and elementary in the first order case which will therefore be studied first.

In Section 23.3 we show that for operators of principal type the correctness of the Cauchy problem requires strict hyperbolicity away from the plane carrying Cauchy data. However, certain types of double roots may occur at that plane. This situation is studied at some length in Section 23.4 since it occurs for the important Tricomi equation. We shall also come across such operators in Section 24.6.

23.1. First Order Operators

In this section we shall study the Cauchy problem in $\mathbb{R}^{n+1}$

(23.1.1) $$\partial u/\partial t+a(t,x,D)u=f,\quad 0<t<T;\quad u=\phi \quad \text{when } t=0.$$

We shall assume

(i) $a_t(x,\xi)=a(t,x,\xi)$ belongs to a bounded set in $S^1(\mathbb{R}^n\times\mathbb{R}^n)$ when $0\leq t\leq T$;

(ii) $t \to a_t$ is continuous with values in $\mathscr{D}'(\mathbb{R}^{2n})$ (or equivalently in $C^\infty(\mathbb{R}^{2n})$);

(iii) $\operatorname{Re} a(t,x,\xi) \geqq -C$, $0 \leqq t \leqq T$.

The assumption (iii) is natural for if a is constant and $f=0$ then the solution of (23.1.1) is $u = e^{-at}\phi$.

It follows from (i), (ii) that $t \mapsto a(t,x,D)u$ is continuous with values in $\mathscr{S}$ if $u \in \mathscr{S}$, for $a(t,x,D)u$ is bounded in $\mathscr{S}$ by Theorem 18.1.6 and continuity with values in C^∞ is obvious. In view of Theorem 18.1.13 it follows that $a_t(x,D)$ is a strongly continuous function of t with values in the set of bounded operators from $H_{(s)}(\mathbb{R}^n)$ to $H_{(s-1)}(\mathbb{R}^n)$, for every s.

The following basic estimate is proved by the energy integral method which we already encountered in the proof of Lemma 17.5.4.

Lemma 23.1.1. *If $s \in \mathbb{R}$ and if $\lambda \in \mathbb{R}$ is larger than some number depending on s, we have for every $u \in C^1([0,T]; H_{(s)}) \cap C^0([0,T]; H_{(s+1)})$ and $p \in [1,\infty]$*

$$(23.1.2)\qquad \Big(\tfrac{1}{2}\int_0^T \|e^{-\lambda t}u(t,.)\|_{(s)}^p \lambda\, dt\Big)^{1/p} \leqq \|u(0,.)\|_{(s)} + 2\int_0^T e^{-\lambda t}\|\partial u/\partial t + a(t,x,D)u\|_{(s)}\,dt,$$

with the obvious interpretation as a maximum when $p=\infty$.

Proof. First assume that $s=0$. By Theorem 18.1.14 it follows from conditions (i) and (iii) that for some constant c

$$\operatorname{Re}(a_t(x,D)v,v) \geqq -c(v,v), \qquad v \in H_{(1)}.$$

Writing $f = \partial u/\partial t + a(t,x,D)u$ and taking scalar products for fixed t, we obtain

$$\begin{aligned} 2\operatorname{Re}(f(t),u(t))e^{-2\lambda t} &= \frac{\partial}{\partial t}e^{-2\lambda t}\|u(t)\|^2 + 2\operatorname{Re}((a_t(x,D)+\lambda)u(t),u(t))e^{-2\lambda t} \\ &\geqq \frac{\partial}{\partial t}e^{-2\lambda t}\|u(t)\|^2 \end{aligned}$$

provided that $\lambda \geqq c$. If we integrate from 0 to $\tau \leqq t \leqq T$ it follows that

$$M(t)^2 = \sup_{0\leqq\tau\leqq t} e^{-2\lambda\tau}\|u(\tau)\|^2 \leqq \|u(0)\|^2 + 2M(t)\int_0^t e^{-\lambda\tau}\|f(\tau)\|\,d\tau.$$

Hence

$$\Big(M(t) - \int_0^t e^{-\lambda\tau}\|f(\tau)\|\,d\tau\Big)^2 \leqq \Big(\|u(0)\| + \int_0^t e^{-\lambda\tau}\|f(\tau)\|\,d\tau\Big)^2,$$

which for $\lambda = c$ implies

$$e^{-ct}\|u(t)\| \leqq \|u(0)\| + 2\int_0^t e^{-c\tau}\|f(\tau)\|\,d\tau.$$

If we multiply by $e^{(c-\lambda)t}$ it follows that

$$e^{-\lambda t}\|u(t)\| \leqq e^{(c-\lambda)t}\|u(0)\| + 2\int_0^t e^{-\lambda\tau}\|f(\tau)\| e^{(c-\lambda)(t-\tau)}\,d\tau.$$

When $\lambda > 2c$ we have $(c-\lambda) < -\lambda/2$, and the norm of $e^{-\lambda t/2}$ in $L^p(0,\infty)$ is $\leqq (2/\lambda)^{1/p}$ if $\lambda > 0$, which gives (23.1.2) when $s=0$.

To prove (23.1.2) for arbitrary s we set $E_s(D)=(1+|D|^2)^{s/2}$,

$$b(t,x,D)=E_s(D)\,a(t,x,D)\,E_{-s}(D).$$

Since $b(t,.)-a(t,.)$ is bounded in S^0 the properties (i)–(iii) remain valid for b. The estimate (23.1.2) follows from the same estimate with a replaced by b, s replaced by 0, and u replaced by $E_s(D)u$. The proof is complete.

The following existence and uniqueness theorem is suggested by (23.1.2) and it extends the validity of this estimate.

Theorem 23.1.2. *Assume that* (i)–(iii) *above are fulfilled and let* $s\in\mathbb{R}$. *For every* $f\in L^1((0,T);H_{(s)})$ *and* $\phi\in H_{(s)}$ *there is then a unique solution* $u\in C([0,T];H_{(s)})$ *of the Cauchy problem* (23.1.1), *and* (23.1.2) *remains valid for this solution.*

Proof. Uniqueness. Assume that $f=0$, $\phi=0$. Since $a(t,x,D)u$ is a continuous function of t with values in $H_{(s-1)}$ it follows that

$$u\in C^1([0,T];H_{(s-1)})\cap C^0([0,T];H_{(s)}).$$

We can therefore apply (23.1.2) which s replaced by $s-1$ and conclude that $u=0$.

Existence. If a solution u of the Cauchy problem (23.1.1) exists and $v\in C_0^\infty(\{(t,x);\,t<T\})$ then

$$\int_0^T (u,\,-\partial v/\partial t+a(t,x,D)^*v)\,dt=\int_0^T (f,v)\,dt+(\phi,v(0)).$$

If we replace t by $T-t$ and $a(t,x,D)$ by $a(t,x,D)^*$ it follows from Lemma 23.1.1 that for some C

$$\sup_{0\leqq t\leqq T}\|v(t)\|_{(-s)}\leqq C\int_0^T\|g(t)\|_{(-s)}\,dt,\quad g=-\partial v/\partial t+a(t,x,D)^*v.$$

Hence

$$\left|\int_0^T (f,v)\,dt+(\phi,v(0))\right|\leqq C\int_0^T\|g(t)\|_{(-s)}\,dt.$$

Using the Hahn-Banach theorem to extend the anti-linear form

$$g\mapsto\int_0^T (f,v)\,dt+(\phi,v(0))$$

we conclude that there is some $u\in L^\infty((0,T);H_{(s)})$ such that

$$\int_0^T (u,\,-\partial v/\partial t+a(t,x,D)^*v)\,dt=\int_0^T (f,v)\,dt+(\phi,v(0))$$

if $v\in C_0^\infty$ and $t<T$ in $\operatorname{supp} v$. Thus

$$\partial u/\partial t+a(t,x,D)u=f$$

in the distribution sense when $0<t<T$. Assuming for a moment that $f\in\mathscr{S}$ we obtain $\partial u/\partial t\in L^\infty((0,T);H_{(s-1)})$, hence $u\in C([0,T];H_{(s-1)})$. Using the

equation again we find that $u \in C^1([0,T]; H_{(s-2)})$ and that $u(0)=\phi$. If $f \in \mathscr{S}$ and $\phi \in \mathscr{S}$ we obtain if s is replaced by $s+2$ in the argument that there is a solution of (23.1.1) to which (23.1.2) is applicable. Now we can for arbitrary f and ϕ satisfying the assumptions in Theorem 23.1.2 choose sequences $\phi_\nu \in \mathscr{S}(\mathbb{R}^n)$ and $f_\nu \in \mathscr{S}(\mathbb{R}^{n+1})$ such that

$$\|\phi-\phi_\nu\|_{(s)} \to 0, \qquad \int_0^T \|f(t)-f_\nu(t)\|_{(s)}\,dt \to 0.$$

Choose $u_\nu \in C^1([0,T]; H_{(s+1)})$ so that $u_\nu(0)=\phi_\nu$ and

$$\partial u_\nu/\partial t + a(t,x,D)u_\nu = f_\nu.$$

Then it follows from (23.1.2) that u_ν is a Cauchy sequence in $C([0,T]; H_{(s)})$. The limit as $\nu \to \infty$ is the required solution of (23.1.1). This completes the proof.

Corollary 23.1.3. *If u satisfies* (23.1.1) *with $f=0$, and $\phi \in H_{(s)}(\mathbb{R}^n)$ for every s, then $u \in C^1([0,T], H_{(s)}(\mathbb{R}^n))$ for every s if this is true for some s. In particular, $u(t,.) \in C^\infty(\mathbb{R}^n)$ for fixed t.*

We shall now study the wave front set of $u(t,.)$ when $\phi \in H_{(-\infty)} = \bigcup H_{(s)}$ but ϕ is not smooth. In doing so we strengthen the hypothesis (iii) to (iv) $a \in S^1_{\mathrm{phg}}$ and the principal symbol is purely imaginary.

The direction of the time can then be reversed in Theorem 23.1.2. Suppose that $(x_0,\xi_0) \in T^*(\mathbb{R}^n) \setminus 0 \setminus WF(\phi)$. We can then choose $q \sim \sum_0^\infty q_j \in S^0_{\mathrm{phg}}$ so that $q_0(x_0,\xi_0) \neq 0$ but q vanishes outside such a small conic neighborhood of (x_0,ξ_0) that $q(x,D)\phi \in C_0^\infty$. We shall obtain information on the solution u of the Cauchy problem

$$\partial u/\partial t + a(t,x,D)u = 0, \quad u(0,.)=\phi, \tag{23.1.3}$$

if we can find $Q(t,x,D)$ with $Q(0,x,D)=q(x,D)$ so that Q commutes with $iD_t + a(t,x,D)$. Indeed, then we have

$$(iD_t + a(t,x,D))\,Qu = 0; \qquad Qu \in C_0^\infty \quad \text{when } t=0;$$

so it follows from Corollary 23.1.3 that $Qu(t,.) \in C^\infty$. Hence the wave front set of $u(t,.)$ is contained in the characteristic set of $Q(t,x,D)$.

With b_j homogeneous of degree $1-j$ and b_0 real we have

$$a(t,x,\xi)/i \sim \sum_0^\infty b_j(t,x,\xi).$$

The symbol $Q(t,x,\xi)$ should be asymptotic to $\sum Q_j(t,x,\xi)$ where Q_j is homogeneous of degree $-j$. The principal symbol of the commutator $[iD_t + a(t,x,D), Q(t,x,D)]$ is

$$\{\tau + b_0(t,x,\xi), Q_0(t,x,\xi)\} = (\partial/\partial t + H_{b_0})Q_0,$$

and the term of order $-j$ is equal to

$$(\partial/\partial t + H_{b_0})Q_j + R_j$$

where R_j is determined by $Q_0, \dots, Q_{j-1}$. Thus we first have to solve the Cauchy problem

$$(\partial/\partial t + H_{b_0}) Q_0 = 0; \quad Q_0 = q_0 \text{ when } t=0; \tag{23.1.4}$$

and then solve inhomogeneous Cauchy problems of the same kind for Q_j. The equation (23.1.4) means that Q_0 is constant on the orbits of the Hamilton field

$$dx/dt = \partial b_0(t,x,\xi)/\partial\xi, \quad d\xi/dt = -\partial b_0(t,x,\xi)/\partial x. \tag{23.1.5}$$

The uniformity hypothesis (i) guarantees that there is a solution with initial values

$$x = y, \quad \xi = \eta \text{ when } t=0 \tag{23.1.6}$$

in a fixed interval $0 \leqq t \leqq t_0$ if $|\eta| = 1$, and the derivatives of (x, ξ) with respect to (y, η) have fixed bounds then. Because of the homogeneity of b_0 the solution exists for all $\eta \neq 0$ when $0 \leqq t \leqq t_0$, and x, ξ are homogeneous of degree 0 and 1 respectively with respect to η. We can pass from $t=0$ to any $t \leqq T$ in at most $|T/t_0| + 1$ steps which shows that the solutions exist for $0 \leqq t \leqq T$. If we write $(x, \xi) = \chi_t(y, \eta)$ it follows from (6.4.12) that χ_t is a homogeneous canonical transformation. Integrating for decreasing t we find that χ_t is invertible. The solution of (23.1.4) is now

$$Q_0(t, x, \xi) = q_0(\chi_t^{-1}(x, \xi)).$$

The later equations

$$(\partial/\partial t + H_{b_0}) Q_j + R_j = 0$$

can be written in the form

$$\partial Q_j(t, \chi_t(y, \eta))/\partial t + R_j(\chi_t(y, \eta)) = 0.$$

The solution which is equal to q_j when $j=0$ is therefore

$$Q_j(t, \chi_t(y, \eta)) = q_j(y, \eta) - \int_0^t R_j(\chi_s(y, \eta))\, ds.$$

We have uniform bounds for all x, ξ derivatives of the functions Q_j so obtained. If we choose $Q(t, x, \xi) \sim \sum Q_j(t, x, \xi)$ it follows that

$$Q(0, x, \xi) - q(x, \xi) \in S^{-\infty}, \quad [iD_t + a(t, x, D), Q(t, x, D)] = R(t, x, D)$$

where $R(t, .)$ is bounded in $S^{-\infty}$ and continuous as a function of t with values in $C^\infty(\mathbb{R}^{2n})$. If u satisfies (23.1.3) with $\phi \in H_{(-\infty)}(\mathbb{R}^n)$ we conclude that

$$(iD_t + a(t, x, D)) Q(t, x, D) u \in C([0, T]; H_{(s)})$$

for any s, and that $Q(t, x, D) u \in H_{(s)}$ for any s when $t=0$. Hence it follows from Theorem 23.1.2 that this remains true for any t. Since $Q_0(t, x, \xi) \neq 0$ when $(x, \xi) = \chi_t(x_0, \xi_0)$, it follows that $\chi_t(x_0, \xi_0) \notin WF(u(t, .))$. The time direction can be reversed, so we have proved

Theorem 23.1.4. *If a satisfies* (i)–(iv) *and* $u \in C([0,T]; H_{(-\infty)}(\mathbb{R}^n))$ *is the solution of the Cauchy problem* (23.1.3), *with* $\phi \in H_{(-\infty)}(\mathbb{R}^n)$, *then* $WF(u(t,.)) = \chi_t WF(\phi)$ *where* χ_t *is the canonical transformation obtained by integrating the Hamilton equations* (23.1.5) *from* 0 *to* t, *with* b_0 *equal to the principal symbol of* a/i.

In the preceding discussion we have only considered the singularities of u for fixed t. If $a \in S^1(\mathbb{R}^{n+1} \times \mathbb{R}^n)$ we can also discuss the singularities of u as a distribution in $\mathbb{R}^{n+1}$. In fact, although $D_t + a(t,x,D)/i$ is not a pseudo-differential operator it becomes one if we multiply by a pseudo-differential operator in (t,x) with symbol vanishing in a conic neighborhood of $\xi = 0$ (Theorem 18.1.35). Multiplying the equation $(D_t + a(t,x,D)/i)u = 0$ by such an operator we conclude that when $0 < t < T$ we have

$$WF(u) \subset \{(t,x; -b_0(t,x,\xi), \xi)\} \cup \{(t,x,\tau,0)\}.$$

From the proof of Theorem 23.1.4 we also obtain $(t,x;\tau,\xi) \notin WF(u)$ if $\xi \neq 0$ and $Q_0(t,x,\xi) \neq 0$. If for some reason we know that $\xi \neq 0$ when $(t,x,\tau,\xi) \in WF(u)$, then it follows in view of Theorem 8.2.4 that

$$WF(u(t,.)) = \{(x,\xi); (t,x; -b_0(t,x,\xi), \xi) \in WF(u)\}.$$

Thus Theorem 23.1.4 means that $WF(u)$ is a union of bicharacteristics of $\tau + b_0(t,x,\xi)$. This is a result of the same form as Theorem 8.3.3′; we shall return to it in Chapter XXVI.

The solution of the Cauchy problem (23.1.3) defines a map

$$F_t: \phi \mapsto u(t,.)$$

such that by Theorem 23.1.4

$$WF(F_t \phi) = \chi_t WF(\phi).$$

The analytic structure of F_t will be elucidated in Chapter XXV; it is an elliptic Fourier integral operator associated with the canonical transformation χ_t. We could also have used the theory of Fourier integral operators in this section. However, this would have been less elementary, and the discussion above provides a concrete background for the study of Fourier integral operators in Chapter XXV.

23.2. Operators of Higher Order

We shall begin by extending Lemma 23.1.1 and Theorem 23.1.2 to operators of the form

$$P = \sum_0^m P_j(t,x,D_x) D_t^j$$

in $\mathbb{R}^{n+1}$. We shall assume that

(i) $P_m=1$, and $P_j\in S^{m-j}(\mathbb{R}^{n+1}\times\mathbb{R}^n)$ when $j<m$;

(ii) P_j has a principal symbol $p_j(t,x,\xi)$ which is homogeneous in ξ of degree $m-j$; thus $P_j(t,x,\xi)-(1-\chi(\xi))\,p_j(t,x,\xi)\in S^{m-1-j}$ if $\chi\in C_0^\infty(\mathbb{R}^n)$ is equal to 1 in a neighborhood of 0;

(iii) the zeros τ of the principal symbol of P

$$p(t,x,\tau,\xi)=\sum_0^m p_j(t,x,\xi)\,\tau^j$$

are real when $\xi\neq 0$, and they are uniformly simple in the sense that for some positive constant c

$$|\partial p(t,x,\tau,\xi)/\partial\tau|\geqq c|\xi|^{m-1} \quad \text{if } p(t,x,\tau,\xi)=0. \tag{23.2.1}$$

Denote the zeros τ of $p(t,x,\tau,\xi)$ by $\lambda_1(t,x,\xi)<\lambda_2(t,x,\xi)<\ldots<\lambda_m(t,x,\xi)$. They are homogeneous in ξ of degree 1, and

$$|D_\xi^\alpha D_{x,t}^\beta\,\lambda_j(t,x,\xi)|\leqq C_{\alpha\beta}|\xi|^{1-|\alpha|}$$

for this is true when $|\xi|=1$ by the implicit function theorem and (23.2.1). Let $\Lambda_j=(1-\chi(\xi))\,\lambda_j(t,x,\xi)$. By commuting the derivatives D_t^j in P to the left and using the identity $D_t=(D_t-\Lambda_j(t,x,D))+\Lambda_j(t,x,D)$ we can find

$$Q_j=\sum_0^{m-1} Q_{jk}(t,x,D_x)\,D_t^k$$

with $Q_{jk}\in S^{m-1-k}$ such that

$$P-(D_t-\Lambda_j(t,x,D_x))\,Q_j=R_j(t,x,D_x). \tag{23.2.2}$$

It is clear that $R_j\in S^m$. The principal symbol $p-(\tau-\lambda_j)\,q_j$ is independent of τ so it is equal to 0, which proves that in fact $R_j\in S^{m-1}$. Later on we shall improve the crude factorization (23.2.2) but it suffices for the extension of Lemma 23.1.1.

Lemma 23.2.1. *If $s\in\mathbb{R}$ and $T>0$ we have for every u in $\bigcap_0^{m-1} C^j([0,T];$ $H_{(s+m-1-j)})$ with $Pu\in L^1((0,T);H_{(s)})$*

$$\sup_{0\leqq t\leqq T}\sum_{j<m}\|D_t^j u(t,.)\|_{(s+m-1-j)} \tag{23.2.3}$$

$$\leqq C_{s,T}\Big(\sum_{j<m}\|D_t^j u(0,.)\|_{(s+m-1-j)}+\int_0^T\|Pu(t,.)\|_{(s)}\,dt\Big).$$

Proof. By Theorem 23.1.2 we may apply (23.1.2) to $Q_j u$ with $a=-i\Lambda_j$. This gives for large λ and $p\in[1,\infty]$

$$\Big(\tfrac{1}{2}\int_0^T\|e^{-\lambda t}Q_j u(t,.)\|_{(s)}^p\,\lambda\,dt\Big)^{1/p} \tag{23.2.4}$$

$$\leqq\|Q_j u(0,.)\|_{(s)}+2\int_0^T e^{-\lambda t}(\|Pu(t,.)\|_{(s)}+C\|u(t,.)\|_{(s+m-1)})\,dt.$$

If q_j is the principal symbol of Q_j then

$$q_j(t,x,\tau,\xi)=\prod_{\nu\neq j}(\tau-\lambda_\nu)$$

so it follows from Lagrange's interpolation formula that

$$\tau^k=\sum q_j(t,x,\tau,\xi)\,\lambda_j^k/\prod_{\nu\neq j}(\lambda_j-\lambda_\nu),\qquad k<m.$$

Let

$$M_{kj}(t,x,\xi)=(1-\chi(\xi))\,\lambda_j(t,x,\xi)^k/\prod_{\nu\neq j}(\lambda_j(t,x,\xi)-\lambda_\nu(t,x,\xi))$$

which is in $S^{k+1-m}(\mathbb{R}^{n+1}\times\mathbb{R}^n)$. Then we have

$$D_t^k u=\sum M_{kj}(t,x,D)\,Q_j u+\sum_0^{m-1} R_{kj}(t,x,D)\,D_t^j u$$

where $R_{kj}\in S^{k-j-1}$. Hence

$$\sum_{k<m}\|D_t^k u(t,.)\|_{(s+m-1-k)}\leqq C\big(\sum\|Q_j u(t,.)\|_{(s)}+\sum_{k<m}\|D_t^k u(t,.)\|_{(s+m-2-k)}\big).$$

In the last sum we can drop the term with $k=m-1$, for $\|Q_1 u-D_t^{m-1}u\|_{(s-1)}$ can be estimated by the other terms. Thus

$$\sum_{k<m}\Big(\int_0^T\|e^{-\lambda t}D_t^k u(t,.)\|_{(s+m-1-k)}^p\lambda\,dt\Big)^{1/p}\leqq C\Big(\sum\Big(\int_0^T\|e^{-\lambda t}Q_j u(t,.)\|_{(s)}^p\lambda\,dt\Big)^{1/p}$$
$$+\sum_{k<m-1}\Big(\int_0^T\|e^{-\lambda t}D_t^k u(t,.)\|_{(s+m-2-k)}^p\lambda\,dt\Big)^{1/p}\Big).$$

Again by (23.1.2) the last sum can be estimated by

$$2\sum_{k<m-1}\|D_t^k u(0,.)\|_{(s+m-2-k)}+4\sum_{k<m-1}\int_0^T\|e^{-\lambda t}D_t^{k+1}u(t,.)\|_{(s+m-2-k)}\,dt,$$

so we obtain

$$\sum_{k<m}\Big(\int_0^T\|e^{-\lambda t}D_t^k u(t,.)\|_{(s+m-1-k)}^p\lambda\,dt\Big)^{1/p}\leqq C\Big(\sum_{k<m}\|D_t^k u(0,.)\|_{(s+m-1-k)}$$
$$+\int_0^T\|e^{-\lambda t}Pu(t,.)\|_{(s)}\,dt+\sum_{k<m}\int_0^T\|e^{-\lambda t}D_t^k u(t,.)\|_{(s+m-1-k)}\,dt\Big).$$

When $p=1$ and λ is large, the last sum is less than $1/2C$ times the left-hand side. Hence we obtain, first for $p=1$ and then for arbitrary p, with a new constant C independent of s, p, and λ when λ is large

$$\sum_{k<m}\Big(\int_0^T\|e^{-\lambda t}D_t^k u(t,.)\|_{(s+m-1-k)}^p\lambda\,dt\Big)^{1/p}\tag{23.2.4$'$}$$
$$\leqq C\Big(\sum_{k<m}\|D_t^k u(0,.)\|_{(s+m-1-k)}+\int_0^T\|Pu(t,.)\|_{(s)}\,dt\Big),$$

which is a much stronger estimate than (23.2.3).

Next we prove an analogue of Theorem 23.1.2:

Theorem 23.2.2. *Assume that* (i)–(iii) *above are fulfilled. For arbitrary* $f \in L^1((0,T); H_{(s)})$ *and* $\phi_j \in H_{(s+m-1-j)}$, $j<m$, *there is then a unique solution* $u \in \bigcap_0^{m-1} C^j([0,T]; H_{(s+m-1-j)})$ *of the Cauchy problem*

$$Pu = f \quad \text{if } 0<t<T; \qquad D_t^j u = \phi_j \quad \text{for } j<m \quad \text{if } t=0. \tag{23.2.5}$$

Proof. The uniqueness is an immediate consequence of Lemma 23.2.1 which gives $u=0$ if $f=0$ and $\phi_0 = \ldots = \phi_{m-1} = 0$. To prove the existence we note that if u is smooth and v is in $C_0^\infty(\mathbb{R}^{n+1})$, $t<T$ in supp v, then

$$\int_0^T (u, P^* v)\, dt = \int_0^T (Pu, v)\, dt - i \sum_{j+k<m} (D_t^j u(0), D_t^k P_{j+k+1}(t,x,D)^* v(0)). \tag{23.2.6}$$

Now it follows from (23.2.3) applied to P^*, with t replaced by $T-t$, that if $g = P^* v$ then

$$\sum_{k<m} \|D_t^k v(t)\|_{(-s-k)} \leq C \int_0^T \|g(t)\|_{(1-m-s)}\, dt.$$

This implies that
$$\left| \int_0^T (f,v)\, dt - i \sum_{j+k<m} (\phi_j, D_t^k P_{j+k+1}(t,x,D)^* v(0)) \right|$$
$$\leq C' \left(\int_0^T \|f\|_{(s)}\, dt + \sum \|\phi_j\|_{(s+m-1-j)} \right) \int_0^T \|g(t)\|_{(1-m-s)}\, dt.$$

Using the Hahn-Banach theorem to extend the anti-linear form in g on the left-hand side we conclude that there is some $u \in L^\infty((0,t); H_{(s+m-1)})$ such that

$$\int_0^T (f,v)\, dt - i \sum (\phi_j, D_t^k P_{j+k+1}(t,x,D)^* v(0)) = \int_0^T (u, P^* v)\, dt \tag{23.2.6$'$}$$

if $v \in C_0^\infty$ and $t<T$ in supp v. This implies that $Pu=f$ if we take v with $0<t<T$ in supp v; that the Cauchy data (23.2.5) are assumed will follow by comparison with (23.2.6) when we have established enough regularity for u.

In view of the estimate (23.2.3) it suffices as in the proof of Theorem 23.1.2 to prove existence when f and ϕ_j are in $\mathscr{S}$. From the fact that $u \in L^2((0,T); H_{(s+m-1)})$ and $Pu \in \mathscr{S}$ we shall deduce that $D_t^j u \in L^2((0,t); H_{(s+m-1-j)})$ if $0 \leq j \leq m$ (this is in fact true for every j). The proof is a minor variation of that of Theorem B.2.9. First note that Theorem B.2.3 remains valid if we define $\bar{H}_{(m,s)}$ as the set of restrictions of elements in $H_{(m,s)}(\mathbb{R}^{n+1})$ to $(0,T) \times \mathbb{R}^n$. Indeed, if $\chi \in C^\infty(\mathbb{R})$ is equal to 1 in $(-\infty, \frac{1}{3})$ and 0 in $(\frac{2}{3}, \infty)$ then Theorem B.2.3 is applicable to χu and also to $(1-\chi)u$ if t is replaced by $T-t$. Now we know for the solution of $Pu=f$ constructed above that

$$D_t^j u \in \bar{H}_{(k-j, s+m-1-k)}, \qquad j \leq m,$$

if $k=0$. Assuming that this is known for some $k<m$ we obtain

$$D_t^m u = Pu - \sum_{j<m} P_j(t,x,D_x) D_t^j u \in \bar{H}_{(k-m+1, s+m-2-k)}.$$

By the modification of Theorem B.2.3 just discussed it follows by induction for decreasing j that $\quad D_t^j u \in \bar{H}_{(k+1-j, s+m-k-2)}$, which means that k is replaced by $k+1$ in the hypothesis. Thus it is true for $k=m$ which means that

$$D_t^j u \in \bar{H}_{(m-j,s-1)}, \quad j \leqq m,$$

and implies that $D_t^j u \in C([0,T]; H_{(s+m-j-3/2)})$ by Theorem B.2.7. This is enough to make (23.2.6) valid. If we compare (23.2.6) and (23.2.6)′ when $D_t^j v = 0$ for $t=0$ and $j<m-1$, we obtain

$$(u(0)-\phi_0, D_t^{m-1} v(0))=0,$$

hence $u(0)=\phi_0$. Taking $D_t^j v=0$ for $j<m-2$ only we then obtain $D_t u(0)=\phi_1$ and so on until all the Cauchy data (23.2.5) are verified. From the uniqueness already established it follows that the Cauchy problem (23.2.5) with $f\in\mathscr{S}$ and all $\phi_j\in\mathscr{S}$ has a solution $u\in\bigcap_s C^{m-1}([0,T]; H_{(s)})$. Using the equation $Pu=f$ repeatedly we obtain $u\in\bigcap_s C^\infty([0,T]; H_{(s)})\subset C^\infty$. This completes the existence proof.

The main purpose of this section is to study strictly hyperbolic differential operators:

Definition 23.2.3. A differential operator P of order m in the C^∞ manifold X, with principal symbol p, is said to be strictly hyperbolic with respect to the level surfaces of $\phi\in C^\infty(X,\mathbb{R})$ if $p(x,\phi'(x))\neq 0$ and the characteristic equation $p(x,\xi+\tau\phi'(x))=0$ has m different real roots τ for every $x\in X$ and $\xi\in T_x^*\setminus\mathbb{R}\phi'(x)$.

The Cauchy problem for a strictly hyperbolic operator always has semi-global solutions:

Theorem 23.2.4. *Let P be a differential operator of order m with C^∞ coefficients in the C^∞ manifold X, and let $Y\Subset X$ be an open subset. Assume that P is strictly hyperbolic with respect to the level surfaces of $\phi\in C^\infty(X,\mathbb{R})$, and set*

$$X_+=\{x\in X; \phi(x)>0\}, \quad X_0=\{x\in X; \phi(x)=0\}.$$

If $f\in H_{(s)}^{\mathrm{loc}}(X)$ has support in the closure of X_+ one can then find $u\in H_{(s+m-1)}^{\mathrm{loc}}(X)$ with support in the closure of X_+ such that $Pu=f$ in Y. If $s\geqq 0$, v is a vector field with $v\phi=1$, and $f\in\bar{H}_{(s)}^{\mathrm{loc}}(X_+)$, $\psi_j\in H_{(s+m-1-j)}^{\mathrm{loc}}(X_0)$, $j<m$, then there is some $u\in\bar{H}_{(s+m-1)}^{\mathrm{loc}}(X_+)$ such that $Pu=f$ in $X_+\cap Y$ and $v^j u=\psi_j$ in $X_0\cap Y$ when $j<m$.

To prove the theorem we shall first study the case where $X\subset\mathbb{R}^n$ and $\phi(x)=x_n$. By hypothesis the coefficient of D_n^m is never 0 so we can divide by it and assume henceforth that it is equal to 1. Let $\lambda_1(x,\xi')<\ldots<\lambda_m(x,\xi')$, $\xi'=(\xi_1,\ldots,\xi_{n-1})$ be the roots of the equation $p(x,\xi',\lambda)=0$, choose $\chi\in C_0^\infty(X)$

with $0 \leqq \chi \leqq 1$ and $\chi = 1$ in Y, and set

$$\tilde{p}(x, \xi) = \prod_1^m (\xi_n - \tilde{\lambda}_j(x, \xi')),$$

$$\tilde{\lambda}_j(x, \xi') = \chi(x)\, \lambda_j(x, \xi') + (1 - \chi(x))\, j|\xi'|.$$

We can then define $\tilde{P}$ with principal symbol $\tilde{p}$ so that $\tilde{P}(x, \xi) = P(x, \xi)$ when $x \in Y$ and $\tilde{P}_j(x, \xi')$ is independent of x outside a compact subset of X. Then the hypotheses of Theorem 23.2.2 are obviously fulfilled with x_n playing the role of t and $(x_1, \ldots, x_{n-1})$ in the role of x.

Proposition 23.2.5. *Assume that P is strictly hyperbolic with respect to the planes $x_n = $ constant in $X \subset \mathbb{R}^n$, and let $Y \Subset X$. If $f \in \bar{H}_{(s,t)}(\mathbb{R}^n_+)$ and $s \geqq 0$, and if $\psi_j \in H_{(s+t+m-1-j)}(\mathbb{R}^{n-1})$, $j < m$, one can find $u \in \bar{H}_{(s+m-1,t)}(\mathbb{R}^n_+)$ such that $D_n^j u$ is a continuous function of x_n with values in $H_{(s+t+m-1-j)}(\mathbb{R}^{n-1})$ when $x_n \geqq 0$, if $j < m$, and*

(23.2.7) $\quad Pu = f \quad$ *in* $Y \cap \mathbb{R}^n_+, \qquad D_n^j u = \psi_j \quad$ *in* $Y \cap (\mathbb{R}^{n-1} \times \{0\}) \quad$ *if* $j < m$.

Proof. Choose $T > x_n$ for $x \in \bar{Y}$. The hypothesis on f implies that $f \in L^1((0, T), H_{(s+t)}(\mathbb{R}^{n-1}))$ so it follows from Theorem 23.2.2 that one can find $u \in \bigcap_{j<m} C^j([0, T]; H_{(s+t+m-1-j)})$ with $\tilde{P}u = f$ when $0 < x_n < T$ and $D_n^j u = \psi_j$ when $x_n = 0$, $j < m$. Since $u \in \bar{H}_{(m-1,s+t)}$ in the slab $\{x; 0 < x_n < T\}$ and $\tilde{P}u \in H_{(s,t)}$ there, it follows just as in the proof of Theorem 23.2.2 that $u \in \bar{H}_{(m-1+s,t)}$, which completes the proof.

Proposition 23.2.5 gives a local version of the second part of Theorem 23.2.4; we shall now prove a local version of the first part.

Proposition 23.2.6. *Assume that P is strictly hyperbolic with respect to the planes $x_n = $ constant in $X \subset \mathbb{R}^n$, and let $Y \Subset X$. If $f \in \dot{H}_{(s,t)}(\bar{\mathbb{R}}^n_+)$ we can find $u \in \dot{H}_{(s+m-1,t)}(\bar{\mathbb{R}}^n_+)$ such that $Pu = f$ in Y. Here s (or t) may be $+\infty$.*

Proof. If $s \geqq 0$ it follows from Theorem 23.2.2 after x_n is replaced by $x_n - 1$, say, that we can find $u \in \dot{H}_{(m-1,s+t)}$ with $\tilde{P}u = f$ when $x_n < T$. As in the proof of Theorem 23.2.2 it follows that $u \in \dot{H}_{(s+m-1,t)}$ by an analogue of Theorem B.2.9. When proving the theorem for $s < 0$ we may now assume that it is already verified when s is replaced by $s+1$. By the remark after Theorem B.2.4 we can write

$$f = f_0 + D_n f_n$$

where $f_0 \in \dot{H}_{(s+1,t-1)}$ and $f_n \in \dot{H}_{(s+1,t)}$. Choose $u_0 \in \dot{H}_{(s+m,t-1)}$ and $u_n \in \dot{H}_{(s+m,t)}$ such that

$$\tilde{P}u_0 = f_0 \quad \text{and} \quad \tilde{P}u_n = f_n.$$

Then $U = u_0 + D_n u_n \in \dot{H}_{(s+m-1,t)}$, and

$$\tilde{P}U - f = [\tilde{P}, D_n]\, u_n \in \dot{H}_{(s+1,t-1)},$$

since $[\tilde{P}, D_n]$ is of order m but only of order $m-1$ in D_n. By the inductive hypothesis we can find $v \in \dot{H}_{(s+m,t-1)}$ such that $\tilde{P}v = [\tilde{P}, D_n]u_n$, and $u = U - v$ is then the desired solution.

To prove Theorem 23.2.4 by piecing local solutions together we need a uniqueness theorem:

Theorem 23.2.7. *If P is strictly hyperbolic in X with respect to the level surfaces of ϕ, then every $x_0 \in X$ has a fundamental system of neighborhoods V such that $u \in \mathscr{D}'(V)$, $\phi \geqq \phi(x_0)$ in* supp u, *and $Pu = 0$ in V implies $u = 0$ in V.*

Proof. We can take a local coordinate system at x_0 such that $x_0 = 0$ and $\phi(x) = \phi(0) + x_n - |x'|^2$. Set

$$V_\varepsilon = \{x; |x_n| < \varepsilon^2, |x'| < \varepsilon\}.$$

Since $p(0, \xi', \xi_n)$ is of degree m with respect to ξ_n and has m different real zeros if $\xi' \in \mathbb{R}^{n-1} \setminus 0$, the same is true for $p(x, \xi', \xi_n)$ if $x \in V_\varepsilon$, $0 < \varepsilon < \varepsilon_0$, for $p(x, \xi)$ is essentially real (cf. Lemma 8.7.3). If $g \in C_0^\infty(V_\varepsilon)$ it follows from Proposition 23.2.6 applied to P^* that we can find $v \in C^\infty(V_\varepsilon)$ such that $P^* v = g$ in V_ε and $v(x) = 0$ if $x_n > \varepsilon^2 - \delta$ with δ so small that $g(x) = 0$ when $x_n > \varepsilon^2 - \delta$. But then we have

$$0 = (Pu, v) = (u, P^* v) = (u, g)$$

so $u = 0$ in V_ε. Thus the neighborhoods V_ε have the desired property.

Proof of Theorem 23.2.4. We can find a finite number of coordinate patches $X_\nu \subset X$ and open sets $Y_\nu \Subset X_\nu$ such that $\bar{Y} \subset \bigcup Y_\nu$. To prove the first statement in Theorem 23.2.4 we use Proposition 23.2.6 to find for every ν some $u_\nu \in H_{(s+m-1)}^{\text{loc}}(X_\nu)$ vanishing when $\phi < 0$ such that $Pu_\nu = f$ in Y_ν. Choose a covering of $X_0 \cap \bar{Y}$ by open sets V_μ such that $Pv = 0$ in V_μ and $v = 0$ when $\phi < 0$ implies $v = 0$ in V_μ. We choose V_μ so small that for all μ, μ' with $V_\mu \cap V_{\mu'} \neq \emptyset$ we have $V_\mu \cup V_{\mu'} \subset Y_\nu$ for some ν. Now define $u = u_\nu$ in V_μ if $V_\mu \subset Y_\nu$. By Theorem 23.2.7 we have $u_\nu = u_{\nu'}$ in V_μ if $V_\mu \subset Y_\nu \cap Y_{\nu'}$, so the choice of ν does not matter. The definition made in V_μ agrees with that in $V_{\mu'}$ in the overlap, since V_μ and $V_{\mu'}$ are contained in Y_ν for the same ν if the intersection is non-empty. It is clear that $Pu = f$ in $V = \bigcup V_\mu$.

Choose $\chi \in C_0^\infty(V)$ equal to 1 in a neighborhood W of $X_0 \cap \bar{Y}$. Then $f = P(\chi u)$ in W. Hence we can find $g \in H_{(s)}^{\text{loc}}(X)$ and $\varepsilon > 0$ such that $g = 0$ when $\phi(x) < \varepsilon$ and $f = P(\chi u) + g$ in Y. It suffices therefore to prove the statement with f replaced by g and ϕ replaced by $\phi - \varepsilon$. A moment's reflection on the geometrical nature of the proof of Theorem 23.2.7 shows that the same ε can be used when ϕ is replaced by $\phi - t$ for some $t > 0$. Since the statement is trivial when $\phi < 0$ in Y, it follows after a finite number of iterations of the preceding local solution. To prove the second statement in Theorem 23.2.4 we just start by using Proposition 23.2.5 instead of Proposition 23.2.6. The construction then continues as before. The proof is complete.

We shall next discuss the singularities of solutions of the Cauchy problem (23.2.5). As a preparation for the proofs we refine the factorization (23.2.2).

Lemma 23.2.8. *If P satisfies the conditions* (i)–(iii) *at the beginning of the section then the operators Λ_j and Q_{jk} can be modified with terms of order* 0 *and $m-2-k$ so that the right-hand side R_j in* (23.2.2) *is in $S^{-\infty}(\mathbb{R}^{n+1}\times\mathbb{R}^n)$.*

Proof. Suppose that we already have (23.2.2) with $R_j \in S^{m-\mu}$ for some $\mu>0$, as we do for $\mu=1$. We can choose $\lambda \in S^{1-\mu}(\mathbb{R}^{n+1}\times\mathbb{R}^n)$ so that

$$R_j(t,x,\xi)+\lambda(t,x,\xi)\sum_0^{m-1} Q_{jk}(t,x,\xi)\Lambda_j(t,x,\xi)^k \in S^{m-\mu-1},$$

for $R_j \in S^{m-\mu}$ by hypothesis and the sum is elliptic of order $m-1$ by (23.2.1) since the principal symbol is $p'_\tau(t,x,\lambda_j,\xi)$. Now it follows as in our earlier discussion of (23.2.2) that we can choose $q_k \in S^{m-1-\mu-k}$ so that with the notation $\tilde\Lambda_j=\Lambda_j+\lambda$ the symbol of

$$R_j(t,x,D_x)+\lambda(t,x,D_x)\sum_0^{m-1} Q_{jk}(t,x,D_x)D_t^k-(D_t-\tilde\Lambda_j(t,x,D_x))\sum_0^{m-2} q_k(t,x,D_x)D_t^k$$

is independent of τ and belongs to $S^{m-\mu-1}$. If Λ_j is replaced by $\tilde\Lambda_j$ and Q_{jk} is replaced by $Q_{jk}+q_k$ when $k<m-1$, we obtain a new identity of the form (23.2.2), now with $R_j \in S^{m-\mu-1}$. If we repeat the argument for $\mu=1,2,\dots$ and add to Λ_j, Q_{jk} the asymptotic sums of the successive corrections λ^μ, q_k^μ (see Proposition 18.1.3), we obtain (23.2.2) with $R_j \in S^{-\infty}$.

We shall now discuss the singularities of the solution of the Cauchy problem (23.2.5), starting with an interior regularity result similar to Theorem 8.3.3′. It is no restriction that we assume the principal symbol of the hyperbolic operator real.

Theorem 23.2.9. *Let P be a differential operator of order m with C^∞ coefficients in the C^∞ manifold X, and assume that P is strictly hyperbolic with respect to the level surfaces of $\phi \in C^\infty(X,\mathbb{R})$. If $Pu=f$ it follows that $WF(u)\setminus WF(f)$ is a subset of $p^{-1}(0)$ which is invariant under the flow defined by the Hamilton vector field of the principal symbol p.*

Proof. That $WF(u)\setminus WF(f)\subset p^{-1}(0)$ is already a consequence of Theorem 8.3.1. Since $WF(u)$ is closed we just have to show that the intersection with any bicharacteristic arc γ of p is open in the complement of $WF(f)$. This is a purely local statement so we may assume that $X\subset\mathbb{R}^n$, that $\phi(x)=x_n$, and that $\gamma_0=(0,\xi_0)\in WF(u)\setminus WF(f)$; we must prove that $WF(u)$ contains a neighborhood of γ_0 on the bicharacteristic γ of p through γ_0. Since (x,N) is non-characteristic if $x\in X$ and $N=(0,\dots,0,1)$, we have $\xi'_0 \neq 0$. We can therefore choose $a \in S^0(\mathbb{R}^n \times \mathbb{R}^n)$ equal to 1 in a conic neighborhood of γ_0 and equal to 0 at infinity in a conic neighborhood of $X \times \mathbb{R}N$ as well as for all x outside a compact subset of X. Set $v = a(x, D)u$, $g = P(x, D)v$. Then $v \in \mathscr{E}'(X)$, $\gamma_0 \in WF(v) \setminus WF(g)$, and

$$(\mathbb{R}^n \times \mathbb{R}N) \cap WF(v) = \emptyset. \tag{23.2.8}$$

We must prove that this implies that $WF(v)$ contains a neighborhood of γ_0 on γ.

First note that if $R \in S^{-\infty}(\mathbb{R}^n \times \mathbb{R}^{n-1})$ then $R(x, D')v \in C^\infty$. In fact, we can write

$$R(x, D')v = R(x, D')q(D)v + R(x, D')(I - q(D))v$$

where $q(\xi) \in S^0$ has support in such a small conic neighborhood of N that $q(D)v \in C^\infty$, hence $R(x, D')q(D)v \in C^\infty$, but $q = 1$ at infinity in another conic neighborhood. By Theorem 18.1.35 we have $R(x, D')(I - q(D)) \in \operatorname{Op} S^{-\infty}$, hence $R(x, D')(I - q(D))v \in C^\infty$. In particular, we have $R(x, D')v \in C^\infty$ if $R = 0$ in a neighborhood of supp v.

Choose $\tilde{P}$ as in the proof of Proposition 23.2.5. Then we obtain $Pv - \tilde{P}v \in C^\infty$. Using Lemma 23.2.8 we can factor $\tilde{P}$

$$\tilde{P}(x, D) = (D_n - \tilde{\Lambda}_j(x, D'))\tilde{Q}_j(x, D) + \tilde{R}_j(x, D')$$

where $D_n - \tilde{\Lambda}_j$ is characteristic at γ_0 but $\tilde{Q}_j$ is non-characteristic there; $\tilde{R}_j \in S^{-\infty}$. Set $w = \tilde{Q}_j(x, D)v$. In view of Theorem 18.1.35 we have $\gamma_0 \in WF(w)$, and since $\tilde{R}_j v \in C^\infty$ as proved above, we obtain $\gamma_0 \notin WF((D_n - \tilde{\Lambda}_j(x, D'))w)$. By the discussion after Theorem 23.1.4 it follows that $WF(w)$ contains a neighborhood γ_1 of γ_0 on γ. Using Theorem 18.1.35 again we conclude that $\gamma_1 \subset WF(v) \subset WF(u)$, which completes the proof.

Remark. The function ϕ plays no role at all in the conclusion of Theorem 23.2.9. This suggests that hyperbolicity is not essential, and we shall prove later (Theorem 26.1.1) that it suffices to assume that the principal symbol is real.

We shall now discuss the regularity of the solution of the Cauchy problem, assuming that P is a differential operator of order m with C^∞ coefficients in the C^∞ manifold X which is strictly hyperbolic with respect to the level surfaces of $\phi \in C^\infty(X, \mathbb{R})$. The sets X_+ and X_0 are defined as in Theorem 23.2.4 and we choose a C^∞ vector field v with $v\phi = 1$. Assume that $f \in \mathscr{N}(\overline{X}_+)$ (Definition 18.3.30), and assume that $u \in \overline{\mathscr{D}}'(X_+)$, $Pu = f$ in X_+. Then it follows from Corollary 18.3.31 that u is the restriction to X_+ of a unique element $\dot{u} \in \mathscr{N}(\overline{X}_+)$, and so is $v^j u$. This implies that the boundary value ϕ_j of $v^j u$ is well defined. By Theorem 18.3.27 (ii) we have $WF_b(f) \subset WF_b(\dot{u})$, and Theorem 18.3.27 (iv) gives $WF(\phi_j) \subset WF_b(\dot{u}) \cap T^*(X_0)$.

Theorem 23.2.10. *Under the preceding hypotheses we have*

$$WF_b(\dot{u})|_{X_0} = WF_b(f)|_{X_0} \cup \left(\bigcup_{j=0}^{m-1} WF(\phi_j)\right); \tag{23.2.9}$$

hence $WF(u)$ is contained in the union of the forward bicharacteristics starting at $WF(f|_{X_+})$ or at points in the inverse image of the right-hand side of (23.2.9) under the natural map

$$T^*(X)|_{X_0} \to T^*(X_0).$$

Proof. The last statement follows from the first if we note that WF_b is closed and apply Theorem 23.2.9. Thus we only have to verify that the left-hand side of (23.2.9) is contained in the right-hand side. This is a local statement. After multiplying u by a suitable cutoff function we may therefore assume that $X \subset \mathbb{R}^n$, $\phi(x) = x_n$, $v = \partial/\partial x_n$, $u \in \mathscr{E}'(\overline{X}_+)$ and that $(x,\xi) \in WF(u)$, $x_n > 0$ implies $\xi' \neq 0$. (See the proof of Theorem 18.3.32.)

Define $\tilde{P}(x,D)$ as in the proof of Proposition 23.2.5 and form the factorizations

$$(D_n - \tilde{\Lambda}_j(x,D'))\,\tilde{Q}_j(x,D) = \tilde{P}(x,D) + \tilde{R}_j(x,D')$$

where $\tilde{R}_j \in S^{-\infty}(\mathbb{R}^n \times \mathbb{R}^{n-1})$ as in the proof of Theorem 23.2.9. Let $\gamma \in T^*(X_0) \smallsetminus 0$, and assume that γ is not contained in the right-hand side of (23.2.9). We must show that γ is not in $WF_b(u)|_{X_0}$. From the proof of Theorem 23.1.4 we know that $\Phi_j(x,\xi') \in S^0$ can be chosen non-characteristic at γ so that

$$[D_n - \tilde{\Lambda}_j(x,D'), \Phi_j(x,D')] = T_j(x,D')$$

where T_j is of order $-\infty$ and Φ_j is of order $-\infty$ outside any given neighborhood of the bicharacteristic of $\xi_n - \tilde{\lambda}_j(x,\xi')$ passing through γ. Now we have outside X_0

$$\begin{aligned}(D_n - \tilde{\Lambda}_j(x,D'))\,\Phi_j(x,D')\,\tilde{Q}_j(x,D)\,u \\ = T_j(x,D')\,\tilde{Q}_j(x,D)\,u + \Phi_j(x,D')\,\tilde{f} + \Phi_j(x,D')\,\tilde{R}_j(x,D')\,u\end{aligned}$$

where $\tilde{f} = \tilde{P}(x,D)\,u$ is equal to f in a neighborhood of $\operatorname{supp} u$, outside X_0. All terms on the right are in $\mathscr{N}$ by Theorem 18.3.32. Since $\Phi_j(x,D')\,\tilde{R}_j(x,D')$ and $T_j(x,D')\,\tilde{Q}_j(x,D)\,u$ are of order $-\infty$ it follows that the right-hand side is in C^∞ for $0 \leqq x_n < \varepsilon$, say, if the support of $\Phi_j(x',0,\xi')$ is in a sufficiently small conic neighborhood of γ. The boundary values of $\tilde{Q}_j(x,D)\,u$ are given by pseudo-differential operators acting on $\phi_0, \ldots, \phi_{m-1}$, so the boundary values of $\Phi_j(x,D')\,\tilde{Q}_j(x,D)\,u$ are in C^∞ if the support of $\tilde{\Phi}_j$ is in a small conic neighborhood of the bicharacteristic through γ. Hence it follows from Theorem 23.1.4 and the remarks after its proof that $\Phi_j(x,D')\,\tilde{Q}_j(x,D)\,u$ is in C^∞ for $0 \leqq x_n < \varepsilon$. Now the operators $\tilde{Q}_j(x,D) = \sum_0^{m-1} \tilde{Q}_{jk}(x,D')\,D_n^k$, $j = 1, \ldots, m$ can be regarded as a Douglis-Nirenberg system composed with the operator $(1, D_n, \ldots, D_n^{m-1})$ (see Section 19.5). Hence we can find pseudo-differential operators W_j and V_k such that

$$\sum_1^m W_j(x,D')\,\Phi_j(x,D')\,\tilde{Q}_j(x,D)\,u = \sum_0^{m-1} V_k(x,D')\,D_n^k\,u$$

where $V_0(x',0,\xi')$ is non-characteristic at γ but $V_k \in S^{-\infty}$ for $k \neq 0$. Hence $V_0(x,D')u \in C^\infty$ for small $x_n \geqq 0$, and it follows from Theorem 18.3.32 that $\gamma \notin WF_b(\dot{u})$. The proof is complete.

23.3. Necessary Conditions for Correctness of the Cauchy Problem

Definition 23.2.3 was motivated only by the constant coefficient theory developed in Chapter XII. One might therefore suspect that it is too restrictive. However, we shall now prove that a weaker hyperbolicity condition is necessary for an existence theorem like Theorem 23.2.4 to hold.

Theorem 23.3.1. *Let P be a differential operator of order m with C^∞ coefficients in the C^∞ manifold X, and let $Y \Subset X$ be an open subset. Let $\phi \in C^\infty(X, \mathbb{R})$, $\phi' \neq 0$ in X, and set*

$$X_+ = \{x \in X; \phi(x) > 0\}, \quad X_0 = \{x \in X; \phi(x) = 0\}.$$

Assume that for every $f \in C_0^\infty(X)$ with $\operatorname{supp} f \subset \overline{X}_+$ one can find $u \in \mathscr{D}'(X)$ with $\operatorname{supp} u \subset \overline{X}_+$ and $Pu = f$ in Y. If $y \in X_0 \cap Y$ and $\xi \in T_y^*(X)$ it follows then that the characteristic equation $p(y, \xi + \tau\phi'(y)) = 0$ in τ has only real roots unless it is fulfilled for every τ.

Proof. The hypotheses remain valid if we shrink X and Y, so there is no restriction in assuming that $X \subset \mathbb{R}^n$, $\phi(x) = x_n$, and $y = 0$. Repeating an argument in the proof of Theorem 10.6.6 we shall prove that if $K \Subset Y$ is a compact neighborhood of 0 then there exists an integer N such that for some C

$$|(f,v)| \leqq C \sum_{|\alpha| \leqq N} \sup |D^\alpha f| \sum_{|\beta| \leqq N} \sup_{x_n > 0} |D^\beta P^* v|; \tag{23.3.1}$$
$$f \in \dot{C}^\infty(\overline{\mathbb{R}}^n_+), \quad v \in C_0^\infty(K).$$

Here $(f,v) = \int f \bar{v}\, dx$ and P^* is the adjoint of P with respect to this scalar product. To prove (23.3.1) we consider (f,v) as a bilinear form on all $f \in C^\infty$ with bounded derivatives and $\operatorname{supp} f \subset \overline{\mathbb{R}}^n_+$ while $v \in C_0^\infty(K)$. Such functions f form a Fréchet space with the topology defined by the semi-norms $\sup |D^\alpha f|$, and the functions v form a metrizable space with the semi-norms $\sup_{x_n > 0} |D^\beta P^* v|$. When $f = Pu$ in Y we have

$$(f,v) = (u, P^* v).$$

If $x_n \geqq 0$ in $\operatorname{supp} u$ this proves the continuity with respect to v for fixed f, in view of Theorem 2.3.10. Thus (f,v) is separately continuous, hence continuous, which proves (23.3.1).

Assume now that for some $\xi' = (\xi_1, \ldots, \xi_{n-1}) \in \mathbb{R}^{n-1}$ the equation

$$p(0, \xi', \tau) = 0$$

has a root with $\operatorname{Im}\tau \neq 0$ but is not identically valid. Then $\xi' \neq 0$ and replacing (ξ',τ) by $-(\xi',\tau)/\operatorname{Im}\tau$ we may assume that $\operatorname{Im}\tau = -1$. Taking $\langle x',\xi'\rangle + x_n \operatorname{Re}\tau$ as new x_{n-1} coordinate while x_n is unchanged we may also assume that $\tau = -i$ and $\xi' = (0,\dots,0,1)$. Thus

$$p(0,\dots,0,\xi_{n-1},\xi_n) = (\xi_n + i\,\xi_{n-1})^r\, Q(\xi_{n-1},\xi_n) \tag{23.3.2}$$

where r is the multiplicity of the root $-i$, thus $Q(1,-i) \neq 0$.

To show that (23.3.1) cannot be valid we shall choose f and v concentrated close to 0. With an integer $\nu > r$ we therefore introduce new coordinates by

$$y_j = x_j \rho^\nu,\quad j \leqq n-2;\qquad y_j = x_j \rho^{2\nu},\quad j = n-1, n,$$

where ρ is large. This blows up any neighborhood of 0 so that it contains any given compact set when ρ is large. P^* is replaced

$$P^*_\rho(y, D_y) = P^*(y_1/\rho^\nu, \dots, y_{n-2}/\rho^\nu, y_{n-1}/\rho^{2\nu}, y_n/\rho^{2\nu},$$
$$\rho^\nu D_1, \dots, \rho^\nu D_{n-2}, \rho^{2\nu} D_{n-1}, \rho^{2\nu} D_n).$$

From (23.3.1) it follows with $\kappa = \nu(n+2+4N)$ that

$$|(f,v)| \leqq C\rho^\kappa \sum_{|\alpha| \leqq N} \sup |D^\alpha f| \sum_{|\beta| \leqq N} \sup_{y_n \geqq 0} |D^\beta P^*_\rho v|, \tag{23.3.1$'$}$$

when $f \in \dot{C}^\infty(\overline{\mathbb{R}}^n_+)$ and $v \in C_0^\infty(M)$ where M is a fixed compact set. As $\rho \to \infty$ we have

$$\rho^{-2m\nu} P^*_\rho(y, D_y) - \bar{p}(0,\dots,0,D_{n-1},D_n) = O(\rho^{-\nu})$$

where $O(\rho^{-\nu})$ means that the product by ρ^ν is a differential operator with coefficients bounded in C^∞ as $\rho \to \infty$. In the main term given by (23.3.2) the factor $D_n - iD_{n-1} = -(\partial/\partial y_{n-1} + i\,\partial/\partial y_n)$ is essentially the Cauchy-Riemann operator in $y_{n-1} + i\,y_n$. Choose an analytic function $\psi(y_{n-1} + i\,y_n)$ which vanishes at 0 and has positive imaginary part when $y_n \geqq 0$ and $0 < |y| \leqq 1$. We can take

$$\psi = 2(y_{n-1} + i\,y_n) + i(y_{n-1} + i\,y_n)^2.$$

If we put

$$\Psi(y) = \sum_1^{n-2} i\,y_j^2 + \psi(y_{n-1} + i\,y_n)$$

it follows then that $(D_n - iD_{n-1})\,\Psi(y) = 0$ and that $\operatorname{Im}\Psi(y) \geqq |y|^2$ when $|y| \leqq 1$ and $y_n \geqq 0$.

Now we shall construct a formal series

$$u_\rho(y) = \sum_0^\infty e^{i\rho\Psi(y)}\, v_j(y)\, \rho^{-j}$$

where $v_j \in C_0^\infty$, $v_0(0) = 1$, such that $P^*_\rho u_\rho = 0$ in a neighborhood of 0. Since $\nu > r$ the lowest term in $P^*_\rho u_\rho$ is $\rho^{2m\nu + m - r}$ times $a(D_n - iD_{n-1})^r v_0$. Thus we require

$$a(D_n - iD_{n-1})^r v_0 = 0,$$

where $a=\bar{Q}(\partial\Psi/\partial y_{n-1},\partial\Psi/\partial y_n)\neq 0$ in a neighborhood V_0 of 0. Choose $\chi\in C_0^\infty(V_0)$ equal to 1 in another neighborhood V_1 of 0 and set $v_0=\chi$. The subsequent equations are of the form

$$a(D_n-iD_{n-1})^r v_j=F_j \tag{23.3.3}$$

where F_j is determined by the preceding amplitudes $v_0,\dots,v_{j-1}$. Repeated solution of Cauchy-Riemann equations gives a solution of (23.3.3) which we then cut off by multiplication with χ. The equation remains valid in V_1, and $v_j\in C_0^\infty(V_0)$.

If we choose for u a partial sum u_ρ^μ of the series u_ρ of higher order, it is now clear that the second sum on the right-hand side of (23.3.1)′ tends to 0 as $\rho\to\infty$, like any desired power of $1/\rho$. Choose $F\in C_0^\infty(\mathbb{R}^n_+)$ and set

$$f_\rho(y)=F(\rho y)\rho^n.$$

Then the right-hand side of (23.3.1)′ tends to 0 as $\rho\to\infty$ if μ is large enough, and

$$\int\overline{f_\rho(y)}\,u_\rho^\mu(y)\,dy\to\int\overline{F(y)}\,e^{2i(y_{n-1}+iy_n)}\,dy.$$

If F is chosen so that the integral on the right-hand side is not 0, it follows that (23.3.1)′ cannot hold. This completes the proof.

From the constant coefficient case (see Section 12.8) we know that no stronger conclusion on the principal symbol at a point in X_0 would be possible in Theorem 23.3.1. In particular, X_0 may be characteristic. However, in the strictly hyperbolic case it follows from Theorem 23.2.7 that there is a finite propagation speed for the support of the solution to the Cauchy problem, just as in Theorem 12.5.6. We shall now prove that this is only true in the non-characteristic case. To simplify the statement we place ourselves in $\mathbb{R}^n$ right away and write $x'=(x_1,\dots,x_{n-1})$.

Theorem 23.3.2. *Let P be a differential operator of order m with C^∞ coefficients in $X\subset\mathbb{R}^n$, let $0<B\leqq A$, and let $Y\subset X$ be an open subset containing 0. Assume that for every $f\in C_0^\infty(X)$ with $x_n\geqq A|x'|$ in* supp f *one can find $u\in\mathscr{D}'(X)$ with $x_n\geqq B|x'|$ in* supp u *such that $Pu=f$ in Y. Then the plane $x_n=0$ is non-characteristic at 0 provided that $p(0,\xi)$ is not identically 0.*

Proof. Let $K_t=\{x;x_n\geqq t|x'|\}$. The hypothesis implies that if $K\Subset Y$ is a compact neighborhood of 0 then

$$|(f,v)|\leqq C\sum_{|\alpha|\leqq N}\sup|D^\alpha f|\sum_{|\beta|\leqq N}\sup_{K_B}|D^\beta P^*v| \tag{23.3.4}$$

if $v\in C_0^\infty(K)$ and $f\in C^\infty(\mathbb{R}^n)$, supp $f\subset K_A$. This follows by an obvious modification of the proof of (23.3.1) which we leave as an exercise. If $p(0,\dots,0,1)=0$ we can write

$$p(0,\xi)=\sum_r^m q_j(\xi')\xi_n^{m-j}$$

where q_j is homogeneous of degree j, $q_r \not\equiv 0$ and $r>0$. Now we introduce the new coordinates $y=x/\rho^\nu$ where $\nu>r$. We have

$$\rho^{-m\nu} P_\rho^*(y, D_y) = \sum_r^m q_j(D') D_n^{m-j} + O(\rho^{-\nu}),$$

and (23.3.4) is replaced by a similar estimate for P_ρ^* with a power of ρ in the right-hand side. We choose $\Psi(y)=i\,y_n$ now and determine a formal solution

$$u_\rho(y) = \sum_0^\infty e^{i\rho\Psi(y)} v_j(y)\,\rho^{-j}$$

of the equation $P_\rho^* u_\rho = 0$ in a neighborhood of 0. The leading term in the equation is

$$q_r(D')\,v_0 = 0$$

so we can take $v_0 \in C_0^\infty(V_0)$ equal to 1 in a neighborhood V_1 of 0. The subsequent equations are of the form

$$q_r(D')\,v_j = F_j$$

where F_j is determined by the preceding amplitudes. They can be solved successively in V_1 with $v_j \in C_0^\infty(V_0)$. The proof is now completed just as that of Theorem 23.3.1, with $F \in C_0^\infty(K_A)$ of course.

We can now motivate Definition 23.2.3 of strict hyperbolicity better. First recall that by Theorem 10.4.10 the strength of a differential operator with constant coefficients is independent of the terms of lower order if and only if it is of principal type in the sense that the differential of the principal symbol is never 0 outside the origin. We shall devote Chapters XXVI and XXVII entirely to such operators. The following proposition examines the combination of this principal type condition with the necessary condition in Theorem 23.3.1 imposed in all of $\overline{X}_+$.

Proposition 23.3.3. *Let X be a C^∞ manifold, let $\phi \in C^\infty(X)$ and let $p \in C^\infty(T^*(X))$ be a polynomial along the fibers such that*

(i) *$dp(x,\xi) \neq 0$ in $T^*(X) \setminus 0$;*

(ii) *when $\phi(x) \geqq 0$ the equation $p(x, \xi + \tau\phi'(x)) = 0$ in τ has only real roots unless it is satisfied for every τ.*

Then it follows that when $\phi(x) \geqq 0$ the equation $p(x, \xi + \tau\phi'(x)) = 0$ cannot have a root τ of finite order $k>1$ unless $k=2$, $\phi(x)=0$, and

$$(H_p^2\phi)(x, \xi + \tau\phi'(x)) > 0.$$

Proof. There is nothing to prove unless $\phi'(x) \neq 0$. We can then choose local coordinates so that $\phi(x) = x_n$, say. If $\phi(x)>0$ then Lemma 8.7.2 with $\theta = (0, \dots, 0, 1)$ shows that if τ is a zero of order k of $p(x, \xi + \tau\phi'(x))$ then all derivatives of p of order $<k$ vanish at $(x, \xi + \tau\phi'(x))$. (Note that the proof of Lemma 8.7.2 only requires analyticity in the θ direction.) Hence $k=1$ by condition (i). If $\phi(x)=0$ we can only conclude that the derivatives of p of

order $<k$ with respect to the variables other than x_n vanish at $(x, \xi+\tau\phi'(x))$. If $k\neq 1$ it follows from (i) that $\partial p/\partial x_n \neq 0$. When $(s,t)\to 0$ we have

$$p(x', s, \xi', \xi_n+\tau+t) = a s + b t^k + O(s^2+|s t|+|t|^{k+1})$$

where $a=p_{(n)}(x, \xi+\tau\phi'(x))$ and $b=\partial^k p(x, \xi+\tau\phi'(x))/\partial\xi_n^k/k!$. By condition (ii) there must be k small real zeros t for all small $s\geqq 0$. Replacing s by $\varepsilon^k s$ and t by εt we conclude as in the proof of Lemma 8.7.2 when $\varepsilon\to 0$ that the equation $a s+b t^k=0$ has k real roots t for every $s\geqq 0$. Thus $k=2$ and $a b<0$. Now

$$H_p x_n = \partial p/\partial\xi_n, \quad H_p^2 x_n = -\partial p/\partial x_n\, \partial^2 p/\partial\xi_n^2 \quad \text{at } (x, \xi+\tau\phi'(x)),$$

which completes the proof.

In Section 23.4 we shall prove that the results of Section 23.2 remain valid with very small modifications if strict hyperbolicity is relaxed as in Proposition 23.3.3 on the manifold carrying the Cauchy data. However, at genuine double characteristics where $dP_m=0$ one can usually prove necessary conditions on the subprincipal symbol analogous to Corollary 12.4.9. There is also an extensive literature on their sufficiency. It is perhaps particularly interesting that the Cauchy problem is solvable for arbitrary lower order terms for some hyperbolic operators with double characteristics. However, it would take us too far to develop these results here, and we shall content ourselves with references in the notes.

23.4. Hyperbolic Operators of Principal Type

In this section we shall extend the results of Section 23.2 to operators P satisfying the conditions in Theorem 23.3.2 and Proposition 23.3.3. If $\phi(x)=x_n$ and X is a neighborhood of 0 in $\mathbb{R}^n$, this means that the principal symbol is assumed to satisfy the following conditions:

(i) $p(x,\xi)=\sum_0^m p_j(x,\xi')\,\xi_n^j$, $p_m=1$;

(ii) when $x\in X$, $x_n>0$ and $\xi'\in\mathbb{R}^{n-1}\setminus 0$ the equation $p(x,\xi)=0$ has only simple real roots ξ_n;

(iii) when $x\in X$, $x_n=0$ and $\xi'\in\mathbb{R}^{n-1}\setminus 0$, the equation $p(x,\xi)=0$ has only real roots, and $\partial p/\partial x_n\, \partial^2 p/\partial\xi_n^2<0$ at the double roots.

Suppose that $\xi_n=\tau$ is a double root when $x'=x_0'$, $x_n=0$ and $\xi'=\xi_0'$. Then it follows from the Weierstrass preparation theorem and the homogeneity that in a conic neighborhood of $(x_0', 0, \xi')$ we have

$$p(x,\xi)=((\xi_n-a(x,\xi'))^2-b(x,\xi'))\,q(x,\xi)$$

where a, $b\in C^\infty$ are homogeneous of degree 1, 2 with respect to ξ', $b(x,\xi')\geqq 0$ when $x_n\geqq 0$, and q is a polynomial in ξ_n of degree $m-2$. Moreover, $b=0$, a

$=\tau$ and $\partial b/\partial x_n > 0$ at $(x_0', 0, \xi_0')$, so $b(x, \xi') \geqq c x_n |\xi'|^2$. We shall begin by a study parallel to Section 23.1 of such second order operators, which are differential operators in x_n but pseudo-differential operators in x'. As in Lemma 23.2.1 we shall then return to the original operator.

Before proceeding we give an important example of a differential operator satisfying the preceding conditions which it is useful to keep in mind.

Example 23.4.1. The Tricomi operator

$$D_n^2 - x_n \sum_1^{n-1} D_j^2$$

is strictly hyperbolic for $x_n > 0$ and satisfies the conditions (i)–(iii) above when $x_n = 0$. When $x_n < 0$ it is elliptic.

Lemma 23.4.2. *Let* $a \in S^1(\mathbb{R}^n \times \mathbb{R}^{n-1})$, $b \in S^2(\mathbb{R}^n \times \mathbb{R}^{n-1})$, $\operatorname{Im} a \in S^0(\mathbb{R}^n \times \mathbb{R}^{n-1})$, $\operatorname{Im} b \in S^1(\mathbb{R}^n \times \mathbb{R}^{n-1})$, *and assume that*

$$x_n |\xi'|^2 \leqq C_1 \operatorname{Re} b(x, \xi') + C_2(1 + |\xi'|), \quad 0 \leqq x_1 \leqq 1. \tag{23.4.1}$$

Set $P = (D_n - a(x, D'))^2 - b(x, D')$. *If* $0 \leqq t < T \leqq 1$ *and* $u \in C_0^\infty(\mathbb{R}^n)$, $u = D_n u = 0$ *when* $x_n = t$, *it follows that for every* $s \in \mathbb{R}$

$$\lambda \int_t^T (\|u\|_{(s+1)}^2 + \|D_n u\|_{(s)}^2) e^{-2\lambda x_n} dx_n \leqq C \int_t^T e^{-2\lambda x_n} \|Pu\|_{(s)}^2 dx_n. \tag{23.4.2}$$

Here $\| \ \|_{(s)}$ *is the* $H_{(s)}$ *norm in* $\mathbb{R}^{n-1}$ *for fixed* x_n, *and* C *is independent of* s *when* $\lambda > A_s$.

Proof. A change in the lower order terms changes the right-hand side by an amount which is $O(1/\lambda)$ times the left-hand side. The statement is therefore independent of the lower order terms. It is convenient to choose them so that $A = a(x, D')$ and $B = b(x, D')$ are self-adjoint. By the sharp Gårding inequality (Theorem 18.1.14) we can also assume that $b(x', x_n, D') \geqq 0$ for every $x_n \in [0, 1]$, and we may strengthen (23.4.1) to

$$x_n(|\xi'|^2 + 1) \leqq C_1 \operatorname{Re} b(x, \xi'), \quad 0 \leqq x_n \leqq 1. \tag{23.4.1)'$$

As in the proof of Lemma 23.1.1 the independence of lower order terms shows that it suffices to prove (23.4.2) when $s = 0$.

Set $\psi = e^{-2\lambda x_n}/x_n$, and form with scalar products in $L^2(G)$, $G = \{x \in \mathbb{R}^n; t < x_n < T\}$,

$$2 \operatorname{Im}(\psi(D_n - A)u, Pu) = 2 \operatorname{Im}(\psi(D_n - A)u, (D_n - A)^2 u) - 2 \operatorname{Im}(\psi(D_n - A)u, Bu).$$

With $v = (D_n - A)u$ the first term can be written

$$\begin{aligned} 2 \operatorname{Im}(\psi v, (D_n - A) v) &= 2 \operatorname{Im}(\psi v, D_n v) = \int_G \psi \partial_n |v|^2 dx \\ &= -\int_G |v|^2 \psi' dx + \int |v(x', T)|^2 \psi(T) dx'. \end{aligned}$$

Both terms are non-negative and we shall drop the second one. Next we observe that

$$-2\,\mathrm{Im}(\psi(D_n-A)u, Bu)=i([B\psi,(D_n-A)]u,u)+(\psi Bu(.,T),u(.,T)).$$

The last term is non-negative since B is. The symbol of $i[B\psi, D_n-A]$ is, apart from terms of order 1 multiplied by ψ,

$$-\partial(b\psi)/\partial x_n+\{a,b\}\,\psi=2\lambda b\psi+\psi(b-x_n(\partial b/\partial x_n-\{a,b\}))/x_n.$$

If we apply Taylor's formula to $\mathrm{Re}\, b((x,\xi')-sH_{\xi_n-\mathrm{Re}\,a})$, $0\leqq s\leqq x_n$, recalling that $\mathrm{Re}\, b(x',0,\xi')\geqq 0$ by (23.4.1)′, it follows that

$$\mathrm{Re}(b-x_n(\partial b/\partial x_n-\{a,b\}))\geqq -C(1+|\xi'|^2)\,x_n^2-Cx_n(1+|\xi'|).$$

Hence (23.4.1)′ gives

$$\begin{aligned}&\mathrm{Re}\,\psi(2\lambda b+(b-x_n(\partial b/\partial x_n-\{a,b\}))/x_n)\\ &\geqq e^{-2\lambda x_n}(1+|\xi'|^2)(2\lambda/C_1-C)-C(1+|\xi'|)\,\psi.\end{aligned}$$

When $\lambda>CC_1$ we have $2\lambda/C_1-C\geqq\lambda/C_1$. By the sharp Gårding inequality applied to the quotient by ψ' for fixed x_n it follows that

$$\begin{aligned}(23.4.3)\qquad &\mathrm{Re}\, i([B\psi,(D_n-A)]u,u)-\lambda/C_1\int_t^T e^{-2\lambda x_n}\|u\|_{(1)}^2\,dx_n\\ &\geqq -C\int_t^T|\psi'|\,\|u\|_{(1)}\|u\|_{(0)}\,dx_n.\end{aligned}$$

We estimate the right-hand side by the inequality between geometric and arithmetic means. Summing up, we have now proved

$$\begin{aligned}(23.4.4)\qquad &\int_G e^{-2\lambda x_n}(x_n^{-2}+2\lambda/x_n)\,|(D_n-A)u|^2\,dx+\lambda/2C_1\int_t^T e^{-2\lambda x_n}\|u(.,x_n)\|_{(1)}^2\,dx_n\\ &\leqq 2\int_G e^{-2\lambda x_n}|(D_n-A)u|\,|Pu|\,dx/x_n\\ &\quad+C/\lambda\int_G(\lambda/x_n+x_n^{-2})^2 e^{-2\lambda x_n}|u|^2\,dx.\end{aligned}$$

We estimate the first integral on the right-hand side by the inequality between geometric and arithmetic means so that half of the first term on the left is cancelled. Now an integration by parts similar to those above gives

$$\begin{aligned}I^2&=\int_G e^{-2\lambda x_n}(2\lambda^3/x_n+\lambda^2/x_n^2+2\lambda/x_n^3+3/x_n^4)\,|u|^2\,dx\\ &\leqq \mathrm{Im}-\int_G e^{-2\lambda x_n}(\lambda^2/x_n+1/x_n^3)(D_n-A)\,u\bar{u}\,dx\\ &\leqq I\Big(\int_G e^{-2\lambda x_n}(\lambda/x_n+1/x_n^2)\,|(D_n-A)\,u|^2\,dx\Big)^{\frac{1}{2}}\end{aligned}$$

where we have used the Cauchy-Schwartz inequality. Hence I^2 can be estimated by the left-hand side of (23.4.4). We have therefore proved a stronger estimate than (23.4.2) for $s=0$

(23.4.5) $$\int_t^T e^{-2\lambda x_n}((\lambda^3/x_n+1/x_n^4)\|u\|_{(0)}^2+\lambda\|u\|_{(1)}^2+(x_n^{-2}+\lambda/x_n)\|(D_n-A)u\|^2)\,dx_n \leq C\int_t^T e^{-2\lambda x_n}\|Pu\|_{(0)}^2\,dx_n.$$

(A similar estimate is of course valid when $s \neq 0$.) The proof is complete.

We shall also need an estimate for the solution of the Cauchy problem for decreasing x_n:

Lemma 23.4.3. *With P, t and T as in Lemma 23.4.2 it follows if $u \in C_0^\infty(\mathbb{R}^n)$ and $u=D_n u=0$ when $x_n=T$ that*

(23.4.6) $$\int_t^T (\lambda^2\|u\|_{(s)}^2+\lambda x_n^2(\|u\|_{(s+1)}^2+\|D_n u\|_{(s)}^2))\,e^{2\lambda x_n}\,dx_n \leq C\int_t^T e^{2\lambda x_n}x_n^2\|Pu\|_{(s)}^2\,dx_n,$$

when $\lambda>\Lambda_s$.

Proof. We can essentially repeat the proof of Lemma 23.4.2 taking $\psi=-x_n e^{2\lambda x_n}$ now. The independence of lower order terms is again clear, so we can take $a=0$ and A, D self-adjoint, $B \geq 0$. Now

$$2\,\mathrm{Im}(\psi v, (D_n-A)v) = -\int_G |v|^2\psi'\,dx - \int |v(x',t)|^2\psi(t)\,dx'$$

where $-\psi'=(2\lambda x_n+1)e^{2\lambda x_n}$ and the second term is also positive. The next partial integration works as before, and the symbol of $[iB\psi, D_n-A]$ is, apart from terms of order 1 multiplied by ψ,

$$-2\lambda b\psi-\psi(b+x_n(\partial b/\partial x_n-\{a,b\}))/x_n.$$

Using Taylor's formula as before we find that the real part is bounded from below by

$$e^{2\lambda x_n}((2\lambda C_1-C)x_n^2(1+|\xi'|^2)-Cx_n(1+|\xi'|)).$$

When $\lambda>CC_1$ it follows by the Fefferman-Phong inequality (Theorem 18.6.8) that

(23.4.7) $$\int_G e^{2\lambda x_n}(2\lambda x_n+1)|(D_n-A)u|^2\,dx+\lambda/2C_1\int_t^T x_n^2\|u(.,x_n)\|_{(1)}^2 e^{2\lambda x_n}\,dx_n \leq 2\int_G|(D_n-A)u|\,|Pu|\,x_n e^{2\lambda x_n}\,dx+C\lambda\int_t^T\|u\|_{(0)}^2 e^{2\lambda x_n}\,dx_n.$$

We have

$$I=2\lambda\int_t^T e^{2\lambda x_n}\|u\|_{(0)}^2\,dx \leq 2\,\mathrm{Im}\int_G e^{2\lambda x_n}\bar{u}(D_n-A)u\,dx \leq (2I/\lambda)^{\frac12}\left(\int|(D_n-A)u|^2 e^{2\lambda x_n}\,dx\right)^{\frac12},$$

hence

$$I\lambda \leqq 2\int_G |(D_n - A)u|^2 e^{2\lambda x_n}\, dx.$$

Using Cauchy-Schwarz' inequality in the right-hand side of (23.4.7) and estimating the left-hand side from below by means of this estimate for I, we obtain (23.4.6). The proof is complete.

Let us now return to the differential operator P of order m in the open set $X \subset \mathbb{R}^n$, with principal symbol p satisfying conditions (i)–(iii) discussed at the beginning of the section. Let $Y \Subset X$.

Lemma 23.4.4. *If $0 \leqq t < T \leqq 1$ and $u \in C_0^\infty(Y)$ it follows that for every $s \in \mathbb{R}$ we have*

$$\int_t^T \sum_0^{m-1} \|D_n^{m-1-k} u\|_{(s+k)}^2\, dx_n \leqq C_s \int_t^T \|Pu\|_{(s)}^2\, dx_n, \tag{23.4.8}$$

if $D_n^j u = 0$ when $x_n = t$ and $j < m$;

$$\int_t^T \sum_0^{m-1} \|D_n^{m-1-k} u\|_{(s+k)}^2\, x_n^2\, dx_n + \int_t^T \sum_0^{m-2} \|D_n^{m-2-k} u\|_{(s+k)}^2\, dx_n \tag{23.4.9}$$
$$\leqq C_s \int_t^T \|Pu\|_{(s)}^2\, x_n^2\, dx_n$$

if $D_n^j u = 0$ when $x_n = T$ and $j < m$.

Proof. The proofs are quite similar starting from Lemmas 23.4.2 and 23.4.3 respectively in cases (23.4.8) and (23.4.9), so we shall only discuss (23.4.8). By Lemma 23.2.1 **it is sufficient to prove (23.4.8) where t and T are small,** so we may assume that X and Y are products of open sets X_0 and Y_0 in $\mathbb{R}^{n-1}$ and intervals I_X, I_Y on the x_n axis. We can choose a covering of $\bar{Y}_0 \times (\mathbb{R}^{n-1} \setminus 0)$ by finitely many open cones Γ_j such that

$$p(x,\xi) = \prod_1^N q_\nu^j(x,\xi); \quad (x',\xi') \in \Gamma_j, \; x_n \in \bar{I}_Y;$$

where each q_ν^j is either of the form $\xi_n - \Lambda(x,\xi')$ or of the form $(\xi_n - a(x,\xi'))^2 - b(x,\xi')$ with a, b real valued, $b(x,\xi') \geqq c\, x_n |\xi'|^2$. The factors q_ν^j have no zeros ξ_n in common. The functions Λ, a and b are only defined locally at first but can be extended to the whole space. For the corresponding operators we obtain a priori estimates from Lemma 23.1.1 in the first order case and from Lemma 23.4.2 in the second order case. Choose $\chi_j \in C^\infty$ with support in Γ_j and homogeneous of degree 0 in ξ' when $|\xi'| > 1$ so that

$$\sum \chi_j(x',\xi') = 1 \quad \text{when} \quad |\xi'| > 1.$$

In supp χ_j the polynomials

$$\xi_n^\kappa \prod_{\nu \neq \mu} q_\nu^j(x,\xi), \quad \kappa < \deg q_\mu^j, \; \mu = 1, \ldots, N,$$

form a basis for polynomials of degree $m-1$ in ξ_n. Using the equation $P\chi_j(x',D')u=\chi_j(x',D')(Pu)+[P,\chi_j(x',D')]u$ and factoring P on the principal symbol level in Γ_j, we obtain essentially as in the proof of Lemma 23.2.1 after adding the estimates obtained for $\chi_j(x',D')u$

$$(\lambda-C)\int_t^T \sum_0^{m-1} \|D_n^{m-1-k}u\|_{(s+k)}^2 e^{-2\lambda x_n} dx_n$$
$$\leqq C\int_t^T \|Pu\|_{(s)}^2 e^{-2\lambda x_n} dx_n + C\lambda \int_t^T \sum_0^{m-1} \|D_n^{m-1-k}u\|_{(s+k-1)}^2 e^{-2\lambda x_n} dx_n,$$

if $D_n^j u=0$ when $x_n=t$ and $j<m$. The last sum can be estimated by

$$\int_t^T \sum_0^{m-1} \|D_n^{m-k}u\|_{(s+k-1)}^2 e^{-2\lambda x_n} dx_n.$$

In the term with $k=0$ we write $D_n^m u = Pu+(D_n^m-P)u$ and observe that

$$\|D_n^m u - Pu\|_{(s-1)}^2 \leqq C\sum_0^{m-1} \|D_n^{m-k-1}u\|_{(s+k)}^2.$$

The estimate (23.4.6) follows now when λ is large enough. The remaining details are left for the reader to avoid too much repetition of the proof of Lemma 23.2.1 (see also the proof of Theorem 28.1.8).

The preceding lemmas allow us to extend Theorem 23.2.4 and Theorem 23.2.7 as follows:

Theorem 23.4.5. *Let P be a differential operator of order m with C^∞ coefficients in the C^∞ manifold X, and let $Y \Subset X$ be an open subset. Assume that P is strictly hyperbolic with respect to the level surfaces of $\phi\in C^\infty(X,\mathbb{R})$ in*

$$X_+=\{x\in X;\ \phi(x)>0\},$$

that P remains of principal type in

$$X_0=\{x\in X;\ \phi(x)=0\},$$

and that X_0 is non-characteristic with respect to P. If $f\in H_{(s)}^{\mathrm{loc}}(X)$ has support in the closure of X_+ one can then find $u\in H_{(s+m-1)}^{\mathrm{loc}}(X)$ with support in the closure of X_+ so that $Pu=f$ in Y. This determines u uniquely in a neighborhood of $X_0\cap Y$.

Proof. We just have to supplement Theorems 23.2.4 and 23.2.7 with local existence and uniqueness theorems at X_0. We may therefore assume as above that $X\subset\mathbb{R}^n$ and that $\phi(x)=x_n$. Let $Y\Subset X$ and choose M so large that

$$p(x,\xi)\neq 0 \quad \text{when } x\in Y \quad \text{and} \quad |\xi_n|\geqq M|\xi'|,\ \xi\neq 0.$$

If K is a compact subset of $Y\cap X_0$ then

$$K_T=\{x;\ 0\leqq |x_n|\leqq T, |x'-y'|\leqq M|x_n| \text{ for some } y'\in K\}$$

is contained in Y if T is small enough. Let $f \in C_0^\infty$ and $\mathbb{R}_+^n \cap \operatorname{supp} f \subset K_T$. If $0<t<T$ it follows from the existence Theorem 23.2.4 and the uniqueness Theorem 23.2.7 that there is a solution $u_t \in C_0^\infty(Y)$ of the equation $Pu_t = f$ for $t \leqq x_n \leqq T$ such that $D_n^j u_t = 0$ for $j<m$ when $x_n = t$. (Here we use that the supporting planes of the cones $\{x; M(x_n - y_n) < |x' - y'|\}$ are non-characteristic in Y.) For u_t we have the estimate (23.4.6) which is uniform in t when $t \to 0$. Hence we can find a limit u as $t \to 0$ which satisfies the equation $Pu = f$ for $0 \leqq x_n \leqq T$ and has zero Cauchy data when $t=0$. It can be estimated by (23.4.8) with $t=0$. Hence a solution $u \in \bar{H}_{(m-1,s)}$ exists for all $f \in \bar{H}_{(0,s)}$ with support in the interior of K_T. The proof of local *existence* is now completed by repeating that of Proposition 23.2.6.

To prove local *uniqueness* we argue similarly. Set

$$M_T = \{x; 0 \leqq x_n \leqq T; |x' - y'| \leqq M(T - x_n) \text{ for some } y' \in K\}.$$

We have $M_T \subset Y$ if T is small enough. By Theorems 23.2.4 and 23.2.7 we can for any $g \in C_0^\infty(M_T)$ and $t \in (0, T)$ find a solution v of the equation $P^* v = g$ when $t < x_n \leqq T$ with zero Cauchy data for $x_n = T$ and $\operatorname{supp} v \subset M_T$. We can estimate v by means of (23.4.9) with P replaced by P^* and conclude that all derivatives are continuous up to the boundary plane X_0 since $P^* v = g \in C^\infty$ also. If $Pu = 0$ in Y and $\operatorname{supp} u \subset X_+ \cup X_0$, it follows that

$$(u, g) = (Pu, v) = 0.$$

Hence $u=0$ in the interior of M_T, which proves the uniqueness.

Remark. There is no difficulty in discussing the Cauchy problem with inhomogeneous Cauchy data of the form (23.2.7). However, we must require half a derivative more for ψ_j than in Proposition 23.2.5 in order to be able to reduce the proof to Theorem 23.4.5.

We shall now prove an extension of Theorem 23.2.10. The main step is a microlocal version of Lemma 23.4.2.

Lemma 23.4.6. *Let a, b, P be as in Lemma 23.4.2, let $q \in S^0$ be real valued, $q(x, \xi') = 0$ when $x_n > T$, and assume that*

$$|\partial q / \partial x'| + |\partial q / \partial \xi'|(1 + |\xi'|) \leqq -\delta \partial q / \partial x_n. \tag{23.4.10}$$

Set $Q = q(x, D')$. If T and δ are sufficiently small, it follows that for every $s \in \mathbb{R}$ and $u \in \mathscr{S}(\mathbb{R}^n)$ with $u = D_n u = 0$ when $x_n = 0$ we have when $\lambda > \Lambda_s$

$$\lambda \int_0^T (\|Qu\|_{(s+1)}^2 + \|QD_n u\|_{(s)}^2) e^{-2\lambda x_n} dx_n \leqq C_s \int_0^T (\|QPu\|_{(s)}^2 + \|Pu\|_{(s-\frac{1}{2})}^2) e^{-2\lambda x_n} dx_n. \tag{23.4.11}$$

Proof. From Lemma 23.4.2 we know that

$$\lambda \int_0^T (\|u\|_{(s+\frac{1}{2})}^2 + \|D_n u\|_{(s-\frac{1}{2})}^2) e^{-2\lambda x_n} dx_n \leqq C \int_0^T \|Pu\|_{(s-\frac{1}{2})}^2 e^{-2\lambda x_n} dx_n.$$

For large λ the lower order terms in P and in Q can therefore only change the constant in (23.4.11). If we replace u by $(1+|D'|^2)^{-s/2}u$ and observe that

$$(1+|D'|^2)^{s/2}P(1+|D'|^2)^{-s/2}-P, \quad (1+|D'|^2)^{s/2}Q(1+|D'|^2)^{-s/2}-Q$$

are of order 1 and -1 respectively, it is clear that it suffices to prove (23.4.11) when $s=0$. We may also assume that $A=a(x,D')$ and $B=b(x,D')$ are self-adjoint, that (23.4.1)$'$ holds and that $B\geqq 0$. We write $Q=(q(x,D')+q(x,D')^*)/2$. With scalar products in $L^2(\mathbb{R}^n_+)$ and $\psi(x)=e^{-2\lambda x_n}/x_n$ we form

$$2\,\mathrm{Im}\,(\psi Q^2(D_n-A)u, Pu)=2\,\mathrm{Im}\,(\psi Q^2 v,(D_n-A)v)-2\,\mathrm{Im}\,(\psi Q^2(D_n-A)u, Bu)$$

where $v=(D_n-A)u$. Integrating first for $x_n>t>0$ we obtain when $t\to 0$

$$\begin{aligned}(23.4.12)\quad 2\,\mathrm{Im}\,(\psi Q^2 v,(D_n-A)v)&=i((D_n-A)v,\psi Q^2 v)-i(\psi Q^2 v,(D_n-A)v)\\ &=(-\psi' Qv, Qv)+i(\psi[Q^2, D_n-A]v, v).\end{aligned}$$

Since $-\psi'=e^{-2\lambda x_n}(2\lambda/x_n+x_n^{-2})$ the first term gives control of Qv. Next we observe that

$$\begin{aligned}(23.4.13)\qquad &-2\,\mathrm{Im}\,(\psi Q^2(D_n-A)u, Bu)\\ &=i(\psi Q^2(D_n-A)u, Bu)-i(Bu,\psi Q^2(D_n-A)u)\\ &=i(\psi[B,Q^2]v,u)+i([Q^2,D_n-A]\psi Bu,u)\\ &\quad+i(Q^2[\psi B, D_n-A]u,u).\end{aligned}$$

The terms in (23.4.12) and (23.4.13) containing a commutator with Q^2 are

$$(23.4.14)\quad i((\psi[Q^2,D_n-A]v,v)+(\psi[B,Q^2]v,u)+(\psi[Q^2,D_n-A]Bu,u)).$$

With $U=(v,(1+|D'|^2)^{\frac{1}{4}}u)$ this is of the form $(\psi W(x,D')U,U)$ where W is vector valued of order 0 with principal symbol

$$\begin{pmatrix}\{q^2,\xi_n-a\} & \{b,q^2\}/2|\xi'| \\ \{b,q^2\}/2|\xi'| & \{q^2,\xi_n-a\}\,b/|\xi'|^2\end{pmatrix}.$$

It is positive semi-definite if

$$|\{b,q^2\}/2|=|q\{b,q\}|\leqq\{q^2,\xi_n-a\}\,b^{\frac{1}{2}}=2q(-\partial q/\partial x_n-\{q,a\})\,b^{\frac{1}{2}},$$

that is,

$$|\{b,q\}|\leqq 2b^{\frac{1}{2}}(-\partial q/\partial x_n-\{q,a\}).$$

This inequality follows from (23.4.10) if Lemma 7.7.2 is applied to b. Thus the sharp Gårding inequality for systems gives

$$\begin{aligned}\mathrm{Re}\,(\psi W(x',D)U,U)&\geqq -C\int_0^T\psi(\|v\|^2_{(-\frac{1}{2})}+\|u\|^2_{(\frac{1}{2})})\,dx_n\\ &\geqq -C'/\lambda\int_0^T\|Pu\|^2_{(-\frac{1}{2})}e^{-2\lambda x_n}\,dx_n.\end{aligned}$$

The last estimate follows from (23.4.5) (with the order of the norms lowered by one half).

The last term in (23.4.13) can be written

$$i(Q^2[\psi B, D_n - A]u, u) = i([\psi B, D_n - A]Qu, Qu) + i(Q[Q,[\psi B, D_n - A]]u, u).$$

The principal symbol of $iQ[Q,[\psi B, D_n - A]]$ is purely imaginary and of order 1, hence the self-adjoint part is of order 0 and

(23.4.15) $$|\mathrm{Re}\, i(Q,[Q,[\psi B, D_n - A]]u, Qu)| \leqq C\int|\psi'|\,\|u\|^2_{(0)}\,dx_n.$$

Since $\|u\|^2_{(0)} \leqq \|u\|_{(\frac12)}\|u\|_{(-\frac12)}$ we obtain using (23.4.5) with the order of the norms lowered by one half

(23.4.16) $$\lambda^{\frac12}\int|\psi'|\,\|u\|^2_{(0)}\,dx_n \leqq C\int_0^T \|Pu\|^2_{(-\frac12)}\,e^{-2\lambda x_n}\,dx_n.$$

The term $\mathrm{Re}\, i([\psi B, D_n - A]Qu, Qu)$ is estimated from below by (23.4.3) with u replaced by Qu. Summing up, we obtain

$$\begin{aligned}\int_0^T|\psi'|\,\|Q(D_n - A)u\|^2\,dx_n &+ \lambda/2C_1\int_0^T e^{-2\lambda x_n}\|Qu\|^2_{(1)}\,dx_n\\ &\leqq 2\int_0^T e^{-2\lambda x_n}\|Q(D_n - A)u\|\,\|QPu\|\,dx_n/x_n + C\lambda^{-\frac12}\int_0^T\|Pu\|^2_{(-\frac12)}\,e^{-2\lambda x_n}\,dx_n\\ &\quad + C\lambda^{-1}\int_0^T(\lambda/x_n + x_n^{-2})^2\,e^{-2\lambda x_n}\|Qu\|^2\,dx_n.\end{aligned}$$

In the first term on the right-hand side we use the Cauchy-Schwarz inequality so that one term which occurs is at most one half of the first term on the left. In the half remaining after cancellation we use that $Q(D_n - A) = QD_n - AQ + [A,Q] = (D_n - A)Q + [Q, D_n - A]$ where the commutators are of order 0 so that $[A,Q]u$ and $[D_n - A, Q]u$ can be estimated using (23.4.16). The last step of the proof of Lemma 23.4.2 is now applicable and completes the proof.

Lemma 23.4.7. *If a, b, P are as in Lemma 23.4.2 and $f \in \bar H_{(0,s)}(\mathbb{R}^n_+)$ then the equation $Pu = f$ has a unique solution $u \in \bar H_{(2,s-1)} \subset \bar H_{(1,s)}$ when $x_n < \frac12$, such that $u = D_n u = 0$ when $x_n = 0$. If Q satisfies the conditions in Lemma 23.4.6 with $T = \frac12$ then $Qu \in \bar H_{(1,s+\frac12)}$ if $Qf \in \bar H_{(0,s+\frac12)}$.*

Proof. a) Existence. The estimate (23.4.2) extends by continuity to all $u \in \bar H_{(2,s)}(\mathbb{R}^n_+)$, with $t = 0$ and $T = 1$, so it suffices to prove the existence for f in a dense subset of $\bar H_{(0,s)}$, say all $f \in \bar H_{(0,s+1)}$. The equation $Pu = f$ with the homogeneous Cauchy boundary conditions means that with L^2 scalar products in $\mathbb{R}^n_+$

(23.4.17) $$(u, P^*v) = (f, v), \qquad v \in C_0^\infty(X_1),$$

where $X_1 = \{x \in \mathbb{R}^n; x_n < 1\}$. If $f \in \bar H_{(0,s+1)}$ then it follows from Lemma 23.4.3 that

$$|(f,v)| \leqq \|f\|_{(0,s+1)}\|v\|_{(0,-s-1)} \leqq \|f\|_{(0,s+1)}\|P^*v\|_{(0,-s-1)}$$

where the norm on P^*v is taken with respect to the half space $\mathbb{R}^n_+$. Hence the Hahn-Banach theorem shows that (23.4.17) is satisfied by some $u \in \bar{H}_{(0,s+1)}(\mathbb{R}^n_+)$. By Theorem B.2.9 it follows that χu has the required properties if $\chi \in C_0^\infty(-1,1)$ and $\chi = 1$ in $(-\frac{1}{2}, \frac{1}{2})$.

b) The estimate (23.4.2) with s replaced by $s-1$ is valid for all $u \in \bar{H}_{(2,s-1)}$, which gives the uniqueness immediately.

c) Assume that $Qf \in \bar{H}_{(0,s+\frac{1}{2})}$. Choose $\chi \in C_0^\infty(\mathbb{R}^{n-1})$ with $\int \chi\, dx' = 1$ and let $f_\varepsilon = \hat{\chi}(\varepsilon D')f$ be the convolution with $\varepsilon^{1-n}\chi(x'/\varepsilon)$. Then $f_\varepsilon \in \bar{H}_{(0,\sigma)}$ for every σ, so the Cauchy problem $Pu_\varepsilon = f_\varepsilon$ with zero Cauchy data when $x_n = 0$ has a solution when $x_n < \frac{1}{2}$ satisfying (23.4.11) with s replaced by $s + \frac{1}{2}$. Since

$$Qf_\varepsilon = \hat{\chi}(\varepsilon D')\, Qf + [Q, \hat{\chi}(\varepsilon D')]\, f$$

and the commutator is bounded in $\operatorname{Op} S^{-1}$ for $0 < \varepsilon < 1$, it follows that Qf_ε is bounded in $\bar{H}_{(0,s+\frac{1}{2})}$ when $0 < \varepsilon < 1$. Hence (23.4.11) shows that Qu_ε is bounded in $\bar{H}_{(1,s+\frac{1}{2})}$. We have $u_\varepsilon \to u$ in $\bar{H}_{(1,s)}$ when $\varepsilon \to 0$ so it follows that $Qu \in \bar{H}_{(1,s+\frac{1}{2})}$, which completes the proof.

We can now prove an extension of Theorem 23.2.10.

Theorem 23.4.8. *Let P satisfy the hypotheses of Theorem 23.4.5, let $f \in \mathcal{N}(\bar{X}_+)$, $u \in \bar{\mathcal{D}}'(X_+)$, $Pu = f$ in X_+ and $v^j u|_{X_0} = \phi_j$, $j < m$, where v is a C^∞ vector field transversal to X_0. Then the extension $\dot{u} \in \mathcal{N}(\bar{X}_+)$ of u satisfies*

$$WF_b(u)|_{X_0} = WF_b(f)|_{X_0} \cup \Big(\bigcup_{j=0}^{m-1} WF(\phi_j)\Big). \tag{23.4.18}$$

Proof. As in the proof of Theorem 23.2.10 we just have to verify that the left-hand side of (23.4.18) is contained in the set on the right. To prove this local statement we may assume that $X \subset \mathbb{R}^n$, $\phi(x) = x_n$, $v = \partial/\partial x_n$, $\dot{u} \in \mathcal{E}'(\bar{X}_+)$, and that $(x, \xi) \in WF(u)$, $x_n > 0$, implies $\xi' \neq 0$. At first we also assume that $m = 2$ and that P satisfies the hypotheses in Lemma 23.4.2; thus we allow P to be pseudo-differential along the boundary.

Let $(y', \eta') \in T^*(\mathbb{R}^{n-1}) \setminus 0$ belong to the complement of the right-hand side of (23.4.18). Choose a conic neighborhood $W \subset \mathbb{R}^n \times (\mathbb{R}^{n-1} \setminus 0)$ of $(y', 0, \eta')$ such that $\{(x', \xi'); (x', 0, \xi') \in W\}$ does not intersect the right-hand side of (23.4.18) and $(x, \xi') \in W$, $x_n > 0$ implies $(x, \xi', \xi_n) \notin WF(f)$, $\xi_n \in \mathbb{R}$. By a slight change of the proof of Theorem B.2.9 (see the proof of Lemma 24.4.6) it follows from the equation $Pu = f$ that we can find s so that

$$\chi(x, D')\, u \in \bar{H}_{(1,s)}$$

for all $\chi \in S^0$ with cone supp $\chi \subset W$. Set $U = \chi(x, D')\, u$. Then

$$F = PU = \chi(x, D') f + [P, \chi(x, D')]\, u \in \bar{H}_{(0,s)},$$

and $\chi(x, D')\, \phi_j \in C_0^\infty$. If Q satisfies the conditions in Lemma 23.4.6 and $\chi = 1$ in a neighborhood of supp q, then $QF \in \bar{H}_{(1,\sigma)}$ for all σ since $QP\chi(x, D')$

$-Q\chi(x,D')P$ is of order $-\infty$ tangentially. Hence Lemma 23.4.7 gives $QU \in \bar{H}_{(1,s+\frac{1}{2})}$ which implies $Qu \in \bar{H}_{(1,s+\frac{1}{2})}$.

The condition (23.4.10) is satisfied by

$$q(x,\xi') = \psi(\varepsilon - x_n/\varepsilon - |x'-y'|^2 - |\xi'(1+|\xi'|^2)^{-\frac{1}{2}} - \eta'/|\eta'||^2)$$

if ε is small enough and $\psi \in C_0^\infty(-1,1)$. The support lies in W when $|\xi'|$ is large. If we iterate the preceding argument for functions $\chi(x,\xi')$ of this form only we find inductively that $q(x,D')u \in \bar{H}_{(1,s)}$ for all such q and all s. Using partial hypoellipticity again (see Theorem B.2.9 or Lemma 24.4.6) we conclude that $q(x,D')u \in \bar{C}^\infty$, hence that $(y',\eta') \notin WF_b(u)$.

For arbitrary m we can factor P as in the proof of Theorem 23.2.10 as a product of first and second order factors with an error of order $-\infty$ tangentially. Strictly speaking we only have such a factorization in a small conic neighborhood of $(y',0,\eta')$, but that is all one needs to apply the preceding argument to remove the factors successively. No new ideas are required in the proof so we leave the details for the reader.

Notes

The Cauchy problem for strictly hyperbolic second order equations was first solved by Hadamard [1] with his parametrix method. (See Section 17.4 for operators of the form $D_t^2 - a(x,D)$ where a is elliptic. The general case is similar since as explained in Section C.6 geodesic normal coordinates can be introduced in general.) A variant of the method was developed by M. Riesz [1]. Another approach depending on L^2 estimates was introduced by Friedrichs and Lewy [1] who transferred the idea of energy estimates from the theory of Maxwell's equations. Thus the estimates were proved by a partial integration. However, the extension of the estimates to higher order hyperbolic operators found by Petrowsky [1,5] relied on quite complicated Fourier analysis techniques. The simpler energy estimate method was extended to higher order operators much later by Leray [1] and Gårding [3]. Shortly afterwards the Fourier analysis approach of Petrowsky reappeared in the guise of singular integral operators. (See e.g. Calderón [2], Mizohata [2,3].) We have here steered a middle course, using energy estimates for first order operators in Section 23.1 and then a factorization of higher order operators into first order pseudo-differential factors in Section 23.2.

Both in Section 23.1 and in Section 23.2 we study the singularities of the solutions of the Cauchy problem closely. Such studies were first made by Courant and Lax [1], Lax [3] and Ludwig [1] before the introduction of wave front sets; the results were then less precise and harder to prove since they did not have a local character. In Chapter XXVI we shall prove more general results on interior singularities with a more systematic method.

The necessity of weak hyperbolicity proved in Theorem 23.3.1 is due to Lax [3] for simple roots and to Mizohata [5] in general. The simple proof given here is essentially due to Ivrii and Petkov [1]; see also Hörmander [36]. Theorem 23.3.2 also comes from Ivrii and Petkov [1]. These papers are mainly concerned with conditions on the subprincipal symbol at a doubly characteristic point, that is, one where $P_m = dP_m = 0$. The Hessian of $P_m/2$ is there a hyperbolic quadratic form Q by Theorem 23.3.1 and Lemmas 8.7.2, 8.7.3. A symplectic classification is therefore given by Theorem 21.5.3. If no real eigenvalues exist then solvability of the Cauchy problem for P requires that the subprincipal symbol of P lies between $-\sum \mu_j$ and $\sum \mu_j$ where $i\mu_j$ are the eigenvalues of the Hamilton map on the positive imaginary axis. The proof was given in part by Ivrii and Petkov [1] and completed in Hörmander [36]. On the other hand, when the characteristics are at most double and real eigenvalues exist for the Hamilton map at every double characteristic (the effectively hyperbolic case) then the Cauchy problem can be solved for arbitrary lower order terms (see Ivrii [1], Melrose [6], Iwasaki [1], Nishitani [1]). Much work has also been devoted to double characteristics which are not effectively hyperbolic. (See Lascar and Lascar [1] and the references given there.) However, there are no complete results yet in the doubly characteristic case.

Chapter XXIV. The Mixed Dirichlet-Cauchy Problem for Second Order Operators

Summary

In Section 23.1 we introduced energy integral estimates for first order operators only. The passage to higher order operators in Section 23.2 was made by factorization. However, as we saw in Section 23.4 the energy integral method is also applicable directly in the higher order case, and it was originally introduced for second order operators. We shall discuss it in that case in Section 24.1 to derive estimates for mixed problems. For the sake of simplicity only the mixed Dirichlet-Cauchy problem will be discussed in this chapter. References to work on more general mixed problems will be given in the notes. However, the propagation of singularities at the boundary is a very intricate matter also in this special case. It will be discussed in Sections 24.2 to 24.5 by means of a microlocal version of the energy estimates of Section 24.1. Section 24.3 is devoted to the geometrical aspects of the flow of singularities. Some problems remain concerning bicharacteristics which are tangents of infinite order to the Dirichlet boundary. Thus there is a gap in the final results on propagation of singularities in Section 24.5. Two other cases where these fail are discussed in Section 24.6. Both are related to the Tricomi equation. In Section 24.7 finally we discuss estimates for operators depending on a parameter in order to complete the study in Section 17.5 of the asymptotic properties of the spectral function of the Dirichlet problem.

24.1. Energy Estimates and Existence Theorems in the Hyperbolic Case

Let P be a second order differential operator with C^∞ coefficients in a C^∞ manifold X of dimension n, with interior X° and boundary ∂X. We assume

(i) P is strictly hyperbolic with respect to the level surfaces of $\phi \in C^\infty(X, \mathbb{R})$.

(See Definition 23.2.3.) Let p be the principal symbol. Since P may be multiplied to the left by $p(x, \phi'(x))^{-1}$ we may asusume without restriction that $p(x, \phi'(x)) > 0$ for every $x \in X$. The quadratic form $p(x, \xi)$ can be polarized for fixed x to a symmetric bilinear form $p(x,\xi,\eta)$; $\xi,\eta\in T_x^*$. The orthogonal plane of $\phi'(x)$ with respect to this form is then supplementary to $\mathbb{R}\phi'(x)$, and the hyperbolicity means that $p(x,\xi)$ is negative definite there. Thus $p(x,\xi)$ has Lorentz signature: the indices of inertia are $1, n-1$.

With standard local coordinates $x_1,\dots,x_n$, $\xi_1,\dots,\xi_n$ we shall write

$$p(x,\xi)=\sum g^{jk}(x)\xi_j\xi_k$$

to conform with the usual notation in (pseudo-)Riemannian geometry. We identify $\xi\in T_x^*$ and $t\in T_x$ when $p(x,\xi,\eta)=\langle t,\eta\rangle$, $\eta\in T_x^*$, that is, in local coordinates

$$t_j=\sum g^{jk}(x)\xi_k.$$

This carries p over to the dual quadratic form in $T_x(X)$

$$\sum g_{jk}(x)t_j t_k$$

where (g_{jk}) is the inverse of (g^{jk}). This form defines a pseudo-Riemannian geometry in X. A tangent vector t at x is called timelike if $\sum g_{jk}(x)t_jt_k>0$. Since $\langle t,\phi'(x)\rangle=p(x,\xi,\phi'(x))$ if ξ is the cotangent vector corresponding to t, we must have $\langle t,\phi'(x)\rangle\neq 0$ then so the timelike tangent vectors at x form a double cone. Those with $\langle t,\phi'(x)\rangle>0$ will be called forward directed. Their closure is a convex cone in which the pseudo-Riemannian scalar product of any two non-zero vectors is positive apart from the case of two linearly dependent vectors in the boundary. These boundary vectors are isotropic, that is, $\sum g_{jk}(x)t_jt_k=0$. Vectors with $\sum g_{jk}(x)t_jt_k<0$ finally are called spacelike. Hypersurfaces are classified by their conormals ν: A hypersurface is spacelike (timelike) if $p(\nu)>0$ ($p(\nu)<0$).

We shall assume that

(ii) The map $X\ni x\mapsto\phi(x)$ is proper.
(iii) ∂X is timelike.

Thus $X_{ab}=\{x\in X;\ a\leqq\phi(x)\leqq b\}$ is for arbitrary real $a<b$ compact; the boundary consists of the two spacelike surfaces $\phi^{-1}(a)$ and $\phi^{-1}(b)$ and the part of ∂X where $a\leqq\phi\leqq b$. It intersects the spacelike boundary pieces transversally. The condition (iii) means that the pseudo-Riemannian metric remains of Lorentz signature when restricted to ∂X.

One of the main goals of this section is to prove the following existence theorem:

Theorem 24.1.1. *Let $f\in\bar H^{\text{loc}}_{(s)}(X^\circ)$, $u_0\in H^{\text{loc}}_{(s+1)}(\partial X)$ where $s\geqq 0$, and assume that f and u_0 vanish when $\phi<a$. Under the assumptions* (i)–(iii) *above there is then a unique $u\in\bar H^{\text{loc}}_{(s+1)}(X^\circ)$ vanishing when $\phi<a$ such that $Pu=f$ in X° and $u=u_0$ on ∂X.*

Remark. We have only made a statement here concerning the case of vanishing Cauchy data. This is not a serious restriction since we already know how to solve the Cauchy problem. However, the discussion of compatibility conditions between u_0, f and the Cauchy data in their common domain is otherwise somewhat complicated.

The proof of Theorem 24.1.1 like that of Theorem 23.2.4 will follow once we have proved a local existence theorem analogous to Proposition 23.2.6 and obtained a local uniqueness theorem analogous to Theorem 23.2.7. We shall therefore postpone the proof of Theorem 24.1.1 until we have established these local results.

Thus we assume in what follows that

$$P(x,D)=\sum g^{jk}(x)\,D_j D_k+\sum b_j(x)\,D_j+c(x)$$

where the coefficients are in $C^\infty(\mathbb{R}^n)$ and constant outside a compact set, (g^{jk}) is real with Lorentz signature, and $g^{nn}>0$, $g^{11}=-1$. Then we have hyperbolicity with respect to the planes $t=$constant if $t(x)=x_n$, and the plane $x_1=0$ is timelike. The energy estimate in Lemma 23.1.1 was obtained by taking the scalar product of Pu with $\bar{u}$ multiplied by $e^{-\lambda t}$. Here we argue similarly but multiply with a first order derivative of $\bar{u}$. First note that if g^{jk} and f^i are real constants then

$$\begin{aligned}2\,\mathrm{Re}\sum f^i\partial_i\bar{u}\sum g^{jk}\partial_j\partial_k u&=\sum\partial_j\big(2\,\mathrm{Re}\sum g^{jk}\partial_k u\sum f^i\partial_i\bar{u}\big)-\sum\partial_i\sum f^i g^{jk}\partial_j u\,\partial_k\bar{u}\\&=\sum\partial_j\big(T_i^j(u)f^i\big)\end{aligned}$$

where for a cotangent vector v and $f_m=\sum g_{mi}f^i$

(24.1.1) $\quad\sum T_i^j(u)\,f^i v_j=2\,\mathrm{Re}\sum g^{jk}\partial_k u\,v_j\sum g^{im}\partial_i\bar{u}\,f_m-\sum g^{jk}\partial_j u\,\partial_k\bar{u}\sum g^{li}f_l v_i.$

When f^i and g^{jk} are variable there is only a sesquilinear form in the derivatives of u to add on the right-hand side. Since $\sum g^{jk}\partial_j\partial_k u+Pu$ is of first order, we obtain the basic energy identity

(24.1.2) $$\begin{aligned}&\sum\partial_j\big(e^{-\lambda t}\big(\sum T_i^j(u)f^i+f^j|u|^2\big)\big)+\lambda\sum e^{-\lambda t}\big(\sum T_i^j(u)f^i\partial_j t+|u|^2\langle f,t'\rangle\big)\\&\quad=-2\,\mathrm{Re}\,e^{-\lambda t}\langle f,\bar{u}'\rangle Pu+e^{-\lambda t}R(u)\end{aligned}$$

where R is a sesquilinear form in (u,u') which is independent of λ. We recall that $t=x_n$, hence $t'=(0,\dots,0,1)$. We shall integrate the identity over a quarter space $x_1>0$, $x_n<T$. The first term can be integrated out by means of Green's formula then, which gives a surface integral over each of the boundary planes. The essential point is now to examine when the form involving T_i^j is positive definite; we have kept the notation rather more general than necessary to make this discussion clearer.

Lemma 24.1.2. *Let L be a real quadratic form of Lorentz signature in a finite dimensional vector space V, and let a, b be timelike vectors in the same cone. Thus $L(a)$, $L(b)$ and $L(a,b)$ are positive if $L(\xi,\eta)$ is the symmetric bilinear form with $L(\xi,\xi)=L(\xi)$. Then the quadratic form*

(24.1.3) $$V\ni\xi\mapsto 2L(\xi,a)\,L(\xi,b)-L(\xi,\xi)\,L(a,b)$$

is positive definite.

Proof. The right-hand side of (24.1.3) can be written $L(a,\eta)$ where $\eta = 2\xi L(\xi,b) - bL(\xi,\xi)$, thus $L(\eta,\eta) = L(b,b)\,L(\xi,\xi)^2 \geqq 0$. If $\xi = tb + \theta$, $L(b,\theta) = 0$, then

$$L(\eta,b) = 2L(\xi,b)^2 - L(b,b)\,L(\xi,\xi) = (t^2 L(b,b) - L(\theta,\theta))\,L(b,b) > 0$$

if $\xi \neq 0$. Thus η is then in the same closed Lorentz cone as b, which proves that $L(\eta,a) > 0$.

Remark. Conversely, L or $-L$ has Lorentz signature and a, b are timelike vectors in the same Lorentz cone if (24.1.3) is positive definite. In fact, taking $\xi = a$ or $\xi = b$ we obtain $L(a,a)\,L(a,b) > 0$ and $L(b,b)\,L(a,b) > 0$. After a change of sign if necessary it follows that $L(a)$, $L(b)$, $L(a,b)$ are all positive, and when $L(\xi,a) = 0$ we obtain $L(\xi,\xi) < 0$ if $\xi \neq 0$, which proves that the indices of inertia are $1, n-1$.

If f is timelike and directed forward, it follows that the second term in (24.1.2) is positive definite in (u,u'). The surface integral obtained over the plane $x_n = T$ will also have an integrand which is a positive definite form in (u,u') then. With $v = (-1,0,\dots,0)$ the first order terms in the integrand of the surface integral over the plane $x_1 = 0$ will be $e^{-\lambda t}$ times

$$\sum T_i^j(u)\,f^i v_j = 2\,\mathrm{Re}\,p(u',v)\,p(\bar{u}',f_c) - p(u',\bar{u}')\,p(v,f_c)$$

where f_c is the cotangent vector corresponding to f. If u equals 0 when $x_1 = 0$, then $u' = tv$ so this is equal to $|t|^2\,p(v,v)\,p(v,f_c) = |t|^2 f^1$ which is positive if f points toward the half space where $x_1 > 0$. In that case we have for arbitrary $u \in C^1$, if $u' = tv + \zeta$ where $p(v,\zeta) = 0$

$$\begin{aligned}\sum T_i^j(u)\,f^i v_j &= 2\,\mathrm{Re}\,p(tv+\zeta,v)\,p(\bar{t}v+\bar{\zeta},f_c) - p(tv+\zeta,\bar{t}v+\bar{\zeta})\,p(v,f_c)\\ &= |t|^2 f^1 - 2\,\mathrm{Re}\,\bar{t}\,p(\zeta,f_c) + p(\zeta,\bar{\zeta})\,f^1\\ &\geqq |t|^2 f^1/2 - 2|p(\zeta,f_c)|^2/f^1 + p(\zeta,\bar{\zeta})\,f^1.\end{aligned}$$

Using a partition of unity it is of course easy to choose f so that $f^1 > 0$ on ∂X and f lies in the forward cone everywhere; we can take $f \in C^\infty$ and constant outside a compact set. If we integrate (24.1.2) over $Q_T = \{x;\, x_1 > 0, x_n < T\}$, estimate the integral of the first term on the right-hand side by Cauchy-Schwarz' inequality, and cancel a constant times the volume integral of $|u'|^2 + |u|^2$, we obtain if $u \in C_0^2(\mathbb{R}^n)$ and λ is large enough

$$\text{(24.1.4)}\quad \int_{Q_T} \lambda(|u'|^2 + |u|^2)\,e^{-\lambda x_n}\,dx + \int_{\partial Q_T} (|u'|^2 + |u|^2)\,e^{-\lambda x_n}\,dS$$
$$\leqq C\lambda^{-1} \int_{Q_T} |Pu|^2\,e^{-\lambda x_n}\,dx + C \int_{x_1 = 0,\, x_n < T} \Big(|u|^2 + \sum_2^n |\partial_j u|^2\Big)\,e^{-\lambda x_n}\,dS.$$

As a first application we prove a uniqueness theorem.

Lemma 24.1.3. *If $D^\alpha u \in L^2(Q_T)$, $|\alpha| \leqq 1$, and $Pu = 0$ in Q_T while, $u = 0$ in $\{x;\, x_1 = 0, x_n < T\}$, then $u = 0$ in Q_T if u vanishes outside a compact subset of $\overline{Q}_T$.*

Proof. Choose a so that $a+1<x_n$ when $x\in\operatorname{supp} u$. It follows from (24.1.4) that

$$\int_{Q_T} (|v'|^2+|v|^2)\,dx \leqq C_a \int_{Q_T} |Pv|^2\,dx \tag{24.1.4)$'$}$$

when $v\in C^2(\bar{Q}_T)$ and $v=0$ if $x_n<a$ or $x_1=0$ or $|x|$ is large. Let $\psi\in C_0^\infty(\mathbb{R}^{n-1})$, $\int\psi\,dx=1$, and set

$$\psi_\varepsilon=\varepsilon^{1-n}\psi(x_2/\varepsilon,\dots,x_n/\varepsilon)\,\delta(x_1).$$

Since the plane $x_1=0$ is non-characteristic it follows from Theorem B.2.9 that u is locally in $\bar{H}_{(m,1-m)}$ for every m with respect to the half space $x_1>0$. If $x_n>0$ in $\operatorname{supp}\psi$ and ε is small it follows that $v_\varepsilon=u*\psi_\varepsilon$ is in $C^\infty(\bar{Q}_T)$ and equal to 0 when $x_n<a$. The support is compact so we may apply (24.1.4)$'$. Since $Pu=0$ we have

$$\begin{aligned} Pv_\varepsilon=P(u*\psi_\varepsilon)-(Pu)*\psi_\varepsilon=&\sum(g^{jk}(D_jD_ku*\psi_\varepsilon)-(g^{jk}D_jD_ku)*\psi_\varepsilon)\\ &+\sum(b_j(D_ju*\psi_\varepsilon)-(b_jD_ju)*\psi_\varepsilon)+c(u*\psi_\varepsilon)-(cu)*\psi_\varepsilon. \end{aligned}$$

In the first sum the term with $j=k=1$ is equal to 0 since $g^{11}=-1$. To the others we can apply Friedrichs' lemma (Lemma 17.1.5), for fixed x_1 at first, to conclude that the L^2 norm over Q_T tends to 0 with ε. This is obvious for the other terms since $D^\alpha u\in L^2$ for $|\alpha|\leqq 1$. Hence the L^2 norm of Pv_ε over Q_T tends to 0 as $\varepsilon\to 0$, and since $v_\varepsilon\to u$ in L^2 norm it follows from (24.1.4)$'$ that $u=0$ in Q_T. The proof is complete.

From Lemma 24.1.3 we can deduce a local uniqueness theorem which combined with Theorem 23.2.7 implies the uniqueness in Theorem 24.1.1.

Theorem 24.1.4. *Let the hypotheses* (i), (iii) *of Theorem* 24.1.1 *be fulfilled in a neighborhood of* $x_0\in\partial X$. *Then there is a fundamental system of neighborhoods* V *of* x_0 *in* X *such that* $u\in\bar{H}^{\mathrm{loc}}_{(1)}(V\cap X^\circ)$, $u=0$ *in* $V\cap X^\circ$ *when* $\phi<\phi(x_0)$, $Pu=0$ *in* $V\cap X^\circ$, *and* $u=0$ *in* $V\cap\partial X$ *implies* $u=0$ *in* V.

Proof. We can choose local coordinates at x_0 such that $x_0=0$, $x_1=0$ on ∂X, and $\phi(x)=\phi(0)+x_n-|x'|^2$. Let V_ε be defined as in the proof of Theorem 23.2.7. If the coefficients of P are extended smoothly to a full neighborhood of 0 it is clear that P is strictly hyperbolic there. With $\chi\in C_0^\infty(\mathbb{R}^n)$, $0\leqq\chi\leqq 1$ and $\chi=1$ in a neighborhood of 0 it is also obvious that

$$\chi(x/\gamma)\,P(x,D)+(1-\chi(x/\gamma))\,P(0,D)$$

satisfies the hypotheses of Lemma 24.1.3 if γ is chosen small enough. When ε is so small that $\chi(x/\gamma)=1$ in V_ε it follows from Lemma 24.1.3 that the uniqueness statement in the theorem holds.

Our next goal is to derive an existence theorem from (24.1.4). Thus we return to the hypotheses on P made in the proof of (24.1.4). If P^* is the formal adjoint of P, then $Pu=f$ in Q_T and $u=u_0$ when $x_1=0$ implies

$$\int (P^* v)\bar{u}\,dx = \int_{Q_T} v\bar{f}\,dx - \int_{x_1=0,x_n<T} \partial_1 v\bar{u}_0\,dS \tag{24.1.5}$$

provided that $v\in C_0^\infty(\mathbb{R}^n)$, $x_n<T$ in supp v, and $v=0$ when $x_1=0$. (Recall that $g^{11}=-1$.) Conversely, if (24.1.5) is valid with $u,f\in L^2$ for all such v, we first obtain $Pu=f$ by taking $v\in C_0^\infty(Q_T)$. Thus $u\in H_{(2,-2)}$ locally which makes the boundary values of u when $x_1=0$ well defined. From (24.1.5) with general v vanishing for $x_1=0$ it follows that $u=u_0$ when $x_1=0$.

Assuming for example that $x_n\geqq 0$ in supp f and in supp u_0 we want to find u so that $x_n\geqq 0$ in supp u and (24.1.5) is valid. In view of the duality in Theorem B.2.1 this can be done by means of the Hahn-Banach theorem if we can estimate the right-hand side of (24.1.5) by means of the restriction of P^*v to the half space $\mathbb{R}^n_+$.

The operator P^* satisfies the same hypotheses as P. Replacing P by P^* and changing the sign of x_n, we obtain from (24.1.4) with $T=0$

$$\lambda\int_Q (|v'|^2+|v|^2)e^{\lambda x_n}\,dx + \int_{x_1=0<x_n} |\partial_1 v|^2 e^{\lambda x_n}\,dS \leqq C\lambda^{-1}\int_Q |P^*v|^2 e^{\lambda x_n}\,dx, \tag{24.1.4}''$$

if $v\in C_0^\infty(\mathbb{R}^n)$ and $v=0$ when $x_1=0$. Here $Q=\{x;\ x_1>0, x_n>0\}$ and λ is large. If $x_n<T$ in supp v and T is a small positive number we can take $\lambda=1/T$ and obtain

$$T^{-1}\int_Q (|v'|^2+|v|^2)\,dx + \int_{x_1=0<x_n} |\partial_1 v|^2\,dS \leqq CeT\int_Q |P^*v|^2\,dx. \tag{24.1.6}$$

This is already an estimate of the desired form but to obtain the right smoothness properties of the solution u as stated in Theorem 24.1.1 we must first pass to suitable $H_{(s)}$ norms in (24.1.6). Set

$$E_s(D') = ((1+D_2^2+\ldots+D_{n-1}^2)^{\frac12} - iD_n)^s.$$

If $w(x_2,\ldots,x_n)\in C_0^\infty(\mathbb{R}^{n-1})$ then the L^2 norm $\|\ \|_+$ of $E_s(D')w$ in the half space $\mathbb{R}^{n-1}_+ = \{(x_2,\ldots,x_n),\ x_n>0\}$ is by Theorem B.2.4 equal to the norm in $\bar{H}_{(s)}(\mathbb{R}^{n-1}_+)$ of the restriction of w to that half space.

Lemma 24.1.5. *For every real s we have, if $v\in C_0^\infty(\mathbb{R}^n)$ vanishes when $x_1=0$ and for $x_n>T$ where $0<T\leqq T_s$*

$$T^{-1}\int_0^\infty \|E_{s+1}(D')v(x_1,.)\|_+^2\,dx_1 + \|E_s(D')\partial_1 v(0,.)\|_+^2 \leqq CT\int_0^\infty \|E_s P^* v(x_1,.)\|_+^2\,dx_1. \tag{24.1.7}$$

Proof. $w=E_s v$ belongs to $\bar{\mathscr{S}}(\mathbb{R}^n_+)$ and $w=0$ when $x_n>T$. It is therefore clear that (24.1.6) can be applied to w. We have

$$\|E_1 w(x_1,.)\|_+^2 \leqq 2\|(1+D_2^2+\ldots+D_{n-1}^2)^{\frac{1}{2}} w(x_1,.)\|_+^2+2\|D_n w(x_1,.)\|_+^2$$
$$=2\left(\sum_2^n \|D_j w(x_1,.)\|_+^2+\|w(x_1,.)\|_+^2\right),$$

so the left-hand side of (24.1.7) is at most equal to twice the left-hand side of (24.1.6) with v replaced by w. Now

$$P^* w=E_s P^* v-E_s[P^*, E_{-s}]w.$$

Recall that in the principal part $\sum g^{jk} D_j D_k$ of P^* the coefficient g^{11} is a constant. This term drops out in the commutator and in the others we have at most one factor D_1. It follows that

$$E_s[P^*, E_{-s}]=\sum R_k(x, D') D_k+R_0(x, D')$$

where $R_0, \ldots, R_n$ are in $S^0(\mathbb{R}^n \times \mathbb{R}^{n-1})$. The operators $R_k(x, D')$ are sums of products of the operators E_s, E_{-s} and differential operators, so $R_k(x, D') D_k w = R_k(x, D') w_k$ when $x_n>0$ if $w_k=D_k w$ when $x_n>0$ and $w_k=0$ elsewhere. Hence it follows from (24.1.6) with a constant C_s that

$$(T^{-1}-C_s T)\int_0^\infty\left(\sum_1^n \|D_j w(x_1,.)\|_+^2+\|w(x_1,.)\|_+^2\right) dx_1+\|\partial_1 w(0,.)\|_+^2$$
$$\leqq 2 C e T \int_0^\infty \|E_s P^* v(x_1,.)\|_+^2\, dx_1.$$

When $T^2 C_s<\frac{1}{2}$ we obtain the estimate (24.1.7) by the remarks at the beginning of the proof.

By duality we can now prove a local existence theorem:

Lemma 24.1.6. *Let $s \geqq 0$ and $f \in H_{(s)}(\mathbb{R}^n)$, $u_0 \in H_{(s+1)}(\mathbb{R}^{n-1})$, both with support where $x_n \geqq 0$. Then one can find $u \in \bar{H}_{(s+1)}^{\text{loc}}(Q_{T_s})$ (with respect to the half space $H'=\{x; x_1>0\}$) such that $Pu=f$ in Q_{T_s}, $x_n \geqq 0$ in supp u, and $u=u_0$ on the part of the boundary where $x_1=0$.*

Proof. By Theorem B.2.4 we have

$$\|f\|_{(s)}^2 \geqq \int_0^\infty \|E_s^* f(x_1,.)\|^2\, dx_1, \qquad \|u_0\|_{(s+1)}=\|E_{s+1}^* u_0\|$$

where $\| \ \|$ denotes the L^2 norm. If we apply (24.1.7) with s replaced by $-s-1$ and $T=T_s$, it follows that for some C

$$\left|\int_{Q_T} v \bar{f}\, dx-\int_{x_1=0, x_n<T} \partial_1 v\, \bar{u}_0\, dS\right| \leqq C\left(\int_0^\infty \|E_{-s-1} P^* v(x_1,.)\|_+^2\, dx_1\right)^{\frac{1}{2}}$$

if $v\in C_0^\infty$ and $v=0$ when $x_1=0$ and when $x_n>T$. By the Hahn-Banach theorem we can therefore find a linear form L on C_0^∞ such that

$$|L(w)|\leq C\left(\int_0^\infty \|E_{-s-1}w(x_1,.)\|_+^2\,dx_1\right)^{\frac{1}{2}}, \quad w\in C_0^\infty,$$

$$L(P^*v)=\int_{Q_T} v\bar{f}\,dx-\int_{x_1=0,x_n<T}\partial_1 v\bar{u}_0\,dS,$$

if $v\in C_0^\infty$ and $v=0$ when $x_1=0$ and when $x_n>T$. The continuity means that

$$L(w)=\int_0^\infty (w(x_1,.),u(x_1,.))\,dx_1$$

where $u\in L^2((0,\infty);\dot{H}_{(s+1)}(\mathbb{R}^{n-1}))$. Thus u defines an element in $\bar{H}_{(0,s+1)}(H')$ such that $x_n\geq 0$ in $\operatorname{supp} u$ and $Pu=f$ in Q_T, $u=u_0$ when $x_1=0$. Since $f\in\bar{H}_{(s)}(H')\subset\bar{H}_{(s-1)}(H')$ it follows from Theorem B.2.9 that $u\in\bar{H}^{\mathrm{loc}}_{(s+1)}(Q_T)$ (with respect to the half space H') which completes the proof of the lemma.

Proof of Theorem 24.1.1. The proof of Theorem 23.2.4 can be repeated with the reference to Theorem 23.2.7 (Proposition 23.2.6) replaced by a reference to Theorem 24.1.4 (Lemma 24.1.5). The details are left for the reader to repeat.

The proof of the crucial estimate (24.1.4) can also be adapted to the boundary conditions in Example 12.9.14 b) when all inequalities required for E and F there are fulfilled in a strict sense. The main difference is that another partial integration is required in the boundary plane $x_1=0$ and that the choice of the vector field f requires more attention. We refer to Gårding [6] for details and shall discuss the higher order case in the notes only. The arguments used to exploit the energy estimates here are clearly applicable quite generally.

24.2. Singularities in the Elliptic and Hyperbolic Regions

From now on we shall consider an arbitrary second order differential operator P with real principal symbol p and C^∞ coefficients, defined in a C^∞ manifold X of dimension n, with boundary ∂X. We assume that ∂X is non-characteristic with respect to P and write $X^\circ=X\setminus\partial X$. Our purpose is to study the singularities of a solution $u\in\mathcal{N}(X)$ of the mixed problem

$$Pu=f \quad\text{in } X^\circ, \qquad u=u_0 \quad\text{in } \partial X, \tag{24.2.1}$$

where $f\in\mathcal{N}(X)$. In X° we know that $WF(u)\setminus WF(f)$ is contained in the characteristic set $p^{-1}(0)$. By Theorem 23.2.9 this set is invariant under the flow defined by the Hamilton field H_p of the principal symbol p, if p is strictly hyperbolic. Since only microlocal properties of p are important, it is easy to see that hyperbolicity may be replaced by $p'_\xi \neq 0$. (Compare with the proof of Theorem 20.1.14.) We leave the proof for the reader for even this condition will be eliminated in Section 26.1.

The purpose of this section and the following ones is to prove analogous results at the boundary. By Corollary C.5.3 we can introduce local coordinates $(x_1, \dots, x_n)$ such that X is defined by $x_1 \geqq 0$ and, perhaps after a change of sign,

$$p(x,\xi) = \xi_1^2 - r(x,\xi'), \qquad \xi' = (\xi_2, \dots, \xi_n). \tag{24.2.2}$$

Recall that $T^*(X)|_{\partial X}$ is projected on $T^*(\partial X)$ with the conormal bundle as kernel if we restrict cotangent vectors of X to cotangent vectors of ∂X. The set $E \subset T^*(\partial X)$ of points such that $p \neq 0$ in the inverse image is called the *elliptic set* of the mixed problem; in local coordinates

$$E = \{(x',\xi');\ r(0,x',\xi') < 0\}. \tag{24.2.3}$$

By Theorem 20.1.14 it follows from (24.2.1) that

$$E \cap WF_b(u) = E \cap (WF_b(f) \cup WF(u_0)), \tag{24.2.4}$$

so we have complete control of the singularities in the elliptic set. The *hyperbolic set* H is the set of points in $T^*(\partial X)$ such that p has two different zeros in the inverse image, that is, in local coordinates,

$$H = \{(x',\xi');\ r(0,x',\xi') > 0\}.$$

The corresponding zeros of the characteristic equation are defined by $x_1 = 0$ and $\xi_1 = \pm r(x,\xi')^{\frac{1}{2}}$. The Hamilton field

$$H_p = 2\xi_1\, \partial/\partial x_1 - \partial r/\partial \xi'\, \partial/\partial x' + \partial r/\partial x\, \partial/\partial \xi$$

points into X (out of X) for the positive (negative) root. Thus the positive root defines the initial point of a bicharacteristic which goes transversally from ∂X into X°, while the negative root is the end point of a transversally incoming bicharacteristic.

Theorem 24.2.1. *If* (24.2.1) *holds and*

$$(x',\xi') \in (H \cap WF_b(u)) \setminus (WF_b(f) \cup WF(u_0))$$

then the corresponding outgoing and incoming bicharacteristics are both in $WF(u)$ *in a neighborhood of* $(0,x')$.

On the other hand, if $(x',\xi') \notin WF_b(u)$ then these bicharacteristics do not meet $WF(u)$ over a neighborhood of $(0,x')$, since $WF_b(u)$ is closed. Hence they cannot meet $WF(u)$ until they encounter $WF(f)$ (or return to the boundary). The theorem is of course a microlocal form of the reflection law of optics.

Proof of Theorem 24.2.1. We must show that if $(0,\xi'_0) \in H \setminus (WF_b(f) \cup WF(u_0))$ and, say, the corresponding outgoing bicharacteristic is not in $WF(u)$ near 0 then $(0,\xi_0')\notin WF_b(u)$. After multiplying u by a suitable cutoff function we may assume that u has compact support and that $(x,\xi)\in WF(u)$, $x_1>0$, implies $\xi'\neq 0$, which makes Theorem 18.1.36 applicable. We shall use the factorization in Lemma 23.2.8,

$$P(x,D)=(D_1-\Lambda_+(x,D'))(D_1-\Lambda_-(x,D'))+\omega(x,D') \tag{24.2.5}$$

where $\omega\in S^{-\infty}$ and the principal symbol of $\Lambda_\pm$ is $\pm r^{\frac{1}{2}}$ in a conic neighborhood of $(0,\xi_0')$. There is another such factorization

$$P(x,D)=(D_1-\tilde\Lambda_-(x,D'))(D_1-\tilde\Lambda_+(x,D'))+\tilde\omega(x,D') \tag{24.2.6}$$

where the principal symbol of $\tilde\Lambda_\pm$ is $\pm r^{\frac{1}{2}}$ in a conic neighborhood of $(0,\xi_0')$. Assume now that an open interval γ_+ on the corresponding outgoing bicharacteristic, with one end point at $(0,0,r(0,\xi_0')^{\frac{1}{2}},\xi_0')$, is in $T^*(X)\setminus WF(u)$. From the proof of Theorem 23.1.4 we know that $Q(x,D')$ can be chosen of order $-\infty$ outside any desired neighborhood of $\{(x,\xi');\ (x,r(x,\xi')^{\frac{1}{2}},\xi')\in\bar\gamma_+\}$ and non-characteristic at $(0,\xi_0')$ so that $[Q(x,D'),\ D_1-\Lambda_+(x,D')]$ is of order $-\infty$. Since

$$\begin{aligned}(D_1-\Lambda_+(x,D'))\,Q(x,D')(D_1-\Lambda_-(x,D'))u&=Q(x,D')f-Q(x,D')\,\omega(x,D')u\\&\quad-[Q(x,D'),D_1-\Lambda_+(x,D')](D_1-\Lambda_-(x,D'))u\end{aligned}$$

we have $(D_1-\Lambda_+(x,D'))v\in C^\infty$ by Theorem 18.3.32 for small $x_1\ge 0$ if $v=Q(x,D')(D_1-\Lambda_-(x,D'))u$ and Q has sufficiently small support. Hence $\xi_1=r(x,\xi')^{\frac{1}{2}}$ in $WF(v)$ if $x_1>0$ is small enough. This proves that $WF(v)$ is as close to γ_+ as we please if Q is suitably chosen. Since $WF(v)\subset WF(u)$ we obtain $v\in C^\infty$ for small positive x_1 then. Hence it follows from Corollary 23.1.3 (see also the discussion after Theorem 23.1.4) that $v\in C^\infty$ for small $x_1\ge 0$.

Choose $\tilde Q(x,D')$ commuting in the same way with $D_1-\tilde\Lambda_-(x,D')$ and so that $\tilde Q(0,x',\xi')=Q(0,x',\xi')$, and set $\tilde v=\tilde Q(x,D')(D_1-\tilde\Lambda_+(x,D'))u$. For $x_1=0$ we have $\tilde v-v=Q(0,x',D')(\Lambda_-(0,x',D')-\tilde\Lambda_+(0,x',D'))u_0\in C^\infty$ if the support of Q is small enough. As in the discussion of v above we also have $(D_1-\tilde\Lambda_-(x,D'))\,\tilde v\in C^\infty$ for small $x_1\ge 0$. Hence Corollary 23.1.3 yields that $\tilde v\in C^\infty$ for small $x_1\ge 0$. Now the equations

$$Q(x,D')(D_1-\Lambda_-(x,D'))u=v,\qquad \tilde Q(x,D')(D_1-\tilde\Lambda_+(x,D'))u=\tilde v,$$

can be regarded as an elliptic system for D_1u and u. More precisely, if $WF(a(x,D'))$ is contained in a set where Q and $\tilde Q$ are non-characteristic, we obtain, again by Theorem 18.3.32, that $a(x,D')(D_1-\Lambda_-(x,D'))u$ and $a(x,D')(D_1-\tilde\Lambda_+(x,D'))u$ are in C^∞ for small $x_1\ge 0$. Hence $a(x,D')(\tilde\Lambda_+(x,D')-\Lambda_-(x,D'))\,u\in C^\infty$ for small $x_1\ge 0$. Here the principal symbol $2r(x,\xi')^{\frac{1}{2}}$ of $\tilde\Lambda_+-\Lambda_-$ does not vanish at $(0,\xi_0')$. Choosing $a(x,D')$ non-characteristic there we conclude from Theorem 18.3.32 again that $(0,\xi_0')\notin WF_b(u)$. The proof is complete.

Remark. It is not sufficient to prove that $(0,\xi_0')\notin WF(D_1u|_{\partial X})$ in order to make the desired conclusion. (See the remark at the end of Section 9.6.)

Theorem 24.2.1 suggests that we introduce the following terminology:

Definition 24.2.2. A broken bicharacteristic arc of P is a map

$$I\setminus B\ni t\mapsto \gamma(t)\in T^*(X^\circ)\setminus 0,$$

where I is an interval on $\mathbb{R}$ and B is a discrete subset, such that

(i) If J is an interval $\subset I\setminus B$ then $J\ni t\mapsto\gamma(t)$ is a bicharacteristic of P over X°.

(ii) If $t\in B$ then the limits $\gamma(t-0)$ and $\gamma(t+0)$ exist and belong to $T_x^*(X)\setminus 0$ for some $x\in\partial X$, and the projections in $T_x^*(\partial X)\setminus 0$ are the same hyperbolic point.

Thus we have a bicharacteristic arc arriving transversally to ∂X at $\gamma(t-0)$ and another going out transversally from the reflected point $\gamma(t+0)$ with the same projection in $T^*(\partial X)\setminus 0$. Thus γ has an image $\tilde\gamma$ in $\tilde T^*(X)\setminus 0$ which is well defined and continuous also at the points in B. We shall call $\tilde\gamma$ a *compressed broken bicharacteristic* when a distinction between γ and $\tilde\gamma$ is called for.

Theorem 24.2.1 shows that a compressed broken bicharacteristic arc which does not intersect $WF_b(f)\cup WF(u_0)$ is either contained in or disjoint with $WF_b(u)$. We shall devote the rest of this section to constructing examples where the first case occurs and there are no other singularities. The constructions are related to the proof of Theorem 8.3.8 but modifications are required by the presence of the boundary and the variable coefficients.

At first we shall only make a local construction at a boundary point. We choose coordinates there so that the principal symbol has the form (24.2.2). Denote by $(0,\xi_0')$ a hyperbolic point. We shall then construct u in a neighborhood of 0 in X, defined by $x_1\geqq 0$, so that $u=0$ on ∂X, $Pu\in C^\infty(X)$ and $WF_b(u)$ is the compressed broken bicharacteristic passing through $(0,\xi_0')$. Recall that it is defined by the outgoing bicharacteristic γ_+ starting at $(0,0,r(0,\xi_0')^{\frac12},\xi_0')$ and the bicharacteristic γ_- coming in at $(0,0,-r(0,\xi_0')^{\frac12},\xi_0')$. At first we look for oscillatory asymptotic solutions of $Pu=0$ given by formal series

$$u_\lambda(x)=e^{i\lambda\phi(x)}\sum_0^\infty a_j(x)\lambda^{-j}$$

where $\partial\phi(0,0)/\partial x'=\xi_0'$. The leading term in Pu_λ is $\lambda^2p(x,\phi'(x))u_\lambda$ so our first equation is

$$p(x,\phi'(x))=0. \tag{24.2.7}$$

By Theorem 6.4.5 we can solve (24.2.7) near 0 with prescribed initial values

$$\phi(0,x')=\phi_0(x');\quad \phi_0'(0)=\xi_0',\quad \phi_0(0)=0. \tag{24.2.8}$$

The equation $p(0,0,\xi_1,\xi_0')=0$ has two solutions $\xi_1=\pm r(0,0,\xi_0')^{\frac{1}{2}}$ and this gives two corresponding solutions $\phi_\pm(x)$. For these we have $\phi'_\pm(x)=\xi$ if $(x,\xi)\in\gamma_\pm$. Next we obtain the transport equation for the leading term $a_0^\pm$ of the amplitude corresponding to $\phi_\pm$

$$\sum p^{(k)}(x,\phi'_\pm(x))\,D_k a_0^\pm(x)+c(x)\,a_0^\pm(x)=0 \tag{24.2.9}$$

where c depends on the lower order terms of p also. The $j+1$st equation is a similar equation for $a_j^\pm$ with an inhomogeneous term depending on $a_0^\pm,\dots,a_{j-1}^\pm$ added. All these equations can be solved in a fixed neighborhood of 0 with $a_j^\pm$ prescribed when $x_1=0$. The difference between the series $u_\lambda^\pm$ obtained can then be used as in the proof of Theorem 8.3.8 if we take $a_j^\pm$ with support near 0 when $x_1=0$ and take a large number of terms.

However, we want to prove a global result. An obvious difficulty is that long before the bicharacteristics $\gamma_\pm$ leave the local coordinate patch or return to the boundary the functions $\phi_\pm$ may cease to exist. Still the geometrical solutions as Lagrangians given in Theorem 6.4.3 do exist; the problem is that the projection of a Lagrangian in X may not have bijective differential. We shall avoid this difficulty by using the fact that a strictly positive Lagrangian plane in the complexification of a symplectic vector space S always has bijective projection on the complexification of a Lagrangian subspace of S. (Cf. Proposition 21.5.7.)

First we recall the details of the solution of (24.2.7), (24.2.8) given in Section 6.4. Denote by $\chi_\pm(x_1,y',\eta')$ the solution of the Hamilton equations with Hamiltonian $\xi_1\mp r(x,\xi')^{\frac{1}{2}}$,

$$dx'/dx_1=\mp\partial r(x,\xi')^{\frac{1}{2}}/\partial\xi',\qquad d\xi'/dx_1=\pm\partial r(x,\xi')^{\frac{1}{2}}/\partial x',$$

and initial data $(x',\xi')=(y',\eta')$ when $x_1=0$. The maps $\chi_\pm(x_1,.)$ are well defined and symplectic from a neighborhood of $(0,\xi_0')$ to a neighborhood of the point $\chi_\pm(x_1,0,\xi_0')\in\gamma_\pm$ for fixed x_1 (and the variable ξ_1 dropped). The solution of (24.2.7), (24.2.8) is equal to 0 on $\gamma_\pm$ and $G_{x_1}=\{(x',\partial\phi(x_1,x')/\partial x')\}$ is the image of $G_0=\{(x',\partial\phi_0(x')/\partial x'\}$ under $\chi_\pm(x_1,.)$. The solution continues to exist along $\gamma_\pm$ as long as $G_{x_1}\ni(x',\xi')\mapsto x'$ has a bijective differential.

So far ϕ_0 has been real, but we shall now allow ϕ_0 to be complex valued with the Hessian $\operatorname{Im}\phi_0''(0)$ positive definite. Then it follows from Proposition 21.5.9 that the complexified tangent plane of G_0 is a strictly positive Lagrangian plane. Strictly speaking $\chi_\pm(x_1,y',\partial\phi_0(y')/\partial y')$ is not defined for $y'\neq 0$ but the Taylor expansion at $y'=0$ is meaningful and defines a formal Lagrangian at $\chi_\pm(x_1,0,\xi_0')$. The tangent plane at this point is the image of the complexified tangent plane of G_0 at $(0,\xi_0')$ under the complexification of a real linear symplectic map, hence strictly positive. Let $\gamma_\pm$ be defined for $0\leqq x_1\leqq c$. In view of Proposition 21.5.9 it follows that we can find $\phi_\pm\in C^\infty$ such that

$$\phi_\pm(x)=0,\quad \partial\phi_\pm(x)/\partial x'=\xi',$$

$\operatorname{Im} \partial^2 \phi_\pm/\partial x'^2$ is positive definite if $(x, \xi) \in \gamma_\pm$ and $0 \leqq x_1 \leqq c$; moreover, $\partial \phi_\pm/\partial x_1 \mp r(x, \partial \phi_\pm/\partial x')^{\frac{1}{2}}$ vanishes of infinite order on $\gamma_\pm$. Similarly we can solve the transport equation (24.2.9) and the subsequent ones so that $a_0^\pm \neq 0$ on $\gamma_\pm$ and

$$u_{\lambda,N}^\pm(x) = e^{i\lambda\phi_\pm(x)} \sum_0^N a_j^\pm(x) \lambda^{-j}$$

for every N has the property that

$$D^\alpha P u_{\lambda,N}^\pm(x) = O(\lambda^{|\alpha|-N}).$$

If we choose an asymptotic sum $a^\pm(x, \lambda) \in S^0(\mathbb{R}^n \times \mathbb{R})$, it follows that

$$P(e^{i\lambda\phi_\pm(x)} a^\pm(x, \lambda)) \in S^{-\infty}(\mathbb{R}^n \times \mathbb{R}).$$

We can choose the sums so that $a^+(x, \lambda) = a^-(x, \lambda)$ when $x_1 = 0$ and take $a^\pm$ with support so close to the projection $\pi\gamma_\pm$ of $\gamma_\pm$ that $\operatorname{Im} \phi_\pm(x)$ is strictly positive there except on $\pi\gamma_\pm$.

Now we introduce the absolutely convergent integral

$$U^\pm(x) = \int_0^\infty e^{i\lambda\phi_\pm(x)} a^\pm(x, \lambda)\, d\lambda/(\lambda^2+1). \tag{24.2.10}$$

By Theorem 8.1.9 we have $WF(U^\pm) \subset \mathbb{R}_+\gamma_\pm$ when $x_1 > 0$, and $PU^\pm \in C^\infty$ when $x_1 \geqq 0$ so $U^\pm \in \mathcal{N}$. The boundary values when $x_1 = 0$ are equal, so $U = U^+ - U^-$ vanishes when $x_1 = 0$. If $b(x, D')$ is a pseudo-differential operator with symbol of order $-\infty$ in a conic neighborhood of $(0, 0, \xi_0')$ then it follows from Theorem 8.1.9 and Theorem 18.1.36 that $b(x, D') U^\pm \in C^\infty$ for small $x_1 \geqq 0$. Thus $WF_b(U^\pm) \subset \mathbb{R}_+ \tilde{\gamma}_\pm$. We shall prove now that $(0, \xi_0')$ is in the wave front set of

$$D_1(U^+ - U^-)(0, .) = \int_0^\infty e^{i\lambda\phi_0(x')} b(x', \lambda)\, d\lambda/(\lambda^2+1);$$
$$b(x', \lambda) = 2r(0, x', \partial\phi_0/\partial x')^{\frac{1}{2}} \lambda a^\pm(0, x', \lambda) + D_1(a^+(0, x', \lambda) - a^-(0, x', \lambda)).$$

Here $b \in S^1$ and the leading term $2r(0, x', \partial\phi_0/\partial x')^{\frac{1}{2}} \lambda a_0(0, x')$ does not vanish at 0.

When $\operatorname{Im} z > 0$ we have

$$\int_0^\infty \lambda e^{i\lambda z}\, d\lambda/(\lambda^2+1) + \log|z| = O(1) \quad \text{as } z \to 0.$$

In fact, the integral is equal to

$$\int_0^\infty \lambda e^{i\lambda z'}\, d\lambda/(\lambda^2+|z|^2), \quad z' = z/|z|.$$

The difference between this integral and

$$\int_0^1 \lambda\, d\lambda/(\lambda^2+|z|^2) = \tfrac{1}{2}\log(1+|z|^2) - \log|z|$$

is bounded as $z \to 0$ since $|e^{i\lambda z'}-1| \leq |\lambda|$ and the integral from 1 to ∞ is seen to be bounded by a partial integration. Hence

$$D_1(U^+ - U^-)(0,x') + 2r(0, \partial\phi_0(0)/\partial x')^{\frac{1}{2}} a_0(0,0) \log|\phi_0(x')|$$

is bounded as $x' \to 0$ so $D_1(U^+ - U^-)(0,.)$ is not even a continuous function. In particular $(0,\xi_0')$ must be in the wave front set since it can only contain points along this ray by Theorem 8.1.9. This completes the proof that $WF_b(u)$ contains $(0,\xi_0')$, hence also contains the bicharacteristics which are incoming and outgoing there.

We are now ready to prove

Theorem 24.2.3. *Let $[a,b] \ni t \mapsto \tilde{\gamma}(t)$ be a compressed broken bicharacteristic arc which has an injective projection to the compressed cosphere bundle $\tilde{T}^*(X)/\mathbb{R}_+$ and end points $\tilde{\gamma}(a)$, $\tilde{\gamma}(b)$ over X°. Assume that the projection of $H_p(\tilde{\gamma}(t))$ in $T(X^\circ)$ is never 0 when $\tilde{\gamma}(t) \in T^*(X^\circ)$. Then one can find $u \in \mathcal{N}(X)$ so that $WF_b(u)$ is the cone generated by $\tilde{\gamma}([a,b])$, $WF_b(Pu)$ is generated by $\tilde{\gamma}(\{a,b\})$, and $u=0$ on ∂X.*

Proof. Assuming that $\tilde{\gamma}$ contains some point over ∂X we introduce local coordinates there and start the construction as above. (The argument just given could also be used to prove directly that $\gamma_\pm \subset WF(U^\pm)$ when $0 < x_1 < c$, so we could also start with constructing just U^+ or U^- near a point in X° where x_1 can be used as local coordinate on the bicharacteristic.) The construction can be continued from $x_1 = c$ with another set of local coordinates such that x_1 is a good parameter on the bicharacteristic strip. In fact, changing coordinates in the phases ϕ and the amplitudes a_j we get initial data to use for the new coordinate patch, and if the bicharacteristic strip arrives at the boundary we obtain initial data for the construction of a reflected wave. There can never be any interference between different pieces of the constructed solution since the wave front sets are different by the assumed injectivity of the projection to the compressed cosphere bundle.

The only remaining point is to show how the construction can be cut off at $x_1 = c$ if that corresponds to an end point of $\tilde{\gamma}$. To do so we choose $\kappa \in C_0^\infty(\mathbb{R})$ equal to 1 in $(-\infty,-2)$ and 0 on $(-1,\infty)$ and replace $a^+(x,\lambda)$ by $a^+(x,\lambda)\,\kappa((x_1-c)^3\lambda)$, which is in $S^0_{1,\frac{1}{3}}$ (cf. Example 18.1.2). This does not affect the singularities when $x_1 < c$ but makes $U^+ = 0$ when $x_1 > c$. Theorem 8.1.9 is applicable everywhere so $WF(U^+)$ is still generated by γ_+ for $x_1 > 0$. Now $WF(PU^+)$ can only contain the ray defined by the boundary point of γ_+ with $x_1 = c$, and this point cannot be missing for the interior propagation theorem would then imply that γ_+ does not meet $WF(U^+)$. The proof is complete.

Remark. Changing the order of $a^\pm(x,\lambda)$ by multiplication with a power of $\lambda+1$ we could construct u with any desired regularity, as we shall do in the analogous Theorem 26.1.5 below.

The functional analytic arguments in the proof of Theorem 8.3.8 can be used to show that also limits of the arcs $\tilde{\gamma}$ in Theorem 24.2.3 can carry singularities. This will be done in Theorem 24.5.3 below. However, first we shall discuss the properties of such limits in Section 24.3 and prove converse results analogous to Theorem 24.2.1 in Sections 24.4 and 24.5.

In Theorem 24.2.3 we have assumed that H_p is not a tangent to the fiber of $T^*(X)$, but this is not essential. To remove this hypothesis one can use the theory of Fourier integral operators as in the proof of Theorem 26.1.5. Alternatively one can interpret U as a Lagrangian distribution associated with a positive Lagrangian. In either approach the methods and notions of Chapter XXV are required.

24.3. The Generalized Bicharacteristic Flow

In Section 24.2 we have seen that singularities of solutions of the Dirichlet problem arriving at the boundary on a transversal bicharacteristic will leave again on the reflected bicharacteristic. To prepare for the study of tangentially incoming singularities we shall now examine the geometrical properties of broken bicharacteristics nearby. In doing so it is natural to start from a general symplectically invariant description of the situation and then introduce suitable local coordinates.

Thus we start from a symplectic manifold S of dimension $2n$ with boundary ∂S, and choose $\phi \in C^\infty(S)$ so that $\phi=0$ on ∂S, $\phi>0$ in $S \setminus \partial S$, and $d\phi \neq 0$ on ∂S. Let $p \in C^\infty(V)$ where V is an open neighborhood in S of a point $s_0 \in \partial S$, and assume that

(i) $$p=\{\phi,p\}=0, \quad \{\phi,\{\phi,p\}\}\neq 0 \quad \text{at } s_0.$$

Usually we shall also require that

(ii) $$dp|_{\partial S} \neq 0 \quad \text{at } s_0.$$

Condition (i) means that p restricted to the fiber of the natural fibration of ∂S (by the integral curves of H_ϕ) has a double zero at s_0. Another interpretation of the condition $\{\phi,p\}=0$ is that the bicharacteristic of p starting from s_0 is tangent to ∂S at s_0. We shall study the bicharacteristic foliation of $p^{-1}(0)$, broken by reflection in ∂S as in Definition 24.2.2.

Shrinking V if necessary we may assume that $V \subset T^*(\overline{\mathbb{R}}^n_+)$, $s_0=(0,0)$ and $\phi=x_1$. Then (i) means by the Malgrange preparation theorem that with $g\neq 0$ and $a=r=0$ at 0

$$p(x,\xi)=g(x,\xi)((\xi_1+a(x,\xi'))^2-r(x,\xi')), \qquad \xi'=(\xi_2,\dots,\xi_n).$$

We can take new symplectic coordinates with $y_1=x_1$, $\eta_1=\xi_1+a(x,\xi')$. Since $0=H_{x_1}r=H_{y_1}r$ it follows that with the new coordinates p has the same form but with $a=0$, as we assume from now on. The factor g does not affect the set $p^{-1}(0)$ or its foliation, just the parameter on the bicharacteristics, so we shall assume that $g=1$, thus

$$p(x,\xi)=\xi_1^2-r(x,\xi'),\quad \xi'=(\xi_2,\dots,\xi_n), \tag{24.3.1}$$

where $d_{x',\xi'}r(0,0)\neq 0$ if condition (ii) holds. This is the same form of p as the one which we obtained in Section 24.2 for the principal symbol of a second order differential operator in a manifold X with non-characteristic boundary ∂X.

Definition 24.2.2 has a symplectically invariant meaning in a neighborhood of s_0: $\gamma(t)$ shall be a bicharacteristic of p contained in $S\setminus\partial S$ when $t\notin B$, and when $t\in B$ the limits $\gamma(t\pm 0)$ shall exist and belong to the same leaf of the natural foliation of ∂S.

We shall write $\Sigma=p^{-1}(0)$ for the characteristic set and $\tilde\Sigma$ for the compressed characteristic set obtained by identifying points on the same leaf of the foliation of ∂S. Thus x, ξ are continuous functions on Σ and define the topology there while x, $x_1\xi_1$, ξ' are continuous on $\tilde\Sigma$ and define its topology. The subset of $\tilde\Sigma$ where $x_1=0$ is denoted by $\tilde\Sigma_0$; it is the image of all (x,ξ) with $x_1=0$ and $\xi_1^2=r(x,\xi')\geqq 0$. The set $G\subset\Sigma$ where $x_1=\xi_1=0$ is called the *glancing set*; it is the set of all points in $\phi^{-1}(0)\cap p^{-1}(0)$ where H_p is tangent to ∂S. It is mapped topologically into $\tilde\Sigma_0$ so it can be identified with its image there, and $H=\tilde\Sigma_0\setminus G$ is the hyperbolic set with the terminology of Section 24.2. We can also define invariantly the glancing set G^k of order at least k, $k=2,3,\dots$, by the equations

$$p=0 \quad\text{and}\quad H_p^j\phi=0 \quad \text{for } 0\leqq j<k. \tag{24.3.2}$$

Thus $G=G^2\supset G^3\supset\dots\supset G^\infty$ are closed sets. The definition depends only on $p^{-1}(0)$ and ∂S, for p and ϕ are determined by $p^{-1}(0)$ and ∂S up to smooth factors $\neq 0$ which only affect the parametrization on the bicharacteristics and not the order of the zero of ϕ. When $\phi(x)=x_1$ and p is given by (24.3.1) we set $r_j(x',\xi')=\partial^j r(0,x',\xi')/\partial x_1^j$, $j=0,1$, and shall prove:

Lemma 24.3.1. *Let k be an integer $\geqq 0$. Then $\gamma_0=(0,x',0,\xi')\in G^{k+2}$ if and only if*

$$r_0=0 \quad\textit{and}\quad H_{r_0}^j r_1=0,\quad 0\leqq j<k, \tag{24.3.3}$$

at γ_0, and then we have at γ_0

$$H_p^{k+2}\phi=2(-H_{r_0})^k r_1. \tag{24.3.4}$$

Proof. With H_r^0 denoting the Hamilton field $\partial r/\partial\xi'\,\partial/\partial x'-\partial r/\partial x'\,\partial/\partial\xi'$ of r for fixed x_1, we have

$$H_p=2\xi_1\partial/\partial x_1+r_{(1)}\partial/\partial\xi_1-H_r^0,\qquad r_{(1)}=\partial r/\partial x_1.$$

Hence $H_p x_1=2\xi_1$, $H_p^2 x_1=2r_{(1)}$ which proves the lemma when $k=0$. By an obvious inductive argument it remains only to show that (24.3.4) is valid if

$k>0$, $\gamma_0 \in G^{k+2}$ and (24.3.3) holds. By hypothesis $H_p^j\psi=0$ at γ_0 for $j<k+1$ if $\psi=0$ when $x_1=0$. Thus we have at γ_0

$$H_p^{k+2}\phi = H_p^k 2r_1 = 2(2\xi_1\partial/\partial x_1 + r_1\partial/\partial\xi_1 - H_{r_0})^k r_1,$$

for working from right to left we can successively put $x_1=0$ in each of the factors $r_{(1)}$ and H_p since at most k factors H_p act from the left. We can now drop $\partial/\partial x_1$ since nothing depends on x_1, and then $\partial/\partial\xi_1$ because ξ_1 no longer occurs. This completes the proof of (24.3.4) and of the lemma.

The bicharacteristics of p starting at a point γ_0 with $\phi=p=H_p\phi=0$ but $H_p^2\phi>0$ will lie in the interior defined by $\phi>0$ in a deleted neighborhood of the starting point γ_0. On the other hand, if p is extended to negative values of x_1 then the bicharacteristic will immediately leave the half space $\phi\geqq 0$ if $H_p^2\phi<0$. This convexity (concavity) of ∂X with respect to tangential bicharacteristics leads to drastic differences in the behavior of the broken bicharacteristic flow. We begin the discussion by summing up and extending our terminology:

Definition 24.3.2. The glancing set G^k of order at least $k\geqq 2$ is defined by (24.3.2); $G=G^2$ is just called the glancing set. The glancing set $G^2\setminus G^3$ of order precisely 2 is the union $G_d\cup G_g$ of the diffractive part G_d where $H_p^2\phi>0$ and the gliding part G_g where $H_p^2\phi<0$.

With the notation in Lemma 24.3.1 the diffractive (resp. gliding) sets are defined by $x_1=\xi_1=r_0=0$ and $r_1>0$ (resp. $r_1<0$). To investigate the broken bicharacteristics near G_d and G_g it is by Theorem 21.4.8 sufficient to study the following example provided that (i) and (ii) hold.

Example 24.3.3. If $p(x,\xi)=\xi_1^2\pm x_1+\xi_n$, $\phi(x)=x_1$, then the glancing set is defined by $x_1=\xi_1=\xi_n=0$, and $H_p^2\phi=\mp 2$ so it is gliding for the upper sign and diffractive for the lower sign. The invariants of the Hamilton flow are

$$x_2,\ldots,x_{n-1},\xi_2,\ldots,\xi_n,\xi_1\pm x_n \quad\text{and}\quad \xi_1^2\pm x_1.$$

The broken bicharacteristics are displayed in the x_1x_n plane for the diffractive and gliding cases in Fig. 3 and Fig. 4. We have $\xi_1=0$ when $dx_1/dt=0$. In the *diffractive* case the only limits of broken bicharacteristics are the tangent parabolas defined by $\xi_1^2=x_1$, $\xi_n=0$, ξ_1-x_n, $x_2,\ldots,x_{n-1}$, $\xi_2,\ldots,\xi_{n-1}$ all constant. In the *gliding* case $\xi_1^2+x_1$ is equal to the maximum value c^2 of x_1. Thus $|\xi_1|\leqq c$, $0\leqq x_1\leqq c^2$, $\xi_n=-c^2$ and the different parabolas in Fig. 4 differ only by a translation.

When $c\to 0$ the broken bicharacteristic degenerates to a line $dx/dt=(0,\ldots,0,1)$, $d\xi/dt=0$ in the gliding set, defined by the Hamilton field of ξ_n. This is called a *gliding ray*.

Instead of using the fairly deep result in Theorem 21.4.8 and in order to avoid using condition (ii) we shall now show directly that the conclusions in

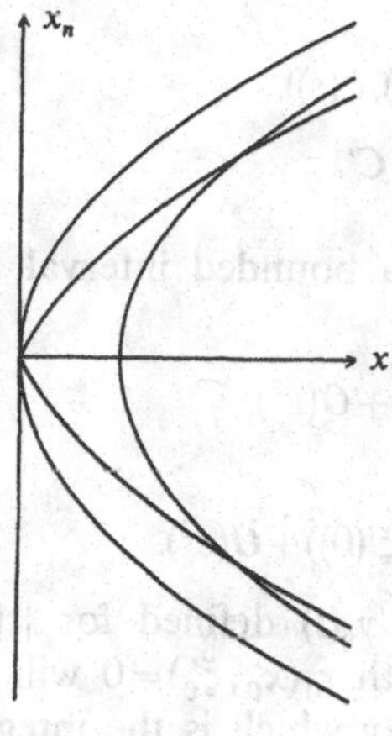

Fig. 3

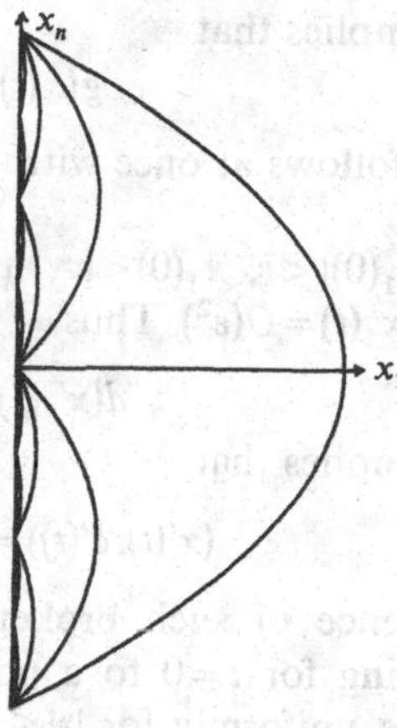

Fig. 4

the example are valid quite generally. As usual V is an open set in the half space defined by $x_1 \geqq 0$ such that $\overline{V}$ is compact, and p is defined in a neighborhood of $\overline{V}$ in this half space.

Lemma 24.3.4. *Assume that $\partial r/\partial x_1 \geqq c > 0$ in $\overline{V}$ and that $t \mapsto \gamma(t)$, $0<|t|<T$, is a broken bicharacteristic reflected when $t=0$. Then $x_1 \geqq ct^2$ so γ cannot return to ∂S when maximally extended.*

Proof. Since $d\xi_1/dt = \partial r/\partial x_1 \geqq c$ and $\xi_1(\pm 0) \gtrless 0$, it follows that $\xi_1(t) > ct$ if $t>0$, $\xi_1(t) < ct$ if $t<0$. Now $dx_1/dt = 2\xi_1$ and $x_1(0)=0$ so it follows that $x_1(t) \geqq ct^2$.

From the lemma it follows that a limit of broken bicharacteristics is either a broken bicharacteristic or else a bicharacteristic starting from a point in the glancing (thus diffractive) set.

Lemma 24.3.5. *Assume that $\partial r/\partial x_1 \leqq -c < 0$ in $\overline{V}$ and that γ is a broken bicharacteristic. Then there are constants C_0 and C such that*

$$\xi_1(t)^2 + x_1(t) \leqq C_0 e^{C|s-t|}(\xi_1(s)^2 + x_1(s)) \tag{24.3.5}$$

for any two points on γ.

Proof. For $g(x,\xi) = \xi_1^2 - x_1\, \partial r/\partial x_1$ we have

$$\xi_1^2 + c x_1 \leqq g(x,\xi) \leqq \xi_1^2 + C' x_1$$

if $\partial r/\partial x_1 \geqq -C'$ also. At the points of reflection $g(x,\xi)$ is continuous, and elsewhere we have

$$dg(x(t),\xi(t))/dt = 2\xi_1\, \partial r/\partial x_1 - 2\xi_1\, \partial r/\partial x_1 - x_1\, d(\partial r/\partial x_1)/dt.$$

Hence

$$|dg(x(t),\xi(t))/dt| \leqq C g(x(t),\xi(t)),$$

which implies that

$$g(x(t),\xi(t)) \leqq e^{C|s-t|} g(x(s),\xi(s)).$$

(24.3.5) follows at once with $C_0 = C'/c$ if $c < 1 < C'$.

If $|\xi_1(0)| < \varepsilon$, $x_1(0) < \varepsilon^2$, it follows for t in a bounded interval that $\xi_1(t) = O(\varepsilon)$, $x_1(t) = O(\varepsilon^2)$. Thus

$$d(x',\xi')/dt = -H_r^0 = -H_{r_0} + O(\varepsilon^2),$$

which implies that

$$(x'(t),\xi'(t)) = \exp(-tH_{r_0})(x'(0),\xi'(0)) + O(\varepsilon^2).$$

A sequence of such broken bicharacteristics $\gamma_\nu(t)$ defined for $|t| < T$ and converging for $t=0$ to a point $(0,x_0',0,\xi_0')$ with $r_0(x_0',\xi_0')=0$ will therefore converge uniformly for $|t| \leqq T$ to the *gliding ray* which is the integral curve of $-H_{r_0}$. (These give a foliation of the glancing set defined by $x_1 = \xi_1 = r_0 = 0$ where $H_{r_0} \neq 0$. When $H_{r_0}=0$ the gliding ray does not move.) By the differential equation $(x_\nu'(t),\xi_\nu'(t))$ is in fact C^1 convergent.

In the glancing set we have

$$H_p = r_1\partial/\partial\xi_1 - H_{r_0}, \qquad H_\phi = -\partial/\partial\xi_1.$$

Thus $-H_{r_0} = H_p + \lambda H_\phi$ with λ chosen so that this vector field is tangential to the glancing set $x_1 = \xi_1 = r_0 = 0$. Invariantly this set is defined by $p = \phi = \{p,\phi\} = 0$, so λ must be chosen so that $H_p^2\phi + \lambda H_\phi H_p\phi = 0$, that is, $\lambda = H_p^2\phi/H_\phi^2 p$.

Definition 24.3.6. The vector field

$$H_p^G = H_p + (H_p^2\phi/H_\phi^2 p)H_\phi$$

tangent to G is called the gliding vector field.

An alternative interpretation is that H_p^G is the Hamilton field of p restricted to the symplectic space where $\phi = \{p,\phi\} = 0$. In fact, with our local coordinates this is defined by $x_1 = \xi_1 = 0$, and p restricts to $-r_0$. Note also that $H_p^G = H_p$ in G^3 so the vector field which is H_p^G in $G \setminus G_d$ and H_p on G_d is continuous. H_p^G is independent of the choice of ϕ, as the notation suggests, and if p is replaced by gp, $g \neq 0$, then H_p^G like H_p is just multiplied by g. This gives a consistent change of parametrization in the following definition suggested by Lemmas 24.3.4 and 24.3.5.

Definition 24.3.7. A generalized bicharacteristic arc of p is a map

$$I \setminus B \ni t \mapsto \gamma(t) \in (S \setminus \partial S) \cup G$$

where I is an interval on $\mathbb{R}$ and B a subset of I, such that $p \circ \gamma = 0$ and

(i) $\gamma(t)$ is differentiable and $\gamma'(t) = H_p(\gamma(t))$ if $\gamma(t) \in S \setminus \partial S$ or $\gamma(t) \in G_d$.

(ii) $\gamma(t)$ is differentiable and $\gamma'(t) = H_p^G(\gamma(t))$ if $\gamma(t) \in G \setminus G_d$.

(iii) Every $t \in B$ is isolated, $\gamma(s) \in S \setminus \partial S$ if $s \neq t$ and $|s-t|$ is small enough, the limits $\gamma(t \pm 0)$ exist and are different points in the same (hyperbolic) fiber of ∂S.

The continuous curve $\tilde{\gamma}$ obtained by mapping γ into $\tilde{\Sigma}$ is called a *compressed generalized bicharacteristic.*

The definition has been stated in an invariant form, but condition (iii) assumes tacitly that p can be given the form (24.3.1). From now on we assume that we have such coordinates and set $\gamma(t) = (x(t), \xi(t))$. Then $x'(t)$, $\xi'(t)$ are continuously differentiable in I,

$$dx'/dt = -\partial r/\partial \xi', \quad d\xi'/dt = \partial r/\partial x'.$$

x_1 is continuous and left and right differentiable with derivative $2\xi_1(t \mp 0)$ also if $t \in B$. Even then the right and left derivatives

$$\lim_{\varepsilon \to +0} (\xi_1(t \pm \varepsilon) - \xi_1(t \pm 0))/(\pm \varepsilon)$$

exist and are equal to $\partial r/\partial x_1$ except in G_g where the derivative is 0. In particular we obtain a uniform Lipschitz condition for $x(t)$, $x_1(t)\xi_1(t)$ and $\xi'(t)$ when $\gamma(t)$ remains in a fixed compact set. Since $\xi_1^2 = r(x, \xi')$ we also have a uniform Lipschitz condition for ξ_1^2 and a Hölder condition of order $\frac{1}{2}$ for $|\xi_1|$.

If $\gamma(0) \in (S \setminus \partial S) \cup G_d$ then $\gamma(t)$ remains in this set for small t. In fact, if $\gamma(0) \in G_d$ then there can be no reflection point $\gamma(t)$ for small t, by Lemma 24.3.4, and $\gamma(t) \notin G \setminus G_d$ since this is a closed set. Hence it follows from condition (i) that $t \mapsto \gamma(t)$ is for small t just the orbit of H_p through $\gamma(0)$.

If $\gamma(0) \in G_g$ then $\gamma(t)$ is for small t the gliding ray through $\gamma(0)$. In fact, there is a neighborhood of $\gamma(0)$ where (24.3.5) is valid. If $x_1(t)$ and $\xi_1(t)$ do not vanish identically for small t it follows that $\gamma(t)$ is a broken bicharacteristic and we get a contradiction when $t \to 0$. Thus $\gamma(t) \in G_g$ and the assertion follows from condition (ii) in Definition 24.3.7.

We shall now study the generalized bicharacteristic when $\gamma(0)$ is close to G^3. Let $(y'(t), \eta'(t))$ be the solution of

$$d(y', \eta')/dt = -H_{r_0}(y', \eta'); \quad y'(0) = x'(0), \quad \eta'(0) = \xi'(0);$$

that is, the orbit of the gliding vector field with the same initial data. Subtracting from the Hamilton equations for x', ξ' we obtain if $\gamma(t)$ is contained in a fixed compact set for $0 \leq t \leq T$

$$|d(x' - y', \xi' - \eta')/dt| \leq C_0(|x' - y'| + |\xi' - \eta'|) + C_1 x_1.$$

With $f(t) = |x'(t) - y'(t)| + |\xi'(t) - \eta'(t)|$ it follows that

$$df(t)/dt \leq C_0 f(t) + C_1 x_1(t), \quad f(0) = 0.$$

Set $e(t)=r_1(y'(t),\eta'(t))=r_{(1)}(0,y'(t),\eta'(t))$ where $r_{(1)}=\partial r/\partial x_1$. Then

$$|r_{(1)}(x(t),\xi'(t))-e(t)|\leqq C_2 f(t)+C_3 x_1(t),$$

which means that for both left and right derivatives

$$d|\xi_1(t)|/dt\leqq|e(t)|+C_2 f(t)+C_3 x_1(t).$$

Furthermore we have $dx_1(t)/dt\leqq 2|\xi_1(t)|$.

Consider now a solution X,Ξ,F of the differential equations

$$dX/dt=2\Xi,\quad d\Xi/dt=|e(t)|+C_2F+C_3X,\quad dF/dt=C_0F+C_1X, \tag{24.3.6}$$

with $X(0)>x_1(0)$, $\Xi(0)>|\xi_1(0)|$, $F(0)>0$. We claim that $x_1<X$, $|\xi_1|<\Xi$ and $f<F$ in $[0,T]$. In fact, if s is the largest number $\leqq T$ such that this is true in $[0,s)$ then $X-x_1$, $\Xi-|\xi_1|$, $F-f$ are increasing in $[0,s)$, hence positive at s, which proves that $s=T$ and that the inequality is valid in $[0,T]$. Hence we can also estimate x_1, $|\xi_1|$, f by the solution of the equations (24.3.6) with $X(0)=x_1(0)$, $\Xi(0)=|\xi_1(0)|$, $F(0)=0$.

Let F, X, Ξ be the solution of

$$dX/dt=2\Xi,\quad d\Xi/dt=C_2F+C_3X,\quad dF/dt=C_0F+C_1X$$

with $X(0)=F(0)=0$ and $\Xi(0)=1$. Then $X(t)=2t+O(t^2)$, $F(t)=C_1t^2+O(t^3)$ as $t\to 0$, and $(X(t),\ \Xi(t),\ F(t))=O(e^{\mu t})$ as $t\to\infty$. Differentiation gives a solution of the same system with initial data $(2,0,0)$, and the two solutions allow us to solve (24.3.6) with initial data $x_1(0)$, $|\xi_1(0)|$, 0. This proves that for $0\leqq t\leqq T$

$$\begin{aligned} x_1(t)&\leqq Ce^{\mu t}\left(\int_0^t(t-s)|e(s)|\,ds+x_1(0)+t|\xi_1(0)|\right),\\ |\xi_1(t)|&\leqq Ce^{\mu t}\left(\int_0^t|e(s)|\,ds+tx_1(0)+|\xi_1(0)|\right),\\ f(t)&\leqq Ce^{\mu t}\left(\int_0^t(t-s)^2|e(s)|\,ds+tx_1(0)+t^2|\xi_1(0)|\right).\end{aligned}\tag{24.3.7}$$

Let us now assume that $\gamma(0)\in G^k$, $k>2$. By Lemma 24.3.1 this means that $e(t)=O(t^{k-2})$ as $t\to 0$. From (24.3.7) we then obtain that $x_1(t)=O(t^k)$ and that $\xi_1(t)=O(t^{k-1})$, just as we would have for the orbit of H_p starting at $\gamma(0)$ if no boundary interfered. This is far more than differentiability of $x_1(t)$ and $\xi_1(t)$ at 0.

Assume for a moment that $\gamma(0)\in G^k\smallsetminus G^{k+1}$ for some $k>2$. Then it follows from Lemma 24.3.1 that

$$e(t)=at^{k-2}+O(t^{k-1}),\qquad e'(t)=a(k-2)t^{k-3}(1+O(t))$$

where $2a=H_p^k\phi(\gamma(0))/(k-2)!\neq 0$. Hence e is monotonic for $t\neq 0$ small enough. This monotonicity alone suffices to determine $\gamma(t)$:

Proposition 24.3.8. *If $\gamma(0)\in G^3$ and $e(t)$ is increasing for small positive t then $\gamma(t)$ is for such t the orbit of H_p starting at $\gamma(0)$. If $\gamma(0)\in G^3$ and $e(t)$ is decreasing for $0\leqq t\leqq T$ then $\gamma(t)$ is a gliding ray for $0\leqq t\leqq T$.*

Proof. Assume that $e(t)$ is increasing for $0\leqq t\leqq T$. With the notation above it follows that for $0\leqq t\leqq T$

$$x_1(t)=O(t^2 e(t)),\quad \xi_1(t)=O(te(t)),\quad f(t)=O(t^3 e(t)),$$
$$r_{(1)}(x(t), \xi'(t))-e(t)=O(t^2 e(t)).$$

Hence $d\xi_1/dt=r_{(1)}(x(t),\xi'(t))\geqq 0$ (for left and right derivatives) if t is small enough. Since jumps of ξ_1 at reflections must be positive it follows that ξ_1 is increasing. If e is equal to 0 for $0\leqq t\leqq T$ then there can be no jumps and $x_1=\xi_1=0$ so we have a gliding ray in G^3 which is also an orbit of H_p. On the other hand, if e is not equal to 0 for all small positive t then $\xi_1(t)>0$ for $t>0$ so $x_1(t)$ is strictly increasing. Thus $\gamma(t)\in S\setminus\partial S$ so $\gamma(t)$ must be an orbit of H_p.

Now assume that $e(t)$ is decreasing for $0\leqq t\leqq T$. In view of the discussion above we may assume that $e(t)$ does not vanish for all small t, thus $e(t)<0$ for $0<t\leqq T$. Then

$$r_{(1)}(x(t),\xi'(t))=e(t)+O(t^2 e(t))<0$$

if t is small enough. As in the proof of Lemma 24.3.5 we introduce

$$g(t)=\xi_1(t)^2-x_1(t)\,r_{(1)}(x(t),\xi'(t))$$

and observe that

$$g'(t)=-x_1(t)\,dr_{(1)}(x(t),\xi'(t))/dt.$$

Since

$$e'(t)=dr_{(1)}(0,y'(t),\eta'(t))/dt,$$
$$x_1(t)=O(t^2 e(t)),\quad f(t)=O(t^3 e(t)),\quad dx_1(t)/dt-2\xi_1(t)=O(te(t)),$$
$$d(x'(t)-y'(t),\xi'(t)-\eta'(t))/dt=O(t^2 e(t)),$$

in view of the Hamilton equations, we obtain

$$dr_{(1)}(x(t),\xi'(t))/dt-e'(t)=O(te(t)).$$

By the obvious estimate $g(t)\geqq -x_1(t)\,r_{(1)}(x(t),\xi'(t))$ we have

$$g'(t)\leqq g(t)(-e'(t)+O(te(t)))/(-e(t)+O(t^2 e(t)))$$
$$\leqq g(t)(2e'(t)/e(t)+Ct)$$

when t is so small that $|O(t^2 e(t))|<|e(t)|/2$. Hence

$$d(g(t)/e(t)^2)/dt=(g'(t)-2g(t)\,e'(t)/e(t))/e(t)^2\leqq Ct\,g(t)/e(t)^2.$$

Since $g(t)/e(t)^2=O(t^2)\to 0$ when $t\to 0$ it follows by integration of this inequality that $g(t)/e(t)^2=0$ identically. The proof is complete.

We can of course argue in the same way for $t<0$. In particular, we have therefore proved:

Theorem 24.3.9. *Let $t\mapsto\gamma(t)$ be a generalized bicharacteristic, let $\gamma(t_0)\in G\setminus G^\infty$ and denote by γ_g the gliding ray with $\gamma_g(t_0)=\gamma(t_0)$. Then it follows that in a one sided deleted neighborhood of t_0 we have either $\gamma_g(t)\in G_g$ and $\gamma(t)=\gamma_g(t)$ or else $\gamma_g(t)\in G_d$ and $\gamma(t)=\exp(tH_p)\gamma(t_0)$.*

Corollary 24.3.10. *A generalized bicharacteristic with no point in G^∞ is uniquely determined by any one of its points. A generalized bicharacteristic is constant if it contains a point in $G\setminus G_d$ where $H_p^G=0$.*

The following example shows that there is not always uniqueness when the monotonicity in Proposition 24.3.8 fails:

Example 24.3.11. We shall construct below a C^∞ convex curve $y=f(x)$ in $\mathbb{R}^2$ such that $f(x)=0$ for $x<0$, and an inscribed polygon with vertices converging to 0 which satisfies the reflection law. This condition means precisely that with $X=\{(x,y,t)\in\mathbb{R}^3;\ y\geqq f(x)\}$ the polygon with t defined as arc length and the dual variables ξ, η, τ defined by $\tau=-1$ and $(\xi,\eta)=(dx/dt, dy/dt)$ is a broken bicharacteristic for the differential equation $D_t^2-D_x^2-D_y^2$ which approaches a point $(0,0,t,-1,0,-1)\in G^\infty$ without being equal to the gliding ray. To construct the curve and the polygon we first choose the polygon with the corners

$$(x_k,y_k)=\sum_{j=k}^{\infty} j^{-2}(\cos\alpha_j,\sin\alpha_j)$$

where α_j is decreasing and $\operatorname{tg}((\alpha_j-\alpha_{j+1})/2)=2^{-j}$. Thus the tangents of half the outer angles are 2^{-j}. Choose a C^∞ convex function $g(x)$ such that $g(x)=-x$ for $x<0$ and $g(x)=2(x-1)$ when $x>1$. Then the graph of $(0,k^{-2})\ni x\mapsto 2^{-k}g(k^2x)/k^2$ rotated by the angle α_k and translated to (x_{k+1},y_{k+1}) defines the desired curve if $k\geqq k_0$. We can extrapolate linearly to the right of the first interval. The curve is then in C^∞ for $x>0$ and all derivatives of f tend to 0 when $x\to 0$ so $f\in C^\infty(\mathbb{R})$ if we define $f(x)=0$ when $x\leqq 0$.

We motivated the definition of generalized bicharacteristics by studying limits of broken bicharacteristics at G_d and at G_g. It will now be proved that

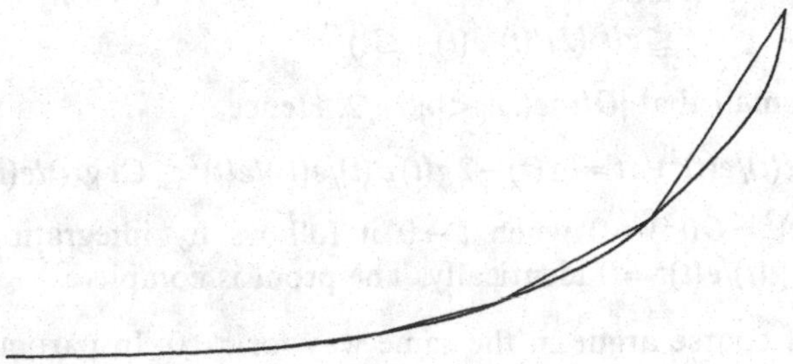

Fig. 5

limits of broken bicharacteristics are indeed generalized bicharacteristics. More generally, assume that we have a sequence of generalized bicharacteristics $[a,b]\ni t\mapsto\gamma^\nu(t)\in K$ where $K\subset S$ is a compact set. If $\gamma^\nu(t)=(x^\nu(t),\xi^\nu(t))$ we have already observed that $x^\nu(t)$, $x_1^\nu(t)\xi_1^\nu(t)$, $\xi_1^\nu(t)^2$ and $\xi^{\nu\prime}(t)$ are uniformly Lipschitz continuous. For a subsequence these functions are therefore uniformly convergent.

Proposition 24.3.12. *Let $[a,b]\ni t\mapsto\gamma^\nu(t)=(x^\nu(t),\xi^\nu(t))$ be generalized bicharacteristics such that $x^\nu(t)$, $x_1^\nu(t)\xi_1^\nu(t)$, $\xi_1^\nu(t)^2$ and $\xi^{\nu\prime}(t)$ are uniformly convergent when $\nu\to\infty$. Then there is a unique generalized bicharacteristic $[a,b]\ni t\mapsto\gamma(t)$ such that $\gamma^\nu(t)\to\gamma(t)$ uniformly for t in any compact subset of $[a,b]$ containing no reflection point of $\gamma(t)$.*

Proof. Let t_0 be a point such that the limit of $x_1^\nu(t_0)$ is not 0. Then $\gamma(t)=\lim\gamma^\nu(t)\in S\setminus\partial S$ exists uniformly in a neighborhood of t_0. From the Hamilton equations it follows that we have C^1 convergence and that $\gamma(t)$ is also an orbit of H_p. Now let t_0 be a point such that $x_1^\nu(t_0)\to 0$ but the limit of $\xi_1^\nu(t_0)^2$ is not 0. Since $dx_1^\nu(t)/dt=2\xi_1^\nu(t)$ it follows from the implicit function theorem that there is a sequence $t_\nu\to t_0$ such that $x_1^\nu(t_\nu)=0$. In a neighborhood of t_0 independent of ν it follows that $\gamma^\nu(t)$ is for $t<t_\nu$ the orbit of H_p which comes in at $\gamma^\nu(t_\nu-0)$ and for $t>t_\nu$ the orbit which goes out at $\gamma^\nu(t_\nu+0)$. This proves that $\gamma(t_0+t)=\lim\gamma^\nu(t_\nu+t)$ exists for fixed small $t\neq 0$, thus with C^1 convergence for small $t>0$ or $t<0$; the limit is a broken bicharacteristic.

Let T be the closed set of all $t\in[a,b]$ such that $x_1^\nu(t)\to 0$ and $\xi_1^\nu(t)\to 0$. Then $\gamma(t)=\lim\gamma^\nu(t)$ exists uniformly for $t\in T$, and $\gamma(t)\in G$. We have already proved the existence of the limit at all points in $[a,b]\setminus T$ except for the countably many reflection points. If $\gamma(t_0)\in G_d$ or G_g for some $t_0\in T$, it follows from Lemmas 24.3.4 and 24.3.5 that $\gamma^\nu(t)$ converges uniformly in a neighborhood of t_0 to the orbit of H_p resp. the gliding ray through $\gamma(t_0)$. If $\gamma(t_0)\in G^3$ we obtain by applying (24.3.7) to $\gamma^\nu(t-t_0)$ and letting $\nu\to\infty$ that $x_1(t)=O((t-t_0)^3)$, $\xi_1(t)=O((t-t_0)^2)$ so $x_1(t)$ and $\xi_1(t)$ are differentiable with derivative 0 at t_0. From the Hamilton equations for $x^{\nu\prime}$, $\xi^{\nu\prime}$ it follows that $x'(t)$, $\xi'(t)$ are everywhere differentiable with derivative $-H_r^0(x(t),\xi(t))$ which proves that γ is a generalized bicharacteristic.

For any $\varepsilon>0$ we have $|\gamma^\nu(t)-\gamma(t)|<\varepsilon$, $t\in T$, if $\nu>\nu(\varepsilon)$. By the uniform continuity of x_1^ν, $|\xi_1^\nu|$, $\xi^{\nu\prime}$ it follows that $|\gamma^\nu(t)-\gamma(t)|<2\varepsilon$ if $\nu>\nu(\varepsilon)$, for all t in a neighborhood V of T independent of ν. In any compact subset of $[a,b]\setminus V$ containing no reflection point of $\gamma(t)$ we have uniform convergence, which completes the proof of the stated uniform convergence of γ^ν.

Our next aim is to show that through every point in Σ there is a generalized bicharacteristic which is a limit of broken bicharacteristics. To do so we must make some preliminary remarks on the bicharacteristic flow. Let $p\in C^\infty(S)$ where S is a symplectic manifold, and let s_0, s_1 be points in S

with $p(s_0)=p(s_1)=0$ and $\exp(t_0 H_p)s_0=s_1$. Then one can choose neighborhoods S_0 and S_1 of s_0 and s_1 such that

$$\{(s',s'')\in S_0\times S_1;\ p(s')=p(s'')=0,\ s''=\exp(tH_p)s' \text{ for some } t \text{ close to } t_0\}$$

is a canonical relation. In fact, we can choose symplectic coordinates y, η vanishing at s_1 such that $p=\eta_1$ in a neighborhood. Then the pullbacks (x,ξ) of (y,η) by $\exp(t_0 H_p)$ are symplectic coordinates at s_0 which vanish at s_0, and $p=0$ is equivalent to $\xi_1=0$ there. (See Section 6.4.) The relation is then

$$\{(x,\xi;y,\eta);\ \xi_1=\eta_1=0,\ x_j=y_j \text{ and } \xi_j=\eta_j \text{ for } j\neq 1\}$$

which is obviously canonical. (See also Proposition 26.1.3.)

Let us now return to the Hamiltonian (24.3.1) in a convex neighborhood of $0\in T^*(\overline{\mathbb{R}}^n_+)$ where we assume that $\partial p/\partial\xi_n>0$. This implies that $dx_n/dt>0$ for the (generalized) bicharacteristics and also that the map

$$\{(x,\xi);\ p(x,\xi)=0\}\ni(x,\xi)\mapsto(x_1,\dots,x_{n-1},\xi_1,\dots,\xi_{n-1})\in T^*(\mathbb{R}^{n-1})$$

is injective and has bijective differential for fixed x_n. The pullback of the symplectic form in $T^*(\mathbb{R}^n)$ by the inverse map is the symplectic form in $T^*(\mathbb{R}^{n-1})$. For fixed s, $t\in\mathbb{R}$ let χ_s^t map $(x_1,\dots,x_{n-1},\ \xi_1,\dots,\xi_{n-1})$ to $(y_1,\dots,y_{n-1},\eta_1,\dots,\eta_{n-1})$ if there exist some ξ_n and η_n such that p vanishes at the point $(x_1,\dots,x_{n-1},s,\xi_1,\dots,\xi_n)\in S$ and the orbit of the Hamilton field H_p starting there arrives at $(y_1,\dots,y_{n-1},\ t,\eta_1,\dots,\eta_n)$ without reaching the boundary of S. This is a canonical transformation for it is obtained by restricting the canonical relation described above to fixed x_n and y_n. For the same reason the map $\chi_s(x_1,\dots,x_{n-1},\ \xi_1,\dots,\xi_{n-1})=(y',\eta')$ defined when the bicharacteristic first meets the boundary transversally at $(0,y',\pm r(0,y',\eta')^{\frac12},\eta')$ is also a canonical transformation. By composing such in and outgoing canonical transformations we obtain

Proposition 24.3.13. *Let p be given by (24.3.1) in a convex neighborhood S of $0\in T^*(\overline{\mathbb{R}}^n_+)$ where $\partial p/\partial\xi_n>0$. For $s,t\in\mathbb{R}$ define*

$$\tilde\chi_s^t(x_1,\dots,x_{n-1},\xi_1,\dots,\xi_{n-1})=(y_1,\dots,y_{n-1},\eta_1,\dots,\eta_{n-1})$$

if there exist some ξ_n and η_n such that $(x_1,\dots,x_{n-1},s,\xi_1,\dots,\xi_n)$ and $(y_1,\dots,y_{n-1},t,\eta_1,\dots,\eta_n)$ are zeros of p in S connected by a broken bicharacteristic arc in S. If s and t are small enough then $\tilde\chi_s^t$ is a symplectic transformation defined in an open set which contains a neighborhood of 0 in $T^*(\mathbb{R}^{n-1})$ apart from a set of measure 0.

Proof. When $\tilde\chi_s^t(x_1,\dots,x_{n-1},\xi_1,\dots,\xi_{n-1})$ is defined, the broken bicharacteristic considered has a finite number N of (transversal) reflection points. By the implicit function theorem we get such a broken bicharacteristic for all $(x_1',\dots,\xi_{n-1}')$ in a neighborhood of $(x_1,\dots,\xi_{n-1})$; as the composition of $N+1$ C^∞ symplectic maps in $T^*(\mathbb{R}^{n-1})$ the map $\tilde\chi_s^t$ then defined is C^∞ and symplectic. By the uniform continuity of x, $|\xi_1|$, ξ' on generalized bicharacteristics we can find $\delta>0$ such that for $|s|+|t|<\delta$ the map $\tilde{\chi}_s^t$ is defined and uniformly bounded in the set A of all $(x_1,\dots,\xi_{n-1})$ with $|x_1|+\dots+|\xi_{n-1}|<\delta$ unless the broken bicharacteristic γ starting there comes arbitrarily close to a point in G before $x_n=t$. Choose ε so that $(t-s)/\varepsilon$ is a large positive integer, and let A_ε be the set of points in A such that $x_1\leqq\varepsilon^2$ and $|\xi_1|\leqq\varepsilon$ at some point $\gamma(\tau)$. Then it follows from (24.3.7) that $x_1\leqq C\varepsilon^2$, $|\xi_1|\leqq C\varepsilon$ at $\gamma(v\varepsilon)$ if v is the largest integer with $v\varepsilon\leqq\tau$. The measure of all $(x_1,\dots,\xi_{n-1})$ with $0\leqq x_1\leqq C\varepsilon^2$, $|\xi_1|\leqq C\varepsilon$ and fixed bounds for $x_2,\dots,\xi_{n-1}$ is $O(\varepsilon^3)$ and it is equal to the measure of the inverse image under the symplectic map $\tilde{\chi}_s^{v\varepsilon}$. If we sum for all positive integers $v\leqq(t-s)/\varepsilon$ it follows that the measure of A_ε is $O(\varepsilon^2)$, hence $\bigcap_{\varepsilon>0}A_\varepsilon$ is of measure 0, which completes the proof.

Corollary 24.3.14. *Under the hypotheses in Proposition* 24.3.13 *there is some* $\delta>0$ *and a neighborhood* V *of* 0 *in* $T^*(\overline{\mathbb{R}}^n_+)$ *such that for every* $\gamma_0\in V\cap p^{-1}(0)$ *there is a generalized bicharacteristic* $(-\delta,\delta)\ni t\mapsto\gamma(t)\in S$ *with* $\gamma(0)=\gamma_0$. *(If* $\gamma(t)$ *has a reflection point for* $t=0$ *this means that* $\gamma(+0)=\gamma_0$ *or* $\gamma(-0)=\gamma_0$.*)*

Proof. By Proposition 24.3.13 one can choose V and δ so that there is such a broken bicharacteristic for every $\gamma_0\in V\cap p^{-1}(0)$ outside a null set in the manifold $p^{-1}(0)$, even for fixed x_n. Since these broken bicharacteristics are uniformly continuous it follows from Proposition 24.3.12 that there is a generalized bicharacteristic for every γ_0.

Remark. From this corollary and the standard existence theorem for the vector field H_p in the interior of S one can conclude the same global existence theorems for generalized bicharacteristics as for bicharacteristics on an open manifold. The proof can also be used to give generalized bicharacteristics which are globally limits of broken bicharacteristics. In view of Corollary 24.3.10 it follows that every generalized bicharacteristic which does not intersect G^∞ is a limit of broken bicharacteristics. However, it is an open problem if every generalized bicharacteristic can be obtained as such a limit.

We shall now prove a stronger existence theorem for generalized bicharacteristics which will be essential in the proof of an extension of Theorem 24.2.3.

Lemma 24.3.15. *Let the hypotheses of Proposition* 24.3.13 *be fulfilled, and let* F *be a closed subset of* $S\cap p^{-1}(0)$ *such that*

(i) *If* $\gamma\in F\setminus(G_g\cup G^3)$ *then* F *contains a neighborhood of* γ *on the broken or diffractive bicharacteristic through* γ.

(ii) *For every compact subset* K *of* $F\cap(G_g\cup G^3)$ *and* $\varepsilon>0$ *there exists some* $\delta>0$ *and* $\gamma_\varepsilon(t,\gamma_0)\in F$, $-\delta\leqq t\leqq\delta$, $\gamma_0\in K$, *such that*

$$|(\gamma_\varepsilon(t,\gamma_0)-\gamma_0)/t-H_p^G(\gamma_0)|'<\varepsilon,\quad 0<|t|\leqq\delta. \tag{24.3.8}$$

Here $H_p^G(\gamma_0)$ is the tangent of the gliding ray, and $|v|'$ denotes for $v \in T^*(\mathbb{R}^n) = \mathbb{R}^{2n}$ the Euclidean length when the ξ_1 component is dropped. Then there exists for every $\gamma_0 \in F$ a generalized bicharacteristic $\gamma(t)$, $t_- < t < t_+$, such that $t_- < 0 < t_+$, $\gamma(0) = \gamma_0$, $\gamma((t_-, t_+)) \subset F$, and $\gamma(t)$ leaves every compact subset of F when $t \to t_\pm$.

Proof. Since $dx_n/dt = \partial p/\partial \xi_n > 0$ on a generalized bicharacteristic it is sufficient to show that for every γ_0 in a compact set $K \subset F$ there is a generalized bicharacteristic through γ_0 with end points on the boundary of K in F. To do so we first choose δ as in condition (ii) for some small $\varepsilon > 0$, and then define an approximate generalized bicharacteristic as follows. If $\gamma_0 \notin G_g \cup G^3$ we define $\gamma_\varepsilon(t)$ as the (broken) bicharacteristic of p with $\gamma(0) = \gamma_0$ as long as it exists and is contained in K. We are then through for positive t unless we get $\gamma_\varepsilon(t)$ defined just in an interval $0 \leqq t < \tau$. The limit of $\gamma_\varepsilon(t)$ as $t \to \tau$ must then exist and belong to $G_g \cup G^3$. Now we define $\gamma_\varepsilon(t)$ for $\tau \leqq t \leqq \tau + \delta$ by

$$\gamma_\varepsilon(t) = \gamma_\varepsilon(\tau) + (\gamma_\varepsilon(\delta, \gamma_\varepsilon(\tau)) - \gamma_\varepsilon(\tau))(t - \tau)/\delta$$

where we have used the function in condition (ii). Continuing in this way also for negative t we arrive at the boundary of K in F at some bounded time t since we have a positive lower bound for dx_n/dt on the curve $[t_-(\varepsilon), t_+(\varepsilon)] \ni t \mapsto \gamma_\varepsilon(t)$ obtained. The end points of the curve are thus on the boundary of K in F. In view of the uniform Lipschitz continuity of the coordinates other than ξ_1 we can choose a sequence $\varepsilon_j \to 0$ such that a limit $\gamma(t) \in F$ exists as in Proposition 24.3.12. Since $G_g \cup G^3$ is closed it follows as there that γ is a broken bicharacteristic elsewhere. If $\gamma(t) \in G_g$ then $\gamma(s)$ must be the gliding ray through $\gamma(t)$ for small $|s-t|$. In fact, if $\gamma(s) \notin G_g$ for some s in any neighborhood of t then it follows from Lemma 24.3.5 that we must have a broken bicharacteristic there. Thus $\gamma(s) \in G_g$ for small $|s-t|$ so $x_1(s) = \xi_1(s) = 0$ then, and

$$d(x'(s), \xi'(s))/ds = -H_{r_0}(x'(s), \xi'(s))$$

since for γ_ε the derivative of these components is $-H_r^0(\gamma_\varepsilon(s))$ or $-H_{r_0}(\gamma_\varepsilon(s) + O(\delta))$ except at finitely many points. Finally, if $\gamma(t) \in G^3$ we obtain in the same way that $(x(s), \xi'(s))$ is differentiable at t with the derivative $(0, -H_{r_0}(x'(t), \xi'(t)))$. It remains to prove that $\xi_1(s)$ is differentiable with derivative 0 at t, that is, that $\xi_1(s)^2 = r(x(s), \xi'(s)) = O((s-t)^2)$. Now

$$|r(x(s), \xi'(s))| + |r_{(1)}(x(s), \xi'(s))| = O(s-t)$$

since the left-hand side is Lipschitz continuous and vanishes at t. The right and left derivatives of $r(x(s), \xi'(s))$ are $r_{(1)}(x(s), \xi'(s))\,dx_1/ds$ since $d(x'(s), \xi'(s))/ds = -H_r^0(x(s), \xi'(s))$ which makes the other terms drop out. Here the left and right derivatives dx_1/ds are equal to $\pm\xi_1(s) = \pm r(x(s), \xi'(s))^{\frac{1}{2}}$ so the derivatives of $r(x(s), \xi'(s))$ are $O(|s-t|^{\frac{3}{2}})$. Thus

$$r(x(s), \xi'(s)) = O(|s-t|^{\frac{5}{2}})$$

which completes the proof.

In Sections 24.4 and 24.5 we shall respectively verify conditions (i) and (ii) in Lemma 24.3.15 for $WF_b(u)$ when u satisfies a second order differential equation with smooth right-hand side and has smooth boundary values; condition (i) is then already proved in part by Theorem 24.2.1.

24.4. The Diffractive Case

As in Section 24.2 we assume that P is a second order differential operator with real principal symbol p and C^∞ coefficients defined in a C^∞ manifold X of dimension n, with non-characteristic boundary ∂X. The present section is devoted to the proof of the following analogue of Theorem 24.2.1. Recall that the diffractive set $G_d \subset T^*(\partial X)$ can be identified with a subset of $T^*(X)|_{\partial X}$.

Theorem 24.4.1. *Let $u, f \in \mathcal{N}(X)$, $u_0 \in \mathcal{D}'(\partial X)$, and assume that*

$$Pu=f \quad \text{in } X \smallsetminus \partial X, \qquad u=u_0 \quad \text{in } \partial X. \tag{24.4.1}$$

If $\gamma \in (WF_b(u) \cap G_d) \smallsetminus (WF_b(f) \cup WF(u_0))$ then a neighborhood of γ on the orbit of H_p through γ is in $WF_b(u)$.

In the proof we may again assume that $X \subset \overline{\mathbb{R}}^n_+ = \{x \in \mathbb{R}^n; x_1 \geqq 0\}$ and that $p(x,\xi) = \xi_1^2 - r(x,\xi')$. Thus

$$P = D_1^2 - r(x,D') - \sum_1^n R_j(x) D_j - R_0(x).$$

If c is a C^∞ function then the operator $u \mapsto e^{-c} P(e^c u)$ has the same principal symbol but R_1 is replaced by $R_1 - 2D_1 c$. We can therefore eliminate the D_1 term and assume in what follows that

$$P = D_1^2 - R(x,D') \tag{24.4.2}$$

where R is a differential operator in the variables $x' = (x_2, \dots, x_n)$ with principal symbol $r(x,\xi')$. We may assume that the coefficients have compact support.

The first step is now to extend the energy identity (24.1.2) by substituting for the first order differential operator $\sum f^i \partial_i$ an operator which is pseudo-differential along the boundary but a first order differential operator in x_1. It is convenient later to have the total order equal to 0 so we set

$$Q(x,D) = Q_1(x,D') D_1 + Q_0(x,D'), \tag{24.4.3}$$

where Q_j is a pseudo-differential operator in the x' variables of order $-j$ and principal symbol q_j. We shall assume that Q is self adjoint,

$$Q^* = Q, \quad \text{that is, } Q_1^* = Q_1 \quad \text{and} \quad Q_0^* = Q_0 - [D_1, Q_1], \tag{24.4.4}$$

which implies that q_0 and q_1 are real. Let $u \in C_0^\infty(X)$ and set $u_j = D_1^j u$. With $(\ ,\ )_X$ and $(\ ,\ )_{\partial X}$ denoting the usual sesquilinear scalar products in X and in ∂X, we shall prove

Lemma 24.4.2. *With P of the form (24.4.2) and Q of the form (24.4.3) satisfying (24.4.4), we have if $u \in C_0^\infty(\overline{\mathbb{R}}^n_+)$*

$$2\operatorname{Im}(Pu, Qu)_X = \sum_0^1 (B_{jk}(x', D')u_k, u_j)_{\partial X} + \sum_0^1 (C_{jk}(x, D')u_k, u_j)_X. \tag{24.4.5}$$

Here $B_{11} = Q_1$, $B_{01}^* = B_{10} = Q_0$, $B_{00} = Q_1(R+R^*)/2$, for $x_1 = 0$, the principal symbol c_{jk} (of order $1-j-k$) of C_{jk} is real, $c_{01} = c_{10}$ and

$$\sum c_{jk}(x, \xi')\xi_1^{j+k} = \{p, q\} + 2q \operatorname{Im} p_s \tag{24.4.6}$$

where

$$p_s(x, \xi') = -\tfrac{1}{2} i \sum_2^n r_{(j)}^{(j)}(x, \xi') - \sum_2^n R_j(x)\xi_j$$

is the subprincipal symbol of P.

Proof. First we compute the contributions from the term D_1^2,

$$\begin{aligned}((D_1^2 u, Qu)_X - (Qu, D_1^2 u)_X)/i &= (D_1 u, Qu)_{\partial X} + (Qu, D_1 u)_{\partial X}\\ &+ ((D_1 u, QD_1 u)_X + (D_1 u, [D_1, Q]u)_X - (QD_1 u, D_1 u)_X - ([D_1, Q]u, D_1 u)_X)/i.\end{aligned}$$

Another integration by parts gives in view of (24.4.4)

$$((D_1 u, QD_1 u)_X - (QD_1 u, D_1 u)_X)/i = -(Q_1 D_1 u, D_1 u)_{\partial X}.$$

The boundary terms add up to those in (24.4.5) with $j+k \neq 0$. The operator $[D_1, Q]/i$ is self adjoint with principal symbol $-\partial q/\partial x_1$, so we get the contribution $2\xi_1 \partial q/\partial x_1 = \{\xi_1^2, q\}$ to the sum in (24.4.6) from the terms in the second sum (24.4.5) now obtained.

It remains to study

$$((Qu, Ru)_X - (Ru, Qu)_X)/i.$$

In doing so we set $R = R' + iR''$ where R' and R'' are self adjoint. Both operate in the x' variables and R'' is of order 1. Now

$$\begin{aligned}&((Qu, R'u)_X - (R'u, Qu)_X)/i = (R'u, Q_1 u)_{\partial X} + ([R', Q]u, u)_X/i,\\ &-(R''u, Qu)_X - (Qu, R''u)_X\\ &\quad = -(Q_1 R''u, D_1 u)_X - (R''Q_1 D_1 u, u)_X - ((Q_0^* R'' + R''Q_0)u, u)_X\end{aligned}$$

which gives the term $B_{00} = Q_1 R'$ in (24.2.5) and contributes to the sum in (24.4.6) the terms $-\{r, q\} - 2qr''$, where $r'' = -\operatorname{Im} p_s$ is the principal symbol of R''. The proof is complete.

Theorem 24.4.1 will be proved by means of an estimate derived from (24.4.5). To see what conditions Q ought to satisfy we first observe that if $u=0$ on ∂X then $(B_{11}u_1, u_1)_{\partial X}$ is the only boundary term in (24.4.5). In view of the sharp Gårding inequality (Theorem 18.1.14) it can be estimated above by $C\|u_1(0,.)\|^2_{(-1)}$ if $q_1\leqq 0$. If $\sum c_{jk}(x,\xi')\xi_1^{j+k}\leqq 0$ then

$$\sum c_{jk}(x,\xi')z_k\bar{z}_j\leqq 0, \quad (z_0,z_1)\in\mathbb{C}^2,$$

since (c_{jk}) is real and symmetric. Hence the sharp Gårding inequality for systems (see the remark after Theorem 18.1.14 or Theorem 18.6.14) shows that the real part of the second sum in (24.4.5) can be estimated above by $C\sum\|u_j\|^2_{(0,-j)}$, that is, $C\|u\|^2_{(1,-1)}$ by Theorem B.2.3. (In Appendix B the half space $\overline{\mathbb{R}}^n_+$ is defined by $x_n\geqq 0$ while it is defined by $x_1\geqq 0$ here; this change of notation should cause no confusion.) Actually we are using a slightly modified version of the sharp Gårding inequality with different degrees assigned to u_1 and u_0 just as in Theorem 19.5.3; it follows from the standard version by the proof of that theorem.

To derive an estimate from (24.4.5) we shall of course have to make $\sum c_{jk}(x,\xi')\xi_1^{j+k}$ strictly negative in some important region. This will be achieved by choosing q so that $H_p q\leqq 0$ with strict inequality in some important region. However, we shall only be able to do so in the set where $p(x,\xi)=\xi_1^2-r(x,\xi')=0$, that is, we shall have q decreasing along the (broken) bicharacteristic flow. The following lemma will allow us to use this weaker positivity condition.

Lemma 24.4.3. *Let Y be an open subset of $\overline{\mathbb{R}}^\nu_+=\{y\in\mathbb{R}^\nu;\, y_1\geqq 0\}$, and let $r\in C^\infty(Y)$. We assume that r is real valued, that $dr\neq 0$ when $r=0$ and that $\partial r/\partial y_1>0$ when $r=y_1=\partial r/\partial y_j=0$ for $j\neq 1$. Let*

$$A(t,y)=\sum_0^2 a_j(y)t^j$$

be a quadratic polynomial in t with coefficients in $C^\infty(Y)$ such that

$$A(t,y)=-\psi(t,y)^2 \quad \text{when } t^2=r(y), \tag{24.4.7}$$

where $\psi\in C^\infty(\mathbb{R}\times Y)$. Then one can find $\psi_0,\psi_1,g\in C^\infty(Y)$ with $\psi_0(y)=\psi(0,y)$, $\psi_1(y)=\partial\psi(0,y)/\partial t$ when $r(y)=0$, and

$$A(t,y)+(\psi_0(y)+\psi_1(y)t)^2\leqq g(y)(t^2-r(y)); \quad t\in\mathbb{R},\ y\in Y. \tag{24.4.8}$$

Proof. If $r(y)>0$ we set $r(y)=s^2$ and observe that

$$\psi_0(y)+\psi_1(y)t=\psi(t,y) \quad \text{when } t=\pm s$$

provided that

$$\psi_0(y)=(\psi(s,y)+\psi(-s,y))/2, \quad \psi_1(y)=(\psi(s,y)-\psi(-s,y))/2s.$$

In view of Theorem C.4.4 we can choose $\Psi_j\in C^\infty(\mathbb{R}\times Y)$ such that

$$(\psi(s,y)+\psi(-s,y))/2=\Psi_0(s^2,y), \quad (\psi(s,y)-\psi(-s,y))/2s=\Psi_1(s^2,y)$$

in $\mathbb{R}\times Y$. Thus $\psi_j(y)=\Psi_j(r(y),y)\in C^\infty(Y)$ and

$$B(t,y)=A(t,y)+(\psi_0(y)+\psi_1(y)\,t)^2-(a_2(y)+\psi_1(y)^2)(t^2-r(y))$$

is linear in t and vanishes when $t^2=r(y)$. (We could also have applied the Malgrange preparation theorem here.) If we write

$$B(t,y)=b_1(y)\,t+b_0(y)$$

it follows that $b_j(y)=0$ when $r(y)\geqq 0$. We shall have

$$B(t,y)\leqq f(y)\,(t^2-r(y))$$

if $f(y)=0$ when $r(y)>0$ and $f(y)\geqq F(y)$ where

$$F(y)=|b_1(y)|/|r(y)|^{\frac{1}{2}}+|b_0(y)|/|r(y)|.$$

Here $F=O(|r|^N)$ for any N in any compact subset K of Y. In fact, by the implicit function theorem we can for every $y\in K$ such that $r(y)$ is negative and small enough find $\tilde{y}\in Y$ with $r(\tilde{y})=0$ and $|\tilde{y}-y|\leqq C|r(y)|$. Hence Lemma 24.4.3 is a consequence of

Lemma 24.4.4. *Let Y and r satisfy the hypotheses in Lemma* 24.4.3, *and let F be a non-negative function vanishing for $r>0$ such that $F=O(r^N)$ for any N on any compact subset of Y. Then one can find $f\in C^\infty(Y)$ such that $F\leqq f$ and $f=0$ when $r\geqq 0$.*

Proof. Local solutions of this problem can be pieced together by a partition of unity, so we may work in a compact set K where $r\leqq 0$ at some point. Set

$$F_1(t)=\sup\{F(x);\, r(x)\geqq -t\}\quad \text{if } t>0;\qquad F_1(t)=0\quad \text{if } t\leqq 0.$$

Then F_1 is an increasing function of t which vanishes of infinite order when $t=0$. Set

$$f_1(t)=\int F_1(t\,s)\,\chi(s)\,ds/s=\int F_1(s)\,\chi(s/t)\,ds/s$$

where $\chi\in C_0^\infty((1,2))$ is non-negative and $\int\chi(s)\,ds/s=1$. Then $f_1\geqq F_1$ since F_1 is increasing, $f_1=0$ when $t\leqq 0$, and

$$t^k f_1^{(k)}(t)=\int F_1(s)\,\chi_k(s/t)\,ds/s=\int F_1(s\,t)\,\chi_k(s)\,ds/s$$

where $\chi_0=\chi$ and $\chi_{k+1}(t)=-k\,\chi_k(t)-t\,\chi_k'(t)\in C_0^\infty$. The right-hand side can be estimated by

$$F_1(2t)\int|\chi_k(s)|\,ds/s=O(t^{k+1})$$

so all derivatives of f_1 tend to 0 when $t\to 0$. This proves that $f_1\in C^\infty$, so $f=f_1(-r)$ has the required properties. The proof is complete.

Lemma 24.4.5. *Let $A_{jk}\in S^{1-j-k}(\mathbb{R}^n\times\mathbb{R}^{n-1})$; $j,k=0,1$; have real homogeneous principal symbols $a_{jk}(x,\xi')$ vanishing for large $|x|$, $a_{01}=a_{10}$, and assume that*

$$\sum a_{jk}(x,\xi')\,\xi_1^{j+k}=-\psi(x,\xi)^2\quad \textit{when}\ \ \xi_1^2=r(x,\xi'),$$

where $\psi \in C^\infty(\mathbb{R}^n \times (\mathbb{R}^n \setminus 0))$ *is homogeneous of degree* $\frac{1}{2}$. *Further assume that* $\partial r/\partial x_1 > 0$ *or that* $\partial r/\partial(x', \xi') \neq 0$ *in* $\bigcup \operatorname{supp} a_{jk}$. *Then one can choose* $\Psi_0(x, \xi') \in S^{\frac{1}{2}}(\mathbb{R}^n \times \mathbb{R}^{n-1})$ *and* $\Psi_1(x, \xi') \in S^{-\frac{1}{2}}(\mathbb{R}^n \times \mathbb{R}^{n-1})$ *with principal symbols equal to* $\psi(x, \xi)$ *and* $\partial\psi(x, \xi)/\partial\xi_1$ *when* $\xi_1 = r(x, \xi') = 0$, *such that for any* $u \in \bar{C}_0^\infty(\mathbb{R}^n_+)$ *and any* $\kappa \in \mathbb{R}$

$$\text{(24.4.9)} \quad \operatorname{Re} \sum (A_{jk}(x, D') D_1^k u, D_1^j u)_X + \|\Psi_0(x, D') u + \Psi_1(x, D') D_1 u\|_X^2 \\ \leqq C_\kappa(\|u\|^2_{(1,-1)} + \|D_1 u(0,.)\|_{(-\kappa-1)} \|u(0,.)\|_{(\kappa)} + \|Pu\|^2_{(0,-1)}).$$

Proof. Choose C^∞ functions $\psi_0(x, \xi')$ and $\psi_1(x, \xi')$ homogeneous of degree $\frac{1}{2}$ and $-\frac{1}{2}$ respectively so that $\psi_j(x, \xi') = \partial^j \psi(x, \xi)/\partial\xi_1^j$ when $\xi_1 = r(x, \xi') = 0$ and so that

$$\text{(24.4.10)} \quad \sum a_{jk}(x, \xi') \xi_1^{j+k} + (\psi_0(x, \xi') + \psi_1(x, \xi') \xi_1)^2 \\ \leqq g(x, \xi')(\xi_1^2 - r(x, \xi'))$$

for some $g \in C^\infty$ homogeneous of degree -1. This is possible by Lemma 24.4.3, with $y = (x, \xi')$ and $t = \xi_1$; the homogeneity is obtained by taking the restriction to the set where $|\xi'| = 1$ and extending by homogeneity. We can take ψ_j and g equal to 0 for large $|x|$. Choose Ψ_j and G with principal symbols ψ_j and g, and consider

$$\text{(24.4.11)} \quad \operatorname{Re} \sum (A_{jk}(x, D') u_k, u_j)_X + \sum (\Psi_j^*(x, D') \Psi_k(x, D') u_k, u_j)_X \\ (G(x, D') u_1, u_1)_X + (G(x, D') R(x, D') u_0, u_0)_X.$$

The principal symbol is negative semi-definite by (24.4.10), so it follows from the sharp Gårding inequality that (24.4.11) can be estimated above by $C(\|u_0\|^2_{(0,0)} + \|u_1\|^2_{(0,-1)})$, as pointed out above. We take $u_k = D_1^k u$ and observe that the sum of the last two terms in (24.4.11) differs from $(-G(x, D') Pu, u)_X$ by

$$(G(x, D') D_1^2 u, u)_X - (G(x, D') D_1 u, D_1 u)_X \\ = i(G(0, x', D') D_1 u(0,.), u(0,.))_{\partial X} - ([D_1, G(x, D')] D_1 u, u)_X.$$

Since $[D_1, G(x, D')]$ is of order -1, the last term can be estimated by $C\|u\|_{(0,0)} \|D_1 u\|_{(0,-1)}$, and the boundary term can be estimated by $C_\kappa \|D_1 u(0,\cdot)\|_{(-\kappa-1)} \|u(0,.)\|_{(\kappa)}$. In view of Theorem B.2.3, the proof is complete.

Remark. We shall sometimes refer to $\sum a_{jk} \xi_1^{j+k}$ as the principal symbol of the form $\sum (A_{jk}(x, D') D_1^k u, D_1^j u)_X$.

We are now prepared to start the proof of Theorem 24.4.1. We assume as before that $X \subset \overline{\mathbb{R}}^n_+$, that P is of the form (24.4.2) with the coefficients in $C_0^\infty(\overline{\mathbb{R}}^n_+)$, and that $\gamma = (0, \xi_0)$, $\xi_0 = (0, \xi_0')$. Replacing X by a smaller neighborhood $X = X_0 \times [0, c)$ of 0 we may assume that $WF(f)$ and therefore $WF(u)$ contains no element (x, ξ) over $X_0 \times (0, c)$ with $\xi' = 0$. These hypotheses are not affected if u is multiplied by a cutoff function which is equal to 1 near 0, so we assume that $u\in\dot{\mathscr{E}}'(X)$. Note that by Theorem 18.1.36 and Theorem 18.3.32 it follows then that $a(x,D')u\in C^\infty(X)$ resp. $a(x,D')f\in C^\infty(X)$ if $a\in S^\infty(\mathbb{R}^n\times\mathbb{R}^{n-1})$ is of order $-\infty$ in a conic neighborhood of $WF_b(u)$ (resp. $WF_b(f)$) projected to $\mathbb{R}^n\times(\mathbb{R}^{n-1}\smallsetminus 0)$.

Now choose an open conic neighborhood W of $(0,\xi_0')$ in $X\times(\mathbb{R}^{n-1}\smallsetminus 0)$ such that

$$\begin{aligned}&(x,\xi')\in W, \quad x_1\neq 0\Rightarrow(x,\xi)\notin WF(f);\\ &(0,x',\xi')\in W\Rightarrow(x',\xi')\notin WF_b(f)\cup WF(u_0);\\ &(x,\xi')\in W\Rightarrow \partial r(x,\xi')/\partial x_1>0.\end{aligned}\tag{24.4.12}$$

This is possible since $WF_b(f)$ and $WF(u_0)$ are closed and the hypothesis $\gamma\in G_d$ means that $r=0$, $\partial r/\partial x_1>0$ at γ. Assuming that there is some $\gamma_0\neq\gamma$ on the bicharacteristic of p through γ such that the bicharacteristic interval $[\gamma_0,\gamma]$ lies over W and $\gamma_0\notin WF(u)$, we must prove that $(0,\xi_0')\notin WF_b(u)$. Let γ lie after γ_0 on the bicharacteristic; the proof is completely analogous when γ_0 lies after γ. Choose an open conic neighborhood Γ_0 of γ_0 in $T^*(X^\circ)\smallsetminus 0$ with $\Gamma_0\cap WF(u)=\emptyset$, and let W_0 be the set of all $(x,\xi')\in W$ such that if $r(x,\xi')\geqq 0$ there are bicharacteristic intervals, possibly broken by one reflection or tangency at ∂X, which lie over W, have initial point in Γ_0 and end point $(x,\pm r(x,\xi')^{\frac12},\xi')$. For reasons of continuity W_0 is an open subset of W; it is clear that W_0 is conic and contains $(0,\xi_0')$. We shall prove that $\chi(x,D')u\in C^\infty$ if $\chi\in S^0$ and $\operatorname{cone\,supp}\chi\subset W_0$; by Theorem 18.3.32 this implies that $(0,\xi_0')\notin WF_b(u)$. (By $\operatorname{cone\,supp}\chi$ we mean the smallest closed conic set containing $\operatorname{supp}\chi$.)

By Theorem 24.2.1 and (24.2.4) we know that

$$\{(x',\xi');(0,x',\xi')\in W_0,\quad (x',\xi')\in WF_b(u)\}\subset G_d\tag{24.4.13}$$

if G_d is regarded as a subset of $T^*(\partial X_0)$, and that

$$\{(x,\xi)\in WF(u);\, x_1>0,(x,\xi')\in W_0\}$$

is contained in the forward H_p flowout of G_d over W_0; thus $\xi_1>0$ there. (Recall that ξ_1 is strictly increasing on the (broken) bicharacteristics since $\partial r/\partial x_1>0$, and that $dx_1/dt=2\xi_1$ there.) In fact, if (x,ξ) is not in this flowout then the definition of W_0 shows that the bicharacteristic going backwards from (x,ξ) must reach Γ_0 after at most a transversal reflection while remaining over W.

The proof of Theorem 24.2.1 gives more,

$$\chi(x,D')(D_1-\tilde{\Lambda}_+(x,D'))u\in C^\infty(X)\tag{24.4.14}$$

if $\chi\in S^0$ and $r>0$ in $\operatorname{cone\,supp}\chi\subset W_0$. Here the operator $D_1-\tilde{\Lambda}_+(x,D')$ in the factorization (24.2.6) is defined on the symbol level when $r>0$ so the product by $\chi(x,D')$ is defining up to an operator of order $-\infty$. For the proof we just have to observe that the bicharacteristic going backwards from a point $(x, -r(x,\xi')^{\frac{1}{2}}, \xi')$ with $(x,\xi')\in$cone supp χ remains over W until it has entered Γ_0, and that $r(x,\xi')=\xi_1^2>0$ there because ξ_1 decreases in the backward direction on a bicharacteristic.

We shall also need a microlocal form of Theorem B.2.9.

Lemma 24.4.6. *Under the hypotheses above, if $\chi(x,D')u\in\overline{H}_{(m,s)}$ for all $\chi\in S^0$ with* cone supp $\chi\subset W_0$ *then $\chi(x,D')u\in\overline{H}_{(m+1,s-1)}$ for all such χ.*

Proof. Since

$$D_1^2\chi(x,D')u=\chi(x,D')f+R(x,D')\chi(x,D')u+[P,\chi(x,D')]u$$

and $[P,\chi(x,D')]=\chi_1(x,D')D_1+\chi_0(x,D')$ with χ_j of order $1-j$ and order $-\infty$ outside cone supp χ, we have $D_1^2\chi(x,D')u\in\overline{H}_{(m-1,s-1)}$. From Theorem B.2.3 we now obtain first that $D_1\chi(x,D')u\in\overline{H}_{(m,s-1)}$ and then that $\chi(x,D')u\in\overline{H}_{(m+1,s-1)}$.

Since $u\in\overline{H}_{(m,s)}$ for some m, s we conclude by repeated use of Lemma 24.4.6 that for some s

$$(24.4.15)_s\quad \chi(x,D')u\in\overline{H}_{(1,s-1)},\quad \chi(x,D')D_1u|_{x_1=0}\in H_{(s-1)}(\mathbb{R}^{n-1})$$

for all $\chi\in S^0$ with cone supp $\chi\subset W_0$. The proof of Theorem 24.4.1 will be completed if we show that $(24.4.15)_s\Rightarrow(24.4.15)_{s+\frac{1}{2}}$, for by Lemma 24.4.6 we can then conclude that $\chi(x,D')u\in C^\infty$ for all $\chi\in S^0$ with cone supp $\chi\subset W_0$. The decisive step in the induction is the following

Lemma 24.4.7. *Assume that $q(x,\xi)=q_1(x,\xi')\xi_1+q_0(x,\xi')$ where the coefficients $q_j\in C^\infty(\mathbb{R}^n\times(\mathbb{R}^{n-1}\setminus 0))$ have support in W_0 and are homogeneous of degree $-j$, and let $q_1(0,x',\xi')=-t(x',\xi')^2$ for some $t\in C^\infty$. Further assume that for some constant M*

$$(24.4.16)\quad \{p,q\}+qM|\xi'|=-\psi^2+\rho(\xi_1-r^{\frac{1}{2}})\text{ and }\quad q=v^2\quad\text{when } p=0,$$

where $\psi, v\in C^\infty(\mathbb{R}^n\times(\mathbb{R}^n\setminus 0))$ have support over W_0 and are homogeneous of degree $\frac{1}{2}$ and 0, while $\rho\in C^\infty(\mathbb{R}^n\times(\mathbb{R}^{n-1}\setminus 0))$ is homogeneous of degree 0 and $r>0$ in supp ρ, *which is contained in W_0. If u satisfies (24.4.1), $(24.4.15)_s$ and M is larger than some number depending on P and on s, it follows that $T(x',D')D_1u|_{x_1=0}\in H_{(s)}(\mathbb{R}^{n-1})$ if t is the principal symbol of T, and $\Psi_0(x,D')u+\Psi_1(x,D')D_1u\in\overline{H}_{(0,s)}$ for some $\Psi_j\in S^{\frac{1}{2}-j}$ with* cone supp $\Psi_j\subset W_0$ *and principal symbols $\partial^j\psi(0,x',0,\xi')/\partial\xi_1^j$ when $x_1=r(x,\xi')=0$.*

Proof. First note that for reasons of continuity the energy identity (24.4.5) is valid for all $u\in\overline{H}_{(2)}$. Choose $\chi\in C^\infty$ with cone supp $\chi\subset W_0$ homogeneous of degree 0 for $|\xi'|>1$, so that χ is then equal to 1 in a neighborhood of supp q_j for $j=0,1$. We shall apply (24.4.5) to

$$u_\varepsilon=(1+\varepsilon^2|D'|^2)^{-1}(1+|D'|^2)^{s/2}\chi(x,D')u,$$

taking $Q=Q_1(x,D')D_1+Q_0(x,D')$ where $Q_j(x,\xi')=q_j(x,\xi')$ when $|\xi'|>1$. By $(24.4.15)_s$ and Lemma 24.4.6 we have $u_\varepsilon\in\bar{H}_{(3,-1)}\subset\bar{H}_{(2)}$ when $\varepsilon>0$, and u_ε is bounded in $\bar{H}_{(1,-1)}$ as $\varepsilon\to 0$. The lemma will follow when $\varepsilon\to 0$ after appropriate estimates have been given for the terms in (24.4.5).

We start with studying Pu_ε. Writing

$$\Lambda_\varepsilon=(1+\varepsilon^2|D'|^2)^{-1}(1+|D'|^2)^{s/2}$$

we have

$$Pu_\varepsilon=f_\varepsilon+[P,\Lambda_\varepsilon\chi(x,D')]u.$$

Here $f_\varepsilon=\Lambda_\varepsilon\chi(x,D')f$ is bounded in $\bar{H}_{(0,\infty)}$ since $\chi(x,D')f\in\bar{H}_{(0,\infty)}$, and the second term is bounded in $\bar{H}_{(0,-1)}$ by $(24.4.15)_s$. To study $\operatorname{Im}(Pu_\varepsilon,Qu_\varepsilon)$ we must make a more detailed analysis so we write

$$([P,\Lambda_\varepsilon\chi(x,D')]u,Qu_\varepsilon)=(\Lambda_\varepsilon[P,\chi(x,D')]u,Qu_\varepsilon)$$
$$+([P,\Lambda_\varepsilon]\Lambda_\varepsilon^{-1}u_\varepsilon,Qu_\varepsilon).$$

The first term is bounded as $\varepsilon\to 0$ since $Q_j^*\Lambda_\varepsilon[P,\chi(x,D')]$ is of order $-\infty$ in D'. To deal with the second term we write

$$G^\varepsilon(x,D')=[P,\Lambda_\varepsilon]\Lambda_\varepsilon^{-1}=\Lambda_\varepsilon[\Lambda_\varepsilon^{-1},P]=(1+|D'|^2)^{s/2}[(1+|D'|^2)^{-s/2},P]$$
$$+\varepsilon^2(1+\varepsilon^2|D'|^2)^{-1}(1+|D'|^2)^{s/2}[|D'|^2,P](1+|D'|^2)^{-s/2}.$$

$G^\varepsilon(x,\xi')$ is bounded in S^1 since $(\varepsilon^{-2}+|\xi'|^2)^{-1}$ is bounded in S^{-2} when $0<\varepsilon<1$. The leading term is purely imaginary. Let

$$M+2\operatorname{Im}(G^\varepsilon(x,\xi')-p^s(x,\xi'))(1+|\xi'|^2)^{-\frac{1}{2}}\geqq 1$$

for $0<\varepsilon<1$ and all (x,ξ'), and set

$$S^\varepsilon(x,\xi')=(1+|\xi'|^2)^{\frac{1}{4}}(M+2\operatorname{Im}(G^\varepsilon(x,\xi')-p^s(x,\xi'))(1+|\xi'|^2)^{-\frac{1}{2}})^{\frac{1}{2}}$$

which is bounded in $S^{\frac{1}{2}}$ as $0<\varepsilon<1$. Since the symbol of

$$M(1+|D'|^2)^{\frac{1}{2}}+2G^\varepsilon(x,D')/i-2\operatorname{Im}p^s(x,D')-S^\varepsilon(x,D')^*S^\varepsilon(x,D')$$

is bounded in S^0, we have

$$2\operatorname{Im}(G^\varepsilon(x,D')u_\varepsilon,Qu_\varepsilon)\geqq\operatorname{Re}((2\operatorname{Im}p^s(x,D')-M(1+|D'|^2)^{\frac{1}{2}})u_\varepsilon,Qu_\varepsilon)$$
$$+\operatorname{Re}(S^\varepsilon(x,D')u_\varepsilon,QS^\varepsilon(x,D')u_\varepsilon)-C_1$$

for some C_1. Here we have also used that $[S^\varepsilon(x,D'),Q]$ is bounded of order $-\frac{1}{2}$. With $v_\varepsilon=(1+|D'|^2)^{-\frac{1}{4}}S^\varepsilon(x,D')u_\varepsilon$ the term involving S^ε can be written

$$\operatorname{Re}\sum(A_{jk}(x,D')D_1^k v_\varepsilon,D_1^j v_\varepsilon)$$

where

$$A_{11}=0,\qquad A_{00}=(1+|D'|^2)^{\frac{1}{4}}Q_0(x,D')(1+|D'|^2)^{\frac{1}{4}},$$
$$A_{10}=A_{01}=\tfrac{1}{2}(1+|D'|^2)^{\frac{1}{4}}Q_1(x,D')(1+|D'|^2)^{\frac{1}{4}}.$$

The principal symbol is $|\xi'|v^2$ when $p=0$, so we can apply Lemma 24.4.5. As already observed $\|u_\varepsilon\|_{(1,-1)}+\|Pu_\varepsilon\|_{(0,-1)}$ is bounded as $\varepsilon\to 0$. Hence

$\|v_\varepsilon\|_{(1,-1)}+\|Pv_\varepsilon\|_{(0,-1)}$ is bounded and so are the other terms in the right-hand side of (24.4.9) when u is replaced by v_ε and κ is large. In fact, $v_\varepsilon(0,.)$ is bounded in $H_{(\infty)}$. Summing up, we have proved that

$$2\operatorname{Im}(Pu_\varepsilon, Qu_\varepsilon) \geqq \operatorname{Re}((2\operatorname{Im} p^s(x,D') - M(1+|D'|^2)^{\frac{1}{2}})u_\varepsilon, Qu_\varepsilon) - C_2. \tag{24.4.17}$$

Next we shall derive an upper bound for the right-hand side of (24.4.5) applied to u_ε by using Lemma 24.4.5 to estimate

$$\begin{aligned} &\operatorname{Re}\sum (C_{jk}(x,D') D_1^k u, D_1^j u) \\ &+\operatorname{Re}((M(1+|D'|^2)^{\frac{1}{2}} - 2\operatorname{Im} p^s(x,D'))u_\varepsilon, Qu_\varepsilon) \\ &-\operatorname{Re}(\phi(x,D')(D_1 - \tilde{\Lambda}_+(x,D'))u_\varepsilon, \rho^0(x,D')u_\varepsilon). \end{aligned} \tag{24.4.18}$$

Here C_{jk} comes from Lemma 24.4.2, $\rho^0(x,\xi')\in C^\infty$ is equal to $\rho(x,\xi')$ when $|\xi'|>1$, and $\phi\in S^0$ is equal to 1 in a neighborhood of $\operatorname{supp}\rho^0$, cone supp $\phi \subset W_0$, and $r>0$ in cone supp ϕ. When $p(x,\xi)=0$ the corresponding principal symbol defined as in Lemma 24.4.5 is

$$\{p,q\} + qM|\xi'| - \rho(\xi_1 - r^{\frac{1}{2}}) = -\psi^2.$$

Hence the estimates for u_ε and Pu_ε already used in the proof of (24.4.17) show that (24.4.18) can be estimated above by

$$C_3 - \|\Psi_0(x,D')u_\varepsilon + \Psi_1(x,D')D_1 u_\varepsilon\|_X^2$$

where Ψ_j is related to ψ as in Lemma 24.4.5. Using (24.4.14) and the fact that $[\phi(x,D')(D_1-\tilde{\Lambda}_+(x,D')), \Lambda_\varepsilon\chi(x,D')]$ is bounded in S^s, we conclude that

$$\operatorname{Re}(\phi(x,D')(D_1 - \tilde{\Lambda}_+(x,D'))u_\varepsilon, \rho^0(x,D')u_\varepsilon) \leqq C_4. \tag{24.4.19}$$

Summing up, we have therefore proved that

$$\begin{aligned} &\|\Psi_0(x,D')u_\varepsilon + \Psi_1(x,D')D_1 u_\varepsilon\|_X^2 \\ &\leqq C_2 + C_3 + C_4 + \operatorname{Re}\sum (B_{jk}(x',D')D_1^k u_\varepsilon(0,.), D_1^j u_\varepsilon(0,.))_{\partial X}. \end{aligned} \tag{24.4.20}$$

Since $B_{11} + T(x',D')^* T(x',D')$ is of order -2, we have

$$\operatorname{Re}(B_{11}(x',D')u_{\varepsilon 1}, u_{\varepsilon 1})_{\partial X} + \|Tu_{\varepsilon 1}\|_{\partial X}^2 \leqq C\|u_{\varepsilon 1}(0,.)\|_{(-1)}^2 \leqq C_5.$$

The other boundary terms in (24.4.5) are bounded when $\varepsilon\to 0$ since $u_\varepsilon(0,.)$ is even bounded in C^∞ by the second part of (24.4.12). Thus

$$\|\Psi_0(x,D')u_\varepsilon + \Psi_1(x,D')D_1 u_\varepsilon\|_X^2 + \|Tu_{\varepsilon 1}\|_{\partial X}^2 \leqq C_6,$$

and the lemma follows when $\varepsilon\to 0$.

From Lemma 24.4.7 it follows that $D_1 u|_{x_1=0}$ is in $H_{(s-\frac{1}{2})}$ at (x',ξ') if $t(x',\xi')\neq 0$. If using different choices of q we obtain $\Psi_0(x,D')u + \Psi_1(x,D')D_1 u \in \bar{H}_{(0,s)}$ for two different systems Ψ_0, Ψ_1 of order $\frac{1}{2}$, $-\frac{1}{2}$ forming an elliptic system in cone supp χ, where $\chi\in S^0$, then it will follow that $\chi(x,D')u\in\bar{H}_{(0,s+\frac{1}{2})}$, $\chi(x,D')D_1 u\in\bar{H}_{(0,s-\frac{1}{2})}$ which is a local form of $(24.4.15)_{s+\frac{1}{2}}$. Thus the main point which remains is to construct operators satisfying the conditions in Lemma 24.4.7. This is fairly easy since (24.4.16) is only a condition on the restriction of q to the surface $p(x,\xi)=0$. At first we can therefore ignore the condition that q shall be linear in ξ_1. Before giving the construction we prove two technical lemmas which will enable us to take the square root t.

Lemma 24.4.8. *If $\chi \in C^\infty(\mathbb{R}^N)$ is non-negative and*

$$|D^\alpha \chi| \leqq C_{|\alpha|,k}\chi^k \tag{24.4.21}$$

for all multi-indices α and all $k<1$, it follows that $\chi^{\frac{1}{2}} \in C^\infty(\mathbb{R}^N)$. The set of all χ satisfying (24.4.21) with fixed $C_{|\alpha|,k}$ is convex and contains the origin; the product of two such functions has the same property with $C_{|\alpha|,k}$ replaced by $2^{|\alpha|}C_{|\alpha|,k}$.

Proof. In the set where $\chi \neq 0$ we can write $D^\alpha \chi^{\frac{1}{2}}$ as a linear combination of terms of the form

$$\chi^{\frac{1}{2}-j} D^{\alpha_1}\chi \dots D^{\alpha_j}\chi;$$

this follows at once by induction. Hence

$$|D^\alpha \chi^{\frac{1}{2}}| \leqq C'_{|\alpha|,k}\chi^{\frac{1}{2}-j(1-k)} = C'_{|\alpha|,k}\chi^{k'/2}$$

where $k'=1-2j(1-k)>0$ if $k>1-1/(2j)$. By Corollary 1.1.2 this proves that $\chi^{\frac{1}{2}} \in C^\infty$; in fact, estimates of the form (24.4.21) with other constants are valid for $\chi^{\frac{1}{2}}$. If $\chi=\lambda_1\chi_1+\lambda_2\chi_2$ where χ_j satisfy (24.4.21) and λ_j are non-negative reals with sum 1, then

$$|D^\alpha \chi| \leqq C_{|\alpha|,k}(\lambda_1\chi_1^k+\lambda_2\chi_2^k) \leqq C_{|\alpha|,k}(\lambda_1\chi_1+\lambda_2\chi_2)^k = C_{|\alpha|,k}\chi^k$$

by Hölder's inequality. This proves the convexity, and the last statement follows from Leibniz' formula.

Lemma 24.4.9. *If $\psi \in C^\infty(\mathbb{R}^N)$ and $\chi = \partial_1\psi$ satisfies (24.4.21) then*

$$(\psi(t_1+t_0,t_2,\dots,t_N)-\psi(t_1-t_0,t_2,\dots,t_N))/2t_0 = \phi(t_0^2,t_1,\dots,t_N)^2$$

where $\phi \in C^\infty(\mathbb{R}^{N+1})$.

Proof. The difference quotient can be written in the form

$$F(t_0,\dots,t_N) = \int_{-1}^{1} \chi(t_1+st_0,t_2,\dots,t_N)\,ds/2.$$

The integrand obviously satisfies (24.4.21) for fixed s with no change of the constants, for differentiation can only bring out powers of s. In view of the convexity proved in Lemma 24.4.8 it follows that F has the same property, hence $F^{\frac{1}{2}} \in C^\infty$. By Corollary C.4.4 this even function of t_0 can be written in the form $\phi(t_0^2,t_1,\dots,t_N)$ with $\phi \in C^\infty(\mathbb{R}^{N+1})$. The proof is complete.

An example of a function satisfying (24.4.21) is

$$\chi_0(t)=\exp(-1/t) \quad \text{if } t>0; \qquad \chi_0(t)=0 \quad \text{if } t\leqq 0.$$

We shall use χ_0 as a cutoff function below. Also χ_0' satisfies (24.4.21).

Proof of Theorem 24.4.1. We shall now construct functions q satisfying the conditions in Lemma 24.4.7. Let $(0,y',\eta')\in W_0$, $(y',\eta')\in G_d$, $|\eta'|=1$, and set

(24.4.22) $$\phi(x,\xi)=\phi_1(x,\xi')\,\xi_1+\phi_0(x,\xi'),$$

(24.4.23) $$\phi_1(x,\xi')=1/|\xi'|, \quad \phi_0(x,\xi')=x_1^2+|x'-y'|^2+|\xi'/|\xi'|-\eta'|^2.$$

Then ϕ is an increasing function of ξ_1 and

$$H_p\phi=\{\xi_1^2-r,\phi_1\,\xi_1+\phi_0\}=\xi_1(2\,\partial\phi_0/\partial x_1-\{r,\phi_1\})+\{\xi_1,r\}\,\phi_1-\{r,\phi_0\}.$$

In a conic neighborhood $W_{y',\eta'}\subset W_0$ of $(0,y',\eta')$ we have by (24.4.12) for some positive constants c and C

(24.4.24) $$\{\xi_1,r\}=\partial r/\partial x_1\geqq 4c|\xi'|^2,$$

$$\{r,\phi_0\}\leqq c|\xi'|, \quad |2\partial\phi_0/\partial x_1-\{r,\phi_1\}|\leqq C.$$

Hence $H_p\phi>0$ when $|\xi_1|\,C\leqq c|\xi'|$.

Choose δ so small that $(x,\xi')\in W_{y',\eta'}$ if $\phi_0(x,\xi')\leqq 3\delta$, and consider as a preliminary choice of q

$$f=\chi_2(\phi_0/\delta)^2\,\chi_0(1-\phi/\delta)$$

where χ_0 is defined above and $\chi_2\in C_0^\infty((-3,3))$ is non-negative and equal to 1 in $(-2,2)$. Later on we shall also need a cutoff function $\chi_1\in C_0^\infty((-2,2))$ which is equal to 1 in $(-1,1)$. In $\operatorname{supp} f$ we have $\phi\leqq\delta$, hence $\xi_1\leqq\delta|\xi'|$, and in $\operatorname{supp} f\cap\operatorname{supp} d\chi_2(\phi_0/\delta)$ we have $\phi_0+\phi_1\,\xi_1\leqq\delta$, $\phi_0\geqq 2\delta$, hence $\xi_1\leqq-\delta|\xi'|$. This implies $r\geqq\delta^2|\xi'|^2$ if $p(x,\xi)=0$. Thus

$$H_pf+fM|\xi'|=-\psi^2+\rho(\xi_1-r^{\frac12}) \quad \text{when} \quad p(x,\xi)=0$$

where

(24.4.25) $$\psi=\chi_1(\xi_1/\delta|\xi'|)\,N^{\frac12},$$

(24.4.26) $$-2r^{\frac12}\rho=-(1-\chi_1(\xi_1/\delta|\xi'|)^2)\,N+\chi_0(1-\phi/\delta)\,H_p\chi_2(\phi_0/\delta)^2,$$

evaluated for $\xi_1=-r^{\frac12}$ if $r>0$, and $\rho=0$ if $r\leqq 0$. Here we have written

(24.4.27) $$N=\chi_2(\phi_0/\delta)^2(\chi_0'(1-\phi/\delta)\,H_p\phi/\delta-\chi_0(1-\phi/\delta)\,M|\xi'|).$$

In the support of ψ we have $|\phi_1\xi_1|\leqq 2\delta$ and $\phi_0\leqq 3\delta$, hence $1-\phi/\delta\leqq 3$. When $t\leqq 3$ and $|10a|\leqq 1$, it is clear that

$$(1-a\chi_0(t)/\chi_0'(t))^{\frac12}=(1-at^2)^{\frac12}$$

is a C^∞ function $G(a,t)$. We can write

$$N^{\frac12}=\chi_2(\phi_0/\delta)(\chi_0'(1-\phi/\delta)\,H_p\phi/\delta)^{\frac12}\,G(M|\xi'|\,\delta/H_p\phi,1-\phi/\delta)$$

and conclude for small δ that $\psi\in C^\infty$. In the support of $(1-\chi_1^2)N$ we have $\xi_1\leqq-\delta|\xi'|$, hence $r\geqq\delta^2|\xi'|^2$, so it follows at once that ρ is also in C^∞ and that $r\geqq\delta^2|\xi'|^2$ in the support. With $v=\chi_1(\xi_1/|\xi'|)\chi_2(\phi_0/\delta)\chi_0(1-\phi/\delta)^{\frac{1}{2}}$ we have $v\in C^\infty$ and $v^2=f$ when $p=0$ if δ is small enough.

Now we define $q(x,\xi)$ as the remainder when f is divided by p using the Malgrange preparation theorem,

$$f(x,\xi)=p(x,\xi)\,g(x,\xi)+q_1(x,\xi')\,\xi_1+q_0(x,\xi').$$

Since $\{p,f\}=\{p,q\}$ when $p(x,\xi)=0$ the condition (24.4.16) follows. When $r(x,\xi')>0$ we have

$$q_1(x,\xi')=(f(x,\xi_1,\xi')-f(x,-\xi_1,\xi'))/2\xi_1,\quad \text{if } \xi_1^2=r(x,\xi').$$

Since $-\partial f/\partial\xi_1=\chi_2(\phi_0/\delta)^2\,\chi_0'(1-\phi/\delta)\,\phi_1/\delta$, it follows from Lemma 24.4.9 that the right-hand side is equal to $-F(\xi_1^2,x,\xi')^2$ where $F\in C^\infty$, so $q_1(x,\xi')=-F(r,x,\xi')^2$ when $r>0$. Without loss of (24.4.16) we may change q_1 to $-t^2$ where $t=F(r(x,\xi'),x,\xi')$, and have then verified all the hypotheses of Lemma 24.4.7. If ψ_j is the principal symbol of Ψ_j we have at (y',η') resp. $(0,y',\eta')$

$$t=(\chi_0'(1)\,\phi_1/\delta)^{\frac{1}{2}},\qquad \psi_0^2=\chi_0'(1)\,\partial r/\partial x_1\,\phi_1/\delta-\chi_0(1)\,M,$$
$$2\psi_0\,\psi_1=-\chi_0''(1)\,\partial r/\partial x_1\,\phi_1^2/\delta^2-\chi_0'(1)\,\{r,\phi_1\}/\delta+\chi_0'(1)\,M\phi_1/\delta.$$

Thus t is non-characteristic at (y',η') and ψ_1/ψ_0 is close to $-\phi_1\,\chi_0''(1)/2\chi_0'(1)\,\delta$ at $(0,y',\eta')$ if δ is small. This ratio is different for two different small choices of δ. By the discussion preceding Lemma 24.4.8 we have therefore proved $(24.4.15)_{s+\frac{1}{2}}$ if the support of χ is sufficiently close to (y',η').

To finish the proof we must establish a similar result at the H_p flowout of $G_d\cap W_0$ in W_0. This can be done by using the proof of Theorem 24.2.1 again. First we choose an operator $Q(x,D')$ of order 0 with support close to the bicharacteristic Γ going out from a point in $G_d\cap W_0$ and commuting approximately with $D_1-\Lambda_+$ there, except when x_1 is so small that one has entered the set already controlled; Q is then cut down to 0. Then

$$(D_1-\Lambda_+(x,D'))\,(Q(x,D')\,(D_1-\Lambda_-(x,D'))\,u\in\bar{H}_{(0,s-\frac{1}{2})}$$

by the result already proved. From Theorem 23.1.2 it follows therefore that $Q(x,D')(D_1-\Lambda_-(x,D'))\,u\in\bar{H}_{(0,s-\frac{1}{2})}$. Together with the already established fact that $\chi(x,D')(D_1-\tilde{\Lambda}_+(x,D'))\,u\in C^\infty$ when $\chi\in S^0$ and $r>0$ in cone supp $\chi\subset W_0$ this implies by standard elliptic theory that $(24.4.15)_{s+\frac{1}{2}}$ is valid when supp χ is in a sufficiently small conic neighborhood of a point on Γ. The proof of the theorem is now complete.

Remark. It is clear that the proof yields a more precise result on microlocal $H_{(s)}$ regularity. We have omitted it for the sake of brevity. (Cf. Theorem 26.1.4 for the corresponding interior regularity theorem.)

24.5. The General Propagation of Singularities

We shall begin by proving property (ii) in Lemma 24.3.15 for $WF_b(u)$ when u is a solution of the Dirichlet problem. Note that (ii) is a consequence of (i) in the diffractive set, so we can work in the entire glancing set omitting the subset where the gliding vector is zero since the generalized bicharacteristic through such a non-diffractive point is trivial by Corollary 24.3.10.

As in Section 24.4 we assume that P is of the form (24.4.2) with the coefficients of R of compact support. Let $u \in \mathcal{N}(X)$ be a solution of the boundary problem

$$Pu = f \quad \text{in} \quad X \setminus \partial X, \qquad u = u_0 \quad \text{in} \quad \partial X \tag{24.5.1}$$

where X is an open subset of $\overline{\mathbb{R}}^n_+$ and $\partial X = X \cap \partial \mathbb{R}^n_+$. We may assume that u has compact support in X and that $(x, \xi) \in WF(u)$, $x_1 > 0$, implies $\xi' \neq 0$. Let W be an open conic set in $X \times (\mathbb{R}^{n-1} \setminus 0)$ such that

$$\begin{gathered}(x, \xi') \in W, \quad x_1 \neq 0 \Rightarrow (x, \xi) \notin WF(f),\\ (0, x', \xi') \in W \Rightarrow (x', \xi') \notin WF_b(f) \cup WF(u_0).\end{gathered} \tag{24.5.2}$$

By K we denote a compact subset of $\{(x', \xi') \in G, (0, x', \xi') \in W\}$ such that

$$H_{r_0}(x', \xi') \quad \text{is not radial when } (x', \xi') \in K. \tag{24.5.3}$$

It will be convenient to assume that $|\xi'| = 1$ in K. For every point in K we can choose $N \in C^\infty(\mathbb{R}^{n-1} \times (\mathbb{R}^{n-1} \setminus 0))$ homogeneous of degree 0 so that $H_{r_0} N = |\xi'|$ in a conic neighborhood. Splitting K into a finite number of compact sets we may therefore assume that

$$\{r_0, N\} = |\xi'| \quad \text{in a conic neighborhood of } K. \tag{24.5.4}$$

From Lemma 24.4.6 it follows that for suitable s

$$\chi(x, D') u \in \bar{H}_{(0,s)} \quad \text{if } \chi \in S^0, \text{ cone supp } \chi \subset W. \tag{24.5.5$_s$}$$

Our purpose is to improve this at a point where there are fewer singularities near the backward gliding ray than asserted by condition (ii) in Lemma 24.3.15; the proof for the forward gliding ray is similar and is left as an exercise for the reader. (Recall that the gliding ray is the orbit of $-H_{r_0}$ where $r_0(x', \xi') = r(0, x', \xi')$.) In the following statement we use the notation

$$\begin{aligned} WF'(u) = \{(x, \xi');\ &(x', \xi') \in WF_b(u) \quad \text{if } x_1 = 0 \quad \text{and}\\ &(x, \xi_1, \xi') \in WF(u) \quad \text{for some } \xi_1 \quad \text{if } x_1 > 0\}.\end{aligned}$$

Proposition 24.5.1. *Assume that* (24.5.1)–(24.5.4) *are valid and that* $|\eta'| = 1$ *if* $(y', \eta') \in K$, *If* $\varepsilon > 0$ *it follows that there is some* $\delta_\varepsilon > 0$ *such that if* $(y', \eta') \in K$ *and for some* $\delta \in (0, \delta_\varepsilon)$

$$x_1 < \varepsilon\delta,\ |(x', \xi') - (y', \eta') - \delta H_{r_0}(y', \eta')| < \varepsilon\delta \Rightarrow (x, \xi') \notin WF'(u) \tag{24.5.6}$$

then $(y', \eta') \notin WF_b(u)$.

Since $\exp(\delta H_{r_0})(y',\eta')=(y',\eta')+\delta H_{r_0}(y',\eta')+O(\delta^2)$ it follows from (24.5.6) that for small enough δ

$$(24.5.6)' \qquad x_1<\varepsilon\delta,\ |(x',\xi')-\exp(\delta H_{r_0})(y',\eta')|<\varepsilon\delta/2 \Rightarrow (x,\xi')\notin WF'(u).$$

This form of the condition will be more convenient below.

The main point in the proof is a slight variation of Lemma 24.4.7.

Lemma 24.5.2. *Let $q, v, b, e\in C^\infty(\mathbb{R}^n\times(\mathbb{R}^{n-1}\setminus 0))$ be homogeneous functions of degree $0, 0, \frac{1}{2}, \frac{1}{2}$ respectively, all with support in W, satisfying*

$$(24.5.7) \qquad \{p,q\}+qM|\xi'|=-\psi^2-b^2+e^2,\ q=v^2 \quad \text{when} \quad p=0.$$

Here $\psi\in C^\infty(\mathbb{R}^n\times(\mathbb{R}^n\setminus 0))$ is homogeneous of degree $\frac{1}{2}$ and has support over W, and M is a constant. Choose $B, E\in C^\infty$ equal to b and e outside a compact set. If (24.5.1)–(24.5.4), (24.5.5)$_s$ *hold and M is larger than some constant depending on P and s, it follows that $B(x,D')u\in\bar{H}_{(0,s)}$ if $E(x,D')u\in\bar{H}_{(0,s)}$.*

Proof. We follow the proof of Lemma 24.4.7 closely, with q independent of ξ_1 now. First we define u_ε as there. The proof of (24.4.17) requires no change. Instead of (24.4.18) we consider the form

$$\operatorname{Re}\sum(C_{jk}(x,D')D_1^k u_\varepsilon, D_1^j u_\varepsilon)+\operatorname{Re}((M(1+|D'|^2)^{\frac{1}{2}}-2\operatorname{Im} p^s(x,D'))u_\varepsilon, Qu_\varepsilon) + \|B(x,D')u_\varepsilon\|^2-\|E(x,D')u_\varepsilon\|^2$$

which has the principal symbol $-\psi(x,\xi)^2$ when $p(x,\xi)=0$. Since $u_\varepsilon(0,.)$ is bounded in $H_{(\kappa)}$ for every κ we conclude as before that this is bounded above as $\varepsilon\to 0$. (We drop the positive contributions which could be obtained from ψ as in Lemma 24.4.7, for the term involving B is now the essential one.) The boundary terms are all bounded since $Q_1=0$. A new point is that

$$E(x,D')u_\varepsilon=\Lambda_\varepsilon\chi(x,D')E(x,D')u+[E(x,D'),\Lambda_\varepsilon\chi(x,D')]u$$

is bounded in L^2 as $\varepsilon\to 0$ because $E(x,D')u\in\bar{H}_{(0,s)}$ by hypothesis and $[E(x,D'),\Lambda_\varepsilon\chi(x,D')]$ is bounded in $S^{s-\frac{1}{2}}$ as $\varepsilon\to 0$. Thus we have proved that $\|B(x,D')u_\varepsilon\|$ is bounded which by the same argument implies that $\|\Lambda_\varepsilon\chi(x,D')B(x,D')u\|$ is bounded when $\varepsilon\to 0$. Hence $\chi(x,D')B(x,D')u\in\bar{H}_{(0,s)}$ which proves the lemma if χ is taken equal to 1 in $\operatorname{supp} B$.

Proof of Proposition 24.5.1. For every $(y',\eta')\in K$ we choose a function $\omega(x',\xi')=\omega(x',\xi';y',\eta')$ in a conic neighborhood by solving the Cauchy problem

$$(24.5.8) \quad H_{r_0}\omega=0;\ \omega(x',\xi')=|x'-y'|^2+|\xi'/|\xi'|-\eta'|^2 \quad \text{for } N(x',\xi')=N(y',\eta').$$

By (24.5.4) there is a unique solution when $|x'-y'|^2+|\xi'/|\xi'|-\eta'|^2$ is small enough. It vanishes of second order on the bicharacteristic of r_0 through

(y', η'), hence on the cone it generates, but the Hessian is positive in directions transversal to it since this is true at (y', η'). Since r_0 vanishes on the bicharacteristic we obtain

$$|r_0| \leqq C\omega^{\frac{1}{2}}|\xi'|^2. \tag{24.5.9}$$

Similarly we have

$$|\partial\omega/\partial x'|^2 + |\partial\omega/\partial\xi'|^2|\xi'|^2 \leqq C\omega, \tag{24.5.10}$$

which is also a consequence of Lemma 7.7.2 since $\omega \geqq 0$. With

$$\phi(x, \xi') = N(y', \eta') - N(x', \xi') + x_1/\varepsilon + \omega(x', \xi')/\delta\varepsilon^2 \tag{24.5.11}$$

we obtain

$$H_p\phi = H_r N + 2\xi_1/\varepsilon - H_r\omega/\delta\varepsilon^2.$$

By (24.5.8) and (24.5.10)

$$H_r\omega = (H_r - H_{r_0})\omega = O(x_1)O(|\xi'|\omega^{\frac{1}{2}}),$$

and on the characteristic set we have

$$\xi_1^2 = r \leqq r_0 + O(x_1|\xi'|^2) = O(\omega^{\frac{1}{2}} + x_1)|\xi'|^2.$$

When $N(x', \xi') - N(y', \eta') \leqq 2\delta$ and $\phi(x, \xi') \leqq 2\delta$, we have $x_1 \leqq 4\delta\varepsilon$ and $\omega(x', \xi') < 4\delta^2\varepsilon^2$. For small δ it follows that

$$H_p\phi = \{r, N\} + O((\delta/\varepsilon)^{\frac{1}{2}})|\xi'|, \quad \text{if } p(x, \xi) = 0,\ \phi(x, \xi') \leqq 2\delta \text{ and } N(x', \xi') - N(y', \eta') \leqq 2\delta. \tag{24.5.12}$$

The first term is larger than $|\xi'|/2$ by (24.5.4) if δ is small so it dominates then.

As in the proof of Theorem 24.4.1 we set

$$\chi_0(t) = \exp(-1/t) \quad \text{if } t > 0; \ \chi_0(t) = 0 \quad \text{if } t \leqq 0.$$

Let $\chi_1(t) = 0$ for $t < 0$, $\chi_1(t) = 1$ for $t > 1$ and $0 \leq \chi_1' \in C_0^\infty((0, 1))$. We choose χ_1' satisfying (24.4.21) and conclude that χ_1 does so too. For $(y', \eta') \in K$, a small ε, $\delta \ll \varepsilon$, and $0 \leq t \leq 1$, we now introduce

$$q_t(x, \xi') = \chi_0(1 + t - \phi/\delta)\chi_1((N(y', \eta') - N(x', \xi') + \delta)/\varepsilon\delta + t). \tag{24.5.13}$$

It is clear that q_t is homogeneous of degree 0. The support of q_t is in the interior of that of q_T when $t < T$ because χ_0 and χ_1 are increasing. In the support of the first factor we have $\phi \leqq \delta(1 + t) \leqq 2\delta$, and in the support of the second one we have $N(y', \eta') - N(x', \xi') + \delta \geqq -t\varepsilon\delta$. It is equal to 1 when $N(y', \eta') - N(x', \xi') + \delta \geqq (1 - t)\varepsilon\delta$ so it constant except in a narrow strip. Set

$$e_t = (\chi_0(1 + t - \phi/\delta)\chi_1'((N(y', \eta') - N(x', \xi') + \delta)/\varepsilon\delta + t)\{r, N\}/\varepsilon\delta)^{\frac{1}{2}}. \tag{24.5.14}$$

This is a C^∞ function homogeneous of degree $\frac{1}{2}$, for $\chi_0^{\frac{1}{2}}$ and $\chi_1'^{\frac{1}{2}}$ are C^∞ functions and $\{r, N\} > 0$ by (24.5.4). If $(x, \xi') \in \operatorname{supp} e_t$ then we have $|N(y', \eta') - N(x', \xi') + \delta| \leqq \varepsilon\delta$, hence $N(x', \xi') - N(y', \eta') \leqq 2\delta$, so $x_1 \leqq 4\varepsilon\delta$ and

$\omega(x',\xi') \leqq 4\varepsilon^2\delta^2$. If $(y'(\delta),\eta'(\delta)) = \exp(\delta H_{r_0})(y',\eta')$ it follows from (24.5.4) that

$$|N(y'(\delta),\eta'(\delta)) - N(x',\xi')| \leqq \varepsilon\delta + O(\delta^2).$$

For small enough δ we obtain

$$x_1^2 + |x'-y'(\delta)|^2 + |\xi'|\eta'(\delta)|/|\xi'| - \eta'(\delta)|^2 \leqq C^2(\varepsilon\delta)^2 \tag{24.5.15}$$

since $|N(x',\xi') - N(y'(\delta),\eta'(\delta))|^2 + \omega(x',\xi')$ is equivalent to the square of the distance from $(x',\xi'/|\xi'|)$ to the ray through $(y'(\delta),\eta'(\delta))$. Assuming that (24.5.6)$'$ is valid with ε replaced by $2C\varepsilon$, we conclude that $\operatorname{supp} e_t \cap WF'(u) = \emptyset$, hence $E_t(x,D')u \in C^\infty$.

To apply Lemma 24.5.2 we must also choose b_t and ψ_t so that

$$\chi_1(\chi_0'(1+t-\phi/\delta) H_p\phi/\delta - M|\xi'|\chi_0(1+t-\phi/\delta)) = \psi_t^2 + b_t^2. \tag{24.5.16}$$

Here the argument of χ_1 is the same as in (24.5.13). Recalling (24.5.12) we set

$$b_t(x,\xi') = (\chi_1\chi_0'(1+t-\phi/\delta)|\xi'|/2\delta)^{\frac{1}{2}}, \tag{24.5.17}$$

$$\psi_t(x,\xi) = \chi_1^{\frac{1}{2}}(\chi_0'(1+t-\phi/\delta)(H_p\phi - |\xi'|/2)/\delta - \chi_0(1+t-\phi/\delta) M|\xi'|)^{\frac{1}{2}}. \tag{24.5.18}$$

It is clear that $b_t \in C^\infty$. In the support of q_t we have $1+t-\phi/\delta \leqq 4$, and $H_p\phi \geqq 2|\xi'|/3$ in the intersection with the characteristic set by (24.5.12). Since $\chi_0(s)/\chi_0'(s) = s^2$ we obtain for small δ that $\psi_t \in C^\infty$ in a conic neighborhood of the characteristic set, for we can first break out the smooth factor $\chi_0'(1+t-\phi/\delta)^{\frac{1}{2}}|\xi'/\delta|^{\frac{1}{2}}$. We can multiply ψ_t by a cutoff function with support in this neighborhood which is 1 in the characteristic set.

Having verified all the hypotheses of Lemma 24.5.2 we conclude that $B_t(x,D')u \in \bar{H}_{(s+\frac{1}{2})}$. Let

$$W_t = \{(x,\xi'); b_t(x,\xi') > 0\}.$$

This is a conic neighborhood of $\operatorname{supp} q_{t'}$ if $t' < t$, and we have $(24.5.5)_{s+\frac{1}{2}}$ when cone $\operatorname{supp}\chi \subset W_t$. For any $t<1$ and any integer k we can apply this argument k times with t replaced by $t_1,\dots,t_k$ where $t<t_k<\dots<t_1<1$. Then we obtain $(24.5.5)_{s+\frac{1}{2}k}$ when cone $\operatorname{supp}\chi \subset W_t$, so $\chi(x,D')u \in C^\infty$ then in view of Lemma 24.4.6. By Theorem 18.3.32 it follows that $(y',\eta') \notin WF_b(u)$, which completes the proof.

Using Lemma 24.3.15 we can now sum up (24.2.4), Theorems 24.2.1 and 24.4.1, and Proposition 24.5.1 as follows:

Theorem 24.5.3. *Let P be a second order differential operator with real principal symbol p and C^∞ coefficients defined in a C^∞ manifold X with non-characteristic boundary ∂X. Let $u \in \mathcal{N}(X)$ be a solution of the boundary problem (24.5.1) with $f \in \mathcal{N}(X)$ and $u_0 \in \mathscr{D}'(\partial X)$. Every $\gamma_0 \in WF_b(u) \setminus (WF_b(f) \cup WF(u_0))$ is then either a characteristic of P in the interior of X or else a point in the hyperbolic or the glancing set $\subset T^*(\partial X)$. An open interval*

$(-T, T)\ni t\mapsto\tilde{\gamma}(t)$ with $\tilde{\gamma}(0)=\gamma_0$ on a compressed generalized bicharacteristic is contained in $WF_b(u)$.

Note that the last statement is empty if $H_p(\gamma_0)$ is radial or if $H_p^G(\gamma_0)$ is radial and γ_0 is non-diffractive. The generalized bicharacteristic is then just a ray. Otherwise singularities continue to move along a generalized bicharacteristic as long as it stays in a compact set, does not reach the singularities of the data f, u_0 and does not converge to a singular point of the type just described. To show that this result is optimal one should extend Theorem 24.2.3 to generalized bicharacteristics. Only a weaker result is known though:

Theorem 24.5.4. *Let P be a second order differential operator with real principal symbol p and C^∞ coefficients defined in a C^∞ manifold X with non-characteristic boundary ∂X. Let $[a,b]\ni t\mapsto\tilde{\gamma}_\nu(t)$ be compressed broken bicharacteristic arcs nowhere tangential to the fibers of $T^*(X)$, converging as $\nu\to\infty$ to a compressed generalized bicharacteristic $[a,b]\ni t\mapsto\tilde{\gamma}(t)$ which has injective projection to the compressed cosphere bundle and end points $\tilde{\gamma}(a)$, $\tilde{\gamma}(b)$ over X°. Then one can find $u\in\mathcal{N}(X)$ so that $WF_b(u)$ is the cone generated by $\tilde{\gamma}([a,b])$, $WF_b(Pu)$ is generated by $\tilde{\gamma}(\{a,b\})$, and $u=0$ on ∂X.*

Proof. It is enough to find $u\in C(X)$ vanishing on ∂X such that $WF_b(Pu)$ is contained in the cone Γ' generated by $\tilde{\gamma}(a)$ and $\tilde{\gamma}(b)$ while $WF_b(u)$ is contained in the cone Γ generated by $\tilde{\gamma}([a,b])$ but not in Γ'. Indeed, then $u\in\mathcal{N}(X)$ by Corollary 18.3.31, and it follows from Theorem 24.5.3 that $WF_b(u)=\Gamma$, $WF_b(Pu)=\Gamma'$. The functional analytic setup in the proof of Theorem 8.3.8 can still be used here. Thus we denote by $\mathcal{F}$ the Fréchet space of all $u\in C_0(K)$ vanishing on ∂X with $WF_b(Pu)\subset\Gamma'$ and $WF_b(u)\subset\Gamma$, where $K\subset X$ is a compact neighborhood of the projection of $\tilde{\gamma}([a,b])$ in X. The topology can be defined by semi-norms of the following types:

$$u\mapsto\sup|u|; \tag{24.5.19}$$

(24.5.20) $u\mapsto\sup|APu|$ where $A\in\Psi^\infty(\mathbb{R}^n_+)$ is properly supported and $\tilde{\gamma}(\{a,b\})\cap WF(A)=\emptyset$;

(24.5.21) $u\mapsto\sup|Au|$ where $A\in\Psi^\infty(\mathbb{R}^n_+)$ is properly supported and $\tilde{\gamma}([a,b])\cap WF(A)=\emptyset$;

(24.5.22) $u\mapsto\sup_{K'}|D^\alpha Pu|$ where α is any multi-index and K' is a compact subset of K with the projections of $\tilde{\gamma}(a)$ and $\tilde{\gamma}(b)$ removed;

(24.5.23) $u\mapsto\sup|a(x,D')u|$ where $a\in S^\infty(\mathbb{R}^n\times\mathbb{R}^{n-1})$ vanishes for large $|x|$ and is of order $-\infty$ in a conic neighborhood of the projection of $\tilde{\gamma}([a,b])$ in $\mathbb{R}^n\times\mathbb{R}^{n-1}$.

Let $K_0\subset K$ be a closed ball with center at a point x_0 in the projection of Γ such that K_0 and the projection of Γ' are disjoint. The theorem will be proved if we show that there is some $u\in\mathcal{F}$ which is not in $C^1(K_0)$. If every

$u \in \mathscr{F}$ is in $C^1(K_0)$ it follows by the closed graph theorem that when $|\alpha|=1$

$$\sup_{K_0} |D^\alpha u| \leqq C \sum_1^5 N_j(u), \quad u \in \mathscr{F}, \tag{24.5.24}$$

where N_j are semi-norms in $\mathscr{F}$ of the type (24.5.18+j). We shall complete the proof by showing that (24.5.24) cannot be valid.

Choose ν so large that the conditions imposed on the seminorms of type (24.5.20)–(24.5.23) occurring in (24.5.24) remain valid with $\tilde{\gamma}$ replaced by $\tilde{\gamma}_\nu$. We shall apply (24.5.24) to functions U_ε constructed as in the proof of Theorem 24.2.3 for $\tilde{\gamma}_\nu$, but with a factor $e^{-\varepsilon\lambda}$ inserted in (24.2.10). (We may assume that $a^\pm(x,\lambda)$ has support in K for every λ.) This new factor is bounded in S^0 as $\varepsilon \to 0$. The proof of Theorem 24.2.3 therefore yields uniform bounds for $N_j(U_\varepsilon)$ as $\varepsilon \to 0$ if $2 \leqq j \leqq 5$. It is also obvious that $\sup|U_\varepsilon|$ is uniformly bounded. Assuming that x_0 is only the projection of a single point in $\tilde{\gamma}_\nu([a,b])$ we obtain from the discussion of $D_1(U^+ - U^-)$ in the proof of Theorem 24.2.3 (preceding the statement) that for some α with $|\alpha| = 1$ we have $|D^\alpha U_\varepsilon(x_0)| \to \infty$ as $\varepsilon \to 0$. Thus (24.5.24) cannot hold, and the theorem is proved.

Not every generalized bicharacteristic is a limit of broken bicharacteristics. Indeed, if $(y',\eta') \in G_d$ and $r_0 = 0$ in a neighborhood, then there can be no reflection point nearby. Since $r(x,\xi') > 0$ when $x_1 > 0$ in a neighborhood of $(0,y',\eta')$ it follows that $\xi_1 \neq 0$ on a bicharacteristic so x_1 cannot have a minimum other than 0 on a bicharacteristic. In Section 24.6 we shall prove by a special argument that nevertheless there is a solution which is singular on a given bicharacteristic of this type. It is not known if there are other generalized bicharacteristics which are not limits of broken bicharacteristics and give rise to a gap between Theorems 24.5.3 and 24.5.4. However, in Example 24.3.11 it is clear that both the polygonal and the gliding bicharacteristic constructed through the point in G^∞ are limits of broken bicharacteristics. Hence there is no certain way of predicting the future path of a singularity in general.

24.6. Operators Microlocally of Tricomi's Type

In this section we shall discuss situations where the results of Section 24.5 are inadequate. Both are related to the Tricomi operator

$$D_1^2 - x_1(D_2^2 + \ldots + D_n^2)$$

in the hyperbolic or elliptic half spaces defined respectively by $x_1 > 0$ and $x_1 < 0$. In the first case Theorem 24.5.4 is not applicable since all the bicharacteristics have cusps when $x_1 = 0$ (see Fig. 6). In the second case the

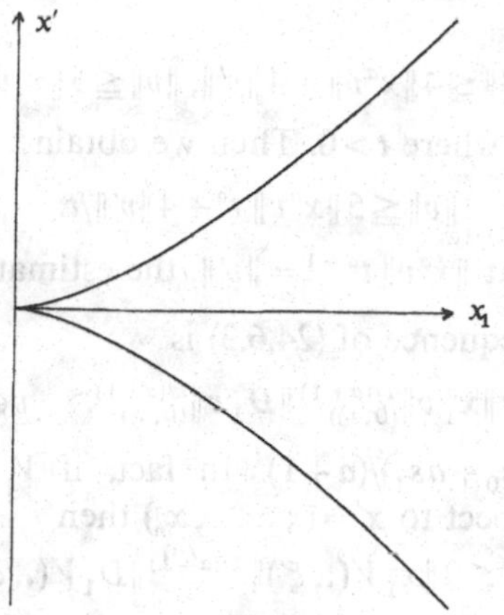

Fig. 6

Propagation Theorem 24.5.3 is empty but we shall see that there is hypoellipticity. We shall discuss this situation first.

Definition 24.6.1. The hypoelliptic set $HE \subset T^*(\partial X)\setminus 0$ is the intersection of the gliding set and the complement of the closure of the hyperbolic set.

If X as usual is defined by $x_1 \geqq 0$ and the principal symbol p of P is of the form $\xi_1^2 - r(x, \xi')$ then

(24.6.1) $$HE = \{(x', \xi');\, \partial r(0, x', \xi')/\partial x_1 < 0,\ r(0, y', \eta') \leqq 0 \text{ in a neighborhood of } (x', \xi')\}.$$

The terminology is justified by the following analogue of (24.2.4)

Theorem 24.6.2. *Let $u, f \in \mathcal{N}(X)$ and $Pu = f$, $u = u_0$ on ∂X. Then*

(24.6.2) $$HE \cap WF_b(u) = HE \cap (WF_b(f) \cup WF(u_0)).$$

The proof of the estimates leading to Theorem 24.6.2 will require the following lemma:

Lemma 24.6.3. *If $v \in \bar{C}_0^\infty(\mathbb{R}_+)$ and $a > 0$ then*

(24.6.3) $$\|v\| \leqq 9\, \|x^a v\|^{1/(a+1)}\, \|v'\|^{a/(a+1)}$$

where $\|\ \|$ is the L^2 norm.

Proof. Let $\chi(x) = 1$ for $x \leqq 1$, $\chi(x) = 2 - x$ for $1 \leqq x \leqq 2$ and $\chi(x) = 0$ for $x \geqq 2$, and set $w = \chi v$. Then $\|v - w\| \leqq \|x^a v\|$ and

$$\|w'\| \leqq \|\chi' v\| + \|\chi v'\| \leqq \|x^a v\| + \|v'\|.$$

Since

$$\|w\|^2 = \int_0^\infty |w|^2\, dt = -2 \operatorname{Re} \int_0^\infty t w \bar{w}'\, dt \leqq 4 \|w\|\, \|w'\|$$

it follows that

$$\|w\| \leqq 4\|w'\| \leqq 4\|x^a v\| + 4\|v'\|, \ \|v\| \leqq 5\|x^a v\| + 4\|v'\|.$$

Now replace v by $v(./t)$ where $t>0$. Then we obtain

$$\|v\| \leqq 5\|x^a v\| \, t^a + 4\|v'\|/t.$$

When t is chosen so that $\|x^a v\| \, t^{a+1} = \|v'\|$, the estimate (24.6.3) follows.

An immediate consequence of (24.6.3) is

$$\|v\|_{(0,s)} \leqq 9\|x_1^a v\|_{(0,s_0)}^{1/(a+1)} \|D_1 v\|_{(0,s_1)}^{a/(a+1)}, \quad v \in \bar{C}_0^\infty(\mathbb{R}^n_+), \tag{24.6.4}$$

where $s_j \in \mathbb{R}$ and $s=(s_0+as_1)/(a+1)$. In fact, if V is the partial Fourier transform of v with respect to $x'=(x_2,\dots,x_n)$ then

$$\|V(.,\xi')\| \leqq 9\|x_1^a V(.,\xi')\|^{1/(a+1)} \|D_1 V(.,\xi')\|^{a/(a+1)}$$

by (24.6.3). If we square, multiply by

$$(1+|\xi'|^2)^s = (1+|\xi'|^2)^{s_0/(a+1)} (1+|\xi'|^2)^{s_1 a/(a+1)}$$

and integrate with respect to ξ', then (24.6.4) follows by Hölder's inequality. Similarly we obtain

$$\|v(0,.)\|_{(s)}^2 \leqq 2\|v\|_{(0,s_0)} \|D_1 v\|_{(0,s_1)}, \quad s=(s_0+s_1)/2, \tag{24.6.5}$$

if we start from the estimate

$$|v(0)|^2 = -2\,\mathrm{Re} \int_0^\infty v\bar{v}'\,dt \leqq 2\|v\| \, \|v'\|, \quad v \in \bar{C}_0^\infty(\mathbb{R}_+).$$

Lemma 24.6.4. *Let $p(x,\xi)=\xi_1^2 - r(x,\xi')$ where $r \in S^2$ and*

$$r(x,\xi') \leqq -c x_1 |\xi'|^2, \ 0<x_1<1, \tag{24.6.6}$$

for some $c>0$. If $u \in \bar{C}_0^\infty(\mathbb{R}^n_+)$ and $x_1<1$ in $\operatorname{supp} u$, it follows that

$$\sum_0^2 \|D_1^{2-j} u\|_{(0,2j/3)} \leqq C(\|p(x,D)u\|_{(0,0)} + \|u\|_{(0,0)} + \|u(0,.)\|_{(5/3)}). \tag{24.6.7}$$

Proof. An integration by parts gives

$$(p(x,D)u, D_1^2 u) = \|D_1^2 u\|^2 - (r(x,D')D_1 u, D_1 u) + ([r(x,D'),D_1]u, D_1 u) - (r(0,x',D')u, \partial_1 u),$$

where the last scalar product is in L^2 on the boundary plane defined by $x_1=0$. Since it follows from (24.6.5) that $\|\partial_1 u(0,.)\|_{(\frac{1}{3})}$ can be estimated in terms of the left-hand side of (24.6.7), we estimate this term by $\|u(0,.)\|_{(\frac{5}{3})} \|\partial_1 u(0,.)\|_{(\frac{1}{3})}$. We have $[r(x,D'),D_1] = C(x,D')$ where C is of second order and $\mathrm{Re}\, C$ is of first order. Thus

$$(C(x,D')u, D_1 u) + (D_1 u, C(x,D')u) = ([D_1, C(x,D')]u, u) + ((C(x,D') + C(x,D')^*) D_1 u, u) - i(C(0,x',D')u, u)$$

where the last scalar product is in the boundary. Since $[D_1, C]$ is of second order and $C+C^*$ is of first order we obtain when taking real parts

(24.6.8) $$\|D_1^2 u\|^2 - 2\operatorname{Re}(r(x,D')D_1 u, D_1 u) \leqq \|p(x,D)u\|^2 + C((\|u\|_{(0,1)} + \|D_1 u\|_{(0,0)})\|u\|_{(0,1)} + \|u(0,.)\|_{(\frac{2}{3})}\|D_1 u(0,.)\|_{(\frac{1}{3})}).$$

Now we use the hypothesis (24.6.6). By the sharp Gårding inequality (Theorem 18.1.14) applied for fixed x_1 we have

$$c\|x_1^{\frac{1}{2}} D_1 u\|^2_{(0,1)} \leqq -\operatorname{Re}(r(x,D')D_1 u, D_1 u) + C\|D_1 u\|^2_{(0,\frac{1}{2})}.$$

Hence

(24.6.9) $$\|D_1^2 u\|^2 + c\|x_1^{\frac{1}{2}} D_1 u\|^2_{(0,1)} \leqq \|p(x,D)u\|^2 + C(\|u\|^2_{(0,1)} + \|D_1 u\|^2_{(0,\frac{1}{2})} + \|u(0,.)\|_{(\frac{2}{3})}\|D_1 u(0,.)\|_{(\frac{1}{3})}).$$

Now (24.6.4) shows that we can estimate $\|D_1 u\|^2_{(0,\frac{2}{3})}$ by means of the left-hand side, and (24.6.5) allows us to estimate $\|D_1 u(0,.)\|_{(\frac{1}{3})}$ by $\|D_1^2 u\| + \|D_1 u\|_{(0,\frac{2}{3})}$. Hence it follows with a new constant C that

(24.6.10) $$\|D_1^2 u\|^2 + \|D_1 u\|^2_{(0,\frac{2}{3})} \leqq C(\|p(x,D)u\|^2 + \|u\|^2_{(0,1)} + \|D_1 u\|^2 + \|u(0,.)\|^2_{(\frac{2}{3})}).$$

To estimate the term with $j=2$ in the left-hand side of (24.6.7) we shall use the equation $r(x,D')u = D_1^2 u - p(x,D)u$. The sharp Gårding inequality gives in view of (24.6.6) if we now regard x_1 as a parameter

$$c\|x_1^{\frac{1}{2}} u\|^2_{(2)} \leqq \operatorname{Re}(-r(x,D')u, |D'|^2 u) + C\|u\|^2_{(\frac{3}{2})}.$$

If we multiply by x_1, integrate with respect to x_1 also, and use the inequality $ab \leqq \frac{1}{2}(ca^2 + b^2/c)$, it follows that

$$c\|x_1 u\|^2_{(0,2)} \leqq \frac{1}{c}\|r(x,D')u\|^2_{(0,0)} + 2C\|x_1^{\frac{1}{2}} u\|^2_{(0,\frac{3}{2})}.$$

By Cauchy-Schwarz' inequality

$$\|x_1^{\frac{1}{2}} u\|^2_{(0,\frac{3}{2})} \leqq \|x_1 u\|_{(0,2)}\|u\|_{(0,1)}$$

so we have with another constant C

$$\|x_1 u\|^2_{(0,2)} \leqq C(\|r(x,D')u\|^2_{(0,0)} + \|u\|^2_{(0,1)}).$$

Since $r(x,D')u = D_1^2 u - p(x,D)u$ it follows that we may add $\|x_1 u\|^2_{(0,2)}$ on the left-hand side of (24.6.10), with an appropriate change of C. Since by (24.6.4)

$$\|u\|^2_{(0,\frac{4}{3})} \leqq C(\|D_1 u\|^2_{(0,\frac{2}{3})} + \|x_1 u\|^2_{(0,2)})$$

we have finally proved that

$$\sum_0^2 \|D_1^{2-j} u\|^2_{(0,2j/3)} \leqq C(\|p(x,D)u\|^2 + \|u\|^2_{(0,1)} + \|D_1 u\|^2 + \|u(0,.)\|^2_{(\frac{2}{3})}).$$

Now $\|u\|_{(0,1)} \leqq \|u\|^{\frac{1}{4}}_{(0,0)}\|u\|^{\frac{3}{4}}_{(0,\frac{4}{3})}$, and $\|D_1 u\|^2 \leqq 2\|D_1^2 u\|\,\|u\|$ by the proof of Lemma 17.5.2. The middle terms on the right-hand side can thus be replaced by a small constant times the left-hand side plus a large constant times $\|u\|^2$. This completes the proof of Lemma 24.6.4.

For any $s \in \mathbb{R}$ we have more generally

$$(24.6.7)' \quad \sum_0^2 \|D_1^{2-j} u\|_{(0,s+2j/3)} \leqq C(\|p(x,D)u\|_{(0,s)} + \|u\|_{(0,s)} + \|u(0,.)\|_{(s+\frac{5}{3})}).$$

In fact, if we apply (24.6.7) to $(1+|D'|^2)^{s/2} u$ and use that

$$[p(x,D),(1+|D'|^2)^{s/2}] = [-r(x,D'),(1+|D'|^2)^{s/2}]$$

is of order $s+1$, we obtain (24.6.7)' with an additional term $\|u\|_{(0,s+1)}$ in the right-hand side. Since

$$\|u\|_{(0,s+1)} \leqq \|u\|^{\frac{3}{4}}_{(0,s+\frac{4}{3})} \|u\|^{\frac{1}{4}}_{(0,s)}$$

the estimate (24.6.7)' follows.

Proof of Theorem 24.6.2. If $(x_0', \xi_0') \in HE$ then the principal symbol $p(x,\xi) = \xi_1^2 - r(x,\xi')$ of P satisfies (24.6.6) in a neighborhood of $(0, x_0', \xi_0')$. Assume for a moment that (24.6.6) is valid when $0 < x_1 < 1$, with $r \in S^2$ now, and that $u_0 \in C^\infty$,

$$u \in \overline{H}^{\mathrm{comp}}_{(2,s-1)}(\mathbb{R}^n_+), \qquad Pu \in \overline{H}^{\mathrm{comp}}_{(0,s)}(\mathbb{R}^n_+).$$

Let $\phi \in C_0^\infty(\mathbb{R}^{n-1})$, $\int \phi \, dx' = 1$, and let $A_\varepsilon = \hat{\phi}(\varepsilon D')$ be convolution with $\phi(x'/\varepsilon)\varepsilon^{1-n}$ in the x' variables. Then $A_\varepsilon u$ has support in a fixed compact set for small ε, and $A_\varepsilon u \in \overline{H}^{\mathrm{comp}}_{(2,\infty)}(\mathbb{R}^n_+)$ so (24.6.7)' can be applied to u for reasons of continuity. We have $p(x,D)u \in \overline{H}^{\mathrm{comp}}_{(0,s)}(\mathbb{R}^n_+)$ and

$$p(x,D)(A_\varepsilon u) = A_\varepsilon p(x,D)u - [r(x,D), A_\varepsilon] u$$

where the commutator is uniformly bounded in $\operatorname{Op} S^1$. Hence $p(x,D)(A_\varepsilon u)$ is uniformly bounded in $\overline{H}_{(0,s)}$ and it follows from (24.6.7)' when $\varepsilon \to 0$ that $u \in \overline{H}^{\mathrm{comp}}_{(2,s-\frac{2}{3})}$, which improves the hypothesis made on u.

The arguments used to prove Lemma 20.1.13 and Theorem 20.1.14 can now be copied with no change to complete the proof of the theorem. We leave the repetition as an exercise for the reader.

Theorem 24.5.3 shows that singularities always propagate from diffractive points. Theorem 24.5.4 gives a solution singular on the bicharacteristic through $(0, x', \xi') \in G_d$ if (x', ξ') is in the closure of the hyperbolic or the elliptic set. In fact, if $(y', \eta') \in E$ is close to (x', ξ') then we can find a small $y_1 > 0$ so that $r(y, \eta') = 0$, for $\partial r(0, x', \xi')/\partial x_1 > 0$ and $r(0, y', \eta') < 0$. The bicharacteristic through $(y, 0, \eta')$ lies in the half space $x_1 \geqq y_1$ for a fixed amount of time and converges to the bicharacteristic through $(0, x', 0, \xi')$ when $(y', \eta') \to (x', \xi')$ (cf. Lemma 24.3.4).

However, when (x', ξ') is not in the closure of $H \cup E$, that is, $r(0, y', \eta') = 0$ in a neighborhood of (x', ξ'), then all bicharacteristics passing close to $(0, x', 0, \xi')$ turn with a cusp at the plane $x_1 = 0$, as pointed out in the beginning of the section. We shall now show that nevertheless there is a solution which is singular precisely on a given bicharacteristic.

Theorem 24.6.5. *If $(y', \eta') \in G_d \cap \complement \overline{E}$ then one can find $u \in C(X)$ so that $u = 0$ on ∂X, $Pu \in C^\infty$. in a neighborhood of $(0, y')$, and $WF_b(u)$ is precisely generated there by the compressed bicharacteristic through (y', η').*

Proof. We may assume that $X \subset \mathbb{R}^n$ and that P has the usual form (24.4.2). For the principal symbol r we have by hypothesis $r(0, x', \xi') \geq 0$ and $\partial r(0, x', \xi')/\partial x_1 > 0$ in a neighborhood of (y', η'). Hence

$$r(x, \xi') \geq c x_1 |\xi'|^2$$

in a conic neighborhood of $(0, y', \eta')$. Choose $b \in S^2$ equal to R in a conic neighborhood of $(0, y', \eta')$ so that b satisfies (23.4.1). Set $B = b(x, D')$.

By Lemma 23.4.7 we can then find $u \in \bar{H}_{(1, n-1)}(\mathbb{R}^n_+)$ so that $(D_1^2 - B(x, D'))u = 0$ when $x_1 < \frac{1}{2}$ and $u = 0$, $D_n u = \phi$ when $x_n = 0$, where $\phi \in H_{(n-\frac{1}{2})}(\mathbb{R}^{n-1})$ is given. In fact, we just have to choose some $u_0 \in \bar{H}_{(2, n-1)}$ with these Cauchy data (Theorem B.1.9) and can then find $u - u_0 \in \bar{H}_{(1, n-1)} \subset \bar{C}^0$ with Cauchy data 0 such that

$$(D_1^2 - B)(u - u_0) = (B - D_1^2) u_0 \in \bar{H}_{(0, n-1)}.$$

If we choose ϕ with $WF(\phi)$ generated by (y', η') it follows from (the proof of) Theorem 23.4.8 and the interior propagation theorem that $WF(u)$ for $0 < x_n < \frac{1}{2}$ is contained in the cone Γ generated by the bicharacteristic through $(0, y', 0, \eta')$. Hence $Pu \in C^\infty$ in a neighborhood of $(0, y')$. The wave front set of u must there be equal to Γ, for Theorem 24.5.3 would otherwise show that $(y', \eta') \notin WF_b(u)$, hence that $(y', \eta') \notin WF(\phi)$. The proof is complete.

Remark. For the Tricomi equation itself an explicit construction by means of Airy functions is easily made.

24.7. Operators Depending on Parameters

In this section we shall complete the arguments of Section 17.5 by proving Lemma 17.5.14. To do so we must take another look at Theorem 24.5.3 for operators depending on parameters. First we rephrase Theorem 24.5.3 as an estimate.

Proposition 24.7.1. *Let P and X be as in Theorem 24.5.3 and let Γ, Γ_1, Γ_2 be open cones in $\tilde{T}^*(X) \setminus 0$ such that every compressed generalized bicharacteristic $\subset \Gamma$ which intersects Γ_1 must also intersect Γ_2 if maximally extended in Γ. Let B be a Banach space containing $\bar{C}_0^\infty(X)$ as a dense subset such that B is continuously embedded in $\bar{\mathscr{D}}'(X)$ and the restriction to ∂X extends from $\bar{C}_0^\infty(X)$ to a continuous map $B \to \mathscr{D}'(\partial X)$. For every compactly supported $A_1 \in \Psi_b^\infty(X)$ of order $-\infty$ outside a compactly generated cone $\subset \Gamma_1$ it is then possible to find compactly supported A, $A_2 \in \Psi_b^\infty(X)$ of order $-\infty$ outside compactly generated cones $\subset \Gamma$ resp. $\subset \Gamma_2$ such that*

$$\|A_1 u\| \leq \|APu\| + \|A_2 u\| + C\|u\|_B \tag{24.7.1}$$

if $u \in \bar{C}_0^\infty(X)$ and $u = 0$ on ∂X. ($\| \ \|$ denotes L^2 norms.)

Proof. The space

$$\mathscr{F} = \{u \in B;\ \Gamma \cap WF_b(Pu) = \emptyset, \quad u = 0 \ \text{ on } \ \partial X,\ \Gamma_2 \cap WF_b(u) = \emptyset\}$$

is a Fréchet space with the topology defined by $\|u\|_B$ and the semi-norms $\|APu\|$ and $\|A_2 u\|$ with A and A_2 as in the proposition. By hypothesis and Theorem 24.5.3 we have $\Gamma_1 \cap WF_b(u) = \emptyset$ if $u \in \mathscr{F}$. Hence the map

$$\mathscr{F} \ni u \mapsto A_1 u \in L^2(X)$$

is everywhere defined, and it is closed since it is continuous in the $\mathscr{D}'$ topology for $A_1 u$. Hence the map is continuous by the closed graph theorem, which proves the proposition.

We shall now allow P to depend on a parameter z in a neighborhood Z of 0 in $\bar{\mathbb{R}}^n_+$. We assume that the coefficients are in $C^\infty(X \times Z)$ and that P_0 satisfies the hypotheses of Proposition 24.7.1. We shall first prove the stability of this condition.

Lemma 24.7.2. *Assume that the hypotheses in Proposition* 24.7.1 *are fulfilled by* P_0. *For every compact set* $K_1 \subset \Gamma_1$ *one can then choose compact sets* $K_2 \subset \Gamma_2$ *and* $K \subset \Gamma$ *and a neighborhood* $Z_0 \subset Z$ *of* 0 *such that on every maximally extended compressed generalized bicharacteristic of* P_z *intersecting* K_1 *there is an interval* $\subset K$ *with one end point in* K_1 *and one in* K_2, *provided that* $z \in Z_0$.

Proof. Choose increasing sequences of compact sets $K_2^\nu \subset \Gamma_2$ and $K^\nu \subset \Gamma$, each in the interior of the following one and with union Γ_2 and Γ respectively. If the statement is false we can find a sequence $z_\nu \in Z$ converging to 0 and compressed generalized bicharacteristics $I_\nu \ni t \mapsto \beta_\nu(t)$ of P_{z_ν} such that

(i) I_ν is a neighborhood of 0 on $\mathbb{R}$ and $\beta_\nu(0) \in K_1$;
(ii) $K_2^\nu \cap \beta_\nu(I_\nu) = \emptyset$;
(iii) $\beta_\nu(I_\nu) \subset K^\nu$;
(iv) If $t \in I_\nu \cap \partial I_\nu$ then $\beta_\nu(t) \in \partial K^\nu$;
(v) If $t \in \complement I_\nu \cap \partial I_\nu$ then β_ν is maximally extended at t.

We shall derive a contradiction by studying the limit as $\nu \to \infty$.

When studying $\beta_\nu(t)$ for $t \geqq 0$ we distinguish two cases:

a) $\beta_\nu(I_\nu \cap \mathbb{R}_+) \subset K^\mu$ for a fixed μ and infinitely many indices ν. After passage to a subsequence the right end point a_ν of I_ν has a limit $a > 0$ as $\nu \to \infty$ and $\beta(t) = \lim \beta_\nu(t)$ exists if $0 \leqq t < a$, by the uniform Lipschitz condition observed after Definition 24.3.7. The proof of Proposition 24.3.12 shows that $\beta(t)$ is a compressed generalized bicharacteristic of P_0. We have $\beta(0) \in K_1$ by (i), $\beta(t) \notin \Gamma_2$ when $0 \leqq t < a$ by (ii), and we claim that $\beta(t)$ cannot be extended beyond a. If this is not true then the limit $\beta(a-0)$ exists, H_{p_0} is non-radial there, and if $\beta(a-0)$ is a glancing non-diffractive point then the gliding vector is not 0 there. But then it follows from Corollary 24.3.14 that $\beta_\nu(s)$ can be extended to $s \leqq a + \varepsilon$ for some $\varepsilon > 0$ and all large ν, which is a contradiction.

b) Assume now that for every μ we have $\partial K^\mu \cap \beta_\nu(I_\nu \cap \mathbb{R}_+) \neq \emptyset$ for large ν. For fixed μ we have a uniform Lipschitz condition for the restriction of β_ν to the component of 0 in $\{t \in I_\nu \cap \bar{\mathbb{R}}_+;\ \beta_\nu(t) \in K^\mu\}$. Passing to a subsequence we thus obtain a limit $\beta(t)$ defined for $0 \leqq t < a$ such that $\beta(0) \in K_1$, $\beta(t) \notin \Gamma_2$ if $0 \leqq t < a$ and $\beta([0,a)) \cap \partial K^n \neq \emptyset$ for all μ. Hence β is maximally extended at a, as a compressed generalized bicharacteristic in Γ.

If we argue in the same way for $t<0$ we obtain a compressed generalized bicharacteristic of P_0 intersecting K_1 but not Γ_2 which is maximally extended in Γ. This is a contradiction proving the lemma.

The analytical statement in Proposition 24.7.1 also has an analogue involving parameters:

Proposition 24.7.1′. *Let P_z have C^∞ coefficients in $X \times Z$ and assume that P_0 satisfies the hypotheses in Proposition 24.7.1. For every $\Gamma_0 \Subset \Gamma_1$ one can find $\delta>0$ such that if $A_1 \in \Psi_b^\infty(X)$ is compactly supported and of order $-\infty$ outside Γ_0 then one can find A, $A_2 \in \Psi_b^\infty(X)$ of order $-\infty$ outside compactly generated cones $\subset \Gamma$ and Γ_2 respectively, such that (24.7.1) is valid with P replaced by P_z when $z \in Z$ and $|z|<\delta$.*

The proof of Proposition 24.7.1 was obtained by invoking the closed graph theorem to change the qualitative statement in Theorem 24.5.3 to a quantitative one. However, we could equally well have inspected all the steps in the proof of Theorem 24.5.3 to prove the estimate which would then have been given a more precise form. Moreover, a proof of that kind remains valid for z so small that Lemma 24.7.2 is applicable. We leave for the reader to satisfy himself that this is true and turn to our application of Proposition 24.7.1′, the proof of Lemma 17.5.14.

As a preparation we shall give an explicit interpretation of the geometric condition in Proposition 24.7.1 in the simplest case. Thus we consider the constant coefficient wave operator $\partial^2/\partial t^2 - \Delta$ in $\{(t,x) \in \mathbb{R}^{n+1}; x_1 \geqq 0\}$ and determine the generalized bicharacteristic flow backward in time from points in the forward light cone $\{(t,x); t>|x|, x_1 \geqq 0\}$. This is identical to the bicharacteristic flow backward in time in $\mathbb{R}^{n+1}$ from the full open forward light cone, that is, the union W of the conormal bundles of the backward light cones with vertex in this set. We claim that if $-|x|<t \leqq |x|$ then $(t,x,\tau,\xi) \in W$ if and only if

$$\tau(\langle x,\xi\rangle + t\tau) > 0. \tag{24.7.2}$$

For the proof we first observe that (t,x,τ,ξ) is in the conormal bundle of the light cone with vertex at (t_0,x_0) if

$$(t-t_0)^2 = |x-x_0|^2, \quad \tau(x-x_0)+(t-t_0)\xi = 0$$

(cf. (17.4.11)). This implies $|\tau| = |\xi| \neq 0$. The condition can be written

$$(\tau x + t\xi)/t_0 - \xi = \tau x_0/t_0$$

and since x_0 is any point with $|x_0|<t_0$ in $\mathbb{R}^n$ it is equivalent to $|y-\xi|<|\tau|$ for some $t_0>0$ if $y=(\tau x + t\xi)/t_0$. Since $|\xi|=|\tau|$ this is true if and only if $\langle \tau x + t\xi, \xi\rangle > 0$, which is condition (24.7.2).

Proof of Lemma 17.5.14. We shall establish the required estimates for z close to a point $z_0 \in K_0$ where we may assume that $P_{z_0} = -\Delta$. We shall therefore first consider solutions of the wave equation

$$(\partial^2/\partial t^2 - \Delta)v = 0 \quad \text{in } \Omega = \{(t,x); |t|<3, x_1 \geqq 0\}$$

which vanish when $x_1=0$ and in

$$\Omega_2=\{(t,x);\ |x|>|t|+\tfrac{1}{2},\ |t|<3,\ x_1\geqq 0\}.$$

By Lemma 17.5.12 the last condition follows if $|y|<\frac{1}{2}$ when $(0,y)$ is in the support of the Cauchy data. (The same is true for solutions of $(\partial^2/\partial t^2+P_z)v_z=0$ satisfying (i), (ii) in Lemma 17.5.14 if $|z-z_0|$ is small enough.) The Green's kernel with pole at such a point is only singular in the union of the cones $\{(t,x);\ t^2=|x-y|^2\}$ when $|y|<\frac{1}{2}$, so it is C^∞ in

$$\Omega_1=\{(t,x);\ t>|x|+\tfrac{1}{2},\ t<3,\ x_1\geqq 0\}.$$

Let $\Gamma_j=\tilde{T}^*(\Omega_j)\setminus 0$ for $j=1,2$, and let Γ be the union of Γ_1, the complement of the compressed characteristic set $\tilde{\Sigma}$ in $\tilde{T}^*(\Omega_3)\setminus 0$ and the compressed generalized bicharacteristic flow backwards in time over Ω from Γ_1. Here

$$\Omega_3=\{(t,x);\ t>\tfrac{1}{2}-|x|,\ |t|<3,\ x_1\geqq 0\}.$$

It is clear that $\Gamma\subset\tilde{T}^*(\Omega_3)$ and that the intersection with $\tilde{\Sigma}$ is defined over $\Omega_3\setminus\Omega_1$ by (24.7.2) with t replaced by $t-\frac{1}{2}$. Hence Γ is an open cone. Note that $\Omega_2\subset\Omega_3$ and that $\Omega_3\cap\{(t,x);\ t\leqq 0\}\subset\Omega_2$. The hypotheses of Proposition 24.7.1 are thus fulfilled by the wave operator and Γ, Γ_1, Γ_2; we may also replace Γ_1 by Γ.

Let $M\Subset\Omega_1$ be a compact neighborhood of

$$M_0=\{(t,x);\ 1\leqq t\leqq 2,\ |x|\leqq\tfrac{2}{5}\}.$$

Our problem is to estimate the derivatives of the functions v_z in Lemma 17.5.14 uniformly in M. Let K_1 be the unit sphere bundle in $\tilde{T}^*$ restricted to M, and choose, using Lemma 24.7.2, a compact neighborhood $Z_0\subset Z$ of z_0 and open cones

$$\Gamma_1'\Subset\Gamma_1,\quad \Gamma_2'\Subset\{(t,x)\in\Gamma_2;\ t<-1\},\quad \Gamma'\Subset\Gamma,$$

such that $K_1\subset\Gamma_1'$ and the hypotheses of Proposition 24.7.1 are fulfilled for these cones and any P_z, $z\in Z_0$. Fix z arbitrarily in Z_0 now, and let K_1^z be the

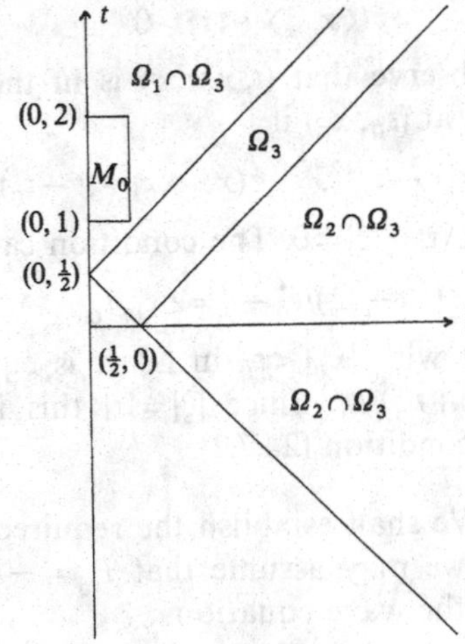

Fig. 7

union of K_1 and the compressed generalized bicharacteristics for P_z backward in time from K_1 to the plane $t=-1$. Let K_2^z be the intersection with that plane. We can then choose open conic neighborhoods $\Gamma^{(\nu)}$ of K_1^z such that

$$\Gamma' \supset \Gamma^{(1)} \supset \Gamma^{(2)} \supset \ldots \supset K_1^z$$

and the hypotheses in Proposition 24.7.1 are fulfilled for P_z with Γ_1 replaced by $\Gamma^{(\nu+1)}$ and Γ replaced by $\Gamma^{(\nu)}$ when $\nu=1,2,\ldots$. In fact, when $\Gamma^{(\nu)}$ has already been chosen it follows from Lemma 24.7.2 that the compressed generalized bicharacteristics going backwards from a sufficiently small neighborhood of K_1^z must be contained in $\Gamma^{(\nu)}$ until $t=-1-\varepsilon$ for a suitably small ε.

From Proposition 24.7.1′ we obtain if ζ is sufficiently close to z that there is a uniform bound for $\|A^{(1)}v_\zeta\|$ for any $A^{(1)} \in \Psi_b^\infty(\Omega)$ which is compactly supported and of order $-\infty$ outside a compactly generated cone $\subset \Gamma^{(1)}$; the bound depends on $A^{(1)}$ of course. Differentiation of the equation $(\partial^2/\partial t^2 + P_\zeta)v_\zeta = 0$ with respect to ζ_j gives

$$(\partial^2/\partial t^2 + P_\zeta)\,\partial v_\zeta/\partial \zeta_j = -(\partial P_\zeta/\partial \zeta_j)\,v_\zeta = f.$$

Since we already control $A^{(1)}f$ we now obtain from Proposition 24.7.1′ a bound for $\|A^{(2)}\,\partial v_\zeta/\partial \zeta_j\|$ when ζ is in some possibly smaller neighborhood of z, independent of $A^{(2)}$, and $A^{(2)}$ is of order $-\infty$ outside a compactly generated cone $\subset \Gamma^{(2)}$. Continuing in this way we obtain a bound for $\|A^{(\nu)} D_\zeta^\alpha v_\zeta\|$ if $A^{(\nu)}$ is of order $-\infty$ outside a compactly generated cone $\subset \Gamma^{(\nu)}$ and ζ is sufficiently close to z, $|\alpha|<\nu$. In particular we obtain uniform bounds for any derivative $D_{t,x,\zeta}^\alpha v_\zeta(t,x)$ when $(t,x)\in M$ and ζ is in a neighborhood of $z \in Z_0$ depending on α. By the Borel-Lebesgue lemma this implies a uniform bound for all $z \in Z_0$ which completes the proof.

Remark. Note that over Ω_3 there are many singularities for the Green's functions. It is only thanks to the microlocal point of view that we have been able to avoid difficulties from them. That is why it was not possible to prove Lemma 17.5.14 in Section 17.5.

Notes

The energy integral method was already used by Krzyżański-Schauder [1] to derive estimates for the mixed Dirichlet-Cauchy problem for second order operators. Much later Gårding [6] determined the general first order boundary conditions which can be used instead of the Dirichlet condition to derive fairly weak estimates. For higher order operators energy estimates were first given by Agmon [6]. In addition to what has later become known as the uniform Lopatinski condition he required that the real roots of the characteristic polynomial in the normal direction should be at most double. This condition was removed by Sakamoto [2]; similar results were proved for systems by Kreiss [1]. (See also Chazarain-Piriou [1].) For a discussion of correctness in L^2 we also refer to Sakamoto [3].

We have restricted ourselves here to the mixed Dirichlet-Cauchy problem for second order operators in order to be able to cover the boundary regularity theory fully in that case. As seen in Section 24.2 the discussions of the elliptic and the hyperbolic regions are essentially microlocal versions of elliptic boundary problems and strictly hyperbolic Cauchy problems. (See also Chazarain [1] for the hyperbolic case.) However, the glancing case poses essential new problems. These were first overcome in the diffractive case by Melrose [3] and Taylor [2] using parametrix constructions. (See also Eškin [2].) The work of Melrose was inspired by the explicit solution given by Friedlander [1] for the wave equation in a layered medium; his example is the model operator in Theorem 21.4.12 here. Taylor on the other hand was led by the asymptotic expansions of geometrical optics as given by Ludwig [2]. The gliding case was then treated by Andersson-Melrose [1] using the equivalence of glancing hypersurfaces proved by Melrose [2]. Another proof based on a parametrix construction was given by Eškin [2]. Later on Melrose and Sjöstrand [1] gave the geometrical discussion of generalized bicharacteristics which we have followed here in Section 24.3 with minor modification. They proved Theorem 24.5.3 using only energy integral arguments independent of the Melrose equivalence theorem except in the diffractive case. For that case Ivrii [2] derived the propagation theorem by energy integral arguments, thus eliminating the need for the parametrix constructions of Melrose and Taylor. We have followed his paper to the extent that it has been possible to do so. A new result may be the observation that in the diffractive set it is not necessary to assume that the gliding vector does not vanish. For results on more general boundary conditions the reader should consult Melrose-Sjöstrand [1] and Eškin [3, 4].

The interesting Example 24.3.11 has been taken from Taylor [2]. Further information on the non-uniqueness of generalized bicharacteristics at G^∞ can be found in Melrose-Sjöstrand [1]. For real analytic boundaries this problem does not occur. However, if one is interested in analytic singularities then singularity propagation along the gliding ray occurs also in the diffractive case, so singularities will in general spread from bicharacteristics to sets of higher dimension. (See Friedlander-Melrose [1] for a special case and Sjöstrand [4] for general results.)

The construction of solutions with given singularities in Sections 24.2 and 24.5 is a modification of the constructions in the interior case which seems to have been given first in Ralston [1]. The simplification given here by means of positive Lagrangians has been used by many authors for similar problems concerning interior singularities (see e.g. Hörmander [29]). In the present context it has been used recently by Ralston [2].

The results in Section 24.7 are taken from the simplification by Melrose [7] of the work of Ivrii [3] on the asymptotic properties of the eigenvalues of the Dirichlet problem. Melrose also used the idea to derive the qualitative part of Theorem 17.5.5 and applied it to similar questions for systems as well.

Appendix B. Some Spaces of Distributions

The purpose of this appendix is to collect the definitions and basic properties of the spaces of distributions used in the text. To a large extent they are special cases of spaces discussed in Sections 10.1 and 14.1. However, we shall give an essentially self contained exposition since this requires only minor repetitions.

B.1. Distributions in $\mathbb{R}^n$ and in an Open Manifold

Already in Section 7.9 we introduced the important Sobolev space $H_{(s)}(\mathbb{R}^n)$ consisting of all $u \in \mathscr{S}'(\mathbb{R}^n)$ with $\hat{u} \in L^2_{\text{loc}}$ and

$$\|u\|_{(s)} = ((2\pi)^{-n} \int |\hat{u}(\xi)|^2 (1+|\xi|^2)^s \, d\xi)^{\frac{1}{2}} < \infty. \tag{B.1.1}$$

When s is a non-negative integer then $H_{(s)}$ consists of all u such that $D^\alpha u \in L^2(\mathbb{R}^n)$, $|\alpha| \leqq s$. (Using the pseudo-differential operators introduced in Chapter XVIII one can define $H_{(s)}$ for any s as the set of distributions mapped to L^2 by all operators of order s.) Thus $H_{(s)}$ is the inverse Fourier transform of $L^2(\mathbb{R}^n, (1+|\xi|^2)^s d\xi/(2\pi)^n)$. In Section 14.1 we introduced modifications of these L^2 spaces. We shall now consider their inverse Fourier transforms which are modifications of $H_{(s)}$ spaces. As in (14.1.2) we set

$$X_j = \{\xi \in \mathbb{R}^n;\, R_{j-1} < |\xi| < R_j\}, \quad \text{where } R_0 = 0,\ R_j = 2^{j-1} \text{ if } j > 0. \tag{B.1.2}$$

Recall that by (14.1.8)

$$R_j^2/4 < (1+|\xi|^2) < 2R_j^2, \quad \xi \in X_j. \tag{B.1.3}$$

For any $u \in \mathscr{S}'$ with $\hat{u} \in L^2_{\text{loc}}$ we set

$$\pi_j u(x) = (2\pi)^{-n} \int_{X_j} e^{i\langle x,\xi\rangle} \hat{u}(\xi) \, d\xi,$$

that is, we multiply the Fourier transform by the characteristic function of X_j. Then it follows from (B.1.3) that

$$2^{-|s|} \|\pi_j u\|_{(0)} \leqq \|\pi_j u\|_{(s)} R_j^{-s} \leqq 2^{|s|} \|\pi_j u\|_{(0)}, \tag{B.1.4}$$

and we have

$$\|u\|_{(s)}^2 = \sum \|\pi_j u\|_{(s)}^2.$$

This suggests the following definition of more general *Besov* spaces:

Definition B.1.1. If $1 \leqq p \leqq \infty$ and $s \in \mathbb{R}$, then ${}^pH_{(s)}$ is the Banach space of all $u \in \mathscr{S}'$ with $\hat{u} \in L^2_{\text{loc}}$ and the norm

$$ {}^p\|u\|_{(s)} = (\sum \|\pi_j u\|^p_{(s)})^{1/p} \tag{B.1.5}$$

finite. (This should be read as $\sup_j \|\pi_j u\|_{(s)}$ if $p = \infty$.)

It is clear that ${}^pH_{(s)}$ is a Banach space and that ${}^2H_{(s)} = H_{(s)}$. By (B.1.4) we have

$$ 2^{-|s|}\, {}^p\|u\|_{(s)} \leqq (\sum \|R_j^s \pi_j u\|^p_{(0)})^{1/p} \leqq 2^{|s|}\, {}^p\|u\|_{(s)} \tag{B.1.6}$$

which describes ${}^pH_{(s)}$ and its norm in terms of the L^2 norm $\| \ \|_{(0)}$ only. Since $l^p \subset l^q$ if $p \leqq q$ it follows at once that ${}^pH_{(s)}$ increases with p. We have

$$ {}^\infty H_{(s_1)} \subset {}^1H_{(s_2)} \quad \text{if } s_1 > s_2. \tag{B.1.7}$$

In fact,

$$ {}^1\|u\|_{(s_2)} \leqq 2^{|s_2|} \sum \|R_j^{s_2} \pi_j u\|_{(0)} \leqq 2^{|s_2|+|s_1|} \sum R_j^{s_2 - s_1}\, {}^\infty\|u\|_{(s_1)}.$$

Summing up, we have found

Proposition B.1.2. ${}^{p_1}H_{(s_1)} \subset {}^{p_2}H_{(s_2)}$ *if (and only if)* $s_1 > s_2$ *or* $s_1 = s_2$ *and* $p_1 \leqq p_2$.

If $p < \infty$ then

$$ {}^p\|\sum_1^N \pi_j u - u\|_{(s)} \to 0, \quad N \to \infty, \quad \text{if } u \in {}^pH_{(s)}.$$

Since C_0^∞ is dense in L^2_{comp} it follows that C_0^∞ is dense in the Fourier transform of ${}^pH_{(s)}$. Hence $\mathscr{S}$ is dense in ${}^pH_{(s)}$ which implies that C_0^∞ is dense in ${}^pH_{(s)}$ because C_0^∞ is dense in $\mathscr{S}$. If $1/p + 1/p' = 1$ we have

$$ |(u, v)| = |u(\bar{v})| \leqq {}^{p'}\|u\|_{(-s)}\, {}^p\|v\|_{(s)}; \quad u \in {}^{p'}H_{(-s)},\ v \in C_0^\infty;$$

which proves that ${}^{p'}H_{(-s)}$ is continuously embedded in the antidual space of ${}^pH_{(s)}$. From the converse of Hölder's inequality it follows that the embedding is an isomorphism:

Proposition B.1.3. *If* $1 \leqq p < \infty$ *then the antidual of* ${}^pH_{(s)}$ *is* ${}^{p'}H_{(-s)}$ *where* $1/p + 1/p' = 1$.

Example. The spaces B and B^* of (14.1.1) and (14.1.3) are the Fourier transforms of ${}^1H_{(\frac{1}{2})}$ and ${}^\infty H_{(-\frac{1}{2})}$ respectively.

We shall now discuss how ${}^pH_{(s)}$ can be obtained as an interpolation space which is a convenient way of deriving various estimates. (See also

Lemmas 14.1.3 and 14.1.4.) If $u \in {}^pH_{(s)}$ and $u_k = \pi_k u$ then we know that $u = \sum u_k$ with convergence in $\mathscr{S}'$ and that

$$\|u_k\|_{(t)} \leqq R_k^t 2^{|t|} M_k, \quad t \in \mathbb{R},$$

where $M_k = \|\pi_k u\|_{(0)}$, thus

$$(\sum (R_k^s M_k)^p)^{1/p} \leqq 2^{|s|}\, {}^p\|u\|_{(s)}.$$

The following proposition gives a converse.

Proposition B.1.4. *Assume that $u_k \in H_{(s_0)} \cap H_{(s_1)}$, $k=1,2,\dots$ and that*

(B.1.8) $$\|u_k\|_{(s_\nu)} \leqq R_k^{s_\nu} M_k; \quad \nu = 1,2;\ k=1,2,\dots.$$

If $s_0 < s < s_1$ and $R_k^s M_k \in l^p$, then $\sum u_k$ converges in $\mathscr{S}'$, the sum belongs to ${}^pH_{(s)}$, and

(B.1.9) $${}^p\|\sum u_k\|_{(s)} \leqq C(s) (\sum (R_j^s M_j)^p)^{1/p}.$$

If $p < \infty$ the series converges in ${}^pH_{(s)}$ norm.

Proof. It suffices to prove (B.1.9) when the sum is finite and then apply it to the difference between partial sums with s replaced by a smaller number if $p = \infty$. Then $\pi_j u = \sum \pi_j u_k$, and (B.1.8) gives

$$\|R_j^{s_\nu} \pi_j u_k\|_{(0)} \leqq C \|u_k\|_{(s_\nu)} \leqq R_k^{s_\nu} M_k.$$

With $\nu = 0$ when $k \geqq j$ and $\nu = 1$ when $k < j$ we obtain

$$R_j^s \|\pi_j u\|_{(0)} \leqq \sum a_{jk} R_k^s M_k,$$

$$a_{jk} = R_k^{s_0 - s} R_j^{s - s_0} \quad \text{if } k \geqq j; \quad a_{jk} = R_k^{s_1 - s} R_j^{s - s_1} \quad \text{if } k < j.$$

Since R_k is a geometric progression we have

$$\sum_j a_{jk} \leqq C(s), \quad \sum_k a_{jk} \leqq C(s) \quad \text{if } C(s) = (1 - 2^{s_0 - s})^{-1} + (2^{s_1 - s} - 1)^{-1}.$$

The estimate (B.1.9) follows now from a classical lemma of Schur:

Lemma B.1.5. *If $\sum_j |a_{jk}| \leqq C$ and $\sum_k |a_{jk}| \leqq C$ for all k and j respectively, then*

$$(\sum_j |\sum_k a_{jk} x_k|^p)^{1/p} \leqq C (\sum |x_k|^p)^{1/p}; \quad 1 \leqq p \leqq \infty,\ x \in l^p.$$

Proof. The statement is obvious if $p=1$ or $p=\infty$. Otherwise we obtain if $1/p + 1/p' = 1$ using Hölder's inequality

$$|\sum_k a_{jk} x_k|^p \leqq (\sum_k |a_{jk}|)^{p/p'} \sum_k |a_{jk}|\, |x_k|^p \leqq C^{p/p'} \sum_k a_{jk} |x_k|^p.$$

Hence

$$\sum_j |\sum_k a_{jk} x_k|^p \leqq C^{1+p/p'} \sum |x_k|^p = C^p \sum |x_k|^p$$

which completes the proof.

Corollary B.1.6. *If T is a continuous linear map $H_{(s_0)} \to H_{(s_0-m)}$ which restricts to a continuous linear map $H_{(s_1)} \to H_{(s_1-m)}$ for some $s_1 > s_0$ then T restricts to a continuous linear map ${}^pH_{(s)} \to {}^pH_{(s-m)}$ if $s_0 < s < s_1$ and $1 \leqq p \leqq \infty$.*

Proof. If $u \in {}^pH_{(s)}$ then $Tu = \sum T\pi_k u$ with convergence in $H_{(s_0-m)}$ and for $\nu = 0, 1$ we have

$$\|T\pi_k u\|_{(s_\nu - m)} \leqq C\|\pi_k u\|_{(s_\nu)} \leqq C' R_k^{s_\nu} \|\pi_k u\|_{(0)},$$

$$(\sum \|R_k^s \pi_k u\|_{(0)}^p)^{1/p} \leqq 2^{|s|\,p} \|u\|_{(s)}.$$

The statement follows now from Proposition B.1.4.

Theorem B.1.7. *If $\phi \in C^\infty$ and $D^\alpha \phi$ is bounded for every α, then $u \mapsto \phi u$ is a continuous map in ${}^pH_{(s)}$ for all s and p.*

Proof. The continuity in $H_{(s)}$ is obvious if s is a non-negative integer. Since the adjoint of the map $H_{(-s)} \ni u \mapsto \bar{\phi} u \in H_{(-s)}$ is the map $H_{(s)} \ni u \mapsto \phi u \in H_{(s)}$ if s is a negative integer, the continuity in $H_{(s)}$ follows also in that case. Hence the theorem is a consequence of Corollary B.1.6.

Theorem B.1.8. *Let ψ be a diffeomorphism of an open set $X_1 \subset \mathbb{R}^n$ onto another open set $X_2 \subset \mathbb{R}^n$, and let $\chi \in C_0^\infty(X_2)$. Then the pullback $\psi^*(\chi v) \in \mathscr{E}'(X_1) \subset \mathscr{E}'(\mathbb{R}^n)$ is in ${}^pH_{(s)}(\mathbb{R}^n)$ if $v \in {}^pH_{(s)}(\mathbb{R}^n)$, and we have*

$$^p\|\psi^*(\chi v)\|_{(s)} \leqq C(s)\, {}^p\|v\|_{(s)}.$$

Proof. By Corollary B.1.6 it suffices to prove that

$$\|\psi^*(\chi v)\|_{(s)} \leqq C(s)\, \|v\|_{(s)}$$

if s is an integer. When s is a positive integer this follows immediately by differentiating the definition

$$\psi^*(\chi v) = \chi(\psi)\, v(\psi)$$

and taking $\psi(x)$ as a new integration variable. If s is a negative integer we shall argue by duality. Choose $\chi_1 \in C_0^\infty(X_1)$ equal to 1 near $\operatorname{supp} \psi^* \chi$. Then we have for $u \in C_0^\infty(\mathbb{R}^n)$

$$|(\psi^*(\chi v), u)| = |(\psi^*(\chi v), \chi_1 u)| = |(|\det(\psi^{-1})'|\, \chi v, (\psi^{-1})^*(\chi_1 u))| \leqq C\, \|v\|_{(s)}\, \|u\|_{(-s)}$$

by Theorem B.1.7 and the first part of the proof. Hence

$$\|\psi^*(\chi v)\|_{(s)} \leqq C\, \|v\|_{(s)},$$

which proves the theorem.

Remark. For the spaces $H_{(s)} = {}^2H_{(s)}$, arbitrary $s \in \mathbb{R}$, one can also prove Theorem B.1.8 by means of (7.9.3), (7.9.4).

By Theorem B.1.7 we can use Definition 10.1.18 to define ${}^pH^{\text{loc}}_{(s)}(X)$ if X is an open subset of $\mathbb{R}^n$; this is the set of all $u\in\mathscr{D}'(X)$ such that $\phi u\in{}^pH_{(s)}(\mathbb{R}^n)$ for every $\phi\in C_0^\infty(X)$. More generally, if X is a C^∞ manifold a distribution $u\in\mathscr{D}'(X)$ is said to be in ${}^pH^{\text{loc}}_{(s)}(X)$ if

$$(\kappa^{-1})^* u = u_\kappa \in {}^pH^{\text{loc}}_{(s)}(\tilde{X}_\kappa)$$

for every C^∞ coordinate system $\kappa\colon X_\kappa \to \tilde{X}_\kappa \subset \mathbb{R}^n$. The topology is defined by the semi-norms

$$u \mapsto {}^p\|\phi u_\kappa\|_{(s)}$$

where ϕ is an arbitrary element in $C_0^\infty(\tilde{X}_\kappa)$ and κ is any C^∞ coordinate system. From Theorem B.1.8 it follows as in the proof of Theorem 6.3.4 that it suffices to take κ in an atlas for X and $\phi=\phi_\kappa$ provided that the sets $\{x\in X_\kappa;\ \phi_\kappa(\kappa x)\neq 0\}$ cover X. In particular, if $X\subset\mathbb{R}^n$ and $K\subset X$ is compact, then ${}^pH^{\text{loc}}_{(s)}(X)\cap\mathscr{E}'(K)\subset{}^pH_{(s)}(\mathbb{R}^n)$ and the topology is that defined by ${}^p\|\ \|_{(s)}$.

We shall now discuss restrictions to submanifolds starting with the restriction to a hyperplane in $\mathbb{R}^n$. If $u\in\mathscr{S}(\mathbb{R}^n)$ we denote by γu the restriction to the hyperplane $x_n=0$, identified with $\mathbb{R}^{n-1}$:

$$(\gamma u)(x') = u(x',0), \qquad x'\in\mathbb{R}^{n-1}.$$

Theorem B.1.9. *γ extends to a continuous map from ${}^1H_{(\frac12)}(\mathbb{R}^n)$ to $H_{(0)}(\mathbb{R}^{n-1}) = L^2(\mathbb{R}^{n-1})$, which we shall also denote by γ. For every $s>0$ the restriction of γ maps ${}^pH_{(s+\frac12)}(\mathbb{R}^n)$ continuously onto ${}^pH_{(s)}(\mathbb{R}^{n-1})$ and has a continuous right inverse from ${}^pH_{(s)}(\mathbb{R}^{n-1})$ to ${}^pH_{(s+\frac12)}(\mathbb{R}^n)$.*

Proof. With the notation $\pi_j u$ in (B.1.4) we have $\pi_j u\in C^\infty$ and

$$\begin{aligned}\|\gamma\pi_j u\|^2_{(0)} &= (2\pi)^{1-n}\int\Big|\int_{\xi\in X_j}\hat{u}(\xi)\,d\xi_n/2\pi\Big|^2 d\xi'\\ &\leqq (2\pi)^{-1-n}2R_j\int_{X_j}|\hat{u}(\xi)|^2\,d\xi = \pi^{-1}R_j\|\pi_j u\|^2_{(0)}.\end{aligned}$$

If $\hat{u}\in C_0^\infty$ then $u=\sum\pi_j u$ with only finitely many terms $\neq 0$, and the triangle inequality gives

$$\|\gamma u\|_{(0)} \leqq \sum\|\gamma\pi_j u\|_{(0)} \leqq C\,{}^1\|u\|_{(\frac12)}.$$

Such functions are dense in ${}^1H_{(\frac12)}$ which proves the first statement and that $\gamma u=\sum\gamma\pi_j u$ for every $u\in{}^1H_{(\frac12)}$.

If $u\in{}^pH_{(s+\frac12)}(\mathbb{R}^n)$ we have for $t\geqq 0$

$$\|\gamma\pi_j u\|_{(t)} \leqq 2^t R_j^t\|\gamma\pi_j u\|_{(0)},$$

and

$$\|R_j^s\gamma\pi_j u\|_{(0)} \leqq \|R_j^{s+\frac12}\pi_j u\|_{(0)}$$

where the right hand side is in l^p. Hence Proposition B.1.4 applied to partial sums gives if $s>0$

$${}^p\|\gamma u\|_{(s)} \leqq C(s)\,{}^p\|u\|_{(s+\frac12)}.$$

Thus γ is continuous ${}^pH_{(s+\frac{1}{2})}(\mathbb{R}^n)\to {}^pH_{(s)}(\mathbb{R}^{n-1})$. To construct a right inverse we choose $\psi\in C_0^\infty(\mathbb{R})$ with $\psi(0)=1$ and set for $v\in {}^pH_{(s)}(\mathbb{R}^{n-1})$

$$u_k(x)=v_k(x')\,\psi(R_k x_n)$$

where $x'=(x_1,\dots,x_{n-1})$ and $v_k=\pi_k v$ is defined as before but with $\mathbb{R}^n$ replaced by $\mathbb{R}^{n-1}$. Then

$$\begin{aligned}\hat{u}_k(\xi)&=\hat{v}_k(\xi')\,\hat{\psi}(\xi_n/R_k)/R_k,\\ \|u_k\|_{(t)}^2&=(2\pi)^{-n}\int|\hat{v}_k(\xi')|^2\,|\hat{\psi}(\xi_n/R_k)/R_k|^2(1+|\xi|^2)^t\,d\xi\\ &\leqq C_t R_k^{2t-1}\int|\hat{v}_k(\xi')|^2\,|\hat{\psi}(\xi_n)|^2(1+\xi_n^2)^t\,d\xi\\ &\leqq C_t' R_k^{2t-1}\,\|\pi_k v\|_{(0)}^2.\end{aligned}$$

This gives (B.1.8) with $M_k=C_t'' R_k^{-\frac{1}{2}}\|\pi_k v\|_{(0)}$. Since

$$R_k^{s+\frac{1}{2}}M_k=C_t'' R_k^s\,\|\pi_k v\|_{(0)}\in l^p$$

if $v\in {}^pH_{(s)}$, it follows from Proposition B.1.4 that the map

$$v\to\sum u_k$$

is continuous from ${}^pH_{(s)}$ to ${}^pH_{(s+\frac{1}{2})}$. It is obviously a right inverse of γ when $s>0$. The proof is complete.

Remark. In Chapter XIV we proved that γ is also surjective from ${}^1H_{(\frac{1}{2})}(\mathbb{R}^n)$ to $H_{(0)}(\mathbb{R}^{n-1})$ but the right inverse above is not continuous from $H_{(0)}(\mathbb{R}^{n-1})$ to ${}^1H_{(\frac{1}{2})}(\mathbb{R}^n)$.

In Theorem B.1.9 we have just discussed the restriction to a hyperplane. However, iterating the result k times we obtain that there is a continuous restriction map from ${}^1H_{(k/2)}(\mathbb{R}^n)$ to $H_{(0)}(\mathbb{R}^{n-k})$ if $x'\in\mathbb{R}^{n-k}$ is identified with $(x',0)\in\mathbb{R}^n$. It restricts to a continuous map from ${}^pH_{(s+k/2)}(\mathbb{R}^n)$ to ${}^pH_{(s)}(\mathbb{R}^{n-k})$ for every $s>0$ and integer $k>0$.

Since we have defined the restriction map γ by continuous extension from the smooth case it is clear that it is invariant under a change of variables. If Y is a C^∞ submanifold of codimension k of a C^∞ manifold X **and $u\in {}^pH_{(s)}^{\mathrm{loc}}(X)$ we can therefore define the restriction $\gamma_Y u$ of u to Y if $s>k/2$, and $\gamma_Y u\in {}^pH_{(s-k/2)}^{\mathrm{loc}}$ then.**

The attentive reader may have noticed that the right inverse constructed in the proof of Theorem B.1.9 was defined for all s although we could not assert that it was a right inverse of γ if $s\leqq 0$ since γ is not defined in ${}^pH_{(s+\frac{1}{2})}$ then. This suggests that γ should be defined in some larger spaces. We shall now define such modifications of the $H_{(s)}$ spaces. (For the sake of brevity we no longer consider analogues of the Besov spaces.) In the definition the plane $x_n=0$ plays a distinguished role; we note that the first $n-1$ coordinates $\xi'=(\xi_1,\dots,\xi_{n-1})$ of $\xi\in\mathbb{R}^n$ are coordinates in its dual space.

Definition B.1.10. If $m, s \in \mathbb{R}$ we denote by $H_{(m,s)}(\mathbb{R}^n)$ the set of all $u \in \mathscr{S}'(\mathbb{R}^n)$ with $\hat{u} \in L^2_{\text{loc}}$ and

$$\|u\|_{(m,s)} = ((2\pi)^{-n} \int |\hat{u}(\xi)|^2 (1+|\xi|^2)^m (1+|\xi'|^2)^s \, d\xi)^{\frac{1}{2}} < \infty. \tag{B.1.10}$$

This is the special case of the spaces $B_{2,k}$ of Definition 10.1.6 with

$$k(\xi) = (1+|\xi|^2)^{m/2} (1+|\xi'|^2)^{s/2}.$$

Thus multiplication defines a continuous bilinear map

$$\mathscr{S} \times H_{(m,s)} \ni (\phi, u) \mapsto \phi\, u \in H_{(m,s)}$$

by Theorem 10.1.15, and Theorem 10.1.8 gives

$$H_{(m_1,s_1)}(\mathbb{R}^n) \subset H_{(m_2,s_2)}(\mathbb{R}^n) \Leftrightarrow m_2 \leqq m_1 \quad \text{and} \quad m_2+s_2 \leqq m_1+s_1. \tag{B.1.11}$$

$C_0^\infty(\mathbb{R}^n)$ is a dense subset of $H_{(m,s)}$ for all m, s. If m is a positive integer then $H_{(m,s)}$ consists of all u with $D_n^j u \in H_{(0,s+m-j)}$ for $0 \leqq j \leqq m$, and $v \in H_{(0,s)}$ means that v can be regarded as an L^2 function of x_n with values in $H_{(s)}(\mathbb{R}^{n-1})$.

As in Theorem B.1.9 we denote by γ the restriction map

$$(\gamma u)(x') = u(x', 0), \qquad u \in \mathscr{S}(\mathbb{R}^n).$$

Theorem B.1.11. *If $m > \frac{1}{2}$ there exists a continuous extension of γ to a map $H_{(m,s)}(\mathbb{R}^n) \to H_{(m+s-\frac{1}{2})}(\mathbb{R}^{n-1})$. There is also an extension map e from $\bigcup H_{(s)}(\mathbb{R}^{n-1})$ to $\mathscr{S}'(\mathbb{R}^n)$ such that e is continuous from $H_{(s)}(\mathbb{R}^{n-1})$ to $H_{(m,s-m+\frac{1}{2})}(\mathbb{R}^n)$ for all s, m, and γe is the identity. Moreover, e can be chosen so that $\gamma D_n^k e = 0$ for every positive integer k, and $x_n^k e$ is a continuous map from $H_{(s)}(\mathbb{R}^{n-1})$ to $H_{(m,s+k-m+\frac{1}{2})}(\mathbb{R}^n)$ for all s, m.*

The last statement has a corollary which is analogous to Corollary 1.3.4:

Corollary B.1.12. *If m is a positive integer and $f_k \in H_{(s+m-k-\frac{1}{2})}(\mathbb{R}^{n-1})$, $k = 0, \dots, m-1$, then one can find $u \in H_{(m,s)}(\mathbb{R}^n)$ with $\gamma D_n^k u = f_k$, $k = 0, \dots, m-1$.*

Proof. We just have to take

$$u = \sum_0^{m-1} x_n^k e\, f_k / k!.$$

Proof of Theorem B.1.11. If $u \in \mathscr{S}$ and $v = \gamma u$ then

$$|\hat{v}(\xi')|^2 = |\int \hat{u}(\xi)\, d\xi_n / 2\pi|^2 \leqq \int |\hat{u}(\xi)|^2 (1+|\xi|^2)^m \, d\xi_n \int (1+|\xi|^2)^{-m} \, d\xi_n / (2\pi)^2$$
$$\leqq C_m^2 \int |\hat{u}(\xi)|^2 (1+|\xi|^2)^m (1+|\xi'|^2)^{\frac{1}{2}-m} \, d\xi_n / 2\pi.$$

The last inequality follows by taking $t = \xi_n / (1+|\xi'|^2)^{\frac{1}{2}}$ as a new variable instead of ξ_n. Here it is essential for the convergence that $m > \frac{1}{2}$. If we multiply by $(1+|\xi'|^2)^{s+m-\frac{1}{2}}$ and integrate it follows that

$$\|v\|_{(s+m-\frac{1}{2})} \leqq C_m \|u\|_{(m,s)}.$$

To construct the extension operator e we argue essentially as in the proof of Theorem B.1.9 taking $\psi \in C_0^\infty$ equal to 1 in a neighborhood of 0. With $\lambda = (1+|\xi'|^2)^{\frac{1}{2}}$ we define $e\,v = u$ for $v \in \mathscr{S}$ so that

$$\hat{u}_n(\xi', x_n) = \psi(\lambda x_n)\,\hat{v}(\xi').$$

Here $\hat{u}_n$ denotes the Fourier transform with respect to the x' variables. It is then clear that $u \in \mathscr{S}$, that $u(.,0) = v$ and that $D_n^j u = 0$ for $x_n = 0$ if $j \neq 0$. We have

$$\begin{aligned}\|D_n^j x_n^k u\|^2_{(0,s+k-j+\frac{1}{2})} &= (2\pi)^{1-n} \int |\hat{v}(\xi')|^2 \lambda^{2(s+k-j)+1}\, d\xi' \int |D_n^j x_n^k \psi(\lambda x_n)|^2\, dx_n \\ &= (2\pi)^{1-n} \int |\hat{v}(\xi')|^2 \lambda^{2s}\, d\xi' \int |D_n^j x_n^k \psi(x_n)|^2\, dx_n.\end{aligned}$$

If m is a positive integer we obtain by adding for $0 \leqq j \leqq m$

$$\|x_n^k e\, v\|_{(m,s+k-m+\frac{1}{2})} \leqq C_{m,k} \|v\|_{(s)}$$

which completes the proof.

B.2. Distributions in a Half Space and in a Manifold with Boundary

We shall denote by $\mathbb{R}^n_+$ the open half space of $\mathbb{R}^n$ defined by $x_n > 0$ and by $\mathbb{R}^n_-$ the complement of its closure $\overline{\mathbb{R}}^n_+$. If F is a space of distributions in $\mathbb{R}^n$ we shall use the notation $\bar{F}(\mathbb{R}^n_+)$ for the space of *restrictions* to $\mathbb{R}^n_+$ of elements in F and we shall write $\dot{F}(\overline{\mathbb{R}}^n_+)$ for the set of distributions in F supported by $\overline{\mathbb{R}}^n_+$. Thus $\bar{F}(\mathbb{R}^n_+)$ is a subspace of $\mathscr{D}'(\mathbb{R}^n_+)$ and a quotient space $F/\dot{F}(\overline{\mathbb{R}}^n_-)$ of F; it contains the quotient of $\dot{F}(\overline{\mathbb{R}}^n_+)$ by the subspace of elements with support in $\partial\mathbb{R}^n_+ = \{x \in \mathbb{R}^n; x_n = 0\}$. The notation is meant as a reminder of these facts.

Example. $\bar{C}_0^\infty(\mathbb{R}^n_+)$ consists of the restrictions to $\mathbb{R}^n_+$ of functions in $C_0^\infty(\mathbb{R}^n)$. It follows from Theorem 1.2.6 that it can be identified with the space of functions in $C^\infty(\overline{\mathbb{R}}^n_+)$ vanishing outside a compact set. The space $\dot{C}_0^\infty(\overline{\mathbb{R}}^n_+)$ is the subspace of functions vanishing of infinite order when $x_n = 0$. The dual space of $\bar{C}_0^\infty(\mathbb{R}^n_+)$ is by Theorem 2.3.3 equal to the space $\dot{\mathscr{D}}'(\overline{\mathbb{R}}^n_+)$ of distributions with support in $\overline{\mathbb{R}}^n_+$, and that of $\dot{C}_0^\infty(\overline{\mathbb{R}}^n_+)$ is the space $\bar{\mathscr{D}}'(\mathbb{R}^n_+)$ of extendible distributions, by the Hahn-Banach theorem and Theorem 2.3.3.

We shall mainly discuss the spaces $\bar{H}_{(m,s)}(\mathbb{R}^n_+)$ and $\dot{H}_{(m,s)}(\overline{\mathbb{R}}^n_+)$, with the quotient and induced topology respectively. Thus we set

$$\|u\|_{(m,s)} = \inf \|U\|_{(m,s)}, \qquad u \in \bar{H}_{(m,s)}(\mathbb{R}^n_+),$$

with the infimum taken over all $U \in H_{(m,s)}(\mathbb{R}^n)$ equal to u in $\mathbb{R}^n_+$.

Theorem B.2.1. *$C_0^\infty(\mathbb{R}^n_+)$ is dense in $\dot{H}_{(m,s)}(\overline{\mathbb{R}}^n_+)$, $\bar{C}_0^\infty(\mathbb{R}^n_+)$ is dense in $\overline{H}_{(m,s)}(\mathbb{R}^n_+)$, and the spaces $\dot{H}_{(m,s)}(\overline{\mathbb{R}}^n_+)$, $\overline{H}_{(-m,-s)}(\mathbb{R}^n_+)$ are dual with respect to an extension of the sequilinear form*

$$\int u\bar{v}\,dx;\quad u\in C_0^\infty(\mathbb{R}^n_+),\ v\in\bar{C}_0^\infty(\mathbb{R}^n_+).$$

Proof. Since C_0^∞ is dense in $H_{(m,s)}(\mathbb{R}^n)$ it is clear that $\bar{C}_0^\infty(\mathbb{R}^n_+)$ is dense in $\overline{H}_{(m,s)}(\mathbb{R}^n_+)$. That $C_0^\infty(\mathbb{R}^n_+)$ is dense in $\dot{H}_{(s)}(\overline{\mathbb{R}}^n_+)$ follows from Theorems 10.1.16 and 10.1.17, where we take ϕ with support in $\mathbb{R}^n_+$. Now $\dot{H}_{(m,s)}(\overline{\mathbb{R}}^n_+)$ is a closed subspace of $H_{(m,s)}(\mathbb{R}^n)$ so its dual space is the quotient of $H_{(-m,-s)}(\mathbb{R}^n)$ by the annihilator of the dense subset $C_0^\infty(\mathbb{R}^n_+)$, that is, the subspace of elements vanishing in $\mathbb{R}^n_+$. That is by definition $\overline{H}_{(-m,-s)}(\mathbb{R}^n_+)$.

It follows from Theorem 10.1.8 that

$$\overline{H}_{(m_1,s_1)}(\mathbb{R}^n_+)\subset\overline{H}_{(m_2,s_2)}(\mathbb{R}^n_+)\Leftrightarrow m_2\leqq m_1\quad\text{and}\quad m_2+s_2\leqq m_1+s_1. \tag{B.2.1}$$

Using Theorem 10.1.10 we obtain

Theorem B.2.2. *Let K be a compact subset of $\overline{\mathbb{R}}^n_+$. Then*

$$\{u;\,u\in\overline{H}_{(m_1,s_1)}(\mathbb{R}^n_+),\ \|u\|_{(m_1,s_1)}\leqq 1,\ \operatorname{supp}u\subset K\}$$

is a compact subset of $\overline{H}_{(m_2,s_2)}(\mathbb{R}^n_+)$ if $m_2<m_1$ and $m_2+s_2<m_1+s_1$. If K has interior points, these conditions are also necessary.

Proof. The necessity follows immediately from Theorem 10.1.10 if we shrink K to a compact subset of $\mathbb{R}^n_+$. To prove the sufficiency we take a sequence $u_\nu\in\overline{H}_{(m_1,s_1)}(\mathbb{R}^n_+)$ with $\operatorname{supp}u_\nu\subset K$ and $\|u_\nu\|_{(m_1,s_1)}\leqq 1$. Then we can find $U_\nu\in H_{(m_1,s_1)}(\mathbb{R}^n)$ so that $U_\nu=u_\nu$ in $\mathbb{R}^n_+$ and $\|U_\nu\|_{(m_1,s_1)}\leqq 2$. Choose $\Phi\in C_0^\infty(\mathbb{R}^n)$ so that $\Phi=1$ in a neighborhood of K and set $U'_\nu=\Phi U_\nu$. Then we have

$$\|U'_\nu\|_{(m_1,s_1)}\leqq C\|U_\nu\|_{(m_1,s_1)}\leqq 2C$$

where C is a constant. Since $\operatorname{supp}U_\nu\subset\operatorname{supp}\Phi$ for every ν, it follows from Theorem 10.1.10 that there exists a subsequence U'_{ν_j} converging in $H_{(m_2,s_2)}(\mathbb{R}^n)$ to a limit $U_\nu\in H_{(m_1,s_1)}(\mathbb{R}^n)$. But then the sequence u_{ν_j} of the restrictions to $\mathbb{R}^n_+$ converges in $\overline{H}_{(m_2,s_2)}(\mathbb{R}^n_+)$ to the restriction of U to $\mathbb{R}^n_+$. The proof is complete.

We shall now prove some results which lead to a more direct description of $\overline{H}_{(m,s)}(\mathbb{R}^n_+)$ when m is a non-negative integer.

Theorem B.2.3. *In order that $u\in\overline{H}_{(m,s)}(\mathbb{R}^n_+)$ it is necessary and sufficient that $u\in\overline{H}_{(m-1,s+1)}(\mathbb{R}^n_+)$ and that $D_n u\in\overline{H}_{(m-1,s)}(\mathbb{R}^n_+)$; we have*

$$\tfrac{1}{2}\|u\|^2_{(m,s)}\leqq\|D_n u\|^2_{(m-1,s)}+\|u\|^2_{(m-1,s+1)}\leqq\|u\|^2_{(m,s)}. \tag{B.2.2}$$

Moreover, $u \in \overline{H}_{(m,s)}(\mathbb{R}^n_+)$ *if and only if* $u \in \overline{H}_{(m,s-1)}(\mathbb{R}^n_+)$ *and* $D_j u \in \overline{H}_{(m,s-1)}(\mathbb{R}^n_+)$ *when* $j<n$, *and we have*

$$\|u\|^2_{(m,s)} = \|u\|^2_{(m,s-1)} + \sum_1^{n-1} \|D_j u\|^2_{(m,s-1)}. \tag{B.2.3}$$

Proof. Let $u \in \overline{H}_{(m,s)}(\mathbb{R}^n_+)$ and choose $U \in H_{(m,s)}(\mathbb{R}^n)$ so that $U=u$ in $\mathbb{R}^n_+$ and $\|U\|_{(m,s)} = \|u\|_{(m,s)}$. Then

$$\|D_n U\|^2_{(m-1,s)} + \|U\|^2_{(m-1,s+1)} = \|U\|^2_{(m,s)} = \|u\|^2_{(m,s)}$$

which proves that $D_n u \in \overline{H}_{(m-1,s)}(\mathbb{R}^n_+)$, that $u \in \overline{H}_{(m-1,s+1)}(\mathbb{R}^n_+)$ and that the second inequality in (B.2.2) is valid. The rest of the theorem follows at once from

Theorem B.2.4. *Set* $\xi' = (\xi_1, \dots, \xi_{n-1})$, $\langle \xi' \rangle = (1+|\xi'|^2)^{\frac{1}{2}}$ *and*

$$\Lambda_{m,s}(\xi) = (\langle \xi' \rangle + i\xi_n)^m \langle \xi' \rangle^s. \tag{B.2.4}$$

Then $\Lambda_{m,s}(D)\phi = F^{-1}(\Lambda_{m,s}\hat{\phi})$, $\phi \in \dot{\mathscr{S}}(\overline{\mathbb{R}}^n_+)$, *is an isomorphism of* $\dot{\mathscr{S}}(\overline{\mathbb{R}}^n_+)$ *which extends by continuity to an isomorphism* $L^2(\mathbb{R}^n_+) \to \dot{H}_{(-m,-s)}(\overline{\mathbb{R}}^n_+)$. *The adjoint* $\bar{\Lambda}_{m,s}(D)$ *extends by continuity from* $\bar{C}_0^\infty(\mathbb{R}^n_+)$ *to an isomorphism from* $\overline{H}_{(m,s)}(\mathbb{R}^n_+)$ *to* $L^2(\mathbb{R}^n_+)$. *Thus*

$$\|u\|_{(m,s)} = \|\bar{\Lambda}_{m,s}(D)u\|_{L^2(\mathbb{R}^n_+)}, \quad u \in \overline{H}_{(m,s)}(\mathbb{R}^n_+), \tag{B.2.5}$$

and $\overline{H}_{(m,s)}(\mathbb{R}^n_+)$ *consists of all* $u \in \bar{\mathscr{S}}'(\mathbb{R}^n_+)$ *such that* $\Lambda_{m,s} u \in L^2(\mathbb{R}^n_+)$.

Proof. By Example 7.1.17 the inverse Fourier transform of $\Lambda_{m,s}$ (resp. $\bar{\Lambda}_{m,s}$) has support in $\overline{\mathbb{R}}^n_+$ (resp. $\overline{\mathbb{R}}^n_-$). Hence $\Lambda_{m,s}(D)$ is a continuous map in $\dot{\mathscr{S}}'(\overline{\mathbb{R}}^n_+)$ with inverse $\Lambda_{-m,-s}(D)$, which proves the first statement since $|\Lambda_{m,s}(\xi)|^2 = (1+|\xi|^2)^m (1+|\xi'|^2)^s$. The second statement follows by duality.

Remark. Writing $\Lambda_{m,s} = (\langle D' \rangle + iD_n)\Lambda_{m-1,s}$ we conclude from Theorem B.2.4 that every $u \in \dot{H}_{(-m,-s)}(\overline{\mathbb{R}}^n_+)$ can be written $u = u_0 + D_n u_n$ where $u_0 \in \dot{H}_{(1-m,-s-1)}(\overline{\mathbb{R}}^n_+)$ and $u_n \in \dot{H}_{(1-m,-s)}(\overline{\mathbb{R}}^n_+)$. We shall occasionally need this dual version of Theorem B.2.3.

Corollary B.2.5. *If m and s are non-negative integers, then* $\overline{H}_{(m,s)}(\mathbb{R}^n_+)$ *consists of all* $u \in L^2(\mathbb{R}^n_+)$ *such that* $D^\alpha u \in L^2(\mathbb{R}^n_+)$ *when* $|\alpha| \leqq m+s$ *and* $\alpha_n \leqq m$. *For the norm we have the estimate*

$$C_1 \|u\|^2_{(m,s)} \leqq \sum_{\substack{|\alpha| \leqq m+s \\ \alpha_n \leqq m}} \|D^\alpha u\|^2_{L^2} \leqq \|u\|^2_{(m,s)}, \quad u \in \overline{H}_{(m,s)}(\mathbb{R}^n_+). \tag{B.2.6}$$

Here C_1 *is a positive constant depending on s and m but not on u.*

The proof follows immediately from Theorem B.2.3 by induction with respect to m and the obvious fact that $\bar{H}_{(0,0)}(\mathbb{R}^n_+) = L^2(\mathbb{R}^n_+)$. In the same way we obtain for any m and s

$$\|D^\alpha u\|_{(m-x_n, s+\alpha_n-|\alpha|)} \leqq \|u\|_{(m,s)} \quad \text{if } u \in \bar{H}_{(m,s)}(\mathbb{R}^n_+). \tag{B.2.7}$$

Since the elements in $H_{(0,s)}(\mathbb{R}^n)$ can be considered as L^2 functions of x_n with values in $H_{(s)}(\mathbb{R}^{n-1})$ it is clear that $\overline{H}_{(0,s)}(\mathbb{R}^n_+)=L^2(\mathbb{R}_+, H_{(s)}(\mathbb{R}^{n-1}))$. Hence we also obtain from Theorem B.2.3:

Corollary B.2.6. *If m is a non-negative integer the space $\overline{H}_{(m,s)}(\mathbb{R}^n_+)$ consists of all $u\in\mathscr{S}'(\mathbb{R}^n_+)$ such that $D_n^j u$ is an L^2 function of $x_n\in\mathbb{R}_+$ with values in $H_{(s+m-j)}(\mathbb{R}^{n-1})$ when $0\leqq j\leqq m$, and $\|u\|_{(m,s)}$ is equivalent to*

$$\left(\sum_{j=0}^{m}\int_0^\infty \|D_n^j u(.,x_n)\|^2_{(s+m-j)}\,dx_n\right)^{\frac{1}{2}}.$$

The following theorem is an immediate consequence of Theorem B.1.11.

Theorem B.2.7. *If j is a non-negative integer and $m>j+\frac{1}{2}$, then the mapping*

$$C_0^\infty(\overline{\mathbb{R}}^n_+)\ni u\mapsto D_n^j u(.,x_n)$$

can for fixed $x_n\geqq 0$ be extended continuously to a mapping from $\overline{H}_{(m,s)}(\mathbb{R}^n_+)$ into $H_{(s+m-j-\frac{1}{2})}(\mathbb{R}^{n-1})$. For every $u\in\overline{H}_{(m,s)}(\mathbb{R}^n_+)$ the map

$$\overline{\mathbb{R}}_+\ni x_n\mapsto D_n^j u(.,x_n)\in H_{(s+m-j-\frac{1}{2})}(\mathbb{R}^{n-1})$$

thus defined is continuous, and

$$\sup_{x_n\geqq 0}\|D_n^j u(.,x_n)\|_{(s+m-j-\frac{1}{2})}\leqq C_{m,j}\|u\|_{(m,s)},\qquad u\in\overline{H}_{(m,s)}(\mathbb{R}^n_+).$$

Since $H_{(s)}(\mathbb{R}^{n-1})\subset C^\nu(\mathbb{R}^{n-1})$ if $s>(n-1)/2+\nu$ (Corollary 7.9.4), we obtain

Theorem B.2.8. *If $u\in\overline{H}_{(m,s)}(\mathbb{R}^n_+)$ it follows that $D^\alpha u$ is continuous in $\overline{\mathbb{R}}^n_+$ if $|\alpha|<m+s-n/2$ and $\alpha_n<m-\frac{1}{2}$.*

In the study of boundary problems for differential operators Theorem B.2.7 combined with the following theorem related to Theorem 4.4.8′ often allows one to give a precise interpretation of the boundary conditions. Let X be an open subset of $\overline{\mathbb{R}}^n_+$ and let $X^\circ=X\cap\mathbb{R}^n_+$. Since $u\in\overline{H}_{(m,s)}(\mathbb{R}^n_+)$ and $\phi\in C_0^\infty(\mathbb{R}^n)$ implies $\phi u\in\overline{H}_{(m,s)}(\mathbb{R}^n_+)$ we can as in Theorem 10.1.19 introduce the local space $\overline{H}^{\mathrm{loc}}_{(m,s)}(X^\circ)$ consisting of all $u\in\mathscr{D}'(X^\circ)$ such that $\phi u\in\overline{H}_{(m,s)}(\mathbb{R}^n_+)$ for all $\phi\in C_0^\infty(X)$.

Theorem B.2.9. *Let P be a differential operator of order μ,*

$$P=\sum_{|\alpha|\leqq\mu}a_\alpha(x)D^\alpha$$

where $a_\alpha\in C^\infty(X)$ and the coefficient of D_n^μ is equal to 1. If

$$u\in\overline{H}^{\mathrm{loc}}_{(m_1,s_1)}(X^\circ)\quad\text{and}\quad Pu\in\overline{H}^{\mathrm{loc}}_{(m_2-\mu,s_2)}(X^\circ)$$

it follows that $u\in\overline{H}^{\mathrm{loc}}_{(m,s)}(X^\circ)$ if $m\leqq m_2$ and $m+s\leqq m_j+s_j$, $j=1,2$.

Proof. Choose $\phi \in C_0^\infty(X)$ and set $v=\phi u$. Assume first that $m \leqq m_1+1$. Then we have

$$\text{(B.2.8)} \qquad P v \in \bar{H}_{(m-\mu, s)}(\bar{\mathbb{R}}^n_+),$$

for $\phi P u \in \bar{H}_{(m_2-\mu, s_2)} \subset \bar{H}_{(m-\mu, s)}$ by (B.2.1) since $m \leqq m_2$ and $m+s \leqq m_2+s_2$, and $D^\alpha u \in \bar{H}^{\text{loc}}_{(m_1+1-\mu, s_1)} \subset \bar{H}^{\text{loc}}_{(m-\mu, s)}$ if $|\alpha|<\mu$ since $m \leqq m_1+1$ and $m+s \leqq m_1+s_1$. Also note that

$$\text{(B.2.9)} \qquad D^\alpha v \in \bar{H}_{(m_1-\alpha_n, s_1-|\alpha'|)}$$

since $v \in \bar{H}_{(m_1, s_1)}$. This implies $D^\alpha v \in \bar{H}_{(m-\mu, s)}$ if $\alpha_n<\mu$ and $|\alpha| \leqq \mu$, so it follows from (B.2.8) that $D_n^\mu v \in \bar{H}_{(m-\mu, s)}$ since this is true for all other terms in Pv. We claim that $D_n^j v \in \bar{H}_{(m-j, s)}$, $0 \leqq j \leqq \mu$. Since this is already proved when $j=\mu$ we may assume that $j<\mu$ and that $D_n^{j+1} v \in \bar{H}_{(m-j-1, s)}$. Now our claim follows from Theorem B.2.3 for $D_n^j v \in \bar{H}_{(m_1-j, s_1)} \subset \bar{H}_{(m-j-1, s+1)}$. When $j=0$ we obtain $v \in \bar{H}_{(m, s)}$, that is, $u \in \bar{H}^{\text{loc}}_{(m, s)}$, so the theorem is proved provided that $m \leqq m_1+1$. To prove it in general we may as well assume that $m_1+s_1=m_2+s_2$ and that $m_1=m_2-k$ for some integer $k>1$. Then we have shown that $u \in \bar{H}^{\text{loc}}_{(m_1+1, s_1-1)}$, and repeating this argument we obtain

$$u \in \bar{H}^{\text{loc}}_{(m_1+2, s_1-2)}, \ldots, u \in \bar{H}^{\text{loc}}_{(m_1+k, s_1-k)} = \bar{H}^{\text{loc}}_{(m_2, s_2)}.$$

The proof is complete.

Remark. In the theorem we could equally well allow u to have values in $\mathbb{C}^N$ provided that the coefficients are $N \times N$ matrices and the coefficient of D_n^μ is invertible. After multiplication by the inverse of this matrix the proof is unchanged.

The notion of C^∞ *manifold with boundary* is defined by allowing the sets $\tilde{X}_\kappa$ in Definition 6.3.1 to be open subsets of the closed half space $\bar{\mathbb{R}}^n_+$. Since $\mathbb{R}^n$ is diffeomorphic to the open half space $\mathring{\mathbb{R}}^n_+$ it is clear that every manifold is also a manifold with boundary. Furthermore, if X is a manifold of dimension n with boundary, the set X° of all points $x \in X$ for which $\kappa x \in \mathbb{R}^n_+$ for some (and hence for every) coordinate system κ such that $x \in X_\kappa$, is a manifold of dimension n called the *interior* of X. The set of all $x \in X$ such that κx belongs to the boundary of $\bar{\mathbb{R}}^n_+$ for some (and hence for every) κ with $x \in X_\kappa$ forms a manifold of dimension $n-1$, called the *boundary* of X, and is denoted by ∂X. (The coordinate systems in X° and ∂X are of course the restrictions of those in X.)

If X and Y are open subsets of $\bar{\mathbb{R}}^n_+$ and $\psi\colon Y \to X$ is a C^∞ diffeomorphism, then we can use Theorem 1.2.6 to extend ψ to a C^∞ map $\hat{\psi}\colon \hat{Y} \to \hat{X}$ between neighborhoods of Y and X in $\mathbb{R}^n$. It is a diffeomorphism if $\hat{Y}$ is chosen small enough and $\hat{X}=\hat{\psi}(\hat{Y})$. If $u \in \mathscr{D}'(\hat{X})$ and $\operatorname{supp} u \subset X$ then $\hat{\psi}^* u \in \mathscr{D}'(\hat{Y})$ and $\operatorname{supp} \hat{\psi}^* u \subset Y$. There is an obvious identification between the set of such distributions for different choices of $\hat{X}$; we denote it by $\dot{\mathscr{D}}'(X)$. It follows from Theorem 2.3.3 that $\hat{\psi}^*$ defines a map ψ^*:

$\dot{\mathscr{D}}'(X) \to \dot{\mathscr{D}}'(Y)$ independent of the choice of $\hat{X}, \hat{Y}, \hat{\psi}$. Similarly we define the set $\bar{\mathscr{D}}'(X^\circ)$ as the set of restrictions to X° of distributions in some neighborhood of X in $\mathbb{R}^n$. The pullback ψ^* maps $\bar{\mathscr{D}}'(X^\circ)$ to $\bar{\mathscr{D}}'(Y^\circ)$ since an extension to $\hat{Y}$ is provided by the pullback by $\hat{\psi}$ of an extension to $\hat{X}$.

For any C^∞ manifold X with boundary we can now define the space $\bar{\mathscr{D}}'(X^\circ)$ of extendible distributions in X and the space $\dot{\mathscr{D}}'(X)$ of distributions supported by X by an obvious modification of Definition 6.3.3. In view of Theorem B.1.8 we can also define the spaces $\bar{H}^{\text{loc}}_{(s)}(X^\circ) \subset \bar{\mathscr{D}}'(X^\circ)$ and $\dot{H}^{\text{loc}}_{(s)}(X) \subset \dot{\mathscr{D}}'(X)$ just as in the case of open manifolds.

The notion of differential operator in a C^∞ manifold X with boundary is defined just as in Section 6.4 for the case without boundary. If the order is $\leqq m$ it is clear that it defines a continuous map from $\bar{H}^{\text{loc}}_{(s)}(X^\circ)$ to $\bar{H}^{\text{loc}}_{(s-m)}(X^\circ)$ for every s. A boundary problem in X involves, besides a differential operator in X, a certain number of boundary conditions on the boundary ∂X. Each of these is defined by a boundary differential operator, that is, a linear map p of $C^\infty(X)$ into $C^\infty(\partial X)$ such that for every C^∞ coordinate system κ we have

$$(pu)\circ\kappa^{-1} = p^\kappa(u\circ\kappa^{-1}) \quad \text{in} \quad \tilde{X}_\kappa \cap \mathbb{R}^n_0 \quad \text{if } u \in C^\infty(X).$$

Here we have used the notation $\mathbb{R}^n_0 = \{x \in \mathbb{R}^n; x_n = 0\}$, and p^κ denotes a differential operator

$$p^\kappa(x, D) = \sum_\alpha a^\kappa_\alpha(x)\, D^\alpha$$

with coefficients in $C^\infty(\mathbb{R}^n_0 \cap \tilde{X}_\kappa)$. Besides the order of p, which was invariantly defined in Section 6.4, we shall also consider the *transversal* order of p, which is defined as the smallest t such that $a^\kappa_\alpha = 0$ for every κ when $\alpha_n > t$. Equivalently, it is the smallest t such that $pu = 0$ in X for every u vanishing of order $t+1$ on ∂X. (The transversal order, like the order, may be $+\infty$ if ∂X is not compact.) Note that a boundary differential operator of transversal order 0 can also be considered as a differential operator in the manifold ∂X.

Now assume that p is a boundary differential operator with C^∞ coefficients and of order $\mu < \infty$. If the transversal order is $< m - \frac{1}{2}$, it follows from Theorem B.2.7 that the mapping

$$C^\infty(\tilde{X}_\kappa) \ni v \mapsto p^\kappa v \in C^\infty(\mathbb{R}^n_0 \cap \tilde{X}_\kappa)$$

has a unique extension to a continuous mapping of $\bar{H}^{\text{loc}}_{(m)}(\tilde{X}^\circ_\kappa)$ into $\bar{H}_{(m-\mu-\frac{1}{2})}(\mathbb{R}^n_0 \cap \tilde{X}_\kappa)$; we denote the extension also by p^κ. If $u \in \bar{H}^{\text{loc}}_{(m)}(X^\circ)$ it is clear that the distributions $p^\kappa u_\kappa$ in $\mathbb{R}^n_0 \cap \tilde{X}_\kappa$ define a distribution in the manifold ∂X. We denote this distribution by pu and obtain

Theorem B.2.10. *A boundary differential operator p with C^∞ coefficients and of order μ defines a continuous linear mapping*

$$\bar{H}^{\text{loc}}_{(m)}(X^\circ) \ni u \mapsto pu \in H^{\text{loc}}_{(m-\mu-\frac{1}{2})}(\partial X)$$

if the transversal order of p is $< m - \frac{1}{2}$.

As in Section 6.4 there is no difficulty in extending the preceding discussion to distributions with values in vector bundles and differential operators between them. We shall make use of such extensions freely with no further comments.

Appendix C. Some Tools from Differential Geometry

In this appendix we have collected some facts on maps and vector fields which are required in the main text. For most of the topics a more thorough discussion can be found in many texts on differential geometry.

C.1. The Frobenius Theorem and Foliations

The basic existence theorem for first order systems of differential equations may be stated as follows: If

$$v=\sum_1^n v_j(x)\,\partial/\partial x_j$$

is a C^∞ vector field in a neighborhood of 0 in $\mathbb{R}^n$ with $v(0)\neq 0$ then one can introduce new local coordinates $y_1, \dots, y_n$ such that $v=\partial/\partial y_1$ in a neighborhood of 0. We shall now discuss a similar result where we have several given vector fields. Recall that the commutator of two vector fields v and w is defined by

$$[v, w]\,u=v(wu)-w(vu); \qquad u\in C^\infty.$$

This means if $w=\sum w_j(x)\,\partial/\partial x_j$ that

$$[v, w]=\sum(v_k\,\partial w_j/\partial x_k-w_k\,\partial v_j/\partial x_k)\,\partial/\partial x_j.$$

Theorem C.1.1 *(Frobenius). Let $v_1, \dots, v_r$ be C^∞ vector fields in a neighborhood of 0 in $\mathbb{R}^n$. If*

(C.1.1) $\qquad v_1(0), \dots, v_r(0)$ *are linearly independent,*

(C.1.2) $\qquad [v_i, v_j]=\sum c_{ijk}\, v_k$

then there exist new local coordinates $y_1, \dots, y_n$ such that

$$\partial/\partial y_i=\sum_1^r b_{ij}\, v_j, \qquad i=1, \dots, r$$

where b_{ij} is an invertible matrix.

Proof. First we note that $V_i = \sum a_{ij} v_j$, $i=1,\dots,r$, also have the properties (C.1.1), (C.1.2) if $\det(a_{ij}) \neq 0$, for

$$[\sum a_{ik} v_k, \sum a_{jl} v_l] = \sum a_{ik} a_{jl} [v_k, v_l] + \sum a_{ik} (v_k a_{jl}) v_l - \sum a_{jl} (v_l a_{ik}) v_k.$$

The conditions are also invariant under a change of variables. (This proves already that (C.1.1), (C.1.2) are necessary for the conclusion to be true.) Assume now that Theorem C.1.1 is already proved for smaller values of n, which is legitimate since it is trivial when $n=1$. Changing variables we may then as well assume that $v_1 = \partial/\partial x_1$, and after subtraction of a multiple of v_1 we may also assume that

$$v_j = \sum_2^n b_{jl}\, \partial/\partial x_l, \quad j=2,\dots,r.$$

When $x_1 = 0$ these vector fields in $\mathbb{R}^{n-1}$ satisfy our hypotheses so changing the coordinates $x_2, \dots, x_n$ we may by the inductive hypothesis assume that $b_{jl} = 0$ for $j=2,\dots,r$ and $l>r$ when $x_1=0$. Now

$$\partial b_{jl}/\partial x_1 = v_1 v_j x_l = [v_1, v_j] x_l = \sum_{k=2}^r c_{1jk} v_k x_l$$

$$= \sum_{k=2}^r c_{1jk} b_{kl}; \quad l=2,\dots,n;\ j=2,\dots,r.$$

The uniqueness of the Cauchy problem for first order ordinary systems of differential equations now gives that $b_{jl}=0$ if $l>r$, in a neighborhood of 0, and this completes the proof.

Corollary C.1.2. *Assume that $v_1, \dots, v_r$ satisfy the hypotheses of Theorem C.1.1, and let $f_1, \dots, f_r \in C^\infty$. Then the equations*

(C.1.3) $$v_j u = f_j, \quad j=1,\dots,r,$$

have a C^∞ solution in a neighborhood of 0 if and only if

(C.1.4) $$v_i f_j - v_j f_i = \sum_{k=1}^r c_{ijk} f_k; \quad i,j=1,\dots,r.$$

If S is a C^∞ manifold of codimension k with tangent plane at 0 supplementary to the plane spanned by $v_1, \dots, v_r$ then there is a unique solution in a neighborhood of 0 with prescribed restriction $u_0 \in C^\infty(S)$ to S.

Proof. The conditions (C.1.4) follow immediately from (C.1.3) and (C.1.2), and they are obviously invariant if we change variables or make linear combinations of the equations. By Theorem C.1.1 we may therefore assume that $v_j = \partial/\partial x_j$, $j=1,\dots,r$. Then $c_{ijk}=0$ and (C.1.4) reduce to

(C.1.4)′ $$\partial f_j/\partial x_i - \partial f_i/\partial x_j = 0; \quad i,j=1,\dots,r.$$

If $x' = (x_1, \dots, x_r)$ and $x'' = (x_{r+1}, \dots, x_n)$ then S is defined by $x' = h(x'')$ in a neighborhood of 0, and (C.1.3) means that for fixed x'' the differential of u is the form $\sum_1^r f_j\,dx_j$ in x', which is closed by (C.1.4)'. Thus

$$u(x)=\int_{h(x'')}^{x'} \sum_1^r f_j\,dx_j+u_0(h(x''),x'')$$

is the unique solution of (C.1.3) which is equal to u_0 on S.

The coordinates y in Theorem C.1.1 are of course not uniquely determined. If $\tilde{y}_1,\dots,\tilde{y}_n$ are other coordinates which enjoy the same properties, then

$$\partial/\partial y_j=\sum_{k=1}^n \partial\tilde{y}_k/\partial y_j\,\partial/\partial\tilde{y}_k$$

must be a linear combination of $\partial/\partial\tilde{y}_1,\dots,\partial/\partial\tilde{y}_r$ when $j\leqq r$. Thus $\partial\tilde{y}_k/\partial y_j=0$ when $j\leqq r$ and $k>r$, that is, $(\tilde{y}_{r+1},\dots,\tilde{y}_n)$ is (locally) a function of $(y_{r+1},\dots,y_n)$. This suggests the following global notion:

Definition C.1.3. A C^∞ manifold X is said to be foliated of dimension r if it has a distinguished atlas $\mathscr{F}$ such that for the map

$$\phi=\kappa'\circ\kappa^{-1}\colon \kappa(X_\kappa\cap X_{\kappa'})\to\kappa'(X_\kappa\cap X_{\kappa'})$$

the components $\phi_{r+1},\dots,\phi_n(x)$ are functions of $x_{r+1},\dots,x_n$ only if $\kappa,\kappa'\in\mathscr{F}$.

Here we have kept the notation in Definition 6.3.1. A foliation in X defines for every $x\in X$ a plane $\pi_x\subset T_x(X)$ of dimension r by the equations $dx_{r+1}=\dots=dx_n=0$. By Theorem C.1.1 the additional property of a foliation is that, in a neighborhood of any point, π_x has a basis $v_1(x),\dots,v_r(x)$ such that $v_1,\dots,v_r$ are C^∞ vector fields satisfying the Frobenius conditions (C.1.2). (To see that this is sufficient one should note that every atlas has a refinement for which an arbitrary intersection of coordinate patches is connected.) For every $x_0\in X$ there is locally a unique r dimensional manifold $\Sigma\subset X$, called the *leaf* of the foliation through x_0, such that $T_x(\Sigma)=\pi_x$ for every $x\in\Sigma$. With local coordinates in the distinguished atlas, Σ is defined by $x_{r+1}=$constant, ..., $x_n=$constant. Alternatively, it is the set of points which can be reached from x_0 by moving along integral curves of the vector fields $v_1,\dots,v_r$ inside a suitable neighborhood of x_0. The set of all points which can be reached along such curves is the global leaf of the foliation. With any point it contains the local leaf through it but in any neighborhood there may be a countable number of other leaves corresponding to integral curves which are nearly closed. For further details we refer the reader to Haefliger [1].

C.2. A Singular Differential Equation

Let v be a real C^∞ vector field defined in a neighborhood of 0 in $\mathbb{R}^n$. We assume given a splitting $x=(x',x'')$ of the coordinates in $\mathbb{R}^n$, where x'

$=(x_1, \ldots, x_k)$ and $x''=(x_{k+1}, \ldots, x_n)$, and that $v=v_0+w$ where

(C.2.1) $$v_0=\langle Ax', \partial/\partial x'\rangle,$$

(C.2.2) $$w=\sum_1^n w_j(x)\,\partial/\partial x_j; \quad w_j=0,\ \partial w_j/\partial x=0 \quad \text{when } x'=0.$$

Thus v is singular when $x'=0$ and only then if the real $k\times k$ matrix A is non-singular and $|x'|$ is sufficiently small.

Theorem C.2.1. *Let $v=v_0+w$ be a C^∞ vector field in a neighborhood of 0 in $\mathbb{R}^n$ such that v_0 and w satisfy (C.2.1) and (C.2.2), and $\operatorname{Re}\lambda>0$ if λ is an eigenvalue of A. Let g, h be C^∞ functions with $g(0)>0$. Then the equation*

(C.2.3) $$gu+vu=h$$

has a unique C^∞ solution in a neighborhood of 0.

In the proof we shall first construct u so that $gu+vu-h$ vanishes of infinite order when $x'=0$. This requires the following lemma.

Lemma C.2.2. *Let A be a $k\times k$ complex matrix with eigenvalues $\lambda_1, \ldots, \lambda_k$. Then the eigenvalues of $\langle Az, \partial/\partial z\rangle$ as a linear transformation in the space π_m of homogeneous polynomials of degree m in $(z_1, \ldots, z_k)$ are the sums*

$$\langle\alpha, \lambda\rangle=\sum_1^k \alpha_j\lambda_j; \quad \alpha_j\geqq 0 \quad \text{and} \quad \sum\alpha_j=m.$$

Proof. The matrix of $\langle Az, \partial/\partial z\rangle$ operating on π_1 is the adjoint of A so it is obvious that it has the eigenvalues $\lambda_1, \ldots, \lambda_k$. By a linear change of the z variables we can put A in upper triangular form, thus $A_{jk}=0$ when $j>k$ and $A_{jj}=\lambda_j$. Thus

$$\langle Az, \partial/\partial z\rangle z^\alpha=\langle\alpha, \lambda\rangle z^\alpha+Rz^\alpha, \quad R=\sum_{j<k} A_{jk}z_k\,\partial/\partial z_j.$$

The monomials z^α with $|\alpha|=m$ form a basis in π_m. If we order them lexicographically it is clear that R increases the lexicographical order so $\langle Az, \partial/\partial z\rangle$ as an operator in π_m has upper triangular form and the diagonal elements $\langle\alpha, \lambda\rangle$. The proof is complete.

In particular, the lemma shows that $\langle Az, \partial/\partial z\rangle$ is bijective on π_m for every $m\neq 0$ if 0 is not in the convex hull of the eigenvalues.

To pass from formal to genuine solutions of (C.2.3) we shall study the orbits of v by means of the following lemma.

Lemma C.2.3. *If A is a linear transformation in a real k dimensional vector space V and all the eigenvalues are in the open right half plane, then there is a*

Euclidean scalar product $(\ ,\ )$ *in* V *such that for some* $c>0$

$$(Ax, x) \geqq c(x, x), \qquad x \in V.$$

Proof. Let $|\ |$ be a Euclidean norm in V and set

$$\|x\|^2 = \int_0^\infty |e^{-At}x|^2\, dt, \qquad x \in V.$$

The integral is convergent since $e^{-At}x$ is exponentially decreasing, and it is clear that $\|x\|$ is a Euclidean norm. Now

$$\|e^{As}x\|^2 = \int_{-s}^\infty |e^{-At}x|^2\, dt = \|x\|^2 + \int_{-s}^0 |e^{-At}x|^2\, dt.$$

If $(\ ,\ )$ is the scalar product associated with $\|\ \|$, it follows when $s \to 0$ that

$$2(Ax, x) = |x|^2 \geqq 2c\|x\|^2,$$

which proves the lemma.

Proof of Theorem C.2.1. We shall first find a formal power series

$$\sum_0^\infty u^\mu(x', x''),$$

where u^μ is a homogeneous polynomial of degree μ in x' with coefficients C^∞ in x'', which satisfies (C.2.3) of infinite order when $x'=0$. The term of degree μ in the equation is

$$(g(0, x'') + \langle Ax', \partial/\partial x' \rangle)\, u^\mu(x', x'') + R_\mu(x', x'') = 0$$

where R_μ is determined by the data and $u^0, \ldots, u^{\mu-1}$. Since $g>0$ it follows from Lemma C.2.2 that these linear equations for the coefficients of u^μ have a unique solution. By Theorem 1.2.6 we can find $u_0 \in C^\infty$ with the Taylor expansion $\sum u^\mu(x', x'')$ with respect to x' when $x'=0$. The equation (C.2.3) can be written

$$g(u-u_0) + v(u-u_0) = h - gu_0 - vu_0$$

where the right-hand side vanishes of infinite order when $x'=0$. Hence it remains only to prove Theorem C.2.1 when h vanishes of infinite order for $x'=0$, and we know that if a solution exists it must also vanish of infinite order then.

To solve (C.2.3) exactly we must examine the orbits of the vector field v,

(C.2.4) $$dx'/dt = Ax' + w'; \qquad dx''/dt = w''.$$

By Lemma C.2.3 we may assume that $(Ax', x') \geqq c(x', x')$. Hence

$$(dx'/dt, x') \geqq c_0 \|x'\|^2, \qquad \|dx''/dt\| \leqq c_0 \|x'\|$$

if $0 < c_0 < c$ and $|x'| + |x''|$ is small enough. Thus $d\|x'\|/dt \geqq c_0 \|x'\|$ so there is an orbit $x(t)$, $t \leqq 0$, with $x(0) = y$ given sufficiently small, and

$$\|x'(t)\| \leqq \|y'\|\, e^{c_0 t}, \qquad \|x'(t)\| + \|x''(t)\| \leqq \|y'\| + \|y''\| \qquad \text{if } t \leqq 0.$$

We denote the solution by $x(t, y)$ and observe that (C.2.3) can be written

$$(gu)(x(t, y)) + du(x(t, y))/dt = h(x(t, y)). \tag{C.2.3$'$}$$

Set

$$G(t, y) = \int_0^t g(x(s, y))\, ds, \qquad t<0.$$

Then $|G(t, y)| \leq C|t|$. If the function h in (C.2.3) vanishes of infinite order when $x'=0$ and a C^∞ solution u exists, we have seen that it must also vanish of infinite order when $x'=0$ so we must have

$$u(y) = \int_{-\infty}^{0} h(x(t, y))\, e^{G(t, y)}\, dt. \tag{C.2.5}$$

The integral converges since $h(x(t), y) = O(|y'|^N e^{Nc_0 t})$ for any N, and it is clear that (C.2.5) gives a solution of (C.2.3)$'$. By differentiation of (C.2.4) with respect to the parameters y it follows successively that the derivatives of $x(t, y)$ with respect to y are at most exponentially increasing in t. Hence $u \in C^\infty$ which completes the proof.

Remark. Although the perturbation w plays a small role in the preceding proof it is not always possible to remove it by a change of variables, for there might be resonances among the eigenvalues of A.

C.3. Clean Intersections and Maps of Constant Rank

First recall that two C^∞ submanifolds Y and Z of a C^∞ manifold X are said to intersect transversally at $x_0 \in Y \cap Z$ if $T_x(Y) + T_x(Z) = T_x(X)$ when $x = x_0$. An equivalent condition is of course that $N_x(Y) \cap N_x(Z) = \{0\}$. If Y is defined locally by $f_1 = \ldots = f_k = 0$ where $df_1, \ldots, df_k$ are linearly independent at x_0 and Z is similarly defined by $g_1 = \ldots = g_l = 0$, it follows that $df_1, \ldots, df_k, dg_1, \ldots, dg_l$ are all linearly independent at x_0. Thus the intersection $Y \cap Z$ defined by $f_1 = \ldots = f_k = g_1 = \ldots = g_l = 0$ is a C^∞ manifold in a neighborhood of x_0 and

$$\operatorname{codim}(Y \cap Z) = \operatorname{codim} Y + \operatorname{codim} Z, \tag{C.3.1}$$

$$T_x(Y \cap Z) = T_x(Y) \cap T_x(Z), \qquad x \in Y \cap Z. \tag{C.3.2}$$

Now the intersection $Y \cap Z$ may very well be a manifold satisfying (C.3.2) even if the intersection is not transversal. An example is given by any two linear subspaces Y and Z of a vector space X. In that case (C.3.1) is replaced by

$$\operatorname{codim} Y + \operatorname{codim} Z = \operatorname{codim}(Y \cap Z) + e \tag{C.3.3}$$

where $e = \operatorname{codim}(Y + Z)$. This example is in fact quite general:

Proposition C.3.1. *Let Y, Z and $Y \cap Z$ be C^∞ submanifolds of a C^∞ manifold X, and assume that* (C.3.2) *is valid. For any $x_0 \in Y \cap Z$ one can then choose local coordinates in a neighborhood such that Y and Z are defined there by linear equations in the local coordinates. Thus* (C.3.3) *is valid with an integer $e \geqq 0$ vanishing precisely when the intersection is transversal.*

Proof. We choose local coordinates such that $x_0 = 0$, Y is defined by $x_1 = \ldots = x_\nu = 0$, and $Y \cap Z$ is defined by $x_1 = \ldots = x_{\nu+\mu} = 0$ where $\nu = \operatorname{codim} Y$ and $\nu + \mu = \operatorname{codim}(Y \cap Z)$. The manifold Z is defined by $N = \operatorname{codim} Z$ equations $F_j(x) = 0$ with linearly independent differentials. Since $F_j = 0$ when $x_1 = \ldots = x_{\nu+\mu} = 0$, we can write (Theorem 1.1.9)

$$F_j(x) = \sum_{k=1}^{\nu+\mu} F_{jk}(x)\, x_k$$

where $F_{jk} \in C^\infty$. Thus

$$dF_j(x) = \sum_{k=1}^{\nu+\mu} F_{jk}(x)\, dx_k$$

in $Y \cap Z$. By hypothesis these equations together with $dx_1 = \ldots = dx_\nu = 0$ define the tangent plane of $Y \cap Z$, so the matrix

$$(F_{jk}(x))_{j=1,\ldots,N;\, k=\nu+1,\ldots,\nu+\mu}$$

is of rank μ. Suppose that the determinant with $j = 1, \ldots, \mu$ is not 0. In a neighborhood of 0 we can then by elimination find a C^∞ matrix (a_{ij}) such that

$$G_i(x) = F_i(x) - \sum_{j=1}^{\mu} a_{ij}(x) F_j(x) = \sum_{k=1}^{\nu} G_{ik}(x)\, x_k, \quad i = \mu+1, \ldots, N.$$

In a suitably small neighborhood of 0 the equations $G_i(x) = 0$ with $i = \mu+1, \ldots, N$ define a manifold W of codimension $e = N - \mu$ containing Y and Z. The codimensions of Y, Z and $Y \cap Z$ in W are respectively $\nu - e$, $N - e$ and $\nu + \mu - e = \nu - e + N - e$, so (C.3.1) is fulfilled, that is, Y and Z intersect transversally as submanifolds of W. If we choose new local coordinates $y = (y', y'')$ so that W is defined by $y' = 0$, the statement is now reduced to the transversal case where we have already observed that it is true.

Definition C.3.2. If the intersection $Y \cap Z$ of two C^∞ submanifolds of a C^∞ manifold X is a C^∞ submanifold satisfying (C.3.2) then the intersection is said to be clean and the non-negative integer e in (C.3.3) is called its excess.

When using clean intersections it is useful to choose coordinates according to Proposition C.3.1 which allows one to apply statements from linear algebra.

Now consider two C^∞ manifolds X and Y and a C^∞ map $f\colon Y \to X$. If $y_0 \in Y$ and $f'(y_0)$ is injective, then a neighborhood of y_0 is mapped by f on a submanifold of X of dimension $\dim Y$. On the other hand, if $f'(y_0)$ is surjective, we obtain a local fibration of Y over X with fibers $f^{-1}(x)$. This follows from the implicit function theorem. We shall now discuss the more general case where f' has constant but arbitrary rank.

Proposition C.3.3. *Assume that Y and X are C^∞ manifolds and that $f: Y \to X$ is a C^∞ map such that f' is of constant rank r in a neighborhood of $y_0 \in Y$. Then one can choose local coordinates $x_1, \dots, x_m$ and $y_1, \dots, y_n$ at $f(y_0)$ and y_0 so that $f(y) = (y_1, \dots, y_r, 0, \dots, 0)$. Thus f defines locally a fibration of Y over a submanifold of X.*

Proof. We may assume that $X \subset \mathbb{R}^m$, that $f(y_0) = 0$, and that for $f = (f_1, \dots, f_m)$ the differentials of $f_1, \dots, f_r$ are linearly independent at y_0. Coordinates can then be chosen in Y so that $y_0 = 0$ and $f_j(y) = y_j$, $j = 1, \dots, r$. Then $\partial f_j / \partial y_k = 0$ when $j > r$ and $k > r$, for the rank of $\partial f / \partial y$ would otherwise exceed r. Hence $f_{r+1}, \dots, f_m$ are function of $y_1, \dots, y_r$ only, which means that $f(y) = f(y_1, \dots, y_r, 0, \dots, 0)$. The range of f is therefore a submanifold of X of dimension r. Now we change the coordinates in X so that it is defined by $x_{r+1} = \dots = x_m = 0$ and start the proof over again. Then $f_{r+1} = \dots = f_m = 0$ which proves the statement.

C.4. Folds and Involutions

If Y and X are C^∞ manifolds and $f: Y \to X$ is a C^∞ map, then there is a well defined Hessian map going from $\operatorname{Ker} f'(y) \subset T_y(Y)$ to $\operatorname{Coker} f'(y) = T_{f(y)}(X)/f'(y)\, T_y(Y)$ for every $y \in Y$. Indeed, if $\phi: \mathbb{R} \to Y$ is a C^∞ map with $\phi(0) = y$, $\phi'(0) = \eta \in \operatorname{Ker} f'(y)$ then we have in terms of local coordinates

$$\begin{aligned} f(\phi(s)) - f(y) &= f'(y)(\phi(s) - y) + \langle f''(y)(\phi(s) - y), \phi(s) - y\rangle/2 + O(s^3) \\ &= s^2 (f'(y)\, \phi''(0) + \langle f''(y)\eta, \eta\rangle)/2 + O(s^3). \end{aligned}$$

The term $f'(y)\,\phi''(0)$ drops out when we take the image in $\operatorname{Coker} f'(y)$ which shows that

$$\operatorname{Ker} f'(y) \ni \eta \mapsto \langle f''(y)\eta, \eta\rangle \in \operatorname{Coker} f'(y)$$

is an invariantly defined quadratic form, called the Hessian of f.

In the following definition we introduce the class of maps which fail to be diffeomorphisms in the simplest possible way.

Definition C.4.1. If Y and X are C^∞ manifolds and $f: Y \to X$ is a C^∞ map, then f is said to have a fold at $y_0 \in Y$ if $\dim \operatorname{Ker} f'(y_0) = \dim \operatorname{Coker} f'(y_0) = 1$ and the Hessian of f at y_0 is not equal to 0.

Note that the hypothesis implies that Y and X have the same dimension.

Theorem C.4.2. *If $f: Y \to X$ has a fold at y_0 then one can choose local coordinates $y_1, \dots, y_n$ in Y which vanish at y_0 and local coordinates $x_1, \dots, x_n$ in X which vanish at $f(y_0)$ such that f is defined by*

$$f(y_1, \dots, y_n) = (y_1, \dots, y_{n-1}, y_n^2).$$

Proof. Choose local coordinates in X such that for the resulting representation $f = (f_1, \dots, f_n)$ we have $df_n(y_0) = 0$. Then $df_1, \dots, df_{n-1}$ are linearly independent at y_0 so we can choose local coordinates in Y with $y_j = f_j$, $j < n$. Then

$$f(y) = (y_1, \dots, y_{n-1}, f_n(y)),$$

$df_n(0) = 0$ and $\partial^2 f_n(0)/\partial y_n^2 \neq 0$ by hypothesis. The implicit function theorem shows that the equation

$$\partial f_n / \partial y_n = 0$$

has a unique local solution $y_n = g(y')$, $y' = (y_1, \dots, y_{n-1})$, such that $g(0) = 0$. Replacing the variable y_n by $y_n - g(y')$ we obtain new coordinates such that $\partial f_n / \partial y_n = 0$ when $y_n = 0$, that is,

$$f_n(y) = f_n(y', 0) + y_n^2 F(y)$$

where $F \in C^\infty$ and $F(0) \neq 0$ (Taylor's formula). If we take $x_n - f_n(x', 0)$ as a new variable instead of x_n, we have in the new coordinates

$$f(y) = (y_1, \dots, y_{n-1}, y_n^2 F(y)).$$

Changing the sign of x_n if necessary we may assume that $F > 0$. We can then complete the proof by taking $y_n F(y)^{\frac{1}{2}}$ as a new coordinate instead of y_n.

The theorem explains the term "fold" if we observe that when $n = 1$ the graph $\{(f(y), y)\} \subset X \times Y$ is a parabola folded around the x axis.

Corollary C.4.3. *If $f: Y \to X$ has a fold at y_0 then there is in a neighborhood V of y_0 a unique C^∞ map $i: V \to V$ with $f \circ i = f$ which is not the identity.*

Proof. In the local coordinates of Theorem C.4.2 there is precisely one map with this property, and it is given by $(y', y_n) \mapsto (y', -y_n)$.

The map i is an *involution*, that is, $i \circ i$ is the identity. This follows from the explicit representation or else from the uniqueness: If $f \circ i = f$ then $f \circ i \circ i = f \circ i = f$ so $i \circ i = \text{identity}$ since $i \circ i \neq i$. We shall call i the involution defined by the fold. Its fixed point set is the hypersurface where f' is not bijective, defined by $y_n = 0$ in the local coordinates. The importance of the involution is due to the following result:

Theorem C.4.4. *Let $u \in C^\infty$ in a neighborhood of $y_0 \in Y$, where f has a fold. Then one can find $v \in C^\infty$ in a neighborhood of $f(y_0) \in X$ with $f^* v = u$, if and only if $i^* u = u$.*

Proof. If $u=f^*v$ then $i^*u=(f\circ i)^*v=f^*v=u$, so the necessity is clear. To prove the sufficiency we work in the local coordinates of Theorem C.4.2. The function u is then even in y_n so in the formal Taylor expansion

$$\sum \partial_n^j u(y',0)\, y_n^j/j!$$

the terms with j odd must vanish. Choose $v_0\in C^\infty$ according to Theorem 1.2.6 so that v_0 has the Taylor expansion

$$\sum \partial_n^{2j} u(x',0)\, x_n^j/(2j)!.$$

Then $u_1(y)=u(y)-v_0(y', y_n^2)$ vanishes of infinite order when $y_n=0$. Hence

$$v_1(x)=u_1(x', x_n^{\frac{1}{2}}),\quad x_n>0;\qquad v_1(x)=0,\quad x_n\leqq 0,$$

is a C^∞ function, for all derivatives tend to 0 when $x_n\to 0$ since

$$D^\alpha u_1(x', x_n^{\frac{1}{2}})\, x_n^{-N}\to 0,\qquad x_n\to +0,$$

for all α and N. Now $v=v_0+v_1$ has the required property.

Locally every involution leaving a hypersurface fixed can be defined by a folding map:

Theorem C.4.5. *Let Y be a C^∞ manifold and $i\colon Y\to Y$ a C^∞ involution with fixed point set equal to a hypersurface containing $y_0\in Y$. Then one can choose local coordinates $y_1,\dots,y_n$ at y_0 vanishing at y_0 such that*

$$i(y_1,\dots,y_n)=(y_1,\dots,y_{n-1},-y_n).$$

Proof. If $u\in C^\infty(Y)$ then i^*u+u and i^*u-u are respectively even and odd under the involution. When u runs through a system of local coordinates it follows that we can pick out from the functions obtained a local coordinate system $y_1,\dots,y_n$ where every coordinate is either even or odd, that is, $i^*y_j=\pm y_j$. The codimension of the fixed point set is the number of odd coordinates so it must be equal to 1.

On the hypersurface of fixed points the involution defines a line bundle in the tangent bundle consisting of the eigenvectors of the Jacobian map i' with eigenvalue -1. Such vectors are of course transversal to the hypersurface. We shall refer to the bundle as the *reflection bundle* of the involution. If i is defined by a folding map f then the bundle is the kernel of the differential of f.

We shall now examine when two involutions with the same fixed hypersurface can be simultaneously put into a simple form.

Theorem C.4.6. *Assume that f and g are non-trivial C^∞ involutions in Y with the same hypersurface S of fixed points and with linearly independent reflection bundles at $y_0\in S$. Then there exist local coordinates $y_1,\dots,y_n$ vanish-*

ing at y_0 *such that*

$$f(y', y_n) = (y', -y_n), \quad g(y', y_n) = f(y', y_n) + y_n e_1 \tag{C.4.1}$$

where $e_1 = (1, 0, \dots, 0)$.

We could of course have taken any other vector $\neq 0$ with last component equal to 0 instead of e_1. Note that the reflection bundles for these involutions are generated by e_n and by $e_n - e_1/2$ where $e_n = (0, \dots, 0, 1)$.

Proof of Theorem C.4.6. We choose local coordinates $y_1, \dots, y_n$ such that f has the desired form. Since g is an involution leaving the plane $y_n = 0$ pointwise fixed, the Jacobian matrix g' at $(y', 0)$ must agree with the identity matrix except in the last column which is $(a_1(y'), \dots, a_{n-1}(y'), -1)$. It maps (η', η_n) to $(\eta', 0) + \eta_n(a_1(y'), \dots, a_{n-1}(y'), -1)$ which shows that the reflection bundle is generated by $(a_1(y'), \dots, a_{n-1}(y'), -2)$. By hypothesis some a_j is therefore $\neq 0$. We want to introduce new coordinates preserving the simple form of f but making g come closer to the desired form

$$g_0(y_1, \dots, y_n) = (y_1 + y_n, y_2, \dots, y_{n-1}, -y_n).$$

So far we only know that with the notation $a = (a_1, \dots, a_{n-1})$ and some $A \in C^\infty$

$$g(y) = f(y) + y_n(a(y'), y_n A(y')) + O_2.$$

Here and below the notation O_k means that the first $n-1$ components are $O(y_n^k)$ and the last is $O(y_n^{k+1})$. This classification of errors will turn out to be convenient when the variables are changed so that the plane $y_n = 0$ is respected. Now we introduce new coordinates

$$u(y) = (u_1(y'), \dots, u_{n-1}(y'), y_n u_n(y')).$$

This preserves the form of f, and we have

$$\begin{aligned} g_0 \circ u(y) &= (u_1(y') + y_n u_n(y'), u_2(y'), \dots, u_{n-1}(y'), -y_n u_n(y')), \\ u \circ g(y) &= (u_1(y' + y_n a(y')), \dots, u_{n-1}(y' + y_n a(y')), \\ &\quad (-y_n + y_n^2 A(y')) u_n(y' + y_n a(y'))) + O_2. \end{aligned}$$

These compositions are equal apart from an O_2 error if

$$\langle u_1', a\rangle = u_n; \quad \langle u_j', a\rangle = 0, \quad j = 2, \dots, n-1; \quad \langle u_n', a\rangle = A u_n.$$

These differential equations can be solved locally with initial data $u_1 = 0$, $u_n = 1$ and u_j equal to a system of local coordinates on a hypersurface transversal to a in the plane $y_n = 0$. (In particular this means that we put the foliation of S, defined by the intersection of the tangent space of S and the sum of the reflection bundles, in the form where the leaves are along the y_1 axis.) Taking $u_1, \dots, u_{n-1}, y_n u_n$ as new local coordinates we have achieved that $g = g_0 + O_2$.

Assume more generally that we have succeeded to find local coordinates such that f has the simple form (C.4.1) and

$$g(y)=g_0(y)+O_k \tag{C.4.2}$$

for some $k \geqq 2$. We write

$$g(y)=g_0(y)+y_n^k(h(y'), y_n H(y'))+O_{k+1} \tag{C.4.3}$$

where h has $n-1$ components and H has one. Two cases must be distinguished.

a) k is even. Since g is an involution we obtain after a simple calculation

$$y=g \circ g(y)=g_0 \circ g_0(y)+2y_n^k(h(y'), -y_n H(y'))+O_{k+1}.$$

It follows that $h=0$ and that $H=0$.

b) k is odd. Set

$$u(y)=y+y_n^{k-1}(v(y'), y_n V(y')).$$

Then

$$u^{-1}(y)=y-y_n^{k-1}(v(y'), y_n V(y'))+O_{2k-2},$$

and $2k-2 \geqq k+1$ since $k \geqq 3$. A straightforward computation gives

$$\begin{aligned} g \circ u^{-1}(y)=&\, g_0(y)-y_n^{k-1}(v(y')+y_n V(y')e_1, -y_n V(y')) \\ &+y_n^k(h(y'), y_n H(y'))+O_{k+1}, \\ u \circ g \circ u^{-1}(y)=&\, g_0(y)-y_n^{k-1}(v(y')+y_n V(y')e_1, -y_n V(y'))+y_n^k(h(y'), y_n H(y')) \\ &+y_n^{k-1}(v(y'+y_n e_1), -V(y'+y_n e_1)y_n)+O_{k+1} \\ =&\, g_0(y)+y_n^k(h(y')+\partial v(y')/\partial y_1-V(y')e_1, \\ &\, y_n(H(y')-\partial V(y')/\partial y_1))+O_{k+1}. \end{aligned}$$

Here we have written e_1 for the first unit vector in $\mathbb{R}^{n-1}$. This is equal to $g_0 + O_{k+1}$ if

$$\partial v/\partial y_1 = Ve_1 - h, \qquad \partial V/\partial y_1 = H,$$

and these equations are easily solved.

Iteration of the preceding steps leads to a formal power series solution with respect to y_n of the problem of finding $u(y)$ such that

$$u \circ f \circ u^{-1}=f \quad \text{and} \quad u \circ g \circ u^{-1}=g_0.$$

By Theorem 1.2.6 we can choose a C^∞ function u with this Taylor expansion. Introducing the new coordinates u we have then reduced the proof to the case where f is exactly of the form (C.4.1) and $g-g_0$ vanishes of infinite order when $y_n=0$.

As in the proof of Theorem C.2.1 we switch now to another argument in order to eliminate this small error. First note that if w is one of the desired coordinate functions other than y_1 then it must be even (odd) both with respect to f and with respect to g, thus

$$w \circ f = w \circ g. \tag{C.4.4}$$

Conversely, if we have a function w satisfying this equation then

$$w \pm w \circ f = w \pm w \circ g$$

is even (odd) both with respect to f and g. Thus we look for solutions of (C.4.4) or rather the equivalent equation

$$w \circ \delta = w, \qquad \delta = f \circ g. \tag{C.4.4$'$}$$

From the first part of the proof we have that

$$\delta(y) = y + y_n e_1 + \rho(y)$$

where ρ vanishes of infinite order when $y_n = 0$. Set $w_0 = y_j$ for some $j \neq 1$ and set $w = w_0 - w_1$. Then (C.4.4)$'$ is equivalent to

$$w_1 - w_1 \circ \delta = h; \qquad h = w_0 - w_0 \circ \delta.$$

Here h vanishes of infinite order when $y_n = 0$. Choose $h_1 \in C_0^\infty$ equal to h in a neighborhood of 0. It suffices to solve the equation

$$w_1 - w_1 \circ \delta = h_1$$

which is formally done by the infinite series

$$w_1 = h_1 + h_1 \circ \delta + h_1 \circ \delta \circ \delta + \ldots. \tag{C.4.5}$$

We shall prove below that (C.4.5) converges and defines a C^∞ function vanishing of infinite order when $y_n = 0$, for every $h_1 \in C_0^\infty$ which vanishes of infinite order. Accepting this for a moment we replace the coordinate functions $y_2, \ldots, y_n$ by the modifications just constructed, which are even (odd) with respect to both f and g. With the new coordinates f still has the form (C.4.1), since y_1 is also even with respect to f, and

$$g(y) = g_0(y) + e_1 R(y)$$

where R vanishes of infinite order when $y_n = 0$. Now put $w_0 = y_1$. We have

$$w_0 - w_0 \circ \delta = -y_n + h_2(y)$$

where h_2 vanishes of infinite order when $y_n = 0$. Hence we can solve the equation

$$w_1 - w_1 \circ \delta = h_2$$

as before in a neighborhood of 0. Then $w = w_0 - w_1$ satisfies the equation $w - w \circ \delta = -y_n$, and $w - y_1$ vanishes of infinite order when $y_n = 0$. The equation $w - w \circ f \circ g = -y_n$ can be written $w \circ g - w \circ f = y_n$. Hence $\tilde{w} = (w + w \circ f)/2$ is even with respect to f while $\tilde{w} \circ g = \tilde{w} + y_n$. When $\tilde{w}$ is taken as new y_1 coordinate both f and g have the desired form.

It remains to prove the convergence of the series (C.4.5). Put $\delta^k = \delta \circ \ldots \circ \delta$ with k "factors". Then $\delta^k(y) = y + k y_n e_1 + \varepsilon_k(y)$,

$$\varepsilon_{k+1} = \varepsilon_k + E_k e_1 + \rho(y + k y_n e_1 + \varepsilon_k)$$

where E_k is the nth component of ε_k. Thus $\varepsilon_0 = 0$ and for some constant C

$$|\varepsilon_{k+1} - \varepsilon_k - E_k e_1| \leqq C(|y_n| + |E_k|)^4.$$

If $|E_j| \leqq |y_n|$ when $j \leqq k$ it follows by taking the nth component that

$$|E_{j+1} - E_j| \leqq 16 C y_n^4, \quad j \leqq k,$$

$$|E_{k+1}| = |E_{k+1} - E_0| \leqq 16 C k y_n^4 \leqq y_n^2 \leqq |y_n| \quad \text{if } k < (16 C y_n^2)^{-1}, |y_n| < 1.$$

Since this implies

$$|\varepsilon_{j+1} - \varepsilon_j| \leqq y_n^2 + 16 C y_n^4, \quad j < (16 C y_n^2)^{-1},$$

we obtain

$$|\delta^k(y) - y - k y_n e_1| = |\varepsilon_k| \leqq k y_n^2 (1 + 16 C y_n^2) < 2 \quad \text{if } k < (16 C y_n^2)^{-1},$$

provided that $16C > 1$ and $|y_n| < 1$. Thus the iterates move almost as if there were no error term.

Let

$$\omega = \{y;\, |y_n| < \eta, |y| < 1\}$$

where η is small, and let $|y| \leqq M$ in $\operatorname{supp} h_1$. If $y \in \omega$ we have $|\delta^k(y)| > M$ if $|k y_n| > M + 3$ and $k < (16 C y_n^2)^{-1}$; such a k exists if

$$1 + (M+3)/|y_n| < (16 C y_n^2)^{-1}.$$

This is true for all $y \in \omega$ if we choose η so small that

$$(M+3)\eta + \eta^2 < (16C)^{-1}.$$

Breaking off the series (C.4.5) at the first such integer we still have a C^∞ solution of (C.4.4)′ in ω when $y_n \neq 0$. The number of terms is $O(1/|y_n|)$. It remains to give bounds showing that all derivatives tend to 0 when $y_n \to 0$.

Let N be any positive integer and set

$$|f|_N = \sum_{|\alpha| \leqq N} |\partial^\alpha f|.$$

Then we have with a constant C_N depending on N

$$|\varepsilon_{k+1}|_N \leqq |\varepsilon_k|_N + |E_k|_N + C_N |y_n|^5 (1 + |\varepsilon_k|_N)^N,$$

$$|E_{k+1}|_N \leqq |E_k|_N + C_N |y_n|^5 (1 + |\varepsilon_k|_N)^N.$$

We want to show that $|\varepsilon_j|_N < y_n^2$, $j \leqq k$, if y_n is small and k is the number of terms which we have estimated by $O(1/|y_n|)$. This would imply that $|w_1|_N = O(y_n)$ and prove that $w_1 \in C^\infty(\omega)$ with all derivatives 0 when $y_n = 0$. Assume that we know that $|\varepsilon_j|_N < y_n^2$ for all $j \leqq J < k$. If y_n is small it follows that

$$|E_{j+1}|_N \leqq |E_j|_N + 2 C_N |y_n|^5, \quad j \leqq J,$$

and adding these inequalities we obtain since $E_0=0$

$$|E_j|_N \leqq 2C_N k|y_n|^5 < C'_N y_n^4, \quad j \leqq J+1.$$

But this implies that

$$|\varepsilon_{j+1}|_N \leqq |\varepsilon_j|_N + C''_N y_n^4 \quad \text{if } j \leqq J,$$

and adding these inequalities we obtain if y_n is small

$$|\varepsilon_j|_N \leqq C''_N k y_n^4 < y_n^2 \quad \text{if } j \leqq J+1.$$

By induction it follows that $|\varepsilon_j|_N < y_n^2$ if $j \leqq k$ and y_n is small, and this finishes the proof. Note that the estimates have followed the simple principle that the nth components of the iterations are estimated first since they hardly change at all. The other components are easy to estimate afterwards. This ends the proof of Theorem C.4.6.

Corollary C.4.7. *Let the hypotheses of Theorem C.4.6 be fulfilled and let v be a tangent vector of S at y_0 which is not in the span of the reflection bundles. Then there is a hypersurface $Y_1 \subset Y$ through y_0 which is transversal to v and is invariant under f and g so that they induce involutions in Y_1 with all the properties assumed in Y.*

Proof. With the local coordinates in Theorem C.4.6 the hypothesis on v is that $v=(v_1, \dots, v_{n-1}, 0)$ with $v_j \neq 0$ for some $j \neq 1$. The surface

$$\left\{y;\ \sum_2^{n-1} y_j v_j = 0\right\}$$

is then transversal to v and clearly the involutions f and g have basically the same form there.

It is now easy to prove a homogeneous version of Theorem C.4.6.

Theorem C.4.8. *Assume that f and g are non-trivial homogeneous C^∞ involutions of the conic C^∞ manifold Y having the same fixed point hypersurface Y_0. Assume that the two reflection bundles and the radial vector are linearly independent at $y_0 \in Y_0$. Then $n=\dim Y \geqq 3$ and there exist local coordinates $y_1, \dots, y_n$ in a conic neighborhood of y_0 such that $y_1, \dots, y_{n-1}$ are homogeneous of degree 0, y_n is homogeneous of degree 1, and*

$$f(y_1, \dots, y_n) = (-y_1, y_2, \dots, y_n);$$
$$g(y_1, \dots, y_n) = (-y_1, y_2, \dots, y_{n-1} + y_1, y_n).$$

Proof. The surface Y_0 is conic so the radial vector ρ is a tangent at y_0. By hypothesis it is not in the linear span of the reflection bundles so using Corollary C.4.7 and Theorem C.4.6 we can choose a hypersurface Y_1 through y_0 transversal to $\rho(y_0)$ and local coordinates $y_1, \dots, y_{n-1}$ in Y_1 which vanish at y_0, such that $(y_1, \dots, y_{n-1})$ is mapped to $(-y_1, y_2, \dots, y_{n-1})$

by f and to $(-y_1, y_2, \dots, y_{n-1}+y_1)$ by g. Now extend these coordinates to polar coordinates with respect to Y_1, that is, define $y_n=1$ in Y_1 and extend all y_j to homogeneous functions of the stated degree of homogeneity. By the homogeneity of f and g we then obtain the stated conclusion.

C.5. Geodesic Normal Coordinates

Let X be a C^∞ manifold and $g(x,\xi)$ a real C^∞ quadratic form in $T^*(X)$. If g is non-degenerate then the dual quadratic form in $T(X)$ is a (pseudo-) Riemannian metric and it is well known then how one can introduce geodesic normal coordinates with respect to a submanifold. We shall now give a symplectic derivation of such results using weaker hypotheses on g.

Let $Y\subset X$ be a C^∞ submanifold of X such that g restricted to the conormal bundle $N^*(Y)$ is non-degenerate. This implies that

$$V=\{(y, \partial g(y,\eta)/\partial\eta);\ (y,\eta)\in N^*(Y)\}$$

is a subbundle of $T_Y(X)$ of total dimension equal to $\dim X$. The annihilator in $T_Y^*(X)$ of V is the g orthogonal bundle of $N^*(Y)$. Thus it is disjoint with $N^*(Y)\setminus 0$ so V is dual to $N^*(Y)$, hence isomorphic to the normal bundle $T_Y(X)/T(Y)$. We shall deduce a natural diffeomorphism between a neighborhood of 0 in V and a neighborhood of Y in X by means of this identification and the flowout from the Lagrangian $N^*(Y)$ along the Hamilton field of $g(x,\xi)/2$, defined in local coordinates by

$$H=(\partial g/\partial\xi\,\partial/\partial x-\partial g/\partial x\,\partial/\partial\xi)/2.$$

We shall write ϕ_t for $\exp tH$.

Theorem C.5.1. *If U is a sufficiently small neighborhood of the zero section in $N^*(Y)$ then the Lagrangian $\phi_1 U\subset T^*(X)$ is a section S over a neighborhood U_X of Y in X. If $\pi_S\colon S\to U_X$ is the restriction of the projection to S, then $S=\{(x, dF(x)), x\in U_X\}$ where $F=(\phi_1^{-1}\pi_S^{-1})^* g/2$. One can choose local coordinates (x', x'') in X near any point in Y so that Y is defined by $x'=0$ and with a fixed non-degenerate quadratic form $g_0(\xi')$*

(C.5.1) $$\partial g(x,\xi)/\partial\xi=\partial g_0(\xi')/\partial\xi \quad \text{when } \xi''=0 \text{ and } x'=t\,\partial g_0(\xi')/\partial\xi' \quad \text{for some } t\in\mathbb{R}.$$

In particular, $g(0,x'',\xi)=g_0(\xi')+g_1(x'',\xi'')$ for some quadratic form g_1.

Corollary C.5.2. *When Y is a point we obtain coordinates with the origin at Y and*

$$\partial g(x,\xi)/\partial\xi=\partial g(0,\xi)/\partial\xi \quad \text{when} \quad x=t\,\partial g(0,\xi)/\partial\xi \quad \text{for some } t\in\mathbb{R}.$$

If we write $g(x,\xi)=\langle g(x)\xi,\xi\rangle$ with $g(x)$ symmetric, this means that

(C.5.2) $$x=g(x)\,g(0)^{-1}\,x.$$

Corollary C.5.3. *When Y is a non-characteristic hypersurface, then ξ' is one dimensional and*

$$g(x,\xi)=\pm\xi_1^2+r(x,\xi'')$$

if we write ξ_1 instead of ξ' and $\xi''=(\xi_2,\dots,\xi_n)$.

In the positive definite case Corollary C.5.2 gives standard normal coordinates when we take $g(0)=$ identity. Note that Corollary C.5.3 does not assume that g is non-degenerate.

Proof of Theorem C.5.1. By the homogeneity of g the Hamilton equations

$$2dx/dt=\partial g/\partial\xi, \quad 2d\xi/dt=-\partial g/\partial x \tag{C.5.3}$$

for the orbits of H imply for fixed $s\neq 0$ that

$$2dx/d(st)=g'_\xi(x,\xi/s), \quad d(\xi/s)/d(st)=-g'_x(x,\xi/s).$$

Hence

$$\phi_t(x,\xi)=(x(t),\xi(t)) \text{ implies } \phi_{st}(x,\xi/s)=(x(t),\xi(t)/s). \tag{C.5.4}$$

Choose local coordinates (x',x'') so that Y is defined near 0 by $x'=0$. Since (C.5.4) gives

$$\pi\phi_1(x,\varepsilon\xi)=\pi\phi_\varepsilon(x,\xi)=x+\varepsilon/2\,\partial g/\partial\xi+O(\varepsilon^2)$$

it follows that

$$\pi\phi_1(x,\xi)=x+\partial(g/2)/\partial\xi+O(|\xi|^2), \quad \xi\to 0. \tag{C.5.5}$$

By hypothesis the map $\xi'\mapsto\partial g(0,\xi',0)/\partial\xi'$ is surjective so the implicit function theorem shows that $\pi\phi_1$ is a local diffeomorphism at the zero section of $N^*(Y)=\{(0,x'',\xi',0)\}$. Choose a neighborhood U with convex fibers so small that $\pi\phi_1$ is a diffeomorphism on U. Since ϕ_1 is canonical we know that $S=\phi_1 U$ is Lagrangian. We shall compute the differential $\xi\,dx$ restricted to S in terms of the coordinates in U. To do so we compute

$$\frac{\partial}{\partial t}\phi_t^*(\xi\,dx)=\phi_t^*\frac{\partial}{\partial s}\phi_s^*(\xi\,dx)|_{s=0}=\phi_t^*(d\xi/dt\,dx+\xi\,d(dx/dt)).$$

The parenthesis is equal to

$$\begin{aligned}(-\partial g/\partial x\,dx+\xi\,d(\partial g/\partial\xi))/2&=((\xi\,\partial^2 g/\partial\xi\,\partial x-\partial g/\partial x)\,dx+\xi\,\partial^2 g/\partial\xi^2\,d\xi)/2\\&=(\partial g/\partial x\,dx+\partial g/\partial\xi\,d\xi)/2=dg/2\end{aligned}$$

by Euler's identity for homogeneous functions. Now g is constant on the orbits of H so we obtain $\phi_1^*(\xi\,dx)=dg/2$, for $\phi_0^*(\xi\,dx)=0$ since $\xi\,dx=0$ on $N^*(Y)$. Thus S is the graph of dF as stated.

We can choose our local coordinates so that the vector bundle V is tangent to the fibers $x''=$ constant when $x'=0$. In fact, since V is transversal to $T(Y)$ it is defined by an equation of the form $dx''=f(x'')\,dx'$ where f is a smooth matrix. If we replace the coordinates x'' by $x''-f(x'')x'$ the desired property is obtained. It means that $\partial^2 g/\partial\xi'\partial\xi''=0$ on $N^*(Y)$.

With our local coordinates the quadratic form g restricted to $N^*(Y)$ is a non-degenerate form in ξ' depending on the parameter x''. By completion of squares we obtain a local change of variables $y'=T(x'')x'$, $y''=x''$ which makes it independent of x''. Thus there is a non-degenerate quadratic form $g_0(\eta')$ such that $g(0,y'',\eta)-g_0(\eta')$ is a form in the η'' variables. Now consider the diffeomorphism from V to X

$$(g_0'(\eta')/2, y'')\mapsto \pi\phi_1(0,y'',\eta',0); \quad (y'',\eta')\in U.$$

By (C.5.5) the Jacobian matrix is the identity when $\eta'=0$. We can change the coordinates in X so that the map becomes the identity, without affecting the conditions previously imposed on the coordinates at Y. Then the function F is defined by

$$F(g_0'(\eta')/2, y'')=g_0(\eta')/2.$$

If G_0 is the dual quadratic form of g_0 then $y'=g_0'(\eta')/2$ is equivalent to $\eta'=G_0'(y')/2$, thus $F'(y)=(G_0'(y')/2,0)$ or

$$F'(g_0'(\eta')/2, y'')=(\eta',0) \tag{C.5.6}$$

By (C.5.4) we have with our present coordinates

$$\pi\phi_t(0,y'',\eta',0)=\pi\phi_1(0,y'',t\eta',0)=(t\,g_0'(\eta')/2, y'').$$

Thus the solution of the Hamilton equations (C.5.3) with initial values $(0,y'',\eta',0)$ is $x=(t\,g_0'(\eta')/2, y'')$, and when $t=1$ we have $\xi=F'(x)=(\eta',0)$. Since the Hamilton equation for dx/dt gives for $t=1$

$$(g_0'(\eta')/2,0)=\tfrac{1}{2}\partial g(g_0'(\eta')/2, y'',\xi)/\partial\xi,$$

this completes the proof of Theorem C.5.1.

C.6. The Morse Lemma with Parameters

The advantage of the Morse lemma is that it shows that a C^∞ function is equivalent to a polynomial near a non-degenerate critical point. We need the result with parameters:

Lemma C.6.1. *Let $f(x,y)$ $(x\in\mathbb{R}^n, y\in\mathbb{R}^N)$ be a real valued C^∞ function in a neighborhood of $(0,0)$. Assume that $f_x'(0,0)=0$ and that $A=f_{xx}''(0,0)$ is non-singular. Then the equation $f_x'(x,y)=0$ determines in a neighborhood of 0 a C^∞ function $x(y)$ with $x(0)=0$, and we have in a neighborhood of $(0,0)$*

$$f(x,y)=f(x(y),y)+\langle Az,z\rangle/2$$

where $z=x-x(y)+O(|x-x(y)|(|x|+|y|))$ is a C^∞ function of (x,y) at $(0,0)$.

Proof. By the implicit function theorem the equation $f'_x(x,y)=0$ has a unique solution $x(y)$ near 0. Introducing $x-x(y)$ as a new variable instead of x we may assume that $f'_x(0,y)=0$ for small y. By Taylor's formula we have

$$f(x,y)-f(0,y)=\sum b_{jk}(x,y)\,x_j x_k/2,$$

$$b_{jk}(x,y)=2\int_0^1 (1-t)\,\partial_j\partial_k f(tx,y)\,dt.$$

Thus $B=(b_{jk})$ is a C^∞ function of (x,y) with values in the finite dimensional vector space of symmetric $n\times n$ matrices, and $B(0,0)=A$. Write $z=R(x,y)\,x$ where R is a $n\times n$ matrix to be determined so that $R(0,0)=I$ and

$$R^*AR=B. \tag{C.6.1}$$

The equation is valid when $x=y=0$ and $R=I$, $B=A$. The differential of the map $R\mapsto R^*AR$ is then

$$R\mapsto R^*A+AR.$$

It is surjective for if C is a symmetric matrix we have $R^*A+AR=C$ when $R=A^{-1}C/2$. Hence the inverse function theorem shows that (C.6.1) is fulfilled when $R=F(B)$ where F is a C^∞ function defined for symmetric $n\times n$ matrices close to A with arbitrary $n\times n$ matrix values and $F(A)=I$. The lemma follows with $z=F(B(x,y))\,x$.

Lemma C.6.1 could have been used to derive Theorem 7.7.6 from Lemma 7.7.3 although it is hard to calculate the operators $L_{f,j,y}$ there in that way. Another application occurs in Section 22.3:

Lemma C.6.2. *Let f be a non-negative C^∞ function with $f(0)=0$ defined in a neighborhood of $0\in\mathbb{R}^k$, and let n be the rank of $f''(0)$. Then one can find C^∞ functions $c_1,\dots,c_n$ and g vanishing at 0 such that $dc_1,\dots,dc_n$ are linearly independent at 0, $g\geqq 0$, $g''(0)=0$ and*

$$f=\sum_1^n c_j^2+g \tag{C.6.2}$$

Proof. Since $f(0)=0\leqq f$ we have $f'(0)=0$. By a linear change of coordinates we can make sure that $\det f''_{xx}(0)\neq 0$ if $x=(x_1,\dots,x_n)$ are the first n coordinates. Let $y=(y_1,\dots,y_{k-n})$ be the others. If we apply Lemma C.6.1 and introduce z,y as new variables, we obtain

$$f(x)=g(y)+\langle Az,z\rangle$$

where A is positive definite and $g\geqq 0$. The Hessian of f in these coordinates is the direct sum of A and of $g''(0)$ so $g''(0)=0$ by the definition of n. If we write $\langle Az,z\rangle$ as a sum of squares and return to the original variables, the lemma is proved.

Notes

For Sobolev spaces $H_{(s)}$ the results discussed in Section B.1 are so well known that we shall not try to trace their origins. In order to have appropriate spaces for an intrinsic definition of Lagrangian distributions we have also included a very limited discussion of Besov spaces. For a more systematic presentation we refer to Peetre [4]. The results in Section B.2 for a manifold with boundary have been taken from Peetre [1] via the predecessor of this book.

We shall also refrain from historical comments on the majority of the scattered topics discussed in Appendix C. The only recent result is Theorem C.4.6. It has been isolated from the proof of Theorem 21.4.12 in Melrose [2] as the non-symplectic part of his argument. The rest of it was given in Section 21.4.

Bibliography

Agmon, S.: [1] The coerciveness problem for integro-differential forms. J. Analyse Math. 6, 183–223 (1958).
– [2] Spectral properties of Schrödinger operators. Actes Congr. Int. Math. Nice 2, 679–683 (1970).
– [3] Spectral properties of Schrödinger operators and scattering theory. Ann. Scuola Norm. Sup. Pisa (4) 2, 151–218 (1975).
– [4] Unicité et convexité dans les problèmes différentiels. Sém. Math. Sup. No 13, Les Presses de l'Univ. de Montreal, 1966.
– [5] Lectures on elliptic boundary value problems. van Nostrand Mathematical Studies 2, Princeton, N.J. 1965.
– [6] Problèmes mixtes pour les équations hyperboliques d'ordre supérieur. Coll. Int. CNRS 117, 13–18, Paris 1962.
– [7] Some new results in spectral and scattering theory of differential operators on $\mathbb{R}^n$. Sém. Goulaouic-Schwartz 1978–1979, Exp. II, 1–11.
Agmon, S., A. Douglis and L. Nirenberg: [1] Estimates near the boundary for solutions of elliptic partial differential equations satisfying general boundary conditions. I. Comm. Pure Appl. Math. 12, 623–727 (1959); II. Comm. Pure Appl. Math. 17, 35–92 (1964).
Agmon, S. and L. Hörmander: [1] Asymptotic properties of solutions of differential equations with simple characteristics. J. Analyse Math. 30, 1–38 (1976).
Agranovich, M.S.: [1] Partial differential equations with constant coefficients. Uspehi Mat. Nauk 16:2, 27–94 (1961). (Russian; English translation in Russian Math. Surveys 16:2, 23–90 (1961).)
Ahlfors, L. and M. Heins: [1] Questions of regularity connected with the Phragmén-Lindelöf principle. Ann. of Math. 50, 341–346 (1949).
Airy, G.B.: [1] On the intensity of light in a neighborhood of a caustic. Trans. Cambr. Phil. Soc. 6, 379–402 (1838).
Alinhac, S.: [1] Non-unicité du problème de Cauchy. Ann. of Math. 117, 77–108 (1983).
– [2] Non-unicité pour des opérateurs différentiels à caractéristiques complexes simples. Ann. Sci. École Norm. Sup. 13, 385–393 (1980).
– [3] Uniqueness and non-uniqueness in the Cauchy problem. Contemporary Math. 27, 1–22 (1984).
Alinhac, S. and M.S. Baouendi: [1] Uniqueness for the characteristic Cauchy problem and strong unique continuation for higher order partial differential inequalities. Amer. J. Math. 102, 179–217 (1980).
Alinhac, S. and C. Zuily: [1] Unicité et non-unicité du problème de Cauchy pour des opérateurs hyperboliques à caractéristiques doubles. Comm. Partial Differential Equations 6, 799–828 (1981).
Alsholm, P.K.: [1] Wave operators for long range scattering. Mimeographed report, Danmarks Tekniske Højskole 1975.
Alsholm, P.K. and T. Kato: [1] Scattering with long range potentials. In Partial Diff. Eq., Proc. of Symp. in Pure Math. 23, 393–399. Amer. Math. Soc. Providence, R.I. 1973.

Amrein, W.O., Ph.A. Martin and P. Misra: [1] On the asymptotic condition of scattering theory. Helv. Phys. Acta 43, 313–344 (1970).

Andersson, K.G.: [1] Propagation of analyticity of solutions of partial differential equations with constant coefficients. Ark. Mat. 8, 277–302 (1971).

Andersson, K.G. and R.B. Melrose: [1] The propagation of singularities along glidings rays. Invent. Math. 41, 197–232 (1977).

Arnold, V.I.: [1] On a characteristic class entering into conditions of quantization. Funkcional. Anal. i Priložen. 1, 1–14 (1967) (Russian); also in Functional Anal. Appl. 1, 1–13 (1967).

Aronszajn, N.: [1] Boundary values of functions with a finite Dirichlet integral. Conference on Partial Differential Equations 1954, University of Kansas, 77–94.

– [2] A unique continuation theorem for solutions of elliptic partial differential equations or inequalities of second order. J. Math. Pures Appl. 36, 235–249 (1957).

Aronszajn, N., A. Krzywcki and J. Szarski: [1] A unique continuation theorem for exterior differential forms on Riemannian manifolds. Ark. Mat. 4, 417–453 (1962).

Asgeirsson, L.: [1] Über eine Mittelwerteigenschaft von Lösungen homogener linearer partieller Differentialgleichungen 2. Ordnung mit konstanten Koeffizienten. Math. Ann. 113, 321–346 (1937).

Atiyah, M.F.: [1] Resolution of singularities and division of distributions. Comm. Pure Appl. Math. 23, 145–150 (1970).

Atiyah, M.F. and R. Bott: [1] The index theorem for manifolds with boundary. Proc. Symp. on Differential Analysis, 175–186. Oxford 1964.

– [2] A Lefschetz fixed point formula for elliptic complexes. I. Ann. of Math. 86, 374–407 (1967).

Atiyah, M.F., R. Bott and L. Gårding: [1] Lacunas for hyperbolic differential operators with constant coefficients. I. Acta Math. 124, 109–189 (1970).

– [2] Lacunas for hyperbolic differential operators with constant coefficients. II. Acta Math. 131, 145–206 (1973).

Atiyah, M.F., R. Bott and V.K. Patodi: [1] On the heat equation and the index theorem. Invent. Math. 19, 279–330 (1973).

Atiyah, M.F. and I.M. Singer: [1] The index of elliptic operators on compact manifolds Bull. Amer. Math. Soc. 69, 422–433 (1963).

– [2] The index of elliptic operators. I, III. Ann. of Math. 87, 484–530 and 546–604 (1968).

Atkinson, F.V.: [1] The normal solubility of linear equations in normed spaces. Mat. Sb. 28 (70), 3–14 (1951) (Russian).

Avakumovič, V.G.: [1] Über die Eigenfunktionen auf geschlossenen Riemannschen Mannigfaltigkeiten. Math. Z. 65, 327–344 (1956).

Bang, T.: [1] Om quasi-analytiske funktioner. Thesis, Copenhagen 1946, 101 pp.

Baouendi, M.S. and Ch. Goulaouic: [1] Nonanalytic-hypoellipticity for some degenerate elliptic operators. Bull. Amer. Math. Soc. 78, 483–486 (1972).

Beals, R.: [1] A general calculus of pseudo-differential operators. Duke Math. J. 42, 1–42 (1975).

Beals, R. and C. Fefferman: [1] On local solvability of linear partial differential equations. Ann. of Math. 97, 482–498 (1973).

– [2] Spatially inhomogeneous pseudo-differential operators I. Comm. Pure Appl. Math. 27, 1–24 (1974).

Beckner, W.: [1] Inequalities in Fourier analysis. Ann. of Math. 102, 159–182 (1975).

Berenstein, C.A. and M.A. Dostal: [1] On convolution equations I. In L'anal. harm. dans le domaine complexe. Springer Lecture Notes in Math. 336, 79–94 (1973).

Bernstein, I.N.: [1] Modules over a ring of differential operators. An investigation of the fundamental solutions of equations with constant coefficients. Funkcional. Anal. i Priložen. 5:2, 1–16 (1971) (Russian); also in Functional Anal. Appl. 5, 89–101 (1971).

Bernstein, I.N. and S.I. Gelfand: [1] Meromorphy of the function P^λ. Funkcional. Anal. i Priložen. 3:1, 84–85 (1969) (Russian); also in Functional Anal. Appl. 3, 68–69 (1969).
Bernstein, S.: [1] Sur la nature analytique des solutions des équations aux dérivées partielles du second ordre. Math. Ann. 59, 20–76 (1904).
Beurling, A.: [1] Quasi-analyticity and general distributions. Lectures 4 and 5, Amer. Math. Soc. Summer Inst. Stanford 1961 (Mimeographed).
– [2] Sur les spectres des fonctions. Anal. Harm. Nancy 1947, Coll. Int. XV, 9–29.
– [3] Analytic continuation across a linear boundary. Acta Math. 128, 153–182 (1972).
Björck, G.: [1] Linear partial differential operators and generalized distributions. Ark. Mat. 6, 351–407 (1966).
Björk, J.E.: [1] Rings of differential operators. North-Holland Publ. Co. Math. Library series 21 (1979).
Bochner, S.: [1] Vorlesungen über Fouriersche Integrale. Leipzig 1932.
Boman, J.: [1] On the intersection of classes of infinitely differentiable functions. Ark. Mat. 5, 301–309 (1963).
Bonnesen, T. and W. Fenchel: [1] Theorie der konvexen Körper. Erg. d. Math. u. ihrer Grenzgeb. 3, Springer Verlag 1934.
Bony, J.M.: [1] Une extension du théorème de Holmgren sur l'unicité du problème de Cauchy. C.R. Acad. Sci. Paris 268, 1103–1106 (1969).
– [2] Extensions du théorème de Holmgren. Sém. Goulaouic-Schwartz 1975–1976, Exposé no. XVII.
– [3] Equivalence des diverses notions de spectre singulier analytique. Sém. Goulaouic-Schwartz 1976–1977, Exposé no. III.
Bony, J.M. and P. Schapira: [1] Existence et prolongement des solutions holomorphes des équations aux dérivées partielles. Invent. Math. 17, 95–105 (1972).
Borel, E.: [1] Sur quelques points de la théorie des fonctions. Ann. Sci. École Norm. Sup. 12 (3), 9–55 (1895).
Boutet de Monvel, L.: [1] Comportement d'un opérateur pseudo-différentiel sur une variété à bord. J. Analyse Math. 17, 241–304 (1966).
– [2] Boundary problems for pseudo-differential operators. Acta Math. 126, 11–51 (1971).
– [3] On the index of Toeplitz operators of several complex variables. Invent. Math. 50, 249–272 (1979).
– [4] Hypoelliptic operators with double characteristics and related pseudo-differential operators. Comm. Pure Appl. Math. 27, 585–639 (1974).
Boutet de Monvel, L., A. Grigis and B. Helffer: [1] Parametrixes d'opérateurs pseudo-différentiels à caractéristiques multiples. Astérisque 34–35, 93–121 (1976).
Boutet de Monvel, L. and V. Guillemin: [1] The spectral theory of Toeplitz operators. Ann. of Math. Studies 99 (1981).
Brézis, H.: [1] On a characterization of flow-invariant sets. Comm. Pure Appl. Math. 23, 261–263 (1970).
Brodda, B.: [1] On uniqueness theorems for differential equations with constant coefficients. Math. Scand. 9, 55–68 (1961).
Browder, F.: [1] Estimates and existence theorems for elliptic boundary value problems. Proc. Nat. Acad. Sci. 45, 365–372 (1959).
Buslaev, V.S. and V.B. Matveev: [1] Wave operators for the Schrödinger equation with a slowly decreasing potential. Theor. and Math. Phys. 2, 266–274 (1970). (English translation.)
Calderón, A.P.: [1] Uniqueness in the Cauchy problem for partial differential equations. Amer. J. Math. 80, 16–36 (1958).
– [2] Existence and uniqueness theorems for systems of partial differential equations. Fluid Dynamics and Applied Mathematics (Proc. Symp. Univ. of Maryland 1961), 147–195. New York 1962.

– [3] Boundary value problems for elliptic equations. Outlines of the joint Soviet-American symposium on partial differential equations, 303–304, Novosibirsk 1963.

Calderón, A.P. and R. Vaillancourt: [1] On the boundedness of pseudo-differential operators. J. Math. Soc. Japan 23, 374–378 (1972).

– [2] A class of bounded pseudo-differential operators. Proc. Nat. Acad. Sci. U.S.A. 69, 1185–1187 (1972).

Calderón, A.P. and A. Zygmund: [1] On the existence of certain singular integrals. Acta Math. 88, 85–139 (1952).

Carathéodory, C.: [1] Variationsrechnung und partielle Differentialgleichungen Erster Ordnung. Teubner, Berlin, 1935.

Carleman, T.: [1] Sur un problème d'unicité pour les systèmes d'équations aux dérivées partielles à deux variables indépendentes. Ark. Mat. Astr. Fys. 26B No 17, 1–9 (1939).

– [2] L'intégrale de Fourier et les questions qui s'y rattachent. Publ. Sci. Inst. Mittag-Leffler, Uppsala 1944.

– [3] Propriétés asymptotiques des fonctions fondamentales des membranes vibrantes. C.R. Congr. des Math. Scand. Stockholm 1934, 34–44 (Lund 1935).

Catlin, D.: [1] Necessary conditions for subellipticity and hypoellipticity for the $\bar{\partial}$ Neumann problem on pseudoconvex domains. In Recent developments in several complex variables. Ann. of Math. Studies 100, 93–100 (1981).

Cauchy, A.: [1] Mémoire sur l'intégration des équations linéaires. C.R. Acad. Sci. Paris 8 (1839). In Œuvres IV, 369–426, Gauthier-Villars, Paris 1884.

Cerezo, A., J. Chazarain and A. Piriou: [1] Introduction aux hyperfonctions. Springer Lecture Notes in Math. 449, 1–53 (1975).

Chaillou, J.: [1] Hyperbolic differential polynomials and their singular perturbations. D. Reidel Publ. Co. Dordrecht, Boston, London 1979.

Chazarain, J.: [1] Construction de la paramétrix du problème mixte hyperbolique pour l'équation des ondes. C.R. Acad. Sci. Paris 276, 1213–1215 (1973)

– [2] Formules de Poisson pour les variétés riemanniennes. Invent. Math. 24, 65–82 (1974).

Chazarain, J. and A. Piriou: [1] Introduction à la théorie des équations aux dérivées partielles linéaires. Gauthier-Villars 1981.

Chester, C., B. Friedman and F. Ursell: [1] An extension of the method of steepest descent. Proc. Cambr. Phil. Soc. 53, 599–611 (1957).

Cohen, P.: [1] The non-uniqueness of the Cauchy problem. O.N.R. Techn. Report 93, Stanford 1960.

– [2] A simple proof of the Denjoy-Carleman theorem. Amer. Math. Monthly 75, 26–31 (1968).

– [3] A simple proof of Tarski's theorem on elementary algebra. Mimeographed manuscript, Stanford University 1967, 6 pp.

Colin de Verdière, Y.: [1] Sur le spectre des opérateurs elliptiques à bicharactéristiques toutes périodiques. Comment. Math. Helv. 54, 508–522 (1979).

Cook, J.: [1] Convergence to the Møller wave matrix. J. Mathematical Physics 36, 82–87 (1957).

Cordes, H.O.: [1] Über die eindeutige Bestimmtheit der Lösungen elliptischer Differentialgleichungen durch Anfangsvorgaben. Nachr. Akad. Wiss. Göttingen Math.-Phys. Kl. IIa, No. 11, 239–258 (1956).

Cotlar, M.: [1] A combinatorial inequality and its application to L^2 spaces. Rev. Math. Cuyana 1, 41–55 (1955).

Courant, R. and D. Hilbert: [1] Methoden der Mathematischen Physik II. Berlin 1937.

Courant, R. and P.D. Lax: [1] The propagation of discontinuities in wave motion. Proc. Nat. Acad. Sci. 42, 872–876 (1956).

De Giorgi, E.: [1] Un esempio di non-unicitá della soluzione del problema di Cauchy relativo ad una equazione differenziale lineare a derivate parziali ti tipo parabolico. Rend. Mat. 14, 382–387 (1955).

– [2] Solutions analytiques des équations aux dérivées partielles à coefficients constants. Sém. Goulaouic-Schwartz 1971–1972, Exposé 29.

Deič, V.G., E.L. Korotjaev and D.R. Jafaev: [1] The theory of potential scattering with account taken of spatial anisotropy. Zap. Naučn. Sem. Leningrad Otdel. Mat. Inst. Steklov 73, 35–51 (1977).

Dencker, N.: [1] On the propagation of singularities for pseudo-differential operators of principal type. Ark. Mat. 20, 23–60 (1982).

– [2] The Weyl calculus with locally temperate metrics and weights. Ark. Mat. 24, 59–79 (1986).

Dieudonné, J.: [1] Sur les fonctions continus numériques définies dans un produit de deux espaces compacts. C.R. Acad. Sci. Paris 205, 593–595 (1937).

Dieudonné, J. and L. Schwartz: [1] La dualité dans les espaces ($\mathscr{F}$) et ($\mathscr{L}\mathscr{F}$). Ann. Inst. Fourier (Grenoble) 1, 61–101 (1949).

Dollard, J.D.: [1] Asymptotic convergence and the Coulomb interaction. J. Math. Phys. 5, 729–738 (1964).

– [2] Quantum mechanical scattering theory for short-range and Coulomb interactions. Rocky Mountain J. Math. 1, 5–88 (1971).

Douglis, A. and L. Nirenberg: [1] Interior estimates for elliptic systems of partial differential equations. Comm. Pure Appl. Math. 8, 503–538 (1955).

Duistermaat, J.J.: [1] Oscillatory integrals, Lagrange immersions and unfolding of singularities. Comm. Pure Appl. Math. 27, 207–281 (1974).

Duistermaat, J.J. and V.W. Guillemin: [1] The spectrum of positive elliptic operators and periodic bicharacteristics. Invent. Math. 29, 39–79 (1975).

Duistermaat, J.J. and L. Hörmander: [1] Fourier integral operators II. Acta Math. 128, 183–269 (1972).

Duistermaat, J.J. and J. Sjöstrand: [1] A global construction for pseudo-differential operators with non-involutive characteristics. Invent. Math. 20, 209–225 (1973).

DuPlessis, N.: [1] Some theorems about the Riesz fractional integral. Trans. Amer. Math. Soc. 80, 124–134 (1955).

Egorov, Ju.V.: [1] The canonical transformations of pseudo-differential operators. Uspehi Mat. Nauk 24:5, 235–236 (1969).

– [2] Subelliptic pseudo-differential operators. Dokl. Akad. Nauk SSSR 188, 20–22 (1969); also in Soviet Math. Doklady 10, 1056–1059 (1969).

– [3] Subelliptic operators. Uspehi Mat. Nauk 30:2, 57–114 and 30:3, 57–104 (1975); also in Russian Math. Surveys 30:2, 59–118 and 30:3, 55–105 (1975).

Ehrenpreis, L.: [1] Solutions of some problems of division I. Amer. J. Math. 76, 883–903 (1954).

– [2] Solutions of some problems of division III. Amer. J. Math. 78, 685–715 (1956).

– [3] Solutions of some problems of division IV. Amer. J. Math. 82, 522–588 (1960).

– [4] On the theory of kernels of Schwartz. Proc. Amer. Math. Soc. 7, 713–718 (1956).

– [5] A fundamental principle for systems of linear differential equations with constant coefficients, and some of its applications. Proc. Intern. Symp. on Linear Spaces, Jerusalem 1961, 161–174.

– [6] Fourier analysis in several complex variables. Wiley-Interscience Publ., New York, London, Sydney, Toronto 1970.

– [7] Analytically uniform spaces and some applications. Trans. Amer. Math. Soc. 101, 52–74 (1961).

– [8] Solutions of some problems of division V. Hyperbolic operators. Amer. J. Math. 84, 324–348 (1962).

Enqvist, A.: [1] On fundamental solutions supported by a convex cone. Ark. Mat. 12, 1–40 (1974).

Enss, V.: [1] Asymptotic completeness for quantum-mechanical potential scattering. I. Short range potentials. Comm. Math. Phys. 61, 285–291 (1978).

– [2] Geometric methods in spectral and scattering theory of Schrödinger operators.

In Rigorous Atomic and Molecular Physics, G. Velo and A. Wightman ed., Plenum, New York, 1980–1981 (Proc. Erice School of Mathematical Physics 1980).

Eškin, G.I.: [1] Boundary value problems for elliptic pseudo-differential equations. Moscow 1973; Amer. Math. Soc. Transl. of Math. Monographs 52, Providence, R.I. 1981.

– [2] Parametrix and propagation of singularities for the interior mixed hyperbolic problem. J. Analyse Math. 32, 17–62 (1977).

– [3] General initial-boundary problems for second order hyperbolic equations. In Sing. in Boundary Value Problems. D. Reidel Publ. Co., Dordrecht, Boston, London 1981, 19–54.

– [4] Initial boundary value problem for second order hyperbolic equations with general boundary conditions I. J. Analyse Math. 40, 43–89 (1981).

Fedosov, B.V.: [1] A direct proof of the formula for the index of an elliptic system in Euclidean space. Funkcional. Anal. i Priložen. 4:4, 83–84 (1970) (Russian); also in Functional Anal. Appl. 4, 339–341 (1970).

Fefferman, C.L.: [1] The uncertainty principle. Bull. Amer. Math. Soc. 9, 129–206 (1983).

Fefferman, C. and D.H. Phong: [1] On positivity of pseudo-differential operators. Proc. Nat. Acad. Sci. 75, 4673–4674 (1978).

– [2] The uncertainty principle and sharp Gårding inequalities. Comm. Pure Appl. Math. 34, 285–331 (1981).

Fredholm, I.: [1] Sur l'intégrale fondamentale d'une équation différentielle elliptique à coefficients constants. Rend. Circ. Mat. Palermo 25, 346–351 (1908).

Friedlander, F.G.: [1] The wave front set of the solution of a simple initial-boundary value problem with glancing rays. Math. Proc. Cambridge Philos. Soc. 79, 145–159 (1976).

Friedlander, F.G. and R.B. Melrose: [1] The wave front set of the solution of a simple initial-boundary value problem with glancing rays. II. Math. Proc. Cambridge Philos. Soc. 81, 97–120 (1977).

Friedrichs, K.: [1] On differential operators in Hilbert spaces. Amer. J. Math. 61, 523–544 (1939).

– [2] The identity of weak and strong extensions of differential operators. Trans. Amer. Math. Soc. 55, 132–151 (1944).

– [3] On the differentiability of the solutions of linear elliptic differential equations. Comm. Pure Appl. Math. 6, 299–326 (1953).

– [4] On the perturbation of continuous spectra. Comm. Pure Appl. Math. 1, 361–406 (1948).

Friedrichs, K. and H. Lewy: [1] Über die Eindeutigkeit und das Abhängigkeitsgebiet der Lösungen beim Anfangswertproblem linearer hyperbolischer Differentialgleichungen. Math. Ann. 98, 192–204 (1928).

Fröman, N. and P.O. Fröman: [1] JWKB approximation. Contributions to the theory. North-Holland Publ. Co. Amsterdam 1965.

Fuglede, B.: [1] A priori inequalities connected with systems of partial differential equations. Acta Math. 105, 177–195 (1961).

Gabrielov, A.M.: [1] A certain theorem of Hörmander. Funkcional. Anal. i Priložen. 4:2, 18–22 (1970) (Russian); also in Functional Anal. Appl. 4, 106–109 (1970).

Gårding, L.: [1] Linear hyperbolic partial differential equations with constant coefficients. Acta Math. 85, 1–62 (1951).

– [2] Dirichlet's problem for linear elliptic partial differential equations. Math. Scand. 1, 55–72 (1953).

– [3] Solution directe du problème de Cauchy pour les équations hyperboliques. Coll. Int. CNRS, Nancy 1956, 71–90.

– [4] Transformation de Fourier des distributions homogènes. Bull. Soc. Math. France 89, 381–428 (1961).

– [5] Local hyperbolicity. Israel J. Math. 13, 65–81 (1972).

– [6] Le problème de la dérivée oblique pour l'équation des ondes. C.R. Acad. Sci. Paris 285, 773–775 (1977). Rectification C.R. Acad. Sci. Paris 285, 1199 (1978).

– [7] On the asymptotic distribution of the eigenvalues and eigenfunctions of elliptic differential operators. Math. Scand. 1, 237–255 (1953).

Gårding, L. and J.L. Lions: [1] Functional analysis. Nuovo Cimento N. 1 del Suppl. al Vol. (10)14, 9–66 (1959).

Gårding, L. and B. Malgrange: [1] Opérateurs différentiels partiellement hypoelliptiques et partiellement elliptiques. Math. Scand. 9, 5–21 (1961).

Gask, H.: [1] A proof of Schwartz' kernel theorem. Math. Scand. 8, 327–332 (1960).

Gelfand, I.M. and G.E. Šilov: [1] Fourier transforms of rapidly increasing functions and questions of uniqueness of the solution of Cauchy's problem. Uspehi Mat. Nauk 8:6, 3–54 (1953) (Russian); also in Amer. Math. Soc. Transl. (2) 5, 221–274 (1957).

– [2] Generalized functions. Volume 1: Properties and operations. Volume 2: Spaces of fundamental and generalized functions. Academic Press, New York and London 1964, 1968.

Gevrey, M.: [1] Démonstration du théorème de Picard-Bernstein par la méthode des contours successifs; prolongement analytique. Bull. Sci. Math. 50, 113–128 (1926).

Glaeser, G.: [1] Etude de quelques algèbres Tayloriennes. J. Analyse Math. 6, 1–124 (1958).

Godin, P.: [1] Propagation des singularités pour les opérateurs pseudo-différentiels de type principal à partie principal analytique vérifiant la condition (*P*), en dimension 2. C.R. Acad. Sci. Paris 284, 1137–1138 (1977).

Gorin, E.A.: [1] Asymptotic properties of polynomials and algebraic functions of several variables. Uspehi Mat. Nauk 16:1, 91–118 (1961) (Russian); also in Russian Math. Surveys 16:1, 93–119 (1961).

Grubb, G.: [1] Boundary problems for systems of partial differential operators of mixed order. J. Functional Analysis 26, 131–165 (1977).

– [2] Problèmes aux limites pseudo-différentiels dépendant d'un paramètre. C.R. Acad. Sci. Paris 292, 581–583 (1981).

Grušin, V.V.: [1] The extension of smoothness of solutions of differential equations of principal type. Dokl. Akad. Nauk SSSR 148, 1241–1244 (1963) (Russian); also in Soviet Math. Doklady 4, 248–252 (1963).

– [2] A certain class of hypoelliptic operators. Mat. Sb. 83, 456–473 (1970) (Russian); also in Math. USSR Sb. 12, 458–476 (1970).

Gudmundsdottir, G.: [1] Global properties of differential operators of constant strength. Ark. Mat. 15, 169–198 (1977).

Guillemin, V.: [1] The Radon transform on Zoll surfaces. Advances in Math. 22, 85–119 (1976).

– [2] Some classical theorems in spectral theory revisited. Seminar on sing. of sol. of diff. eq., Princeton University Press, Princeton, N.J., 219–259 (1979).

– [3] Some spectral results for the Laplace operator with potential on the *n*-sphere. Advances in Math. 27, 273–286 (1978).

Guillemin, V. and D. Schaeffer: [1] Remarks on a paper of D. Ludwig. Bull. Amer. Math. Soc. 79, 382–385 (1973).

Guillemin, V. and S. Sternberg: [1] Geometrical asymptotics. Amer. Math. Soc. Surveys 14, Providence, R.I. 1977.

Gurevič, D.I.: [1] Counterexamples to a problem of L. Schwartz. Funkcional. Anal. i Priložen. 9:2, 29–35 (1975) (Russian); also in Functional Anal. Appl. 9, 116–120 (1975).

Hack, M.N.: [1] On convergence to the Møller wave operators. Nuovo Cimento (10) 13, 231–236 (1959).

Hadamard, J.: [1] Le problème de Cauchy et les équations aux dérivées partielles linéaires hyperboliques. Paris 1932.

Haefliger, A.: [1] Variétés feuilletées. Ann. Scuola Norm. Sup. Pisa 16, 367–397 (1962).

Hanges, N.: [1] Propagation of singularities for a class of operators with double characteristics. Seminar on singularities of sol. of linear partial diff. eq., Princeton University Press, Princeton, N.J. 1979, 113–126.

Hardy, G.H. and J.E. Littlewood: [1] Some properties of fractional integrals. (I) Math. Z. 27, 565–606 (1928); (II) Math. Z. 34, 403–439 (1931–32).

Hausdorff, F.: [1] Eine Ausdehnung des Parsevalschen Satzes über Fourierreihen. Math. Z. 16, 163–169 (1923).

Hayman, W.K. and P.B. Kennedy: [1] Subharmonic functions I. Academic Press, London, New York, San Francisco 1976.

Hedberg, L.I.: [1] On certain convolution inequalities. Proc. Amer. Math. Soc. 36, 505–510 (1972).

Heinz, E.: [1] Über die Eindeutigkeit beim Cauchyschen Anfangswertproblem einer elliptischen Differentialgleichung zweiter Ordnung. Nachr. Akad. Wiss. Göttingen Math.-Phys. Kl. IIa No. 1, 1–12 (1955).

Helffer, B.: [1] Addition de variables et applications à la régularité. Ann. Inst. Fourier (Grenoble) 28:2, 221–231 (1978).

Helffer, B. and J. Nourrigat: [1] Caractérisation des opérateurs hypoelliptiques homogènes invariants à gauche sur un groupe de Lie nilpotent gradué. Comm. Partial Differential Equations 4:8, 899–958 (1979).

Herglotz, G.: [1] Über die Integration linearer partieller Differentialgleichungen mit konstanten Koeffizienten I–III. Berichte Sächs. Akad. d. Wiss. 78, 93–126, 287–318 (1926); 80, 69–114 (1928).

Hersh, R.: [1] Boundary conditions for equations of evolution. Arch. Rational Mech. Anal. 16, 243–264 (1964).

– [2] On surface waves with finite and infinite speed of propagation. Arch. Rational Mech. Anal. 19, 308–316 (1965).

Hirzebruch, F.: [1] Neue topologische Methoden in der algebraischen Geometrie. Springer Verlag, Berlin-Göttingen-Heidelberg 1956.

Hlawka, E.: [1] Über Integrale auf konvexen Körpern. I. Monatsh. Math. 54, 1–36 (1950).

Holmgren, E.: [1] Über Systeme von linearen partiellen Differentialgleichungen. Öfversigt af Kongl. Vetenskaps-Akad. Förh. 58, 91–103 (1901).

– [2] Sur l'extension de la méthode d'intégration de Riemann. Ark. Mat. Astr. Fys. 1, No 22, 317–326 (1904).

Hörmander, L.: [1] On the theory of general partial differential operators. Acta Math. 94, 161–248 (1955).

– [2] Local and global properties of fundamental solutions. Math. Scand. 5, 27–39 (1957).

– [3] On the regularity of the solutions of boundary problems. Acta Math. 99, 225–264 (1958).

– [4] On interior regularity of the solutions of partial differential equations. Comm. Pure Appl. Math. 11, 197–218 (1958).

– [5] On the division of distributions by polynomials. Ark. Mat. 3, 555–568 (1958).

– [6] Differentiability properties of solutions of systems of differential equations. Ark. Mat. 3, 527–535 (1958).

– [7] Definitions of maximal differential operators. Ark. Mat. 3, 501–504 (1958).

– [8] On the uniqueness of the Cauchy problem I, II. Math. Scand. 6, 213–225 (1958); 7, 177–190 (1959).

– [9] Null solutions of partial differential equations. Arch. Rational Mech. Anal. 4, 255–261 (1960).

– [10] Differential operators of principal type. Math. Ann. 140, 124–146 (1960).

– [11] Differential equations without solutions. Math. Ann. 140, 169–173 (1960).

– [12] Hypoelliptic differential operators. Ann. Inst. Fourier (Grenoble) 11, 477–492 (1961).

– [13] Estimates for translation invariant operators in L^p spaces. Acta Math. 104, 93–140 (1960).
– [14] On the range of convolution operators. Ann. of Math. 76, 148–170 (1962).
– [15] Supports and singular supports of convolutions. Acta Math. 110, 279–302 (1963).
– [16] Pseudo-differential operators. Comm. Pure Appl. Math. 18, 501–517 (1965).
– [17] Pseudo-differential operators and non-elliptic boundary problems. Ann. of Math. 83, 129–209 (1966).
– [18] Pseudo-differential operators and hypoelliptic equations. Amer. Math. Soc. Symp. on Singular Integrals, 138–183 (1966).
– [19] An introduction to complex analysis in several variables. D. van Nostrand Publ. Co., Princeton, N.J. 1966.
– [20] Hypoelliptic second order differential equations. Acta Math. 119, 147–171 (1967).
– [21] On the characteristic Cauchy problem. Ann. of Math. 88, 341–370 (1968).
– [22] The spectral function of an elliptic operator. Acta Math. 121, 193–218 (1968).
– [23] Convolution equations in convex domains. Invent. Math. 4, 306–317 (1968).
– [24] On the singularities of solutions of partial differential equations. Comm. Pure Appl. Math. 23, 329–358 (1970).
– [25] Linear differential operators. Actes Congr. Int. Math. Nice 1970, 1, 121–133.
– [26a] The calculus of Fourier integral operators. Prospects in math. Ann. of Math. Studies 70, 33–57 (1971).
– [26] Fourier integral operators I. Acta Math. 127, 79–183 (1971).
– [27] Uniqueness theorems and wave front sets for solutions of linear differential equations with analytic coefficients. Comm. Pure Appl. Math. 24, 671–704 (1971).
– [28] A remark on Holmgren's uniqueness theorem. J. Diff. Geom. 6, 129–134 (1971).
– [29] On the existence and the regularity of solutions of linear pseudo-differential equations. Ens. Math. 17, 99–163 (1971).
– [30] On the singularities of solutions of partial differential equations with constant coefficients. Israel J. Math. 13, 82–105 (1972).
– [31] On the existence of real analytic solutions of partial differential equations with constant coefficients. Invent. Math. 21, 151–182 (1973).
– [32] Lower bounds at infinity for solutions of differential equations with constant coefficients. Israel J. Math. 16, 103–116 (1973).
– [33] Non-uniqueness for the Cauchy problem. Springer Lecture Notes in Math. 459, 36–72 (1975).
– [34] The existence of wave operators in scattering theory. Math. Z. 146, 69–91 (1976).
– [35] A class of hypoelliptic pseudo-differential operators with double characteristics. Math. Ann. 217, 165–188 (1975).
– [36] The Cauchy problem for differential equations with double characteristics. J. Analyse Math. 32, 118–196 (1977).
– [37] Propagation of singularities and semiglobal existence theorems for (pseudo-) differential operators of principal type. Ann. of Math. 108, 569–609 (1978).
– [38] Subelliptic operators. Seminar on sing. of sol. of diff. eq. Princeton University Press, Princeton, N.J., 127–208 (1979).
– [39] The Weyl calculus of pseudo-differential operators. Comm. Pure Appl. Math. 32, 359–443 (1979).
– [40] Pseudo-differential operators of principal type. Nato Adv. Study Inst. on Sing. in Bound. Value Problems. Reidel Publ. Co., Dordrecht, 69–96 (1981).
– [41] Uniqueness theorems for second order elliptic differential equations. Comm. Partial Differential Equations 8, 21–64 (1983).
– [42] On the index of pseudo-differential operators. In Elliptische Differentialgleichungen Band II, Akademie-Verlag, Berlin 1971, 127–146.
– [43] L^2 estimates for Fourier integral operators with complex phase. Ark. Mat. 21, 297–313 (1983).
– [44] On the subelliptic test estimates. Comm. Pure Appl. Math. 33, 339–363 (1980).

Hurwitz, A.: [1] Über die Nullstellen der Bessel'schen Funktion. Math. Ann. 33, 246–266 (1889).

Iagolnitzer, D.: [1] Microlocal essential support of a distribution and decomposition theorems – an introduction. In Hyperfunctions and theoretical physics. Springer Lecture Notes in Math. 449, 121–132 (1975).

Ikebe, T.: [1] Eigenfunction expansions associated with the Schrödinger operator and their applications to scattering theory. Arch. Rational Mech. Anal. 5, 1–34 (1960).

Ikebe, T. and Y. Saito: [1] Limiting absorption method and absolute continuity for the Schrödinger operator. J. Math. Kyoto Univ. 12, 513–542 (1972).

Ivrii, V.Ja: [1] Sufficient conditions for regular and completely regular hyperbolicity. Trudy Moskov. Mat. Obšč. 33, 3–65 (1975) (Russian); also in Trans. Moscow Math. Soc. 33, 1–65 (1978).

– [2] Wave fronts for solutions of boundary value problems for a class of symmetric hyperbolic systems. Sibirsk. Mat. Ž. 21:4, 62–71 (1980) (Russian); also in Sibirian Math. J. 21, 527–534 (1980).

– [3] On the second term in the spectral asymptotics for the Laplace-Beltrami operator on a manifold with boundary. Funkcional. Anal. i Priložen. 14:2, 25–34 (1980) (Russian); also in Functional Anal. Appl. 14, 98–106 (1980).

Ivrii, V.Ja and V.M. Petkov: [1] Necessary conditions for the correctness of the Cauchy problem for non-strictly hyperbolic equations. Uspehi Mat. Nauk 29:5, 3–70 (1974) (Russian); also in Russian Math. Surveys 29:5, 1–70 (1974).

Iwasaki, N.: [1] The Cauchy problems for effectively hyperbolic equations (general case). J. Math. Kyoto Univ. 25, 727–743 (1985).

Jauch, J.M. and I.I. Zinnes: [1] The asymptotic condition for simple scattering systems. Nuovo Cimento (10) 11, 553–567 (1959).

Jerison D. and C.E. Kenig: [1] Unique continuation and absence of positive eigenvalues for Schrödinger operators. Ann. of Math. 121, 463–488 (1985).

John, F.: [1] On linear differential equations with analytic coefficients. Unique continuation of data. Comm. Pure Appl. Math. 2, 209–253 (1949).

– [2] Plane waves and spherical means applied to partial differential equations. New York 1955.

– [3] Non-admissible data for differential equations with constant coefficients. Comm. Pure Appl. Math. 10, 391–398 (1957).

– [4] Continuous dependence on data for solutions of partial differential equations with a prescribed bound. Comm. Pure Appl. Math. 13, 551–585 (1960).

– [5] Linear partial differential equations with analytic coefficients. Proc. Nat. Acad. Sci. 29, 98–104 (1943).

Jörgens, K. and J. Weidmann: [1] Zur Existenz der Wellenoperatoren. Math. Z. 131, 141–151 (1973).

Kashiwara, M.: [1] Introduction to the theory of hyperfunctions. In Sem. on microlocal analysis, Princeton Univ. Press, Princeton, N.J., 1979, 3–38.

Kashiwara, M. and T. Kawai: [1] Microhyperbolic pseudo-differential operators. I. J. Math. Soc. Japan 27, 359–404 (1975).

Kato, T.: [1] Growth properties of solutions of the reduced wave equation with a variable coefficient. Comm. Pure Appl. Math. 12, 403–425 (1959).

Keller, J.B.: [1] Corrected Bohr-Sommerfeld quantum conditions for nonseparable systems. Ann. Physics 4, 180–188 (1958).

Kitada, H.: [1] Scattering theory for Schrödinger operators with long-range potentials. I: Abstract theory. J. Math. Soc. Japan 29, 665–691 (1977). II: Spectral and scattering theory. J. Math. Soc. Japan 30, 603–632 (1978).

Knapp, A.W. and E.M. Stein: [1] Singular integrals and the principal series. Proc. Nat. Acad. Sci. U.S.A. 63, 281–284 (1969).

Kohn, J.J.: [1] Harmonic integrals on strongly pseudo-convex manifolds I, II. Ann. of Math. 78, 112–148 (1963); 79, 450–472 (1964).

– [2] Pseudo-differential operators and non-elliptic problems. In Pseudo-differential operators, CIME conference, Stresa 1968, 157–165. Edizione Cremonese, Roma 1969.

Kohn, J.J. and L. Nirenberg: [1] On the algebra of pseudo-differential operators. Comm. Pure Appl. Math. 18, 269–305 (1965).

– [2] Non-coercive boundary value problems. Comm. Pure Appl. Math. 18, 443–492 (1965).

Kolmogorov, A.N.: [1] Zufällige Bewegungen. Ann. of Math. 35, 116–117 (1934).

Komatsu, H.: [1] A local version of Bochner's tube theorem. J. Fac. Sci. Tokyo Sect. I-A Math. 19, 201–214 (1972).

– [2] Boundary values for solutions of elliptic equations. Proc. Int. Conf. Funct. Anal. Rel. Topics, 107–121. University of Tokyo Press, Tokyo 1970.

Kreiss, H.O.: [1] Initial boundary value problems for hyperbolic systems. Comm. Pure Appl. Math. 23, 277–298 (1970).

Krzyźański, M. and J. Schauder: [1] Quasilineare Differentialgleichungen zweiter Ordnung vom hyperbolischen Typus. Gemischte Randwertaufgaben. Studia Math. 6, 162–189 (1936).

Kumano-go, H.: [1] Factorizations and fundamental solutions for differential operators of elliptic-hyperbolic type. Proc. Japan Acad. 52, 480–483 (1976).

Kuroda, S.T.: [1] On the existence and the unitary property of the scattering operator. Nuovo Cimento (10) 12, 431–454 (1959).

Lascar, B. and R. Lascar: [1] Propagation des singularités pour des équations hyperboliques à caractéristiques de multiplicité au plus double et singularités Masloviennes II. J. Analyse Math. 41, 1–38 (1982).

Lax, A.: [1] On Cauchy's problem for partial differential equations with multiple characteristics. Comm. Pure Appl. Math. 9, 135–169 (1956).

Lax, P.D.: [2] On Cauchy's problem for hyperbolic equations and the differentiability of solutions of elliptic equations. Comm. Pure Appl. Math. 8, 615–633 (1955).

– [3] Asymptotic solutions of oscillatory initial value problems. Duke Math J. 24, 627–646 (1957).

Lax, P.D. and L. Nirenberg: [1] On stability for difference schemes: a sharp form of Gårding's inequality. Comm. Pure Appl. Math. 19, 473–492 (1966).

Lebeau, G.: [1] Fonctions harmoniques et spectre singulier. Ann. Sci. École Norm. Sup. (4) 13, 269–291 (1980).

Lelong, P.: [1] Plurisubharmonic functions and positive differential forms. Gordon and Breach, New York, London, Paris 1969.

– [2] Propriétés métriques des variétés définies par une équation. Ann. Sci. École Norm. Sup. 67, 22–40 (1950).

Leray, J.: [1] Hyperbolic differential equations. The Institute for Advanced Study, Princeton, N.J. 1953.

– [2] Uniformisation de la solution du problème linéaire analytique de Cauchy près de la variété qui porte les données de Cauchy. Bull. Soc. Math. France 85, 389–429 (1957).

Lerner, N.: [1] Unicité de Cauchy pour des opérateurs différentiels faiblement principalement normaux. J. Math. Pures Appl. 64, 1–11 (1985).

Lerner, N. and L. Robbiano: [1] Unicité de Cauchy pour des opérateurs de type principal. J. Analyse Math. 44, 32–66 (1984/85).

Levi, E.E.: [1] Caratterische multiple e problema di Cauchy. Ann. Mat. Pura Appl. (3) 16, 161–201 (1909).

Levinson, N.: [1] Transformation of an analytic function of several variables to a canonical form. Duke Math. J. 28, 345–353 (1961).

Levitan, B.M.: [1] On the asymptotic behavior of the spectral function of a self-adjoint differential equation of the second order. Izv. Akad. Nauk SSSR Ser. Mat. 16, 325–352 (1952).

– [2] On the asymptotic behavior of the spectral function and on expansion in eigenfunctions of a self-adjoint differential equation of second order II. Izv. Akad. Nauk SSSR Ser. Mat. 19, 33–58 (1955).

Lewy, H.: [1] An example of a smooth linear partial differential equation without solution. Ann. of Math. 66, 155–158 (1957).
– [2] Extension of Huyghen's principle to the ultrahyperbolic equation. Ann. Mat. Pura Appl. (4) 39, 63–64 (1955).
Lions, J.L.: [1] Supports dans la transformation de Laplace. J. Analyse Math. 2, 369–380 (1952–53).
Lions, J.L. and E. Magenes: [1] Problèmes aux limites non homogènes et applications I–III. Dunod, Paris, 1968–1970.
Łojasiewicz, S.: [1] Sur le problème de division. Studia Math. 18, 87–136 (1959).
Lopatinski, Ya.B.: [1] On a method of reducing boundary problems for a system of differential equations of elliptic type to regular integral equations. Ukrain. Mat. Ž. 5, 123–151 (1953). Amer. Math. Soc. Transl. (2) 89, 149–183 (1970).
Ludwig, D.: [1] Exact and asymptotic solutions of the Cauchy problem. Comm. Pure Appl. Math. 13, 473–508 (1960).
– [2] Uniform asymptotic expansions at a caustic. Comm. Pure Appl. Math. 19, 215–250 (1966).
Luke, G.: [1] Pseudodifferential operators on Hilbert bundles. J. Differential Equations 12, 566–589 (1972).
Malgrange, B.: [1] Existence et approximation des solutions des équations aux dérivées partielles et des équations de convolution. Ann. Inst. Fourier (Grenoble) 6, 271–355 (1955–56).
– [2] Sur une classe d'opérateurs différentiels hypoelliptiques. Bull. Soc. Math. France 85, 283–306 (1957).
– [3] Sur la propagation de la régularité des solutions des équations à coefficients constants. Bull. Math. Soc. Sci. Math. Phys. R.P. Roumanie 3 (53), 433–440 (1959).
– [4] Sur les ouverts convexes par rapport à un opérateur différentiel. C. R. Acad. Sci. Paris 254, 614–615 (1962).
– [5] Sur les systèmes différentiels à coefficients constants. Coll. CNRS 113–122, Paris 1963.
– [6] Ideals of differentiable functions. Tata Institute, Bombay, and Oxford University Press 1966.
Mandelbrojt, S.: [1] Analytic functions and classes of infinitely differentiable functions. Rice Inst. Pamphlet 29, 1–142 (1942).
– [2] Séries adhérentes, régularisations des suites, applications. Coll. Borel, Gauthier-Villars, Paris 1952.
Martineau, A.: [1] Les hyperfonctions de M. Sato. Sém. Bourbaki 1960–1961, Exposé No 214.
– [2] Le "edge of the wedge theorem" en théorie des hyperfonctions de Sato. Proc. Int. Conf. Funct. Anal. Rel. Topics, 95–106. University of Tokyo Press, Tokyo 1970.
Maslov, V.P.: [1] Theory of perturbations and asymptotic methods. Moskov. Gos. Univ., Moscow 1965 (Russian).
Mather, J.: [1] Stability of C^∞ mappings: I. The division theorem. Ann. of Math. 87, 89–104 (1968).
Melin, A.: [1] Lower bounds for pseudo-differential operators. Ark. Mat. 9, 117–140 (1971).
– [2] Parametrix constructions for right invariant differential operators on nilpotent groups. Ann. Global Analysis and Geometry 1, 79–130 (1983).
Melin, A. and J. Sjöstrand: [1] Fourier integral operators with complex-valued phase functions. Springer Lecture Notes in Math. 459, 120–223 (1974).
– [2] Fourier integral operators with complex phase functions and parametrix for an interior boundary value problem. Comm. Partial Differential Equations 1:4, 313–400 (1976).
Melrose, R.B.: [1] Transformation of boundary problems. Acta Math. 147, 149–236 (1981).

– [2] Equivalence of glancing hypersurfaces. Invent. Math. 37, 165–191 (1976).
– [3] Microlocal parametrices for diffractive boundary value problems. Duke Math. J. 42, 605–635 (1975).
– [4] Local Fourier-Airy integral operators. Duke Math. J. 42, 583–604 (1975).
– [5] Airy operators. Comm. Partial Differential Equations 3:1, 1–76 (1978).
– [6] The Cauchy problem for effectively hyperbolic operators. Hokkaido Math. J. To appear.
– [7] The trace of the wave group. Contemporary Math. 27, 127–167 (1984).
Melrose, R.B. and J. Sjöstrand: [1] Singularities of boundary value problems I, II. Comm. Pure Appl. Math. 31, 593–617 (1978); 35, 129–168 (1982).
Mihlin, S.G.: [1] On the multipliers of Fourier integrals. Dokl. Akad. Nauk SSSR 109, 701–703 (1956) (Russian).
Mikusiński, J.: [1] Une simple démonstration du théorème de Titchmarsh sur la convolution. Bull. Acad. Pol. Sci. 7, 715–717 (1959).
– [2] The Bochner integral. Birkhäuser Verlag, Basel and Stuttgart 1978.
Minakshisundaram, S. and Å. Pleijel: [1] Some properties of the eigenfunctions of the Laplace operator on Riemannian manifolds. Canad. J. Math. 1, 242–256 (1949).
Mizohata, S.: [1] Unicité du prolongement des solutions des équations elliptiques du quatrième ordre. Proc. Jap. Acad. 34, 687–692 (1958).
– [2] Systèmes hyperboliques. J. Math. Soc. Japan 11, 205–233 (1959).
– [3] Note sur le traitement par les opérateurs d'intégrale singulière du problème de Cauchy. J. Math. Soc. Japan 11, 234–240 (1959).
– [4] Solutions nulles et solutions non analytiques. J. Math. Kyoto Univ. 1, 271–302 (1962).
– [5] Some remarks on the Cauchy problem. J. Math. Kyoto Univ. 1, 109–127 (1961).
Møller, C.: [1] General properties of the characteristic matrix in the theory of elementary particles. I. Kongl. Dansk. Vidensk. Selsk. Mat.-Fys. Medd. 23, 2–48 (1945).
Morrey, C.B.: [1] The analytic embedding of abstract real-analytic manifolds. Ann. of Math. 68, 159–201 (1958).
Morrey, C.B. and L. Nirenberg: [1] On the analyticity of the solutions of linear elliptic systems of partial differential equations. Comm. Pure Appl. Math. 10, 271–290 (1957).
Moyer, R.D.: [1] Local solvability in two dimensions: Necessary conditions for the principle-type case. Mimeographed manuscript, University of Kansas 1978.
Müller, C.: [1] On the behaviour of the solutions of the differential equation $\Delta U = F(x, U)$ in the neighborhood of a point. Comm. Pure Appl. Math. 7, 505–515 (1954).
Münster, M.: [1] On A. Lax's condition of hyperbolicity. Rocky Mountain J. Math. 8, 443–446 (1978).
– [2] On hyperbolic polynomials with constant coefficients. Rocky Mountain J. Math. 8, 653–673 (1978).
von Neumann, J. and E. Wigner: [1] Über merkwürdige diskrete Eigenwerte. Phys. Z. 30, 465–467 (1929).
Nirenberg, L.: [1] Remarks on strongly elliptic partial differential equations. Comm. Pure Appl. Math. 8, 648–675 (1955).
– [2] Uniqueness in Cauchy problems for differential equations with constant leading coefficients. Comm. Pure Appl. Math. 10, 89–105 (1957).
– [3] A proof of the Malgrange preparation theorem. Liverpool singularities I. Springer Lecture Notes in Math. 192, 97–105 (1971).
– [4] On elliptic partial differential equations. Ann. Scuola Norm. Sup. Pisa (3) 13, 115–162 (1959).
– [5] Lectures on linear partial differential equations. Amer. Math. Soc. Regional Conf. in Math., 17, 1–58 (1972).

Nirenberg, L. and F. Treves: [1] Solvability of a first order linear partial differential equation. Comm. Pure Appl. Math. 16, 331–351 (1963).
– [2] On local solvability of linear partial differential equations. I. Necessary conditions. II. Sufficient conditions. Correction. Comm. Pure Appl. Math. 23, 1–38 and 459–509 (1970); 24, 279–288 (1971).
Nishitani, T.: [1] Local energy integrals for effectively hyperbolic operators I, II. J. Math. Kyoto Univ. 24, 623–658 and 659–666 (1984).
Noether, F.: [1] Über eine Klasse singulärer Integralgleichungen. Math. Ann. 82, 42–63 (1921).
Olejnik, O.A.: [1] On the Cauchy problem for weakly hyperbolic equations. Comm. Pure Appl. Math. 23, 569–586 (1970).
Olejnik, O.A. and E.V. Radkevič: [1] Second order equations with non-negative characteristic form. In Matem. Anal. 1969, ed. R.V. Gamkrelidze, Moscow 1971 (Russian). English translation Plenum Press, New York-London 1973.
Oshima, T.: [1] On analytic equivalence of glancing hypersurfaces. Sci. Papers College Gen. Ed. Univ. Tokyo 28, 51–57 (1978).
Palamodov, V.P.: [1] Linear differential operators with constant coefficients. Moscow 1967 (Russian). English transl. Grundl. d. Math. Wiss. 168, Springer Verlag, New York, Heidelberg, Berlin 1970.
Paley, R.E.A.C. and N. Wiener: [1] Fourier transforms in the complex domain. Amer. Math. Soc. Coll. Publ. XIX, New York 1934.
Pederson, R.: [1] On the unique continuation theorem for certain second and fourth order elliptic equations. Comm. Pure Appl. Math. 11, 67–80 (1958).
– [2] Uniqueness in the Cauchy problem for elliptic equations with double characteristics. Ark. Mat. 6, 535–549 (1966).
Peetre, J.: [1] Théorèmes de régularité pour quelques classes d'opérateurs différentiels. Thesis, Lund 1959.
– [2] Rectification à l'article "Une caractérisation abstraite des opérateurs différentiels". Math. Scand. 8, 116–120 (1960).
– [3] Another approach to elliptic boundary problems. Comm. Pure Appl. Math. 14, 711–731 (1961).
– [4] New thoughts on Besov spaces. Duke Univ. Math. Series I, Durham, N.C. 1976.
Persson, J.: [1] The wave operator and P-convexity. Boll. Un. Mat. Ital. (5) 18-B, 591–604 (1981).
Petrowsky, I.G.: [1] Über das Cauchysche Problem für Systeme von partiellen Differentialgleichungen. Mat. Sb. 2 (44), 815–870 (1937).
– [2] Über das Cauchysche Problem für ein System linearer partieller Differentialgleichungen im Gebiete der nichtanalytischen Funktionen. Bull. Univ. Moscow Sér. Int. 1, No. 7, 1–74 (1938).
– [3] Sur l'analyticité des solutions des systèmes d'équations différentielles. Mat. Sb. 5 (47), 3–70 (1939).
– [4] On the diffusion of waves and the lacunas for hyperbolic equations. Mat. Sb. 17 (59), 289–370 (1945).
– [5] Some remarks on my papers on the problem of Cauchy. Mat. Sb. 39 (81), 267–272 (1956). (Russian.)
Pham The Lai: [1] Meilleures estimations asymptotiques des restes de la fonction spectrale et des valeurs propres relatifs au laplacien. Math. Scand. 48, 5–31 (1981).
Piccinini, L.C.: [1] Non surjectivity of the Cauchy-Riemann operator on the space of the analytic functions on $\mathbb{R}^n$. Generalization to the parabolic operators. Bull. Un. Mat. Ital. (4) 7, 12–28 (1973).
Pliš, A.: [1] A smooth linear elliptic differential equation without any solution in a sphere. Comm. Pure Appl. Math. 14, 599–617 (1961).
– [2] The problem of uniqueness for the solution of a system of partial differential equations. Bull. Acad. Pol. Sci. 2, 55–57 (1954).

— [3] On non-uniqueness in Cauchy problem for an elliptic second order differential equation. Bull. Acad. Pol. Sci. 11, 95–100 (1963).

Poincaré, H.: [1] Sur les propriétés du potentiel et les fonctions abeliennes. Acta Math. 22, 89–178 (1899).

Povzner, A.Ya.: [1] On the expansion of arbitrary functions in characteristic functions of the operator $-\Delta u + cu$. Mat. Sb. 32 (74), 109–156 (1953).

Radkevič, E.: [1] A priori estimates and hypoelliptic operators with multiple characteristics. Dokl. Akad. Nauk SSSR 187, 274–277 (1969) (Russian); also in Soviet Math. Doklady 10, 849–853 (1969).

Ralston, J.: [1] Solutions of the wave equation with localized energy. Comm. Pure Appl. Math. 22, 807–823 (1969).

— [2] Gaussian beams and the propagation of singularities. MAA Studies in Mathematics 23, 206–248 (1983).

Reed, M. and B. Simon: [1] Methods of modern mathematical physics. III. Scattering theory. Academic Press 1979.

Rempel, S. and B.-W. Schulze: [1] Index theory of elliptic boundary problems. Akademie-Verlag, Berlin (1982).

de Rham, G.: [1] Variétés différentiables. Hermann, Paris, 1955.

Riesz, F.: [1] Sur l'existence de la dérivée des fonctions d'une variable réelle et des fonctions d'intervalle. Verh. Int. Math. Kongr. Zürich 1932, I, 258–269.

Riesz, M.: [1] L'intégrale de Riemann-Liouville et le problème de Cauchy. Acta Math. 81, 1–223 (1949).

— [2] Sur les maxima des formes bilinéaires et sur les fonctionnelles linéaires. Acta Math. 49, 465–497 (1926).

— [3] Sur les fonctions conjuguées. Math. Z. 27, 218–244 (1928).

— [4] Problems related to characteristic surfaces. Proc. Conf. Diff. Eq. Univ. Maryland 1955, 57–71.

Rothschild, L.P.: [1] A criterion for hypoellipticity of operators constructed from vector fields. Comm. Partial Differential Equations 4:6, 645–699 (1979).

Saitō, Y.: [1] On the asymptotic behavior of the solutions of the Schrödinger equation $(-\Delta + Q(y) - k^2)V = F$. Osaka J. Math. 14, 11–35 (1977).

— [2] Eigenfunction expansions for the Schrödinger operators with long-range potentials $Q(y) = O(|y|^{-\varepsilon})$, $\varepsilon > 0$. Osaka J. Math. 14, 37–53 (1977).

Sakamoto, R.: [1] E-well posedness for hyperbolic mixed problems with constant coefficients. J. Math. Kyoto Univ. 14, 93–118 (1974).

— [2] Mixed problems for hyperbolic equations I. J. Math. Kyoto Univ. 10, 375–401 (1970), II. J. Math. Kyoto Univ. 10, 403–417 (1970).

Sato, M.: [1] Theory of hyperfunctions I. J. Fac. Sci. Univ. Tokyo I, 8, 139–193 (1959).

— [2] Theory of hyperfunctions II. J. Fac. Sci. Univ. Tokyo I, 8, 387–437 (1960).

— [3] Hyperfunctions and partial differential equations. Proc. Int. Conf. on Funct. Anal. and Rel. Topics, 91–94, Tokyo University Press, Tokyo 1969.

— [4] Regularity of hyperfunction solutions of partial differential equations. Actes Congr. Int. Math. Nice 1970, 2, 785–794.

Sato, M., T. Kawai and M. Kashiwara: [1] Hyperfunctions and pseudodifferential equations. Springer Lecture Notes in Math. 287, 265–529 (1973).

Schaefer, H.H.: [1] Topological vector spaces. Springer Verlag, New York, Heidelberg, Berlin 1970.

Schapira, P.: [1] Hyperfonctions et problèmes aux limites elliptiques. Bull. Soc. Math. France 99, 113–141 (1971).

— [2] Propagation at the boundary of analytic singularities. Nato Adv. Study Inst. on Sing. in Bound. Value Problems. Reidel Publ. Co., Dordrecht, 185–212 (1981).

— [3] Propagation at the boundary and reflection of analytic singularities of solutions of linear partial differential equations. Publ. RIMS, Kyoto Univ., 12 Suppl., 441–453 (1977).

Schechter, M.: [1] Various types of boundary conditions for elliptic equations. Comm. Pure Appl. Math. 13, 407–425 (1960).
– [2] A generalization of the problem of transmission. Ann. Scuola Norm. Sup. Pisa 14, 207–236 (1960).
Schwartz, L.: [1] Théorie des distributions I, II. Hermann, Paris, 1950–51.
– [2] Théorie des noyaux. Proc. Int. Congr. Math. Cambridge 1950, I, 220–230.
– [3] Sur l'impossibilité de la multiplication des distributions. C. R. Acad. Sci. Paris 239, 847–848 (1954).
– [4] Théorie des distributions à valeurs vectorielles I. Ann. Inst. Fourier (Grenoble) 7, 1–141 (1957).
– [5] Transformation de Laplace des distributions. Comm. Sém. Math. Univ. Lund, Tome suppl. dédié à Marcel Riesz, 196–206 (1952).
– [6] Théorie générale des fonctions moyenne-périodiques. Ann. of Math. 48, 857–929 (1947).
Seeley, R.T.: [1] Singular integrals and boundary problems. Amer. J. Math. 88, 781–809 (1966).
– [2] Extensions of C^∞ functions defined in a half space. Proc. Amer. Math. Soc. 15, 625–626 (1964).
– [3] A sharp asymptotic remainder estimate for the eigenvalues of the Laplacian in a domain of $\mathbb{R}^3$. Advances in Math. 29, 244–269 (1978).
– [4] An estimate near the boundary for the spectral function of the Laplace operator. Amer. J. Math. 102, 869–902 (1980).
– [5] Elliptic singular integral equations. Amer Math. Soc. Symp. on Singular Integrals, 308–315 (1966).
Seidenberg, A.: [1] A new decision method for elementary algebra. Ann. of Math. 60, 365–374 (1954).
Shibata, Y.: [1] E-well posedness of mixed initial-boundary value problems with constant coefficients in a quarter space. J. Analyse Math. 37, 32–45 (1980).
Siegel, C.L.: [1] Zu den Beweisen des Vorbereitungssatzes von Weierstrass. In Abhandl. aus Zahlenth. u. Anal., 299–306. Plenum Press, New York 1968.
Sjöstrand, J.: [1] Singularités analytiques microlocales. Prépublications Université de Paris-Sud 82-03.
– [2] Analytic singularities of solutions of boundary value problems. Nato Adv. Study Inst. on Sing. in Bound. Value Prob., Reidel Publ. Co., Dordrecht, 235–269 (1981).
– [3] Parametrices for pseudodifferential operators with multiple characteristics. Ark. Mat. 12, 85–130 (1974).
– [4] Propagation of analytic singularities for second order Dirichlet problems I, II, III. Comm. Partial Differential Equations 5:1, 41–94 (1980), 5:2, 187–207 (1980), and 6:5, 499–567 (1981).
– [5] Operators of principal type with interior boundary conditions. Acta Math. 130, 1–51 (1973).
Sobolev, S.L.: [1] Méthode nouvelle à résoudre le problème de Cauchy pour les équations linéaires hyperboliques normales. Mat. Sb. 1 (43), 39–72 (1936).
– [2] Sur un théorème d'analyse fonctionnelle. Mat. Sb. 4 (46), 471–497 (1938). (Russian; French summary.) Amer. Math. Soc. Transl. (2) 34, 39–68 (1963).
Sommerfeld, A.: [1] Optics. Lectures on theoretical physics IV. Academic Press, New York, 1969.
Stein, E.M.: [1] Singular integrals and differentiability properties of functions. Princeton Univ. Press 1970.
Sternberg, S.: [1] Lectures on differential geometry. Prentice-Hall Inc. Englewood Cliffs, N.J., 1964.
Stokes, G.B.: [1] On the numerical calculation of a class of definite integrals and infinite series. Trans. Cambridge Philos. Soc. 9, 166–187 (1850).

Svensson, L.: [1] Necessary and sufficient conditions for the hyperbolicity of polynomials with hyperbolic principal part. Ark. Mat. 8, 145–162 (1968).

Sweeney, W.J.: [1] The D-Neumann problem. Acta Math. 120, 223–277 (1968).

Szegö, G.: [1] Beiträge zur Theorie der Toeplitzschen Formen. Math. Z. 6, 167–202 (1920).

Täcklind, S.: [1] Sur les classes quasianalytiques des solutions des équations aux dérivées partielles du type parabolique. Nova Acta Soc. Sci. Upsaliensis (4) 10, 1–57 (1936).

Tarski, A.: [1] A decision method for elementary algebra and geometry. Manuscript, Berkeley, 63 pp. (1951).

Taylor, M.: [1] Gelfand theory of pseudodifferential operators and hypoelliptic operators. Trans. Amer. Math. Soc. 153, 495–510 (1971).

– [2] Grazing rays and reflection of singularities of solutions to wave equations. Comm. Pure Appl. Math. 29, 1–38 (1976).

– [3] Diffraction effects in the scattering of waves. In Sing. in Bound. Value Problems, 271–316. Reidel Publ. Co., Dordrecht 1981.

– [4] Pseudodifferential operators. Princeton Univ. Press, Princeton, N.J., 1981.

Thorin, O.: [1] An extension of a convexity theorem due to M. Riesz. Kungl Fys. Sällsk. Lund. Förh. 8, No 14 (1939).

Titchmarsh, E.C.: [1] The zeros of certain integral functions. Proc. London Math. Soc. 25, 283–302 (1926).

Treves, F.: [1] Solution élémentaire d'équations aux dérivées partielles dépendant d'un paramètre. C. R. Acad. Sci. Paris 242, 1250–1252 (1956).

– [2] Thèse d'Hörmander II. Sém. Bourbaki 135, 2e éd. (Mai 1956).

– [3] Relations de domination entre opérateurs différentiels. Acta Math. 101, 1–139 (1959).

– [4] Opérateurs différentiels hypoelliptiques. Ann. Inst. Fourier (Grenoble) 9, 1–73 (1959).

[5] Local solvability in L^2 of first order linear PDEs. Amer. J. Math. 92, 369–380 (1970).

– [6] Fundamental solutions of linear partial differential equations with constant coefficients depending on parameters. Amer. J. Math. 84, 561–577 (1962).

– [7] Un théorème sur les équations aux dérivées partielles à coefficients constants dépendant de paramètres. Bull. Soc. Math. France 90, 473–486 (1962).

– [8] A new method of proof of the subelliptic estimates. Comm. Pure Appl. Math. 24, 71–115 (1971).

– [9] Introduction to pseudodifferential and Fourier integral operators. Volume 1: Pseudodifferential operators. Volume 2: Fourier integral operators. Plenum Press, New York and London 1980.

Vauthier, J.: [1] Comportement asymptotique des fonctions entières de type exponentiel dans $\mathbb{C}^n$ et bornées dans le domaine réel. J. Functional Analysis 12, 290–306 (1973).

Vekua, I.N.: [1] Systeme von Differentialgleichungen erster Ordnung vom elliptischen Typus und Randwertaufgaben. Berlin 1956.

Veselič, K. and J. Weidmann: [1] Existenz der Wellenoperatoren für eine allgemeine Klasse von Operatoren. Math. Z. 134, 255–274 (1973).

– [2] Asymptotic estimates of wave functions and the existence of wave operators. J. Functional Analysis. 17, 61–77 (1974).

Višik, M.I.: [1] On general boundary problems for elliptic differential equations. Trudy Moskov. Mat. Obšč. 1, 187–246 (1952) (Russian). Also in Amer. Math. Soc. Transl. (2) 24, 107–172 (1963).

Višik, M.I. and G.I. Eškin: [1] Convolution equations in a bounded region. Uspehi Mat. Nauk 20:3 (123), 89–152 (1965). (Russian.) Also in Russian Math. Surveys 20:3, 86–151 (1965).

– [2] Convolution equations in a bounded region in spaces with weighted norms. Mat. Sb. 69 (111), 65–110 (1966) (Russian). Also in Amer. Math. Soc. Transl. (2) 67, 33–82 (1968).
– [3] Elliptic convolution equations in a bounded region and their applications. Uspehi Mat. Nauk 22:1 (133), 15–76 (1967). (Russian.) Also in Russian Math. Surveys 22:1, 13–75 (1967).
– [4] Convolution equations of variable order. Trudy Moskov. Mat. Obšč. 16, 25–50 (1967). (Russian.) Also in Trans. Moskov. Mat. Soc. 16, 27–52 (1967).
– [5] Normally solvable problems for elliptic systems of convolution equations. Mat. Sb. 74 (116), 326–356 (1967). (Russian.) Also in Math. USSR-Sb. 3, 303–332 (1967).
van der Waerden, B.L.: [1] Einführung in die algebraische Geometrie. Berlin 1939.
– [2] Algebra I–II. 4. Aufl. Springer Verlag, Berlin-Göttingen-Heidelberg 1959.
Wang Rou-hwai and Tsui Chih-yung: [1] Generalized Leray formula on positive complex Lagrange-Grassmann manifolds. Res. Report, Inst. of Math., Jilin Univ. 8209, 1982.
Warner, F.W.: [1] Foundations of differentiable manifolds and Lie Groups. Scott, Foresman and Co., Glenview, Ill., London, 1971.
Weinstein, A.: [1] The order and symbol of a distribution. Trans. Amer. Math. Soc. 241, 1–54 (1978).
– [2] Asymptotics of eigenvalue clusters for the Laplacian plus a potential. Duke Math. J. 44, 883–892 (1977).
– [3] On Maslov's quantization condition. In Fourier integral operators and partial differential equations. Springer Lecture Notes in Math. 459, 341–372 (1974).
Weyl, H.: [1] The method of orthogonal projection in potential theory. Duke Math. J. 7, 411–444 (1940).
– [2] Die Idee der Riemannschen Fläche. 3. Aufl., Teubner, Stuttgart, 1955.
– [3] Über gewöhnliche Differentialgleichungen mit Singularitäten und die zugehörigen Entwicklungen willkürlicher Funktionen. Math. Ann. 68, 220–269 (1910).
– [4] Das asymptotische Verteilungsgesetz der Eigenwerte linearer partieller Differentialgleichungen (mit einer Anwendung auf die Theorie der Hohlraumstrahlung). Math. Ann. 71, 441–479 (1912).
Whitney, H.: [1] Analytic extensions of differentiable functions defined in closed sets. Trans. Amer. Math. Soc. 36, 63–89 (1934).
Widom, H.: [1] Eigenvalue distribution in certain homogeneous spaces. J. Functional Analysis 32, 139–147 (1979).
Yamamoto, K.: [1] On the reduction of certain pseudo-differential operators with non-involution characteristics. J. Differential Equations 26, 435–442 (1977).
Zeilon, N.: [1] Das Fundamentalintegral der allgemeinen partiellen linearen Differentialgleichung mit konstanten Koeffizienten. Ark. Mat. Astr. Fys. 6, No 38, 1–32 (1911).
Zerner, M.: [1] Solutions de l'équation des ondes présentant des singularités sur une droite. C. R. Acad. Sci. Paris 250, 2980–2982 (1960).
– [2] Solutions singulières d'équations aux dérivées partielles. Bull. Soc. Math. France 91, 203–226 (1963).
– [3] Domaine d'holomorphie des fonctions vérifiant une équation aux dérivées partielles. C. R. Acad. Sci. Paris 272, 1646–1648 (1971).
Zuily, C.: [1] Uniqueness and non-uniqueness in the Cauchy problem. Progress in Math. 33, Birkhäuser, Boston, Basel, Stuttgart 1983.
Zygmund, A.: [1] On a theorem of Marcinkiewicz concerning interpolation of operators. J. Math. Pures Appl. 35, 223–248 (1956).

Index

to Volumes III and IV (page numbers in *italics* refer to Volume IV)

Index of Notation

to Volumes III and IV (page numbers in *italics* refer to Volume IV)

Spaces of Functions and Distributions, and Their Norms

Spaces of Symbols

Some Special Notation

Printing: Krips bv, Meppel
Binding: Stürtz, Würzburg

CIM

M. Aigner Combinatorial Theory ISBN 978-3-540-61787-7
A. L. Besse Einstein Manifolds ISBN 978-3-540-74120-6
N. P. Bhatia, G. P. Szegő Stability Theory of Dynamical Systems ISBN 978-3-540-42748-3
J. W. S. Cassels An Introduction to the Geometry of Numbers ISBN 978-3-540-61788-4
R. Courant, F. John Introduction to Calculus and Analysis I ISBN 978-3-540-65058-4
R. Courant, F. John Introduction to Calculus and Analysis II/1 ISBN 978-3-540-66569-4
R. Courant, F. John Introduction to Calculus and Analysis II/2 ISBN 978-3-540-66570-0
P. Dembowski Finite Geometries ISBN 978-3-540-61786-0
A. Dold Lectures on Algebraic Topology ISBN 978-3-540-58660-9
J. L. Doob Classical Potential Theory and Its Probabilistic Counterpart ISBN 978-3-540-41206-9
R. S. Ellis Entropy, Large Deviations, and Statistical Mechanics ISBN 978-3-540-29059-9
H. Federer Geometric Measure Theory ISBN 978-3-540-60656-7
S. Flügge Practical Quantum Mechanics ISBN 978-3-540-65035-5
L. D. Faddeev, L. A. Takhtajan Hamiltonian Methods in the Theory of Solitons ISBN 978-3-540-69843-2
I. I. Gikhman, A. V. Skorokhod The Theory of Stochastic Processes I ISBN 978-3-540-20284-4
I. I. Gikhman, A. V. Skorokhod The Theory of Stochastic Processes II ISBN 978-3-540-20285-1
I. I. Gikhman, A. V. Skorokhod The Theory of Stochastic Processes III ISBN 978-3-540-49940-4
D. Gilbarg, N. S. Trudinger Elliptic Partial Differential Equations of Second Order ISBN 978-3-540-41160-4
H. Grauert, R. Remmert Theory of Stein Spaces ISBN 978-3-540-00373-1
H. Hasse Number Theory ISBN 978-3-540-42749-0
F. Hirzebruch Topological Methods in Algebraic Geometry ISBN 978-3-540-58663-0
L. Hörmander The Analysis of Linear Partial Differential Operators I – Distribution Theory and Fourier Analysis ISBN 978-3-540-00662-6
L. Hörmander The Analysis of Linear Partial Differential Operators II – Differential Operators with Constant Coefficients ISBN 978-3-540-22516-4
L. Hörmander The Analysis of Linear Partial Differential Operators III – Pseudo-Differential Operators ISBN 978-3-540-49937-4
L. Hörmander The Analysis of Linear Partial Differential Operators IV – Fourier Integral Operators ISBN 978-3-642-00117-8
K. Itô, H. P. McKean, Jr. Diffusion Processes and Their Sample Paths ISBN 978-3-540-60629-1
T. Kato Perturbation Theory for Linear Operators ISBN 978-3-540-58661-6
S. Kobayashi Transformation Groups in Differential Geometry ISBN 978-3-540-58659-3
K. Kodaira Complex Manifolds and Deformation of Complex Structures ISBN 978-3-540-22614-7
Th. M. Liggett Interacting Particle Systems ISBN 978-3-540-22617-8
J. Lindenstrauss, L. Tzafriri Classical Banach Spaces I and II ISBN 978-3-540-60628-4
R. C. Lyndon, P. E Schupp Combinatorial Group Theory ISBN 978-3-540-41158-1
S. Mac Lane Homology ISBN 978-3-540-58662-3
C. B. Morrey Jr. Multiple Integrals in the Calculus of Variations ISBN 978-3-540-69915-6
D. Mumford Algebraic Geometry I – Complex Projective Varieties ISBN 978-3-540-58657-9
O. T. O'Meara Introduction to Quadratic Forms ISBN 978-3-540-66564-9
G. Pólya, G. Szegő Problems and Theorems in Analysis I – Series. Integral Calculus. Theory of Functions ISBN 978-3-540-63640-3
G. Pólya, G. Szegő Problems and Theorems in Analysis II – Theory of Functions. Zeros. Polynomials. Determinants. Number Theory. Geometry ISBN 978-3-540-63686-1
W. Rudin Function Theory in the Unit Ball of $\mathbb{C}^n$ ISBN 978-3-540-68272-1
S. Sakai C^*-Algebras and W^*-Algebras ISBN 978-3-540-63633-5
C. L. Siegel, J. K. Moser Lectures on Celestial Mechanics ISBN 978-3-540-58656-2
T. A. Springer Jordan Algebras and Algebraic Groups ISBN 978-3-540-63632-8
D. W. Stroock, S. R. S. Varadhan Multidimensional Diffusion Processes ISBN 978-3-540-28998-2
R. R. Switzer Algebraic Topology: Homology and Homotopy ISBN 978-3-540-42750-6
A. Weil Basic Number Theory ISBN 978-3-540-58655-5
A. Weil Elliptic Functions According to Eisenstein and Kronecker ISBN 978-3-540-65036-2
K. Yosida Functional Analysis ISBN 978-3-540-58654-8
O. Zariski Algebraic Surfaces ISBN 978-3-540-58658-6